The 1st International Electronic Conference on Chemical Sensors and Analytical Chemistry

The 1st International Electronic Conference on Chemical Sensors and Analytical Chemistry

Editor

Nicole Jaffrezic-Renault

MDPI • Basel • Beijing • Wuhan • Barcelona • Belgrade • Manchester • Tokyo • Cluj • Tianjin

Editor
Nicole Jaffrezic-Renault
Institute of Analytical
Sciences, University of Lyon
France

Editorial Office
MDPI
St. Alban-Anlage 66
4052 Basel, Switzerland

This is a reprint of articles from the Proceedings published online in the open access journal *Chemistry Proceedings* (ISSN 2673-4583) (available at: https://www.mdpi.com/2673-4583/5/1).

For citation purposes, cite each article independently as indicated on the article page online and as indicated below:

LastName, A.A.; LastName, B.B.; LastName, C.C. Article Title. *Journal Name* **Year**, *Volume Number*, Page Range.

ISBN 978-3-0365-4115-0 (Hbk)
ISBN 978-3-0365-4116-7 (PDF)

Contents

About the Editor xv

Preface to "The 1st International Electronic Conference on Chemical Sensors and Analytical Chemistry" xvii

Guzel Ziyatdinova and Ekaterina Guss
Sensitive and Selective Voltammetric Sensors for the Simultaneous Quantification of Natural Phenolic Antioxidants in Cognac and Brandy
Reprinted from: *Chem. Proc.* **2021**, *5*, 1, doi:10.3390/CSAC2021-10420 1

Ana Carolina P. Afonso and Luís Pinto da Silva
Carbon Dots as a Fluorescence pH Nanosensor by Application of an Active Surface Preservation Strategy
Reprinted from: *Chem. Proc.* **2021**, *5*, 2, doi:10.3390/CSAC2021-10638 9

Guadalupe Yoselin Aguilar-Lira, Prisciliano Hernandez, Giaan Arturo Álvarez-Romero and Juan Manuel Gutiérrez
Simultaneous Quantification of Four Principal NSAIDs through Voltammetry and Artificial Neural Networks Using a Modified Carbon Paste Electrode in Pharmaceutical Samples
Reprinted from: *Chem. Proc.* **2021**, *5*, 3, doi:10.3390/CSAC2021-10450 11

Franc Paré, Rebeca Castro, Xavier Guimera, Gemma Gabriel and Mireia Baeza
An Inkjet-Printed Amperometric H_2S Sensor for Environmental Applications
Reprinted from: *Chem. Proc.* **2021**, *5*, 4, doi:10.3390/CSAC2021-10462 19

Anastasiia Surkova, Aleksandra Paderina, Andrey Legin, Elena Grachova and Dmitry Kirsanov
Molecular Emitters as a Tunable Light Source for Optical Multisensor Systems
Reprinted from: *Chem. Proc.* **2021**, *5*, 5, doi:10.3390/CSAC2021-10611 25

Nadezhda S. Komova, Ksenya V. Serebrennikova, Anna N. Berlina, Svetlana M. Pridvorova, Anatoly V. Zherdev and Boris B. Dzantiev
Gold Nanoparticles Functionalized with Mercaptosuccinic Acidas a Means for Detecting Fe(III) Ions
Reprinted from: *Chem. Proc.* **2021**, *5*, 6, doi:10.3390/CSAC2021-10624 31

Agnieszka Królicka, Jerzy Zarębski and Andrzej Bobrowski
Implementation of Metallic Film Electrodes for Catalytic Adsorptive Stripping Voltammetric Determination of Germanium(IV)
Reprinted from: *Chem. Proc.* **2021**, *5*, 7, doi:10.3390/CSAC2021-10484 37

Laia L. Fernández, Julio Bastos-Arrieta, Cristina Palet and Mireia Baeza
Tunable Electrochemical Sensors Based on Carbon Nanocomposite Materials towards Enhanced Determination of Cadmium, Lead and Copper in Water
Reprinted from: *Chem. Proc.* **2021**, *5*, 8, doi:10.3390/CSAC2021-10456 43

Amer Charbaji, Hojat Heidari-Bafroui, Nasim Rahmani, Constantine Anagnostopoulos and Mohammad Faghri
Colorimetric Determination of Nitrate after Reduction to Nitrite in a Paper-Based Dip Strip
Reprinted from: *Chem. Proc.* **2021**, *5*, 9, doi:10.3390/CSAC2021-10459 51

André Carvalho, Maria Freitas, Henri P. A. Nouws and Cristina Delerue-Matos
A Voltammetric Nanodiamond-Coated Screen-Printed Immunosensor for The Determination of a Peanut Allergen in Commercial Food Products
Reprinted from: *Chem. Proc.* **2021**, *5*, 10, doi:10.3390/CSAC2021-10458 . . . . **61**

Edward B. Ogugu, Ross N. Gillanders and Graham A. Turnbull
Thermal Desorption of Explosives Vapour from Organic Fluorescent Sensors
Reprinted from: *Chem. Proc.* **2021**, *5*, 11, doi:10.3390/CSAC2021-10559 . . . . **67**

Bogdan-Catalin Serban, Octavian Buiu, Marius Bumbac, Roxana Marinescu, Niculae Dumbravescu, Viorel Avramescu, Cornel Cobianu, Cristina Mihaela Nicolescu, Mihai Brezeanu, Cristiana Radulescu and Florin Comanescu
Ternary Oxidized Carbon Nanohorns/TiO_2/PVP Nanohybrid as Sensitive Layer for Chemoresistive Humidity Sensor
Reprinted from: *Chem. Proc.* **2021**, *5*, 12, doi:10.3390/CSAC2021-10616 . . . . **73**

Annalisa Scroccarello, Flavio Della Pelle, Qurat Ul Ain Bukhari, Filippo Silveri, Daniele Zappi, Enrico Cozzoni and Dario Compagnone
Eucalyptus Biochar as a Sustainable Nanomaterial for Electrochemical Sensors
Reprinted from: *Chem. Proc.* **2021**, *5*, 13, doi:10.3390/CSAC2021-10618 . . . . **81**

Filippo Silveri, Flavio Della Pelle, Daniel Rojas and Dario Compagnone
Graphene Nanoflakes Incorporating Natural Phytochemicals Containing Catechols as Functional Material for Sensors
Reprinted from: *Chem. Proc.* **2021**, *5*, 14, doi:10.3390/CSAC2021-10619 . . . . **83**

Ashutosh Kumar, Hojat Heidari-Bafroui, Amer Charbaji, Nasim Rahmani, Constantine Anagnostopoulos and Mohammad Faghri
Numerical and Experimental Modeling of Paper-Based Actuators
Reprinted from: *Chem. Proc.* **2021**, *5*, 15, doi:10.3390/CSAC2021-10468 . . . . **85**

Vardan Galstyan, Nicola Poli and Elisabetta Comini
Study of Gas-Sensing Properties of Titania Nanotubes for Health and Safety Applications
Reprinted from: *Chem. Proc.* **2021**, *5*, 16, doi:10.3390/CSAC2021-10625 . . . . **99**

Diana M. A. Crista, Joaquim C. G. Esteves da Silva and Luís Pinto da Silva
Validation of Spent Coffee Grounds as Precursors for the Development of Sustainable Carbon Dot-Based for Fe^{3+} Optical Sensing
Reprinted from: *Chem. Proc.* **2021**, *5*, 17, doi:10.3390/CSAC2021-10452 . . . . **103**

Mafalda Reis-Pereira, Rui C. Martins, Aníbal Filipe Silva, Fernando Tavares, Filipe Santos and Mário Cunha
Unravelling Plant-Pathogen Interactions: Proximal Optical Sensing as an Effective Tool for Early Detect Plant Diseases
Reprinted from: *Chem. Proc.* **2021**, *5*, 18, doi:10.3390/CSAC2021-10560 . . . . **105**

Cátia Magro, Marcelo Morais, Paulo A. Ribeiro, Susana Sério, Pedro Vieira and Maria Raposo
Development of a Gas Sensor for *Eucalyptol* Supervision:A Supporting Tool for ExtremeWildfire Management
Reprinted from: *Chem. Proc.* **2021**, *5*, 19, doi:10.3390/CSAC2021-10432 . . . . **115**

Bernabe Nuñez-Estevez, Tiane C. Finimundy, Maria Carpena, Marta Barral-Martinez, Ricardo Calhelha, Tânia C. S. P. Pires, Paz Otero, Pascual Garcia-Perez, Jesus Simal-Gandara, Isabel C. F. R. Ferreira, Miguel A. Prieto and Lillian Barros
Bioactive Compound Profiling and Nutritional Composition of Three Species from the Amaranthaceae Family
Reprinted from: *Chem. Proc.* **2021**, *5*, 20, doi:10.3390/CSAC2021-10563 . . . . **121**

Edgar G. Mendez-Lopez, Jersson X. Leon-Medina and Diego A. Tibaduiza
Development of a Pattern Recognition Tool for the Classification of Electronic Tongue Signals Using Machine Learning
Reprinted from: *Chem. Proc.* **2021**, *5*, 21, doi:10.3390/CSAC2021-10447 . . . . **127**

Heena Tyagi, Emma Daulton, Ayman S. Bannaga, Ramesh P. Arasaradnam and James A. Covington
Electronic Nose for Bladder Cancer Detection
Reprinted from: *Chem. Proc.* **2021**, *5*, 22, doi:10.3390/CSAC2021-10438 . . . . **133**

George-Octavian Buica, Georgiana-Luiza Tatu (Arnold), Eleonora-Mihaela Ungureanu and Gabriela Geanina Vasile
Voltammetric Detection of Mercury Ions at Poly(azulene-EDTA)-like Screen Printed Modified Electrodes
Reprinted from: *Chem. Proc.* **2021**, *5*, 23, doi:10.3390/CSAC2021-10630 . . . . **139**

Esther Sánchez-Tirado, Sara Guerrero, Araceli González-Cortés, Lourdes Agüí, Paloma Yáñez-Sedeño and José Manuel Pingarrón
Electrochemical Immunosensor for Simultaneous Determination of Emerging Autoimmune Disease Biomarkers in Human Serum
Reprinted from: *Chem. Proc.* **2021**, *5*, 24, doi:10.3390/CSAC2021-10437 . . . . **141**

Meijuan Liang, Qi Zhang and Peiwu Li
Advances in Visual Immunoassays for Sensitive Detection of Mycotoxins in Food—A Review
Reprinted from: *Chem. Proc.* **2021**, *5*, 25, doi:10.3390/CSAC2021-10443 . . . . **143**

Giulia Zambotti and Andrea Ponzoni
Reproductivity Study of Metal Oxide Gas Sensors Using Two Different Temperature Setups
Reprinted from: *Chem. Proc.* **2021**, *5*, 26, doi:10.3390/CSAC2021-10613 . . . . **157**

Amall A. Ramanathan
First Principles Investigation of the Optoelectronic Properties of Molybdenum Dinitride for Optical Sensing Applications
Reprinted from: *Chem. Proc.* **2021**, *5*, 27, doi:10.3390/CSAC2021-10429 . . . . **163**

Anton Soria-Lopez, Maria Carpena, Bernabe Nuñez-Estevez, Paula Garcia-Oliveira, Nicolas Collazo,Paz Otero, Pascual Garcia-Perez, Hui Cao, Jianbo Xiao, Márcio Carocho, Lillian Barros, Jesus Simal-Gandara and Miguel A. Prieto
Essential Oils as Possible Candidates to Be Included in Active Packaging Systems and the Use of Biosensors to Monitor the Quality of Foodstuff
Reprinted from: *Chem. Proc.* **2021**, *5*, 28, doi:10.3390/CSAC2021-10485 . . . . **171**

Matteo Tonezzer, Franco Biasioli and Flavia Gasperi
From Single Nanowires to Smart Systems: Different Ways to Assess Food Quality
Reprinted from: *Chem. Proc.* **2021**, *5*, 29, doi:10.3390/CSAC2021-10605 . . . . **179**

Coral Salvo-Comino, Clara Perez-Gonzalez, Fernando Martin-Pedrosa, Cristina Garcia-Cabezon and Maria Luz Rodriguez-Mendez
Silver Nanomaterials as Electron Mediators in a Bio-Electronic Tongue Dedicated to the Analysis of Milks. The Role of the Aspect Ratio of Nanoparticles vs. Nanowires
Reprinted from: *Chem. Proc.* **2021**, *5*, 30, doi:10.3390/CSAC2021-10554 . . . . **181**

Clara Pérez-González, Coral Salvo-Comino, Fernando Martin-Pedrosa, Cristina García-Cabezón and María Luz Rodríguez Méndez
Development of a Bioelectronic Tongue Modified with Gold Nanoparticles for Dairy Analysis
Reprinted from: *Chem. Proc.* **2021**, *5*, 31, doi:10.3390/CSAC2021-10553 . . . . **185**

Karolis Norvaiša and Mathias Otto Senge
Core Modulation of Porphyrins for Chemical Sensing
Reprinted from: *Chem. Proc.* **2021**, *5*, 32, doi:10.3390/CSAC2021-10417 . . . . **189**

Mallikarjun Madagalam, Federica Catania, Mattia Bartoli, Alberto Tagliaferro and Sandro Carrara
Nanostructured Bismuth Electrodes for Non-Enzymatic Paracetamol Sensing: Development, Testing, and Computational Approach
Reprinted from: *Chem. Proc.* **2021**, *5*, 33, doi:10.3390/CSAC2021-10427 . . . . **191**

Bruno Miranda, Rosalba Moretta, Selene De Martino, Principia Dardano, Ilaria Rea, Carlo Forestiere and Luca De Stefano
Plasmonic Hydrogel Nanocomposites with Combined Optical and Mechanical Properties for Biochemical Sensing
Reprinted from: *Chem. Proc.* **2021**, *5*, 34, doi:10.3390/CSAC2021-10467 . . . . **199**

Uttam Narendra Thakur, Radha Bhardwaj and Arnab Hazra
Statistical Analysis for Selective Identifications of VOCs by Using Surface Functionalized MoS_2 Based Sensor Array
Reprinted from: *Chem. Proc.* **2021**, *5*, 35, doi:10.3390/CSAC2021-10451 . . . . **201**

Ítala M. G. Marx, Nuno Rodrigues, Ana C. A. Veloso, José A. Pereira and António M. Peres
Evaluation of the Effect of Extracted Time Conditions on the Phenolic Content of Olive Pastes from *cv.* Arbequina and Discrimination Using a Lab-Made Potentiometric Electronic Tongue
Reprinted from: *Chem. Proc.* **2021**, *5*, 36, doi:10.3390/CSAC2021-10556 . . . . **209**

Antonio Fotia, Patrizia Frontera, Lucio Bonaccorsi and Angela Malara
Conductive Electrospun Nanofibers for Multifunctional Portable Devices
Reprinted from: *Chem. Proc.* **2021**, *5*, 37, doi:10.3390/CSAC2021-10634 . . . . **215**

Praskoviya Boltovets, Sergii Kravchenko, Oleksiy Kovalenko and Borys Snopok
Polysaccharide-Based Organic Frameworks with Embedded Nanoparticles: Advanced SPR Study on the Antiviral Activity of Gold Composites Derived from Glucuronoxylomannan
Reprinted from: *Chem. Proc.* **2021**, *5*, 38, doi:10.3390/CSAC2021-10475 . . . . **217**

Aleksandra Tobolska, Nina E. Wezynfeld, Urszula E. Wawrzyniak, Wojciech Bal and Wojciech Wróblewski
Metal–Peptide Complexes—A Novel Class of Molecular Receptors for Electrochemical Phosphate Sensin
Reprinted from: *Chem. Proc.* **2021**, *5*, 39, doi:10.3390/CSAC2021-10449 . . . . **223**

Marco Grossi, Enrico Valli, Alessandra Bendini, Tullia Gallina Toschi and Bruno Riccò
Evaluation of Olive Oil Quality Grade Using a Portable Battery-Operated Sensor System
Reprinted from: *Chem. Proc.* **2021**, *5*, 40, doi:10.3390/CSAC2021-10614 . . . . **227**

Katerina Lazarova, Silvia Bozhilova, Sijka Ivanova, Darinka Christova and Tsvetanka Babeva
Optical Characterization of Acetone-Sensitive Thin Films of poly(vinyl alcohol)-g-poly(methyl acrylate)
Reprinted from: *Chem. Proc.* **2021**, *5*, 41, doi:10.3390/CSAC2021-10416 . . . . **233**

Bernardo Dias, João P. Mendes, José M. M. M. de Almeida and Luís C. C. Coelho
Characterization and Comparison of the Relative Humidity Response of Hydromorphic Polymers in Long-Period Fiber Grating Structures
Reprinted from: *Chem. Proc.* **2021**, *5*, 42, doi:10.3390/CSAC2021-10461 . . . . **235**

Paula Garcia-Oliveira, Antia G. Pereira, Maria Fraga-Corral, Catarina Lourenço-Lopes,Franklin Chamorro, Aurora Silva, Pascual Garcia-Perez, Fatima Barroso, Lillian Barros, Isabel C. F. R. Ferreira, Jesus Simal-Gandara and Miguel A. Prieto
Identification, Quantification, and Method Validation of Anthocyanins
Reprinted from: *Chem. Proc.* **2021**, *5*, 43, doi:10.3390/CSAC2021-10680 . . . . **243**

Marc Parrilla, Amorn Slosse, Robin Van Echelpoel, Noelia Felipe Montiel, Filip Van Durme and Karolien De Wael
Portable Electrochemical Detection of Illicit Drugs in Smuggled Samples: Towards More Secure Borders
Reprinted from: *Chem. Proc.* **2021**, *5*, 44, doi:10.3390/CSAC2021-10612 . . . . **251**

Iryna Makarchuk, Anton Nikolaev, Alexander Thesseling, Lisa Dejon, Daniel Lamberty, Laura Stief, Thorsten Friedrich, Petra Hellwig, Hamid R. Nasiri and Frederic Melin
The Inhibition Study of Cytochrome bd Oxidase Using the Enzyme-Based Electrochemical Sensor
Reprinted from: *Chem. Proc.* **2021**, *5*, 45, doi:10.3390/CSAC2021-10555 . . . . **259**

Diana A. Toriz-Gutiérrez, Humberto Ramírez-Gasca, Luis E. Cárdenas-Galindo and Eloisa Gallegos-Arellano
Electrochemical Measurement System for Chlorides in Drinking and Wastewater
Reprinted from: *Chem. Proc.* **2021**, *5*, 46, doi:10.3390/chemproc2021005046 . . . . **261**

Anastasiya Zhupanova and Guzel Ziyatdinova
Poly(bromocresol purple)-Based Voltammetric Sensor for the Simultaneous Quantification of Ferulic Acid and Vanillin
Reprinted from: *Chem. Proc.* **2021**, *5*, 47, doi:10.3390/CSAC2021-10441 . . . . **267**

João P. Mendes, Luís C. C. Coelho, Viviana P. Pereira, Manuel A. Azenha, Pedro A. S. Jorge and Carlos M. Pereira
Label-Free Anti-Human IgG Biosensor Based on Chemical Modification of a Long Period Fiber Grating Surface
Reprinted from: *Chem. Proc.* **2021**, *5*, 48, doi:10.3390/CSAC2021-10454 . . . . **271**

Alice Mieting, Sitao Wang, Mia Schliephake, Daniela Franke, Margarita Guenther, Stefan Odenbach and Gerald Gerlach
Precipitation of Iron Oxide in Hydrogel with Superparamagnetic and Stimuli-Responsive Properties
Reprinted from: *Chem. Proc.* **2021**, *5*, 49, doi:10.3390/chemproc2021005049 . . . . **279**

Anxo Carreira-Casais, Maria Carpena, Antia G. Pereira, Franklin Chamorro, Anton Soria-Lopez, Pascual Garcia Perez, Paz Otero, Hui Cao, Jianbo Xiao, Jesus Simal-Gandara and Miguel A. Prieto
Critical Variables Influencing the Ultrasound-Assisted Extraction of Bioactive Compounds—A Review
Reprinted from: *Chem. Proc.* **2021**, *5*, 50, doi:10.3390/CSAC2021-10562 . . . . **287**

Marta Barral-Martinez, Paula Garcia-Oliveira, Bernabe Nuñez-Estevez, Aurora Silva, Tiane C. Finimundy, Ricardo Calhelha, Marija Nenadic, Marina Sokovic, Fatima Barroso, Jesus Simal-Gandara, Isabel C. F. R. Ferreira, Lillian Barros and Miguel A. Prieto
Plants of the Family Asteraceae: Evaluation of Biological Properties and Identification of Phenolic Compounds
Reprinted from: *Chem. Proc.* **2021**, *5*, 51, doi:10.3390/CSAC2021-10486 . . . . **293**

Mohammed Moufid, Carlo Tiebe, Nezha El Bari, Matthias Bartholmai and Benachir Bouchikhi
Characterization of Unpleasant Odors in Poultry Houses Using Metal Oxide Semiconductor-Based Gas Sensor Arrays and Pattern Recognition Methods
Reprinted from: *Chem. Proc.* **2021**, *5*, 52, doi:10.3390/CSAC2021-10481 . . . . **301**

Ali Othman, Akhtar Hayat and Silvana Andreescu
Europium-Doped Ceria Nanocrystals as Nanozyme Fluorescent Probes for Biosensing
Reprinted from: *Chem. Proc.* **2021**, *5*, 53, doi:10.3390/CSAC2021-10549 . . . . **309**

Elvira Yakupova and Guzel Ziyatdinova
Electrode Modified with Tin(IV) Oxide Nanoparticles and Surfactants as Sensitive Sensor for Hesperidin
Reprinted from: *Chem. Proc.* **2021**, *5*, 54, doi:10.3390/CSAC2021-10615 . . . . **311**

Helena Vasconcelos, Ana Matias, Pedro Jorge, Cristina Saraiva, João Mendes, João Araújo, Bernardo Dias, Paulo Santos, José M. M. M. Almeida and Luís C. C. Coelho
Optical Biosensor for the Detection of Hydrogen Peroxide in Milk
Reprinted from: *Chem. Proc.* **2021**, *5*, 55, doi:10.3390/CSAC2021-10466 . . . . **319**

Jersson X. Leon-Medina, Maribel Anaya and Diego A. Tibaduiza
Locally Linear Embedding as Nonlinear Feature Extraction to Discriminate Liquids with a Cyclic Voltammetric Electronic Tongue
Reprinted from: *Chem. Proc.* **2021**, *5*, 56, doi:10.3390/CSAC2021-10426 . . . . **323**

Paulo M. Zagalo, Cátia Magro, Paulo A. Ribeiro and Maria Raposo
Applied Voltage Effect in Lbl Sensors While Detecting 17α-Ethinylestradiol in Water Samples
Reprinted from: *Chem. Proc.* **2021**, *5*, 57, doi:10.3390/CSAC2021-10460 . . . . **333**

Cátia Magro, Tiago Moura, Paulo A. Ribeiro, Maria Raposo and Susana Sério
Smart Sensing for Antibiotic Monitoring in Mineral and SurfaceWater: Development of an Electronic Tongue Device
Reprinted from: *Chem. Proc.* **2021**, *5*, 58, doi:10.3390/CSAC2021-10606 . . . . **341**

Cecilia Lete, Mariana Marin, Francisco Javier del Campo, Ioana Diaconu and Stelian Lupu
Antimony Tin Oxide—Prussian Blue Screen-Printed Electrodes for Electrochemical Sensing of Potassium Ions
Reprinted from: *Chem. Proc.* **2021**, *5*, 59, doi:10.3390/CSAC2021-10639 . . . . **347**

Eleonora Pargoletti, Francesca Tessore, Gabriele Di Carlo, Gian Luca Chiarello and Giuseppe Cappelletti
Towards Low Temperature VOCs Chemoresistors: Graphene Oxide Versus Porphyrin-Based Materials
Reprinted from: *Chem. Proc.* **2021**, *5*, 60, doi:10.3390/CSAC2021-10418 . . . . **349**

Rachid Laref, Etienne Losson, Alexandre Sava and Maryam Siadat
Field Nitrogen Dioxide and Ozone Monitoring Using Electrochemical Sensors with Partial Least Squares Regression
Reprinted from: *Chem. Proc.* **2021**, *5*, 61, doi:10.3390/CSAC2021-10622 . . . . **355**

Ricarda Torre, Maria Freitas, Estefanía Costa-Rama, Henri P. A. Nouws and Cristina Delerue-Matos
Tropomyosin Analysis in Foods Using an Electrochemical Immunosensing Approach
Reprinted from: *Chem. Proc.* **2021**, *5*, 62, doi:10.3390/CSAC2021-10471 . . . . **361**

Hafsa El Youbi, Alassane Diouf, Benachir Bouchikhi and Nezha El Bari
Simultaneous Sensing of Codeine and Diclofenac in Water Samples Using an Electrochemical Bi-MIP Sensor and a Voltammetric Electronic Tongue
Reprinted from: *Chem. Proc.* **2021**, *5*, 63, doi:10.3390/CSAC2021-10483 . . . . **367**

Youssra Aghoutane, Nezha El Bari, Zoubida Laghrari and Benachir Bouchikhi
Electrochemical Detection of Fenthion Insecticide in Olive Oils by a Sensitive Non-Enzymatic Biomimetic Sensor Enhanced with Metal Nanoparticles
Reprinted from: *Chem. Proc.* **2021**, *5*, 64, doi:10.3390/CSAC2021-10773 . . . . **375**

Sarra Takita, Alexei Nabok, David Smith and Anna Lishchuk
Spectroscopic Ellipsometry Detection of Prostate Cancer Bio-Marker PCA3 Using Specific Non-Labeled Aptamer: Comparison with Electrochemical Detection
Reprinted from: *Chem. Proc.* **2021**, *5*, 65, doi:10.3390/CSAC2021-10453 . . . . **383**

Roussin Lontio Fomekong and Bilge Saruhan
Titanium Based Materials for High-Temperature Gas Sensor inHarsh Environment Application
Reprinted from: *Chem. Proc.* **2021**, *5*, 66, doi:10.3390/CSAC2021-10480 . . . . **391**

Javier Echave, Catarina Lourenço-Lopes, Anxo Carreira-Casais, Franklin Chamorro, Maria Fraga-Corral, Paz Otero, Pascual Garcia-Perez, Sergio Baamonde, Fermín Fernández-Saa, Hui Cao, Jianbo Xiao, Miguel A. Prieto and Jesus Simal-Gandara
Nutritional Composition of the Atlantic Seaweeds *Ulva rigida, Codium tomentosum, Palmaria palmata* and *Porphyra purpurea*
Reprinted from: *Chem. Proc.* **2021**, *5*, 67, doi:10.3390/CSAC2021-10681 . . . . **399**

Anastasiia Shuba, Tatiana Kuchmenko and Dariya Menzhulina
Drift Compensation of the Electronic Nose in the Developmentof Instruments for Out-of-Laboratory Analysis
Reprinted from: *Chem. Proc.* **2021**, *5*, 68, doi:10.3390/CSAC2021-10464 . . . . **407**

Mashuni Mashuni, Halimahtussaddiyah Ritonga, Muhammad Jahiding, La Ode Ahmad Nur Ramadhan, Desy Kurniawati and Fitri Handayani Hamid
The Performance of Organophosphate Pesticides Determination Using Biosensor Based on Small Device Potentiometer as a Transducer
Reprinted from: *Chem. Proc.* **2021**, *5*, 69, doi:10.3390/CSAC2021-10604 . . . . **415**

Franklin Chamorro, Lucia Cassani, Catarina Lourenço-Lopes, Anxo Carreira-Casais, Maria Carpena, Javier Echave, Sergio Baamonde, Fermin Fernández-Saa, Paz Otero, Pacual Garcia-Perez, Jesus Simal-Gandara and Miguel Angel Prieto
Optimization of Bioactive Compounds with Antioxidant Activity of *Himanthalia elongata* by Microwave-Assisted Extraction Using Response Surface Methodology
Reprinted from: *Chem. Proc.* **2021**, *5*, 70, doi:10.3390/CSAC2021-10478 . . . . **423**

Ambra Fioravanti, Sara Morandi and Maria Cristina Carotta
Semiconductor Oxide Gas Sensors: Correlation between Conduction Mechanisms and Their Sensing Performances
Reprinted from: *Chem. Proc.* **2021**, *5*, 71, doi:10.3390/CSAC2021-10472 . . . . **429**

Ekaterina Yuskina, Kirill Tugashov, Vladimir B. Shur, Irina A. Tikhonova, Vasily Babain and Dmitry Kirsanov
Cross-Sensitive Potentiometric Sensors Based on Anti-Crown $(C_6HgF_4)_3$
Reprinted from: *Chem. Proc.* **2021**, *5*, 72, doi:10.3390/CSAC2021-10424 . . . . **433**

María Elena Martínez-Hernández, Xabier Sandua, Pedro J. Rivero, Javier Goicoechea and Francisco J. Arregui
An Optical Fiber Sensor for Hg^{2+} Detection Based on the LSPR of Silver and Gold Nanoparticles Embedded in a Polymeric Matrix as an Effective Sensing Material
Reprinted from: *Chem. Proc.* **2021**, *5*, 73, doi:10.3390/CSAC2021-10633 . . . . **439**

Teena Gakhar and Arnab Hazra
Development of Graphene-Doped TiO_2-Nanotube Array-Based MIM-Structured Sensors and Its Application for Methanol Sensing at Room Temperature
Reprinted from: *Chem. Proc.* **2021**, *5*, 74, doi:10.3390/CSAC2021-10620 . . . . **447**

Andrea Ponzoni
Morphological Effects in SnO_2 Chemiresistors for Ethanol Detection: A Systematic Statistical Analysis of Results Published in the Last 5 Years
Reprinted from: *Chem. Proc.* **2021**, *5*, 75, doi:10.3390/CSAC2021-10474 . . . . **455**

Maha Wajeeh Aqra and Amall Ahmed Ramanathan
Review of the Recent Advances in Nano-Biosensors and Technologies for Healthcare Applications
Reprinted from: *Chem. Proc.* **2021**, *5*, 76, doi:10.3390/CSAC2021-10473 . . . . **461**

Teresa Guerra Barroso, Lénio Ribeiro, Hugo Gregório, Filipe Santos and Rui Costa Martins
Feasibility of Total White Blood Cells Counts by Visible-Near Infrared Spectroscopy
Reprinted from: *Chem. Proc.* **2021**, *5*, 77, doi:10.3390/CSAC2021-10434 . . . . **471**

Teresa Guerra Barroso, Lénio Ribeiro, Hugo Gregório, Filipe Santos and Rui Costa Martins
Visible–Near-Infrared Platelets Count: Towards Thrombocytosis Point-of-Care Diagnosis
Reprinted from: *Chem. Proc.* **2021**, *5*, 78, doi:10.3390/CSAC2021-10435 . . . . **477**

Roberto Di Chio, Monica Galtieri, Nicola Donato and Giovanni Neri
Development of an Integrated In-Vehicle Driver Breath EthanolSystem Based on α-Fe_2O_3 Sensing Material
Reprinted from: *Chem. Proc.* **2021**, *5*, 79, doi:10.3390/CSAC2021-10476 . . . . **483**

Carlos Torres-Méndez, Jayendra Ellamathy, Maria Ines Mascarenhas, Yifan Liu, Georgia-Vasiliki Gkountana, Patrizia Kühne, Javier Sebastián, Ivana Jovanovic, David Bern, Sharmilee Nandi, Maike Lüftner, Viktoria Langwallner, Maria Lysandrou, Sam Taylor, Klara Martinovic, Abdul-Raouf Atif, Ehsan Manouchehri Doulabi, Masood Kamali-Moghaddam and Gemma Mestres
Developing an Electrochemical Biosensor for the Detection of Hemagglutinin Protein of Influenza A Virus Subtype H1N1 in Artificial Saliva
Reprinted from: *Chem. Proc.* **2021**, *5*, 80, doi:10.3390/CSAC2021-10477 . . . . **491**

Qing Wang and Manel del Valle
Determination of Chemical Oxygen Demand (COD) Using Nanoparticle-Modified Voltammetric Sensors and Electronic Tongue Principles
Reprinted from: *Chem. Proc.* **2021**, *5*, 81, doi:10.3390/CSAC2021-10442 . . . . **499**

Amall Ahmed Ramanathan
New Half Metal Perovskite $NbScO_3$ for Spintronic Sensing Applications
Reprinted from: *Chem. Proc.* **2021**, *5*, 82, doi:10.3390/CSAC2021-10628 . . . . **501**

Helena J. Spikes, Shelby J. Jarrett-Noland, Stephan M. Germann, Wendy Olivas, Janet Braddock-Wilking and Cynthia M. Dupureur
Group 14 Metallafluorenes for Lipid Structure Detection and Cellular Imaging
Reprinted from: *Chem. Proc.* **2021**, *5*, 83, doi:10.3390/CSAC2021-10455 . . . . **507**

Nadezhda Vladimirova, Julia Ashina and Dmitry Kirsanov
QSPR Modelling of Potentiometric HCO_3^-/Cl^- Selectivity forPolymeric Membrane Sensors
Reprinted from: *Chem. Proc.* **2021**, *5*, 84, doi:10.3390/CSAC2021-10621 . . . . **513**

Mohammed Majeed Alkhabet, Saad Hayatu Girei, Abdul Hadi Ismail, Suriati Paiman, Norhana Arsad, Mohd Adzir Mahdi and Mohd Hanif Yaacob
Room Temperature Hydrogen Sensing Based on Tapered Optical Fiber Coated with Polyaniline (PANI)
Reprinted from: *Chem. Proc.* **2021**, *5*, 85, doi:/10.3390/CSAC2021-10415 . . . . **519**

Rana Dalapati and Ling Zang
Aqueous Medium Fluoride Anion Sensing by Fluorophore Encapsulated UiO-66 Type ZirconiumMetal–Organic Framework
Reprinted from: *Chem. Proc.* **2021**, *5*, 86, doi:10.3390/CSAC2021-10551 . . . . **527**

Karina Torres-Rivero, Clara Pérez-Ràfols, Julio Bastos-Arrieta, Núria Serrano, Vicenç Martí and Antonio Florido
Customized Screen-Printed Electrodes Based on Ag-Nanoseeds for Enhanced Electroanalytical Response towards Cd(II), Pb(II) and As(V) in Aqueous Samples
Reprinted from: *Chem. Proc.* **2021**, *5*, 87, doi:10.3390/CSAC2021-10469 . . . . **533**

Aníbal Filipe Silva, Klara Löfkvist, Mikael Gilbertsson, Erik Van Os, Geert Franken, Jos Balendonck, Tatiana M. Pinho, José Boaventura-Cunha, Luís Coelho, Pedro Jorge and Rui Costa Martins
Hydroponics Monitoring through UV-Vis Spectroscopy and Artificial Intelligence: Quantification of Nitrogen, Phosphorous and Potassium
Reprinted from: *Chem. Proc.* **2021**, *5*, 88, doi:10.3390/CSAC2021-10448 . . . . **541**

Achraf El Mohajir, Jean-Baptiste Sanchez, Mohammad Arab Pour Yazdi, Olivier Heintz and Nicolas Martin
Detection of Indoor Air Pollutants Using Reactive Sputtering/GLAD of Tin Oxide Thin Films
Reprinted from: *Chem. Proc.* **2021**, *5*, 89, doi:10.3390/CSAC2021-10548 . . . . **549**

Aleksandra Kalinowska, Patrycja Matusiak, Sandra Skorupska, Ilona Grabowska-Jadach and Patrycja Ciosek-Skibińska
Influence of the Type and Amount of Plasticizer on the Sensory Properties of Microspheres Sensitive to Lipophilic Ions
Reprinted from: *Chem. Proc.* **2021**, *5*, 90, doi:10.3390/CSAC2021-10487 . . . . **551**

Patricia Peredo-Guzmán, Miriam Trigo-López, Saúl Vallejos, Félix García and José Miguel García
Intrinsically Coloured Red Aromatic Polyamides
Reprinted from: *Chem. Proc.* **2021**, *5*, 91, doi:10.3390/CSAC2021-10421 . . . . **559**

About the Editor

Nicole Jaffrezic-Renault

Received her engineering degree from the Ecole Nationale Supérieure de Chimie, Paris, in 1971, and her Doctorat d'Etat és Sciences Physiques from the University of Paris in 1976. She is the Emeritus Director of Research at the Centre National de la Recherche Scientifique, the past President of the Chemical Micro-Sensor Club (CMC2), the President of the Analytical Division, and a Member of the Administrative Council of the French Chemical Society. She was decorated as a Knight of the Legion of Honor and an Officer of the National Order of Merit. Her research activities in the Institute of Analytical Sciences include the conception and design of (bio)chemical sensors and their integration into microsystems. She coordinates several European and national projects for the development of microsystems for biomedical and environmental monitoring, as well as for food safety. She has published more than 630 papers with more than 14190 citations (h-index: 58).

Preface to "The 1st International Electronic Conference on Chemical Sensors and Analytical Chemistry"

The 1st International Electronic Conference on Chemical Sensors and Analytical Chemistry was held on 1–15 July 2021. The scope of this online conference was to gather experts that are well-known worldwide who are currently working in chemical sensor technologies and to provide an online forum for the presention and discussion of new results. Throughout this event, topics of interest included, but were not limited to, the following: electrochemical devices and sensors; optical chemical sensors; mass-sensitive sensors; materials for chemical sensing; nano- and micro-technologies for sensing; chemical assays and validation; chemical sensor applications; analytical methods; gas sensors and apparatuses; electronic noses; electronic tongues; microfluidic devices; lab-on-a-chip; single-molecule sensing; nanosensors; and medico-diagnostic testing.

Nicole Jaffrezic-Renault
Editor

chemistry proceedings

MDPI

Proceeding Paper

Sensitive and Selective Voltammetric Sensors for the Simultaneous Quantification of Natural Phenolic Antioxidants in Cognac and Brandy †

Guzel Ziyatdinova * and Ekaterina Guss

Analytical Chemistry Department, Kazan Federal University, Kremleyevskaya, 18, 420008 Kazan, Russia; kozlova.ekaterina1992@mail.ru

* Correspondence: Ziyatdinovag@mail.ru

† Presented at the 1st International Electronic Conference on Chemical Sensors and Analytical Chemistry, 1–15 July 2021; Available online: https://csac2021.sciforum.net/.

Abstract: Aged distilled beverages (cognac and brandy) contain phenolic antioxidants as their quality markers. Voltammetric sensors based on the carbon nanotubes and electropolymerized pyrocatechol violet (PCV) or *p*-aminobenzoic acid (ABA) were developed for the simultaneous determination of the phenolic antioxidants in cognac and brandy. The polymerization conditions of PCV and ABA were optimized. Sensors allow for the simultaneous sensitive determination of gallic and ellagic acids as well as syringaldehyde and vanillin. The analytical characteristics are improved vs. other modified electrodes. The sensors show selectivity in the presence of typical interferences and other natural phenolics. The sensors that were developed were tested on cognac and brandy samples.

Keywords: electrochemical sensors; carbon nanotubes; electropolymerization; gallic acid; ellagic acid; syringaldehyde; vanillin; cognac and brandy

Citation: Ziyatdinova, G.; Guss, E. Sensitive and Selective Voltammetric Sensors for the Simultaneous Quantification of Natural Phenolic Antioxidants in Cognac and Brandy. *Chem. Proc.* **2021**, *5*, 1. https://doi.org/10.3390/CSAC2021-10420

Academic Editor: Huangxian Ju

Published: 30 June 2021

1. Introduction

Aged distilled beverages, namely cognac and brandy, contain natural phenolic antioxidants that are often considered as quality markers for these beverages [1]. Phenolic acid (gallic and ellagic) and aromatic aldehydes (vanillin and syringaldehyde) are the major contributors to the antioxidant properties of aged distilled beverages [1–3]. Their simultaneous determination is of practical interest.

Different types of chromatography and capillary electrophoresis are usually applied for their quantification [2,4,5]. Nevertheless, less tedious and simpler methods are encouraged. Electrochemical sensors are simple, reliable, and cost-effective and could be an effective alternative tool for these purposes [6–8]. The only problem facing electrochemical approaches is their low selectivity of when determining the phenolic content due to the structural similarity of the analytes. This limitation can be overcome using chemically modified electrodes. Thus, electrodes based on a Printex L6 carbon-silver hybrid nanomaterial [9], layer-by-layer combination of multi-walled carbon nanotubes (MWCNT), and electropolymerized quercetin [10] or gallic acid [11], reduced graphene oxide, and poly(glutamic acid) [12] were developed for the determination of gallic acid. An MWCNT-based sensor is described for the ellagic acid quantification [3]. Syringaldehyde is not usually considered for electroanalysis, while many electrochemical sensors have been developed for vanillin determination, including ones that are based on a combination of carbon nanomaterials and electropolymerized films [13,14].

Nevertheless, the simultaneous voltammetric detection of gallic and ellagic acids as well as aromatic aldehydes not often considered. Currently, there is just one example of gallic and ellagic acid quantification in a mixture using an MWCNT-modified carbon paste electrode [6]. The possibility of the simultaneous determination of syringaldehyde and vanillin has been demonstrated on a glassy carbon electrode (GCE) that was modified with carbon nanofibers and cationic surfactant cetylpyridinium bromide [7]. The analytical characteristics presented in Table 1 can be improved further. The disadvantage of both approaches is the absence of data on the selectivity of the electrode response to the target analytes. Furthermore, the use of this application in real samples was not realized.

Table 1. Analytical characteristics of simultaneous voltammetric determination of natural phenolic acids and aromatic aldehydes.

Electrode	Methods	Analyte	Detection Limit, µM	Linear Dynamic Range, µM	Ref.
MWCNT-carbon paste electrode	AdDPV [1]	Gallic acid Ellagic acid	– –	2.50–21.25 0.0075–0.019	[6]
Cetylpyridinium bromide/Carbon nanofibers/GCE	DPV [2]	Syringaldehyde Vanillin	0.53 1.17	2.5–30 5.0–40	[7]

[1] Adsorptive differential pulse voltammetry. [2] Differential pulse voltammetry.

Thus, novel voltammetric sensors that are based on carbon nanotubes and electropolymerized pyrocatechol violet (PCV) or *p*-aminobenzoic acid (ABA) have been developed for the simultaneous determination of phenolic antioxidants in cognac and brandy for the first time.

2. Materials and Methods

PCV and 99% ABA from Sigma-Aldrich (India and Germany, respectively) were used as monomers to obtain polymeric coverage. Their standard solutions (10 mM for PCV and 25 mM for ABA) were prepared in distilled water. Gallic (99%) and ellagic (95%) acids, vanillin (99%) from Sigma (Germany), and syringaldehyde (98%) from Aldrich (Germany) were used as standards. Their 10 mM (0.86 mM for ellagic acid) stock solutions in c.p. grade methanol (ethanol (rectificate) in the case of aromatic aldehydes) were prepared in 5.0 mL flasks. The exact dilution with the corresponding solvent was used for the preparation of less concentrated solutions.

Britton–Robinson buffer with a pH of 2.0 was used as a supporting electrolyte for the quantification of the natural phenolics. It was prepared from a mixture of 0.04 M boric acid, 0.04 M phosphoric acid, and 0.04 M acetic acid. The pH value was adjusted using a 0.2 M NaOH solution on a pH meter.

MWCNT (outer diameter 40–60 nm, inner diameter 5–10 nm and 0.5–500 µm length) and polyaminobenzene sulfonic acid functionalized single-walled carbon nanotubes (f-SWCNT) ($d \times l$ is 1.1 nm × 0.5–1.0 µm) were purchased from Aldrich and Sigma-Aldrich (Steinheim, Germany), respectively. MWCNT (0.5 mg mL^{-1} homogeneous suspension in 1% sodium dodecylsulfate (Panreac, Barcelona, Spain) was prepared by sonication for 30 min in an ultrasonic bath (WiseClean WUC-A03H (DAIHAN Scientific Co., Ltd., Wonju-si, Korea). A homogeneous 1.0 mg mL^{-1} suspension of f-SWCNT was obtained by ultrasonic dispersion for 30 min in dimethylformamide.

All reagents were chemical grade purity. Double distilled water was used for the measurements. The experiments were conducted at laboratory temperature (25 ± 2 °C).

Voltammetric measurements were conducted on the potentiostat/galvanostat Autolab PGSTAT 12 (Eco Chemie B.V., Utrecht, The Netherlands) with GPES software, version 4.9.005. Electrochemical impedance spectroscopy (EIS) was performed on the potentiostat/galvanostat Autolab PGSTAT 302N with the FRA 32M module (Eco Chemie B.V., Utrecht, The Netherlands) and NOVA 1.10.1.9 software. The 10 mL glassy electrochemical cell consisted of the working GCE with a 7.07 mm^2 geometric surface area (CH Instruments, Inc., Bee Cave, TX, USA), or a modified electrode, a silver-silver chloride saturated KCl reference electrode, and a platinum wire as the counter electrode was used.

An "Expert-001" pH meter (Econix-Expert Ltd., Moscow, Russian Federation) equipped with a glassy electrode was applied for the pH measurements.

Scanning electron microscopy (SEM) was conducted on a high-resolution field emission scanning electron microscope MerlinTM (Carl Zeiss, Oberkochen, Germany) at the accelerating voltage of 5 kV and at the emission current of 300 pA.

3. Results and Discussion

3.1. Polymer-Based Sensors Creation and Their Characteristics

A carbon nanotube-modified GCE was used as a substrate for polymeric coverage and electrochemical deposition. For this reason, f-SWCNT/GCE and MWCNT/GCE were obtained by drop-casting 2.0 or 4.0 μL of the f-SWCNT or MWCNT suspension, respectively, at the GCE surface. The application of the carbon nanotubes provided high surface area and conductivity for the electrode. Then, the electropolymerization of PCV and ABA in potentiodynamic mode was performed. Both PCV and ABA form conductive polymers, the formation of which were confirmed by the appearance of quasi-reversible redox peaks on the cyclic voltammograms. The redox currents are increased when the number of cycles is grown. The polymerization conditions of PCV and ABA were optimized based on the voltammetric response of the pairs of target analytes (gallic and ellagic acids on the polyPCV/f-SWCNT/GCE and syringaldehyde and vanillin on the polyABA/MWCNT/GCE). The peak potential separation for both types of analytes did not change, regardless of whether it was on the polymer-based sensors vs. the carbon nanotube-modified electrodes (Figure 1), while oxidation currents showed a statistically significant increase. Thus, the electropolymerization of PCV should be performed using a 50 μM monomer in 0.1 M H_2SO_4 with 10 cycles ranging from −0.2 to 1.1 V at 50 mV s^{-1}. The optimum conditions for ABA electropolymerization are 20-fold potential cycling from −0.5 to 2.0 V at 100 mV s^{-1} from a 100 μM monomer in Britton–Robinson buffer pH 2.0.

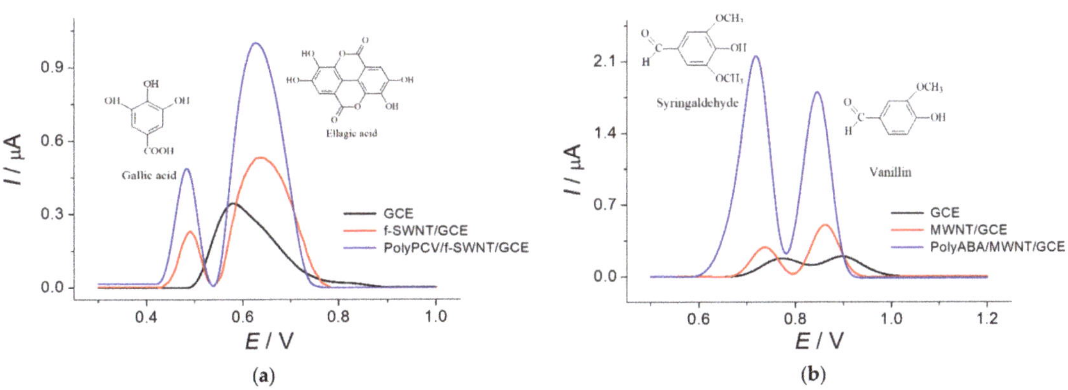

Figure 1. Background-subtracted differential pulse voltammograms of the 10 μM mixtures of (**a**) gallic and ellagic acids and (**b**) syringaldehyde and vanillin in Britton–Robinson buffer with a pH of 2.0. ΔE_{pulse} = 50 mV, t_{pulse} = 50 ms, υ = 10 mV s^{-1}.

The sensors that were developed have been characterized with scanning electron microscopy (SEM), cyclic voltammetry, chronoamperometry, and electrochemical impedance spectroscopy (EIS). According to SEM data, the polymeric coverages exhibit a porous structure, with the shape of the particles and their aggregates deposited on the surface of the carbon nanomaterials confirming that electropolymerization was successful (Figure 2).

(a)

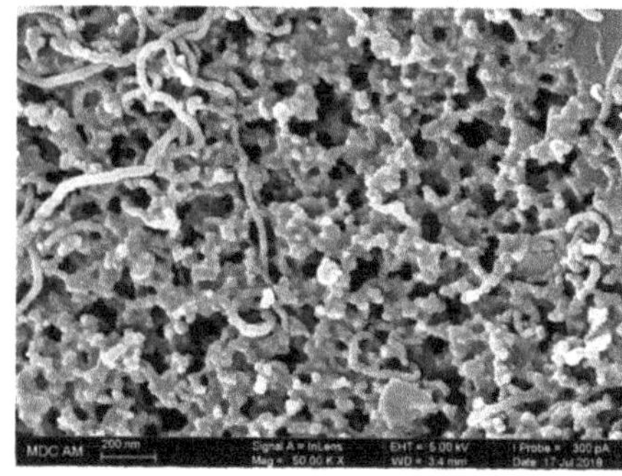

(b)

Figure 2. SEM images of (**a**) polyPCV/f-SWCNT/GCE and (**b**) polyABA/MWCNT/GCE.

The electroactive surface area of the electrodes was calculated on the basis of cyclic voltammetry and chronoamperometry (for GCE) in the presence of 1.0 mM $[Fe(CN)_6]^{4-}$. A statistically significant increase in the effective surface area (49.0 ± 0.2 mm^2 for polyPCV/f-SWCNT/GCE and 89 ± 4 mm^2 for polyABA/MWCNT/GCE vs. 38.9 ± 0.6 mm^2 for f-SWCNT/GCE, 75 ± 3 mm^2 for MWCNT/GCE, and 8.2 ± 0.3 mm^2 for GCE) explain the increase in the analyte oxidation currents well. The EIS data show that the polymer-based sensors demonstrate lower charge transfer resistance (26.0 ± 0.4 and 4.9 ± 0.3 kΩ for polyPCV/f-SWCNT/GCE and polyABA/MWCNT/GCE) compared to the GCE (68 ± 4 and 72 ± 3 kΩ) and the GCE modified with carbon nanotubes (19.2 ± 0.8 kΩ for f-SWCNT/GCE and 12.1 ± 0.9 kΩ for MWCNT/GCE). Thus, polymer-based sensors are characterized by a higher electron transfer rate. SEM and electrochemical method data confirm the effectivity of the developed modifiers.

3.2. Simultaneous Quantification of Natural Phenolic Antioxidants

For the quantification of the phenolics under consideration, the sensors that were created were used under differential pulse voltammetry conditions in Britton–Robinson buffer medium with a pH of 2.0 in order to provide the highest possible oxidation currents for the analytes (phenolic acids and aromatic aldehydes). The linear dynamic ranges of 0.75–10 and 10–100 μM for gallic acid and 0.75–7.5 and 7.5–100 μM for ellagic acid on the polyPCV-based sensor were obtained. The detection limits were equal to 0.12 μM for gallic acid and 0.11 μM for ellagic acid. The PolyABA-based sensor provided linear response in the ranges of 0.075–7.5 and 7.5–100 μM for syringaldehyde and 0.50–7.5 and 7.5–100 μM for vanillin when the detection limits were 0.018 and 0.19 μM, respectively. The detection limits were calculated as $3SD_a/b$, where SD_a is the standard deviation of the intercept of the calibration graph and where b is the slope of the calibration graph. The analytical characteristics that were obtained are improved vs. those of other modified electrodes [6,7]. The proven independent electrooxidation of the analyte pairs allows the application of calibration graphs for the equimolar mixtures, which is another advantage of the approaches developed here. The high accuracy of the sensors has been confirmed by recovery values (97.1–101%) that were determined using model solutions for the studied phenolics. The relative standard deviation values of 0.41–4.8% indicate that the sensor response has high reproducibility as long as a new electrode is prepared before each measurement.

The sensor selectivity in the presence of typical interferences and other natural phenolics was obtained, which is an important advantage. The 1000-fold excess of inorganic ions (K^+, Mg^{2+}, Ca^{2+}, NO_3^-, Cl^-, and SO_4^2), glucose, rhamnose, and sucrose as well as ascorbic acid (1000- and 100-fold excess in the case of phenolic acids and aromatic aldehydes, respectively) does not show an interference effect. It has been proven that there is an absence of an interference effect caused by syringaldehyde and vanillin (<2.5 and <10 μM, respectively) on the oxidation peaks of phenolic acids. On contrary, gallic and ellagic acid are the major potential causes of interference for the determination of the aromatic aldehydes that are under consideration. A 10-fold excess of gallic acid and <1.0 μM of ellagic acid do not interfere with the determination of vanillin and syringaldehyde. The interference effect of ellagic acid caused by its high concentration in cognac and brandy can be excluded via sample dilution. Thus, the sensors that have been developed here can be applied for the analysis of aged distilled beverages.

3.3. Real Samples Analysis

The sensors that were developed were successfully tested on cognac and brandy samples. The phenolic acids and aromatic aldehydes show resolved oxidation peaks on the differential pulse voltammograms for cognac and brandy that is confirmed by the standard addition method (Figure 3). Due to the low concentration of vanillin, the sample volume varied for the quantification of syringaldehyde and vanillin (40 and 500 μL, respectively). The recovery rate of 98.9–102% indicates the absence of matrix effects in these determinations.

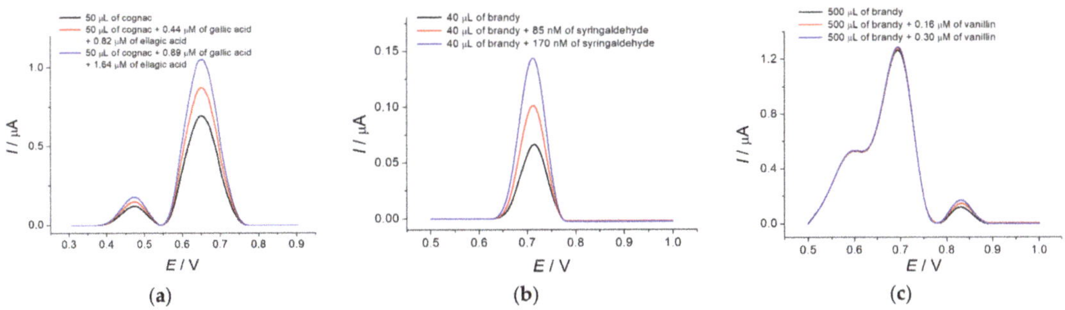

Figure 3. Background-subtracted differential pulse voltammograms of (**a**) 50 μL of cognac with the addition of phenolic acids at the polyPCV-based sensor; (**b**) 40 μL and (**c**) 500 μL of brandy with the addition of aromatic aldehydes at the polyABA-based sensor. Supporting electrolyte was Britton–Robinson buffer with a pH of 2.0. ΔE_{pulse} = 75 mV (50 mV for aromatic aldehydes), t_{pulse} = 25 ms, υ = 10 mV s^{-1}.

The quantification of natural phenolic antioxidants in cognac and brandy has been performed and compared using an independent chromatographic determination (Figure 4). The results that were obtained agree with each other well. *t*-Test data (0.0100–2.19) are less than the critical value of 2.45, which indicates the absence of systematic errors in the determination. Similarly, the *F*-test results (1.00–17.36) achieved the less than critical value of 19.25, indicating that the methods that were used in this study demonstrate uniform precision.

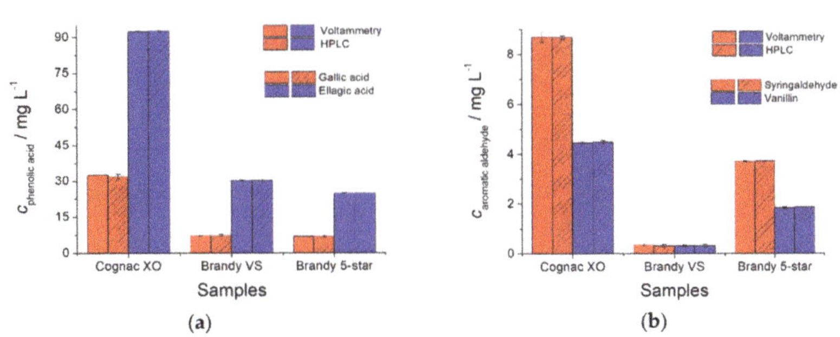

Figure 4. Quantification of (**a**) gallic and ellagic acids and (**b**) syringaldehyde and vanillin in cognac and brandy (n = 5; p = 0.95).

Thus, the novel highly sensitive and selective voltammetric sensors that were developed in this study for the simultaneous determination of the structurally related phenolic antioxidants in cognac and brandy are characterized by the simplicity of their fabrication as well as their reliability and cost-efficiency, and they can be applied for routine analysis as an alternative to chromatographic methods.

Author Contributions: Conceptualization, G.Z.; methodology, G.Z. and E.G.; investigation, E.G.; writing—original draft preparation, G.Z.; writing—review and editing, G.Z.; visualization, G.Z. and E.G. All authors have read and agreed to the published version of the manuscript.

Funding: This research received no external funding.

Institutional Review Board Statement: Not applicable.

Informed Consent Statement: Not applicable.

Data Availability Statement: The data presented in this study are available upon request from the corresponding author.

Acknowledgments: The authors thank Rustam Davletshin (Department of High Molecular and Organoelement Compounds, Kazan Federal University) for the chromatographic measurements and Aleksei Rogov (Laboratory of the Scanning Electron Microscopy, Interdisciplinary Center for Analytical Microscopy, Kazan Federal University) for the SEM measurements.

Conflicts of Interest: The authors declare no conflict of interest.

References

1. Ziyatdinova, G.; Salikhova, I.; Skorobogatova, N.; Chibisova, M.; Budnikov, H. New electrochemistry based approaches to brandy quality evaluation using antioxidant parameters. *Food Anal. Methods* **2015**, *8*, 1794–1803. [CrossRef]
2. Goldberg, D.M.; Hoffman, B.; Yang, J.; Soleas, G.J. Phenolic constituents, furans, and total antioxidant status of distilled spirits. *J. Agric. Food Chem.* **1999**, *47*, 3978–3985. [CrossRef] [PubMed]
3. Ziyatdinova, G.; Salikhova, I.; Budnikov, H. Chronoamperometric estimation of cognac and brandy antioxidant capacity using MWNT modified glassy carbon electrode. *Talanta* **2014**, *125*, 378–384. [CrossRef] [PubMed]
4. Canas, S.; Belchior, A.P.; Spranger, M.I.; Bruno-de-Sousa, R. High-performance liquid chromatography method for analysis of phenolic acids, phenolic aldehydes, and furanic derivatives in brandies. Development and validation. *J. Sep. Sci.* **2003**, *26*, 496–502. [CrossRef]
5. Panossian, A.; Mamikonyan, G.; Torosyan, M.; Gabrielyan, E.; Mkhitaryan, S. Analysis of aromatic aldehydes in brandy and wine by high-performance capillary electrophoresis. *Anal. Chem.* **2001**, *73*, 4379–4383. [CrossRef] [PubMed]
6. Ghoreishi, S.M.; Behpour, M.; Khayatkashani, M.; Motaghedifard, M.H. Simultaneous determination of ellagic and gallic acid in *Punica granatum*, *Myrtus communis* and Itriphal formulation by an electrochemical sensor based on a carbon paste electrode modified with multi-walled carbon nanotubes. *Anal. Methods* **2011**, *3*, 636–645. [CrossRef] [PubMed]
7. Ziyatdinova, G.K.; Kozlova, E.V.; Ziganshina, E.R.; Budnikov, H.C. Application of electrode modified with carbon nanofibers and cationic surfactant for simultaneous voltammetric determination of syringaldehyde and vanillin. *Butlerov Commun.* **2015**, *42*, 132–137.
8. Ziyatdinova, G.; Guss, E.; Morozova, E.; Budnikov, H.; Davletshin, R.; Vorobev, V.; Osin, Y. Simultaneous voltammetric determination of gallic and ellagic acids in cognac and brandy using electrode modified with functionalized SWNT and poly(pyrocatechol violet). *Food Anal. Methods* **2019**, *12*, 2250–2261. [CrossRef]
9. Raymundo-Pereira, P.A.; Campos, A.M.; Prado, T.M.; Furini, L.N.; Boas, N.V.; Calegaro, M.L.; Machado, S.A.S. Synergy between Printex nano-carbons and silver nanoparticles for sensitive estimation of antioxidant activity. *Anal. Chim. Acta* **2016**, *926*, 88–98. [CrossRef] [PubMed]
10. Ziyatdinova, G.; Kozlova, E.; Budnikov, H. Polyquercetin/MWNT-modified electrode for the determination of natural phenolic antioxidants. *Electroanalysis* **2017**, *29*, 2610–2619. [CrossRef]
11. Abdel-Hamid, R.; Newair, E.F. Voltammetric determination of polyphenolic content in pomegranate juice using a poly(gallic acid)/multiwalled carbon nanotube modified electrode. *Beilstein J. Nanotechnol.* **2016**, *7*, 1104–1112. [CrossRef] [PubMed]
12. Feminus, J.J.; Manikandan, R.; Narayanan, S.S.; Deepa, P.N. Determination of gallic acid using poly(glutamic acid): Graphene modified electrode. *J. Chem. Sci.* **2019**, *131*, 11. [CrossRef]

13. Wu, W.; Wang, H.; Yang, L.; Zhao, F.; Zeng, B. Sensitively voltammetric determination of vanillin with a molecularly imprinted ionic liquid polymer-carboxyl single-walled carbon nanotubes composite electrode. *Int. J. Electrochem. Sci.* **2016**, *11*, 6009–6022. [CrossRef]
14. Deng, P.; Xu, Z.; Zeng, R.; Ding, C. Electrochemical behavior and voltammetric determination of vanillin based on an acetylene black paste electrode modified with graphene–polyvinylpyrrolidone composite film. *Food Chem.* **2015**, *180*, 156–163. [CrossRef] [PubMed]

chemistry proceedings

MDPI

Abstract

Carbon Dots as a Fluorescence pH Nanosensor by Application of an Active Surface Preservation Strategy †

Ana Carolina P. Afonso * and Luís Pinto da Silva

Chemistry Research Unit (CIQUP), Faculty of Sciences, University of Porto, R. Campo Alegre 687, 4169-007 Porto, Portugal; luis.silva@fc.up.pt

* Correspondence: up201207846@edu.fc.up.pt

† Presented at the 1st International Electronic Conference on Chemical Sensors and Analytical Chemistry, 1–15 July 2021; Available online: https://csac2021.sciforum.net/.

Abstract: In the environmental, industrial, and biomedical fields, pH monitorization is of the upmost importance. However, the most used type of pH sensors, glass pH-electrodes, still present limitations in their application in low volume samples and in cellular pH sensing, due to their size and invasive nature. Fluorescence-based sensors present a solution to such issues, providing a non-invasive solution to pH sensing. Herein, we report the rational development of carbon dots (CDs) as a pH nanosensor via an active surface preservation (ASP) method. Carbon dots (CDs) are carbon-based fluorescent nanoparticles with valuable properties such as high aqueous solubility, low cost and good biocompatibility, with remarkable fluorescence performance, been increasingly used as fluorescent nanosensors. Namely, these nanomaterials present advantages over molecular probes in terms of (photo)stability and water solubility, among others. By employing ASP strategies, the CDs will be prepared by using precursors with known active functional features. The ASP method allows the nanoparticles to retain the structural features of precursors, thus retaining their properties, without the need for costly and time-consuming post-synthesis functionalization procedures. In this work, we intend to provide a proof-of-concept of this type of strategy by utilizing the known pH-sensitivity of fluorescein to provide a pH-based response to CDs. The resulting CDs presented reversible response by fluorescence enhancement in the range of pH from 4 to 12. The nanoparticles exhibited excellent photostability, in different pH solutions. The studied CDs were also unaffected by, either variation of ionic strength or the presence of interferent species, while being compatible with human cancer cells. Finally, CDs were able to determine the pH of real samples. Thus, a selective pH fluorescent CDs-based nanosensor was developed.

Keywords: carbon dots; pH sensing; active surface preservation; nanosensor; fluorescence; biocompatibility

Citation: Afonso, A.C.P.; Pinto da Silva, L. Carbon Dots as a Fluorescence pH Nanosensor by Application of an Active Surface Preservation Strategy. *Chem. Proc.* **2021**, *5*, 2. https://doi.org/10.3390/CSAC2021-10638

Academic Editor: Elena Benito-Peña

Published: 7 July 2021

Funding: Acknowledgment to projects PTDC/QUI-QFI/2870/2020, UIDB/00081/2020 and UIDB/05748/2020.

chemistry proceedings

MDPI

Proceeding Paper

Simultaneous Quantification of Four Principal NSAIDs through Voltammetry and Artificial Neural Networks Using a Modified Carbon Paste Electrode in Pharmaceutical Samples †

Guadalupe Yoselin Aguilar-Lira [1], Prisciliano Hernandez [2], Giaan Arturo Álvarez-Romero [1] and Juan Manuel Gutiérrez [3,*]

1 Laboratorio de Química Analítica, Área Académica de Química, Instituto de Ciencias Básicas e Ingeniería, Universidad Autónoma del Estado de Hidalgo, Carretera Pachuca-Tulancingo km. 4.5, Mineral de la Reforma 42076, Mexico; guadalupe_aguilar@uaeh.edu.mx (G.Y.A.-L.); giaan@uaeh.edu.mx (G.A.Á.-R.)

2 Laboratorio de Ingeniería y Energía, Área de Energía, Universidad Politécnica de Francisco I. Madero, Domicilio Conocido, Tepatepec 42640, Mexico; phernandez@upfim.edu.mx

3 Sección de Bioelectrónica, Departamento de Ingeniería Eléctrica, Centro de Investigación y de Estudios Avanzados del I.P.N., Col San Pedro Zacatenco, Av. IPN 2508, Ciudad de Mexico 07360, Mexico

* Correspondence: mgutierrez@cinvestav.mx

† Presented at the 1st International Electronic Conference on Chemical Sensors and Analytical Chemistry, 1–15 July 2021; Available online: https://csac2021.sciforum.net/.

Abstract: This work describes the development of a novel and low-cost methodology for the simultaneous quantification of four main nonsteroidal anti-inflammatory drugs (NSAIDs) in pharmaceutical samples using differential pulse voltammetry coupled with an artificial neural network model (ANN). The working electrode used as a detector was a carbon paste electrode (CPE) modified with multi-wall carbon nanotubes (MWCNT-CPE). The specific voltammetric determination of the drugs was performed by cyclic voltammetry (CV). Some characteristic anodic peaks were found at potentials of 0.446, 0.629, 0.883 V related to paracetamol, diclofenac, and aspirin. For naproxen, two anodic peaks were found at 0.888 and 1.14 V and for ibuprofen, an anodic peak was not observed at an optimum pH of 10 in 0.1 mol L^{-1} Britton–Robinson buffer. Since these drug's oxidation process turned out to be irreversible and diffusion-controlled, drug quantification was carried out by differential pulse voltammetry (DPV). The Box Behnken design technique's optimal parameters were: step potential of 5.85 mV, the amplitude of 50 mV, period of 750 ms, and a pulse width of 50 ms. A data pretreatment was carried out using the Discrete Wavelet Transform using the db4 wavelet at the fourth decomposition level applied to the voltammetric records obtained. An ANN was built to interpret the obtained approximation coefficients of voltammograms generated at different drug concentrations to calibrate the system. The ANN model's architecture is based on a Multilayer Perceptron Network (MLP) that employed a Bayesian regularization training algorithm. The trained MLP achieves significant R values for the test data to simultaneous quantification of the four drugs in the presence of aspirin.

Keywords: carbon paste electrode; voltammetry; artificial neural network; quantification; nonsteroidal anti-inflammatory

Citation: Aguilar-Lira, G.Y.; Hernandez, P.; Álvarez-Romero, G.A.; Gutiérrez, J.M. Simultaneous Quantification of Four Principal NSAIDs through Voltammetry and Artificial Neural Networks Using a Modified Carbon Paste Electrode in Pharmaceutical Samples. *Chem. Proc.* **2021**, *5*, 3. https://doi.org/10.3390/CSAC2021-10450

Academic Editor: Frederic Melin

Published: 30 June 2021

1. Introduction

Nonsteroidal anti-inflammatory analgesics NSAIDs are important drugs worldwide due to their low cost and easy accessibility. Most of these drugs can be purchased without a prescription; they are widely used to relieve pain, reduce inflammation and reduce high temperatures. Standing out from this large group of NSAIDs are paracetamol, diclofenac, naproxen, aspirin, and ibuprofen, which are the most frequently used. The pharmacological action of these drugs is that they block the enzyme cyclooxygenase and break down the prostaglandins produced by the cells of the body that increase inflammation, pain, and fever. Although NSAIDs are commonly used, they are not suitable for everyone and can sometimes cause adverse side effects if their use is constant: peptic ulceration, digestive disorders, temporary deafness. Recent studies mention that they may be related to heart attacks [1–3]. Due to the high demand for NSAIDs in pharmaceutical samples, many analytical methods have been proposed for their quantification, the most common being liquid chromatography (HPLC) [4]. This method has some disadvantages, such as the need for sample preparation by chemical reaction or extraction. Some cases include previous derivatization, long analysis times, and a high cost associated with the use and maintenance of the equipment.

An alternative to traditional analysis methods in areas such as the food industry, pharmaceuticals, and environmental monitoring is known as Electronic Tongues (ETs) [5,6]. These systems combine electrochemical techniques (e.g., potentiometry, voltammetry, and impedance spectroscopy) with sophisticated multivariate analysis tools to classify or quantify samples [7]. Their main advantage compared to traditional methods is that they allow quick and low-cost measurements, avoiding the pretreatment of samples in most cases. Although ETs using potentiometric and voltammetric techniques have been reported in the literature [8,9], voltammetric methods are usually the most widespread due to their advantages such as short analysis time and high sensitivity [10,11]. In addition to this, data processing techniques based on artificial neural networks (ANNs), principal component analysis (PCA), and partial least squares (PLS) are popular for decoding the acquired voltammograms of aqueous solutions containing mixtures of different chemical species giving results favorable in the quantification of these species [12,13]. In this work, a methodology based on voltammetric methods is proposed together with ANNs as a modeling and calibration tool to quantify NSAIDs simultaneously.

2. Methods and Materials

2.1. Instrumentation and Reagents

The chemical reagents used are of an analytical grade. All solutions used were prepared using high purity deionized water (18.2 MΩ cm). The experiments were carried out using a potentiostat Autolab PGSTAT302N (Metrohm, Utrecht, The Netherlands) connected to a computer. A three-electrode system was used consisting of a reference electrode with saturated Ag/AgCl (BASi), a graphite bar as the auxiliary electrode, and a carbon paste electrode (CPE) modified with multi-wall carbon nanotubes (MWCNT-CPE) OD × L 6–9 nm × 5 µm> 95% (Sigma-Aldrich, CAS 308068-56-6, Mexico) as working electrode. The pH measurements were performed on a Corning pH/Ion meter 450 digital pH meter. A Sartorius CPA224S brand analytical balance was used. All the potentials referred to in this work are referred to the Ag/AgCl electrode. Standard solutions of diclofenac sodium, naproxen sodium, paracetamol, aspirin, and ibuprofen (Sigma-Aldrich, México) were flushed with high purity nitrogen. A Britton–Robinson (BR) 0.1 mol·L^{-1} buffer solution was used in a range of pH 7–11. Buffer BR was prepared by mixing appropriate volumes of acids (phosphoric acid, boric acid, and acetic acid) and adjusted with concentrated NaOH to the desired pH.

2.2. Electrochemical Characterization

2.2.1. Electrode Preparation

The paste mixture for the proposed working electrode consisted of a 3:2 ratio for mineral oil and multi-walled carbon nanotube graphite powder (MWCNT). The graphite mix is made up of 30% graphite powder and 10% MWCNT. The paste obtained is placed with a spatula in a 1 mL syringe tube (30 mm long by 6 mm wide) and compacted with the syringe's plunger, placing it on a flat surface until excess air is eliminated. At one end of the syringe with the paste, the electrical contact (copper wire) is placed. Finally, the working electrode is built.

2.2.2. Electrochemical Analysis of the NSAIDs in the Proposed Working Electrode

Cyclic voltammetry (CV) is performed for the supporting electrolyte (0.1 $molL^{-1}$ BR buffer at pH 7) and the NSAIDs-BR system (5×10^{-4} $mol{\cdot}L^{-1}$ for each drug), starting at the zero-current potential, in anodic direction and cycling in a potential window from 0 to 1.3 V considering a scan rate of 0.1 Vs^{-1}. At different scan rates, anodic and cathodic CV peaks were analyzed to determine the mechanism that controls the oxidative processes. Moreover, a pH study for the NSAIDs-BR system was performed to choose the maximum anodic current intensity for analytical quantification.

Quantification of the drug is carried out by DPV. The optimization of the parameters of the technique is carried out, with the Box Behnken (BBD) four-factor design, step potential "X_1" (V), interval time "X_2" (s), modulation time "X_3" (s) y modulation amplitude "X_4" (V) and three levels for each factor so that the highest intensity of the anodic current is obtained. The design matrix considers 27 experimental units at random that include the three replicas of the central point. Using a polynomial regression, the response variable was predicted as a function of the independent variables and their interactions. The prediction of the model is described in Equation (1).

$$Y = \beta_0 + \sum_{i=1}^{K} \beta_i X_i + \sum_{i=1}^{k} \beta_{ii} X_i^2 + \sum_{i=1}^{k} \sum_{j=1}^{k} \beta_{ij} X_i X_j + \varepsilon, \tag{1}$$

where Y is the response variable (maximum anode current intensity), X_i and X_j are coded independent variables, and β_0, β_i, β_{ii}, and β_{ij} are coefficients of intercept, linear, quadratic, and interaction terms, respectively. k is the number of independent variables (k = 4 in this study), ε is an experimental error [14]. Determination and regression coefficients were estimated using Minitab® Statistical software version 18.

2.3. *Quantification of NSAIDs by ANN*

2.3.1. Data Processing

Having the optimal parameters of the DPV, the DPV's are carried out at different concentrations of the NSAIDs (ranging from 5×10^{-7} to 7×10^{-5} $mol{\cdot}L^{-1}$). A matrix of peak intensities of dimensions [189 × 27] (intensities x number of samples) was built with the recorded voltammogram records. The pretreatment of the data was carried out using the Discrete Wavelet Transform using the 4th level wavelet decomposition of a Daubechies function (db4). In this pretreatment, only approximation coefficients were chosen considering the degree of similarity between the original voltammogram and the one recovered with these coefficients [13]. Thus, the final input matrix to feed the ANN model has a dimension of [18 × 27] (approximation coefficients x number of samples). The ANN calibration model is based on a Multilayer Perceptron Network (MLP). The MLP input layer was established considering the number of approximation wavelet coefficients. In contrast, the output layer was defined using one neuron for each of the analytes to be quantified since it is associated with the matrix of concentrations of dimension [4 × 27] (i.e., paracetamol, diclofenac, naproxen, and ibuprofen), aspirin was not quantified in the model as it was considered only as an interferer. The hidden layers were established through a trial-and-error process, modifying the number of neurons in the layers until an appropriate number of neurons were found that favored obtaining a satisfactory linear regression coefficient. In this way, the final MLP model was 18 × 10 × 8 × 4 (18 input neurons, 10 neurons in the first hidden layer, 8 neurons in the second hidden layer and 4 output neurons).

2.3.2. ANN Modeling

The described data set of 27 samples, were selected for the training set. The testing set was conformed using an external set of 10 additional samples randomly generated within the concentration range described above. All data sets were normalized in the interval of [−1, 1] to favor the training process. The activation functions established were: purelin for the input layer, tansing for the two hidden layers, and purelin for the output layer. In the same way, the chosen training algorithm was Bayesian regularization, with a training error set at a value of 0.001, together with a learning rate of 0.01. The MLP models were programmed on the MATLAB® R2021a (MathWorks, Natick, MA, USA) platform using the Deep Learning and Wavelet Toolboxes.

3. Results and Discussion

3.1. Electrochemical Characterization

The electrochemical study of the NSAIDs was carried out by CV in a solution of 4×10^{-4} mol·L^{-1} of the drug standards, in BR buffer at pH 7 (Figure 1a). In the window from 0 to 1.3 V in the first sweep in the anodic direction, the oxidation peak corresponding to paracetamol, diclofenac, at a potential of 0.446 and 0.629, respectively, was observed. Naproxen presented two oxidation peaks at a potential of 0.888 and 1.14 V. In the case of ibuprofen, no anodic signal was observed, but the baseline was modified compared to the blank, so we believe that if it is carried out by the oxidation of the drug. Reversing the sweep in the cathodic direction, only one reduction peak was observed with a maximum peak potential at 0.263 V, which corresponds to the reduction of paracetamol. Moreover, a CV was performed in a mixture of the five NSAIDS and the BR Buffer (blank) using the proposed working electrode (Figure 1b), under the same conditions of the systems shown in Figure 1a. In Figure 1b, only three oxidation peaks can be observed in the anodic sense, since some signals overlap in the drugs, and it should also be noted that in the cathodic sweep the reduction signal of paracetamol is not observed, and this could be because the product that was reduced reacted chemically with some oxidation products of the other drugs and the product is no longer electroactive in reduction.

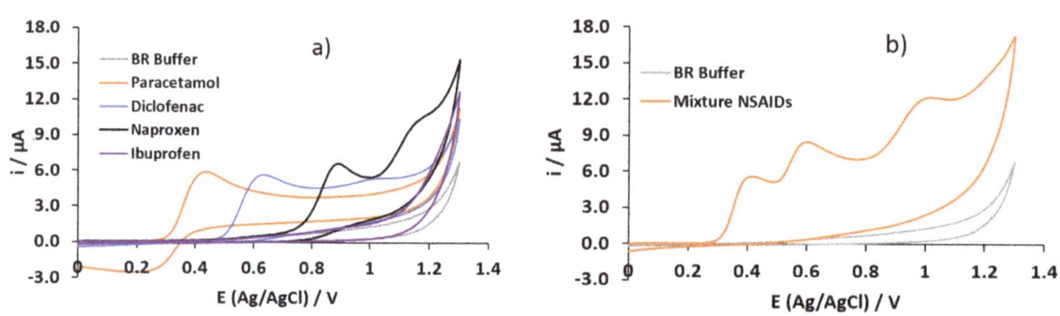

Figure 1. CVs obtained for the systems containing the NSAIDs and the supporting electrolyte, using the proposed working electrode and in presence of aspirin. Potential window from 0 to 1.3 V and at a scan rate of 100 mVs^{-1}, (**a**) individual for each NSAIDs, (**b**) mixture of the four NSAIDs.

Considering the anodic wave corresponding to the oxidation of the NSAIDs for the mixture of the drugs. A pH study was carried out by CV, in a 0.1 mol L^{-1} BR buffer with a concentration of 5×10^{-4} mol L^{-1} for each drug, in Figure 2 the anodic sweep is shown. It can be observed that the highest current intensity is obtained at pH 10 for the mixture of drugs; it can also be observed that as the pH increases the anodic peak potentials of the drugs shift to lower values. Different CV scan rates (10 to 300 mV·s^{-1}) were studied, and the maximum anodic peak current was plotted vs. the square root of the scan rate for each NSAID. A correlation coefficient greater than 0.99 was obtained after the proper statistical analysis, which suggests that the diffusion of the electroactive species to the surface of the electrode governs the oxidation processes.

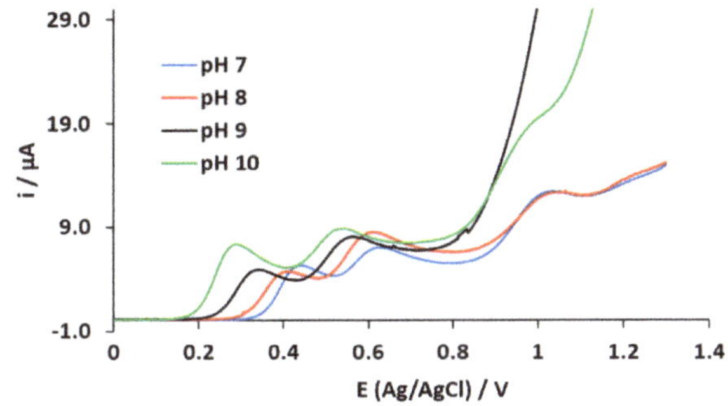

Figure 2. CVs obtained for the NSAIDs mixture at different pHs (range of 7–10) in a 0.1 mol·L^{-1} BR buffer, using a potential window of 0 to 1.3 V and a scan rate of 100 mV·s^{-1}, and a concentration of 4×10^{-4} mol·L^{-1} for the NSAIDs.

A BBD with three levels was used for the optimization of the four variables related to the DPV technique to maximize the anodic current peak of paracetamol (this NSAID has the highest peak current intensity). Twenty-seven experiments were carried out, generating the corresponding voltammograms of NSAIDs using the MWCNT-CPE at pH 10. The parameters of the DPV technique were chosen considering the capabilities of the potentiostat used. The proposed second-order model regression that correlates the current response and the DPV factors is shown in Equation (2).

$$Y = 1.747 + 0.22X_1 + 0.352X_2 - 0.935X_3 + 1.082X_4 - 0.476X_2^2 - 0.431X_2^2 - 0.308X_3^2 - 0.012X_4^2 \\ -0.309X_1X_2 + 0.211X_1X_3 + 0.209X_1X_4 + 0.297X_2X_3 + 0.66X_2X_4 + 1.507X_3X_4. \quad (2)$$

Table 1 shows the theoretical response (Y) after the optimization of the DPV parameters, obtained by using the Response Optimizer function in the Minitab® V.18 software. The maximum anodic current for paracetamol was determined as 5.24 μA under optimal conditions.

Table 1. Optimal DPV parameters found with the Box Behnken design.

X_1 (V)	X_2 (s)	X_3 (s)	X_4 (V)	Y (μA)
0.00585	0.75	0.05	0.05	5.24

3.2. Quantification of NSAIDs Using ANN

Using the optimal parameters of the DPV to analyze the 27 samples considering different concentrations of the NSAIDs, and they were performed using a 3^{5-2} fractional facto-rial design. The trained MLP was used to determine the performance in the quantification task of drugs and the relationship between the concentrations obtained and those expected was evaluated, both for the training and the test phases. In this sense, the linear regression obtained from the comparison was a measure of the model's goodness. Given ideal conditions, the line must have a slope equal to 1 and its intersection equal to 0. The comparative graphs between the real concentrations of paracetamol, diclofenac, naproxen, and ibuprofen and those predicted with the MLP model for the training and test data set (Figures 3 and 4, respectively). The high level of linearity allows a linear regression coefficient of the data obtained very close to one (R = 0.98) for paracetamol and diclofenac, while for the naproxen and ibuprofen, the correlation value was 0.87 and 0.78 respectively; aspirin was present as an interfering agent in the mixture.

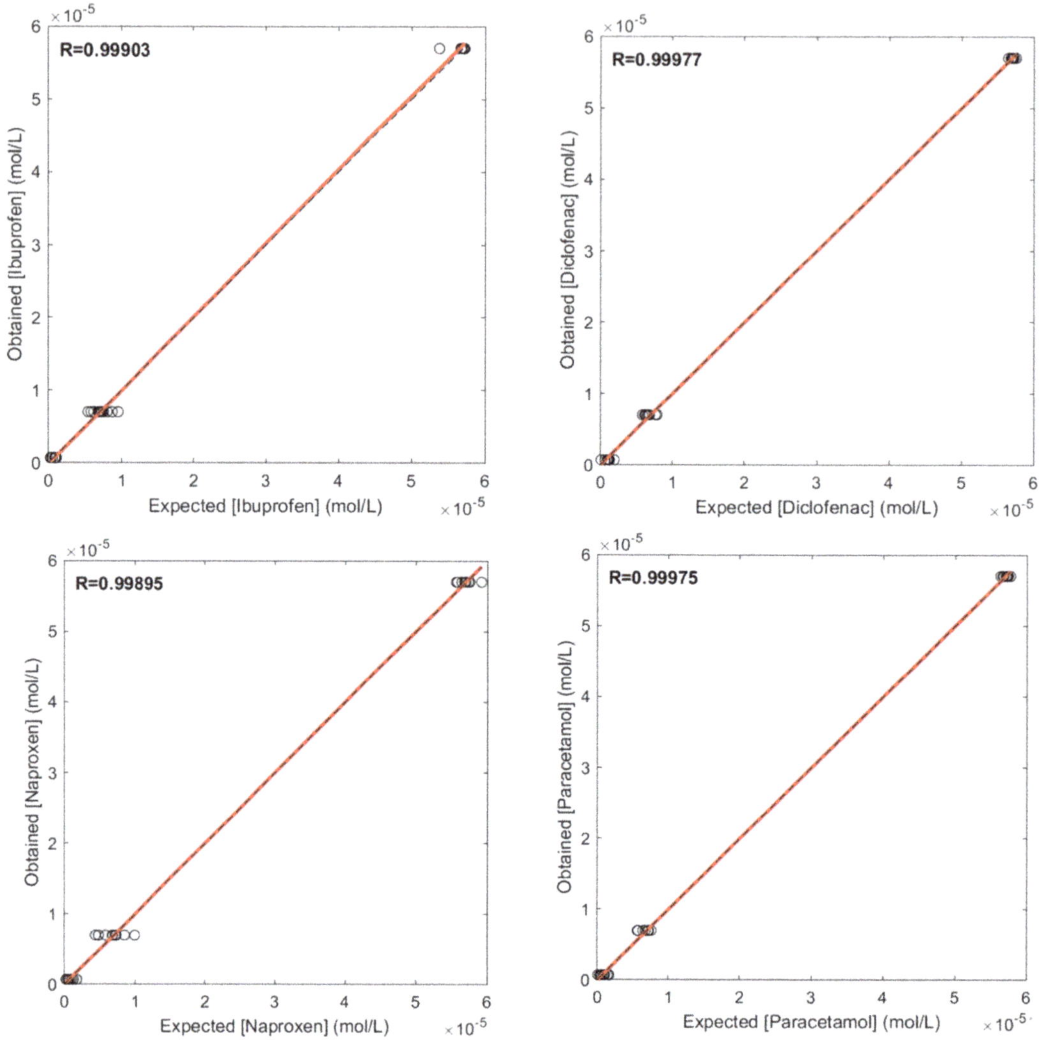

Figure 3. Comparison between the expected NSAIDs concentration and those obtained after MLP training phase.

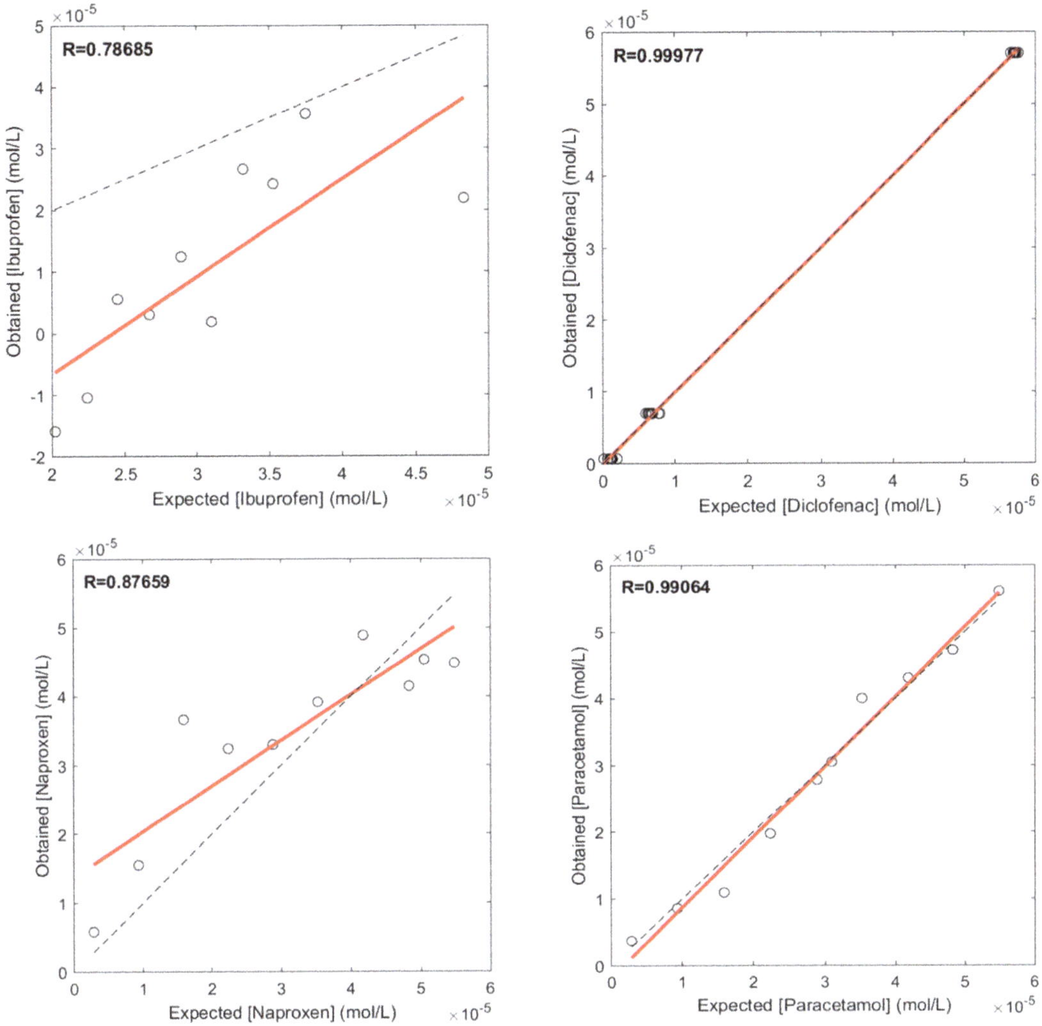

Figure 4. Comparison between the expected NSAIDs concentration and those obtained during MLP test phase.

4. Conclusions

In this work, a potential tool for voltammetric determination is presented. A combination of DPV and DWT-ANN allowed us to obtain satisfactory results for quantifying paracetamol, diclofenac, naproxen, and ibuprofen in the presence of aspirin. The use of DWT is helpful to compact voltammograms, preserving the analytical information of the original records. Multivariable models created with ANN correctly describe the complexity in voltammograms caused by overlapping peaks without the need for a pretreatment step on the samples. Finally, carbon paste electrodes with nanotubes are low-cost and easy-to-make devices that allow us to determine the drugs in the order of microgram per liter.

Supplementary Materials: The following are available online at https://www.mdpi.com/article/10.3390/CSAC2021-10450/s1.

Institutional Review Board Statement: Not applicable.

Informed Consent Statement: Not applicable.

Data Availability Statement: Not applicable.

Acknowledgments: The authors thank the Mexican National Council of Science and Technology for financial support through the National System of Researchers (SNI) and the distinction of their memberships.

Conflicts of Interest: The authors declare no conflict of interest.

References

1. Villoria, C.M. *Tratamiento de Dolor Crónico, Diagnóstico, Clínica y Tratamiento*; Arán Ediciones, S.L.: Madrid, Spain, 2007.
2. Aguilar-Lira, G.Y.; Rojas-Hernández, A.; Rodriguez, J.A.; Páez-Hernández, M.E.; Álvarez-Romero, G.A. Optimized quantification of naproxen based on DPV and a multiwalled MWCNT-carbon paste electrode. *J. Electrochem. Soc.* **2020**, *167*, 166510. [CrossRef]
3. Sondergaard, K.B.; Weeke, P.; Wissenberg, M.; Olsen, A.-M.S.; Fosbol, E.L.; Lippert, F.K.; Torp-Pedersen, C.; Gislason, G.H.; Folke, F. Non-steroidal anti-inflammatory drug use is associated with increased risk of out-of-hospital cardiac arrest: A nationwide case–time–control study. *Eur. Heart J. Cardiovasc. Pharmacother.* **2017**, *3*, 100–107. [CrossRef] [PubMed]
4. Aguilar-Arteaga, K.; Rodriguez, J.; Miranda, J.M.; Medina, J.; Barrado, E. Determination of non-steroidal anti-inflammatory drugs in wastewaters by magnetic matrix solid phase dispersion-HPLC. *Talanta* **2010**, *80*, 1152. [CrossRef] [PubMed]
5. Alcañiz, M.; Vivancos, J.-L.; Masot, R.; Ibañez, J.; Raga, M.; Soto, J.; Martínez-Máñez, R. Design of an electronic system and its application to electronic tongues using variable amplitude pulse voltammetry and impedance spectroscopy. *J. Food Eng.* **2012**, *111*, 122–128. [CrossRef]
6. Cosio, M.S.; Ballabio, D.; Benedetti, S.; Gigliotti, C. Evaluation of different storage conditions of extra virgin olive oils with an innovative recognition tool built by means of electronic nose and electronic tongue. *Food Chem.* **2007**, *101*, 485–491. [CrossRef]
7. Vlasov, Y.G.; Legin, A.V.; Rudnitskaya, A.M.; D'Amico, A.; Natale, C.D. Sens. *Actuators B* **2000**, *65*, 235–236. [CrossRef]
8. Martínez-Máñez, R.; Soto, J.; Garcia-Breijo, E.; Gil, L.; Ibáñez, J.; Llobet, E. An "electronic tongue" design for the qualitative analysis of natural waters. *Sens. Actuators B* **2005**, *104*, 302–307. [CrossRef]
9. Lvova, L.; Martinelli, E.; Mazzone, E.; Pede, A.; Paolesse, R.; Natale, C.D.; D'Amico, A. Electronic tongue based on an array of metallic potentiometric sensors. *Talanta* **2006**, *70*, 833–839. [CrossRef] [PubMed]
10. Bessant, C.; Saini, S. Simultaneous Determination of Ethanol, Fructose, and Glucose at an Unmodified Platinum Electrode Using Artificial Neural Networks. *Anal. Chem.* **1999**, *71*, 2806–28013. [CrossRef]
11. Martina, V.; Ionescu, K.; Pigani, L.; Terzi, F.; Ulrici, A.; Zanardi, C.; Seeber, R. Development of an electronic tongue based on a PEDOT-modified voltammetric sensor. *Anal. Bioanal. Chem.* **2007**, *387*, 2101–2110. [CrossRef] [PubMed]
12. Garcia, B.; Olguin, P.; Masot, P.; Alcañiz, F.; Martínez-Máñeza, R.; Soto, C. TNT detection using a voltammetric electronic tongue based on neural networks. *Sens. Actuator A Phys.* **2013**, *192*, 1–8. [CrossRef]
13. Aguilar-Lira, G.Y.; Gutiérrez-Salgado, J.M.; Rojas-Hernández, A.; Rodríguez-Ávila, J.A.; Páez-Hernández, M.E.; Álvarez-Romero, G.A. Artificial neural network for the voltamperometric quantification of diclofenac in presence of other nonsteroidal anti-inflammatory drugs and some commercial excipients. *J. Electroanal. Chem.* **2017**, *801*, 527–535. [CrossRef]
14. Montgomery, D.C. *Design and Analysis of Experiments*, 8th ed.; Wiley: Hoboken, NJ, USA, 2012.

chemistry proceedings

MDPI

Proceeding Paper

An Inkjet-Printed Amperometric H_2S Sensor for Environmental Applications †

Franc Paré [1,2], Rebeca Castro [3], Xavier Guimera [3], Gemma Gabriel [4,5] and Mireia Baeza [1,2,*]

1 Department of Chemistry, Faculty of Science, Edifici C-Nord, Universitat Autònoma de Barcelona, Carrer dels Til·lers, Bellaterra, 08193 Barcelona, Spain; franc.pare@uab.cat
2 GENOCOV Research Group, Universitat Autònoma de Barcelona, Bellaterra, 08193 Barcelona, Spain
3 Department of Mining, Industrial and ICT Engineering, Universitat Politècnica de Catalunya, Avinguda de les Bases de Manresa 61-73, 08240 Manresa, Spain; rebeca.ignacia.castro@upc.edu (R.C.); xavier.guimera@upc.edu (X.G.)
4 Instituto de Microelectrónica de Barcelona, IMB-CNM (CSIC), Esfera UAB, Campus Universitat Autònoma de Barcelona, Bellaterra, 08193 Barcelona, Spain; gemma.gabriel@imb-cnm.csic.es
5 CIBER, de Bioingeniería, Biomateriales y Nanomedicina (CIBER-BBN), 28029 Madrid, Spain
* Correspondence: mariadelmar.baeza@uab.cat

† Presented at the 1st International Electronic Conference on Chemical Sensors and Analytical Chemistry, 1–15 July 2021; Available online: https://csac2021.sciforum.net/.

Abstract: Hydrogen sulfide (H_2S) is a highly toxic chemical capable of causing severe health issues. Due to its environmental impact, it is critical to create effective methods for its monitoring. Inkjet printing technology has become an alternative for sensor fabrication because it is an economic, fast, and reproducible method for mass producing micro-electrodes. Herein, a miniaturized 25 mm^2 inkjet-printed amperometric sensor is presented. A gold electrode coupled with a silver track was modified with two inks: single-walled carbon nanotubes (SWCNTs) and a mixture of SCWCNTs and poly(vinyl alcohol) (PVA). Morphological and electrochemical properties were studied, as well as H_2S sensor performance. This approach is a suitable option for environmental H_2S tracking.

Keywords: electrochemical sensor; amperometric sensor; H_2S sensor; single-walled carbon nanotubes

Citation: Paré, F.; Castro, R.; Guimera, X.; Gabriel, G.; Baeza, M. An Inkjet-Printed Amperometric H_2S Sensor for Environmental Applications. *Chem. Proc.* **2021**, *5*, 4. https://doi.org/10.3390/CSAC2021-10462

Academic Editor: Núria Serrano

Published: 30 June 2021

1. Introduction

Environmental equilibrium is a hard-to-preserve resource, dangerously impacted by human heavy industrial activities [1,2]. It is naturally regulated through biogeochemical cycles. Among these, the sulfur cycle is of crucial importance, since it is vital for maintaining the composition of both the atmosphere and soils, as well as most living beings. Nonetheless, even more dangerous is an excess of highly toxic compounds such as water-soluble hydrogen sulfide (H_2S) gas. This is a poisonous, inflammable, and corrosive chemical, hazardous to human health at concentrations as low as 20 ppm (1.1 μM) for prolonged exposure [3]. Even though it generally appears as a gas, it has labile hydrogens, meaning it coexists as different species in aqueous media. Hydrogen sulfide can appear as different species depending on the pH of its medium, being capable of losing both its protons and transitioning from a gas to ions. It has a pka_1 of approximately 7 and a pka_2 of about 13.5, meaning that HS^- predominates between pH 7.5 and 13. Due to the dangerous nature of H_2S and its frequent appearance in gas streams, a great need has recently arisen for many biotechnological processes to remove it [4], which require adequate systems for quick and easy tracking.

Nonetheless, many of these removal processes occur in aqueous media, requiring consideration of the environmental pH for adequate quantification. Thus, it is appropriate to incorporate a simultaneous pH measurement with the H_2S tracking.

In recent years, printed electronics have steadily replaced more traditional electrode fabrication methods. Among these, inkjet printing has the main advantage of not requiring any mask preparation, greatly reducing the time and cost of device fabrication.

Moreover, its low drop volume and precision, added to its capacity for printing metal-based inks, allow for the fabrication of highly reproducible micro-electrodes.

For biotechnological applications, microsensors can be printed on different substrates and their designs can be adapted to the shapes of bioreactors with little cost impact [5]. Therefore, inkjet printing technology is an interesting alternative for the development of a microsensors platform for H_2S and pH measurements.

Moreover, electrochemical sensors have the advantages of high sensitivity, in situ application, and a broad range of applicable materials. However, H_2S determination using electrochemical sensors has many design and implementation challenges to solve, such as the pH influence on measurements [6] and electrode passivation by accumulation of S^0 produced from H_2S oxidation [7].

Among the authors who have developed H_2S sensors, Yang et al. (2018) [8] fabricated a sensor using Nafion for H_2S measurement in gaseous samples. This membrane was added to carbon fibers modified with platinum and rhodium nanoparticles. The H_2S was adsorbed on the platinum, releasing protons that crossed the Nafion membrane while electrons moved through an external circuit. The H_2S concentration was proportional to the circulating charge. This sensor had a linear range from 2.9 μM to 5.9 mM, and the minimum detectable signal was 2.9 μM. Brown et al. (2019) [9] deposited a thin layer of S^0 on a glassy carbon electrode (GCE) and covered it with electro-polymers to avoid passivation of the electrode via accumulation of S^0. The H_2S was measured by constant potential amperometry (CPA), using a potential of 0.3 V vs. Ag/AgCl. This sensor exhibited a high degree of selectivity and a linear range between 0 μM and 15 μM, with lowest and highest detection limits of (9 ± 6) nM and (79 ± 51) nM, respectively.

Furthermore, new carbon materials with excellent electronic properties have become suitable options for electrode development. As an example, Lawrence et al. (2004) [10] modified a GCE with carbon nanotubes (CNTs) deposited on the surface by drop casting a solution of CNTs in dimethylformamide (DMF). The main advantage found was the catalytic capacity that decreased the oxidation potential from 0.4 V to −0.3 V (vs. Ag/AgCl). This allowed amperometric measurements at 0.1 V (vs. Ag/AgCl) in a range between 1.25 μM and 112.5 μM, with a detection limit of 0.3 μM. Li et al. (2017) [6] used cobalt to magnetically attach MoS_2 monolayer sheets to CNTs, which were deposited on glassy carbon electrodes using a Nafion membrane as an adhesive. Analysis was performed by amperometry obtaining a linear range of application from 0.05 μM to 0.6 μM, with a detection limit of 7.6 nM.

Despite authors applying carbon materials for H_2S sensor development, there are no reports of the use of SWCNTs ink for an inkjet-printed H_2S sensor. Moreover, several studies have reported poly(vinyl alcohol) (PVA) addition to conductive materials to improve its mechanical and adhesive properties [11–13].

In this study, the fabrication of a H_2S amperometric microsensor has been studied using both a SWCNTs ink and a SWCNTs–PVA ink for Au electrode modification. Morphological and electrochemical characterizations were carried out for electrode performance analysis.

2. Materials and Methods

2.1. Inkjet-Printed H_2S Sensor Fabrication

The printing process for the electrode was performed using Ag NPs ink (SI-J20x Nanosilver Inkjet Printing Ink from Agfa, Mortsel, Belgium), Au NPs ink (Drycure Au-JB 1010B from C-ink, Japan), and SWCNTs ink (Carbon nanotube, single-walled, conductive aqueous ink, SWCNT 1.00 mg/mL from Sigma-Aldrich, Spain). To passivate the electrode, SU8 ink (XP PriElex SU-8 1.0 Inkjettable Dielectric from Kayaku, Westborough, MA, USA) was used. All inks were printed over polyethylene teraphtalate (PET) sheets (Q65HA,

Du Pont Teijin Films, Dumfries, UK), using a Dimatix printer (DMP-2831 from FUJIFILM Dimatix, Santa Clara, CA, USA). The SWCNTs–PVA composite ink was prepared by mixing a commercial SWCNTs ink, and a 5 wt.% PVA solution, both acquired from Sigma-Aldrich. For SWCNTs ink and SWCNTs–PVA ink deposition, the drop casting technique was performed on a thermal plate. The applied temperature was a studied parameter. The final electrode dimensions were a 1 mm diameter gold disk with a total length of 26 mm and surface of 25 mm^2, and an approximately 2 mm diameter SWCNTs disk.

2.2. Sensor Characterization

Morphological characterization of the electrodes was carried out via optical microscopy. Images were obtained with a digital microscope (USB microscope AM4815ZTL from DinoLite, Alemere, The Netherlands).

Electrochemical characterization was performed with a PalmSens potentiostat–galvanostat (PalmSens4 from PalmSens, The Netherlands). A three-electrode configuration was used for the electrochemical cell. An Ag/AgCl (1 M KCl) reference electrode (reference electrode with Ag/AgCl in aqueous KCl from ItalSens, PalmSens, Houten, The Netherlands) and a platinum wire counter electrode (counter electrode made of platinum wire from ItalSens, PalmSens, Houten, The Netherlands) were used. The fabricated electrode functioned as the working electrode. Cyclic voltammetry (CV) measurements were carried out using the redox pair $K_3[Fe(CN)_6]/K_4[Fe(CN)_6]$ (0.01 M) at 0.01 V/s from −0.1 V to 0.5 V. Intensity peaks (Ip) resulting from redox reactions allowed for electrochemical characterization.

To study the effect of deposition temperature in the SWCNTs–PVA ink, ten layers of PVA were drop casted over a gold inkjet-printed electrode at 90 °C and at room temperature. Intensity-peak values were obtained.

For H_2S calibration, a stock H_2S solution (0.1 M) was prepared by dissolving $NaS{\cdot}9H_2O$ and NaOH (both from Sigma-Aldrich, Spain) in deionized water (Milli-Q from Millipore Corporation, Burlington, MA, USA). Standardization was carried out according to standard methods [14]. The stock solution was diluted in phosphate-buffered saline solution (PBS) (from Sigma-Aldrich, Madrid, Spain) to obtain a 0.02 M H_2S solution.

3. Results and Discussion

3.1. Inkjet-Printed H_2S Miniaturized Sensor Fabrication

Inkjet-printed micro-electrodes were fabricated by first printing the Ag tracks and pads, drying at 100 °C in an oven, then printing the Au surfaces and connecting tracks, which were later dried at 120 °C. Subsequently, inks were sintered at 150 °C for 60 min to improve their conductivity. Finally, micro-electrodes were passivated by printing an SU8 layer over the tracks, preventing the short circuiting of the electrodes.

Both SWCNTs and SWCNTs–PVA inks were drop casted on a gold electrode, covering it completely (Figure 1). Both inks showed good adhesion to gold, allowing morphological and electrochemical characterization of the microsensors.

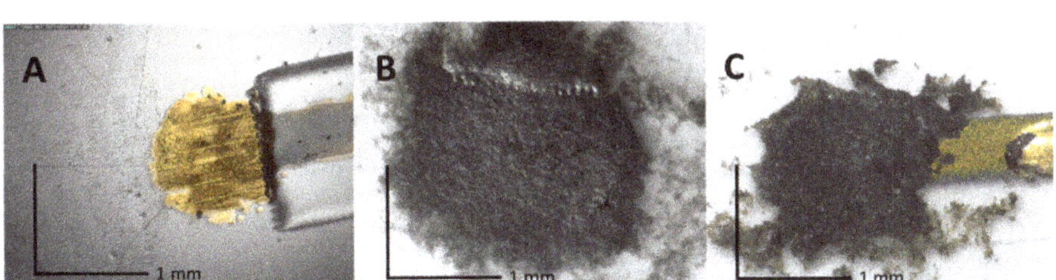

Figure 1. Photographs of (**A**) inkjet-printed Au electrode, (**B**) SWCNTs drop casted over Au electrode, and (**C**) drop-casted SWCNTs–PVA over Au electrode.

3.2. Microsensor Electrochemical Characterization

Cyclic voltammetry measurements of Au, SWCNTs, and SWCNTs–PVA electrodes were carried out. These allowed verification of whether the applied modifications yielded an electrode with properties of interest that could be used for an H_2S sensor. The main interest was in fabricating a sensor capable of oxidizing H_2S at low voltages and with a high tolerance for sulfur poisoning since it is well known that S^0 is highly insoluble in water [7] and has a high affinity for Au atoms. Thus, the electrode would gradually deteriorate by accumulating non-conductive layers of sulfur atoms.

Modified electrodes presented a similar current density and a smaller potential gap than bare Au electrodes (Figure 2). The smaller peak separation, specifically due to reducing the potential necessary for ferrocyanide oxidation, meant that SWCNTs and SWCNTs–PVA are both favorable for use as an H_2S sensor, with the added benefit of lowered rates of sulfur deposition on their surfaces due to the less-favorable S–C interaction compared with S–Au.

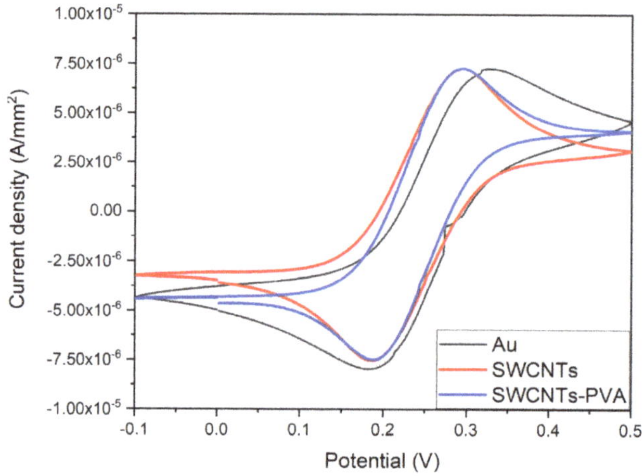

Figure 2. Cyclic voltammetry of Au, SWCNTs, and SWCNTs-PVA electrodes in hexacyanoferrate/hexacyanoferrate (0.01 M).

3.3. PVA Deposition Temperature Study

The effect of PVA deposition temperature was studied through the drop casting of 10 layers of PVA on an Au inkjet-printed electrode at 90 °C and at room temperature. Cyclic voltammetry was performed, and current-peak (Ip) values were obtained. The results showed that at room temperature, PVA deposition had a smaller passivation effect than at 90 °C, in which case the electrode's current values dropped significantly (Table 1).

Table 1. Effect of polymer deposition temperature on Au inkjet-printed electrodes.

Temperature	Ip Au Electrode (μA)	Ip Au Electrode + PVA (μA)	% Ip Reduction
25 °C	8.12	7.46	8.1
90 °C	3.92	$1.22 \cdot 10^{-5}$	100.0

3.4. Microsensor Calibration and Analytical Response

Calibration of SWCNTs and SWCNTs–PVA microsensors was performed by adding different volumes of H_2S standard to a PBS solution, measuring a concentration range between 0 and 600 μM. The chronoamperometry method was used for H_2S oxidation (Equation (1)) at a polarization voltage of 50 mV, and the resulting current was measured.

$$H_2S \rightarrow S^0 + 2e^- + 2H^+ \quad (1)$$

Calibrations were performed for both SWCNTs and SWCNTs–PVA under the same conditions (Figure 3). The results showed good correlation between the measured current and the analyte concentration, with a similar slope for both sensors but different linear ranges.

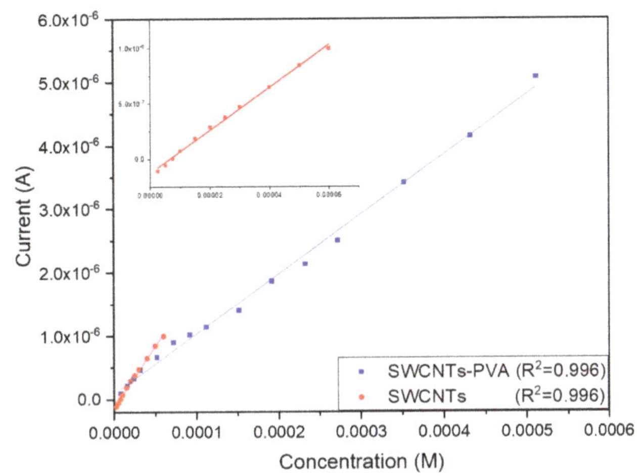

Figure 3. Calibrations of two different sensors made of Au electrodes modified with SWCNTs and SWCNTs-PVA by drop casting. Inset shows a reduced range of H_2S concentration for the SWCNTs sensor calibration.

While the SWCNTs sensor showed a higher sensitivity (19.3 ± 0.4 mA/M) compared to SWCNTs–PVA (9.4 ± 0.2 mA/M), it also had a reduced working range. SWCNTs can measure H_2S concentrations from 8 μM to 60 μM, with a limit of detection (LD) of 4.3 μM. On the other hand, SWCNTs–PVA is capable of measuring from 52 μM to 512 μM, with a LD of 34 μM.

4. Conclusions

We demonstrated a novel miniaturized inkjet-printed amperometric H_2S sensor fabricated by modification of a gold electrode with SWCNTs ink. The results showed that the addition of a stabilizing polymer allowed for an increased range of H_2S concentration measurements. Due to the low versatility of the SWCNTs sensor, the addition of PVA was considered, to improve the H_2S sensor performance. However, it was necessary to adapt the fabrication temperature conditions to avoid electrode passivation, limiting the SWCNTs–PVA ink-deposition temperature to 25 °C. This sensor offers an approach for H_2S tracking with environmental and biotechnological applications.

Supplementary Materials: The following are available online at https://www.mdpi.com/article/10.3390/CSAC2021-10462/s1.

Author Contributions: Conceptualization, M.B. and G.G.; methodology, F.P.; validation, M.B. and X.G.; formal analysis, F.P. and R.C.; investigation, F.P. and R.C.; resources, G.G. and M.B.; data curation, F.P. and R.C.; writing—original draft preparation, F.P. and R.C.; writing—review and editing, M.B., G.G., and X.G.; visualization, G.G. and M.B.; supervision, M.B., G.G., and X.G.; project administration, M.B.; funding acquisition, M.B. and G.G. All authors have read and agreed to the published version of the manuscript.

Funding: This research was funded by the Spanish Government, Ministerio de Economía y Competitividad, through projects RTI2018-099362-B-C21 and RTI2018-099362-B-C22 MINECO/FEDER, EU.

Institutional Review Board Statement: Not applicable.

Informed Consent Statement: Not applicable.

Data Availability Statement: The data presented in this study are available on request from the corresponding author. The data are not publicly available due to the repository that is used to keep the data is a private one provided by the University.

Acknowledgments: The authors wish to acknowledge David Gabriel and Xavier Gamisans for their coordinated direction of the ENSURE project.

Conflicts of Interest: The authors declare no conflict of interest.

References

1. Li, M.; Li, C.; Zhang, M. Exploring the spatial spillover effects of industrialization and urbanization factors on pollutants emissions in China's Huang-Huai-Hai region. *J. Clean. Prod.* **2018**, *195*, 154–162. [CrossRef]
2. Aikawa, M.; Hiraki, T.; Suzuki, M.; Tamaki, M.; Kasahara, M. Separate chemical characterizations of fog water, aerosol, and gas before, during, and after fog events near an industrialized area in Japan. *Atmos. Environ.* **2007**, *41*, 1950–1959. [CrossRef]
3. Guidotti, T.L. Hydrogen Sulfide. *Int. J. Toxicol.* **2010**, *29*, 569–581. [CrossRef] [PubMed]
4. Syed, M.; Soreanu, G.; Falletta, P.; Béland, M. Removal of Hydrogen Sulfide from Gas Streams Using Biological Processes—A Review. *Can. Biosyst. Eng.* **2006**, *48*, 2.
5. Moya, A.; Gabriel, G.; Villa, R.; del Campo, F.J. Inkjet-printed electrochemical sensors. *Curr. Opin. Electrochem.* **2017**, *3*, 29–39. [CrossRef]
6. Li, C.; Zhang, D.; Wang, J.; Hu, P.; Jiang, Z. Magnetic MoS_2 on multiwalled carbon nanotubes for sulfide sensing. *Anal. Chim. Acta* **2017**, *975*, 61–69. [CrossRef] [PubMed]
7. Hall, J.R.; Schoenfisch, M.H. Direct Electrochemical Sensing of Hydrogen Sulfide without Sulfur Poisoning. *Anal. Chem.* **2018**, *90*, 5194–5200. [CrossRef] [PubMed]
8. Yang, X.; Zhang, Y.; Hao, X.; Song, Y.; Liang, X.; Liu, F.; Liu, F.; Sun, P.; Gao, Y.; Yan, X.; et al. Nafion-based amperometric H2S sensor using Pt-Rh/C sensing electrode. *Sens. Actuators B Chem.* **2018**, *273*, 635–641. [CrossRef]
9. Brown, M.D.; Hall, J.R.; Schoenfisch, M.H. A direct and selective electrochemical hydrogen sulfide sensor. *Anal. Chim. Acta* **2019**, *1045*, 67–76. [CrossRef] [PubMed]
10. Lawrence, N.S.; Deo, R.P.; Wang, J. Electrochemical determination of hydrogen sulfide at carbon nanotube modified electrodes. *Anal. Chim. Acta* **2004**, *517*, 131–137. [CrossRef]
11. Samsudin, A.M.; Hacker, V. Preparation and Characterization of PVA/PDDA/Nano-Zirconia Composite Anion Exchange Membranes for Fuel Cells. *Polymers* **2019**, *11*, 1399. [CrossRef] [PubMed]
12. Ben, J.; Song, Z.; Liu, X.; Lü, W.; Li, X. Fabrication and Electrochemical Performance of PVA/CNT/PANI Flexible Films as Electrodes for Supercapacitors. *Nanoscale Res. Lett.* **2020**, *15*, 151. [CrossRef] [PubMed]
13. Chen, W.; Tao, X.; Xue, P.; Cheng, X. Enhanced mechanical properties and morphological characterizations of poly(vinyl alcohol)–carbon nanotube composite films. *Appl. Surf. Sci.* **2005**, *252*, 1404–1409. [CrossRef]
14. American Public Health Association. *Standard Methods: For the Examination of Water and Wastewater*, 22th ed.; American Public Health Association: Washington, DC, USA, 2012.

chemistry proceedings

MDPI

Proceeding Paper

Molecular Emitters as a Tunable Light Source for Optical Multisensor Systems †

Anastasiia Surkova [1,2,*], Aleksandra Paderina [1], Andrey Legin [1], Elena Grachova [1] and Dmitry Kirsanov [1]

1 Institute of Chemistry, St. Petersburg State University, Universitetskii pr. 26, 198504 St. Petersburg, Russia; ksaniasha@list.ru (A.P.); andrey.legin@gmail.com (A.L.); e.v.grachova@mail.ru (E.G.); d.kirsanov@gmail.com (D.K.)

2 Department of Analytical and Physical Chemistry, Samara State Technical University, Molodogvardeyskaya Street 244, 443100 Samara, Russia

* Correspondence: a.surkova@spbu.ru

† Presented at the 1st International Electronic Conference on Chemical Sensors and Analytical Chemistry, 1–15 July 2021; Available online: https://csac2021.sciforum.net/.

Abstract: In this study, optical multisensor systems based on molecular emitters as a light source are introduced. To obtain such light sources, cyclometalated Ir(III) complexes and Cu(I)-based complexes were synthetized and investigated. Since each complex has its own emission spectrum in the visible range, it is possible to choose an appropriate set of emitters for specific analytical tacks. The developed analytical device was successfully applied for fluoride and phosphate quantification in surface water.

Keywords: optical multisensor systems; molecular emitters; iridium complexes; copper complexes; photoluminescence; chemometrics; water analysis; fluoride; phosphate

Citation: Surkova, A.; Paderina, A.; Legin, A.; Grachova, E.; Kirsanov, D. Molecular Emitters as a Tunable Light Source for Optical Multisensor Systems. *Chem. Proc.* **2021**, *5*, 5. https://doi.org/10.3390/CSAC2021-10611

Academic Editor: Elena Benito-Peña

Published: 6 July 2021

1. Introduction

The development of portable and inexpensive analyzers allowing fast determination of the integral sample characteristics is a current trend in analytical chemistry. Optical spectroscopy in the visible and near infrared (NIR) range has a great potential due to the advances in modern optical engineering. Optical multisensor systems (OMS) are devices working on the principle of optical spectroscopy but optimized for a specific analytical task and composed of cheaper elements: light-emitting diodes (LEDs), optical fibers, 3D-printed parts, stamped optics, etc. Such specialization enables essential reduction of analyzers' price, size, and weight, thus making the analysis widely available for both real-time application and in-field measurements. There are many examples of OMS applications for various analytical problems in the recent literature [1–5].

In the present work, a novel platform for construction of OMS was suggested. The idea is to use a combination of molecular emitters as a multichannel light source with tunable intensity and wavelength range. Cyclometalated Ir(III) complexes [6] and Cu(I)-based complexes [7] were synthetized and tested in order to obtain such a light source. Each individual complex has its own emission spectrum in the visible range. This enables the selection and optimization of the light source for a specific analytical application. Several optical setup designs of OMS were developed. The proposed prototype was tested to analyze the metal ions in aqueous mixtures. The practical application of the OMS was demonstrated for the quantification of fluoride and phosphate in real surface and tap waters.

2. Methods and Materials

To obtain light sources for OMS, two sets of molecular emitters were synthesized. One of them consisted of eight cyclometalated Ir(III) compounds: firstly synthesized

$[Ir(dfppy)_2(bpbpy)]PF_6$ (**1**), $[Ir(ppy)_2(bpbpy)]PF_6$ (**2**), $[Ir(pybt)_2(bpbpy)]PF_6$ (**3**), and synthesized according to the standard procedure: (**4**) [8], (**5**) [9], (**6**) [10,11], (**7**) [12], and (**8**) [13]. The second set consists of six Cu(I)-based complexes: $[Cu(MePPy_3)I]_2[Cu_2I_4]$ (**9**) [14], $[Cu_4I_4(py)_4]$ (**10**) [15], [Cu(Tpdp)I] (**11**) [16], $[CuCl(PPh_3)_2(py)]$ (**12**) [17], $[Cu(PPh_3)_3$(4-Mepy)]Br (**13**), and $[CuI(PPh_3)_2$(4-Mepy)] (**14**) [18]. The mixture of each series gives multiband light source in the region from about 400 to 800 nm.

A different optical setup was constructed for each set of emitters. A homogeneous mixture of molecular emitter powders (mixture of iridium(III) or copper(I) complexes) was placed on a glass substrate under the sample. The solution under study was placed either in a glass cup (1 cm in diameter) for the setup with Ir(III)-complexes or in a polystyrene Petri dish (3.5 cm in diameter) for the setup with Cu(I)-complexes. Initially, photoluminescence of the molecular emitters was initiated using the laser diode (λ_{exct} = 365 nm and λ_{exct} = 385 nm for the first and the second experimental setup, respectively). Further, the laser diode was replaced by a UV flashlight with λ_{exct} = 365 nm. The light that passed through the sample was recorded by a fiber-optic UV−vis spectrometer AvaSpec-ULS2048CL-EVO (Avantes, Apeldoorn, the Netherlands). The optimal geometry of the device and the required amount and placement of sample solution were chosen experimentally to obtain a stable and reproducible analytical signal.

To compare Ir(III)-based and Cu(I)-based OMS, the optical setups were tested on two separate calibration series prepared from aqueous solutions of Co(II) and Cu(II) nitrates. Each calibration set consisted of seven samples with different concentrations ranging from 0.01 to 0.1 M with a 0.015 M step for Ir(III)-based setup and ten samples with different concentrations ranging from 0.01 to 0.1 M with a 0.01 M step for Cu(I)-based setup.

The practical application of the Cu(I)-based OMS was demonstrated for quantitative determination of fluoride and phosphate in surface and tap waters. A total of 5 samples for analysis were collected from the tap, rivers, and a lake. The standard photometric methods were used as a reference for the quantification of phosphate [19] and fluoride [20] in water samples. Sample preparation for analysis was carried out in accordance with the procedures described in [19,20]. The mass concentrations of phosphate and fluoride in the calibration solutions were 0–0.96 mg/L with a 0.12 mg/L step and 0–0.4 mg/L with a 0.04 mg/L step, respectively.

To relate the analytical signal response of OMS to the analyte content, the calibration models were built using partial least-squares (PLS) regression [21]. The model performance was estimated by full cross-validation (CV) and validation with a test set. Root mean-square error (RMSE) of calibration (RMSEC), prediction (RMSEP, for the test set) or cross-validation (RMSECV) and the respective coefficients of determination (R^2) were used to compare the models.

3. Results and Discussion

In order to choose a set of appropriate emitters for multiband light sources, several requirements were established: the absorption spectrum must not overlap with the emission spectrum; excitation radiation at 365/385 nm must fit into the absorption maximum; the emission should be intense enough to be employed as the light source in the solid phase. Ir(III) luminescent complexes that have bright controlled emission fit these requirements. Another set of molecular emitters based on Cu(I) complexes is a cost-effective alternative to Ir(III) complexes and their application as a light source in OMS was also tested. The resulting mixture of each set of complexes upon excitation with the laser diode has shown emission spectra in the range of 450–800 nm (Figure 1). Emission of the mixture of eight Ir(III) compounds is higher at the right end of the spectrum from 600 to 800 nm (black line in Figure 1), while emission of the mixture of six Cu(I) complexes is higher at the left end of the spectrum (red line in Figure 1). Despite this difference, both sets are suitable as a multiband light source in the visible region.

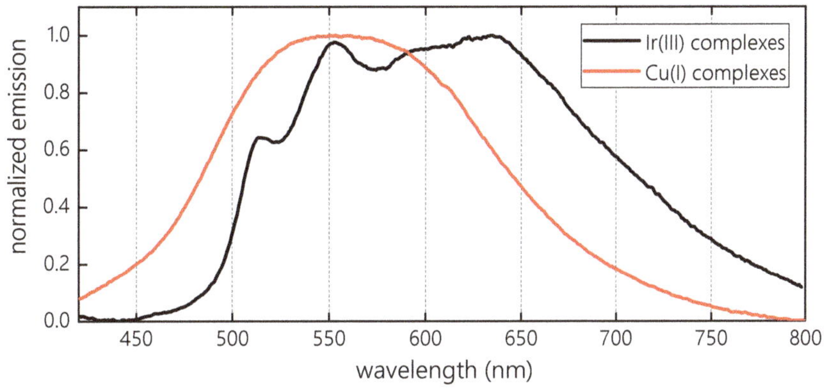

Figure 1. Normalized emission spectra of: mixture of eight Ir(III) complexes is marked in black and mixture of six Cu(I) complexes is marked in red.

The block diagram of the experimental setup is shown in Figure 2. In accordance with the proposed optical setup, the emission of molecular emitters is excited by the laser diode. The light passing through a sample is registered with a fiber-optic cable connected with a spectrometric detector. The laser diode can be replaced by inexpensive UV flashlight. Another advantage of UV flashlight is that its power is higher than that of the laser diode. Therefore, two optical setups were constructed and compared: with laser diode and with UV flashlight as excitation light.

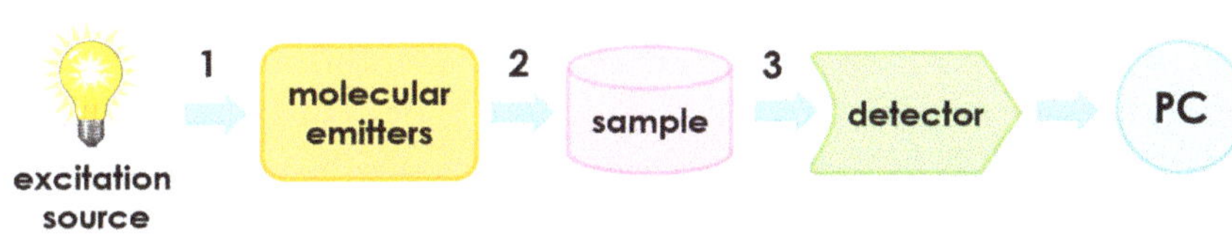

Figure 2. Block diagram of the experimental setup, where 1—exciting light; 2—emitted light; 3—transmitted light.

In order to compare Ir(III) and Cu(I) compounds as a light source in OMS, the series of colored aqueous solutions of copper and cobalt nitrates were chosen because these solutions absorb the light in the visible range ($Co(NO_3)_2$ in the range of 400–600 nm, $Cu(NO_3)_2$—700–900 nm). Individual PLS models were built to relate the registered optical signals with copper and cobalt content, and the respective modeling statistics are presented in Table 1. Despite the simplicity of the developed analytical platform, PLS models for copper and cobalt exhibit a good performance for both OMSs (with Ir(III) and Cu(I) complexes). The parameters of the model for copper quantification are somewhat worse in Cu(I)-based OMS than in Ir(III)-based OMS, while for the cobalt model they are better. A set of copper(I)-based molecular emitters was chosen for further research because it is more efficient both economically and environmentally.

At the next stage, the laser diode was replaced by a UV flashlight. The experimental results showed that the recorded optical density of the initial emission of Cu(I) complexes upon excitation with the UV flashlight is much higher than that upon the laser diode. The PLS results for OMS with UV flashlight are better for both Cu and Co models (Table 1).

Table 1. PLS modeling and validation statistics.

	Emission Source	Light Source	Calibration		CV	
			RMSE	R^2	RMSE	R^2
Co [a]	Laser (365 nm)	Ir(III) complexes [c]	0.005	0.98	0.009	0.94
Co [b]	Laser (385 nm)	Cu(I) complexes [d]	0.002	1	0.005	0.98
Co [b]	UV-flashlight	Cu(I) complexes [d]	0.002	0.99	0.004	0.99
Cu [a]	Laser (365 nm)	Ir(III) complexes [c]	0.002	1	0.003	0.99
Cu [b]	Laser (385 nm)	Cu(I) complexes [d]	0.006	0.96	0.013	0.84
Cu [b]	UV-flashlight	Cu(I) complexes [d]	0.003	0.99	0.004	0.98
PO_4^{3-}	UV-flashlight	Cu(I) complex (9)	0.006	1	0.031	0.99
F^-	UV-flashlight	Cu(I) complex (10)	0.020	0.97	0.029	0.96

[a] 7 samples, [b] 10 samples, [c] mixture of 8 Ir(III) complexes, [d] mixture of 6 Cu(I) complexes. The interval range for the modeling: 450–800 nm for Co, Cu, and F^-, 600–750 nm for PO_4^{3-}.

The practical application of the developed Cu(I)-based OMS was demonstrated for the determination of fluoride and phosphate content in real surface and tap waters. Fluoride and phosphate are essential components for living cells; however, excess concentrations in surface water can lead to various human diseases and a general reduction in water quality. The evaluation of fluoride and phosphate by simple and inexpensive analytical methods is an important task for timely environmental monitoring.

The colored complexes of fluoride and phosphate absorb light in the regions of 450–800 (absorption maximum at around 590 nm) and 550–800 nm (absorption maximum at around 700 nm), respectively. Since these analytes are determined using an individual calibration sample set, a single molecular emitter was chosen for each of the anions based on its the emission properties: complex (**9**) with emission maximum of 659 nm for fluoride determination and complex (**10**) with emission maximum of 584 nm for phosphate determination.

The RMSE and R^2 for the full cross-validation are similar for both phosphate and fluoride PLS models (Table 1 and Figure S1 in Supporting Information). Further, the calibration models were used to predict the content of fluoride and phosphate in five water samples. The content of anions evaluated by a standard spectrophotometric technique was employed for model precision assessment. The prediction performance of the phosphate model is pretty good: RMSEP = 0.073 mg/L with R^2 = 0.97. The PLS model for fluoride has also relatively low RMSEP (0.074 mg/L), but R^2 is 0.7. This can be caused by the fact that water samples were taken in different regions and from different sources (several from rivers, one from a lake, and one from a tap). Each sample may contribute strongly to the model, and more samples are needed to make more accurate predictions. Mean squared errors (MSE) that show the average squared difference between the estimated by reference method values and predicted by OMS values were 0.34% and 0.54% for phosphate and fluoride, respectively.

4. Conclusions

The proposed approach to OMS development allows reducing analysis time and does not require additional sample preparation. Moreover, OMS based on molecular emitters can be adopted for the particular analytical task by selecting the appropriate wavelength region. Despite the relative technical simplicity of OMS, its application in combination with modern chemometric methods provides high accuracy of analysis, comparable with that of full-featured spectrometers. Both synthetized sets of Ir(III) and Cu(I)-based complexes are suitable as a light source in OMS. However, Cu(I) complexes are easier to produce, cheaper, and environmentally friendly. The demonstrated application of OMS based on molecular emitters for the determination of fluorides and phosphates in surface water proves their high practical significance.

Supplementary Materials: The following are available online at https://www.mdpi.com/article/10.3390/CSAC2021-10611/s1, Figure S1: Predicted versus measured values of cross-validation for quantification of (A) PO_4^{3-} and (B) F^-.

Author Contributions: Conceptualization, D.K.; methodology, A.S.; formal analysis, A.S.; investigation, A.S. and A.P.; writing—original draft preparation, A.S.; writing—review and editing, E.G.; supervision, E.G. and D.K.; project administration, A.L. All authors have read and agreed to the published version of the manuscript.

Funding: This research was funded by the RSF, grant number 19-79-00076.

Conflicts of Interest: The authors declare that they have no known competing financial interests or personal relationships that could have appeared to influence the work reported in this paper.

References

1. Fu, X.; Wang, X.; Rao, X. An LED-based spectrally tuneable light source for visible and near-infrared spectroscopy analysis: A case study for sugar content estimation of citrus. *Biosyst. Eng.* **2017**, *163*, 87–93. [CrossRef]
2. Lastra-Mejias, M.; Villa-Martinez, A.; Izquierdo, M.; Aroca-Santos, R.; Cancilla, J.C.; Torrecilla, J.S. Combination of LEDs and cognitive modeling to quantify sheep cheese whey in watercourses. *Talanta* **2019**, *203*, 290–296. [CrossRef] [PubMed]
3. Müller-Maatsch, J.; Alewijn, M.; Wijtten, M.; Weesepoel, Y. Detecting fraudulent additions in skimmed milk powder using a portable, hyphenated, optical multi-sensor approach in combination with one-class classification. *Food Control* **2021**, *121*, 107744. [CrossRef]
4. Bogomolov, A.; Zabarylo, U.; Kirsanov, D.; Belikova, V.; Ageev, V.; Usenov, I.; Galyanin, V.; Minet, O.; Sakharova, T.; Danielyan, G.; et al. Development and testing of an led-based near-infrared sensor for human kidney tumor diagnostics. *Sensors* **2017**, *17*, 1914. [CrossRef] [PubMed]
5. De Lima, K.M.G. A portable photometer based on LED for the determination of aromatic hydrocarbons in water. *Microchem. J.* **2012**, *103*, 62–67. [CrossRef]
6. Gitlina, A.Y.; Surkova, A.; Ivonina, M.V.; Sizov, V.V.; Petrovskii, S.K.; Legin, A.; Starova, G.L.; Koshevoy, I.O.; Grachova, E.V.; Kirsanov, D.O. Cyclometalated Ir(III) complexes as tuneable multiband light sources for optical multisensor systems: Feasibility study. *Dyes Pigments* **2020**, *180*, 108428. [CrossRef]
7. Surkova, A.A.; Paderina, A.V.; Legin, A.V.; Grachova, E.V.; Kirsanov, D.O. Cu(I)-based molecular emitters for quantification of fluoride and phosphate in surface waters. *Measurement* **2021**, *184*, 109976. [CrossRef]
8. Gitlina, A.Y.; Ivonina, M.V.; Sizov, V.V.; Starova, G.L.; Pushkarev, A.P.; Volyniuk, D.; Tunik, S.P.; Koshevoy, I.O.; Grachova, E.V. A rare example of a compact heteroleptic cyclometalated iridium(III) complex demonstrating well-separated dual emission. *Dalton Trans.* **2018**, *47*, 7578–7586. [CrossRef] [PubMed]
9. Henwood, A.F.; Bansal, A.K.; Cordes, D.B.; Slawin, A.M.Z.; Samuel, I.D.W.; Zysman-Colman, E. Solubilised bright blue-emitting iridium complexes for solution processed OLEDs. *J. Mater. Chem. C* **2016**, *4*, 3726–3737. [CrossRef] [PubMed]
10. Wu, C.; Chen, H.-F.; Wong, K.-T.; Thompson, M.E. Study of ion-paired iridium complexes (soft salts) and their application in organic light emitting diodes. *J. Am. Chem. Soc.* **2010**, *132*, 3133–3139. [CrossRef] [PubMed]
11. Di Censo, D.; Fantacci, S.; De Angelis, F.; Klein, C.; Evans, N.; Kalyanasundaram, K.; Bolink, H.J.; Grätzel, M.; Nazeeruddin, M.K. Synthesis, characterization, and DFT/TDDFT calculations of highly phosphorescent blue light-emitting anionic iridium complexes. *Inorg. Chem.* **2008**, *47*, 980–989. [CrossRef] [PubMed]
12. Nazeeruddin, M.K.; Humphry-Baker, R.; Berner, D.; Rivier, S.; Zuppiroli, L.; Graetzel, M. Highly phosphorescence iridium complexes and their application in organic lightemitting devices. *J. Am. Chem. Soc.* **2003**, *125*, 8790–8797. [CrossRef] [PubMed]
13. Baranoff, E.; Curchod, B.F.E.; Frey, J.; Scopelliti, R.; Kessler, F.; Tavernelli, I.; Rothlisberger, U.; Grätzel, M.; Nazeeruddin, M.K. Acid-induced degradation of phosphorescent dopants for OLEDs and its application to the synthesis of trisheteroleptic iridium(III) bis-cyclometalated complexes. *Inorg. Chem.* **2012**, *51*, 215–224. [CrossRef] [PubMed]
14. Artem'ev, A.V.; Pritchina, E.A.; Rakhmanova, M.I.; Gritsan, N.P.; Bagryanskaya, I.Y.; Malysheva, S.F.; Belogorlova, N.A. Alkyl-dependent self-assembly of the first red-emitting zwitterionic $\{Cu_4I_6\}$ clusters from [alkyl-P(2-Py)$_3$]$^+$ salts and CuI: When size matters. *Dalton Trans.* **2019**, *48*, 2328–2337. [CrossRef]
15. Kirakci, K.; Fejfarová, K.; Martinčík, J.; Nikl, M.; Lang, K. Tetranuclear copper (I) iodide complexes: A new class of X-ray phosphors. *Inorg. Chem.* **2017**, *56*, 4609–4614. [CrossRef] [PubMed]
16. Zhang, J.; Duan, C.; Han, C.; Yang, H.; Wei, Y.; Xu, H. Balanced dual emissions from tridentate phosphine-coordinate copper(I) complexes toward highly efficient yellow OLEDs. *Adv. Mater.* **2016**, *28*, 5975–5979. [CrossRef] [PubMed]
17. Ohara, H.; Ogawa, T.; Yoshida, M.; Kobayashi, A.; Kato, M. Reversible luminescent colour changes of mononuclear copper(I) complexes based on ligand exchange reactions by N-heteroaromatic vapours. *Dalton Trans.* **2017**, *46*, 3755–3760. [CrossRef] [PubMed]
18. Ohara, H.; Kobayashi, A.; Kato, M. Simple and extremely efficient blue emitters based on mononuclear Cu(I)-halide complexes with delayed fluorescence. *Dalton Trans.* **2014**, *43*, 17317–17323. [CrossRef] [PubMed]
19. GOST. GOST 18309-2014 Water. Methods for Determination of Phosphorus-Containing Matters. 2016. Available online: https://docs.cntd.ru/document/1200115799 (accessed on 26 November 2021).

20. GOST. GOST 4386-89. Drinking Water. Methods for Determination of Fluorides Mass Concentration. 2001. Available online: https://docs.cntd.ru/document/1200012569 (accessed on 26 November 2021).
21. Wold, S.; Sjöström, M.; Eriksson, L. PLS-regression: A basic tool of chemometrics. *Chemometr. Intell. Lab. Syst.* **2001**, *58*, 109–130. [CrossRef]

chemistry proceedings

MDPI

Proceeding Paper

Gold Nanoparticles Functionalized with Mercaptosuccinic Acid as a Means for Detecting Fe(III) Ions †

Nadezhda S. Komova, Ksenya V. Serebrennikova, Anna N. Berlina, Svetlana M. Pridvorova, Anatoly V. Zherdev and Boris B. Dzantiev *

A.N. Bach Institute of Biochemistry, Research Center of Biotechnology of the Russian Academy of Sciences, Leninsky Prospect 33, 119071 Moscow, Russia; nad4883@yandex.ru (N.S.K.); ksenijasereb@mail.ru (K.V.S.); anberlina@yandex.ru (A.N.B.); sh-p_s@mail.ru (S.M.P.); zherdev@inbi.ras.ru (A.V.Z.)

* Correspondence: dzantiev@inbi.ras.ru; Tel.: +7-495-954-3142

† Presented at the 1st International Electronic Conference on Chemical Sensors and Analytical Chemistry, 1–15 July 2021. Available online: https://csac2021.sciforum.net/.

Abstract: The application of mercaptosuccinic acid-capped gold nanoparticles as a sensing probe for the colorimetric detection of Fe(III) is reported. The well-dispersed gold nanoparticles (AuNPs) with a diameter of around 20 nm were obtained by a one-step reaction of tetrachloroauratic acid with mercaptosuccinic acid (MSA) as a reducing and capping agent, respectively. Fe(III) reportedly causes the aggregation of prepared MSA-capped AuNPs followed by a change in color and a shift to long wavelengths in the absorbance spectra. The resulting method allows for a visual and spectrophotometric Fe(III) determination with detection limits of 30 ng/mL and 23 ng/mL, respectively. MSA-capped AuNPs have been used as sensing probes for the detection of Fe(III) in drinking water samples with a detection limit that is much lower than the maximum permissible level of Fe(III) specified by official regulations (300 ng/mL).

Keywords: mercaptosuccinic acid; gold nanoparticles; Fe(III) ions; colorimetry; aggregation; drinking water

Citation: Komova, N.S.; Serebrennikova, K.V.; Berlina, A.N.; Pridvorova, S.M.; Zherdev, A.V.; Dzantiev, B.B. Gold Nanoparticles Functionalized with Mercaptosuccinic Acid as a Means for Detecting Fe(III) Ions. *Chem. Proc.* **2021**, *5*, 6. https://doi.org/10.3390/CSAC2021-10624

Academic Editor: Run Zhang

Published: 7 July 2021

1. Introduction

Currently, the control of the quality and composition of consumed drinking water has become extremely important. One of the main problems with a centralized water supply is the almost ubiquitous increased concentration of heavy metals in water, the main concentration of which is iron. This is primarily due to the widespread occurrence of this chemical element in various soils [1]; however, significant amounts of Fe(III) can also come from wastewater from the metallurgical, metalworking, textile, paint, and varnish industries as well as from agricultural wastewater. Considering that a high consumption of Fe(III) can cause toxic effects, the determination of Fe(III) content in water resources is of great importance for human life.

Various analytical methods, such as atomic absorption spectrometry [2], inductively couple plasma mass spectrometry [3], liquid chromatography [4], and spectrophotometric [5] and fluorescent [6] methods, are successfully applied for Fe(III) determination. Despite the high sensitivity of these methods, they are complex and time-consuming, and usually require expensive equipment operated by skilled personnel. On the contrary, colorimetric systems offer a promising approach for the detection of various metal ions largely due to their simplicity and rapidity as well as due to the opportunity to estimate results visually. To date, a number of colorimetric sensors, based on the aggregation of nanomaterials induced by metal ions through registration of color changes (visually), and a shift of the plasmon resonance peak (spectrophotometrically) have been reported [7]. A widespread approach used to increase the selectivity of these techniques is the functionalization of the surface of nanomaterials with various ligands [8]. Among these, pyrophosphate [9], hydroxamic acid [10], oxamic and p-aminobenzoic acids [11], casein [12], and (glycol)chitosan [13,14] have been employed for the colorimetric detection of Fe(III) in various environmental and biological samples.

A colorimetric system based on gold nanoparticles (AuNPs) functionalized with mercaptosuccinic acid (MSA) for the simple, rapid, selective, and cost-effective detection of trace Fe(III) ions in water was developed. The choice of the functionalizing agent was determined by previously described binding properties of mercaptosuccinic acid [15–17]. Moreover, the one-step preparation and functionalization of the proposed AuNPs greatly simplifies and accelerates the preparation of the sensor system. To the best of our knowledge, this is the first report of using mercaptosuccinic acid-functionalized AuNPs as a colorimetric sensor for the detection of trace levels of Fe(III) in aqueous media.

2. Experimental Design

2.1. Chemicals and Materials

An aqueous solution of Fe(III) (1 g/L) was obtained from the Center of Standardization of Samples and High-Purity Substances (St. Petersburg, Russia). Salts of Cd^{2+}, As^{3+}, Cu^{2+}, Zn^{2+}, Pb^{2+}, Sn^{2+}, and Cr^{3+} were also obtained from the Center of Standardization of Samples and High-Purity Substances. 2-Mercaptosuccinic acid (MSA) and tetrachloroauric acid (HAuCl4) were purchased from Sigma-Aldrich (St. Louis, MO, USA). Milli-Q-purified water was obtained using a Milli-Q Simplicity system from Millipore (Bedford, MA, USA) and used to prepare all aqueous solutions.

2.2. Synthesis of MSA-Functionalized AuNPs

Gold nanoparticles were synthesized through the reduction of $HAuCl_4$ by mercaptosuccinic acid [18]. First, 100 mL of the 0.01% $HAuCl_4$ solution was heated to boiling temperature and stirred using a magnet stirrer. After that, 12.5 mL of the 1 mM aqueous solution of MSA was added to the reaction mixture. The MSA solution was preliminarily neutralized with a sodium hydroxide in a stoichiometric ratio of 1:2. Thereafter, the reaction mixture was incubated with continuous stirring for 15 min and cooled to room temperature. The synthesized MSA-functionalized AuNPs were concentrated 10 times by centrifugation, resuspended in Milli-Q water with an adjustment to pH 3–4, and stored at 4–6 $^{\circ}$ C until analysis.

2.3. Transmission Electron Microscopy

The prepared MSA-functionalized AuNPs were applied to 300 mesh grids (Pelco International, Redding, CA, USA) coated with a support film of polyvinyl formal deposited from chloroform. A JEM CX-100 electron microscope (JEOL, Tokyo, Japan) operating at 80 kV was used to obtain the images. The digital images were analyzed using Image Tool software (University of Texas Health Science Center, San Antonio, TX, USA).

2.4. Detection of Fe(III) Ions

To determine the Fe(III) ions, 5 μL of a concentrated MSA-AuNP colloidal solution was added to an aqueous solution (pH 5) containing various amounts of analyte. After 5 min, the absorption spectra were measured by the EnSpire Multi-mode Plate Reader (PerkinElmer Inc., Waltham, MA, USA). When applying this technique for real water samples, a preliminary dilution (2–15 times depending on the water source) of the samples was used. To test the selectivity of the developed technique and the interference from other heavy metal ions, the solutions containing Hg^{2+}, Cd^{2+}, As^{3+}, Cu^{2+}, Zn^{2+}, Pb^{2+}, Sn^{2+}, or Cr^{3+} (100 ng/mL) were examined.

3. Results and Discussion

3.1. Synthesis and Characterization of MSA-Capped AuNPs

Mercaptosuccinic acid was chosen as a reducing agent due to the presence of two carboxyl groups providing functionalization by a chelate structure (Figure 1). The procedure of the MSA-capped AuNPs' synthesis includes mixing a $HAuCl_4$ and MSA solution at an optimized molar ratio of 2:1, which ensures a spherical form of particles with a size distribution within 20–25 nm [18]. Figure 2a,b shows the TEM image of MSA-functionalized AuNPs and the size distribution of the nanoparticles. The obtained nanoparticles have a spherical morphology with an average diameter of 19.9 ± 7.1 nm (number of particles was 195). In addition, the protection layer on the gold particles was observed in a TEM image. The obtained absorbance spectra of the MSA-AuNPs in the presence and absence of Fe(III) are given at Figure 2c. A strong peak at 530 nm was observed for MSA-AuNPs.

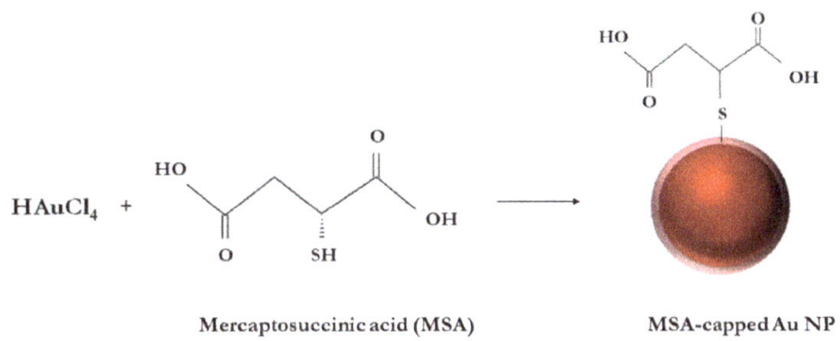

Figure 1. Scheme of MSA-functionalized AuNPs' synthesis.

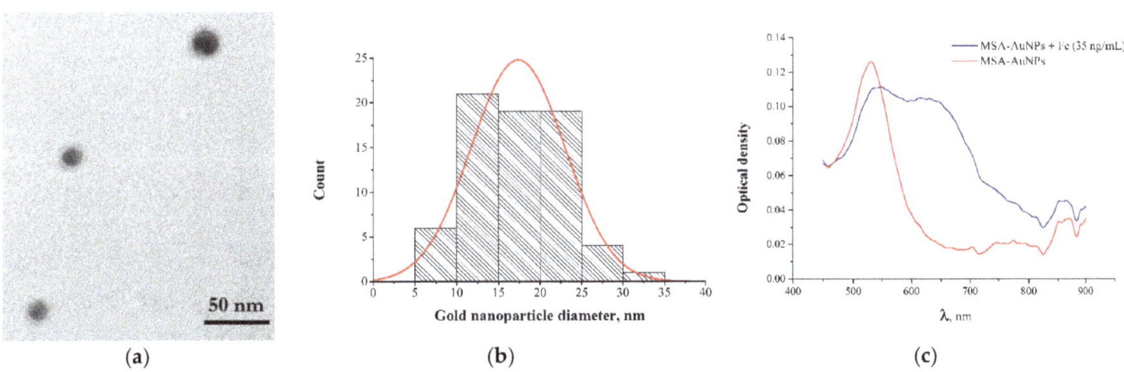

Figure 2. (**a**) TEM image of MSA-capped AuNPS. (**b**) Histogram of MSA-AuNPs particles' diameter distribution. (**c**) Absorption spectrum of MSA-AuNPs before (red) and after (blue) the addition of Fe(III) ions.

3.2. Colorimetric Determination of Fe(III) Ions

The proposed assay scheme is illustrated in Figure 3. In the presence of Fe(III), the coordination of MSA on the particle surface induced the critical convergence and loss of stability of the colloidal solution, resulting in a color change from red to blue. When the Fe(III) concentration reached 20 ng/mL, particle aggregation occurred, as registered by a decrease in the absorption peak at 530 nm and an increase in the peak at 660 nm. Consequently, the A_{530}/A_{660} ratio was selected as an analytical signal.

To find the optimal conditions for the determination of Fe(III) ions, we studied some factors affecting the analytical signal, such as the pH of the medium and the volume ratio of the reacting components (MSA-AuNPs and Fe(III)). The effect of pH on the changes in the analytical signal in the presence of Fe(III) ions was studied; the maximum detection sensitivity was achieved at pH 4.5.

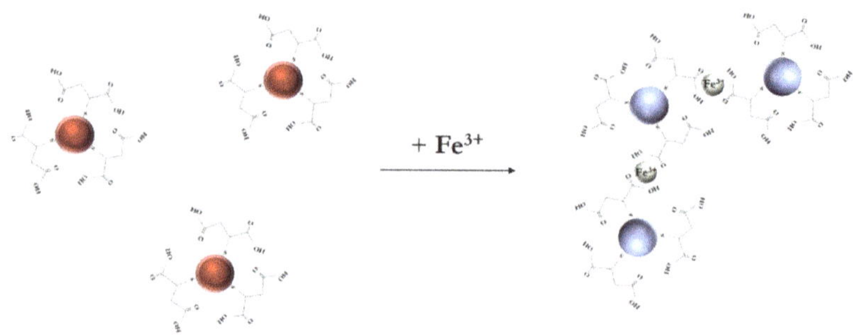

Figure 3. Scheme of the colorimetric detection of Fe(III) ions using MSA-capped gold nanoparticles.

To investigate the optimum volume ratio of the reaction components (MSA-AuNPs and Fe(III) solution), various ratios of the volumes of MSA-AuNPs and Fe(III) ion containing solutions (1:1, 1:3, and 1:30) were tested. The best sensitivity of the analysis was achieved with the volume ratio of MSA-AuNPs/Fe(III) ion solutions equal to 1:30.

The aggregation of MSA-AuNPs in the presence of Fe(III) was proceeded by complexation with the chelating ligands of mercaptosuccinic acid, which was accompanied by a blue shift in the absorption spectrum. The detection limit using MSA-AuNPs as the sensing probe was 23 ng/mL, at least 10 times lower than the maximum permissible concentration of Fe(III) for drinking water, which is 300 ng/mL by official regulations. The linear range of the calibration curve was 20–30 ng/mL, with an approximation factor of R2 0.98 (Figure 4). In addition, the developed sensing system allowed for the detection of Fe(III) by the naked eye at a concentration of 30 ng/mL, which corresponded to the blue color of the reaction mixture. The effectiveness of the developed system for the analysis of real samples was confirmed by the determination of Fe(III) ions in water samples without a matrix influence.

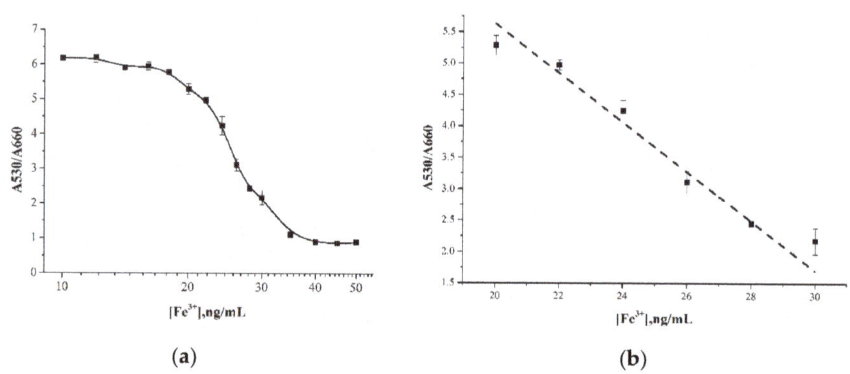

Figure 4. (**a**) Calibration curve for Fe(III) ion detection using AuNP-MSA. (**b**) Linear range of the calibration curve.

3.3. Selectivity of Fe(III) Ions' Detection

To check for the possible influence of other metal ions on the analytical signals for the detection of Fe(III) ions, solutions of Hg(II), Cd(II), As(II), Cu(II), Zn(II), Pb(II), Sn(II), and Cr(II) were tested under the chosen optimal conditions (Figure 5). It was found that the tested ions did not influence the analytical signal; the color change was achieved only for Fe(III) ions. The experimental data showed that the analytical signal (A_{530}/A_{660}) induced by Fe(III) significantly exceeded the signals for other metal ions.

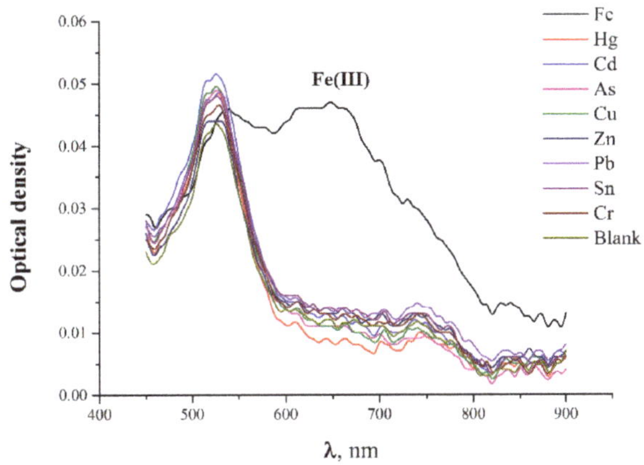

Figure 5. Absorbance spectra of MSA-capped AuNPs after the addition of 100 ng/mL of metal ions.

4. Conclusion

A highly sensitive and selective colorimetric system for Fe(III) ion detection based on mercaptosuccinic acid-functionalized gold nanoparticles was developed. The sensing mechanism is the aggregation of gold nanoparticles due to the selective complexation of mercaptosuccinic acid with Fe(III), resulting in a change in color and in the absorbance spectrum. Under optimal conditions, this system showed a good linear correlation between Fe(III) concentrations and the colorimetric signal, with visual and instrumental detection limits of 30 and 23 ng/mL, respectively. The system demonstrated high selectivity against other metal ions. The assay combined simple fabrication and operation with the possibility of sensitive on-site monitoring. The effectiveness of the developed system was confirmed by the determination of Fe(III) ions in the water samples without a matrix influence.

Supplementary Materials: The following are available online at https://www.mdpi.com/article/10.3390/CSAC2021-10624/s1.

Author Contributions: Conceptualization, A.N.B. and A.V.Z.; methodology, A.N.B., N.S.K. and K.V.S.; software, N.S.K.; validation, N.S.K., K.V.S. and A.N.B.; formal analysis, N.S.K., K.V.S. and S.M.P.; investigation, A.N.B. and K.V.S.; resources, S.M.P.; data curation, A.V.Z. and B.B.D.; writing—original draft preparation, N.S.K., K.V.S. and A.N.B.; writing—review and editing, A.V.Z. and B.B.D.; visualization, N.S.K.; supervision, A.V.Z. and B.B.D.; project administration, A.V.Z.; funding acquisition, A.N.B. All authors have read and agreed to the published version of the manuscript.

Funding: This work was financially supported by the Russian Science Foundation (project number 19-44-02020, assay development) and the Ministry of Science and Higher Education of the Russian Federation (study of assay selectivity).

Institutional Review Board Statement: Not applicable.

Informed Consent Statement: Informed consent was obtained from all subjects involved in the study.

Acknowledgments: This work was financially supported by the Russian Science Foundation (project number 19-44-02020, assay development) and the Ministry of Science and Higher Education of the Russian Federation (study of assay selectivity).

Conflicts of Interest: The authors declare no conflict of interest.

References

1. Xu, T.T.; Yang, J.X.; Song, J.M.; Chen, J.S.; Niu, H.L.; Mao, C.J.; Zhang, S.Y.; Shen, Y.H. Synthesis of high fluorescence graphene quantum dots and their selective detection for Fe^{3+} in aqueous solution. *Sens. Actuators B Chem.* **2017**, *243*, 863–872. [CrossRef]
2. Bagheri, H.; Afkhami, A.; Saber-Tehrani, M.; Khoshsafar, H. Preparation and characterization of magnetic nanocomposite of Schiff base/silica/magnetite as a preconcentration phase for the trace determination of heavy metal ions in water, food and biological samples using atomic absorption spectrometry. *Talanta* **2012**, *97*, 87–95. [CrossRef] [PubMed]
3. Wu, J.; Boyle, E.A. Determination of iron in seawater by high-resolution isotope dilution inductively coupled plasma mass spectrometry after $Mg(OH)_2$ coprecipitation. *Anal. Chim. Acta* **1998**, *367*, 183–191. [CrossRef]
4. Ariga, T.; Ito, K.; Imura, Y.; Yoshimura, E. High-performance liquid chromatography method for ferric iron chelators using a post-column reaction with Calcein Blue. *J. Chromatogr. B* **2015**, *985*, 48–53. [CrossRef] [PubMed]
5. Kuljanin, J.; Janković, I.; Nedeljković, J.; Prstojević, D.; Marinković, V. Spectrophotometric determination of alendronate in pharmaceutical formulations via complex formation with Fe(III) ions. *J. Pharm. Biomed. Anal.* **2002**, *28*, 1215–1220. [CrossRef]
6. Hirayama, T. Fluorescent probes for the detection of catalytic Fe(II) ion. *Free Radic. Biol. Med.* **2019**, *133*, 38–45. [CrossRef]
7. Piriya, V.S.A.; Joseph, P.; Daniel, S.C.G.K.; Lakshmanan, S.; Kinoshita, T.; Muthusamy, S. Colorimetric sensors for rapid detection of various analytes. *Mater. Sci. Eng. C* **2017**, *78*, 1231–1245. [CrossRef]
8. Alex, S.; Tiwari, A. Functionalized Gold Nanoparticles: Synthesis, Properties and Applications. A Review. *J. Nanosci. Nanotechnol.* **2015**, *15*, 1869–1894. [CrossRef]
9. Wu, S.P.; Chen, Y.P.; Sung, Y.M. Colorimetric detection of Fe^{3+} ions using pyrophosphate functionalized gold nanoparticles. *Analyst* **2011**, *136*, 1887–1891. [CrossRef] [PubMed]
10. Karami, C.; Alizadeh, A.; Taher, M.A.; Hamidi, Z.; Bahrami, B. UV-Visible Spectroscopy Detection of Iron(III) Ion on Modified Gold Nanoparticles With a Hydroxamic Acid. *J. Appl. Spectrosc.* **2016**, *83*, 687–693. [CrossRef]
11. Buduru, P.; Reddy, B.C.S.R. Oxamic acid and p-aminobenzoic acid functionalized gold nanoparticles as a probe for colorimetric detection of Fe^{3+} ion. *Sens. Actuators B Chem.* **2016**, *237*, 935–943. [CrossRef]
12. Kim, D.Y.; Shinde, S.; Saratale, R.; Syed, A.; Ameen, F.; Ghodake, G. Spectrophotometric determination of Fe(III) by using casein-functionalized gold nanoparticles. *Microchim. Acta* **2017**, *184*, 4695–4704. [CrossRef]
13. Li, J.; Wang, X.; Huo, D.; Hou, C.; Fa, H.; Yang, M.; Zhang, L. Colorimetric measurement of Fe^{3+} using a functional paper-based sensor based on catalytic oxidation of gold nanoparticles. *Sens. Actuators B Chem.* **2017**, *242*, 1265–1271. [CrossRef]
14. Kim, K.; Nam, Y.S.; Lee, Y.; Le, K.B. Highly Sensitive Colorimetric Assay for Determining Fe^{3+} Based on Gold Nanoparticles Conjugated with Glycol Chitosan. *J. Anal. Methods Chem.* **2017**, *2017*, 3648564. [CrossRef] [PubMed]
15. Cheney, G.E.; Fernando, Q.; Freiser, H. Some Metal Chelates of Mercaptosuccinic Acid. *J. Phys. Chem.* **1959**, *63*, 2055–2057. [CrossRef]
16. Larkworthy, L.F.; Sattari, D. Some complexes of thiomalate with bivalent transition metal ions and gold (I). *J. Inorg. Nucl. Chem.* **1980**, *42*, 551–559. [CrossRef]
17. Pawelec, M.; Stochel, G.; Eldik, R. Mechanistic information on the copper-catalysed autoxidation of mercaptosuccinic acid in aqueous solution. *Dalton Trans.* **2004**, *2*, 292–298. [CrossRef] [PubMed]
18. Vasilev, K.; Zhu, T.; Glasser, G.; Knoll, W.; Kreiter, M. Preparation of gold nanoparticles in an aqueous medium using 2-mercaptosuccinic acid as both reduction and capping agent. *J. Nanosci. Nanotechnol.* **2008**, *8*, 2062–2068. [CrossRef] [PubMed]

chemistry proceedings

MDPI

Proceeding Paper

Implementation of Metallic Film Electrodes for Catalytic Adsorptive Stripping Voltammetric Determination of Germanium(IV) †

Agnieszka Królicka *, Jerzy Zarębski and Andrzej Bobrowski

Department of Building Materials Technology, Faculty of Materials Science and Ceramics, AGH University of Science and Technology, Mickiewicza 30, 30-059 Krakow, Poland; zarebski.jerzy@gmail.com (J.Z.); abobrow@agh.edu.pl (A.B.)

* Correspondence: krolicka@agh.edu.pl

† Presented at the 1st International Electronic Conference on Chemical Sensors and Analytical Chemistry, 1–15 July 2021; Available online: https://csac2021.sciforum.net/.

Abstract: In the present work, it has been shown that bismuth film electrodes deposited on screen-printed carbon supports could be successfully used to provide well-shaped, sensitive and reproducible catalytic adsorptive stripping signals of Ge(IV) in the presence of catechol and V(IV)-HEDTA (HEDTA-N-hydroxyethyl-ethylene diamine-triacetic acid) complex.

Keywords: film electrodes; germanium; stripping voltammetry

Citation: Królicka, A.; Zarębski, J.; Bobrowski, A. Implementation of Metallic Film Electrodes for Catalytic Adsorptive Stripping Voltammetric Determination of Germanium(IV). *Chem. Proc.* **2021**, *5*, 7. https://doi.org/10.3390/CSAC2021-10484

Academic Editor: Núria Serrano

Published: 30 June 2021

1. Introduction

Germanium represents a unique group of elements known as Critical Raw Materials (EU). Although such elements typically constitute only a small percentage of a material by weight, they are essential to its performance [1]. Among the numerous analytical techniques available, voltammetry seems to have much to offer in this regard, as voltammetric techniques are insensitive to the presence of inorganic salts and, at the same time, they offer low detection limits. Among voltammetric methods, catalytic adsorptive stripping voltammetry (CAdSV) plays an essential role in trace analysis due to its remarkable sensitivity. To induce a catalytic effect which gives the method its outstanding sensitivity, ions with oxidizing properties must be added to the examined solution, e.g., nitrate, nitrite, bromate or chlorate. Unfortunately, many electrode materials, both metallic sensing layers as well as auxiliary polymers, are damaged under the influence of these oxidants.

As was reported earlier, using HMDE or silver-amalgam working electrodes and the supporting electrolyte containing V(IV)-HEDTA (HEDTA-N-hydroxyethyl-ethylene diamine-triacetic acid) complex (Figure 1), catechol or its derivatives, the well-developed germanium signals could be recorded at nM level [2,3]. To develop a workable analytical procedure for the detection of trace amounts of germanium, and to meet the current guidelines that impose limitations on the application of mercury in chemical and analytical laboratories, an environmentally friendly alternative should be employed instead. The metallic film electrodes, such as bismuth (BiFE) and lead (PbFE) electrodes are among the most widely used sensors in the field of stripping voltammetry [4].

This work aims to assess the applicability of lead and bismuth film electrodes deposited electrochemically or by physical deposition from gaseous phase on different supports to provide catalytic adsorptive stripping signals of Ge(IV) in the presence of catechol and V(IV)-HEDTA complex.

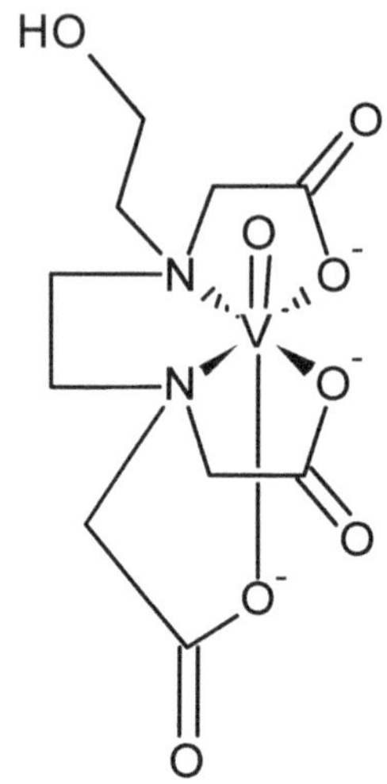

Figure 1. The structure of V(IV)-HEDTA complex.

2. Materials and Methods

Electrochemical study was performed on a Autolab 204 analyzer (Metrohm Autolab B.V., Utrecht, The Netherlands). Disposable screen-printed electrodes (4 mm diameter) (DropSens, Oviedo, Spain) or disc electrodes (3 mm diameter) made of glassy carbon or gold (Mineral, Łomianki-Sadowa, Poland) were used as supports for bismuth films. Platinum wire and Ag/AgCl (3 M KCl) were applied as the anode and reference electrodes. All applied reagents were analytical grade.

Lead and bismuth films were plated just prior to use by means of potentiostatic deposition. Before plating, the disc substrates were polished using an Al_2O_3 suspension (0.3 and 0.05 μm) applied onto a polishing cloth. Screen-printed electrodes did not require any preparation or processing other than 2 min of soaking in the plating solution immediately prior to electrolysis. The plating process performed in quiescent 0.34 M $HClO_4$ containing 0.043 M of Bi(III) or stirred 0.2 M acetate buffer containing 0.003 M Pb(II) was monitored by recording chronoamperometric curves and stopped when the charge reached the defined threshold (E_{plat} = −0.9 V, Q_{plat} = 0.8 mC per mm^2). Pre-plated electrodes were rinsed with 0.1 M acetate buffer (PbFE) or 0.34 M $HClO_4$ (BiFE) and water. The PVD deposition followed the protocol described in the previous work [5].

The supporting electrolyte contained 0.05 M acetate buffer (pH of 4.4), 1 mM of catechol, 1 mM of V(IV) and 1.5 mM of HEDTA. CAdSV voltammograms were recorded after 30 s of accumulation performed at the potential of −0.6 V (PbFE) or −0.4 V (BiFE) by differential pulse mode.

3. Results and Discussion

3.1. Support Selection

To deposit the metallic sensing layers, the following supports were considered: (1) carbon-based electrodes: glassy carbon, carbon paste, impregnated carbon, screen printed carbon and (2) gold-based electrodes: bulk disc, PDV deposited gold, gold screen printed. The sensing layers were deposited electrochemically by ex situ plating or by PVD deposition. The preliminary tests of freshly prepared films rinsed with water were performed by microscopic inspection. In the case of carbon-based supports and bismuth films, the visual changes were not pronounced, as the plating only gave the surface a black, velvety appearance. In the case of lead deposits, the electrode surface was turned gray. In the case of gold supports, the results were more complex. While the expected finding was that the type of the support (gold monolithic disc, gold screen printed layer, or PVD deposited gold) does not play the key role, this study showed that the method of support preparation plays an important role. The films deposited on the gold disc electrode were very stable and adhesive. The lead layers deposited on the SPE or PVD gold were oxidized within minutes once removed from the plating solution (Figure 2). In the case of bismuth, the oxidation at AuSPE was slower but unrelenting. Information on the mechanism of oxidation of the metallic layer was provided by analysis of the bismuth films deposited on the microscratched Au PVD layer (Figure 3a–c). The microscopic images revealed the bismuth crystallites formed near microscratches are susceptible to oxidation. Gradually the gold support was exposed, but after 10 min, further changes were not observed as the bismuth crystallites adjacent to the scratches were depleted. The color histograms of microscopic images of Bi deposited on AuPVD shown in Figure 3d have not demonstrated any further changes for pictures taken 10 and 12.5 min after removal from the plating solution. The microanalysis performed by SEM XRF, and XRD studies of AuSPE as well as AuPVD electrodes did not reveal other elements than gold. Some insight was provided by contact angle measurements, displaying a significant difference in wettability of PVD gold (68.0 ± 0.2°) and SPE (92.0 ± 0.3°) electrodes. Since the surface microstructure of materials correlates closely with the apparent contact angle at the boundary between the liquid and the surface, the different surface properties of AuSPE and AuPVD were confirmed. The AuPVD contact angle value corresponds well to earlier reported values [6], while those obtained for AuSPE are substantially higher, placed on the threshold of hydrophobicity. The high contact angle can be explained by the rough three-dimensional structure of the gold layer, preventing the access of water molecules to the electrode surface by trapped air or by organic compounds of the ink used for screen printing. Regardless of cause, the gold screen printed electrodes cannot be used for bismuth or lead ex situ plating.

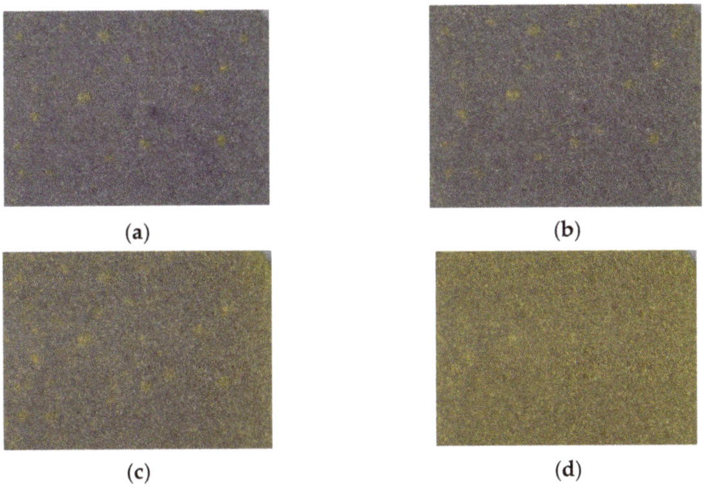

Figure 2. Microscopic images of Pb plated gold SPE recorded after 0 (**a**), 2.5 (**b**), 5 (**c**) and 7.5 (**d**) min after plating. Gray regions represent the lead layer while yellow the exposed gold support.

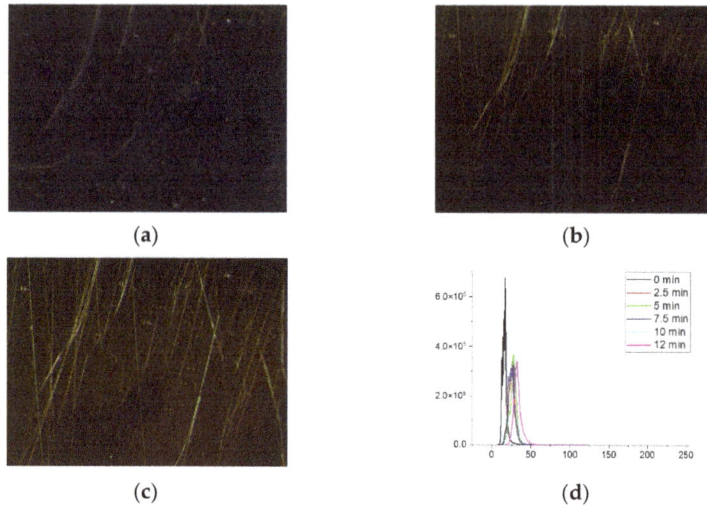

Figure 3. Microscopic images of Bi plated gold sputtered electrode recorded 0 (**a**), 2.5 (**b**) and 12 (**c**) min after plating (150x; images brightened to make the details more visible). Color histogram analysis of the bismuth layer plated on the sputtered gold electrode (**d**) (https://www.dcode.fr/image-histogram, accessed on 8 June 2021).

3.2. Stability Studies of Metallic Films in Contact with V(IV)-HEDTA Solution

The externally plated bismuth and lead films as well as PVD deposited bismuth and lead layers were exposed to the supporting electrolyte containing the V(IV)-HEDTA complex, catechol and acetate buffer. Employing bismuth film electrodes of any type, it was possible to record well-shaped germanium signals in solutions containing from a few to several hundred nM of germanium. In the case of lead film electrodes, only PbFE deposited on glassy carbon, carbon paste and impregnated carbon provided measurable germanium signals. Lead layers deposited on SPE supports of any type did not deliver any germanium signals. Although the signals recorded by the most promising lead electrode, namely, PbFE/GC, in the solution containing 30 nM of Ge(IV), were initially quite pronounced, there was no Ge(IV) signal after recording ten or so voltammograms. In this way, the unsuitability of PbFEs for their intended use in stripping voltammetry became apparent.

3.3. Analytical Performance of Bismuth Plated Screen Printed Electrodes

It was shown [2] that when the HMDE electrode was used as the working electrode, the catalytic activity of HEDTA vanadium complexes towards the germanium complex was correlated with the redox behavior of V(IV)-HEDTA. In the case of pyrogallol germanium complexes, the intensity of the germanium signal correlated with the characteristics of the voltammetric signal representing the reduction of the V(IV) complex [3]. The cyclic voltammograms shown in Figure 4a were recorded by means of bismuth plated glassy carbon and carbon SPE electrodes and compared with the curves obtained when CGMDE and GC were used. Both CGMDE and GC delivered well-shaped reversible voltammograms of V(IV)-HEDTA complex, while at BiFE electrodes only reduction signals were observed. Such behavior indicates that the catalytic process at BiFEs proceeds following a different mechanism than that observed at CGMDE.

The catalytic signals of Ge(IV) recorded at BiFE/SPE in the presence of V(IV)-HEDTA were pronounced and highly reproducible, as it is shown in Figure 4b. The calibration curve is given by the equation $y = (0.083 \pm 0.001)x + (0.06 \pm 0.01)$ ($r^2 = 0.9989$), where y and x denote the peak current (μA) and Ge(IV) concentration (nM). The $I_p = f(c_{Ge(IV)})$ is linear within the range from 2 to 30 nM (LOD = 1.5 nM). Finally, BiFE/SPEs were applied for Ge(IV) determination in Ge(IV) spiked snow water (10 nM) via the standard addition method and the concentration of 10.05 ± 0.11 nM was determined.

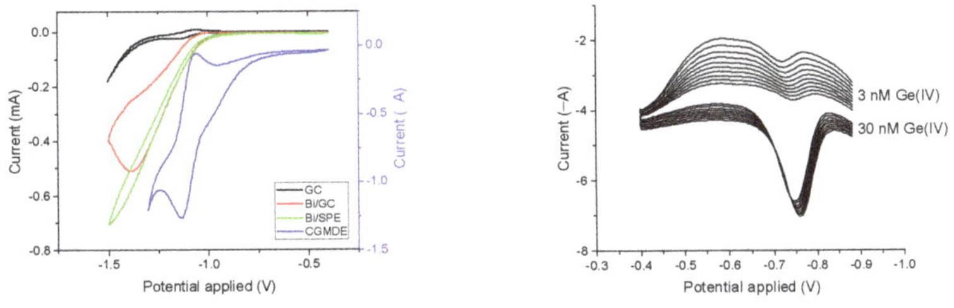

(**a**) (**b**)

Figure 4. CV voltammograms recorded in an 0.05 M acetate buffer containing 1 mM of V(IV) and 2 mM of HEDTA using glassy carbon (GC), bismuth plated glassy carbon (BiFE/GC), bismuth plated screen printed carbon electrode (BiFE/SPE) and a controlled-growth mercury drop electrode (CGMDE). Scan rate = 50 mVs^{-1} (**a**). Ten consecutive differential pulse voltammograms recorded by Bi/SPE electrode in the solution containing 3 or 30 nM of Ge(IV) and 0.05 M acetate buffer (pH of 4.4), 1 mM of catechol, 1 mM of V(IV) and 1.5 mM of HEDTA (**b**).

4. Conclusions

The results reported here demonstrate that the properties of lead and bismuth film electrodes differ considerably and only bismuth plated electrodes enable germanium analytical signals to be obtained when Ge(IV)-catechol-V(IV)-HEDTA system is employed.

Author Contributions: Conceptualization, A.K.; methodology, A.K.; software, A.K.; writing—original draft preparation, A.K.; writing—review, A.B. and J.Z.; visualization, A.K. All authors have read and agreed to the published version of the manuscript.

Funding: This work was supported by the subsidy of the Ministry of Education and Science for the AGH University of Science and Technology in Kraków (Project No 16.16.160.557).

Institutional Review Board Statement: Not applicable.

Informed Consent Statement: Not applicable.

Data Availability Statement: The data are available on request from the corresponding author, (A.K.).

Conflicts of Interest: The authors declare no conflict of interest.

References

1. Eggert, R.G. Minerals go critical. *Nat. Chem.* **2011**, *3*, 688–691. [CrossRef] [PubMed]
2. Zarębski, J.; Bobrowski, A.; Gonciarczyk, J.; Królicka, A. Extremely sensitive germanium stripping voltammetric determination with the use of a new Ge (IV)-catechol-V(IV)-HEDTA catalytic adsorptive system. *Electrochim. Acta* **2019**, *324*, 134859. [CrossRef]
3. Zarębski, J.; Bobrowski, A.; Gonciarczyk, J.; Królicka, A. Selection of Optimal Ligand and Vanadium (IV) Complexonate for Sensitive Catalytic Adsorptive Stripping Voltammetric Quantification of Germanium. *Electroanalysis* **2020**, *32*, 2213–2219. [CrossRef]
4. Economou, A. Screen-printed electrodes modified with "green" metals for electrochemical stripping analysis of toxic elements. *Sensors* **2018**, *18*, 1032. [CrossRef] [PubMed]
5. Bobrowski, A.; Królicka, A.; Śliwa, J.; Zarębski, J.; Januś, M.; Kyzioł, K. PVD fabrication of lead film electrodes and their catalytic adsorptive stripping voltammetric performance in the presence of oxidants. *Electrochem. Commun.* **2018**, *94*, 49–54. [CrossRef]
6. Canning, J.; Tzoumis, N.; Beattie, J.K.; Gibsonc, B.C.; Ilagand, E. Water on Au sputtered films. *Chem. Commun.* **2014**, *50*, 9172–9175. [CrossRef] [PubMed]

chemistry proceedings

MDPI

Proceeding Paper

Tunable Electrochemical Sensors Based on Carbon Nanocomposite Materials towards Enhanced Determination of Cadmium, Lead and Copper in Water †

Laia L. Fernández [1,2], Julio Bastos-Arrieta [3], Cristina Palet [1] and Mireia Baeza [2,*]

1 Grup de Tècniques de Separació en Química (GTS), Departament de Química, Universitat Autònoma de Barcelona, Carrer dels Til·lers, 08193 Bellaterra, Catalunya, Spain; Laia.Lopez@uab.cat (L.L.F.); cristina.palet@uab.cat (C.P.)
2 Group of Biological Treatment and of Liquid and Gaseous Effluents, Nutrient Removal, and Odors and Volatile Organic Compounds (GENOCOV), Departament Química, Universitat Autònoma de Barcelona, Carrer dels Til·lers, 08193 Bellaterra, Catalunya, Spain
3 Grup de Biotecnologia Molecular i Industrial, Departament d'Enginyeria Química, Universitat Politècnica de Catalunya, Rambla Sant Nebridi, 22, 08222 Terrassa, Catalunya, Spain; julio.bastos@upc.edu
* Correspondence: mariadelmar.baeza@uab.cat
† Presented at the 1st International Electronic Conference on Chemical Sensors and Analytical Chemistry, 1–15 July 2021; Available online: https://csac2021.sciforum.net/.

Abstract: Many carbon materials are well-known conductive materials, widely used in the fabrication of composite electrodes. In this work, diverse allotropic forms of carbon such as graphite, MWCNTs and rGO were tested. Furthermore, these materials allow the construction of cheaper, smaller, portable, reliable and easy-to-use devices, which can be easily modified. The above-mentioned composite electrodes were developed for metal analysis in water such as Cu, Cd and Pb that, at a high concentration, can have consequences on human health. SWASV is the selected technique. It would be ideal to exploit the potential properties of mercury for metal detection by tuning the electrode's surface. Due to mercury's hazardous properties and to reduce the amount of this substance used in polarography, the use of nanoparticles is a good option due to their properties. Mercury nanoparticles were used to modify the surface of the composite electrodes to improve electroanalytical sensor response. For this reason, using these modified composite electrodes can lower detection limits and widen the linear range that can be achieved for Cd (0.05–1 mg·L^{-1}) and Pb (0.045–1 mg·L^{-1}). However, for Cu (0.114–1.14 mg·L^{-1}), meaningful variations were not observed compared to the bare electrode.

Keywords: electrochemistry; Hg nanoparticles; graphite; composite electrodes; metal analysis; SWASV

check for updates

Citation: Fernández, L.L.; Bastos-Arrieta, J.; Palet, C.; Baeza, M. Tunable Electrochemical Sensors Based on Carbon Nanocomposite Materials towards Enhanced Determination of Cadmium, Lead and Copper in Water. *Chem. Proc.* **2021**, *5*, 8. https://doi.org/10.3390/CSAC2021-10456

Academic Editor: Núria Serrano

Published: 30 June 2021

Publisher's Note: MDPI stays neutral with regard to jurisdictional claims in published maps and institutional affiliations.

1. Introduction

Water is fundamental for all Earth's living forms, and a key issue for social and economic development. Currently, water analysis is a vital topic, for because monitoring some parameters is important to prevent some health problems. One of the parameters that has become important involves determining the concentration of heavy metals in water. To do this, several analysis techniques are used, such as atomic absorption spectroscopy (AAS) [1], inductively coupled plasma (ICP) [2], high-performance liquid chromatography (HPLC) [3], etc. Some of the metals that can be found in water are Cu, Cd and Pb and, at high concentrations, can have consequences on human health [4–6].

In this work, a voltametric technique has been chosen, known as square-wave anodic stripping voltammetry (SWASV) [7,8]. SWASV consists of two steps: first, applying a potential to preconcentrate the analyte on the surface of the electrode; second, taking a measurement by applying staircase potential to record the current generated.

To use this technique, composite electrodes were construct using different carbon materials and a non-conductor epoxy. The behavior of graphite, reduced graphene oxide (rGO) and carbon nanotubes (CNTs) were tested in the detection of Cd, Pb and Cu. However, we work with the bare electrode; the modification of their surface with mercury nanoparticles (Hg-NPs) was also tested [9]. Mercury was used, a long time ago, in polarography, and it is well-known for its ability to form amalgams with some metals, reducing the potential where they appear [10,11]. Hence, taking advantage of these properties, the aim of this work is to reduce the amount of mercury used in polarography for the determination of Cd, Pb, and Cu.

2. Composite Electrodes Construction, Characterization, and Modification

2.1. Composite Electrode Construction

Composites were constructed using three different carbon materials: graphite, CNTs and rGO. The first step is to weld a copper sheet to a commercial connector; after that, it is placed in a PVC tube. A mixture of one of the carbon materials and Epotek H77 is prepared, and the PVC tube (2.1 cm, ⌀6 mm) is filled with this mixture. Then, it is cured for 2 days at 80 °C. Then, the surface must be polished.

The percentages tested of carbon materials are shown in Table 1. These percentages were optimized previously, and they are related to their respective improvement in the electroanalytical properties of developed sensors, in terms of detection limit and sensitivity [12].

Table 1. Percentages used in the construction of each electrode.

Material	% Carbon Material	% Epotek H77
Graphite	15	85
	20	80
CNTs	10	90
rGO	15	85

2.2. Composite Electrode Chacaterization

Electrodes were characterized using Cyclic Voltammetry (CV) and Electrochemical Impedance Spectroscopy (EIS) using a computer-controlled Multi-AUTOLAB M101 (Eco Chemie, Utrecht, The Netherlands) with a three-electrode cell: a platinum-based electrode 53–671 (Crison Instruments, Alella, Barcelona, Spain) as a counter electrode, an Ag/AgCl handmade electrode as a reference electrode and the constructed composite electrodes as working electrodes. The characterization was performed in solution composed of 0.01 M $K_4Fe(CN_6)$, 0.01 M $K_3Fe(CN_6)$ and 0.1 M KCl. For CV, the scan rate was 10 $mV{\cdot}s^{-1}$ and the rate of frequencies used in EIS was 0.01 to 10^4 Hz.

The behavior of the 15% rGO electrode was unusual, possibly related to the orientation of the layers in the Epotek H77 matrix, and its characterization using CV and EIS was not successful. In Figure 1, the characterization of the rest of the carbon electrodes, with graphite or CNTs, can be observed. The most notable difference is showed in EIS, where the 20% graphite presents the lower charge transference resistance. Thus, a highly conductive surface is then available for the preconcentration of cationic metals.

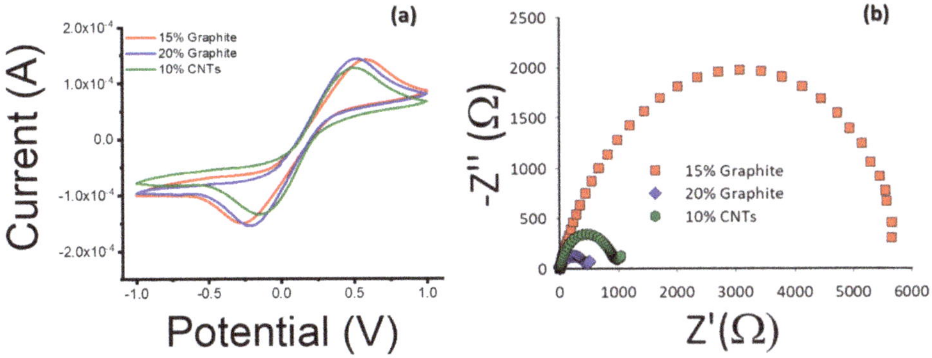

Figure 1. CV (**a**) and EIS (**b**) characterization of the different electrodes.

2.3. Composite Electrode Modification with Hg-Nps

After electrode characterization, the surface of the electrode is modified with mercury nanoparticles (Hg-NPs) following the synthesis from [9]. In the synthesis, 78 mg $Hg_2(NO_3)_2 \cdot 2H_2O$ is used, 1 mL 1 M HNO_3 is added and then 0.5 mL of a solution of 3.5 g of PVA (Polyvinyl Alcohol) added to 16 mL of Milli-Q water. All the steps of the synthesis were performed at 25 °C and under stirring conditions.

A total of 20 µL of the nanoparticle solution is drop casted on the electrode surface and dried in the oven at 80 °C for 2 h. The modified electrodes were characterized using scanning electron microscopy (SEM) (MerlinFe-SEM, Carl Zeiss, Germany) and the Hg-NPs were characterized using transmission electron microscopy (TEM) (JEM-2011 200 kV, Jeol, Peabody, MA, USA) (see Figure 2).

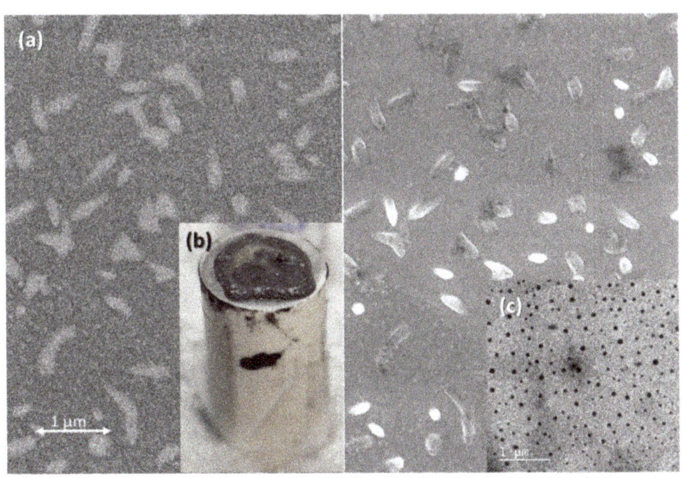

Figure 2. (**a**) Retrodispersive (**left**) and secondary electron (**right**) SEM images; (**b**) 20% graphite electrode drop casted with Hg-NPs image; (**c**) TEM image of the Hg-NPs.

2.4. Metal Solution Preparation and Determination

The metal solutions were prepared using certified stock standards of 37 mg·L^{-1} $Pb(NO_3)_2$ (≥99%, supplied from Sigma-Aldrich), 11,438 mg·L^{-1} $Cu(NO_3)_2$ (99.5 %, purchased from Merck) and 1000 mg·L^{-1} $Cd(NO_3)_2$ (99 %, obtained from Panreac). They were added to a 0.1 M acetic acid (CH_3COOH, 99.9% acquired from J.T.Baker, HPLC reagent)/0.1 M Ammonium acetate (NH_4CH_3COO, 97 % purchased from Panreac) buffer with Milli-Q water at pH 4.6 [13].

2.5. Bare Composite Electrodes

For metal determination, the technique chosen was SWASV. This consists of applying a potential (−1.4 V) for 7 min that reduces the metal ions on the electrode surface; then, staircase potential is applied and the current generated is recorded. This process is performed under N_2 bubbling. Moreover, a modification in the electrochemical cell is used. Instead of using a handmade reference electrode, the one used for the measurements is Orion 900 electrode (Thermo Scientific, Beverly, MA, USA).

Firstly, the bare electrodes were used for the electrochemical detection of Cd, Pb and Cu. The results for all electrodes studied are shown in Figure 3.

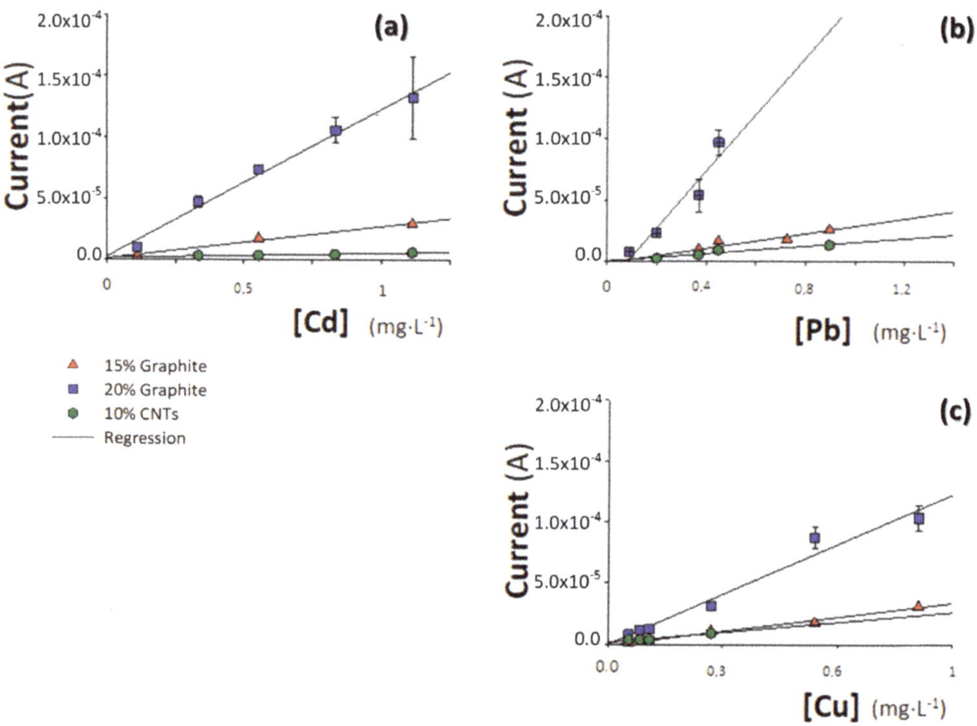

Figure 3. Calibration curves for Cd (**a**), Pb (**b**) and Cu (**c**) for each raw material.

As can be seen, 20% graphite electrodes showed the best response, as it has a better sensitivity compared with 15% graphite and 10% CNTs composite electrodes for three metal cations analyzed.

2.6. Hg-NPs Drop Casted Electrodes

The next step is to modify the surface of the 20% graphite electrode with Hg-NPs, as mentioned above. Once the surface is modified, the electrode is tested for Cd, Pb and Cu determination using SWASV. The corresponding results are shown in Figure 4.

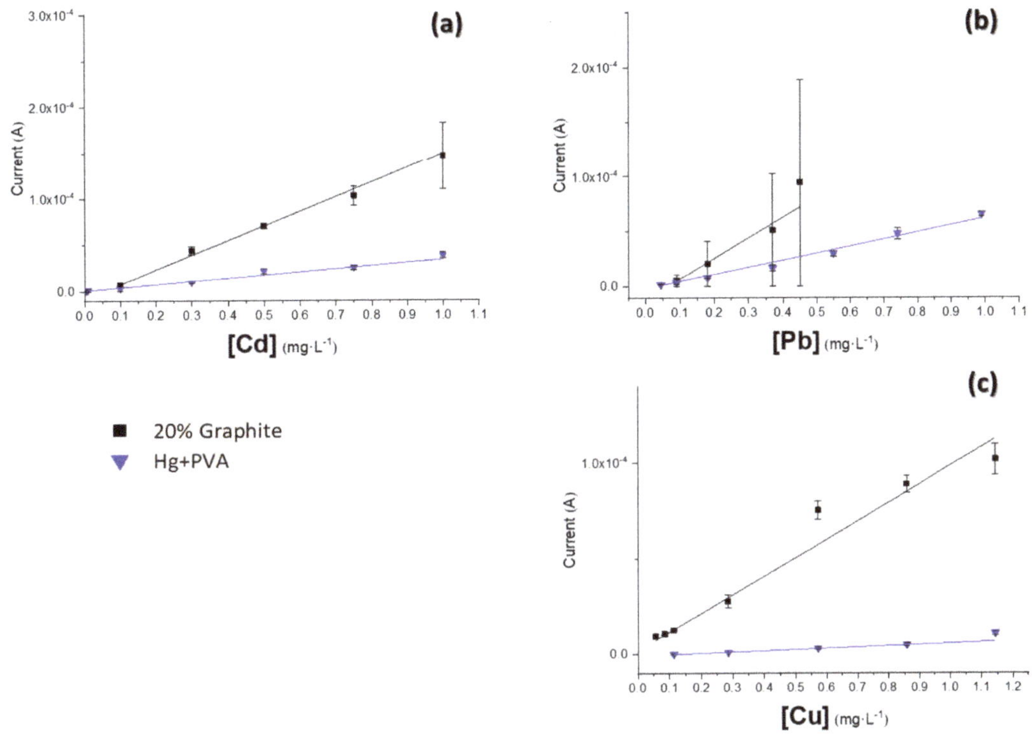

Figure 4. Calibration curves for Cd (**a**), Pb (**b**) and Cu (**c**) for 20% graphite (black) and 20% graphite plus Hg-NPs (blue).

With this modified 20% graphite electrode, lower quantification limits can be achieved. In Table 2, all the parameters of the calibration curves are summarized.

Table 2. Feature parameters: sensitivity, r^2 and linear range of each cationic metal detected separately.

	$[Cd^{2+}]$		
Electrode (20% graphite)	**Sensitivity** $[A{\cdot}(mg{\cdot}L^{-1})^{-1}]$	r^2 **(n)**	**Linear Range** $(mg{\cdot}L^{-1})$
Bare	$(1.6 \pm 0.1) \times 10^{-4}$	0.995 (n = 5)	0.1–1
plus Hg-NPs	$(3.4 \pm 0.2) \times 10^{-5}$	0.98 (n = 6)	0.05–1
	$[Pb^{2+}]$		
Electrode (20% graphite)	**Sensitivity** $[A{\cdot}(mg{\cdot}L^{-1})^{-1}]$	r^2 **(n)**	**Linear Range** $(mg{\cdot}L^{-1})$
Bare	$(1.9 \pm 0.2) \times 10^{-4}$	0.95 (n = 4)	0.09–0.45
plus Hg-NPs	$(6.4 \pm 0.3) \times 10^{-5}$	0.98 (n = 7)	0.045–1
	$[Cu^{2+}]$		
Electrode (20% graphite)	**Sensitivity** $[A{\cdot}(mg{\cdot}L^{-1})^{-1}]$	r^2 **(n)**	**Linear Range** $(mg{\cdot}L^{-1})$
Bare	$(9.7 \pm 0.9) \times 10^{-5}$	0.95 (n = 7)	0.057–1.14
plus Hg-NPs	$(7 \pm 1) \times 10^{-6}$	0.90 (n = 5)	0.114–1.14

3. Conclusions

Carbon composite electrodes are very versatile, robust, and reliable electrodes to work with for Cd, Pb and Cu detection. The well-known properties of mercury to form an amalgam with other metals can be taken advantage of to modify the surface of the carbon composite electrode in order to decrease the limit detection of the bare electrode. To emulate the polarography, the use of Hg-NPs reduces the amount of mercury used without losing its properties. In this case, Cd and Pb form an amalgam with Hg, reducing the detection limit (Cd = 0.05 mg·L^{-1}; Pb = 0.045 mg·L^{-1}) in comparison with the bare electrode. The Cu metallic cation does not exhibit this behavior. Although the bare electrode has higher sensitivity because its electroactive area is not modified, when the electrode was modified with Hg-NPs, its electroactive area decreases. We added a polymer (from the synthesis of the NPs) over the electrode's surface that is not as good as a conductor as graphite. On the other hand, we improved the detection limit due to the specific interaction of mercury with metals cations.

Supplementary Materials: The following are available online at https://www.mdpi.com/article/10.3390/CSAC2021-10456/s1.

Author Contributions: Conceptualization, M.B. and C.P.; methodology, L.L.F.; validation, M.B. and C.P.; formal analysis, L.L.F.; investigation, L.L.F.; resources, M.B.; data curation, L.L.F. and C.P.; writing—original draft preparation, L.L.F.; writing—review and editing, M.B., C.P. and J.B.-A.; visualization, M.B and C.P.; supervision, M.B., C.P. and J.B.-A.; project administration, M.B.; funding acquisition, M.B. All authors have read and agreed to the published version of the manuscript.

Funding: Authors are thankful for the financial support from the RTI2018-099362-B-C21 research project from the Spanish Ministerio de Economía y Competitividad y Fondo Europeo de Desarrollo Regional (MINECO/FEDER, UE).

Institutional Review Board Statement: Not applicable.

Informed Consent Statement: Not applicable.

Data Availability Statement: The data presented in this study are available on request from the corresponding author. The data are not publicly available due to the repository that is used to keep the data is a private one provided by the University.

Acknowledgments: Laia L. Fernández acknowledges Universitat Autònoma de Barcelona (UAB) for the PIF grant. She thanks Servei de Microscopia from UAB for the assistance in electron microscopy characterization.

Conflicts of Interest: The authors declare no conflict of interest.

References

1. Hashim, R.; Song, T.H.; Muslim, N.Z.M.; Yen, T.P. Determination of heavy metal levels in fishes from the lower reach of the kelantan river, Kelantan, Malaysia. *Trop. Life Sci. Res.* **2014**, *25*, 21–39. [PubMed]
2. Pérez-Ràfols, C.; Trechera, P.; Serrano, N.; Díaz-Cruz, J.M.; Ariño, C.; Esteban, M. Determination of Pd(II) using an antimony film coated on a screen-printed electrode by adsorptive stripping voltammetry. *Talanta* **2017**, *167*, 1–7. [CrossRef] [PubMed]
3. Okano, G.; Igarashi, S.; Yamamoto, Y.; Saito, S.; Takagai, Y.; Ohtomo, T.; Kimura, S.; Ohno, O.; Oka, Y. HPLC-spectrophotometric detection of trace heavy metals via 'cascade' separation and concentration. *Int. J. Environ. Anal. Chem.* **2015**, *95*, 135–144. [CrossRef]
4. Wani, A.L.; Ara, A.; Usmani, J.A. Lead toxicity: A review. *Interdiscip. Toxicol.* **2015**, *8*, 55–64. [CrossRef] [PubMed]
5. Azaman, F.; Juahir, H.; Yunus, K.; Azid, A.; Kamarudin, M.K.; Toriman, M.E.; Mustafa, A.D.; Amran, M.A.; Hasnam, C.N.; Saudi, A.S. Heavy metal in fish: Analysis and human health—A review. *J. Teknol.* **2015**, *77*, 61–69. [CrossRef]
6. Godt, J.; Scheidig, F.; Grosse-Siestrup, C.; Esche, V.; Brandenburg, P.; Reich, A.; Groneberg, D.A. The toxicity of cadmium and resulting hazards for human health. *J. Occup. Med. Toxicol.* **2006**, *1*, 1–6. [CrossRef] [PubMed]
7. Nor, Y.M.; McNeal, B.L. Square wave anodic stripping voltammetry: Determination of trace metals in aqueous extracts of soil samples. *Chem. Speciat. Bioavailab.* **1993**, *5*, 135–140. [CrossRef]
8. Scarponi, G. Optimization of Square Wave Anodic Stripping Voltammetry (SWASV) for the Simultaneous Determination of Cd, Pb, and Cu in Seawater and Comparison with Differential Pulse Anodic Stripping Voltammetry (DPASV). April 2002. Available online: http://citeweb.info/20020474074 (accessed on 23 June 2021).
9. Ramesh, G.V.; Prasad, M.D.; Radhakrishnan, T.P. Mercury nanodrops and nanocrystals. *Chem. Mater.* **2011**, *23*, 5231–5236. [CrossRef]
10. Ross, J.W.; Demars, R.D.; Shain, I. Analytical Applications of the Hanging Mercury Drop Electrode. *Anal. Chem.* **1956**, *28*, 1768–1771. [CrossRef]

11. Pujol, L.; Evrard, D.; Groenen-Serrano, K.; Freyssinier, M.; Ruffien-Cizsak, A.; Gros, P. Electrochemical sensors and devices for heavy metals assay in water: The French groups' contribution. *Front. Chem.* **2014**, *2*, 1–24. [CrossRef] [PubMed]
12. Muñoz, J.; Montes, R.; Baeza, M. Trends in electrochemical impedance spectroscopy involving nanocomposite transducers: Characterization, architecture surface and bio-sensing. *TrAC Trends Anal. Chem.* **2017**, *97*, 201–215. [CrossRef]
13. Dropsens, S.L. Serie: Prácticas de laboratorio Ref.: DRP-PL4, "Determinación de cobre en agua de grifo". Available online: https://www.dropsens.com/pdfs_productos/new_brochures/pl1_pl2_pl3_pl4_pl5_pl6_pl7.pdf (accessed on 13 September 2020).

chemistry proceedings

MDPI

Proceeding Paper

Colorimetric Determination of Nitrate after Reduction to Nitrite in a Paper-Based Dip Strip †

Amer Charbaji *, Hojat Heidari-Bafroui, Nasim Rahmani, Constantine Anagnostopoulos and Mohammad Faghri *

Microfluidics Laboratory, Department of Mechanical, Industrial and Systems Engineering, University of Rhode Island, 2 East Alumni Avenue, Kingston, RI 02881, USA; h_heidari@uri.edu (H.H.-B.); nara7@uri.edu (N.R.); anagnostopoulos@uri.edu (C.A.)

* Correspondence: charbaji@uri.edu (A.C.); faghrim@uri.edu (M.F.)

† Presented at the 1st International Electronic Conference on Chemical Sensors and Analytical Chemistry, 1–15 July 2021; Available online: https://csac2021.sciforum.net/.

Abstract: Paper-based microfluidic technology is a relatively new field of research that provides low-cost platforms and sensors for point-of-care diagnostics. While the majority of research in this field has been for biomedical applications, more and more paper-based devices and platforms are being designed and developed for environmental applications, such as water quality monitoring and assessment. One such application is the detection of nitrate in water samples. Colorimetric detection of nitrate by paper-based devices using the Griess assay requires the reduction of nitrate to nitrite before undergoing the reaction. In this paper, we measured the performance of a paper-based dip strip for detecting nitrate and nitrite by calculating its limit of detection and limit of quantification. We also calculated the reduction efficiency of vanadium (III) chloride in the dip strip for detecting nitrate. Our results show that the reduction time of nitrate via vanadium (III) chloride is much longer than that when using zinc microparticles. Our results also show that the performance of the dip strip using vanadium (III) chloride for nitrate detection is not as good as more intricate paper-based devices that have a separate reaction zone with zinc microparticles. The limits of detection and quantification calculated were 3.352 and 7.437 ppm, and the nitrate reduction efficiency varied over the range of nitrate concentrations tested.

Keywords: nitrate reduction; zinc microparticles; vanadium (III) chloride; materials for chemical sensing; nitrate detection; Griess reaction; colorimetric assay; paper-based devices; paper microfluidics; point-of-care diagnostics

Citation: Charbaji, A.; Heidari-Bafroui, H.; Rahmani, N.; Anagnostopoulos, C.; Faghri, M. Colorimetric Determination of Nitrate after Reduction to Nitrite in a Paper-Based Dip Strip. *Chem. Proc.* **2021**, *5*, 9. https://doi.org/10.3390/CSAC2021-10459

Academic Editor: Manel del Valle

Published: 30 June 2021

1. Introduction

Paper-based microfluidic technology has been gaining a lot of attention over the past several years for the many advantages it provides. Most importantly, paper-based microfluidic technology allows the development of low-cost, portable and easy-to-use devices and sensors that can be easily disposed of. These devices can also provide qualitative or quantitative results and data at the point of care without the need for specialized equipment or power sources. Several paper-based devices have been developed for various applications, such as for water analysis [1–4], biomedical applications [5,6], food analysis [7–10], soil analysis [11] and many other miscellaneous applications [12–15]. The field of paper-based microfluidics is expected to continue garnering greater attention as more applications are sought after or the performance improved for the ones already developed [16].

Paper-based devices are generally made up of several different sections that serve different purposes. While more complex devices may include valves and actuators to manipulate fluids and perform multistep reactions [17,18], simpler devices generally include a sample port, transport channels, reactions zones and a detection zone [19]. The majority of paper-based devices use colorimetric detection since it is the simplest technique to produce a quantifiable signal [20,21]. Properties of the material used in paper-based devices influence assay performance and have a substantial impact on the development of paper-based sensors [22]. Therefore, proper material selection and optimization is critical to enhancing the performance of assays in paper-based devices [23]. This is usually an iterative and an ongoing process to learn and adapt different advancements in the field of paper-based technology to check for the possibility of improving the output and performance of paper-based sensors. An example is the selection of a suitable reducing agent to be used in a paper-based device meant for detecting nitrate in water.

Nitrate is part of the nitrogen cycle [24] and is an essential nutrient needed for plant growth; however, it plays a significant role in water nutrient pollution when present in excessive amounts [25,26]. Nitrate is also the most stable form of nitrogen in oxygenated systems, and all other forms of nitrogen-containing compounds can become a source for it [27,28]. Ingesting nitrate has been linked to colorectal cancer, thyroid disease and central nervous system birth defects [29]. Therefore, it is important to measure nitrate levels in water for environmental monitoring purposes and to ensure its safety for consumption. Different techniques are readily available to measure nitrate concentrations in water but are either costly, time-consuming or may require trained personnel [30,31]. Several paper-based sensors have been developed for the rapid and inexpensive detection of nitrate in water, food and human saliva, and their limits of detection (LOD) and limits of quantification (LOQ) are given in Table 1.

Table 1. Performance of paper-based sensors developed for detecting nitrate in different media.

Reference	Media	LOD (ppm)	LOQ (ppm)
[32]	Water	0.533	1.765
[33]	Water	1.178	2.976
[9]	Food Sample	3.6	12
[34]	Food Sample	0.4	NA [1]
[35]	Food Sample	0.4	1.4
[36]	Human Saliva	4.96	16.74

[1] NA, not available.

All of the paper-based devices developed thus far for measuring nitrate levels have used the Griess assay for detection since it is the most commonly used spectrophotometric method for quantifying concentrations of nitrate and nitrite [37,38]. However, this assay is specific to nitrite molecules and, therefore, nitrate molecules have to be reduced to nitrite first before detection. There are several different reducing agents that can reduce nitrate to nitrite, such as cadmium, copperized cadmium, zinc, nitrate reductase, irradiation by ultraviolet light, hydrazine sulfate, titanium (III) chloride, vanadium (III), hydroxylamine, tin chloride or ascorbic acid [36,39,40]. Some of these reducing agents are not suitable for use in paper-based devices, while others have been tested and used in this type of sensors.

Nitrate reductase, irradiation by ultraviolet and hydrazine require lengthy reduction times [41], which may not be suitable for paper-based sensors due to concerns of sample evaporation. Titanium (III) chloride is violet in color and absorbs light in the same range as the azo dye product of the Griess assay [41]. Ferreira et al. [36] tested tin chloride, hydroxylamine, ascorbic acid and zinc microparticles. They used zinc microparticles in their paper-based nitrate sensor since the other agents tested did not extensively reduce nitrate to nitrite. Experimental results by Jayawardane et al. [33] showed that cadmium and zinc microparticles produced similar results for nitrate reduction in their paper-based device. They opted for zinc microparticles due to the higher toxicity of cadmium. Thongkam et al. [35] developed a very simple paper-based device for measuring nitrate and nitrite concentrations in food samples, and they used vanadium (III) chloride to reduce nitrate before detection.

We had previously developed a sensitive paper-based nitrate sensor by testing different device architectures and optimizing the different components of the device [32]. The final device adopted a folding architecture with part of the detection chemistry immobilized at the detection zone. This improved the quality and uniformity of the signal developed. The device also incorporated a new composite material made-up of zinc microparticles and cellulose fibers to enhance nitrate reduction. A nitrate conversion efficiency of 27% was achieved using this new composite material called Zinculose [42]. However, the results obtained by Thongkam et al. [35] for nitrate detection in food samples by using vanadium (III) chloride as a reducing agent are very promising. In this paper, we measure the performance of a dip strip using vanadium (III) chloride for reducing nitrate by calculating its limits of detection and quantifications. We also calculate the nitrate reduction efficiency of vanadium (III) chloride and compare the results to those obtained when using Zinculose.

2. Methods

Thongkam et al. [35] studied the effect of the different parameters on nitrate detection. They tested different concentrations of sulfanilic acid and N-(1-Naphthyl)ethylenediamine dihydrochloride used in the Griess assay for detection. They also examined the effect of different concentrations of vanadium (III) chloride and reaction times on the intensity of the color produced in the detection zone. In this paper, we use the optimum concentrations they have found when preparing the reagents to be used in our experiments.

2.1. Materials

The items below were used in preparing and running the experiments presented in this paper. Whatman grade 1 filter paper (GE Healthcare Whatman 1-1001824), backing cards (DCN Dx MIBA-050), sulfanilamide (98%, Alfa Aesar-A1300136), N-(1-Naphthyl)ethylenediamine dihydrochloride (Alfa Aesar-J6321414), hydrochloric acid (Fisher Chemical-A142-212), sodium nitrate (≥99.5%, Honeywell Fluka-31440), sodium nitrite (≥99%, Honeywell Fluka-31443) and ASTM Type 1 deionized water (resistivity > 18 MΩ/cm, LabChem-LC267405).

2.2. Methods

Strips 1 × 8 cm were cut out from a 30 × 8 cm backing card using a guillotine paper cutter. Three circles, 6 mm in diameter each, were punched out using a tissue biopsy from the Whatman filter paper and stuck onto the backing card, Figures 1 and S1.

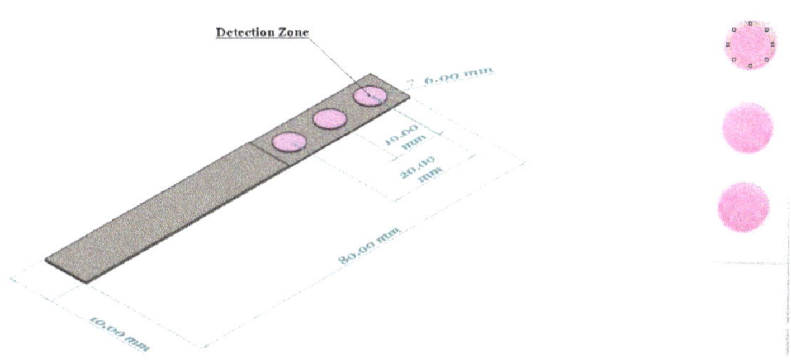

Figure 1. (**a**) Schematic showing the components and dimensions of the dip strip used. (**b**) The yellow circle shows the color analysis zone used in ImageJ to quantify the color intensity of one of the detection zones; the diameter of the circle is about 125 pixels, which is approximately 5.3 mm.

Nitrate and nitrite solutions at concentrations of 1000 ppm were freshly prepared on the day of testing by dissolving the required amount of nitrate or nitrite salt in deionized water. These solutions were then diluted using deionized water into the following concentrations 0.5, 1, 2.5, 5, 10, 20 and 40 ppm. We followed the procedure outlined by Thongkam et al. [35] in preparing the detection chemistry for nitrate and nitrite. For nitrite detection, the solution was called reagent "A" and consisted of equal parts (1:1 ratio) volume of sulfanilic acid and NED solution. For nitrate detection, the solution used was called reagent "B" and consisted of equal parts (1:1:1 ratio) volume of the above sulfanilic acid, NED solution and the reducing reagent solution. The sulfanilic acid used in reagents "A" and "B" was prepared by dissolving 0.1 g of sulfanilamide in 100 mL of 2 mol L^{-1} hydrochloric acid. The NED solution used in reagents "A" and "B" was prepared by dissolving 0.1 g of N-(1-Naphthyl)ethylenediamine dihydrochloride in 100 mL of deionized water. The reducing reagent solution used in reagent "B" was prepared by dissolving 3 g of vanadium (III) chloride in 100 mL of 6 mol L^{-1} hydrochloric acid. 2 µL of reagent A or B was pipetted onto each circle and allowed to air-dry for at least 30 min, Figure S2. Each dip strip was then submerged into the appropriate nitrate or nitrite solution for 1 s, shaken to remove excess fluid and then scanned using a desktop scanner (Canon TS6020) at a resolution of 600 DPI. The nitrate dip strips were scanned after 10 min, and the nitrite dip strips were scanned after 5 min following the optimized scan times previously found by Thongkam et al. [35]. The detection zones were analyzed using ImageJ in RGB mode, similar to how they analyzed their results. We have previously shown that the green component of the measured color intensity shows the largest difference in value over the concentration of nitrate or nitrite for paper-based devices using the Griess assay [43]. Therefore, the data for the different color intensities were provided in the supplementary file, Tables S1–S4. A MATLAB code was used to fit the data to an exponential decay function of the form y = a × exp (−x/b) + c, and the symbolic toolbox was used to calculate the limits of detection and quantification. The limits of detection and quantification were obtained by finding the analyte concentrations corresponding to y_{LOD} or y_{LOQ} on the calibration curves developed. y_{LOD} or y_{LOQ} were calculated using the following equations [44]:

$$y_{LOD} = \overline{y}_B - 3\,\sigma_B$$

$$y_{LOQ} = \overline{y}_B - 10\,\sigma_B$$

where $\overline{y}_B$ corresponds to the mean color intensity of the blank solution (0 ppm) and σ_B is its respective standard deviation.

3. Results and Discussion

The detection zones of the nitrate dip strips showed little to no color change after 10 min, Figure S3, but color started to form after a much longer wait time, so the dip strips were scanned after 1 h as well, Figure S4. The following section shows the results obtained for the nitrate and nitrite dip strips.

3.1. Nitrate and Nitrite Analysis

Figure 2 shows the calibration curves developed for the detection of nitrate in deionized water after a reaction time of 10 min and 1 h. The limits of detection and quantification for nitrate after 10 min are 37.03 and 121 ppm, respectively. The limits of detection and quantification for nitrate after 1 h are 3.352 and 7.437 ppm, respectively.

Figure 3 shows the calibration curves developed for the detection of nitrite in deionized water after a reaction time of 5 min and 1 h. The limits of detection and quantification for nitrite after 5 min are 0.522 and 0.854 ppm, respectively. The limits of detection and quantification for nitrite after 1 h are 0.889 and 1.823 ppm, respectively.

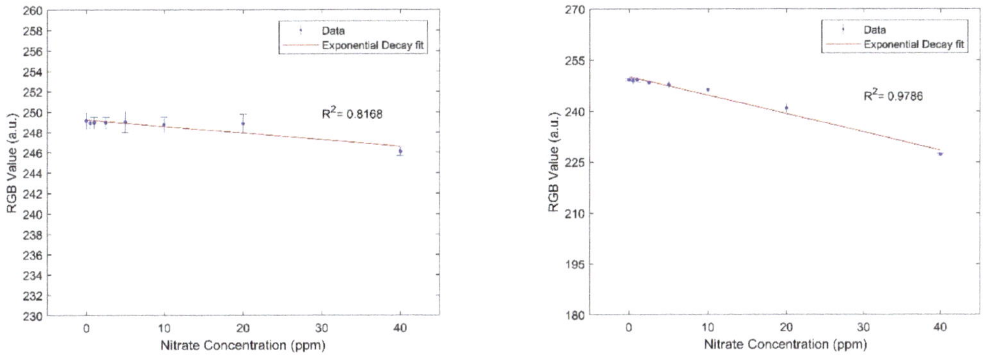

(a) (b)

Figure 2. (**a**) An exponential decay calibration curve in the form y = a × exp (−x/b) + c, where a = 2741, b = 41,430 and c = −2492 was established for nitrate after a reaction time of 10 min. (**b**) An exponential decay calibration curve in the form y = a × exp (−x/b) + c, where a = 80,230, b = 147,700 and c = −79,980 was established for nitrate after a reaction time of 1 h. The error bars represent the standard deviation.

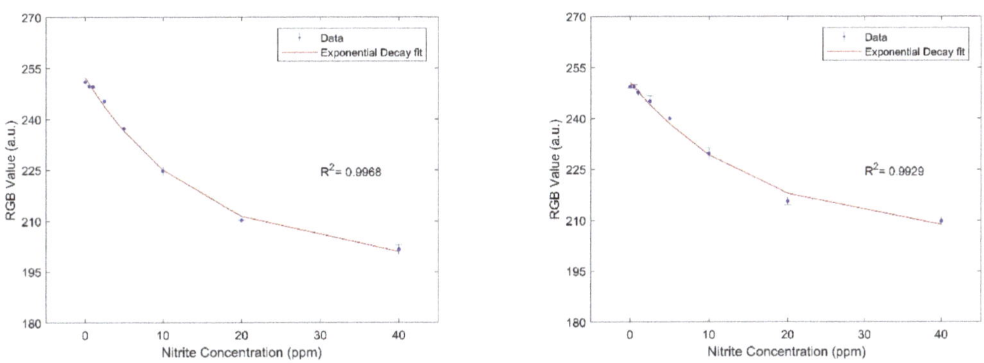

(a) (b)

Figure 3. (**a**) An exponential decay calibration curve in the form y = a × exp (−x/b) + c, where a = 54.94, b = 14.57 and c = 197.4 was established for nitrite after a reaction time of 5 min. (**b**) An exponential decay calibration curve in the form y = a × exp (−x/b) + c, where a = 45.56, b = 15.65 and c = 205.2 was established for nitrite after a reaction time of 1 h. The error bars represent the standard deviation.

3.2. Reduction Efficiency

The reduction efficiency of vanadium (III) chloride was calculated using the data obtained in the above experiments used to calculate the LOD and LOQ for nitrate and nitrite. First, the results obtained from the nitrite experiment after 1 h were used to establish the calibration curve using the method outlined in Section 2.2. Then the results obtained from the nitrate experiment after 1 h were used to calculate the intersection of the measured result with the calibration established for nitrite using the symbolic toolbox. Table 2 gives the nitrate conversion efficiency calculated. As can be seen from the table, the conversion efficiency varies between almost 0% and 27%.

Table 2. Calculated nitrate conversion efficiency.

Nitrate Concentration (ppm)	Normalized Nitrite Concentration Calculated (ppm) [1]	Reduction Efficiency (%)
0	0	0
0.5	0.098	19.54 ± 0.80
1	0.006	0.61 ± 0.23
2.5	0.300	12 ± 0. 05
5	0.524	10.48 ± 0.08
10	1.086	10.86 ± 0.01
20	3.296	16.48 ± 0.03
40	10.896	27.24 ± 0.01

[1] This concentration is normalized by subtracting the intensity calculated for 0 ppm from all other concentrations.

3.3. Discussion

The limits of detection and quantification obtained for nitrate and nitrite in our analysis were much higher than those obtained by commercial dip strips using the Griess assay. This can be attributed to one or more of the following reasons: using the RGB mode in data analysis, not depositing enough reagent volume for reaction or using hydrochloric acid since it evaporates completely without producing acidic conditions when rewet. The reaction with the Griess assay should take place under acidic conditions [45].

A maximum reduction efficiency of 27% was obtained by vanadium (III) chloride. This is similar to the reduction efficiency obtained by Zinculose (27%). However, this reduction efficiency was only obtained for a high nitrate concentration of 40 ppm, while lower concentrations resulted in a much lower reduction efficiency. This raises the question of repeatability and uniformity of vanadium (III) chloride nitrate reduction when used in paper-based devices.

Each of the two reducing agents, zinc microparticles and vanadium (III) chloride, has its own set of advantages and should be used in specific applications with an appropriate device design. Zinculose is a composite material that can be incorporated into any paper-based device. The zinc microparticles in Zinculose are held in place by the matrix, which allows the passage of more sample volume through the material and the reduction of more molecules as they pass through it. This allows for signal amplification as more molecules become available to be captured and detected. However, vanadium (III) chloride is not immobilized and would wash away in any lateral flow paper-based device design. Nitrate reduction using vanadium (III) chloride takes much longer than that by zinc microparticles. That is why commercial dip strips generally use zinc microparticles in the detection zone to reduce nitrate to nitrite before detection, Figure S5. Vanadium (III) chloride allows for the development of simple dip strips since the reducing reagent can be mixed with the detection chemistry and easily deposited in the detection zone. However, the limits of detection and quantification achieved by dip strips utilizing vanadium (III) chloride are not as good as those obtained in more intricate designs using zinc microparticles.

4. Conclusions

Paper-based microfluidic technology is a relatively new field of research that is gaining a lot of attention and is producing a lot of innovation. In this paper, we measured the performance of a dip strip utilizing vanadium (III) chloride to reduce nitrate before detection. We observed that vanadium (III) chloride has some drawbacks that make it impractical for use in paper-based devices meant for detecting nitrate. These include long reduction times required and low limits of detection and quantification obtained. Therefore, we recommend using zinc microparticles as the reducing agent for nitrate detection in paper-based devices. Future work will include developing a suitable lightbox, similar to [46], that emits green light for measuring nitrate and nitrite concentrations using paper-based devices utilizing the Griess assay in the field.

Supplementary Materials: The following are available online at https://www.mdpi.com/article/10.3390/CSAC2021-10459/s1, Figure S1. Dip strip used in experiments, Figure S2. Dip strip used in experiments after the solutions are dried on the detection zones, Figure S3. Color formed in the detection zone vs. nitrate or nitrite concentrations after several minutes, Figure S4. Color formed in the detection zone vs. nitrate or nitrite concentrations after 1 h, Table S1. ImageJ analysis of nitrate detection zones after 10 min. Test order was randomized, Table S2. ImageJ analysis of nitrate detection zones after 1 h. Test order was randomized, Table S3. ImageJ analysis of nitrite detection zones after 5 min. Test order was randomized, Table S4. ImageJ analysis of nitrite detection zones after 1 h. Test order was randomized, Figure S5. Zinc microparticles observed using an electron scanning microscope with EDS analysis in the nitrate test fields of commercial dip strips (a) Quantofix 91313 (b) Quantofix 91351.

Author Contributions: Conceptualization, A.C., N.R., C.A. and M.F.; methodology, A.C.; software, A.C. and H.H.-B.; validation, A.C.; formal analysis, A.C.; investigation, A.C.; resources, C.A. and M.F.; data curation, A.C., C.A. and M.F.; writing—original draft preparation, A.C.; writing— review and editing, A.C., H.H.-B., N.R., C.A. and M.F.; visualization, A.C. and H.H.-B.; supervision, N.R., C.A. and M.F.; project administration, C.A. and M.F.; funding acquisition, C.A. and M.F. All authors have read and agreed to the published version of the manuscript.

Funding: This research was funded by the National Science Foundation under EPSCoR Cooperative Agreement #OIA-1655221.

Institutional Review Board Statement: Not applicable.

Informed Consent Statement: Not applicable.

Data Availability Statement: Data is contained within the article or supplementary material. Additional data not presented in this article is available on request from the corresponding author.

Acknowledgments: The authors would like to acknowledge the support from the Rhode Island EPSCoR, which is funded by the National Science Foundation under Award #OIA-1655221. The authors would also like to acknowledge the students, research scientists and visiting scholars at the Microfluidics Laboratory at the University of Rhode Island for their help and support. SEM and EDS data were acquired at the RI Consortium for Nanoscience and Nanotechnology, a URI College of Engineering core facility partially funded by the National Science Foundation EPSCoR, Cooperative Agreement #OIA-1655221.

Conflicts of Interest: The authors declare no conflict of interest. The funders had no role in the design of the study; in the collection, analyses, or interpretation of data; in the writing of the manuscript; or in the decision to publish the results.

References

1. Bouhoun, M.; Blondeau, P.; Louafi, Y.; Andrade, F. A Paper-Based Potentiometric Platform for Determination of Water Hardness. *Chemosensors* **2021**, *9*, 96. [CrossRef]
2. Firdaus, M.L.; Aprian, A.; Meileza, N.; Hitsmi, M.; Elvia, R.; Rahmidar, L.; Khaydarov, R. Smartphone Coupled with a Paper-Based Colorimetric Device for Sensitive and Portable Mercury Ion Sensing. *Chemosensors* **2019**, *7*, 25. [CrossRef]
3. Oyewunmi, O.D.; Safiabadi-Tali, S.H.; Jahanshahi-Anbuhi, S. Dual-Modal Assay Kit for the Qualitative and Quantitative Determination of the Total Water Hardness Using a Permanent Marker Fabricated Microfluidic Paper-Based Analytical Device. *Chemosensors* **2020**, *8*, 97. [CrossRef]
4. Heidari-Bafroui, H.; Charbaji, A.; Anagnostopoulos, C.; Faghri, M. A Colorimetric Dip Strip Assay for Detection of Low Concentrations of Phosphate in Seawater. *Sensors* **2021**, *21*, 3125. [CrossRef] [PubMed]
5. Kim, H.; Hyung, J.; Noh, H. Rationalization of In-Situ Synthesized Plasmonic Paper for Colorimetric Detection of Glucose in Ocular Fluids. *Chemosensors* **2020**, *8*, 81. [CrossRef]
6. Islam, N.; Ahmed, I.; Anik, M.I.; Ferdous, S.; Khan, M.S. Developing Paper Based Diagnostic Technique to Detect Uric Acid in Urine. *Front. Chem.* **2018**, *6*, 496. [CrossRef] [PubMed]
7. Deroco, P.; Junior, D.W.; Kubota, L. Silver Inkjet-Printed Electrode on Paper for Electrochemical Sensing of Paraquat. *Chemosensors* **2021**, *9*, 61. [CrossRef]
8. Migliorini, F.L.; Dos Santos, D.M.; Soares, A.C.; Mattoso, L.H.C.; Oliveira, J.O.N.; Correa, D.S. Design of A Low-Cost and Disposable Paper-Based Immunosensor for the Rapid and Sensitive Detection of Aflatoxin B1. *Chemosensors* **2020**, *8*, 87. [CrossRef]
9. Teepoo, S.; Arsawiset, S.; Chanayota, P. One-Step Polylactic Acid Screen-Printing Microfluidic Paper-Based Analytical Device: Application for Simultaneous Detection of Nitrite and Nitrate in Food Samples. *Chemosensors* **2019**, *7*, 44. [CrossRef]

10. Rosati, G.; Cunego, A.; Fracchetti, F.; Del Casale, A.; Scaramuzza, M.; De Toni, A.; Torriani, S.; Paccagnella, A. Inkjet Printed Interdigitated Biosensor for Easy and Rapid Detection of Bacteriophage Contamination: A Preliminary Study for Milk Processing Control Applications. *Chemosensors* **2019**, *7*, 8. [CrossRef]
11. Shriver-Lake, L.C.; Zabetakis, D.; Dressick, W.J.; Stenger, D.A.; Trammell, S.A. Paper-Based Electrochemical Detection of Chlorate. *Sensors* **2018**, *18*, 328. [CrossRef]
12. Zikulnig, J.; Khalifa, M.; Rauter, L.; Lammer, H.; Kosel, J. Low-Cost Inkjet-Printed Temperature Sensors on Paper Substrate for the Integration into Natural Fiber-Reinforced Lightweight Components. *Chemosensors* **2021**, *9*, 95. [CrossRef]
13. Vargas, A.; Gámez, F.; Roales, J.; Lopes-Costa, T.; Pedrosa, J. A Paper-Based Ultrasensitive Optical Sensor for the Selective Detection of H_2S Vapors. *Chemosensors* **2021**, *9*, 40. [CrossRef]
14. Li, Z. Nanoporous Silica-Dye Microspheres for Enhanced Colorimetric Detection of Cyclohexanone. *Chemosensors* **2018**, *6*, 34. [CrossRef]
15. Singh, A.T.; Lantigua, D.; Meka, A.; Taing, S.; Pandher, M.; Camci-Unal, G. Paper-Based Sensors: Emerging Themes and Applications. *Sensors* **2018**, *18*, 2838. [CrossRef]
16. Charbaji, A.; Heidari-Bafroui, H.; Anagnostopoulos, C.; Faghri, M. Literature Review of the Use of Zinc and Zinc Compounds in Paper-Based Microfluidic Devices. *J. Miner. Mater. Charact. Eng.* **2021**, *9*, 257–270. [CrossRef]
17. Charbaji, A.; Heidari-Bafroui, H.; Kumar, A.; Ragmani, N. Characterization and Modeling of Paper-based Bi-Material Actuator Cantilever; Application in Phosphate Detection. In Proceedings of the Innovations in Microfluidics and Single Cell Analysis, Boston, MA, USA, 18–19 March 2021.
18. Smith, W.; Rahmani, N.; Charbaji, A.; Lemos, N.; Anagnostopoulos, C.; Fanghri, M.; Hong, C. A Fluidically Controlled Bi-Material Actuator for Automation of Paper-Based Assays. In Proceedings of the International Symposium on Thermal Effects in Gas Flows in Microscale, Ettlingen, Germany, 24–25 October 2019.
19. Charbaji, A.; Heidari-Bafroui, H.; Anagnostopoulos, C.; Faghri, M. Sensitive Detection of Nitrate using a Paper-based Microfluidic Device. In Proceedings of the Innovations in Microfluidics and Single Cell Analysis, Boston, MA, USA, 17–18 August 2020.
20. Xu, Y.; Liu, M.; Kong, N.; Liu, J. Lab-on-paper micro- and nano-analytical devices: Fabrication, modification, detection and emerging applications. *Microchim. Acta* **2016**, *183*, 1521–1542. [CrossRef]
21. Fernandes, G.M.; Silva, W.R.; Barreto, D.N.; Lamarca, R.S.; Lima Gomes, P.C.F.; Flávio, D.S.; Petruci, J.; Batista, A.D. Novel approaches for colorimetric measurements in analytical chemistry—A review. *Anal. Chim. Acta* **2020**, *1135*, 187–203. [CrossRef]
22. Fernandes, S.C.; Walz, J.A.; Wilson, D.; Brooks, J.; Mace, C.R. Beyond Wicking: Expanding the Role of Patterned Paper as the Foundation for an Analytical Platform. *Anal. Chem.* **2017**, *89*, 5654–5664. [CrossRef] [PubMed]
23. Kasetsirikul, S.; Shiddiky, M.; Nguyen, N.-T. Wicking in Paper Strips under Consideration of Liquid Absorption Capacity. *Chemosensors* **2020**, *8*, 65. [CrossRef]
24. World Health Organization. Guidelines for Drinking—Water Quality, 4th Edition, Incorporating the 1st Addendum. 2017. Available online: https://www.who.int/publications/i/item/9789241549950 (accessed on 22 October 2020).
25. US Geological Survey Nitrogen and Water. Available online: https://www.usgs.gov/special-topic/water-science-school/science/nitrogen-and-water?qt-science_center_objects=0#qt-science_center_objects (accessed on 22 October 2020).
26. US EPA Nutrient Pollution. Available online: https://www.epa.gov/nutrientpollution/issue (accessed on 13 May 2021).
27. Silva, C.G.; Pereira, M.F.; Órfão, J.J.M.; Faria, J.L.; Soares, S. Catalytic and Photocatalytic Nitrate Reduction Over Pd-Cu Loaded Over Hybrid Materials of Multi-Walled Carbon Nanotubes and TiO2. *Front. Chem.* **2018**, *6*, 632. [CrossRef] [PubMed]
28. Gupta, S.; Gupta, R.C.; Gupta, A.B.; Eskiocak, S.; Rao, P.; Puttanna, K.; Singhvi, A. Pathophysiology of Nitrate Toxicity in Human and its Mitigation Measures. *Bull. Reg. Assess React. Nitrogen* **2010**, *20*, 1–78.
29. Ward, M.H.; Jones, R.R.; Brender, J.D.; De Kok, T.M.; Weyer, P.J.; Nolan, B.T.; Villanueva, C.M.; Van Breda, S.G. Drinking Water Nitrate and Human Health: An Updated Review. *Int. J. Environ. Res. Public Health* **2018**, *15*, 1557. [CrossRef] [PubMed]
30. Wierzbicka, E. Novel methods of nitrate and nitrite determination—A review. *J. Elem.* **2020**, *25*, 97–106. [CrossRef]
31. Kapoor, A.; Balasubramanian, S.; Muthamilselvi, P.; Vaishampayan, V.; Prabhakar, S. Lab-on-a-Chip Devices for Water Quality Monitoring. In *Nanotechnology in the Life Sciences*; Springer: Cham, Switzerland, 2020; pp. 455–469.
32. Charbaji, A.; Heidari-Bafroui, H.; Anagnostopoulos, C.; Faghri, M. A New Paper-Based Microfluidic Device for Improved Detection of Nitrate in Water. *Sensors* **2020**, *21*, 102. [CrossRef]
33. Jayawardane, B.M.; Wongwilai, W.; Grudpan, K.; Kolev, S.; Heaven, M.; Nash, D.M.; McKelvie, I. Evaluation and Application of a Paper-Based Device for the Determination of Reactive Phosphate in Soil Solution. *J. Environ. Qual.* **2014**, *43*, 1081–1085. [CrossRef]
34. Ratnarathorn, N.; Dungchai, W. Paper-based Analytical Device (PAD) for the Determination of Borax, Salicylic Acid, Nitrite, and Nitrate by Colorimetric Methods. *J. Anal. Chem.* **2020**, *75*, 487–494. [CrossRef]
35. Thongkam, T.; Hemavibool, K. An environmentally friendly microfluidic paper-based analytical device for simultaneous colorimetric detection of nitrite and nitrate in food products. *Microchem. J.* **2020**, *159*, 105412. [CrossRef]
36. Ferreira, F.T.S.M.; Mesquita, R.B.R.; Rangel, A.O.S.S. Novel microfluidic paper-based analytical devices (μPADs) for the determination of nitrate and nitrite in human saliva. *Talanta* **2020**, *219*, 121183. [CrossRef]
37. Wang, Q.-H.; Yu, L.-J.; Liu, Y.; Lin, L.; Lu, R.-G.; Zhu, J.-P.; He, L.; Lu, Z.-L. Methods for the detection and determination of nitrite and nitrate: A review. *Talanta* **2017**, *165*, 709–720. [CrossRef]
38. Mahmud, M.A.P.; Ejeian, F.; Azadi, S.; Myers, M.; Pejcic, B.; Abbassi, R.; Razmjou, A.; Asadnia, M. Recent progress in sensing nitrate, nitrite, phosphate, and ammonium in aquatic environment. *Chemosphere* **2020**, *259*, 127492. [CrossRef]

39. Alahi, M.E.E.; Mukhopadhyay, S.C. Detection methods of nitrate in water: A review. *Sens. Actuators A Phys.* **2018**, *280*, 210–221. [CrossRef]
40. Jaywant, S.A.; Arif, K.M. A Comprehensive Review of Microfluidic Water Quality Monitoring Sensors. *Sensors* **2019**, *19*, 4781. [CrossRef]
41. Ellis, P.S.; Shabani, A.M.H.; Gentle, B.S.; McKelvie, I.D. Field measurement of nitrate in marine and estuarine waters with a flow analysis system utilizing on-line zinc reduction. *Talanta* **2011**, *84*, 98–103. [CrossRef]
42. Charbaji, A.; Smith, W.; Anagnostopoulos, C.; Faghri, M. Zinculose: A new fibrous material with embedded zinc particles. *Eng. Sci. Technol. Int. J.* **2021**, *24*, 571–578. [CrossRef]
43. Charbaji, A.; Heidari-Bafroui, H.; Anagnostopoulos, C.; Faghri, M. A Practical System for the Quantitative Determination of Nitrate and Nitrite in the Field. In Proceedings of the 6th International Microfluidics Conference, Las Vegas, NV, USA, 26 March 2021.
44. James, N.; Miller, J.C. *Statistics and Chemometrics for Analytical Chemistry*, 5th ed.; Pearson: London, UK, 2005.
45. Weng, C.-H.; Chen, M.-Y.; Shen, C.-H.; Yang, R.-J. Colored wax-printed timers for two-dimensional and three-dimensional assays on paper-based devices. *Biomicrofluidics* **2014**, *8*, 066502. [CrossRef] [PubMed]
46. Heidari-Bafroui, H.; Ribeiro, B.; Charbaji, A.; Anagnostopoulos, C.; Faghri, M. Portable infrared lightbox for improving the detection limits of paper-based phosphate devices. *Meas. J. Int. Meas. Conf.* **2021**, *173*, 108607. [CrossRef]

chemistry proceedings

MDPI

Proceeding Paper

A Voltammetric Nanodiamond-Coated Screen-Printed Immunosensor for The Determination of a Peanut Allergen in Commercial Food Products †

André Carvalho, Maria Freitas, Henri P. A. Nouws * and Cristina Delerue-Matos

REQUIMTE/LAQV, Instituto Superior de Engenharia do Porto, Instituto Politécnico do Porto, Rua Dr. António Bernardino de Almeida, 431, 4249-015 Porto, Portugal; andcarv16@gmail.com (A.C.); maria.freitas@graq.isep.ipp.pt (M.F.); cmm@isep.ipp.pt (C.D.-M.)

* Correspondence: han@isep.ipp.pt

† Presented at the 1st International Electronic Conference on Chemical Sensors and Analytical Chemistry, 1–15 July 2021; Available online: https://csac2021.sciforum.net/.

Abstract: A voltammetric immunosensor was developed to quantify a major peanut allergen, Ara h 1, using screen-printed carbon electrodes (SPCE) as transducers. A sandwich-type immunoassay was performed on nanodiamond-coated SPCEs using an alkaline phosphatase-labeled detection antibody and a mixture containing an enzymatic substrate (3-indoxyl phosphate) and silver nitrate. The immunological interaction was detected through the (linear sweep) voltammetric stripping of the enzymatically deposited silver. The immunosensor's applicability was evaluated by analyzing breakfast cereals, cookies, and energy and cereal bars. Ara h 1 was successfully tracked in these commercial food products.

Keywords: peanut allergy; Ara h 1; food allergy; electrochemical immunosensor; screen-printed carbon electrode; nanodiamond

Citation: Carvalho, A.; Freitas, M.; Nouws, H.P.A.; Delerue-Matos, C. A Voltammetric Nanodiamond-Coated Screen-Printed Immunosensor for The Determination of a Peanut Allergen in Commercial Food Products. *Chem. Proc.* **2021**, *5*, 10. https://doi.org/10.3390/CSAC2021-10458

Academic Editor: Núria Serrano

Published: 30 June 2021

1. Introduction

Peanuts are integrated into the Mediterranean dietary pattern, and their consumption has been recommended worldwide. Despite its noteworthy nutritional value, reported cases of peanut allergy have increased, and therefore commercial food tracking is essential since in extreme cases peanut intake causes anaphylaxis [1].

Detection of peanut traces in food samples can be achieved using electrochemical immunosensors that benefit from their advantageous features such as rapid detection and high selectivity and sensitivity [2]. Because SPCEs can be connected to portable devices, they can be used for in situ allergen analysis. Few electrochemical immunosensors were reported for the determination of Ara h 1: a sandwich-type gold nanoparticle-coated SPCE [3], a reagentless label-free single-walled carbon nanotube-based biosensor [4] and an impedimetric immunosensor using a gold electrode functionalized with 11-mercaptoundecanoic acid self-assembled monolayer [5].

Among the distinct carbon-based nanomaterials, nanodiamonds (NDs) have not yet been used in the analysis of allergens. Nevertheless, due to their 3D configuration efficient electrode nanostructuration can improve the analytical signal [6].

The present work reports the development of an electrochemical immunosensor for the analysis of the peanut allergen Ara h 1 using SPCEs/NDs. In a sandwich-type assay, the antibody–antigen interaction was detected through Linear Sweep Voltammetry (LSV).

2. Materials and Methods

2.1. Materials and Solutions

Electrochemical measurements were performed using an Autolab PGSTAT101 potentiostat–galvanostat controlled by the NOVA software package v.1.10 (Metrohm Autolab). Screen-printed carbon electrodes (SPCE, DRP-110) and the connector to interface the electrodes (DRP-CAC) were supplied by Metrohm DropSens.

Albumin from bovine serum (BSA), 3-indoxyl phosphate (3-IP), nanodiamonds (NDs, nanopowder), nitric acid (HNO_3), streptavidin-alkaline phosphatase (S-AP), silver nitrate and tris(hydroxymethyl)aminomethane (Tris) were purchased from Sigma-Aldrich. The capture antibody (CAb, anti-Ara h 1), purified natural Ara h 1 and the detection antibody (DAb; biotin, anti-Ara h 1) were obtained from Indoor Biotechnologies.

Solutions of BSA and CAb were prepared in Tris-HNO_3 (0.1 M, pH 7.2, Tris buffer); Ara h 1, DAb, S-AP solutions were prepared in Tris buffer containing 1.0% BSA (m/V). The solution containing 3-IP and silver nitrate was prepared in Tris 0.1 M (pH 9.8 + $Mg(NO_3)_2$ 2×10^{-2} M). A Tris buffer (0.1 M, pH 8.5) was used to extract Ara h 1 from the food samples. The evaluation of the accuracy of the sensor's results was performed using a commercial ELISA kit (Indoor Biotechnologies).

2.2. Methods

The SPCEs were nanostructurated with NDs (0.10 ng/mL) and the CAb (10 μg/mL) was immobilized overnight. The electrochemical immunoassay consisted of the following incubation steps: (i) Ara h 1 standard solution/food sample extract (30 min); (ii) DAb (250× dilution, 60 min); (iii) S-AP (20,000× dilution, 30 min) and (iv) enzymatic reaction (3-IP (1.0×10^{-3} M) and silver nitrate (4.0×10^{-4} M), 20 min). Washing between the incubation steps was performed using Tris buffer. The electrochemical analysis of the deposited silver was carried out by Linear Sweep Voltammetry (LSV; voltammograms were recorded using the following parameters: potential range from −0.03 to +0.4 V, scan rate: 50 mV/s). A schematic representation of the sandwich-type immunosensor is presented in Figure 1a.

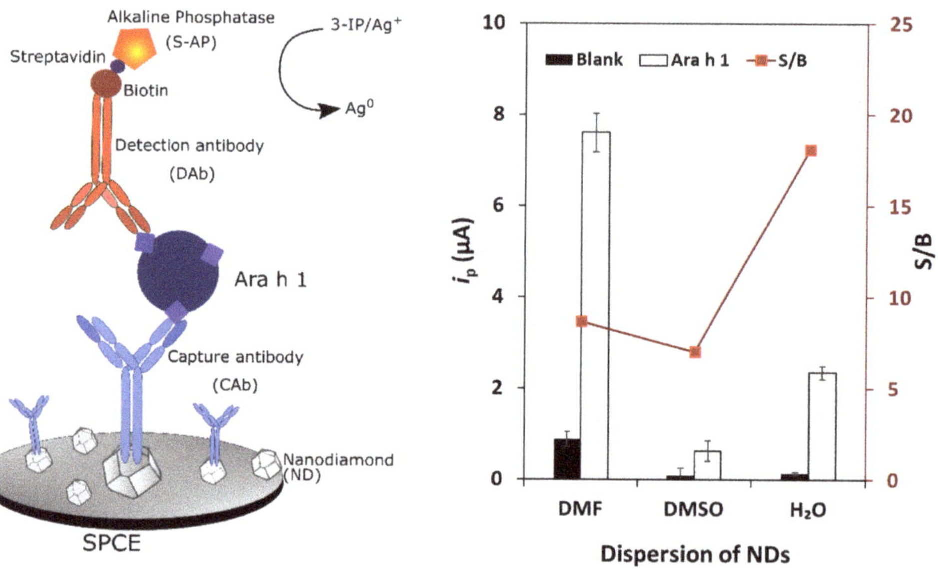

(a) (b)

Figure 1. Nanodiamond-based voltammetric immunosensor. (**a**) Schematic representation of the immunosensor construction; (**b**) results obtained for NDs 1 mg/mL dispersed in DMF, DMSO and H_2O. Experimental parameters: CAb 10 μg/mL, Ara h 1 (0 and 250 ng/mL), DAb 250× dilution, S-AP 200,000× dilution, 3-IP (1.0×10^{-3} M), Ag^+ (4.0×10^{-4} M).

A set of food products was bought in local supermarkets. The extraction procedure was performed as recommended by the Ara h 1 standard supplier. Briefly, 1 g of the food sample was mixed with 10 mL of the extraction buffer, vortexed for 5 s, incubated for 15 min at 60 °C, centrifuged at 2500 rpm for 20 min and stored at −20 °C until use.

3. Results and Discussion

3.1. Optimization of the Experimental Parameters

NDs were used for the SPCE's nanostructuration. Several solvents (DMF, DMSO and H_2O) that are typically employed for the dispersion of carbon-based nanomaterials were tested. The obtained i_p values are presented in Figure 1b, and the signal-to-blank (S/B) ratio in the presence and absence of Ara h 1 (0 and 250 ng/mL) was used to select the best solvent. As can be observed, the dispersion of NDs in H_2O provided the optimum condition to proceed the studies. Then, several ND concentrations were tested, from 1.0 to 0.03 mg/mL, and the best performance was obtained for 0.10 mg/mL.

The electrode surface was biofunctionalized with CAb and distinct concentrations between 5.0 and 25 µg/mL were studied. A 10 µg/mL concentration was selected and used to study the adequate DAb dilution (tested range: from 250× to 1000×). In this case a 250× dilution was found to provide the best performance.

Additionally, several dilutions of the streptavidin-alkaline phosphatase conjugate (S-AP) were tested, from 100,000× to 250,000×, and the selected value was 200,000×. To conclude the optimization process and reduce the assay time, the antigen incubation time was tested (30 min and 60 min), and it was verified that a 30 min incubation allowed the appropriate detection of the allergen.

3.2. Analytical Performance

The analytical responses toward different Ara h 1 concentrations using the nanostructured SPCE/NDs were evaluated. A linear concentration range was established between 25.0 and 500 ng/mL (i_p = (0.027 ± 0.001) [Ara h 1] + (1.41 ± 0.31), r = 0.994, n = 5) with a sensitivity of 0.342 $\mu A \cdot mL \cdot ng^{-1} \cdot cm^{-2}$. The limits of detection (LOD) and quantification (LOQ) were 0.78 and 2.6 ng/mL, respectively (calculated using the equations: LOD = 3 S_{blank}/m and LOQ = 10 S_{blank}/m where S_{blank} is the standard deviation of the blank signal and m is the slope of the calibration plot).

3.3. Precision, Recovery and Stability Studies

The precision of the results provided by the immunosensor was tested using different electrodes on distinct days. A 250 ng/mL Ara h 1 solution was analyzed in triplicate and a relative standard deviation (RSD) of 7.3% was obtained.

To evaluate the accuracy of the results, recovery studies were performed using spiked cookie samples. The result for three replicates of added 250 ng/mL was found to be 75%.

The storage stability of the optimized SPCE/NDs platform was evaluated during several weeks using 0 and 250 ng/mL Ara h 1 solutions. It was verified that the sensing phase was stable for up to two weeks.

3.4. Applicability and Method Validation

The Ara h 1 content in raw peanuts of unknown variety was quantified. The obtained amount (4.29 ± 0.16 mg/g) was in accordance with previously reported results [3].

The immunosensor's applicability was evaluated by analyzing several commercial products: (1) cereal bar (no peanut); (2) energy bar containing peanut; (3) cookie that "may contain peanut"; (4) granola that "may contain peanut"; (5) pineapple cookie containing 8% of peanut. Examples of the obtained LSV voltammograms are shown in Figure 2a. The results were compared with the ones obtained using a commercial ELISA kit (Figure 2b). The good correlation indicated the accuracy of the results. The results of these analyses are presented in Table 1.

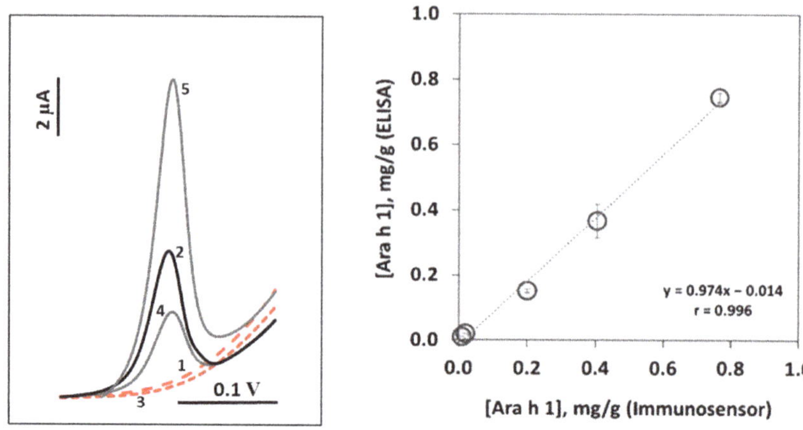

(a) (b)

Figure 2. Analysis of food products. (**a**) LSV voltammograms (solid lines—presence of Ara h 1; dashed lines—absence of Ara h 1); (**b**) correlation between the obtained results for the analysis of food products using the developed immunosensor and the commercial ELISA kit. (1) Cereal bar (no peanut); (2) energy bar containing peanut; (3) cookie that "may contain peanut"; (4) granola that "may contain peanut"; (5) pineapple cookie containing 8% of peanut.

Table 1. Results of the quantification of Ara h 1 (mg/g) in food products using an ELISA kit and the developed voltammetric immunosensor.

Product	ELISA (mg/g)	Immunosensor (mg/g)
Cereal bar (no peanut)	ND	ND
Energy bar containing peanut	0.40 ± 0.04	0.37 ± 0.05
Cookie that "may contain peanut"	ND	ND
Granola that "may contain peanut"	0.20 ± 0.01	0.15 ± 0.01
Pineapple cookie containing 8% of peanut	0.77 ± 0.03	0.75 ± 0.01

* ND: not detected.

Ara h 1 was not detected in the (1) cereal bar (no peanut) and the (3) cookie that "may contain peanut". On the latter product's label the warning may intend to protect the producer and consumers due to possible line-production cross-contaminations. On the other hand, the presence of Ara h 1 was confirmed and quantified in the following products: the (2) energy bar containing peanut (0.37 ± 0.05 mg/g), (4) granola that "may contain peanut" (0.15 ± 0.01 mg/g) and the (5) pineapple cookie containing 8% of peanut (0.75 ± 0.01 mg/g).

4. Conclusions

A nanodiamond-coated SPCE immunosensor was developed to track the major peanut allergen Ara h 1 in commercial food products. Within a total assay time of 2 h 20 min, a limit of detection (LOD) of 0.78 ng/mL was achieved. A set of breakfast products were analyzed and the presence and/or absence of Ara h 1 was confirmed and quantified in the peanut-containing products.

Supplementary Materials: The following are available online at https://www.mdpi.com/article/10.3390/CSAC2021-10458/s1.

Author Contributions: Conceptualization, M.F. and H.P.A.N.; methodology, M.F. and H.P.A.N.; validation, H.P.A.N. and C.D.-M.; formal analysis, M.F. and H.P.A.N.; investigation, A.C. and M.F.; resources, H.P.A.N. and C.D.-M.; data curation, M.F.; writing—original draft preparation, A.C. and M.F.; writing—review and editing, H.P.A.N.; visualization, M.F.; supervision, M.F. and H.P.A.N.; project administration, H.P.A.N.; funding acquisition, H.P.A.N. and C.D.-M. All authors have read and agreed to the published version of the manuscript.

Funding: This research was funded by the European Union (FEDER funds through COMPETE POCI-01-0145-FEDER-030735) and National Funds (Fundação para a Ciência e a Tecnologia) through the project PTDC/QUI-QAN/30735/2017—TracAllerSens—Electrochemical sensors for the detection and quantification of trace amounts of allergens in food products.

Institutional Review Board Statement: Not applicable.

Informed Consent Statement: Not applicable.

Data Availability Statement: Not applicable.

Acknowledgments: This work was also supported by Portuguese national funds (FCT/MCTES, Fundação para a Ciência e a Tecnologia and Ministério da Ciência, Tecnologia e Ensino Superior) through the project UIDB/50006/2020.

Conflicts of Interest: The authors declare no conflict of interest.

References

1. Lieberman, J.A.; Gupta, R.S.; Knibb, R.C.; Haselkorn, T.; Tilles, S.; Mack, D.P.; Pouessel, G. The global burden of illness of peanut allergy: A comprehensive literature review. *Allergy* **2021**, *76*, 1367–1384. [CrossRef] [PubMed]
2. Alves, R.C.; Barroso, M.F.; González-García, M.B.; Oliveira, M.B.; Delerue-Matos, C. New Trends in Food Allergens Detection: Toward Biosensing Strategies. *Crit. Rev. Food Sci. Nutr.* **2016**, *56*, 2304–2319. [CrossRef] [PubMed]
3. Alves, R.C.; Pimentel, F.B.; Nouws, H.P.A.; Marques, R.C.B.; González-García, M.B.; Oliveira, M.B.P.P.; Delerue-Matos, C. Detection of Ara h 1 (a major peanut allergen) in food using an electrochemical gold nanoparticle-coated screen-printed immunosensor. *Biosens. Bioelectron.* **2015**, *64*, 19–24. [CrossRef] [PubMed]
4. Sobhan, A.; Oh, J.-H.; Park, M.-K.; Kim, S.W.; Park, C.; Lee, J. Single walled carbon nanotube-based biosensor for detection of peanut allergy-inducing protein Ara h 1. *Korean J. Chem. Eng.* **2018**, *35*, 172–178. [CrossRef]
5. Huang, Y.; Bell, M.C.; Suni, I.I. Impedance Biosensor for Peanut Protein Ara h 1. *Anal. Chem.* **2008**, *80*, 9157–9161. [CrossRef] [PubMed]
6. Gomes, F.; Freitas, M.; Nouws, H.; Morais, S.; Delerue-Matos, C. Graphene as a material for bioelectrochemistry. In *Encyclopedia of Interfacial Chemistry Surface Science and Electrochemistry*; Wandelt, K., Ed.; Elsevier: Oxford, UK, 2018; pp. 235–240. [CrossRef]

chemistry proceedings

Proceeding Paper

Thermal Desorption of Explosives Vapour from Organic Fluorescent Sensors †

Edward B. Ogugu, Ross N. Gillanders and Graham A. Turnbull *

Organic Semiconductor Centre, School of Physics & Astronomy, SUPA, University of St Andrews, Fife KY16 9SS, UK; ebo3@st-andrews.ac.uk (E.B.O.); rg89@st-andrews.ac.uk (R.N.G.)
* Correspondence: gat@st-andrews.ac.uk
† Presented at the 1st International Electronic Conference on Chemical Sensors and Analytical Chemistry, 1–15 July 2021; Available online: https://csac2021.sciforum.net/.

Abstract: Organic semiconductors can be used as highly sensitive fluorescent sensors for the detection of trace-level vapours of nitroaromatic explosives. This involves fluorescence quenching of the sensors and indicates the presence of explosives in the surrounding environment. However, for many organic fluorescent sensors, the quenching of fluorescence is irreversible and imposes a limitation in terms of the reusability of the sensors. Here, we present a study of thermal desorption of 2,4-DNT from thin-film explosives sensors made from the commercial fluorescent polymers Super Yellow and poly(9-vinyl carbazole). Thermal cycling of the sensor results in recovery of fluorescence, thereby making them reusable.

Keywords: organic semiconductors; nitroaromatic explosives; fluorescence quenching; thermal desorption

Citation: Ogugu, E.B.; Gillanders, R.N.; Turnbull, G.A. Thermal Desorption of Explosives Vapour from Organic Fluorescent Sensors. *Chem. Proc.* **2021**, *5*, 11. https://doi.org/10.3390/CSAC2021-10559

Academic Editor: Ling Zang

Published: 1 July 2021

1. Introduction

Detection of explosives is critical for homeland security, humanitarian demining, and military operations. Various detection technologies [1] are being used complementarily to mitigate explosive hazards which may arise from terrorist attacks. For example, X-ray [2] can be used in conjunction with other sensing technologies, such as ion mobility spectroscopy [3], canines for airport security [4], or radar imaging together with metal detectors for demining operations [5].

Most of these technologies are either expensive and bulky, or complex, requiring a well-experienced operator for optimum performance. Organic semiconductors are promising when used as explosives sensors [6–8], as they can be used to design thin film explosive sensors that be easily fabricated by solution processing. Interestingly, the sensors are comparably portable, cheaper [9,10], and can detect trace vapours of explosive molecules with high sensitivity and fast response time [11,12].

When trace-level vapours of explosives encounter an organic fluorescent sensor, molecules from the vapour are absorbed into the film and modify its light-emitting properties [13,14]. Specifically, an electron is transferred from a photogenerated exciton in the sensor to a sorbed nitroaromatic molecule, which results in fluorescence quenching and indicates the presence of explosives in the surrounding environment. The response to ppb levels of explosives is very rapid; however, for many organic fluorescent sensors, the quenching of fluorescence is irreversible or has slow reversibility [15], which makes them single-use sensors and imposes a limitation in terms of reusability.

It is crucial to have sensors that are reusable, especially in security or military operations. This implies the sensors can recover their fluorescence after exposure to analytes, thereby minimising the number of sensors that would be carried out for operations. There are limited data in the literature showing fluorescence recovery of polymer films after quenching due to explosives vapours [10,12,16]. Zhoa and Swager [16] showed that poly (p-phenylene ethynylene) (PPEs) films can rapidly recover their fluorescence after exposure to DNT and TNT vapours. The measurement was made using a commercially available explosives trace detector called FIDO [11]. In the same experiment, sensory films made from poly(p-phenylenebutadiynylene)s (PPDs) showed much slower recovery, which was attributed to strong binding interactions between the analytes and polymer films.

A promising approach to obtain a reusable organic fluorescent sensor is the application of heat to the fluorescent film, which can thermally desorb the sorbed analytes and result in recovery of the fluorescence [17,18]. Tang et al. [18] showed the thermal desorption of various analytes from carbazole dendrimers films. The heating of the films from 40 °C to 80 °C under the flow of nitrogen resulted in almost full recovery of the sensors' fluorescence.

Here, we present a study of thermal desorption of 2,4-DNT from thin-film explosives sensors made from the commercial fluorescent polymer Super Yellow (SY). Thermal cycling of the sensor results in a recovery of fluorescence, thereby making them reusable, while additionally providing a route to confirm that the fluorescence quenching arises from the analyte response. To optimise the performance of the sensors in terms of reusability, Super Yellow sensors and blended Super Yellow–Poly(9-vinyl carbazole) (PVK) sensors were fabricated. The sensors were exposed to 2,4-DNT vapour in a custom-made chamber while monitoring their fluorescence, before being heated to desorb the DNT molecules from the sensors. Finally, an improvement of the sensors made from the polymer blend and the effect of temperature on these sensors are discussed. This method can be applied to other organic fluorescent sensors, removing the limitation of single-use sensors.

2. Materials and Methods

The fluorescent polymers Super Yellow (SY) and poly(9-vinyl carbazole) (PVK), and the solvents used were purchased from Sigma Aldrich (Merck) and were used without further purification. See Figure 1 for the chemical structure of the polymers.

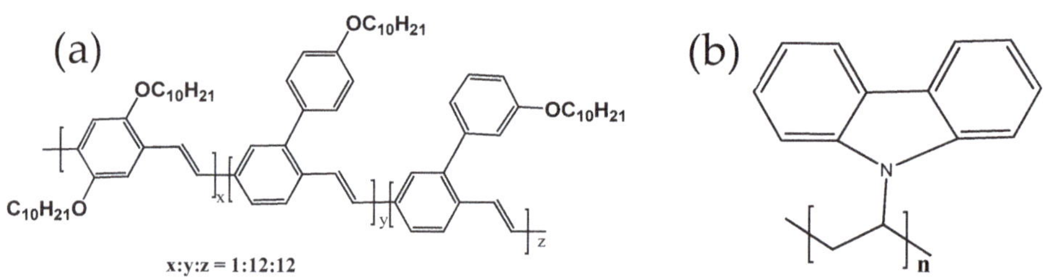

Figure 1. Chemical structures of (**a**) Super Yellow (SY) and (**b**) Poly (9-vinyl carbazole) PVK.

2.1. Sample Preparation

Thin films for sensing were fabricated by first preparing SY and PVK solutions by dissolving in chlorobenzene at 6.5 mg/mL and 20 mg/mL, respectively. The solutions were left on a magnetic stirrer for 24 h at room temperature. For the polymer blend, appropriate mass ratios were measured from the prepared polymer solutions and left for another 24 h on the stirrer for proper mixing of the blend. Thin films for sensing were made by spin-coating the prepared solutions on 25 mm^2 glass substrates at 2000 rpm for 60 s, both for the SY sensor and the blend of SY–PVK. Before spin-coating, the substrates were ultrasonically cleaned in Hellmanex III solution, DI-water three times, acetone and isopropanol for 20, 5×3, 10, and 10 min, respectively, dried in a nitrogen stream, and then plasma ashed in a 100% oxygen plasma (Plasma Technology MiniFlecto) for 3 min. For optical absorbance measurements, fused quartz windows of 12 mm diameter (from UQG optics) were used as substrates.

2.2. Film Characterisation

Optical absorption measurements of the thin films were made using a Cary 300 UV-vis spectrophotometer, and an Edinburgh Instruments FLS980 fluorimeter was used the measure emission spectra. Photoluminescence Quantum Yield (PLQY) measurements were performed in an integrating sphere [19], using a Hamamatsu Photonics C9920-02 measurement system with 444 nm and 340 nm excitation wavelengths. The thickness measurement was made using an Ellipsometer (J. A Woollam M2000U).

2.3. Explosives Vapours Sensing

A custom design vacuum-tight chamber (made of stainless steel) was used for the sensing experiment. The setup for the experiment is the same as that used in our previous work [20], with a slight modification—the sensors were placed in contact with the heater. Excitation of the films was performed using 405 nm continuous wave laser light from a diode laser (Photonic Solutions) after attenuation of the power to 16 μW. Photoluminescence from the sensors was measured using a fibre-coupled CCD spectrometer, taking a measurement every 3 s for 300 s. Explosives vapours were generated using the setup as shown in [21] and nitrogen was used as the carrier gas, and at a flow rate of 6 mL/min. To check for fluorescence recovery, nitrogen gas was used to flush the analyte-exposed sensors, or the sensors heated to 90 °C, followed by a flow of nitrogen gas to flush out the desorbed analytes from the chamber.

3. Results and Discussion

3.1. Photophysical Characterisation

Figure 2a shows the optical absorption of SY, PVK, and a blend of SY and PVK at a 25% to 75% mass ratio. PVK is transparent in the visible wavelength region, SY absorbance peaks at about 443 nm, while the blend retains the properties of both SY and PVK. To understand whether there is a transfer of energy from the PVK to SY molecules in the polymer blend or the blend films phase-separated into regions of SY and PVK, PL measurements of the films were measured (Figure 2b). The films were excited at 340 nm, in a region where both PVK and SY absorb light. PL emissions from both SY and PVK in the blends indicate that there is incomplete transfer of energy between PVK and SY, and there is some level of phase separation of regions of SY and PVK. A blue shift is seen in the peak emission of SY in the blend, which can be attributed to a separation of the SY molecules by the PVK matrix [22].

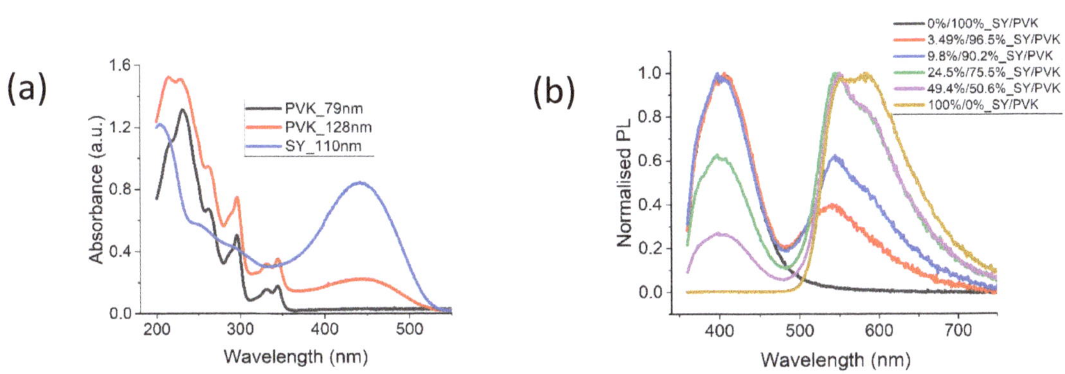

Figure 2. (**a**) Optical absorption of SY, PVK, and a blend of SY and PVK at 25% to 75% mass ratio, respectively; (**b**) Fluorescence spectra of various concentrations of SY–PVK blend at 340 nm excitation wavelength.

3.2. Fluorescence Quenching and Thermal Release of DNT from Sensors

Sensing of DNT vapour using 90 nm thin films of SY and thermal desorption to recover the fluorescence of the quenched SY sensors was demonstrated (Figure 3a). For reference (black line), the PL of the sensor (when only exposed to clean nitrogen gas) remains stable throughout the 300 s. The red line shows the fluorescence from the film when exposed to DNT vapour; fluorescence was collected in clean nitrogen for 60 s, then DNT vapour was introduced in a nitrogen stream (point 1) at 6 mL/min and left to flow for 60 s before cutting off the flow (point 2).

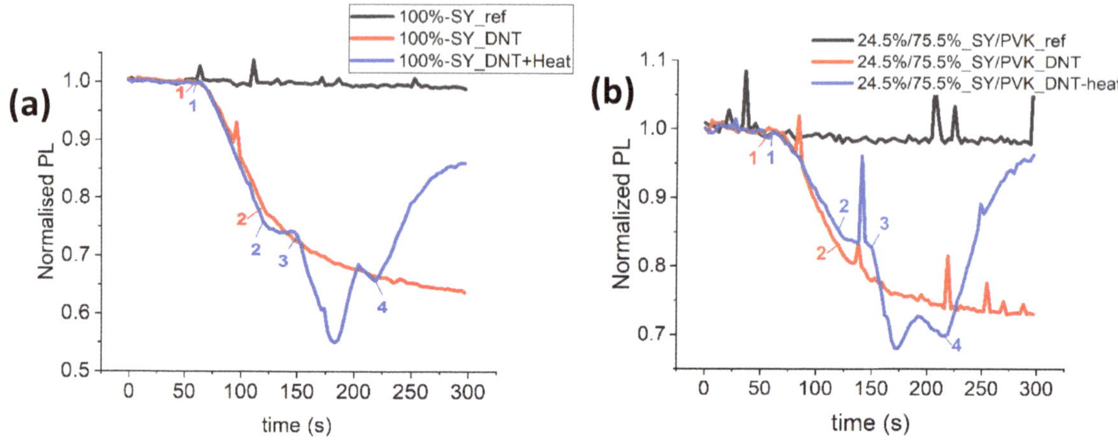

Figure 3. Fluorescence quenching due to DNT vapour and thermal desorption of DNT from sensors made of (**a**) SY film, and (**b**) blend of SY and PVK at 24.5% and 75.5% mass ratio, respectively. Key (Red): 1 = DNT flow on; 2 = DNT flow off and clean nitrogen flow on. Key (Blue): 1 = DNT flow on; 2 = DNT flow off; 3 = heat on; 4 = heat off and clean nitrogen flow on.

Although the source of the DNT vapour was turned off and the system flushed with clean nitrogen, the fluorescence continued to decrease, suggesting that there was a continuing diffusion of DNT vapour into the film [14,21] and that there was a strong binding interaction between the DNT molecules and the SY film [17]. An increase in temperature can weaken the binding strength and release the DNT molecules from the thin film. The blue line shows the thermal desorption of DNT molecules from SY films after fluorescence quenching due to DNT vapour exposure. At point 3, the nitrogen gas flow was temporarily stopped and a heating element was turned on (point 3), which ramped up the temperature of the film from room temperature to 90 °C (point 4). This resulted in an initial thermal degradation [23], then a turning point at around 180 s. We attribute this change to the point when DNT molecules started desorbing from the film, as fluorescence recovery can be seen. When the nitrogen flow was turned back on (point 4), a further increase in the fluorescence was observed. However, the fluorescence could not reach the reference baseline, this may be due either to thermal degradation or there might be some residual DNT molecules left in the film [17].

To optimise the sensor in terms of thermal stability, films fabricated using a blend of SY and PVK were used for similar DNT vapour sensing—see Figure 3b. PVK was chosen because it has a high glass transition temperature and may likely improve the thermal stability of SY sensors [24]. Thermal desorption of the sorbed DNT molecules (point 3 to 4) resulted in a much higher recovery of fluorescence, almost reaching the reference baseline. An investigation of the complex processes during the temperature ramp is in progress.

4. Conclusions

We have shown that commercially available conjugated polymer Super Yellow (SY) can be used as a highly sensitive and reusable sensor for nitroaromatic explosives. An increase in the temperature of the sensors weakens the analyte binding interaction and allows the sorbed analytes to diffuse out of the thin film, which results in the recovery of the PL of the sensors. To optimise the sensors, the high thermally stable polymer PVK was blended with SY, which showed an improvement during thermal desorption of the analytes when resetting the sensors. This method can be applied to other organic fluorescent sensors, removing the limitation of single-use sensors.

Supplementary Materials: The following are available online at https://www.mdpi.com/article/10.3390/CSAC2021-10559/s1.

Author Contributions: Conceptualisation E.B.O., R.N.G., and G.A.T.; methodology, E.B.O., R.N.G., and G.A.T.; software, E.B.O.; formal analysis, E.B.O., R.N.G., and G.A.T.; investigation, E.B.O.; data curation, E.B.O.; writing—original draft preparation, E.B.O.; visualization, E.B.O.; writing—review and editing, R.N.G. and G.A.T.; funding acquisition, R.N.G. and G.A.T.; supervision, G.A.T.; project administration, G.A.T. All authors have read and agreed to the published version of the manuscript.

Funding: This research was funded by the Commonwealth Scholarships Commision, and the NATO Science for Peace & Security Programme under grant agreement MYP G5355.

Institutional Review Board Statement: Not applicable.

Informed Consent Statement: Not applicable.

Data Availability Statement: The research data underpinning this publication can be accessed at https://doi.org/10.17630/b573f7de-553f-4ec4-b907-bfdb62584fc9.

Acknowledgments: We would like to acknowledge useful scientific discussions with I. D. W. Samuel.

Conflicts of Interest: The authors declare no conflict of interest.

References

1. Moore, D.S. Instrumentation for trace detection of high explosives. *Rev. Sci. Instrum.* **2004**, *75*, 2499–2512. [CrossRef]
2. Wells, K.; Bradley, D.A. A review of X-ray explosives detection techniques for checked baggage. *Appl. Radiat. Isot.* **2012**, *70*, 1729–1746. [CrossRef]
3. Perr, J.M.; Furton, K.G.; Almirall, J.R. Solid phase microextraction ion mobility spectrometer interface for explosive and taggant detection. *J. Sep. Sci.* **2005**, *28*, 177–183. [CrossRef]
4. Singh, S.; Singh, M. Explosives detection systems (EDS) for aviation security. *Signal Process.* **2003**, *83*, 31–55. [CrossRef]
5. Daniels, D.J. A review of GPR for landmine detection. *Sens. Imaging Int. J.* **2006**, *7*, 90–123. [CrossRef]
6. Thomas, S.W.; Joly, G.D.; Swager, T.M. Chemical Sensors Based on Amplifying Fluorescent Conjugated Polymers. *Chem. Rev.* **2007**, *107*, 1339–1386. [CrossRef] [PubMed]
7. Swager, T.M.; Wosnick, J.H. Self-Amplifying Semiconducting Polymers for Chemical Sensors. *MRS Bull.* **2002**, *27*, 446–450. [CrossRef]
8. Cumming, C.; Aker, C.; Fisher, M.; Fok, M.; La Grone, M.; Reust, D.; Rockley, M.; Swager, T.; Towers, E.; Williams, V. Using novel fluorescent polymers as sensory materials for above-ground sensing of chemical signature compounds emanating from buried landmines. *IEEE Trans. Geosci. Remote. Sens.* **2001**, *39*, 1119–1128. [CrossRef]
9. Gillanders, R.N.; Samuel, I.D.; Turnbull, G.A. A low-cost, portable optical explosive-vapour sensor. *Sens. Actuators B Chem.* **2017**, *245*, 334–340. [CrossRef]
10. Caron, T.; Guillemot, M.; Montméat, P.; Veignal, F.; Perraut, F.; Prené, P.; Serein-Spirau, F. Ultra trace detection of explosives in air: Development of a portable fluorescent detector. *Talanta* **2010**, *81*, 543–548. [CrossRef]
11. Rose, A.; Zhu, Z.; Madigan, C.F.; Swager, T.M.; Bulovic, V. Chemosensory lasing action for detection of TNT and other analytes. Organic Light Emitting Materials and Devices X. *Proc. SPIE* **2006**, *6333*, 63330Y. [CrossRef]
12. Iii, S.W.T.; Amara, J.P.; Bjork, R.E.; Swager, T.M. Amplifying fluorescent polymer sensors for the explosives taggant 2,3-dimethyl-2,3-dinitrobutane (DMNB). *Chem. Commun.* **2005**, *2005*, 4572–4574. [CrossRef]
13. Shaw, P.E.; Burn, P.L. Real-time fluorescence quenching-based detection of nitro-containing explosive vapours: What are the key processes? *Phys. Chem. Chem. Phys.* **2017**, *19*, 29714–29730. [CrossRef] [PubMed]
14. Ali, M.A.; Geng, Y.; Cavaye, H.; Burn, P.L.; Gentle, I.R.; Meredith, P.; Shaw, P.E. Molecular versus exciton diffusion in fluorescence-based explosive vapour sensors. *Chem. Commun.* **2015**, *51*, 17406–17409. [CrossRef]
15. Bolse, N.; Eckstein, R.; Schend, M.; Habermehl, A.; Eschenbaum, C.; Hernandez-Sosa, G.; Lemmer, U. A digitally printed optoelectronic nose for the selective trace detection of nitroaromatic explosive vapours using fluorescence quenching. *Flex. Print. Electron.* **2017**, *2*, 024001. [CrossRef]
16. Zhao, D.; Swager, T.M. Sensory Responses in Solution vs Solid State: A Fluorescence Quenching Study of Poly(iptycenebutadiynylene)s. *Macromolecules* **2005**, *38*, 9377–9384. [CrossRef]
17. Shaw, P.E.; Cavaye, H.; Chen, S.S.Y.; James, M.; Gentle, I.R.; Burn, P.L. The binding and fluorescence quenching efficiency of nitroaromatic (explosive) vapors in fluorescent carbazole dendrimer thin films. *Phys. Chem. Chem. Phys.* **2013**, *15*, 9845–9853. [CrossRef]
18. Tang, G.; Chen, S.S.Y.; Shaw, P.; Hegedűs, K.; Wang, X.; Burn, P.L.; Meredith, P. Fluorescent carbazole dendrimers for the detection of explosives. *Polym. Chem.* **2011**, *2*, 2360–2368. [CrossRef]
19. Greenham, N.; Samuel, I.; Hayes, G.; Phillips, R.; Kessener, Y.; Moratti, S.; Holmes, A.; Friend, R. Measurement of absolute photoluminescence quantum efficiencies in conjugated polymers. *Chem. Phys. Lett.* **1995**, *241*, 89–96. [CrossRef]

20. Gillanders, R.N.; Glackin, J.M.; Filipi, J.; Kezic, N.; Samuel, I.D.; Turnbull, G.A. Preconcentration techniques for trace explosive sensing. *Sci. Total. Environ.* **2019**, *658*, 650–658. [CrossRef]
21. Campbell, I.A.; Turnbull, G.A. A kinetic model of thin-film fluorescent sensors for strategies to enhance chemical selectivity. *Phys. Chem. Chem. Phys.* **2021**, *23*, 10791–10798. [CrossRef]
22. Summers, M.A.; Buratto, S.K.; Edman, L. Morphology and environment-dependent fluorescence in blends containing a phenylenevinylene-conjugated polymer. *Thin Solid Films* **2007**, *515*, 8412–8418. [CrossRef]
23. Burns, S.; MacLeod, J.; Do, T.T.; Sonar, P.; Yambem, S.D. Effect of thermal annealing Super Yellow emissive layer on efficiency of OLEDs. *Sci. Rep.* **2017**, *7*, srep40805. [CrossRef]
24. Gumyusenge, A.; Tran, D.T.; Luo, X.; Pitch, G.M.; Zhao, Y.; Jenkins, K.A.; Dunn, T.J.; Ayzner, A.L.; Savoie, B.M.; Mei, J. Semiconducting polymer blends that exhibit stable charge transport at high temperatures. *Science* **2018**, *362*, 1131–1134. [CrossRef]

chemistry proceedings

MDPI

Proceeding Paper

Ternary Oxidized Carbon Nanohorns/TiO_2/PVP Nanohybrid as Sensitive Layer for Chemoresistive Humidity Sensor †

Bogdan-Catalin Serban [1,2,*], Octavian Buiu [1,2,*], Marius Bumbac [3,4,*], Roxana Marinescu [1,2], Niculae Dumbravescu [1,2], Viorel Avramescu [1], Cornel Cobianu [1,2,5], Cristina Mihaela Nicolescu [4], Mihai Brezeanu [6], Cristiana Radulescu [3,4] and Florin Comanescu [1]

1 Reliability Department, National Institute for Research and Development in Microtechnologies-IMT Bucharest, 126 A Erou Iancu Nicolae Str., 077190 Voluntari, Romania; roxana.marinescu@imt.ro (R.M.); niculae.dumbravescu@imt.ro (N.D.); viorel.avramescu@imt.ro (V.A.); cornel.cobianu@imt.ro (C.C.); florin.comanescu@imt.ro (F.C.)

2 Research Center for Integrated Systems, Nanotechnologies, and Carbon-Based Nanomaterials (CENASIC)-IMT Bucharest, 077190 Voluntari, Romania

3 Sciences and Advanced Technologies Department, Faculty of Sciences and Arts, Valahia University of Targoviste, 13 Sinaia Alley, 130004 Targoviste, Romania; cristiana.radulescu@valahia.ro

4 Institute of Multidisciplinary Research for Science Technology, Valahia University of Targoviste, 13 Sinaia Alley, 130004 Targoviste, Romania; cristina.nicolescu@valahia.ro

5 Science, and Technology of Information Section, Academy of Romanian Scientists, Str. Ilfov. Nr. 3, 050044 Bucharest, Romania

6 Faculty of Electronics, University Politehnica of Bucharest, Telecommunications and IT, 1-3 Iuliu Maniu Blvd., 6th District, 061071 Bucharest, Romania; scriemiceva@hotmail.com

* Correspondence: bogdan.serban@imt.ro (B.-C.S.); octavian.buiu@imt.ro (O.B.); marius.bumbac@valahia.ro (M.B.)

† Presented at the 1st International Electronic Conference on Chemical Sensors and Analytical Chemistry, 1–15 July 2021; Available online: https://csac2021.sciforum.net/.

Citation: Serban, B.-C.; Buiu, O.; Bumbac, M.; Marinescu, R.; Dumbravescu, N.; Avramescu, V.; Cobianu, C.; Nicolescu, C.M.; Brezeanu, M.; Radulescu, C.; et al. Ternary Oxidized Carbon Nanohorns/TiO_2/PVP Nanohybrid as Sensitive Layer for Chemoresistive Humidity Sensor. *Chem. Proc.* **2021**, *5*, 12. https://doi.org/10.3390/CSAC2021-10616

Academic Editor: Nicole Jaffrezic-Renault

Published: 6 July 2021

Abstract: The relative humidity (RH) sensing response of a chemoresistive sensor using a novel ternary hybrid nanocomposite film as a sensing element is presented. The sensitive layer was obtained by employing the drop-casting technique for depositing a thin film of nanocomposite between the electrodes of an interdigitated (IDT) structure. The sensing support structure consists of an IDT dual-comb structure fabricated on a oSi-SiO_2 substrate. The IDT comprises chromium, as an adhesion layer (10 nm thickness), and a gold layer (100 nm thickness). The sensing capability of a novel thin film based on a ternary hybrid made of oxidated carbon nanohorns–titanium dioxide–polyvinylpyrrolidone (CNHox/TiO_2/PVP) nanocomposite was investigated by applying a direct current with known intensity between the two electrodes of the sensing structure, and measuring the resulting voltage difference, while varying the RH from 0% to 100% in a humid nitrogen atmosphere. The ternary hybrid-based thin film's resistance increased when the sensors were exposed to relative humidity ranging from 0 to 100%. It was found that the performance of the new chemoresistive sensor is consistent with that of the capacitive commercial sensor used as a benchmark. Raman spectroscopy was used to provide information on the composition of the sensing layer and on potential interactions between constituents. Several sensing mechanisms were considered and discussed, based on the interaction of water molecules with each component of the ternary nanohybrid. The sensing results obtained lead to the conclusion that the synergic effect of the p-type semiconductor behavior of the CNHox and the PVP swelling process plays a pivotal role in the overall resistance decrease of the sensitive film.

Keywords: oxidized carbon nanohorns (CNHox); titanium (IV) oxide (TiO_2); polyvinylpyrrolidone (PVP); chemoresistive humidity sensor; swelling

1. Introduction

Relative humidity (RH) sensors have received increasing attention in the last decades due to their importance in many areas of daily life, such as HVAC (heating, ventilation, and air conditioning), food storage, biomedical, climatology, structural health monitoring, agriculture, microelectronics and so forth [1]. The sensing principle, fabrication technology, temperature operating capability, and the sensitive layer play a cardinal role in obtaining superior sensing performances [2]. Due to their excellent sensing properties toward water molecules, abundance and ease of manufacture, low cost, tunable electric properties, and ability to operate under adverse conditions, semiconducting metal oxides (SMOX) have emerged as promising candidates for sensing humidity with high accuracy [3]. TiO_2 is one of the most-used SMOXs and has received increased attention in the last decades due to its fast, linear and sensitive response towards RH changes [4].

At the same time, a lot of recently reported work focused on using carbon-based nanomaterials as sensitive layers within the design of humidity sensors [5]. Among these, in the last years, oxidized carbon nanohorns—single—graphene tubules with oxygen functionalities, mostly carboxylic groups (CNHox) and their nanocomposites, have been extensively explored and have proven to be an attractive option [6–10]. Interestingly, the oxidized carbon nanohorns–TiO_2 nanohybrid was recently used for enhanced photocatalytic hydrogen production [11].

This paper presents, for the first time to our knowledge, the synthesis and characterization of a film based on a ternary nanocomposite comprising oxidized carbon nanohorns–titania–polyvinylpirrolidone CNHox/TiO_2/PVP at 2/1/1 w/w/w ratio. Furthermore, the room temperature RH sensing response of a resistive sensor employing the synthesized sensing film was investigated.

2. Materials and Methods

2.1. Materials

All the materials used in the experiments described below were purchased from Sigma Aldrich (Redox Lab Supplies Com, Bucharest, Romania). CNHox (structure shown in Figure 1a) is characterized by diameters between 2 and 5 nm, lengths between 40 and 50 nm, and a specific surface area around 1300–1400 m^2/g. PVP, with the structure depicted in Figure 1b, has an average molar weight of 40,000 Da. TiO_2 is a nanometric powder (averaged particle size lower than 25 nm), while isopropyl alcohol, $(CH_3)_2CHOH$, is a solution 70% w/w in water.

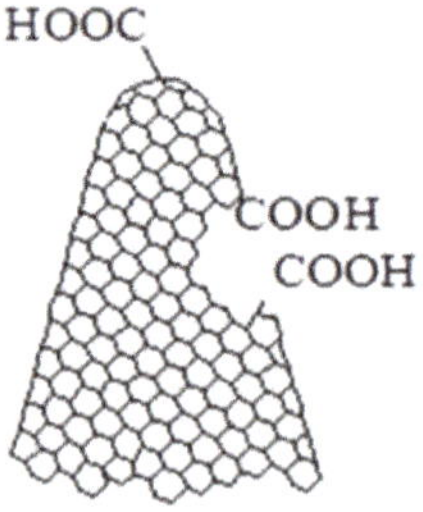

(a)

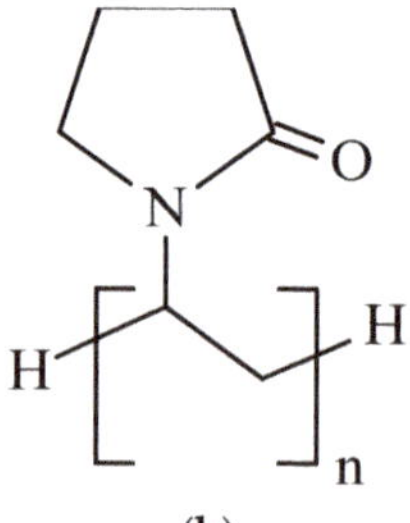

(b)

Figure 1. The structure of: (**a**) CNHox; (**b**) PVP.

2.2. Synthesis of the Ternary Hybrid Nanocomposite Sensing Films and Experimental Setup

The synthesis of the sensitive film based on the ternary nanohybrid CNHox/TiO_2/PVP at 2/1/1 w/w/w ratios is described below.

PVP powder (2 mg) was dissolved in 5 mL isopropyl alcohol solution (70% w/w in water) and subjected to stirring in an ultrasonic bath for one hour at room temperature (RT). CNHox (4 mg) was added to this solution, followed by stirring in the ultrasonic bath for three hours at room temperature. An amount of 2 mg TiO_2 nanopowder was added to the previous dispersion, and continuous stirring was performed in the ultrasound bath for 6 h at RT. The dispersions' homogenization was achieved by employing a mild sonication bath (FS20D Fisher Scientific, Dreieich, Germany) at 42 kHz (output power 70 W). This treatment facilitated a relatively uniform distribution of the CNHox and TiO_2 in the PVP network. The film was annealed in a two-step sequence according to the following procedure:

- Heating for 24 h at 90 °C under low pressure (2 mbar);
- Heating for 24 h at 120 °C under low pressure (2 mbar).

Using the drop-casting method, the sensitive film was obtained by depositing the dispersion of CNHox/TiO_2/PVP in isopropanol solution over the IDT sensing structure (contact area being masked).

The sensing device consists of a metallic IDT dual-comb structure fabricated on a Si substrate (470 μm thickness) covered by SiO_2 (as passivation layer, 1 μm thickness) (Figure 2). The IDT's metal stripes were made by successive deposition of 10 nm chrome (Cr) and 100 nm gold (Au). The sensor has dimensions of 20 × 11 mm^2 [6,7,10].

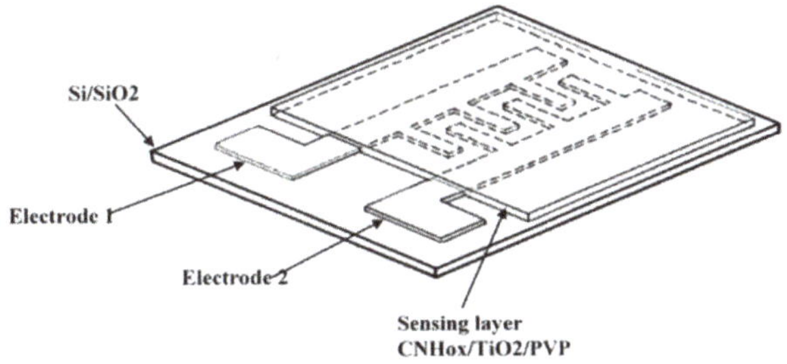

Figure 2. This is a figure. Schemes follow the same formatting.

Raman spectra of the sensing layer were acquired by Lab Ram HR 800 Raman spectrometer (Horiba Europe, Antwerp, Belgium), using a He-Ne laser excitation (633 nm).

Surface topography of the sensing films based on the TiO_2/CNHox/PVP ternary nanohybrid was investigated by scanning electron microscopy (SEM-Nova NanoSEM 630, Thermo Scientific, Waltham, MA, USA). For surface visualization, a field emission gun scanning electron microscope/FEG-SEM-Nova NanoSEM 630 (Thermo Scientific, Waltham, MA, USA) (F.E.I.), with superior low voltage resolution and high surface sensitivity imaging, was used.

The RH sensing measurements were conducted in a suitable experimental setup (Figure 3). The dry nitrogen was purged through two vessels in series containing deionized water for varying the RH in the testing chamber from 0% to 100% RH.

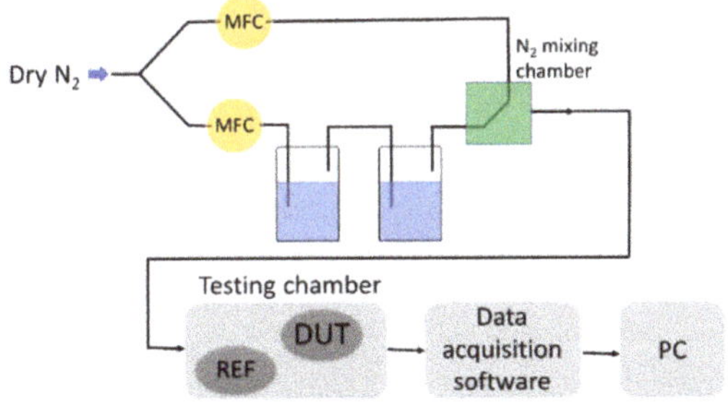

Figure 3. Experimental setup employed for RH measurements.

The experimental chamber included a tandem of sensors: the resistive sensing structure (abbreviated as DUT—device under testing, Figure 3) using the CNHox/TiO_2/PVP-based sensitive film and a capacitive RH commercial sensor (abbreviated as REF, Figure 3). During all measurements, the RH was continuously monitored by the REF sensor. A Keithley 6620 current source (Keithley Instruments GmbH, Germering, Germany), providing current variation between 0.01 and 0.1 A, was employed. The data were collected and analyzed with a PicoLog data logger (PICO Technology, Neots, Cambridgeshire, UK). All experiments were conducted at the ambient temperature [6,7,10].

3. Results

3.1. Raman Spectroscopy

The interaction between CNHox, TiO_2, and PVP was proven using Raman spectroscopy. Figure 4 shows four Raman spectra recorded in four different positions of the CNHox/TiO_2/PVP = 2/1/1 (w/w/w) film, plotted in red, gray, green, and blue color. It can be observed that three active Raman bands (D, G, 2D) were recorded at the wavenumbers of 1318.3, 1591.9, and 2627.8 cm^{-1}, which confirm the presence of the nanocarbonic material (CNHox). One can also identify specific TiO_2 bands as follows: Eg mode at 149.1 cm^{-1} (very sharp and intense), B_{1g} at 397.6 cm^{-1}, A1g at 513.9 cm^{-1}, and Eg at 634.7 cm^{-1}. The peaks associated with PVP are undetectable, most probably being covered by CNHox. The shift of the Raman peak positions of both CNHox and TiO_2 within the ternary nanohybrid compared to the Raman peak positions of each of the two materials considered separately is the most interesting result shown in Figure 4. A plausible explanation for this result can be related to the hydrogen bonds formed between all the components of the synthesized ternary nanohybrid.

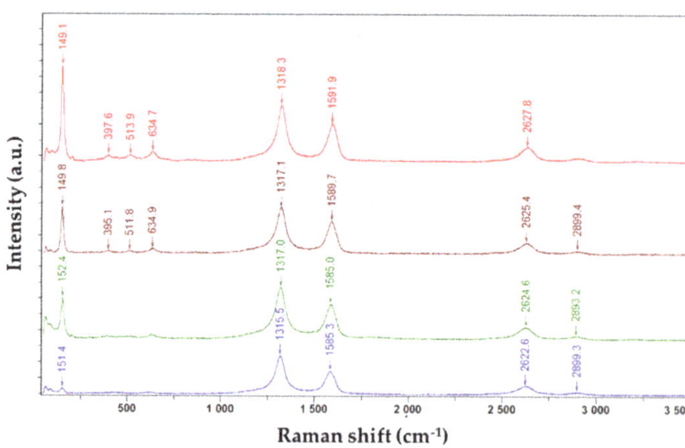

Figure 4. Raman spectra recorded in four different positions of the film deposited from the ternary nanocomposite CNHox/TiO_2/PVP.

3.2. SEM Analysis

Scanning electron micrographs show that the surface morphology of the coating mixture is relatively homogenous in all the cases (Figure 5).

Figure 5. Scanning electron micrographs of the TiO_2/CNHox/PVP at X150,000 magnification.

3.3. *RH Monitoring Capability of the Ternary Hybrid Nanocomposite*

The RH sensing capability of the ternary hybrid nanocomposite-based sensing layer was investigated by applying a current between the two electrodes and measuring the voltage difference when varying the RH from 0% to 100%. A notable characteristic of these sensors is low power consumption, below 2 mW. The behavior of the manufactured sensors is presented in Figure 6.

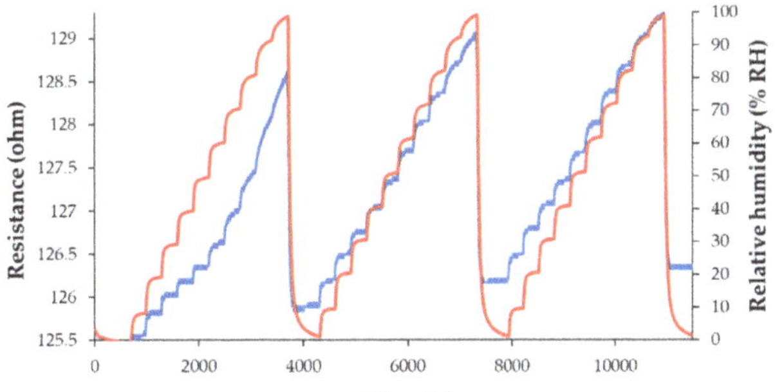

Figure 6. The response of sensor: "RH curve-blue" presented as a function of time for three measurement cycles, when relative humidity was increased in ten steps from 0% RH to 100% RH; "RH curve-red" shows the characteristic measured for a commercial, capacitive sensor.

The resistance of the ternary hybrid-based thin film increases when RH increases. The ternary nanohybrid-based resistive sensors' overall linearity—in humid nitrogen when varying RH from 0% to 100%—is very good, as shown in Figure 7.

An important parameter, such as response time (in seconds), was estimated for both manufactured and commercial RH sensors (Figure 8). It has been observed that, between 10 and 70% RH, comparable response times for either DUT or REH are measured. Differences appear at an RH below 10% and higher than 70%.

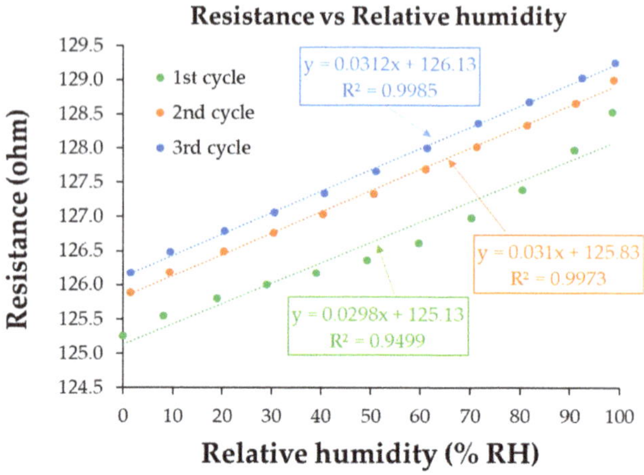

Figure 7. The transfer function of the quaternary nanohybrid-based resistive sensors in humid nitrogen (RH = 0–100%).

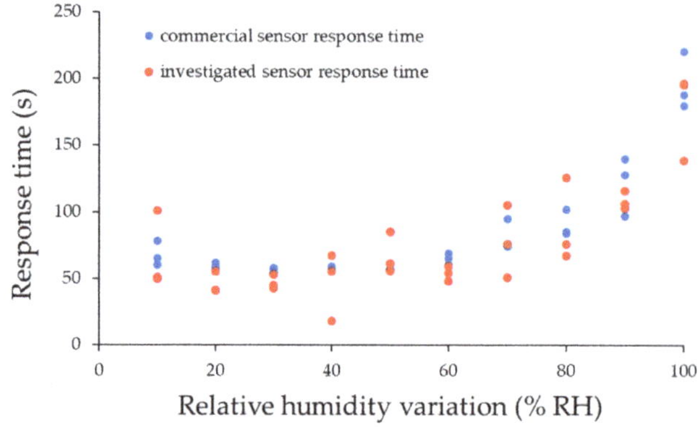

Figure 8. Response time of the tested sensor variation for relative humidity jumps.

3.4. Analysis of the Sensing Mechanism

The most plausible sensing mechanism considers the *p*-type semiconducting material properties of CNHox and the swelling of the hydrophilic polymer. At the interaction with oxidized carbon nanohorns, H_2O molecules donate their electron pairs, decreasing the number of holes in the nanocarbonic materials. Thus, the ternary nanohybrid-based sensing film becomes less conductive. In the same line, the swelling of the hydrophilic polymer PVP increases the distance between the CNHox particles and diminishes the electrically percolating pathways. However, the interaction of water molecules with the surface of TiO_2 yields protonic conduction (Grotthuss mechanism), which should increase the sensing film's conduction. Without completely excluding the later mechanism, one can come to the conclusion that the first two effects prevail and play a pivotal role in the overall increasing resistance of the sensitive film.

4. Conclusions

The RH sensing response of a resistive detector using sensing layers based on a ternary hybrid nanocomposite of CNHox/TiO_2/PVP (2/1/1) was reported. The novel sensitive film used within the design of the chemoresistive sensor exhibited an RT response comparable to that of a commercial capacitive humidity sensor. The ternary nanohybrid-based resistive sensors' overall linearity—in humid nitrogen when varying RH from 0% to 100%—was shown to be excellent. The estimated response times were comparable to those of the commercial sensor. Several sensing mechanism hypotheses were discussed according to the possible chemical interaction between oxidized carbon nanohorn, titania, PVP, and water molecules. Although the Grotthuss mechanism cannot be excluded, the hole conduction ability of CNHox in conjunction with the swelling of hydrophilic polymer prevails and leads to the overall decreasing conduction of the sensing films.

Supplementary Materials: The following are available online at https://www.mdpi.com/article/10.3390/CSAC2021-10616/s1.

Author Contributions: Conceptualization, B.-C.S., C.C. and O.B.; methodology, B.-C.S., C.C., V.A., N.D. and O.B.; software, M.B. (Marius Bumbac), C.M.N. and M.B. (Mihai Brezeanu); validation, B.-C.S., C.C., N.D., O.B. and M.B. (Marius Bumbac); investigation, N.D., B.-C.S., V.A., C.C., R.M. and F.C.; resources, O.B.; writing—original draft preparation, B.-C.S., C.C., O.B. and M.B. (Marius Bumbac); writing—review and editing, all co-authors; visualization, O.B.; supervision, B.-C.S., C.C. and M.B. (Marius Bumbac); project administration, B.-C.S. and O.B.; funding acquisition, C.R., O.B. All authors have read and agreed to the published version of the manuscript.

Funding: The authors would like to acknowledge the financial support provided by: the Romanian Ministry of Research and Education, via the "Nucleu Program" called "MICRO-NANO-SIS PLUS", grant number P.N. 19 16, UEFISCDI contract number 364PED—23 October 2020.

Institutional Review Board Statement: Not applicable.

Informed Consent Statement: Not applicable.

Conflicts of Interest: The authors declare no conflict of interest.

References

1. Yeo, T.L.; Sun, T.; Grattan, K.T.V. Fibre-optic sensor technologies for humidity and moisture measurement. *Sens. Actuators A Phys.* **2008**, *144*, 280–295. [CrossRef]
2. Farahani, H.; Wagiran, R.; Hamidon, M.N. Humidity sensors principle, mechanism, and fabrication technologies: A comprehensive review. *Sensors* **2014**, *14*, 7881–7939. [CrossRef] [PubMed]
3. Ji, H.; Zeng, W.; Li, Y. Gas sensing mechanisms of metal oxide semiconductors: A focused review. *Nanoscale* **2019**, *11*, 22664–22684. [CrossRef] [PubMed]
4. Zakrzewska, K. Gas sensing mechanism of TiO_2-based thin films. *Vacuum* **2004**, *74*, 335–338. [CrossRef]
5. Tulliani, J.M.; Inserra, B.; Ziegler, D. Carbon-based materials for humidity sensing: A short review. *Micromachines* **2019**, *10*, 232. [CrossRef] [PubMed]
6. Serban, B.C.; Buiu, O.; Dumbravescu, N.; Cobianu, C.; Avramescu, V.; Brezeanu, M.; Bumbac, M.; Nicolescu, C.M. Oxidized Carbon Nanohorns as Novel Sensing Layer for Resistive Humidity Sensor. *Acta Chim. Slov.* **2020**, *67*, 469–475. [CrossRef] [PubMed]
7. Serban, B.C.; Buiu, O.; Dumbravescu, N.; Cobianu, C.; Avramescu, V.; Brezeanu, M.; Pachiu, C.; Nicolescu, C.M. Oxidized Carbon Nanohorn-Hydrophilic Polymer Nanocomposite as the Resistive Sensing Layer for Relative Humidity. *Anal. Lett.* **2020**, *54*, 527–540. [CrossRef]
8. Serban, B.C.; Bumbac, M.; Buiu, O.; Cobianu, C.; Brezeanu, M.; Nicolescu, C. Carbon nanohorns and their nanocomposites: Synthesis, properties, and applications. A concise review. *Ann. Acad. Rom. Sci. Ser. Math. Appl.* **2018**, *11*, 5–18.
9. Serban, B.C.; Cobianu, C.; Dumbravescu, N.; Buiu, O.; Bumbac, M.; Nicolescu, C.M.; Cobianu, C.; Serbanescu, M.; Pachiu, C.; Serbanescu, M. Electrical Percolation Threshold and Size Effects in Polyvinylpyrrolidone-Oxidized Single-Wall Carbon Nanohorn Nanocomposite: The Impact for Relative Humidity Resistive Sensors Design. *Sensors* **2021**, *21*, 1435. [CrossRef] [PubMed]
10. Serban, B.-C.; Cobianu, C.; Buiu, O.; Bumbac, M.; Dumbravescu, N.; Avramescu, V.; Nicolescu, C.M.; Brezeanu, M.; Radulescu, C.; Craciun, G.; et al. Quaternary Oxidized Carbon Nanohorns—Based Nanohybrid as Sensing Coating for Room Temperature Resistive Humidity Monitoring. *Coatings* **2021**, *11*, 530. [CrossRef]
11. Jagannatham, M.; Berkmans, A.J.; Haridoss, P.; Reddy, L.; Shankar, M.V. Influence of pre-oxidation, versus post-oxidation of carbon nanohorns in TiO_2 nanohybrid for enhanced photocatalytic hydrogen production. *Mater. Res. Bull.* **2019**, *109*, 34–40.

chemistry proceedings

MDPI

Abstract

Eucalyptus Biochar as a Sustainable Nanomaterial for Electrochemical Sensors †

Annalisa Scroccarello [1,*], Flavio Della Pelle [1], Qurat Ul Ain Bukhari [1], Filippo Silveri [1], Daniele Zappi [2], Enrico Cozzoni [3] and Dario Compagnone [1]

1 Faculty of Bioscience and Technology for Food, Agriculture and Environment University of Teramo, Via Renato Balzarini 1, 64100 Teramo, Italy; fdellapelle@unite.it (F.D.P.); quabukhari@unite.it (Q.U.A.B.); fsilveri@unite.it (F.S.); dcompagnone@unite.it (D.C.)
2 Biosensor, Via Olmetti 44, Formello, 00060 Roma, Italy; daniele.zappi@uniroma1.it
3 GRADO ZERO INNOVATION, Via nove 2/A, Montelupo F.no, 50056 Firenze, Italy; enrico.cozzoni@gradozero.onmicrosoft.com
* Correspondence: ascroccarello@unite.it

† Presented at the 1st International Electronic Conference on Chemical Sensors and Analytical Chemistry, 1–15 July 2021; Available online: https://csac2021.sciforum.net/.

Citation: Scroccarello, A.; Pelle, F.D.; Bukhari, Q.U.A.; Silveri, F.; Zappi, D.; Cozzoni, E.; Compagnone, D. Eucalyptus Biochar as a Sustainable Nanomaterial for Electrochemical Sensors. *Chem. Proc.* **2021**, *5*, 13. https://doi.org/10.3390/CSAC2021-10618

Academic Editor: Elisabetta Comini

Published: 6 July 2021

Abstract: Carbonaceous-based nanomaterials (C-NMs) are the pillar of myriad sensing and catalytic electrochemical applications. In this field, the search for environmentally sustainable C-NMs from renewable sources became a duty in the development of nano-sensors. Herein, water-soluble carbon nanofibers (CF) were produced from eucalyptus scraps-based biochar (BH) through an ultrasound treatment, assisted by sodium cholate used as a stabilizing agent. Noteworthy, thanks to the use of the bio-stabilizing agent, the nanofibers were dispersed in water avoiding the use of organic solvents. The BH-CF was investigated as sensing material onto commercial screen-printed electrodes via drop-casting (BH-SPE) and as thin-film fully integrated into a lab-made flexible electrode. The thin film was produced via BH-CF vacuum filtration followed by the film transferring to a thermo-adhesive plastic substrate through thermal lamination. This approach gave rise to a conductive BH-CF film (BH-Film) easily embodied in a lab-made electrode produced with office-grade instrumentation (i.e., craft-cutter machine, thermal laminator) and materials (i.e., laminating pouches, stencil). The BH-CF amount was optimized and the resulting film morphologically characterized, then, the electrochemical performances were studied. The BH-CF electrochemical features were investigated towards a broad range of analytes containing phenol moieties, discrimination between orto- and mono-phenolic structures were achieved for all the studied compounds. As proof of applicability, the BH-CF-based sensors were challenged for simultaneous determination of mono-phenols and ortho-diphenols in olive oil extracts. LODs ≤ 0.5 µM and ≤ 3.8 µM were obtained for hydroxytyrosol (o-diphenol reference standard) and Tyrosol (m-phenols reference standard), respectively. Moreover, a high inter-sensors precision (RSD calibration-slopes $\leq 7\%$, $n = 3$) and quantitative recoveries in sample analysis (recoveries 91–111%, RSD $\leq 6\%$) were obtained. Here, a solvent-free strategy to obtain water-soluble BH-CF was proposed, and their usability to sensor fabrication and modification proved. This work demonstrated as cost-effective and sustainable renewable sources, rationally used, can lead to obtain useful nanomaterials.

Keywords: biochar; sensor; nanomaterial

Supplementary Materials: The following are available online at https://www.mdpi.com/article/10.3390/CSAC2021-10618/s1.

chemistry proceedings

MDPI

Abstract

Graphene Nanoflakes Incorporating Natural Phytochemicals Containing Catechols as Functional Material for Sensors †

Filippo Silveri *, Flavio Della Pelle, Daniel Rojas and Dario Compagnone

Faculty of Bioscience and Technology for Food, Agriculture and Environment University of Teramo, Via Renato Balzarini 1, 64100 Teramo, Italy; fdellapelle@unite.it (F.D.P.); jdrojastizon@unite.it (D.R.); dcompagnone@unite.it (D.C.)

* Correspondence: fsilveri@unite.it

† Presented at the 1st International Electronic Conference on Chemical Sensors and Analytical Chemistry, 1–15 July 2021; Available online: https://csac2021.sciforum.net/.

Abstract: Phytochemical products start to be employed to assist 2D nanomaterials exfoliation. However, a lack of studies regarding the molecules involved and their capacity to give rise to functional materials is evident. In this work, a novel green liquid-phase exfoliation strategy (LPE) is proposed, wherein a flavonoid namely catechin (CT) exclusively assists the exfoliation of bulk graphite in conductive water-soluble graphene nanoflakes (GF). Physicochemical and electrochemical methods have been employed to characterize the morphological, structural, and electrochemical features of the GF-CT. Surprisingly, the obtained GF-CT integrates well-defined electroactive quinoid adducts. The resulting few-layers graphene flakes intercalated with CT aromatic skeleton ensure strict electrical contact among graphene sheets, whereas the fully reversible quinoid electrochemistry (ΔE = 28 mV, Ip, a/Ip, c = ~1) is attributed to the residual catechol moieties, which work as an electrochemical mediator. The GF-CT intimate electrochemistry is generated directly during the LPE of graphite, not requiring any modification or electro-polymerization steps, resulting in stable (8 months) and reproducible material. The electrocatalytic activity has been proven towards hydrazine (HY) and β-nicotinamide adenine dinucleotide (NADH), a pollutant and a coenzyme, respectively. High sensitivity in extended linear ranges (HY: LOD = 0.1 μM, L.R. 0.5–150 μM; NADH: LOD = 0.6 μM, L.R. 2.5–200 μM) at low overpotential (+0.15 V) was obtained using amperometry, avoiding electrode-fouling. Improved performances, compared with graphite commercial electrodes and graphene exfoliated with a conventional surfactant, were obtained. The GF-CT was successfully used to perform the detection of HY and NADH (recoveries 94–107%, RSD $\leq$ 8%) in environmental and biological matrices, proving the material exploitability even in challenging analytical applications. On course studies aim to combine the intrinsic conductivity of the GF-CT with flexible substrates, in order to construct flexible electrodes/devices able to house GF-CT-exclusively composed conductive films. In our opinion, the proposed GF-CT elects itself as a cost-effective and sustainable material, particularly captivating in the (bio)sensoristics scenario.

Keywords: nanostructured-functional-material; grapheme; 2D-materials; mediator; phytochemicals; catechol-moieties; liquid-phase-exfoliation

---

Supplementary Materials: The following are available at https://www.mdpi.com/article/10.3390/CSAC2021-10619/s1.

Citation: Silveri, F.; Pelle, F.D.; Rojas, D.; Compagnone, D. Graphene Nanoflakes Incorporating Natural Phytochemicals Containing Catechols as Functional Material for Sensors. *Chem. Proc.* **2021**, *5*, 14. https://doi.org/10.3390/CSAC2021-10619

Academic Editor: Elisabetta Comini

Published: 6 July 2021

Publisher's Note: MDPI stays neutral with regard to jurisdictional claims in published maps and institutional affiliations.

chemistry proceedings

MDPI

Proceeding Paper

Numerical and Experimental Modeling of Paper-Based Actuators †

Ashutosh Kumar, Hojat Heidari-Bafroui, Amer Charbaji, Nasim Rahmani, Constantine Anagnostopoulos and Mohammad Faghri *

Microfluidics Laboratory, Department of Mechanical, Industrial and Systems Engineering, University of Rhode Island, Kingston, RI 02881, USA; ashutosh@uri.edu (A.K.); h_heidari@uri.edu (H.H.-B.); charbaji@uri.edu (A.C.); nara7@uri.edu (N.R.); anagnostopoulos@uri.edu (C.A.)

* Correspondence: faghrim@uri.edu

† Presented at the 1st International Electronic Conference on Chemical Sensors and Analytical Chemistry, 1–15 July 2021; Available online: https://csac2021.sciforum.net/.

Abstract: Microfluidic paper-based analytical devices (μPADs) have witnessed a great extent of innovation over the past decade, developing new components and materials assisting the diagnosis of different diseases and sensing of a wide range of biological, chemical, optical, and electrochemical phenomena. The novel paper-based cantilever (PBC) actuator is one the major components that allows autonomous loading and control of multiple fluid reagents required for the accurate operation of paper-based microfluidic devices. This paper provides an extensive overview of numerical and experimental modeling of fluidically controlled PBC actuators for automation of the paper-based assay. The PBC model undergoing hygro-expansion utilizes quasi-static 2D fluid loaded structure governed by the Euler–Bernoulli beam theory for small and moderately large deflections. The solution for the model can avail the response of paper-based actuators for response deflection θ, within 0° to 10° under the assumption of insignificant cross-sectional deformation. The actuation of PBC obtained using a quasi-static theory shows that our results are consistent with quantitative experiments demonstrating the adequacy of models.

Keywords: paper-based cantilever; microfluidic analytical device; paper-based actuator

Citation: Kumar, A.; Heidari-Bafroui, H.; Charbaji, A.; Rahmani, N.; Anagnostopoulos, C.; Faghri, M. Numerical and Experimental Modeling of Paper-Based Actuators. *Chem. Proc.* **2021**, *5*, 15. https://doi.org/10.3390/CSAC2021-10468

Academic Editor: Manel del Valle

Published: 30 June 2021

1. Introduction

The first usage of paper as a substrate material goes back to the early 1800s with litmus paper, the oldest form of pH indicator, for analytical testing of chlorine and carbonic oxide [1]. Developed by Müller and Clegg [2], the first microfluidics channel on filter paper was utilized to elute a mixture of pigment. However, in 2007, the Whiteside's Group of Harvard University gave a new push to the endless possibilities of paper-based microfluidics by introducing a patterned paper as a platform for portable devices [3]. Due to the several benefits of paper for making microfluidic paper-based analytical devices (μPADs), it has attracted extreme attention. Paper is a very cheap and renewable material since it is made of cellulose, the most abundant organic polymer on Earth, and it is also biocompatible and can be used for numerous biological and chemical applications [4,5]; thanks to capillary forces, an external force is not needed for fluid transport in paper.

Paper-based microfluidic devices are made up of different sections that serve different purposes. The simpler devices generally have a sample port, transport channels, reaction zones, and a detection zone. Devices that perform more complex reactions or enzyme-linked immunosorbent assay (ELISA) protocols require proper sample timing and control as it flows into the different reaction zones. Such control is usually achieved by the use of a suitable valving system in the device.

Different valving systems have been developed for use in paper-based microfluidic devices. Some of these valving systems were simple, autonomous and required little-to-no operator input while others required some operator involvement or an external power source to operate. Li et al. [6] reported a method to stop and to promote wicking by manually separating and re-joining two paper channels using a sliding separator or a switch valve. Jahanshahi-Anbuhi et al. [7] utilize the switch valve, which they refer to as a flap, in a paper-based sensor for the detection of pesticides. Han et al. [8] built on the concept and created a 3D slip-PAD in which the operator connects the different fluidic channels by manipulation of the cartridge that holds the paper-based device. Martinez et al. [9] reported on a press valve in a multilayer paper-based device that had small gaps between the paper layers which were created by the finite thickness of the tape in between the layers. The operator would need to press the buttons to mechanically bridge the discrete channels to promote wicking. Rodriguez et al. [10] and Jayawardane et al. [11] utilized discarded separators in their devices that would be removed after a predetermined amount of time to connect appropriate fluidic channels. Noh et al. [12] and Lutz et al. [13] provided designs for metering valve systems that were based on paraffin wax and dissolvable sugar, respectively. While this valving system is cheap, controls the flow rate and requires no operator input, it suffers from the fact that it can only impede the flow rate of the sample without completely stopping it until it is required to move to the next section of the device. Chen et al. [14] and Gerbers et al. [15] developed novel autonomous two- and three-dimensional microfluidic valves involving no external actuation based on altering the hydrophobicity of a multilayered structure by means of a surfactant. With this technique, they were able to control the order and mixing time of the sample and multiple reagents autonomously. However, these valves required long response times and large volumes of actuation fluids. Lai et al. [16] provided a design for a timing valve in their device that consisted of a surfactant and a wax barrier to provide appropriate time delays to sequentially handle the multiple fluidic operations of the device. Koo et al. [17] used an electrowetting valve in which a dielectric material that is normally hydrophobic, is polarized and becomes hydrophilic when the valve is actuated with an applied potential difference. Li et al. [18] developed a magnetic valve with the use of a blend of ferromagnetic nanoparticles and polydimethylsiloxane that magnetized the paper cantilever. Although the operation of the magnetic valve was autonomous, it required an applied potential to actuate the cantilever. Phillips et al. [19] used a thin-film electric heater to thermally actuate a wax valve to pass or block the flow of the sample. Kong et al. [20] reported actuators based on selective wetting of folded paper strips. These strips reduced the actuator's response time to within two seconds from wetting, while utilizing a very small volume of actuation fluid in the order of four microliters. Toley et al. [21] developed a valving toolkit that utilized an expanding element, which was a compressed sponge valve, to control the flow of the fluid in the device. Fu et al. [22] developed a thermally actuated cantilever valve that utilizes a shape-memory polymer.

This paper provides an overview of the behavior of a fluidically controlled paper-based cantilever (PBC) for the automation of a paper-based assay that is exploited in a fluidic circuit to sequentially load several reagents to the analyte detection area. For this purpose, first a simple paper-based cantilever (PBC) will be modeled. Different scenarios for using the cantilever will then be presented. Finally, a mathematical model will be developed for better understanding of the model and for comparing with the experimental results.

2. Methods

A fluidic circuit is used to conduct complex immunoassay to sequentially load two or more additional reagents in addition to the sample fluid. Once a user pipettes a certain amount of sample fluid on the sample port of the paper-based cantilever, as the sample fluid transfers to the free end of the PBC due to the capillary force, and hence, the hygro-expansion of cellulose fibers, the cantilever starts to bend downward and contacts the stationary component within several seconds. This paper incorporates experimental, analytical and numerical methods for analysis and the results from different methods were compared to check for adequacy of the model.

2.1. Materials

The following materials were used in preparing, fabrication, and testing the cantilever valves used in this study. Filter paper (GE Healthcare Whatman grade 4—1004917 & grade 41—1441866), chromatography paper (GE Healthcare Whatman 1—3001878), deionized water, and food coloring (Wilton Icing Colors). The stationary component was a 1 × 1 cm chromatography paper. The dimensions of the cantilevers were using a vector graphics software (CorelDraw X6). The cantilevers were then cut out from paper, in the cross-machine direction, using a laser engraver (Epilog mini 40W). An 8-megapixel video camera with 30 frames per seconds capability was used to record the actuation of the cantilever valve. A media player (Avidemux) was used to playback the recording and collect the data.

2.2. Modeling

2.2.1. Concept of Design

Single Cantilever Design

The design of the single cantilever device comprises a paper-based cantilever, reagent pad, stationary structure and a support structure. Figure 1 demonstrates the single cantilever design of PBC in two conditions (unloaded and loaded). The PBC remains in a normally open position with the separation distance 'd', until the sample fluid is loaded. Upon loading with sample fluid, the PBC absorbs the sample fluid though wicking and the actuator activates and results in deflection, as shown in Figure 1, resulting in the transfer of sample fluid to the reagent pad.

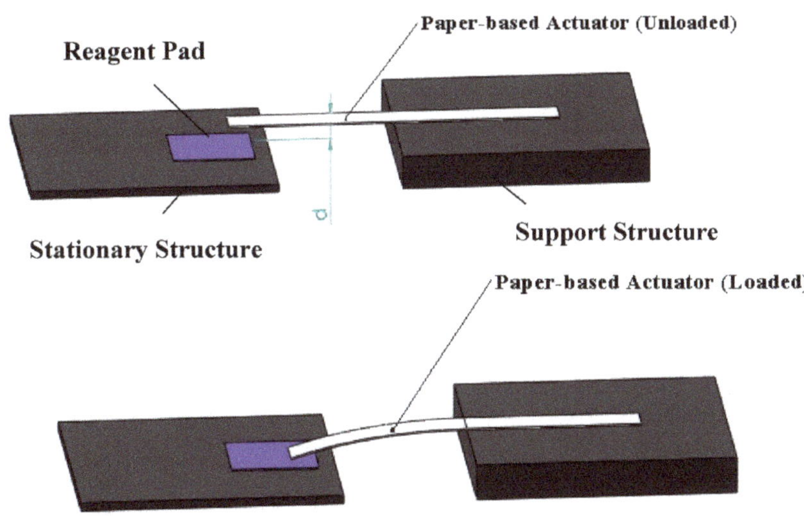

Figure 1. Single cantilever design of PBC actuator device.

Double Cantilever Design

In the double cantilever design, two cantilevers are utilized to allow for the sequential loading of reagents into an area of interest on the paper-based device. Figure 2 shows the diagram of the double cantilever design and Figure 3 shows the sequence of operation for this architecture. The operation of this device is fairly simple; first, the user just needs to pipette a certain amount of activation liquid (e.g., DI water) on the sample pad initiating the actuation of the cantilever on the right, shortly after the addition of the activation fluid, as seen in Figure 3b. This permits the flow of the reagent on the righthand side into the middle section of the device to react with any chemicals dried there. In the meantime, part of the activation fluid is flowing through the timing channel to the left and activates the second cantilever after a predetermined duration of time, as seen in Figure 3d. This connects both zones and thus enables the passage of the reagent on the left into the area of interest to react with the material there, as seen in Figure 3e.

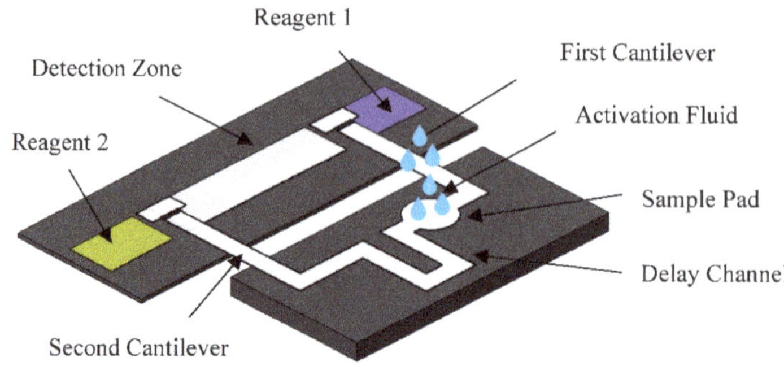

Figure 2. Addition of activation fluid on sample pad for double cantilever design of PBC actuator device.

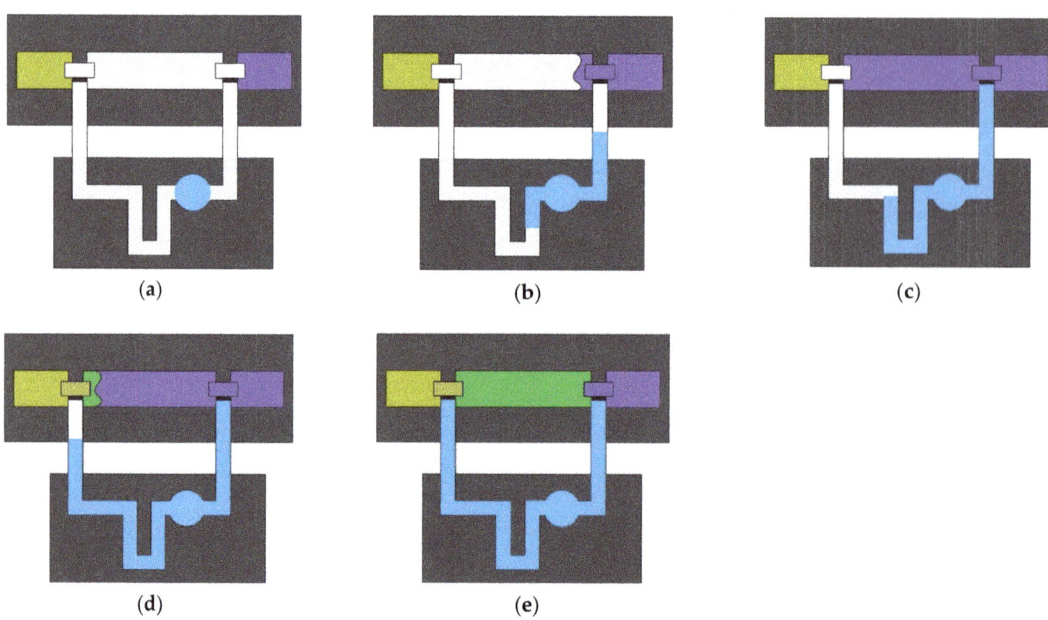

Figure 3. Sequence of operation of the double cantilever design. (**a**) Initial state of device after pipetting activation fluid, (**b**) Loading of the first reagent on-to detection zone upon activation of the first cantilever. (**c**) Activation fluid passing from the delay channel for actuation of the second cantilever. (**d**) Loading of the second reagent on-to the detection zone upon activation of the second cantilever. (**e**) Final state of the device after mixing of two reagents.

2.2.2. Experimental Model

A picture of the experiment is shown in Figure 4, consisting of PBC and a capillary tube to load the fluid and obtain the response deflection.

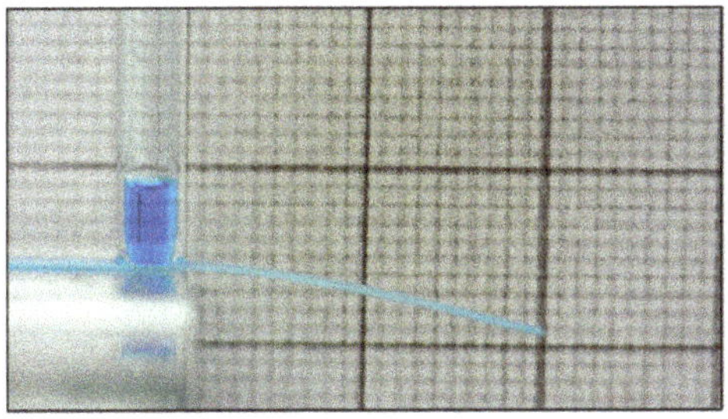

Figure 4. Experimental model for PBC actuator.

In order to measure the deflection of the PBC, Whatman grade 41 filter paper was selected. The filter paper was cut in the cross-machine direction with a 4 mm width and a 40 mm length using an Epilog Mini laser engraver. The picture of samples for PBC can be seen in Figure 5.

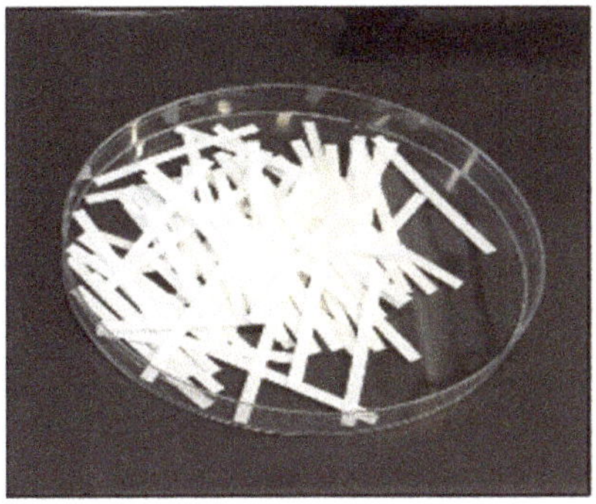

Figure 5. Paper-based cantilevers (PBCs).

A 2 mm in diameter capillary tube was used to introduce fluid into the PBC. A fixture for the position of PBC and capillary was designed and utilized to reduce the uncontrollable error of running experiments (Figure 6).

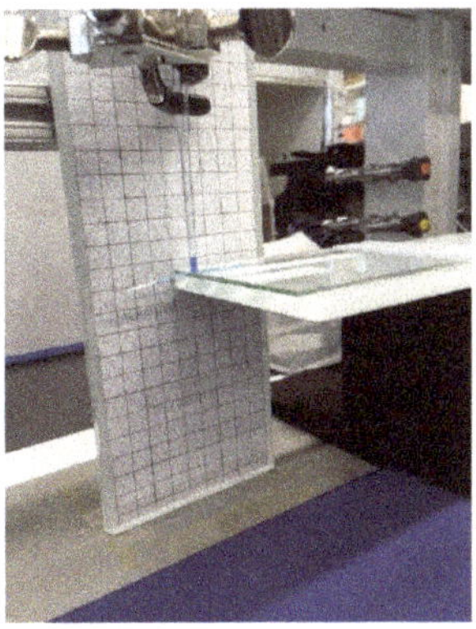

Figure 6. Fixture for experiment.

2.2.3. Mathematical Model

This paper models the response of PBC on the fluidic loading of the paper-based cantilever. The PBC undergoes hygro-expansion when exposed to fluid, which results in the actuation of the cantilever. The mathematical modeling of the PBC determines the response deflection, the Euler–Bernoulli beam theory will be used with the assumption that the cross-section remains normal to the axis of the cantilever deflection and that cross-sectional deformation is not significant.

Modeling of Flow in Paper

The capillary model is adapted to simulate the flow in paper for this experiment. This model will be utilized to develop an analytical expression for fluid imbibition into paper due to capillary action. According to the Washburn equation [23]:

$$L_w = \sqrt{\frac{R\gamma \cos\theta}{2\eta} t} \tag{1}$$

where, L_W—wetted length, R—pore radius, γ—the surface tension of the liquid, t—the time taken for liquid to seep into the capillary, θ—the contact angle of liquid on the capillary walls and η—the viscosity of fluid.

Modeling of PBC

The quasi 2D fluid structure mode is adapted for analysis. Geometry is inspired by the bending of the paper-based cantilever when exposed to fluid. Cantilever actuation is considered as the system output defined by fluid loading, please refer Table 1 for parameters.

Table 1. PBC parameters.

Variables	Description	Dimension
x	Length coordinate	L
z	Height coordinate	L
t	Time	T
w	Deflection	L
q	Transverse loading	MT^{-2}
l	PBC length	L
b	PBC width	L
h	PBC height	L
ρ	Fluid density	ML^{-3}
E	Young's modulus	$ML^{-1}T^{-2}$

Consider the PBC with a uniform rectangular cross-section loaded by a water column, as shown in Figure 7. Moderately large static deflection (Appendix A) of PBC is governed by equation [24],

$$EI\frac{d^4w}{dx^4} + N\frac{d^2w}{dx^2} + q = 0 \tag{2}$$

where, E—Young's modulus of the wet paper, I—second moment of inertia, N—internal (axial) stress in PBC and q—auxiliary loading is the equivalent transverse loading resulting in the same deflection of the PBC.

Internal (axial) stress in PBC is given by,

$$N = EA\epsilon \tag{3}$$

where, A—cross-sectional area of PBC and ϵ– membrane strain (due to hygro-expansion and rotation of PBC) [25].

The auxiliary loading of PBC is given by,

$$q = \rho g\left(\frac{V_f}{l}\right) \tag{4}$$

where, Vf —volume of fluid imbibition into PBC.

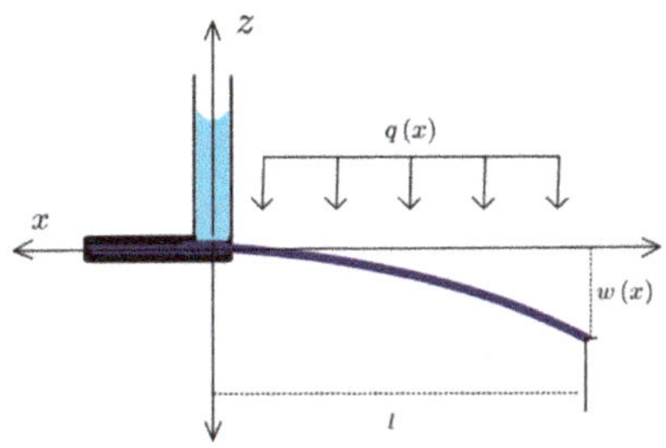

Figure 7. Static actuation model of PBC.

Substituting the above equation to governing Equation (2) we get,

$$w^{IV} + \lambda^2 w^{II} = Q \tag{5}$$

where,

$$\lambda = l\sqrt{\frac{N}{EI}} \tag{6}$$

$$Q = \frac{-\rho g\left(\frac{V_f}{l}\right)}{EI} \tag{7}$$

In the above equations, λ and Q are parameters associated with the type of paper and fluid used for experiment respectively.

In order to solve Equation (2) uniquely, four boundary conditions are required. A four combination of in-plane boundary condition for PBC are given as, deflection: fixed-end x = 0, $w(x) = 0$; slope: fixed-end x = 0, $w^I(x) = 0$, slope; bending moment: free-end x = l, $w^{II}(x) = 0$; shear force: free-end x = l, $w^{III}(x) = 0$.

Non-Dimensional Model of PBC

To get a better idea of the relative size of the terms, we need a non-dimensionalized Equation (5) as per Table 2.

Table 2. PBC parameters.

Variables	Expression	Description
$x*$	$\left(\frac{x}{l}\right)$	Characteristic Length
$w*$	$\left(\frac{w}{R_g}\right)$	Characteristic Deflection
$t*$	$\left(\frac{R_g}{l}\right)t$	Characteristic Time
$\lambda*$	$l\sqrt{\frac{N}{EI}}$	Internal Stress Parameter
$Q*$	$\frac{\rho g V_f l^3}{R_g EI}$	Fluid Loading Parameter
Rg	$\sqrt{\frac{I}{A}}$	Radius of Gyration

Thus, the non-dimensional form of Equation (5) yields to,

$$w^{*IV} + \lambda^{*2} w^{*II} = Q^* \tag{8}$$

The solution to the problems depends on a four combination of in-plane boundary condition for the non-dimensional model given as,

$$\text{Fixed} - \text{end } x^* = 0,\ w^*(x^*) = 0 \tag{9}$$

$$\text{Fixed} - \text{end } x^* = 0,\ w^{*I}(x^*) = 0 \tag{10}$$

$$\text{Free} - \text{end } x^* = 1,\ w^{*II}(x^*) = 0 \tag{11}$$

$$\text{Free} - \text{end } x^* = 1,\ w^{*III}(x^*) = 0 \tag{12}$$

To summarize, Equation (8), together with boundary conditions, related the deflection, hygro-expansive stress and loading of PBC.

Solution for PBC

The problem is solved under the assumption of fixed-free end conditions, and the fluid load is distributed according to the square of wetted length as per the Washburn equation mentioned earlier. There is a considerable elastoplastic effect [26] of PBC response due to fluid imbibition.

In order to determine the deflection shape of the PBC, consider Equation (8). The roots of the characteristic equation are 0, 0, $\pm i\lambda$. Therefore, the general solution of the homogeneous equation is,

$$w_g^*(x^*) = C_0 + C_1 x^* + C_2 cos(\lambda^* x^*) + C_3 sin(\lambda^* x^*) \tag{13}$$

As a particular solution of the inhomogeneous equation,

$$w_p^*(x^*) = Cx^{*2} \tag{14}$$

Substituting the above solution to the governing Equation (8), we get,

$$C = \frac{Q^*}{2\lambda^{*2}} = -\frac{1}{2}\left(\frac{\rho g V_f l}{NR_g}\right) \tag{15}$$

The general solution for Equation (4) is sum of the general solution of the homogeneous equation, and the particular solution of the inhomogeneous equation,

$$w^*(x^*) = w_g^* + w_p^* \tag{16}$$

There are four unknowns, and four boundary conditions as per Equation (4) for transverse deflection. The coefficients determined from boundary conditions are $C_0 = -C/10$, $C_1 = -4C/100$, $C_2 = C/10$, and $C_3 = 9C/1000$.

The general solution for deflection of PBC,

$$w_g^*(x^*) = C_0 + C_1 x^* + C_2 cos(\lambda^* x^*) + C_3 sin(\lambda^* x^*) + Cx^{*2} \tag{17}$$

3. Results

3.1. Analytical Solution

Analytical solution for the problem is obtained by substituting the value of coefficients in Equation (17).

$$w_g^*(x^*) = -\frac{C}{10} - \frac{4C}{100}x^* + \frac{C}{10}cos(\lambda^* x^*) + \frac{9C}{1000}sin(\lambda^* x^*) + Cx^{*2} \tag{18}$$

Eliminating C between Equations (15) and (18), we get

$$w_g^*(x^*) = -\frac{Q^*}{2000\lambda^{*2}}\left(100\cos\lambda^* x^* + 9\sin\lambda^* x^* + 1000x^{*2} - 40x^* - 100\right) \tag{19}$$

For the future reduction of the solution, typical values for the variables can be obtained from Table 3.

An output of MATLAB for the analytical solution for variable values is compatible with the information indicated in Table 3.

Table 3. Variable used to obtain response deflection of PBC.

Variables	Value	Description
ρ_w	$1000\ \mathrm{kg/m^3}$	Density of water
V_f	1.42×10^{-8} L	Volume of fluid moved into PBC
l	20×10^{-3} m	Length of PBC
b	4×10^{-3} m	Width of PBC
h	0.25×10^{-3} m	Height of PBC
N	5.3×10^{-3} N	Axial force
E	20.5 MPa	Young's modulus of wet paper [25]

3.2. Numerical Solution

The numerical solution for the problem is obtained by using d-solve function, an output from MATLAB for the analytical solution for the water and variable values are compatible with the information indicated in Table 3.

3.3. Comparison of Numerical and Analytical Solution

A comparison of the analytical and numerical solution is made in Figure 8, where the solid color and dash lines in plot represent analytical and numerical solutions, respectively.

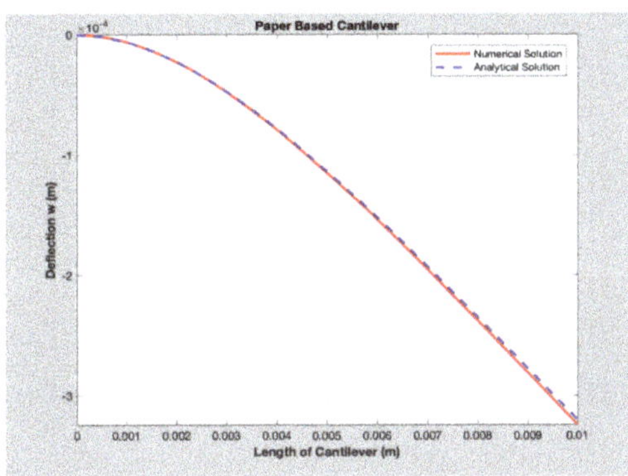

Figure 8. Analytical and numerical solution for PBC.

3.4. Experimental Results

In order to validate the obtained results for the response deflection of the PBC, data for the maximum deflection and change in height of the capillary tube were obtained from the experiment, as presented in Table 4. The experiment was conducted with five replicas of PBC, fabricated at the Microfluidics Laboratory URI, to study the effect of fluidic loading on response deflection (actuation) of PBC.

Table 4. Experimental results for PBC.

Variables	Results (in mm)
Maximum Deflection of PBC	3.04
Change in Height of Capillary Tube	8.96

3.5. Parametric Model

The parametric plots for the fluid (water) loading and internal (axial) stress are shown in Figures 9 and 10, respectively.

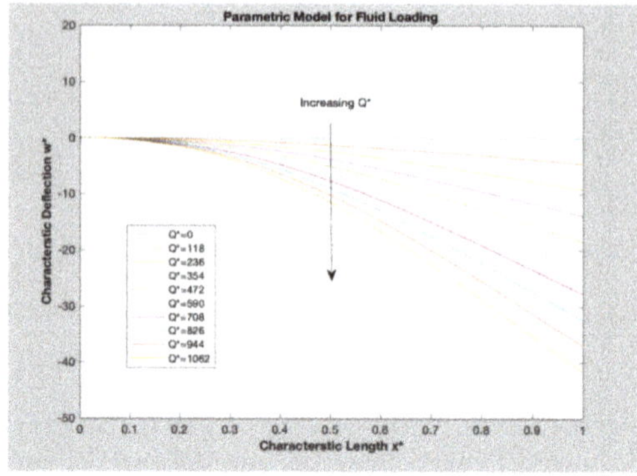

Figure 9. Parametric model for fluid loading $Q*$.

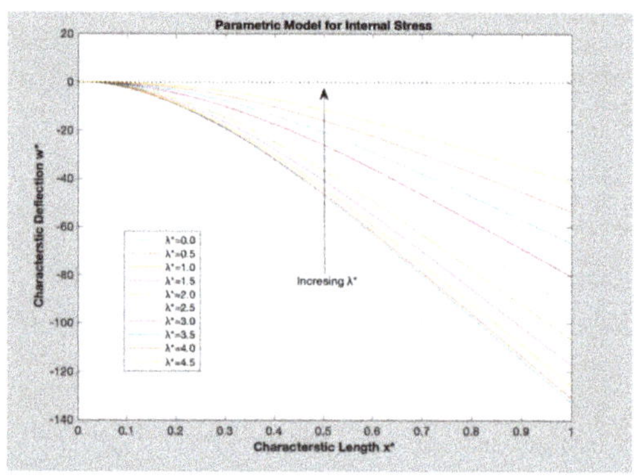

Figure 10. Parametric Model for internal stress parameter $\lambda*$.

Figure 9 displays the relationship between the characteristic length and response deflection of PBC for the increasing value of the fluid loading parameter, $Q*$. Figure 10 displays the relationship between the characteristic length and response deflection of PBC for the increasing value of the internal stress parameter, $\lambda*$.

3.6. *Transverse Displacement of the Free end of PBC*

Figure 11 displays the relationship between the length of the cantilever response deflection of the free end of PBC for different wetted lengths.

3.7. *Model Summary*

The interpretation for the experiment mode is:

- The response deflection of PBC results in identical values both analytically and numerically. The analytical solution is obtained by the use of an experimental value for maximum deflection, whereas the numerical solution is obtained by the use of the material property of PBC, obtained from Table 3;
- Parametric model has been utilized to better understand the effect of fluid loading and internal stress on response deflection of PBC;
- The Washburn flow model is utilized to govern the imbibition of fluid into PBC, and the plots of the response deflection of the free end is demonstrated in Figure 11.

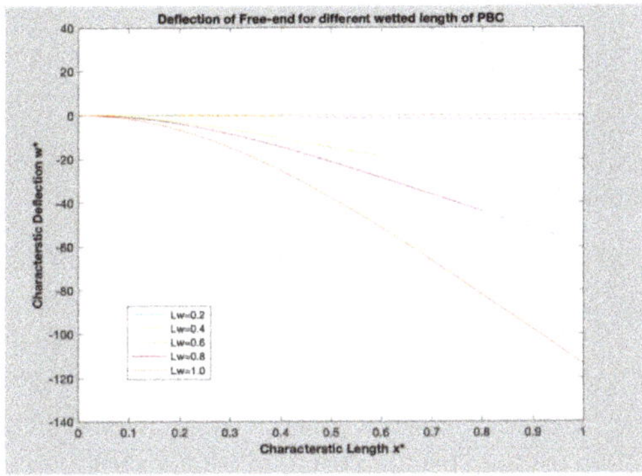

Figure 11. Response deflection of free end of PBC for different wetted length.

4. Discussion

4.1. Summary of Solutions

The maximum deflection of PBC obtained from different models is found to be identical, i.e., 3.04 mm. Thus, the models are adequate and can be used to compare how the response deflection is affected by different fluid loadings.

4.2. Non-Dimensional Model

The non-dimensional model was utilized to compare the magnitude of deflection, fluid loading and internal (axial) stress for PBC. For color water loading, the maximum deflection is found to be around one order of magnitude larger than the internal stress, and more than two orders of magnitude smaller than the fluid loading. The other way to interpret the non-dimensional model is that 1000 units of water loading will induce 250 units of internal stress responsible for membrane strain and 25 units of maximum deflection at the free end of PBC.

4.3. Parametric Model

Comparison for response deflection of PBCs were carried out and the results are shared in Section 3.5.

The increasing value of fluid loading parameter, $Q*$, results in the higher maximum deflection of PBC. The fluid loading parameter is considered as the mass of fluid acting transversely on PBC, the phenomenon of imbibition of fluid from the capillary is carried out until the paper attains its saturation, and at this point the PBC will correspond to maximum deflection. Different paper capillaries possess different amounts of fluid based on their physical properties. The higher value of $Q*$ signifies the high fluid carrying capacity of paper. The graph in Figure 9 demonstrates the increasing response deflection of PBC.

Internal stress parameter $\lambda*$ is another significant factor that needs to be addressed for the analysis of parametric mode. Generally, internal stress results in stress emerging in the structure of loading. In the case of paper, it is comfortable to adapt these resultant stresses in PBC due to the phenomenon of hygro-expansion and bending (rotation). In case of large deflection, these components cancel out others. Figure 10 plots the response deflection of PBC with the increasing value of internal stress parameter $\lambda*$. Contrary to parameter $Q*$, the increasing value of parameter $\lambda*$ will result in a lower maximum deflection of PBC.

4.4. Transverse Displacement of Free End of PBC

It is interesting to analyze the response deflection of the free end for different wetted lengths of PBC. The problem is modeled as a quasi-static 2D fluid structure; however, the fluid flow is governed by Washburn, which makes the model susceptible to dynamic conditions. Instead of analyzing the time dependent solution of the model, the response variable is studied for different wetted lengths. The wetted length of PBC progresses as the square root of the time elapsed, so 5 instants in time are included in the Figure 11, when the value of wetting length is equal to 20, 40, 60, 80 and 100 percent of overall length of PBC.

Due to the fact that Equation (2) is developed for the moderately large static deflection of PBC, it allows for very minimal room to obtain a solution for the axial displacement of the free end of PBC. The model is developed with the limitation of deflection θ, within 0° to 10°. Please refer to Appendix A for details.

5. Conclusions

In this paper, an autonomous paper-based actuation system was introduced in order to sequentially load different reagents into the area of interest. For the purpose of this study, the concluded result will be utilized to model the behavior of PBC on fluidic loading. The solution for the model can be utilized to obtain the maximum deflection of PBC under the assumption, $1^\circ \leq \theta \leq 10^\circ$. For small and large deflection, the model shall be modified under the similar assumption of cross-sectional deformation.

Author Contributions: Conceptualization, A.K., H.H.-B. and A.C.; methodology, A.K., H.H.-B. and A.C.; software, A.K. and H.H.-B.; validation, A.K., H.H.-B., A.C. and N.R.; formal analysis, A.K., H.H.-B.; investigation, A.K., H.H.-B. and A.C.; resources, N.R., C.A. and M.F.; data curation, A.K. and H.H.-B.; writing—original draft preparation, A.K., H.H.-B. and A.C.; writing—review and editing, A.K., H.H.-B., A.C., N.R., C.A. and M.F.; visualization, A.K. and H.H.-B.; supervision, N.R., C.A. and M.F.; project administration, N.R., C.A. and M.F.; funding acquisition, C.A. and M.F. All authors have read and agreed to the published version of the manuscript.

Funding: This research received no external funding.

Acknowledgments: The authors would like to acknowledge the ideas and suggestions that Winfield Smith has shared during the execution of this study in addition to his valuable help and support.

Conflicts of Interest: The authors declare no conflict of interest.

Appendix A

Generalized Relationship [27],
Generalized Stress for cantilever in (x, z) plan,
Bending Moment, $M = \int \sigma_{xx}\, z\, dA$ [Nm]
Normal Stress, $N = \int \sigma_{xx}\, dA$ [N]
Shear Stress, $V = \int \sigma_{xz}\, dA$ [N]

From Euler-Bernoulli hypothesis, and substitution in Axial Force N and Bending Moment M, for constant E and I we get:

$$M = EI\kappa \text{ [Nm]}$$
$$N = EA\,\epsilon^0 \text{ [N]}$$

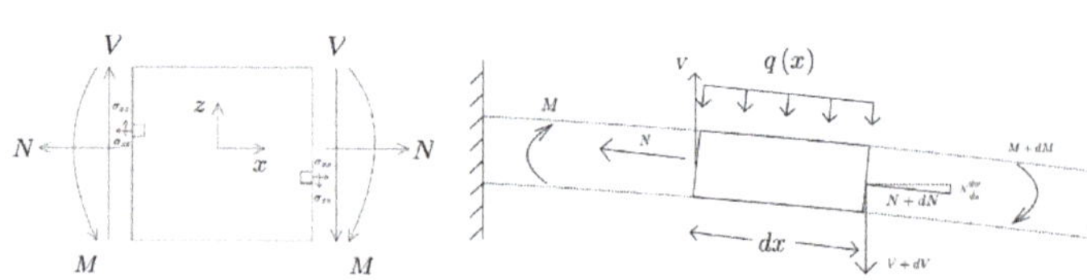

Figure A1. Generalized planar stress for cantilever.

Force equilibrium equation, in x-direction:

$$\frac{dN}{dx} = 0$$

Force equilibrium equation, in z-direction:

$$\frac{d}{dx}\left(V + N\frac{dw}{dx}\right) + q = 0$$

Moment equilibrium equation, about y-axis:

$$\frac{dM}{dx} = V$$

Cantilever equilibrium equation,

$$\frac{d^2M}{dx^2} + N\frac{d^2w}{dx^2} + q(x) = 0$$

Substituting value of M from pervious equation, we get.

$$\frac{d^2}{dx^2}(EI\kappa) + N\frac{d^2w}{dx^2} + q(x) = 0$$

where $\kappa = \frac{d^2w(x)}{dx^2}$

References

1. Davy, J.; Davy, H. LXVIII. On a gaseous compound of carbonic oxide and chlorine. *Philos. Mag.* **1812**, *39*, 443–448. [CrossRef]
2. Müller, R.H.; Clegg, D.L. Automatic Paper Chromatography. *Anal. Chem.* **1949**, *21*, 1123–1125. [CrossRef]
3. Martinez, A.W.; Phillips, S.T.; Butte, M.; Whitesides, G.M. Patterned Paper as a Platform for Inexpensive, Low-Volume, Portable Bioassays. *Angew. Chem. Int. Ed.* **2007**, *46*, 1318–1320. [CrossRef] [PubMed]
4. Böhm, A.; Carstens, F.; Trieb, C.; Schabel, S.; Biesalski, M. Engineering microfluidic papers: Effect of fiber source and paper sheet properties on capillary-driven fluid flow. *Microfluid. Nanofluidics* **2014**, *16*, 789–799. [CrossRef]
5. Carstens, F.; Gamelas, J.; Schabel, S. Engineering microfluidic papers: Determination of fibre source and paper sheet prop-erties and their influence on capillary-driven fluid flow. *Cellulose* **2017**, *24*, 295–309. [CrossRef]
6. Li, X.; Tian, J.F.; Nguyen, T.; Shen, W. Paper-Based Microfluidic Devices by Plasma Treatment. *Anal. Chem.* **2008**, *80*, 9131–9134. [CrossRef]
7. Jahanshahi-Anbuhi, S.; Chavan, P.; Sicard, C.; Leung, V.; Hossain, S.M.Z.; Pelton, R.; Brennan, J.D.; Filipe, C.D. Creating fast flow channels in paper fluidic devices to control timing of sequential reactions. *Lab Chip* **2012**, *12*, 5079–5085. [CrossRef]
8. Han, K.N.; Choi, J.-S.; Kwon, J. Three-dimensional paper-based slip device for one-step point-of-care testing. *Sci. Rep.* **2016**, *6*, 25710. [CrossRef]
9. Martinez, A.W.; Phillips, S.T.; Nie, Z.; Cheng, C.-M.; Carrilho, E.; Wiley, B.; Whitesides, G.M. Programmable diagnostic devices made from paper and tape. *Lab Chip* **2010**, *10*, 2499–2504. [CrossRef]
10. Rodriguez, N.M.; Wong, W.S.; Liu, L.; Dewar, R.; Klapperich, C.M. A fully integrated paperfluidic molecular diagnostic chip for the extraction, amplification, and detection of nucleic acids from clinical samples. *Lab Chip* **2016**, *16*, 753–763. [CrossRef]
11. Jayawardane, B.M.; Wei, S.; McKelvie, I.D.; Kolev, S. Microfluidic Paper-Based Analytical Device for the Determination of Nitrite and Nitrate. *Anal. Chem.* **2014**, *86*, 7274–7279. [CrossRef]
12. Noh, H.; Phillips, S.T. Metering the Capillary-Driven Flow of Fluids in Paper-Based Microfluidic Devices. *Anal. Chem.* **2010**, *82*, 4181–4187. [CrossRef]
13. Lutz, B.; Liang, T.; Fu, E.; Ramachandran, S.; Kauffman, P.; Yager, P. Dissolvable fluidic time delays for programming mul-ti-step assays in instrument-free paper diagnostics. *Lab Chip* **2013**, *13*, 2840–2847. [CrossRef] [PubMed]
14. Chen, H.; Cogswell, J.; Anagnostopoulos, C.; Faghri, M. A fluidic diode, valves, and a sequential-loading circuit fabricated on layered paper. *Lab Chip* **2012**, *12*, 2909. [CrossRef] [PubMed]
15. Gerbers, R.; Foellscher, W.; Chen, H.; Anagnostopoulos, C.; Faghri, M. A new paper-based platform technology for point-of-care diagnostics. *Lab Chip* **2014**, *14*, 4042–4049. [CrossRef] [PubMed]
16. Lai, Y.T.; Tsai, J.S.; Hsu, J.C.; Lu, Y.W. Automated paper-based devices by microfluidic timing-valve for competitive ELI-SA. In Proceedings of the 2017 IEEE 30th International Conference on Micro Electro Mechanical Systems (MEMS), Las Vegas, NV, USA, 22–26 January 2017; pp. 1321–1324.
17. Koo, C.K.W.; He, F.; Nugen, S.R. An inkjet-printed electrowetting valve for paper-fluidic sensors. *Analyst* **2013**, *138*, 4998–5004. [CrossRef]
18. Li, X.; Zwanenburg, P.; Liu, X. Magnetic timing valves for fluid control in paper-based microfluidics. *Lab Chip* **2013**, *13*, 2609–2614. [CrossRef]

19. Phillips, E.A.; Shen, R.; Zhao, S.; Linnes, J.C. Thermally actuated wax valves for paper-fluidic diagnostics. *Lab Chip* **2016**, *16*, 4230–4236. [CrossRef]
20. Kong, T.; Flanigan, S.; Weinstein, M.P.; Kalwa, U.; Legner, C.M.; Pandey, S. A fast, reconfigurable flow switch for paper microfluidics based on selective wetting of folded paper actuator strips. *Lab Chip* **2017**, *17*, 3621–3633. [CrossRef]
21. Toley, B.J.; Wang, J.A.; Gupta, M.; Buser, J.R.; LaFleur, L.K.; Lutz, B.R.; Fu, E.; Yager, P. A versatile valving toolkit for automating fluidic operations in paper microfluidic devices. *Lab Chip* **2015**, *15*, 1432–1444. [CrossRef]
22. Fu, H.; Song, P.; Wu, Q.; Zhao, C.; Pan, P.; Li, X.; Li-Jessen, N.Y.K.; Liu, X. A paper-based microfluidic platform with shape-memory-polymer-actuated fluid valves for automated multi-step immunoassays. *Microsyst. Nanoeng.* **2019**, *5*, 50. [CrossRef]
23. Perez-Cruz, A. Development of Paper-Based Hygro-Mechanical Systems for Liquid Characterization. Ph.D. Thesis, Concordia University, Montréal, QC, Canada, 2017.
24. Yu, H.; Yuan, Y. Analytical Solution for an Infinite Euler-Bernoulli Beam on a Viscoelastic Foundation Subjected to Arbitrary Dynamic Loads. *J. Eng. Mech.* **2014**, *140*, 542–551. [CrossRef]
25. Lee, M.; Kim, S.; Mahadevan, L. Bending and buckling of wet paper. *Phys. Fluids* **2016**, *28*, 042101. [CrossRef]
26. Mäkelä, P.; Östlund, S. Orthotropic elastic–plastic material model for paper materials. *Int. J. Solids Struct.* **2003**, *40*, 5599–5620. [CrossRef]
27. Wang, H.; Qin, Q.-H. *Methods of Fundamental Solutions in Solid Mechanics*; Elsevier BV: Amsterdam, The Netherlands, 2019; pp. 53–90.

chemistry proceedings

MDPI

Proceeding Paper

Study of Gas-Sensing Properties of Titania Nanotubes for Health and Safety Applications †

Vardan Galstyan *, Nicola Poli and Elisabetta Comini

Sensor Lab., Department of Information Engineering, University of Brescia, Via Valotti 9, 25133 Brescia, Italy; nicola.poli@unibs.it (N.P.); elisabetta.comini@unibs.it (E.C.)

* Correspondence: vardan.galstyan@unibs.it

† Presented at the 1st International Electronic Conference on Chemical Sensors and Analytical Chemistry, 1–15 July 2021; Available online: https://csac2021.sciforum.net/.

Abstract: We studied the preparation and gas-sensing performance of a hybrid nanomaterial based on titania nanotubes and graphene derivatives. We fabricated the hybrid structure with tunable chemical-sensing properties, achieved by tailoring the structure and composition of graphene oxide and coupling it with titania nanotubes. The parameters of manufactured sensing structures were investigated for hydrogen and ammonia. Our experimental findings indicate that this research may demonstrate an efficient way to enhance the gas-sensing properties of metal oxide nanomaterials for health and safety applications.

Keywords: nanomaterials; titania; graphene; chemical sensor; gas sensor

Citation: Galstyan, V.; Poli, N.; Comini, E. Study of Gas-Sensing Properties of Titania Nanotubes for Health and Safety Applications. *Chem. Proc.* **2021**, *5*, 16. https://doi.org/10.3390/CSAC2021-10625

Academic Editor: Ye Zhou

Published: 7 July 2021

1. Introduction

Modern gas-sensing systems based on nanotechnology may enable reliable and continuous detection of different gaseous compounds to control atmospheric pollutants and protect human health [1–5]. With their quantum-mechanical properties, wide-bandgap semiconductor nanostructures can affect the characteristics of functional devices [6–8]. Therefore, the application of semiconductor nanomaterials in the development of chemical gas sensors is of great interest [9–11]. Highly ordered transition metal oxide nanostructures have been considered as promising materials for applications in chemical gas sensors due to their good chemical stability and functional properties [12]. In this regard, well-ordered and highly aligned titania nanotubes, with their superior electron transport properties and large surface area, are very attractive structures for the fabrication of gas-sensing systems [13–17]. Herein, we report the preparation and investigation of sensing properties of titania-based nanotubular structures for their application in gas-detection devices. We studied the effect of the additive material on the functionalities of nanotubes to optimize their sensing performance. The morphology, structure, and composition of prepared materials were examined. The sensing properties of materials were studied for hydrogen (H_2) and ammonia (NH_3). We have analyzed the interaction mechanism between the prepared nanotubes and gaseous compounds, considering their structural and compositional modifications. The obtained results demonstrate that the fabricated sensing materials have potential for application in detection systems [17].

2. Materials and Methods

Titania nanotubes were prepared as follows: The metallic titanium films were deposited on alumina substrates by radio frequency magnetron sputtering. Then, metallic films were anodized in a two-electrode system Teflon cell at room temperature. We reported the detailed information on anodization procedure in our previous reports [16,18]. The prepared materials were crystallized via thermal treatment in a tubular furnace at 400 °C for 6 h. We reported the crystallization regimes and analysis of the samples in [19].

The morphological analysis of samples (Figure 1) was performed by means of a LEO 1525 scanning electron microscope (SEM) equipped with a field emission gun. In order to fabricate the hybrid material, we prepared an aqueous dispersion of graphene oxide. Then, we drop-casted the prepared dispersion on the surface of titania nanotubes. To carry out gas-sensing measurements, platinum electrodes and a heater were deposited on the surface of the sensing structures and on the backside of substrates by DC magnetron sputtering. The gas-sensing tests were performed in a test chamber and the measurements were controlled by a computer-controlled gas flow system. The sensor based on pure titania is denoted as S1 and the sensor based on the composite material is denoted as S2 (Figure 2).

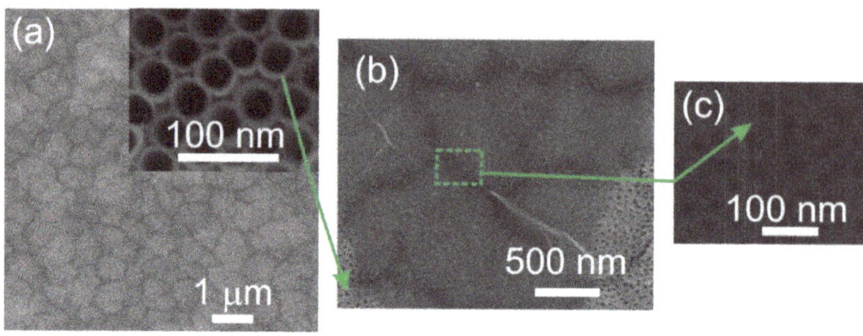

Figure 1. SEM images of the obtained samples. (**a**) The surface morphology of pristine titania nanotubes with different resolutions; (**b**,**c**) the morphologies of the fabricated composite material with different resolutions.

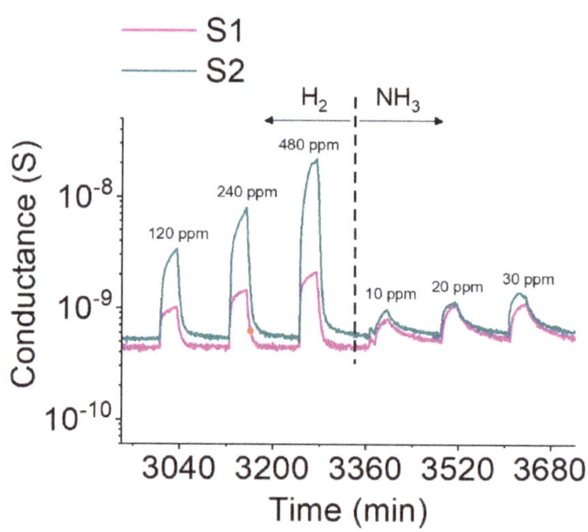

Figure 2. The dynamic response of obtained S1 and S2 sensors for different concentrations of H_2 and NH_3 at 200 °C.

3. Results and Discussion

The results of the morphological analysis of samples are shown in Figure 1. The SEM observations confirmed that highly ordered titania nanotubes were successfully prepared. The tube diameter was 30 nm (Figure 1a). Figure 1b,c shows the surface morphology of the composite structure. As can be seen, the surface of the titania nanotubes was covered by graphene oxide sheets. Figure 2 presents the dynamic response of the fabricated S1 and S2 sensors for concentrations of 120, 240, and 480 ppm of H_2 and 10, 20, and 30 ppm of NH_3. The sensing measurements were carried out at 200 °C. The graphene oxide significantly increased the response of the titania nanotubes towards H_2. Meanwhile, very small differences were observed between the sensing behavior of the S1 and S2 structures towards NH_3. The drastic enhancement in the response of the S2 sensor compared to S1 can be attributed to the depletion layer formed between the titania nanotubes and graphene oxide. In this case, the presence of more active centers improved the adsorption of H_2 on the surface of the hybrid material, which is important for its sensitivity.

4. Conclusions

We fabricated a hybrid structure based on titania nanotubes and graphene oxide. Then, we investigated its gas-sensing performance for H_2 and NH_3. Our experimental findings show that the depletion layer formed between two materials plays a crucial role in tuning the sensing response of the hybrid structure. The hybrid material exhibited a better sensing response towards H_2 compared to pristine titania nanotubes, indicating that this this is an efficient and promising way to enhance the sensing parameters of metal oxide gas sensors. Moreover, a noticeable difference between the responses of the composite structure and pristine nanotubes towards NH_3 was not observed, which indicates an enhancement in the selectivity of the composite.

Funding: This work was partially funded by the NATO Science for Peace and Security Programme under grant No. G5634 "Advanced Electro-Optical Chemical Sensors"; by the "Multi-Messenger and Machine Learning Monitoring of SARS-CoV-2 for occupational health & safety" (4M SARS-CoV-2) project under the Special Integrative Fund for Research (FISR), Ministry of University and Research (MUR), Italy; and by the MIUR "Smart Cities and Communities and social innovation", for the project titled "SWaRM Net/Smart Water Resource Management—Networks".

Institutional Review Board Statement: Not applicable.

Data Availability Statement: The data that support the findings of this study are available from the corresponding author (V.G.), upon reasonable request.

Conflicts of Interest: The authors declare no conflict of interest.

References

1. Ahoulou, S.; Perret, E.; Nedelec, J.-M. Functionalization and characterization of silicon nanowires for sensing applications: A review. *Nanomaterials* **2021**, *11*, 999. [CrossRef] [PubMed]
2. Fioravanti, A.; Marani, P.; Morandi, S.; Lettieri, S.; Mazzocchi, M.; Sacerdoti, M.; Carotta, M.C. Growth mechanisms of ZnO micro-nanomorphologies and their role in enhancing gas sensing properties. *Sensors* **2021**, *21*, 1331. [CrossRef] [PubMed]
3. Cai, Z.; Park, S. Effect of nanoparticle-embedded 1 dimensional-nanostructures to fabricate highly sensitive and selective hydrogen gas sensors. *J. Alloys Compd.* **2021**, *865*, 158929. [CrossRef]
4. Galstyan, V.; Kaur, N.; Zappa, D.; Núñez-Carmona, E.; Sberveglieri, V.; Comini, E. Chemical gas sensors studied at sensor lab, brescia (italy): From conventional to energy-efficient and biocompatible composite structures. *Sensors* **2020**, *20*, 579. [CrossRef] [PubMed]
5. Zhang, G.; Hosomi, T.; Mizukami, W.; Liu, J.; Nagashima, K.; Takahashi, T.; Kanai, M.; Sugiyama, T.; Yasui, T.; Aoki, Y.; et al. A thermally robust and strongly oxidizing surface of WO_3 hydrate nanowires for electrical aldehyde sensing with long-term stability. *J. Mater. Chem. A* **2021**, *9*, 5815–5824. [CrossRef]
6. Lu, Y.; Purwidyantri, A.; Liu, H.; Wang, L.; Shih, C.; Pijanowska, D.G.; Yang, C. Photoelectrochemical detection of β-amyloid peptides by a TiO_2 nanobrush biosensor. *IEEE Sens. J.* **2020**, *20*, 6248–6255. [CrossRef]
7. Agrawal, M.; Prathyusha, E.; Ahmed, H.; Dubey, S.K.; Kesharwani, P.; Singhvi, G.; Naidu, V.G.M.; Alexander, A. Biomaterials in treatment of alzheimer's disease. *Neurochem. Int.* **2021**, *145*, 105008. [CrossRef] [PubMed]
8. Galstyan, V. "Quantum dots: Perspectives in next-generation chemical gas sensors"—A review. *Anal. Chim. Acta* **2021**, 238192. [CrossRef] [PubMed]
9. Milovanov, Y.S.; Skryshevsky, V.A.; Gavrilchenko, I.V.; Kostiukevych, O.M.; Gryn, S.V.; Alekseev, S.A. Ethanol gas sensing performance of electrochemically anodized freestanding porous sic. *Diam. Relat. Mater.* **2019**, *91*, 84–89. [CrossRef]
10. Fioravanti, A.; Carotta, M.C. Year 2020: A snapshot of the last progress in flexible printed gas sensors. *Appl. Sci.* **2020**, *10*, 1741. [CrossRef]
11. Nikolic, M.V.; Milovanovic, V.; Vasiljevic, Z.Z.; Stamenkovic, Z. Semiconductor gas sensors: Materials, technology, design, and application. *Sensors* **2020**, *20*, 6694. [CrossRef] [PubMed]

12. Galstyan, V. Porous TiO_2-based gas sensors for cyber chemical systems to provide security and medical diagnosis. *Sensors* **2017**, *17*, 2947. [CrossRef] [PubMed]
13. Chang, Y.H.; Liu, C.M.; Chen, C.; Cheng, H.E. The heterojunction effects of TiO_2 nanotubes fabricated by atomic layer deposition on photocarrier transportation direction. *Nanoscale Res. Lett.* **2012**, *7*, 231. [CrossRef] [PubMed]
14. Haryński, Ł.; Grochowska, K.; Kupracz, P.; Karczewski, J.; Coy, E.; Siuzdak, K. The in-depth studies of pulsed UV laser-modified TiO_2 nanotubes: The influence of geometry, crystallinity, and processing parameters. *Nanomaterials* **2020**, *10*, 430. [CrossRef] [PubMed]
15. Yadav, S.; Nair, A.; Urs Mb, K.; Kamble, V.B. Protonic titanate nanotube–reduced graphene oxide composites for hydrogen sensing. *ACS Appl. Nano Mater.* **2020**, *3*, 10082–10093. [CrossRef]
16. Galstyan, V.; Comini, E.; Baratto, C.; Ponzoni, A.; Ferroni, M.; Poli, N.; Bontempi, E.; Brisotto, M.; Faglia, G.; Sberveglieri, G. Large surface area biphase titania for chemical sensing. *Sens. Actuators B-Chem.* **2015**, *209*, 1091–1096. [CrossRef]
17. Tian, X.; Cui, X.; Lai, T.; Ren, J.; Yang, Z.; Xiao, M.; Wang, B.; Xiao, X.; Wang, Y. Gas sensors based on TiO_2 nanostructured materials for the detection of hazardous gases: A review. *Nano Mater. Sci.* **2021**, in press. [CrossRef]
18. Galstyan, V.; Vomiero, A.; Comini, E.; Faglia, G.; Sberveglieri, G. TiO_2 nanotubular and nanoporous arrays by electrochemical anodization on different substrates. *RSC Adv.* **2011**, *1*, 1038–1044. [CrossRef]
19. Galstyan, V.; Ponzoni, A.; Kholmanov, I.; Natile, M.M.; Comini, E.; Sberveglieri, G. Highly sensitive and selective detection of dimethylamine through Nb-doping of TiO_2 nanotubes for potential use in seafood quality control. *Sens. Actuators B Chem.* **2020**, *303*, 127217. [CrossRef]

chemistry proceedings

Abstract

Validation of Spent Coffee Grounds as Precursors for the Development of Sustainable Carbon Dot-Based for Fe^{3+} Optical Sensing †

Diana M. A. Crista *, Joaquim C. G. Esteves da Silva and Luís Pinto da Silva

Chemistry Research Unit (CIQUP), Faculty of Sciences of University of Porto, R. Campo Alegre 687, 4169-007 Porto, Portugal; jcsilva@fc.up.pt (J.C.G.E.d.S.); luis.silva@fc.up.pt (L.P.d.S.)

* Correspondence: up200702319@fc.up.pt

† Presented at the 1st International Electronic Conference on Chemical Sensors and Analytical Chemistry, 1–15 July 2021; Available online: https://csac2021.sciforum.net/.

Abstract: Carbon dots (CDs) are fluorescence carbon-based nanomaterials that possess several properties such as photoluminescence, biocompatibility and good water solubility. They can be fabricated from a large variety of precursors; however, most available organic molecules are still expensive and their use or synthesis can lead to significant challenges to the environment and human health. It has become desirable to use biomass waste as alternative precursors in the synthesis of CDs, given that biomass waste material is ubiquitous, nontoxic, cheap and renewable. Spent coffee grounds (SCGs) are the residues of the treatment of coffee powder can be a potential carbon source to a more environmentally sustainable synthesis route. In this work, we fabricated SCG-based CDs via one-pot and solvent-free carbonization at 200 °C of solid samples generating particles with sizes between 2.1 and 3.9 nm. These carbon nanoparticles exhibited blue fluorescence and excitation-dependent emission of carbon dots with moderate quantum yields (2.9–5.8%). The presence of heavy metals in water resources, such as Fe^{3+}, can lead to adverse health effects. SCG-based CDs showed potential for being used as optical Fe^{3+} optical sensors, with Life Cycle Assessment (LCA) studies validating the SCGs as more sustainable precursors than classical precursors, both considering a weight- or function-based functional unit.

Keywords: spent coffee grounds; carbon dots; sustainability; sensing

check for updates

Citation: Crista, D.M.A.; da Silva, J.C.G.E.; da Silva, L.P. Validation of Spent Coffee Grounds as Precursors for the Development of Sustainable Carbon Dot-Based for Fe^{3+} Optical Sensing. *Chem. Proc.* **2021**, *5*, 17. https://doi.org/10.3390/CSAC2021-10452

Academic Editor: Elena Benito-Peña

Published: 30 June 2021

Supplementary Materials: The following are available online at https://www.mdpi.com/article/10.3390/CSAC2021-10452/s1.

Funding: Acknowledgment to projects PTDC/QUI-QFI/2870/2020 and UIDB/00081/2020. Also acknowledge for for funding the PhD grant SFRH/BD/144423/2019 (D.M.A.C.), and funding Scientific Employement Stimulus CEECIND/01425/2017 (L.P.d.S.).

Conflicts of Interest: The authors declare no conflict of interest.

chemistry proceedings

MDPI

Proceeding Paper

Unravelling Plant-Pathogen Interactions: Proximal Optical Sensing as an Effective Tool for Early Detect Plant Diseases †

Mafalda Reis-Pereira [1,2,*], Rui C. Martins [3,*], Aníbal Filipe Silva [1,3], Fernando Tavares [1,4], Filipe Santos [2] and Mário Cunha [1,2,*]

1 Faculty of Sciences of the University of Porto (FCUP), Rua Campo Alegre, s/n, 4169-007 Porto, Portugal; afilipe.silva@fc.up.pt (A.F.S.); ftavares@fc.up.pt (F.T.)
2 Centre of Robotics in Industry and Intelligent Systems, INESC TEC, Dr. Roberto Frias, 4200-465 Porto, Portugal; filipe.n.santos@inesctec.pt
3 Centre for Applied Photonics, INESC TEC, Faculty of Sciences of the University of Porto, Rua do Campo Alegre, s/n, 4169-007 Porto, Portugal
4 Research Centre in Biodiversity and Genetic Resources (CIBIO-InBIO), Rua Padre Armando Quintas, nº 7, 4485-661 Vairão, Portugal
* Correspondence: mafalda.r.pereira@inesctec.pt (M.R.-P.); rui.c.martins@inesctec.pt (R.C.M.); mario.cunha@inesctec.pt (M.C.)

† Presented at the 1st International Electronic Conference on Chemical Sensors and Analytical Chemistry, 1–15 July 2021; Available online: https://csac2021.sciforum.net/.

Abstract: This study analyzed the potential of proximal optical sensing as an effective approach for early disease detection. A compact, modular sensing system, combining direct UV–Vis spectroscopy with optical fibers, supported by a principal component analysis (PCA), was applied to evaluate the modifications promoted by the bacteria *Xanthomonas euvesicatoria* in tomato leaves (cv. cherry). Plant infection was achieved by spraying a bacterial suspension (10^8 CFU mL^{-1}) until run-off occurred, and a similar approach was followed for the control group, where only water was applied. A total of 270 spectral measurements were performed on leaves, on five different time instances, including pre- and post-inoculation measurements. PCA was then applied to the acquired data from both healthy and inoculated leaves, which allowed their distinction and differentiation, three days after inoculation, when unhealthy plants were still asymptomatic.

Keywords: plant disease detection; plant pathology; proximal sensing; spectroscopy; precision agriculture; principal component analysis

Citation: Reis-Pereira, M.; Martins, R.C.; Silva, A.F.; Tavares, F.; Santos, F.; Cunha, M. Unravelling Plant-Pathogen Interactions: Proximal Optical Sensing as an Effective Tool for Early Detect Plant Diseases. *Chem. Proc.* **2021**, 5, 18. https://doi.org/10.3390/CSAC2021-10560

Academic Editor: Elena Benito-Peña

Published: 1 July 2021

1. Introduction

Biotic agents, specifically pests and pathogens, cause significant losses in crop yields, with levels that can range between 20% and 40% [1]. Chemical phytosanitary products are usually applied to prevent and combat these organisms. However, their usage can negatively impact the environment, mainly when applied to treat plant diseases that appear suddenly and spread to large scales [2].

Nowadays, phytopathology methods are considered major challenges because, to be implemented, they often rely on the presence of indicator visible signs of the infection (disease symptoms), which frequently only manifest themselves at the middle to late stages of the process, compromising the effectiveness of phytosanitary measures [3]. An example is the scouting technique, which involves inspecting a crop field to detect and identify infected plant through disease symptoms [4]. Despite being extremely useful, this approach requires specialized trained observers (who must be capable of identifying disease symptoms and distinguishing them from those caused by other abiotic stresses (e.g., nutritional and physiological disorders)), and can be labor-intensive, time-consuming, and expensive [5–11]. Moreover, this approach can be an inefficient in the early stages of the infection and on large areas. Other strategies consist of laboratory-based techniques, namely serological and molecular tests, largely used due to their sensitivity, accuracy, and effectiveness. They include enzyme-linked immunosorbent assay (ELISA) and polymerase chain reaction (PCR) methods, being the first serological approach based on protein in the detection of causative diseases and the second molecular technique based on the DNA sequence of the pathogen. Their development boosted plant disease diagnosis, since they allow the simultaneous processing of several samples and perform a precise pathogen identification. Furthermore, PCR enables the detection of pathogens that have not been cultured. Nevertheless, these procedures present some limitations, especially in the early phase of the infection process, due to the uneven spread of pathogens inside plants, compromising their effectiveness in analyzing asymptomatic samples [9,12–14]. Other drawbacks can also be enumerated. They require several hours to be completed, require the realization of detailed sampling procedures, and destructive sample preparation, not allowing a follow-up of the disease progression [12,13].

Therefore, the necessity of developing fast, accurate, and selective in vivo techniques for plant disease detection arises. These innovative approaches must provide complementary information to the current methods applied in the phytopathology field and combine with them. Several non-invasive methods have been developed in the last decade, which proved to be sensitive, consistent, standardize, rapid, cost-effective, and have high-throughput [15]. Hyperspectral spectroscopy (HS) is one of them and seems to be effective in estimating a wide variety of plant chemical, biophysical, and metabolic traits in living tissue [16–22], namely foliar structure, plant chemical composition, water concentration, and metabolic status [23]. Through spectral measurements in the visible (Vis, 400–700 nm), near-infrared (NIR, 700–1100 nm), and shortwave infrared wavelengths (SWIR, 1100–2500 nm), this approach assesses changes in optical properties of leaves, which derive from interactions between light, chemical bonds, and cellular structure [24]. Briefly, modifications in plants' reflectance in the Vis range are mostly related to pigment concentration and physiological processes, such as photosynthesis. In turn, changes in the NIR are correlated with leaf structure and internal scattering processes. The SWIR region is affected by leaf structural and chemical composition (including lignins and proteins), as well as water content [25–29].

Since phytopathogens induce physiological, biochemical, and structural changes in host plants, HS seems to be promising in plant disease detection, identification, and quantification [30–38]. Hyperspectral sensors can be used alone or mounted in different platforms, allowing the performance of mapping, monitoring, scouting, and application tasks [2]. Their flexibility allows them to assess leaf, single-plant, canopy (proximal sensing), and even plot and regional scales (remote sensing) [2]. Some examples, sorted by measurement scale, include handheld sensors, rail systems, vehicle, and tractor-mounted systems, drones UAVs, as well as aircrafts and satellites [39].

Despite the possibilities provided by these optical devices for simple, rapid, non-destructive disease detection and identification, its application is still very limited, due to the scarcity of extensive agronomic and phytopathological studies aiming to explore their full potential. Their technology readiness levels (TRL) are close to TRL3 (analytical and experimental critical function, and/or characteristic proof-of-concept) [40]. Hence, this study aimed to evaluate the potential of UV–Vis spectroscopy to detect diseased tomato leaves and discriminate between healthy and infected leaves, through a multi-temporal approach. Furthermore, the capability of this technology in detecting changes in the reflectance spectrum of infected leaves was analyzed, before the first symptoms became visible.

2. Materials and Methods

2.1. Experimental Design

Tomato (*Solanum lycopersicum* L.) plants, of the cultivar cherry, were grown in 200 mL pots containing a commercial potting substrate, in a *walk-in* plant growth chamber under controlled conditions (temperature of 25–27 °C, humidity of approximately 60%, and photoperiod of 12/12 h). Plants were divided into two groups, one of them being inoculated with *Xanthomonas euvesicatoria* LMG 905 (Xeu) bacteria and the other being treated with sterile distilled water only (control group, Con). Plants were inoculated in the laboratory, at the growth stage of 5–6 fully expanded leaves, by spraying until they became fully wet, and run-off occurred. The bacterial suspensions used for these inoculation assays consisted of 1×10^8 cells/mL. They were prepared from a 48 hour-old culture, grown on YDC medium (yeast extract, 10.0 g; dextrose, 20.0 g; $CaCO_3$, 20.0 g; agar, 15.0 g; distilled water up to 1.0 L). The inoculated plants were then covered with transparent polythene bags for 48 h to increase the relative humidity that fosters bacterial entry into plant tissues through natural openings, such as stomata [41]. Plants were monitored daily for symptom development for 5 days.

At the same time, to verify if the bacteria cultures used in these inoculation tests were viable, 20 µL of Xeu solution were cultured in different Petri dishes containing YDC media. After 48 h, it was possible to observe the bacteria growth in both nutrient media, proving that bacteria were viable at inoculation.

2.2. Spectral Measurements

Hyperspectral data were collected in vivo from the adaxial side of healthy and infected tomato plant leaves, using a compact benchtop system consisting of a D2 (deuterium) light source (Ocean Optics, model DH-2000-BAL, Ostfildern, Germany), spectrometer (Ocean Optics, model HR4000, Ostfildern, Germany), transmission optical fiber bundle (UV), and stainless-steel slitted reflection probe for sample measurement. The spectrometer operated in the 195–1100 nm wavelength range, with a high spectral response and optical resolution of 0.025 nm (full width at half maximum—FWHM). The measurements were carried out using an experimental setup in the laboratory. A LED light source was placed beneath the leaf and provided homogeneous illumination to its entire surface. The light signal from the sample analyzed was guided to the entrance lens of the spectrometer by the fiber-optic cable placed perpendicularly 1 cm above the measured surface. Specialized software was used for data acquisition and processing. Data acquisition was performed with 10 scans for an integration period of 60 ms, in three leaves per plant, on nine locations on each leaf.

2.3. Data Pre-Processing

Spectral pre-processing was performed, in order to remove possible artifacts, e.g., baseline shifts, Mie and Rayleigh scattering, and stray light. Also, a pretreatment with a fast fourier transform (FFT) was carried out on spectral data to smooth/denoise it. FFT is an algorithm that computes the discrete Fourier transform (DFT) of a sequence or its inverse (IDFT). Fourier analysis converts a signal from its original domain (often time or space) to a representation in the frequency domain and vice versa. The DFT is obtained by decomposing a sequence of values into components of different frequencies [42]. Spectral data pre-processing was performed with RStudio software.

2.4. Data Processing—Analytical Techniques

Spectral data was subjected to a principal component analysis (PCA), a multivariate data analysis technique was used to reduce the dimensionality, while preserving its structure by projecting it into a new coordinate system. This technique allows the preservation of the total variance of the dataset and minimizes the mean square approximate errors. PCA uses eigenvectors and eigenvalues to define the reduced subspace (representing the original coordinate system). It originates principal components (PC), which are linear combinations of interrelated variables. PC1 accounts for the maximum possible proportion of the variance information of the original dataset (explained by the eigenvalue), and subsequent principal components (PC2, PC3, …) account for the maximum proportion of the unexplained residual variance, and so forth [43,44].

Contigous hyperspectral wavebands present redundant information [45]. The application of a PCA allows the transformation of this type of high-dimensional data into a few wavebands that contain most of the information in the original bands. The importance of these hyperspectral bands in each PC is then established based on the magnitude of eigenvectors or factor loadings for crop biophysical and biochemical traits, being that the higher the eigenvector, the higher is the importance of the band. So, PCA allows the selection of the best wavebands to model biophysical and biochemical quantities and the elimination of redundant bands (by highlighting the main bands) [46].

3. Results

The spectral response properties of tomato leaves to the stress caused by *Xanthomonas euvesicatoria* LMG 905 is very important for discriminating bacterial infection levels in precise pest management using hyperspectral proximal sensing data. The averaged raw spectral curves of healthy and diseased tomato leaves were slightly different in some spectral ranges, namely through the visible region of the wavelength spectrum (~480–680 nm) (please refer to Figure 1) Similarly, the spectral measurements assessed on infected leaf tissue presented a decrease in signal intensity throughout the sampling period (24–144 h), which accompanied the appearance of the first visual symptoms of the disease after 72 h.

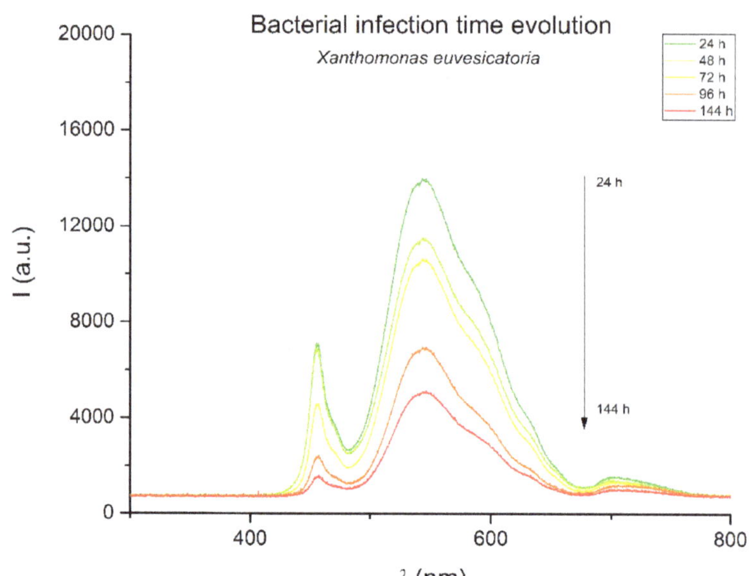

Figure 1. Spectral measurement curve evolution for tomato leaves inoculated with bacteria *Xanthomonas euvesicatoria* LMG 905, within the sampling period (24–144 h). Leaf spectral curves were assessed in vivo on the adaxial side of fully expanded leaves, on the spectral region from 195 to 1100 nm.

Figure 2 presents the principal components (PC) Gabriel plot for the healthy (Con) and diseased (Xeu) leaves spectra, three days after inoculation (before the appearance of the first symptoms). The PCA algorithm has obtained two PCs, accounting for 99.6% of the total variance. PC1 (94.3%) discriminates the effects on the variance of these two types of tomato leaves, which is more evident in PC2 (5.3%).

Figure 2. Gabriel plot of PC1, PC2, and PC3 resulting from the PCA of the dataset three days after inoculation (all leaves were asymptomatic, showing no symptoms of the disease caused by *Xanthomonas euvesicatoria* LMG 905).

The wavelengths that have a higher contribution in these PC are in the interval of ~454–654 nm (visible range of the wavelength spectrum). The ones between ~492–510 nm (essentially the blue region of the electromagnetic spectrum) explain 30% of the variance of the PC1, whereas ~454–461 nm (blue region) explain 40% of the variance of the PC2 and 50% of the PC3. In all the first four dimensions of this analysis, the wavelengths ranging from approximately 445–480 nm (blue) and 580–700 nm (red) were the ones that explain most of the variance of the data.

This evidence can be related to the symptoms caused by Xeu, since these bacteria cause small, brown, angular lesions on leaves (which can be surrounded by a yellow halo with time), affecting the levels of photosynthetic pigments (contributing especially to the reduction of the chlorophyll levels, whose absorption features are more evident in the blue and red ranges of the Vis spectral region), cellular content, and structural arrangement.

4. Discussion

The spectral behavior of tomato plants depends on their biochemical and structural profile. In brief, plants' spectral signature in the visible spectral region (400–700 nm) depends mainly on the content of photosynthetic pigments. These compounds are good absorbers of red and blue wavelengths. Of the major pigments, Chlorophyll a (Chl a) has maximum absorption in the 410–430 and 600–690 nm regions, whereas Chlorophyll b

(Chl b) has maximum absorption in the 450–470 nm range. In healthy plants, chlorophyll concentration is approximately ten times higher than that of other pigments, thus masking out the specific absorption features of these compounds. The green part of the spectrum, on the other hand, is less strongly absorbed, resulting in a reflectance peak in the green domain (at about 550 nm) [25]. Hence, when a light source illuminates healthy plants, they will preferentially absorb red and blue wavelengths, being the green part of the incident light less absorbed and, consequently, more reflected, leading to their green appearance [26]. In the NIR region, the plants' spectral response is related to their structure, structural components, and internal scattering processes. Likewise, the SWIR region is also affected by leaf structural and chemical composition (including the action of lignin's and proteins) and water content [25–29].

Since phytopathogens cause changes in plants' biochemical and structural composition, affecting the levels of photosynthetic pigments and structural elements, tracking changes in plants' spectral behavior can allow an indirect analysis of their phytosanitary status. Generally, unhealthy plants have more reflection in the red region and lower reflectance in the NIR region. Briefly, stress usually promotes an increase of reflectance over the whole spectrum, since it causes a rapid decrease of chlorophylls, which increases reflectance in the Vis range and exposes the absorption characteristics of other pigments, such as carotenoids (responsible for the yellowing of the leaves) and xanthophylls (responsible for the reddening of the leaves). With continuing stress, leaf structures decompose, resulting in extra intra-leaf scattering and an increased NIR signal. At the same time, concentrations of brown pigments, which absorb radiance in the Vis and at the onset of the NIR, can increase leading to a flattening of the red edge. Absorption in the SWIR decreases, due to reduced leaf moisture. With a decay of the leaf tissue, the absorption features characteristics of healthy plants gradually disappear [47].

Our findings seem to be in accord with the previous information, showing evidence that UV–Vis spectroscopy can be suitable for plant disease assessment in laboratory conditions. Data collected in a randomized experimental design, combined with a PCA, allowed the discrimination of healthy and diseased tomato leaves, even at the third day after bacteria inoculation, when no visual symptoms were observable. Most of the variance of the data can be comprised with the first four PCs. In all of them, the wavelengths that explain most of the variance of the data ranged from approximately 445–480 nm (blue) and 580–700 nm (red), which was expected, since *Xanthomonas euvesicatoria* causes tissue lesions, degrading the chlorophylls levels and affecting their absorption features in these spectral regions.

Therefore, our results can be related to those obtained in different research, where sensor-based approaches proved to be capable of assessing modifications in plants' spectral behavior, allowing the detection, identification, and quantification of different types of plant diseases [44,48–51]. They involve the capture and analysis of the optical properties of plants, within different regions of the electromagnetic spectrum and their relationship with modifications in plant physiology, namely alterations in tissue color, structural composition, and transpiration rate [19]. These non-invasive methods have been explored in the last decade, presenting the benefits of being sensitive, consistent, standard, high-throughput, rapid, and cost-effective [47], surpassing the limitations of the current methods used in plant disease detection.

5. Conclusions

The present study suggests that UV–Vis spectroscopy can be a potential tool for the early detection of plant diseases under laboratory conditions, even when unhealthy plants are asymptomatic. Despite these findings, its application is still very limited, due to the scarcity of comprehensive agronomic and phytopathological studies aiming to explore their full potential, as well as the development of applied advanced statistical approaches for data analysis. More research is necessary, especially in field conditions, where more external factors have to surpass, including atmospheric, edaphic, and biotic conditions.

Future research should also include more stress levels to discriminate not only healthy leaves from the diseased ones but also different levels of disease severity.

Author Contributions: Conceptualization, M.R.-P., R.C.M., A.F.S., F.T., F.S. and M.C.; methodology, M.R.-P., R.C.M., A.F.S., F.T., F.S. and M.C.; validation, M.R.-P., R.C.M. and M.C.; formal analysis, M.R.-P., R.C.M. and M.C.; investigation, M.R.-P., R.C.M. and M.C.; resources, R.C.M., F.T., F.S. and M.C.; data curation, M.R.-P. and R.C.M.; writing—original draft preparation, M.R.-P.; writing—review and editing, R.C.M., A.F.S., F.T., F.S. and M.C.; supervision, F.T., F.S. and M.C.; funding acquisition, M.R.-P., R.C.M., A.F.S., F.T., F.S. and M.C. All authors have read and agreed to the published version of the manuscript.

Funding: Mafalda Reis-Pereira and Aníbal Filipe Silva were supported by fellowships from Fundação para a Ciência e a Tecnologia (FCT), with the references SFRH/BD/146564/2019 and DFA/BD/9136/2020, respectively. Rui C. Martins acknowledges Fundação para a Ciência e Tecnologia (FCT) research contract grant (CEEIND/017801/2018). This research was supported by the project 'SpecTOM—Metabolomics Tomography Spectroscopy System', University of Porto, Fundação Amadeus Dias, and Santander-Universities Grant.

Conflicts of Interest: The authors declare no conflict of interest.

References

1. Savary, S.; Ficke, A.; Aubertot, J.-N.; Hollier, C. Crop losses due to diseases and their implications for global food production losses and food security. *Food Secur.* **2012**, *4*, 519–537. [CrossRef]
2. Zhang, N.; Yang, G.; Pan, Y.; Yang, X.; Chen, L.; Zhao, C. A review of advanced technologies and development for hyperspec-tral-based plant disease detection in the past three decades. *Remote Sens.* **2020**, *12*, 3188. [CrossRef]
3. Lowe, A.; Harrison, N.; French, A.P. Hyperspectral image analysis techniques for the detection and classification of the early onset of plant disease and stress. *Plant Methods* **2017**, *13*, 1–12. [CrossRef]
4. Parker, S.R.; Shaw, M.W.; Royle, D.J. The reliability of visual estimates of disease severity on cereal leaves. *Plant Pathol.* **1995**, *44*, 856–864. [CrossRef]
5. Mahlein, A.-K. Plant Disease Detection by Imaging Sensors—Parallels and Specific Demands for Precision Agriculture and Plant Phenotyping. *Plant Dis.* **2016**, *100*, 241–251. [CrossRef]
6. Liaghat, S.; Ehsani, R.; Mansor, S.; Shafri, H.Z.; Meon, S.; Sankaran, S.; Azam, S.H. Early detection of basal stem rot disease (Ganoderma) in oil palms based on hyperspectral reflectance data using pattern recognition algorithms. *Int. J. Remote Sens.* **2014**, *35*, 3427–3439. [CrossRef]
7. Pérez, M.R.V.; Mendoza, M.G.G.; Elías, M.G.R.; González, F.J.; Contreras, H.R.N.; Servin, C.C. Raman Spectroscopy an Option for the Early Detection of Citrus Huanglongbing. *Appl. Spectrosc.* **2016**, *70*, 829–839. [CrossRef]
8. Ali, M.M.; Bachik, N.A.; Muhadi, N.A.; Yusof, T.N.T.; Gomes, C. Non-destructive techniques of detecting plant diseases: A review. *Physiol. Mol. Plant Pathol.* **2019**, *108*. [CrossRef]
9. Sankaran, S.; Mishra, A.; Ehsani, R.; Davis, C. A review of advanced techniques for detecting plant diseases. *Comput. Electron. Agric.* **2010**, *72*, 1–13. [CrossRef]
10. Chen, W.-L.; Lin, Y.-B.; Ng, F.-L.; Liu, C.-Y.; Lin, Y.-W. RiceTalk: Rice Blast Detection Using Internet of Things and Artificial Intelligence Technologies. *IEEE Internet Things J.* **2020**, *7*, 1001–1010. [CrossRef]
11. Khaled, A.Y.; Aziz, S.A.; Bejo, S.K.; Nawi, N.M.; Seman, I.A.; Onwude, D.I. Early detection of diseases in plant tissue using spec-troscopy—applications and limitations. *Appl. Spectrosc. Rev.* **2018**, *53*, 36–64. [CrossRef]
12. Fang, Y.; Ramasamy, R.P. Current and Prospective Methods for Plant Disease Detection. *Biosensors* **2015**, *5*, 537–561. [CrossRef] [PubMed]
13. Martinelli, F.; Scalenghe, R.; Davino, S.; Panno, S.; Scuderi, G.; Ruisi, P.; Villa, P.; Stroppiana, D.; Boschetti, M.; Goulart, L.R.; et al. Advanced methods of plant disease detection. A review. *Agron. Sustain. Dev.* **2015**, *35*, 1–25. [CrossRef]
14. Moghadam, P.; Ward, D.; Goan, E.; Jayawardena, S.; Sikka, P.; Hernandez, E. Plant Disease Detection Using Hyperspectral Imaging. In Proceedings of the 2017 International Conference on Digital Image Computing: Techniques and Applications (DICTA), Sydney, Australia, 29 November–1 December 2017; Volume 2017, pp. 1–8.
15. Golhani, K.; Balasundram, S.K.; Vadamalai, G.; Pradhan, B. A review of neural networks in plant disease detection using hyper-spectral data. *Inf. Process. Agric.* **2018**, *5*, 354–371.
16. Zhao, D.; Reddy, K.R.; Kakani, V.G.; Read, J.J.; Carter, G.A. Corn (*Zea mays* L.) growth, leaf pigment concentration, photosynthesis and leaf hyperspectral reflectance properties as affected by nitrogen supply. *Plant Soil* **2003**, *257*, 205–218. [CrossRef]
17. Delalieux, S.; van Aardt, J.; Keulemans, W.; Schrevens, E.; Coppin, P. Detection of biotic stress (Venturia inaequalis) in apple trees using hyperspectral data: Non-parametric statistical approaches and physiological implications. *Eur. J. Agron.* **2007**, *27*, 130–143. [CrossRef]
18. Jain, N.; Ray, S.S.; Singh, J.P.; Panigrahy, S. Use of hyperspectral data to assess the effects of different nitrogen applications on a potato crop. *Precis. Agric.* **2007**, *8*, 225–239. [CrossRef]

19. Blackburn, G.A.; Ferwerda, J.G. Retrieval of chlorophyll concentration from leaf reflectance spectra using wavelet analysis. *Remote Sens. Environ.* **2008**, *112*, 1614–1632. [CrossRef]
20. Thenkabail, P.S.; Smith, R.B.; De Pauw, E. Hyperspectral vegetation indices and their relationships with agricultural crop char-acteristics. *Remote Sens. Environ.* **2000**, *71*, 158–182. [CrossRef]
21. Abdel-Rahman, E.M.; Ahmed, F.B.; Berg, M.V.D. Estimation of sugarcane leaf nitrogen concentration using in situ spectroscopy. *Int. J. Appl. Earth Obs. Geoinf.* **2010**, *12*, S52–S57. [CrossRef]
22. Couture, J.J.; Serbin, S.P.; Townsend, P.A. Spectroscopic sensitivity of real-time, rapidly induced phytochemical change in response to damage. *New Phytol.* **2013**, *198*, 311–319. [CrossRef]
23. Agrios, G. Plant Pathogens and Disease: General Introduction. *Encycl. Microbiol.* **2009**, 613–646. [CrossRef]
24. Curran, P.J. Remote sensing of foliar chemistry. *Remote Sens. Environ.* **1989**, *30*, 271–278. [CrossRef]
25. Jacquemoud, S.; Baret, F. PROSPECT: A model of leaf optical properties spectra. *Remote Sens. Environ.* **1990**, *34*, 75–91. [CrossRef]
26. Hunt, E.R.; Rock, B.N. Detection of changes in leaf water content using Near- and Middle-Infrared reflectances. *Remote Sens. Environ.* **1989**, *30*, 43–54.
27. Guyot, G. Optical Properties of Vegetation Canopies. *Appl. Remote Sens. Agric.* **1990**, *1990*, 19–43. [CrossRef]
28. Jones, H.G.; Vaughan, R.A. *Remote Sensing of Vegetation: Principles, Techniques, and Applications*; Oxford University Press: Oxford, UK, 2010.
29. Haq, I.U.; Ijaz, S. *Plant Disease Management Strategies for Sustainable Agriculture through Traditional and Modern Approaches*; Springer Nature: Berlin/Heidelberg, Germany, 2020.
30. Arens, N.; Backhaus, A.; Döll, S.; Fischer, S.; Seiffert, U.; Mock, H.-P. Non-invasive Presymptomatic Detection of Cercospora beticola Infection and Identification of Early Metabolic Responses in Sugar Beet. *Front. Plant. Sci.* **2016**, *7*, 1377. [CrossRef]
31. Couture, J.J.; Singh, A.; Charkowski, A.O.; Groves, R.L.; Gray, S.M.; Bethke, P.C.; Townsend, P.A. Integrating Spectroscopy with Potato Disease Management. *Plant. Dis.* **2018**, *102*, 2233–2240. [CrossRef] [PubMed]
32. Oerke, E.-C.; Leucker, M.; Steiner, U. Sensory assessment of Cercospora beticola sporulation for phenotyping the partial disease resistance of sugar beet genotypes. *Plant. Methods* **2019**, *15*, 1–12. [CrossRef]
33. Gold, K.M.; Townsend, P.A.; Herrmann, I.; Gevens, A.J. Investigating potato late blight physiological differences across potato cultivars with spectroscopy and machine learning. *Plant. Sci.* **2020**, *295*, 110316. [CrossRef]
34. Mahlein, A.-K.; Steiner, U.; Dehne, H.-W.; Oerke, E.-C. Spectral signatures of sugar beet leaves for the detection and differentiation of diseases. *Precis. Agric.* **2010**, *11*, 413–431. [CrossRef]
35. Grisham, M.P.; Johnson, R.M.; Zimba, P.V. Detecting sugarcane yellow leaf virus infection in asymptomatic leaves with hyper-spectral remote sensing and associated leaf pigment changes. *J. Virol. Methods.* **2010**, *167*, 140–145. [CrossRef]
36. Krezhova, D.; Petrov, N.; Maneva, S. Hyperspectral remote sensing applications for monitoring and stress detection in cultural plants: Viral infections in tobacco plants. *SPIE Remote Sens.* **2012**, *8531*, 85311H. [CrossRef]
37. Junges, A.H.; Ducati, J.R.; Lampugnani, C.S.; Almanca, M.A.K. Detection of grapevine leaf stripe disease symptoms by hy-perspectral sensor. *Phytopathol. Mediterr.* **2018**, *57*, 399–406.
38. Feng, W.; Qi, S.; Heng, Y.; Zhou, Y.; Wu, Y.; Liu, W.; He, L.; Li, X. Canopy Vegetation Indices from In Situ Hyperspectral Data to Assess Plant Water Status of Winter Wheat under Powdery Mildew Stress. *Front. Plant. Sci.* **2017**, *8*, 1219. [CrossRef]
39. Thomas, S.; Kuska, M.T.; Bohnenkamp, D.; Brugger, A.; Alisaac, E.; Wahabzada, M.; Behmann, J.; Mahlein, A.-K. Benefits of hyperspectral imaging for plant disease detection and plant protection: A technical perspective. *J. Plant. Dis. Prot.* **2017**, *125*, 5–20. [CrossRef]
40. Hirshorn, S.; Jefferies, S. *Final Report of the NASA Technology Readiness Assessment (TRA) Study Team*; 2016. Available online: https://ntrs.nasa.gov/citations/20170005794 (accessed on 8 December 2021).
41. Lamichhane, J.R. Bacterial Diseases of Crops: Elucidation of the factors that lead to differences between field and experimental infections. *Adv. Agron.* **2015**, *134*, 227–246.
42. Heideman, M.T.; Johnson, D.H.; Burrus, C.S. Gauss and the history of the fast Fourier transform. *Granul. Matter* **1985**, *34*, 265–277. [CrossRef]
43. Lee, W.; Alchanatis, V.; Yang, C.; Hirafuji, M.; Moshou, D.; Li, C. Sensing technologies for precision specialty crop production. *Comput. Electron. Agric.* **2010**, *74*, 2–33. [CrossRef]
44. Liu, Z.; Cheng, J.-A.; Huang, W.; Li, C.; Xu, X.; Ding, X. (Eds.) *Hyperspectral Discrimination and Response Characteristics of Stressed Rice Leaves Caused by Rice Leaf Folder*; Springer: Berlin/Heidelberg, Germany, 2012.
45. Thenkabail, P.S.; Smith, R.B.; De Pauw, E. Evaluation of narrowband and broadband vegetation indices for determining optimal hyperspectral wavebands for agricultural crop characterization. *Photogramm. Eng. Rem. S.* **2002**, *68*, 607–621.
46. Zhang, Y.; Migliavacca, M.; Penuelas, J.; Ju, W. Advances in hyperspectral remote sensing of vegetation traits and functions. *Remote Sens. Environ.* **2021**, *252*, 112121. [CrossRef]
47. Nagler, P.; Daughtry, C.; Goward, S. Plant Litter and Soil Reflectance. *Remote Sens. Environ.* **2000**, *71*, 207–215. [CrossRef]
48. Mahlein, A.-K.; Oerke, E.-C.; Steiner, U.; Dehne, H.-W. Recent advances in sensing plant diseases for precision crop protection. *Eur. J. Plant. Pathol.* **2012**, *133*, 197–209. [CrossRef]
49. West, J.S.; Bravo, C.; Oberti, R.; Lemaire, D.; Moshou, D.; McCartney, H.A. The potential of optical canopy measurement for targeted control of field crop diseases. *Annu. Rev. Phytopathol.* **2003**, *41*, 593–614. [CrossRef] [PubMed]

50. Menesatti, P.; Antonucci, F.; Pallottino, F.; Giorgi, S.; Matere, A.; Nocente, F.; Pasquini, M.; D'Egidio, M.G.; Costa, C. Laboratory vs. in-field spectral proximal sensing for early detection of Fusarium head blight infection in durum wheat. *Biosyst. Eng.* **2013**, *114*, 289–293. [CrossRef]
51. Riefolo, C.; Antelmi, I.; Castrignanò, A.; Ruggieri, S.; Galeone, C.; Belmonte, A.; Muolo, M.; Ranieri, N.; Labarile, R.; Gadaleta, G.; et al. Assessment of the Hyperspectral Data Analysis as a Tool to Diagnose *Xylella fastidiosa* in the Asymptomatic Leaves of Olive Plants. *Plants* **2021**, *10*, 683. [CrossRef] [PubMed]

chemistry proceedings

MDPI

Proceeding Paper

Development of a Gas Sensor for *Eucalyptol* Supervision: A Supporting Tool for Extreme Wildfire Management †

Cátia Magro [1,*], Marcelo Morais [1], Paulo A. Ribeiro [2], Susana Sério [2], Pedro Vieira [1] and Maria Raposo [2]

1. Department of Physics, NOVA School of Science and Technology, NOVA University Lisbon, 2829-516 Almada, Portugal; mme.morais@campus.fct.unl.pt (M.M.); pmv@fct.unl.pt (P.V.)
2. Laboratory of Instrumentation, Biomedical Engineering and Radiation Physics (LIBPhys-UNL), Department of Physics, NOVA School of Science and Technology, NOVA University Lisbon, 2829-516 Almada, Portugal; pfr@fct.unl.pt (P.A.R.); susana.serio@fct.unl.pt (S.S.); mfr@fct.unl.pt (M.R.)

* Correspondence: c.magro@fct.unl.pt

† Presented at the 1st International Electronic Conference on Chemical Sensors and Analytical Chemistry, 1–15 July 2021; Available online: https://csac2021.sciforum.net/.

Abstract: Recent research on volatile organic compounds (VOC) released by the heated vegetation has shown that, under specific conditions (e.g., extreme heat, humidity, wind, and topography), VOC might foster wildfire ignition sources and explain sudden changes in fire behavior, particularly in the most susceptible and flammable forests (eucalypt forests). This work aims to develop an electronic nose (e-nose) based on a sensor's array to monitor the concentration of eucalyptol, the major VOC compound of the *Eucalyptus globulus* tree. The detection of this target compound was achieved by measuring the impedance spectra of layer-by-layer developed thin films based on polyethyleneimine, poly(allylamine hydrochloride), and graphene oxide, by injecting the analyte into a custom-made vacuum chamber system. The obtained results were analyzed by the principal component analysis method. The developed e-nose sensor was able to distinguish different concentrations in a range from 411 to 1095 ppm.

Keywords: wildfires; volatile organic compounds; eucalyptol; electronic nose; impedance spectroscopy

Citation: Magro, C.; Morais, M.; Ribeiro, P.A.; Sério, S.; Vieira, P.; Raposo, M. Development of a Gas Sensor for *Eucalyptol* Supervision: A Supporting Tool for Extreme Wildfire Management. *Chem. Proc.* **2021**, *5*, 19. https://doi.org/10.3390/CSAC2021-10432

Academic Editor: Nicole Jaffrezic-Renault

Published: 30 June 2021

1. Introduction

Extreme wildfires cause the loss of various human lives and have a significant impact on the biodiversity of ecosystems. These phenomena are, still, not yet fully understood. Recent studies have proposed a new theory, suggesting that flammable gases generated from heated vegetation, in particular, Volatile Organic Compounds (VOC) common in Mediterranean plants may, under some topographic and wind conditions, accumulate in locations where, after the arrival of the ignition source, they rapidly burst into flames as occurs in explosions [1,2]. VOC can exhibit a flammable nature, enabling fire ignition sources and sudden changes in the fire behavior [3].

The electronic nose (e-nose) system, which comprises an array of sensors with partial specificity and an appropriate pattern recognition system, can recognize complex gases. Moreover, e-noses have shown favorable efficiency in monitoring applications, making them a potential tool for the study of VOC [4–6]. Although the e-nose exhibits high sensitivity, forest environments consist of a complex mixture of gases and therefore the used sensors must be able to detect, classify and quantify the target compound. To improve the sensor's sensitivity, different materials can be used as coatings, thus enhancing the chemical and physical properties of the sensor [7–9].

The purpose of this work was the development of a custom-made measuring system attached to an e-nose system to monitor eucalyptol in a range of concentrations, from 411 to 1095 ppm. The concentration range was chosen based on the concentrations found in

Mediterranean vegetation (including the Eucalyptus tree [10]) and a range that was the lower flammability limit for terpenes (<1% (v/v)) [11], for laboratory safety reasons.

Thus, different layer-by-layer (LbL) thin films were developed in order to attain the best combinations for the monitoring of the target compound. The thin films were produced with polyethyleneimine (PEI), poly(allylamine hydrochloride) (PAH), and graphene oxide (GO), namely (PEI/GO) and (PAH/GO). These films have been already described in [12–14] and, thus, may have an interesting potential to monitor VOC, namely eucalyptol.

2. Materials and Methods

The developed e-nose consists of an array of sensing devices based on ceramic solid supports with deposited gold interdigitated electrodes (IDE), comprising eight "fingers" each, with dimensions of 22.8 × 7.6 × 0.7 mm and each "finger" has 200 μm of width. These solid supports were acquired from DropSens (Oviedo, Asturias, Spain) [15].

The thin films were deposited on the sensors IDE by the layer-by-layer (LbL) technique, which consists of the alternate deposition of polyelectrolytes layers with opposite electrical charges, to obtain several bilayers. The polyelectrolytes used to built-up the thin films layers were polyethyleneimine (PEI), poly(allylamine hydrochloride) (PAH), and graphene oxide (GO), all purchased from Sigma-Aldrich, Steinheim, Germany. The aqueous solutions of the polyelectrolytes were prepared with a 10^{-2} M concentration of each polyelectrolyte. Each of the aqueous solutions were prepared with ultrapure water, obtained in a Milli-Q ultrapure water system (Millipore GmbH, Billerica, MA, USA). This process was carried out by alternated adsorption of the positive PEI or PAH polyelectrolytes and the negatively charged GO molecules.

After each adsorption of the polyelectrolyte layers, the solid support was immersed in water in order to remove any polyelectrolyte molecules that were not completely adsorbed. The immersion time, in which the adsorption of the molecules takes place, was 60 s for each of the polyelectrolytes used and 30 s for the washing process. After the adsorption of each bilayer, the thin film was dried using nitrogen gas stream (99% purity, Air Liquide, Algés, Portugal). Thus, thin films of (PEI/GO) and (PAH/GO), with 5 bilayers each, $(PEI/GO)_5$ and $(PAH/GO)_5$, were produced.

The eucalyptol (99%) used for the experiments was purchased from Sigma-Aldrich, Steinheim, Germany. To test the response of the sensor when exposed to the eucalyptol, a range of concentrations from 411 to 1095 ppm was evaluated.

The measurements of each sensor were performed inside a custom-made vacuum chamber system designed by the author's team, which is depicted in Figure 1.

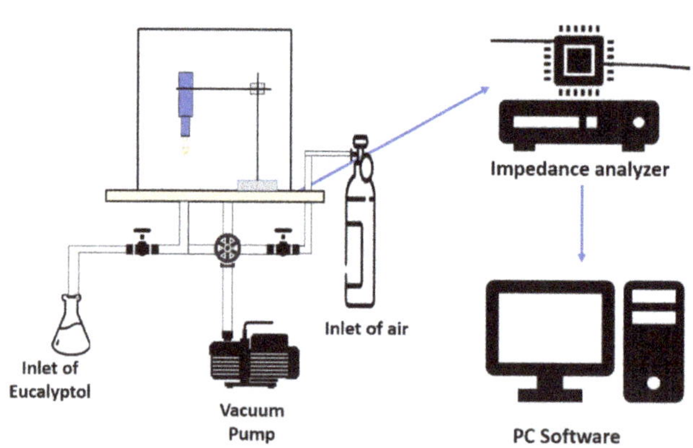

Figure 1. Schematic illustration of the experimental setup, including the custom-made vacuum chamber.

A sample holder containing the sensor was placed inside the chamber which presents an approximate volume of 58 L. A rotatory vacuum pump was also connected so that primary vacuum could be achieved and maintained inside the chamber during the tests, enabling a "clean" environment for the measurements. The sample holder was connected to a Solartron 1260 Impedance Analyzer (Solartron Analytical, AMETEK scientific instruments, Berwyn, PA, USA), in order to measure the impedance spectra at the IDE terminals. The chamber also has two inputs for the inlet of the eucalyptol and compressed air in the chamber, allowing the interaction between the target compound and the sensor.

In order to start the testing process, the sample holder with the sensor was placed inside the chamber and connected to the Impedance Analyzer. After that, the vacuum pump was switched on to achieve a pressure of 1.3×10^{-3} mbar. Following this, the eucalyptol would be evaporated into the chamber by the opening of a needle valve that connects the chamber with the mixture of eucalyptol. Subsequently, and after the evaporation of the compound and reaching a certain pressure level, the inlet was open to inject the compressed air until the pressure of 1.3×10^{-3} mbar was attained. Afterward, the electrical measurements were conducted in a frequency range of 1 Hz to 1 MHz, and an AC signal voltage of 25 mV. This process was then repeated for each eucalyptol concentration.

The electrical impedance spectra data features were assessed with the Principal Component Analysis (PCA) method to reduce the data size and to obtain a new space of orthogonal components in which different concentration patterns can be observed. The principal component analysis (PCA) plots were obtained by performing the normalization (Z-Score normalization (value-μ)/ϑ, μ and ϑ being the mean value and the standard deviation of the samples, respectively) of the impedance spectroscopy data.

3. Results and Discussion

Figure 2 shows the electric impedance spectra of the sensors, coated with thin films of $(PEI/GO)_5$ and $(PAH/GO)_5$, when in contact with different concentrations of eucalyptol, for different frequencies.

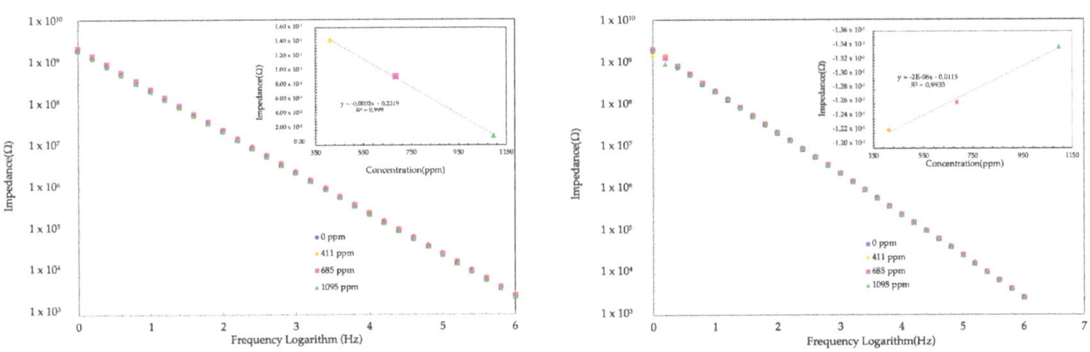

Figure 2. Impedance spectra measured with $(PEI/GO)_5$ (**a**) and $(PAH/GO)_5$ (**b**) LbL films inside of the chamber filled with different eucalyptol concentrations. In the insets are shown the evolution of the impedance at 63 and 10 kHz as a function of concentration, respectively.

In order to have a more thorough analysis, a normalization was performed on the data, using the following equation [12]:

$$\frac{PP(C) - PP(0ppm)}{PP(0ppm)} \quad (1)$$

where $PP(C)$ corresponds to a physical property at a given eucalyptol concentration, and $PP(0ppm)$ to a measure at a reference concentration, in this case when there was 0 ppm of eucalyptol. In each of the cases represented in the insets of Figure 2a,b, the physical property used was the impedance, at a constant frequency, where the effects of the eucalyptol are better represented. In both cases, it may be observed that an increase in the concentration of eucalyptol results in the decrease in the sensors' impedance values.

Furthermore, a preliminary analysis of the electronic nose concept was performed by the mean of the Principal Component Analysis (PCA). Thus, the data of each one of the sensors, $(PEI/GO)_5$ and $(PAH/GO)_5$, when in the presence of different tested concentrations of eucalyptol were plotted, as shown in Figure 3.

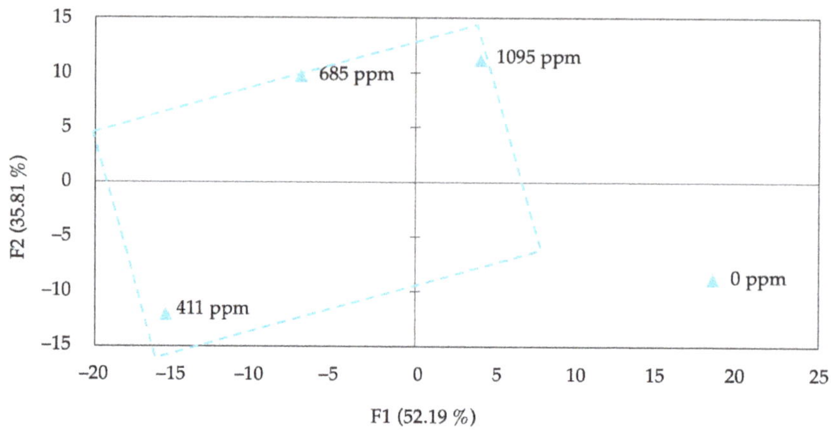

Figure 3. PCA plot for both sensors used with thin films of $(PEI/GO)_5$ and $(PAH/GO)_5$, for a range of eucalyptol from 0 to 1095 ppm.

By the PCA analysis, it can be observed that the e-nose assembled, with two sensors, each one coated with thin films of $(PAH/GO)_5$ and $(PEI/GO)_5$, can distinguish between the blank and the different concentrations of eucalyptol that they interacted with.

In fact, the choice of the GO as the upper layer seems to increase the efficiency and discrimination of the measurements, as the molecules of the thin film may react with the target compound enabling a better adsorption, since it possesses many functional groups.

4. Conclusions

The electronic nose, consisting of an array of two sensors coated with $(PEI/GO)_5$ and $(PAH/GO)_5$ thin films, built-up with the LbL technique, was able to detect eucalyptol and distinguish three different concentrations levels, as the PCA technique has shown. The use of graphene oxide as the thin film bilayer increased the interaction between the custom-made chamber's head space and the thin film coating, and consequently the efficiency of the impedance measurements and capability of the e-nose to distinguish different eucalyptol concentrations.

The present work presents a novel and preliminary study, with an in-deep effort in the build-up of the custom-made chamber. It should be also noted that this project is under development; thus, the build-up of different coatings, e.g., different combination of polyelectrolytes and/or other sensing materials, to improve the e-nose performance, is expected.

Author Contributions: Conceptualization, C.M., M.M., M.R. and P.A.R.; methodology, C.M.; software, M.M.; validation, C.M., M.R. and S.S.; formal analysis, C.M., M.R. and S.S.; investigation, M.M. and C.M.; resources, P.A.R., M.R., S.S. and P.V.; data curation, M.M.; writing—original draft preparation, M.M.; writing—review and editing, C.M., M.R. and S.S.; supervision, C.M., M.R. and S.S.; project administration, P.V.; funding acquisition, P.A.R., M.R. and P.V. All authors have read and agreed to the published version of the manuscript.

Funding: This research was funded by the project PCIF/GFC/0078/2018—Influence of forest VOCs (volatile organic compounds) on extreme fire behaviour from Fundação para a Ciência e a Tecnologia (FCT). The research facilities leading to these results has received support from the Portuguese funding agency FCT—Fundação para a Ciência e a Tecnologia—within projects PTDC/FIS-

NAN/0909/2014, UID/FIS/04559/2020 to LIBPhys-UNL from the FCT/MCTES/PIDDAC, and the Bilateral Project entitled "Deteção de Estrogénio- um Contaminante Emergente em Corpos Hídricos" within the scope of "Cooperação Transnacional_FCT (Portugal)-CAPES (Brazil) 2018".

Institutional Review Board Statement: Not applicable.

Informed Consent Statement: Not applicable.

Acknowledgments: C. Magro acknowledges NOVA.ID of NOVA-FCT for her postdoc fellowship.

Conflicts of Interest: The authors declare no conflict of interest.

References

1. Courty, L.; Chetehouna, K.; Halter, F.; Foucher, F.; Garo, J.P.; Mounaim-Rousselle, C. Flame speeds of α-Pinene/Air and Limonene/Air Mixtures involved in accelerating forest fires. *Combust. Sci. Technol.* **2012**, *184*, 1397–1411. [CrossRef]
2. Centro Ecologia Aplicada. EXTREME: Influence of Forests VOCS (Volatile Organic Compounds) in Extreme Fire Behaviour. Available online: http://www.isa.ulisboa.pt/ceabn/projecto/2/100/extreme-influence-of-forests-vocs-volatile-organic-compounds-in-extreme-fire-behaviour (accessed on 15 June 2021).
3. Chetehouna, K.; Barboni, T.; Zarguili, I.; Leoni, E.; Simeoni, A.; Fernandez-Pello, A.C. Investigation on the emission of volatile organic compounds from heated vegetation and their potential to cause an accelerating forest fire. *Combust. Sci. Technol.* **2009**, *181*, 1273–1288. [CrossRef]
4. Fernández, M.J.; Fontecha, J.L.; Sayago, I.; Aleixandre, M.; Lozano, J.; Gutiérrez, J.; Gràcia, I.; Cané, C.; del Carmen Horrillo, M. Discrimination of volatile compounds through an electronic nose based on ZnO SAW sensors. *Sens. Actuators B Chem.* **2007**, *127*, 277–283. [CrossRef]
5. Kauer, J.S.; White, J. Electronic Nose. In *Encyclopedia of Neuroscience*; Elsevier: Amsterdam, The Netherlands, 2009; pp. 871–877.
6. Gardner, J.W.; Bartlett, P.N. A brief history of electronic noses. *Sens. Actuators B Chem.* **1994**, *18*, 210–211. [CrossRef]
7. Saeed, S.H.; Abbas, Z.; Gopal, B. Experimental use of electronic nose for analysis of volatile organic compound (VOC). In Proceedings of the 2009 International Multimedia, Signal Processing and Communication Technologies, Aligarh, India, 14–16 March 2009; pp. 113–115.
8. Paczkowski, S.; Pelz, S.; Paczkowska, M. Monitoring of Eucalyptus globulus tissue thermal degradation by semi-conductor metal-oxide sensors for early fire detection in eucalypt forests. *Int. J. Wildland Fire* **2019**, *28*, 167–175. [CrossRef]
9. Ahluwalia, A.; De Rossi, D. Artificial Noses and Tongues. In *Encyclopedia of Materials: Science and Technology*; Elsevier: Amsterdam, The Netherlands, 2001; pp. 344–347.
10. Pausas, J.G.; Alessio, G.A.; Moreira, B.; Segarra-Moragues, J.G. Secondary compounds enhance flammability in a Mediterranean plant. *Oecologia* **2016**, *180*, 103–110. [CrossRef] [PubMed]
11. Chatelon, F.-J.; Sauvagnargues, S.; Dusserre, G.; Balbi, J.-H. Generalized Blaze Flash, a "Flashover" Behavior for Forest Fires—Analysis from the Firefighter's Point of View. *Open J. For.* **2014**, *04*, 547–557. [CrossRef]
12. Zhao, Q.; He, Z.; Jiang, Y.; Yuan, Z.; Wu, H.; Su, C.; Tai, H. Enhanced acetone-sensing properties of PEI thin film by GO-NH2 functional groups modification at room temperature. *Front. Mater.* **2019**, *5*, 82. [CrossRef]
13. Graboski, A.M.; Zakrzevski, C.A.; Shimizu, F.M.; Paschoalin, R.T.; Soares, A.C.; Steffens, J.; Paroul, N.; Steffens, C. Electronic Nose Based on Carbon Nanocomposite Sensors for Clove Essential Oil Detection. *ACS Sens.* **2020**, *5*, 1814–1821. [CrossRef] [PubMed]
14. Magro, C.; Zagalo, P.; Pereira-da-Silva, J.; Mateus, E.P.; Ribeiro, A.B.; Ribeiro, P.; Raposo, M. Polyelectrolyte Based Sensors as Key to Achieve Quantitative Electronic Tongues: Detection of Triclosan on Aqueous Environmental Matrices. *Nanomaterials* **2020**, *10*, 640. [CrossRef] [PubMed]
15. Magro, C.; Mateus, E.P.; Paz-Garcia, J.M.; Sério, S.; Raposo, M.; Ribeiro, A.B. Electronic Tongue Coupled to an Electrochemical Flow Reactor for Emerging Organic Contaminants Real Time Monitoring. *Sensors* **2019**, *19*, 5349. [CrossRef] [PubMed]

chemistry proceedings

MDPI

Proceeding Paper

Bioactive Compound Profiling and Nutritional Composition of Three Species from the Amaranthaceae Family †

Bernabe Nuñez-Estevez [1,2], Tiane C. Finimundy [2], Maria Carpena [1], Marta Barral-Martinez [1], Ricardo Calhelha [2], Tânia C. S. P. Pires [2], Paz Otero [1], Pascual Garcia-Perez [1], Jesus Simal-Gandara [1], Isabel C. F. R. Ferreira [2], Miguel A. Prieto [1,2,*] and Lillian Barros [2,*]

[1] Nutrition and Bromatology Group, Faculty of Food Science and Technology, Ourense Campus, University of Vigo, E32004 Ourense, Spain; bernabe.nunez@uvigo.es (B.N.-E.); mcarpena@uvigo.es (M.C.); marta.barral@uvigo.es (M.B.-M.); pazoterofuertes@gmail.com (P.O.); pasgarcia@uvigo.es (P.G.-P.); jsimal@uvigo.es (J.S.-G.)

[2] Centro de Investigação de Montanha (CIMO), Instituto Politécnico de Bragança, Campus de Santa Apolónia, 5300-253 Bragança, Portugal; tcfinimu@hotmail.com (T.C.F.); calhelha@ipb.pt (R.C.); t.pires@live.com.pt (T.C.S.P.P.); iferreira@ipb.pt (I.C.F.R.F.)

* Correspondence: mprieto@uvigo.es (M.A.P.); lillian@ipb.pt (L.B.)

† Presented at the 1st International Electronic Conference on Chemical Sensors and Analytical Chemistry, 1–15 July 2021; Available online: https://csac2021.sciforum.net/.

Abstract: In this work, the chemical and nutritional composition of three Amaranthaceae species (*Alternanthera sessilis*, *Dicliptera chinensis*, and *Dysphania ambrosioides*) was studied. The results showed a differential flavonoid content in the three species: *A. sessilis* and *D. ambrosioides* showed similar flavonoid contents (15.1 ± 0.6 and 15.1 ± 0.1 mg/g extract, respectively), followed by *D. chinensis* (11.4 ± 0.1 mg/g extract). On the other hand, the nutritional results showed a high protein content in all species (16.9–13.9 ± 0.1 g/100 g dw) and revealed the presence of organic acids, such as oxalic and succinic acid. Therefore, bioactive compounds, together with protein and organic acids, could be of great value to the food industry.

Keywords: medicinal plants; phenolic compounds; nutritional value; phytochemistry

check for updates

Citation: Nuñez-Estevez, B.; Finimundy, T.C.; Carpena, M.; Barral-Martinez, M.; Calhelha, R.; Pires, T.C.S.P.; Otero, P.; Garcia-Perez, P.; Simal-Gandara, J.; Ferreira, I.C.F.R.; et al. Bioactive Compound Profiling and Nutritional Composition of Three Species from the Amaranthaceae Family. *Chem. Proc.* **2021**, *5*, 20. https://doi.org/10.3390/CSAC2021-10563

Academic Editor: Giorgio Senesi

Published: 1 July 2021

1. Introduction

Several plant species have played an important role in traditional medicine worldwide, as humans have been using plants as a natural remedy for a multitude of diseases for 60,000 years [1]. In particular, the plants from Amaranthaceae family biosynthesize several bioactive compounds with beneficial biological activities, including essential oils, betalains, terpenoids, and phenolic compounds [2]. Different phytochemical studies have verified the different biological activities associated with the plant extracts belonging to this family, such as antioxidant, antidiabetic, antitumor, antibacterial, anti-inflammatory, among others [3].

Specially, three Amaranthaceae species, namely: *Alternanthera sessilis* (L.) R.Br. ex Dc, *Dicliptera chinensis* (L.) Juss. and *Dysphania ambrosioides* (L.) Mosyakin and Clemants. These species have been little explored in terms of their phytochemical valorization. *A. sessillis* has been used in traditional Malaysian medicine, both as an infusion and as food, while in China, its leaves have even been used for the treatment of eye and skin diseases, snake bites, and wound healing [4]. *D. chinensis* has a major distribution in southern China, Bangladesh, northern India, and Vietnam [5], where it was traditionally used with detoxifying and diuretic purposes, thanks to the production of organic acids, flavonoids, terpenoids, steroids, and polysaccharides [6]. Finally, *D. ambrosioides*, distributed throughout South America, is known to be used in traditional medicine as a remedy for parasitic diseases, and it is still currently used to treat parasitosis because of the presence of ascaridol [7].

Due to the health-enhancing potential attributed to Amaranthaceae species, in this work the nutritional characterization and chemical composition, in terms of phenolic compounds, will be carried out. As a result, this research could be considered as the starting point for a more targeted search for bioactive compounds [8] biosynthesized by these underexplored plant species.

2. Materials and Methods

2.1. Plant Material and Nutritional and Chemical Characterization

The samples proceeding from the Amaranthaceae species involved in this work, *A. sessilis*, *D. chinensis*, and *D. ambrosioides*, were thoroughly washed, air-dried, crushed, and sieved to obtain plant homogenates, which were stored at −80 °C until use.

The nutritional characterization (ashes, proteins, lipids, and carbohydrates, as well as energy) of the three plants was carried out following the methodology adapted previously [9]. The determinations were carried out by duplicate, and results were expressed in terms of percentage of composition for ashes, proteins, lipids, and carbohydrates, whereas energy was expressed as the mean ± standard deviation (SD) in kcal/100 g dry weight (dw).

The chemical composition (total sugars, fatty acids, and organic acids) was evaluated following the methodology also described by Barros et al. (2013) [9]. The determinations were performed in duplicate, and the results were expressed as the mean ± SD in g/100 g dw for total sugars and organic acids composition, whereas fatty acids were expressed as the relative percentage of saturated fatty acids (SFAs), monounsaturated fatty acids (MUFAs), and polyunsaturated fatty acids (PUFAs).

2.2. Sample Extraction and the Determination of Phenolic Compounds

For the determination of phenolic compounds, 1 g of each sample was macerated using 50 mL of ethanol/water (80:20 *v*/*v*) as solvent. This mixture was stirred at room temperature for 1 h and then filtered. This process was repeated twice, and extracts were collected and concentrated at 40 °C in a rotary evaporator, to remove the alcoholic fraction. The aqueous phase was frozen and freeze-dried. For the identification of phenolic compounds, a Dionex Ultimate 3000 UPLC system (Thermo Scientific, San Jose, CA, USA) was used, following a previous methodology [10]. The determination was performed by a diode array detector (DAD) and mass spectrometry (MS) (LTQ XL mass spectrometer, Thermo Finnigan, San Jose, CA, USA) working in negative mode. Once identified and quantified, compounds were grouped by their parental skeleton, being expressed as luteolin derivatives (LD), apigenin derivatives (AD), kaempferol derivatives (KD), quercetin derivatives (QD), and isorhamnetin derivatives (ID), in mg/g dw.

3. Results and Discussion

3.1. Nutritional Characterization

The results for the nutritional composition of Amaranthaceae plants are shown in Figure 1. The inorganic content of all plants, represented by the ashes, showed similar values, being higher than those of other plants of the same family from the genus *Amaranthum* [11] For proteins, *D. chinensis* (16.8 g/100 g dw) and *A. sessilis* (16.1 g/100 g dw) show similar values, higher than those of *D. ambrosioides* (13.9 g/100 g dw). These values are comparable to other species such as *Chenopodium quinoa* (quinoa), and they were also higher than other cereals, such as wheat, maize, or rice [12] With respect to lipids, the values for *A. sessilis* (0.74 g/100 g dw) were very low compared with the other species, which ranged between 1.1–1.8 g/100 g dw, being comparable to the lipid content of fruits and vegetables [12]. Regarding the carbohydrate content, the results were also very similar between the three species: 68.8 g/100 g dw for *D. chinensis*, 73.2 g/100 g dw for *A. sessilis*, and 71.5 g/100 g dw for *D. ambrosioides*, being in accordance with the carbohydrate contents of *C. quinoa* and other cereals, as well as other foods, such as chocolate, flour or bread. [12] Finally, the energy value did not vary much either, ranging from 350–365 kcal/100 g dw with *A. sessilis* being the species with the highest energy intake.

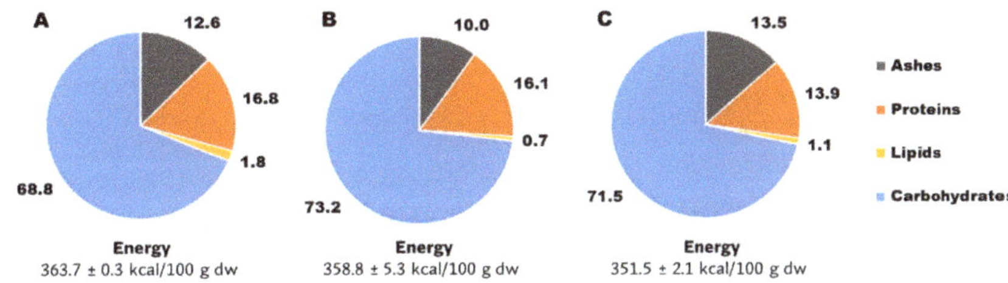

Figure 1. Nutritional characterization (ashes, proteins, lipids, and carbohydrates), and energy determination of three Amaranthaceae plants: (**A**) *A. sessilis*, (**B**) *D. chinensis* and (**C**) *D. ambrosioides*. The results for nutrient content were expressed as relative content in percentage, whereas energy was expressed as mean ± SD, in kcal/100 g dw.

3.2. Chemical Characterization

The results for the chemical characterization of Amarathaceae species, in terms of total sugars, organic acids, and fatty acids contents are shown in Figure 2. The results for free sugars were, in decreasing order, 6.33 g/100 g dw for *D. chinensis*, 4.13 g/100 g dw for *A. sessilis*, and 0.34 g/100 g dw for *D. ambrosioides*. In this case, *A. sessilis* has the most similar content to that estimated for *C. quinoa* (2–3 g/100 g dw) [12].

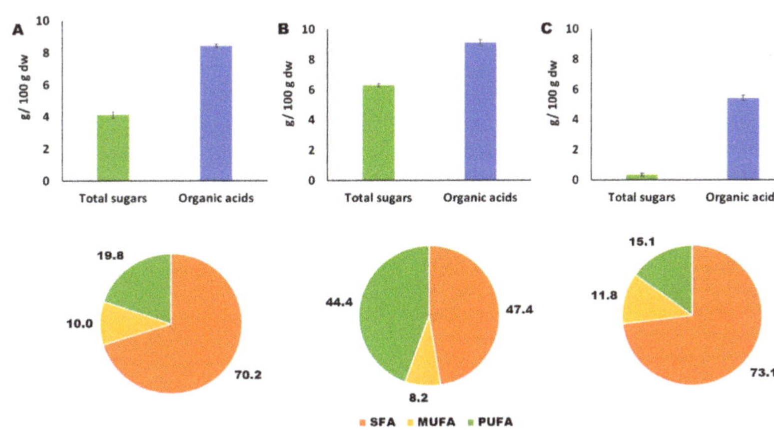

Figure 2. Chemical characterization (total free sugar, organic acids, and fatty acids contents) of three Amaranthaceae plants: (**A**) *A. sessilis*, (**B**) *D. chinensis*, and (**C**) *D. ambrosioides*. Results for total sugars and organic acids were expressed as g/100 g dw, vertical bars indicate standard deviation. The results for fatty acids content were expressed as relative abundance, in percentage. SFA: saturated fatty acids, MUFA: monounsaturated fatty acids, PUFA: polyunsaturated fatty acids.

Organic acids were detected in all three plant species studied, with a total content of ~9.13 g/100 g dw for *D. chinensis*, 8.43 g/100 g dw for *A. sessilis*, and 5.43 g/100 g dw for *D. ambrosioides*. Oxalic acid stood out as the organic acid present in the highest concentrations in all species, especially in the case of *A. sessilis* (data not shown). This acid is associated with reduced dietary Ca^{2+} availability and various kidney diseases [13]. Succinic acid and fumaric acid were also detected in *D. chinensis*, although in lower proportions. Both acids are in high demand by the food, cosmetic and pharmaceutical industries [14]. In addition, in the case of *C. quinoa*, previous data showed the presence of oxalic, citric and fumaric acid [15], revealing a similar profile for this functional food in terms of organic acids content.

Concerning the relative abundance of fatty acids in Amaranthaceae plants (pie charts in Figure 2), SFAs, MUFAs, and PUFAs were detected in these species. According to the data obtained, the plant with the highest amount of SFA was *D. ambrosioides* with 73.1% of the total fatty acids (Figure 2A), followed by *A. sessilis* with 70.2% and *D. chinensis* with 47.4%. As for the MUFA, all plants exhibited very similar values between them, ranging 8.23–11.8%. Finally, the results for PUFAs showed that *D. chinensis* had the highest abundance with 44.4% (Figure 2B), while *D. ambrosioides* and *A. sessilis* presented a similar abundance (19.8% and 15.1%, respectively). According to the data, the most abundant fatty acid in the three species was hexadecanoic acid, with the range of abundance in the percentage of total fatty acids in the three plants being 32–40%. A previous study identified the fatty acids of *A. sessilis*, in which the second most abundant fatty acid was hexadecanoic acid [16]. Regarding the high content in PUFAs for *D. chinensis*, and due to the beneficial properties associated with these bioactive compounds as antioxidant and cardioprotective agents, it is suggested that this species presents the healthiest chemical profile.

3.3. The Determination of Phenolic Compounds

The phenolic profiling of hydroethanolic extracts of Amaranthaceae plants was performed by UPLC-DAD-ESI/MS, revealing that flavonoids were the most abundant family of phenolic compounds in these species. The results for the determination of phenolic compounds are shown in Figure 3. In the case of *A. sessilis* extracts, luteolin derivatives (LD) were the most abundant compounds, with concentrations of 9.7 mg/g extract, with luteolin-8-*C*-(rhamnosyl)ketodeoxyhexoside as the most prevalent derivative (Figure 3A). To a lesser extent, apigenin derivatives (AD), kaempferol derivatives (KD), and quercetin derivatives (QD) were also reported (<1.8 mg/g extract) (Figure 3A). This plant has two varieties distinguished by their colors, red and green, and previous studies have shown that the red variety has a better nutritional composition, a higher content of phenolic compounds, and a greater antioxidant capacity [12]. Thus, the phenolic composition of *A. sessilis* has been seen to be affected by the variety employed.

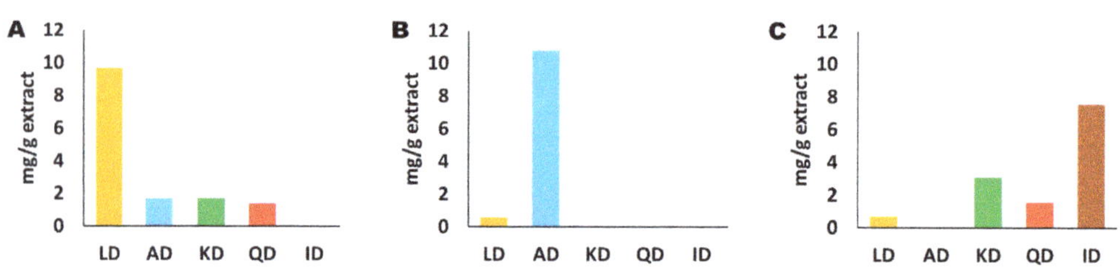

Figure 3. Identification of phenolic compounds (fatty acids, organic acids and tocopherols) of three plants belonging to the Amantharaceae family: (**A**) *A. sessilis*, (**B**) *D. chinensis* and (**C**) *D. ambrosioides*. LD: luteolin derivatives, AD: apigenin derivatives, KD: kaempferol derivatives, QD: quercetin derivatives, and ID: isorhamnetin derivatives.

With respect to *D. chinensis*, the extracts essentially contained different ADs, accounting for 10.82 mg/g extract (Figure 3B), being apigenin 6-C-glucoside-8-C-arabinoside the most prevalent compound. This is the first time, to the best of our knowledge, that the phenolic profiling of *D. chinensis* is determined, being spotted as a potential natural source of apigenin, largely characterized as a bioactive compound [17].

Finally, the results for *D. ambrosioides* (Figure 3C) showed that this was the only species presenting isorhamnetin derivatives (ID), which were the most abundant compounds in the hydroethanolic extracts (7.58 mg/g extract), and isorhamnetin-3-*O*-rutinoside were quantified as the main phenolic compound. Additionally, IDs, LDs, KDs, and QDs were also identified in this species in lower concentrations, as well as a lack of ADs (Figure 3C). In previous studies on *D. ambrosioides*, the highest concentration of phenolic compounds extracted was obtained by methanolic extracts, with 87.7 ± 1.4 µg of gallic acid equivalents/mg extract and 57 ± 1.4 µg quercetin equivalents/mg extract. Moreover, the same authors identified quercetin as the most abundant phenolic compound on *D. ambrosioides* [18], which is in accordance with our results, since quercetin-*O*-rhamnosyl-pentoside was the second most abundant compound in the hydroethanolic extracts. This suggests a critical role of solvent on the extraction of phenolic compounds from Amaranthaceae plants.

4. Conclusions

In this work, the determination of nutritional and chemical characterization, as well as the phenolic profiling of three Amaranthaceae species largely used in traditional medicine was developed. In this regard, similar chemical profiles were obtained for all species, with comparable inorganic, protein, lipid, and carbohydrate contents. The results on fatty acid composition revealed that *D. chinensis* showed the healthier profile with a high proportion of PUFAs. Finally, the determination of phenolic compounds suggested a species-dependent biosynthesis of these compounds, with luteolin derivatives, apigenin derivatives, and isorhamnetin derivatives presenting as the most prevalent phytoconstituents on *A. sessilis*, *D. chinensis*, and *D. ambrosioides*, respectively. Overall, our results shed light on the characterization of these species from a nutritional point of view and suggested that Amaranthaceae species can be considered as sources of bioactive compounds to be applied in the food, cosmetic, and pharmacological industries.

Supplementary Materials: The following are available online at https://www.mdpi.com/article/10.3390/CSAC2021-10563/s1.

Author Contributions: Conceptualization, methodology, software validation, formal analysis, investigation, resources, data curation, writing—original draft preparation, writing—review and editing, visualization, supervision B.N.-E., T.C.F., M.C., M.B.-M., R.C., T.C.S.P.P., P.O., P.G.-P., J.S.-G., I.C.F.R.F., M.A.P. and L.B. All authors have read and agreed to the published version of the manuscript.

Funding: By EcoChestnut Project (Erasmus+ KA202) that supports the work of B.N.-E.; the program Grupos de Referencia Competitiva (GRUPO AA1-GRC 2018) that supports the work of M.B.-M.; Authors are grateful to the Ibero-American Program on Science and Technology (CYTED—AQUA-CIBUS, P317RT0003), to the Bio-Based Industries Joint Undertaking (JU) under grant agreement No 888003, and to the UP4HEALTH Project (H2020-BBI-JTI-2019) that supports the work of Paz Otero. and P. Garcia-Perez. The research leading to these results was funded by Xunta de Galicia supporting the program EXCELENCIA-ED431F 2020/12; to the Ibero-American Program on Science and Technology (CYTED—AQUA-CIBUS, P317RT0003). The JU receives support from the European Union's Horizon 2020 research and innovation program and the Bio-Based Industries Consortium. The project SYSTEMIC Knowledge Hub on Nutrition and Food Security has received funding from national research funding parties in Belgium (FWO), France (INRA), Germany (BLE), Italy (MIPAAF), Latvia (IZM), Norway (RCN), Portugal (FCT), and Spain (AEI) in a joint action of JPI HDHL, JPI-OCEANS and FACCE-JPI launched in 2019 under the ERA-NET ERA-HDHL (No. 696295). The authors are also grateful to FCT, Portugal, for financial support through national funds FCT/MCTES to the CIMO (UIDB/00690/2020); and L.B. and R.C. thank the national funding by FCT, P.I., through the institutional and individual scientific employment program-contract for their contracts.

Institutional Review Board Statement: Not applicable.

Informed Consent Statement: Not applicable.

Data Availability Statement: Not applicable.

Acknowledgments: We would like to thank MICINN for supporting the Ramón y Cajal grant for M.A.P. (RYC-2017-22891) and University of Vigo for supporting the predoctoral grant of M.C. (Uvigo-00VI 131H 6410211).

Conflicts of Interest: The authors declare no conflict of interest.

References

1. Yuan, S.; Chan, H.C.S.; Filipek, S.; Vogel, H. PyMOL and Inkscape Bridge the Data and the Data Visualization. *Structure* **2016**, *24*, 2041–2042. [CrossRef] [PubMed]
2. Mroczek, A. Phytochemistry and bioactivity of triterpene saponins from Amaranthaceae family. *Phytochem. Rev.* **2015**, *14*, 577–605. [CrossRef]
3. Sun, H.X. Adjuvant effect of Achyranthes bidentata saponins on specific antibody and cellular response to ovalbumin in mice. *Vaccine* **2006**, *24*, 3432–3439. [CrossRef] [PubMed]
4. Konda, V.G.R.; Madhavi, E.; Ruckmani, A.; Venkataramana, Y. A review on medicinal plants with potential hypolipidemic activity. *Int. J. Pharma Bio Sci.* **2013**, *4*, 101–107.
5. Ma, J.; Clemants, S. A history and overview of the Flora Reipublicae Popularis Sinicae (FRPS, Flora of China, Chinese edition, 1959–2004). *Taxon* **2006**, *55*, 451–460. [CrossRef]
6. Zhang, K.; Gao, Y.; Zhong, M.; Xu, Y.; Li, J.; Chen, Y.; Duan, X.; Zhu, H. Hepatoprotective effects of Dicliptera chinensis polysaccharides on dimethylnitrosamine-induced hepatic fibrosis rats and its underlying mechanism. *J. Ethnopharmacol.* **2016**, *179*, 38–44. [CrossRef] [PubMed]
7. Monzote, L.; Stamberg, W.; Staniek, K.; Gille, L. Toxic effects of carvacrol, caryophyllene oxide, and ascaridole from essential oil of Chenopodium ambrosioides on mitochondria. *Toxicol. Appl. Pharmacol.* **2009**, *240*, 337–347. [CrossRef] [PubMed]
8. Fabricant, D.S.; Farnsworth, N.R. The Value of Plants Used in Traditional Medicine for Drug Discovery. *Environ. Health Perspect.* **2001**, *109*, 69. [CrossRef] [PubMed]
9. Barros, L.; Pereira, E.; Calhelha, R.C.; Dueñas, M.; Carvalho, A.M.; Santos-Buelga, C.; Ferreira, I.C.F.R. Bioactivity and chemical characterization in hydrophilic and lipophilic compounds of *Chenopodium ambrosioides* L. *J. Funct. Foods* **2013**, *5*, 1732–1740. [CrossRef]
10. Bessada, S.M.F.; Barreira, J.C.M.; Barros, L.; Ferreira, I.C.F.R.; Oliveira, M.B.P.P. Phenolic profile and antioxidant activity of *Coleostephus myconis* (L.) Rchb.f.: An underexploited and highly disseminated species. *Ind. Crops Prod.* **2016**, *89*, 45–51. [CrossRef]
11. Sá, A.G.A.; Moreno, Y.M.F.; Carciofi, B.A.M. Plant proteins as high-quality nutritional source for human diet. *Trends Food Sci. Technol.* **2020**, *97*, 170–184. [CrossRef]
12. Othman, A.; Ismail, A.; Hassan, F.A.; Yusof, B.N.M.; Khatib, A. Comparative evaluation of nutritional compositions, antioxidant capacities, and phenolic compounds of red and green sessile joyweed (*Alternanthera sessilis*). *J. Funct. Foods* **2016**, *21*, 263–271. [CrossRef]
13. Guil, J.L.; Torija, M.E.; Giménez, J.J.; Rodríguez-García, I.; Giménez, A. Oxalic acid and calcium determination in wild edible plants. *J. Agric. Food Chem.* **1996**, *44*, 1821–1823. [CrossRef]
14. Ertugrul, B.; Iplik, E.S.; Cakmakoglu, B. In Vitro Inhibitory Effect of Succinic Acid on T-Cell Acute Lymphoblastic Leukemia Cell Lines. *Arch. Med. Res.* **2020**, in press. [CrossRef] [PubMed]
15. Pereira, E.; Encina-Zelada, C.; Barros, L.; Gonzales-Barron, U.; Cadavez, V.; Ferreira, I.C. Chemical and nutritional characterization of *Chenopodium quinoa* Willd (quinoa) grains: A good alternative to nutritious food. *Food Chem.* **2019**, *280*, 110–114. [CrossRef] [PubMed]
16. Muniandy, K.; Gothai, S.; Tan, W.S.; Kumar, S.S.; Mohd Esa, N.; Chandramohan, G.; Al-Numair, K.S.; Arulselvan, P. In Vitro Wound Healing Potential of Stem Extract of Alternanthera sessilis. *Evid.-Based Complement. Altern. Med.* **2018**, *2018*. [CrossRef] [PubMed]
17. Salehi, B.; Venditti, A.; Sharifi-Rad, M.; Kręgiel, D.; Sharifi-Rad, J.; Durazzo, A.; Lucarini, M.; Santini, A.; Souto, E.B.; Novellino, E.; et al. The therapeutic potential of Apigenin. *Int. J. Mol. Sci.* **2019**, *20*, 1305. [CrossRef] [PubMed]
18. Zohra, T.; Ovais, M.; Khalil, A.T.; Qasim, M.; Ayaz, M.; Shinwari, Z.K. Extraction optimization, total phenolic, flavonoid contents, HPLC-DAD analysis and diverse pharmacological evaluations of *Dysphania ambrosioides* (L.) Mosyakin & Clemants. *Nat. Prod. Res.* **2019**, *33*, 136–142. [CrossRef] [PubMed]

chemistry proceedings

MDPI

Proceeding Paper

Development of a Pattern Recognition Tool for the Classification of Electronic Tongue Signals Using Machine Learning †

Edgar G. Mendez-Lopez [1], Jersson X. Leon-Medina [2,*] and Diego A. Tibaduiza [1]

1 Departamento de Ingeniería Eléctrica y Electrónica, Universidad Nacional de Colombia, Cra 45 No. 26-85, Bogotá 111321, Colombia; egmendezl@unal.edu.co (E.G.M.-L.); dtibaduizab@unal.edu.co (D.A.T.)
2 Departamento de Ingeniería Mecánica y Mecatrónica, Universidad Nacional de Colombia, Cra 45 No. 26-85, Bogotá 111321, Colombia
* Correspondence: jxleonm@unal.edu.co

† Presented at the 1st International Electronic Conference on Chemical Sensors and Analytical Chemistry, 1–15 July 2021; Available online: https://csac2021.sciforum.net/.

Abstract: Electronic tongue type sensor arrays are made of different materials with the property of capturing signals independently by each sensor. The signals captured when conducting electrochemical tests often have high dimensionality, which increases when performing the data unfolding process. This unfolding process consists of arranging the data coming from different experiments, sensors, and sample times, thus the obtained information is arranged in a two-dimensional matrix. In this work, a description of a tool for the analysis of electronic tongue signals is developed. This tool is developed in Matlab® App Designer, to process and classify the data from different substances analyzed by an electronic tongue type sensor array. The data processing is carried out through the execution of the following stages: (1) data unfolding, (2) normalization, (3) dimensionality reduction, (4) classification through a supervised machine learning model, and finally (5) a cross-validation procedure to calculate a set of classification performance measures. Some important characteristics of this tool are the possibility to tune the parameters of the dimensionality reduction and classifier algorithms, and also plot the two and three-dimensional scatter plot of the features after reduced the dimensionality. This to see the data separability between classes and compatibility in each class. This interface is successfully tested with two electronic tongue sensor array datasets with multi-frequency large amplitude pulse voltammetry (MLAPV) signals. The developed graphical user interface allows comparing different methods in each of the mentioned stages to find the best combination of methods and thus obtain the highest values of classification performance measures.

Keywords: electronic tongue; graphical user interface; feature extraction; dimensionality reduction; classification; machine learning

Citation: Mendez-Lopez, E.G.; Leon-Medina, J.X.; Tibaduiza, D.A. Development of a Pattern Recognition Tool for the Classification of Electronic Tongue Signals Using Machine Learning. *Chem. Proc.* **2021**, *5*, 21. https://doi.org/10.3390/CSAC2021-10447

Academic Editor: Nicole Jaffrezic-Renault

Published: 30 June 2021

1. Introduction

The data set obtained from an MLAPV (multifrequency large amplitude pulse voltammetry) electronic tongue device comes from various types of sensors and their magnitudes can have different scales [1]. These signals are characterized by having high dimensionality [2]. This can cause problems in Machine Learning models, both in pattern recognition and in the accuracy of data classification [3]. Due to this, it is necessary to perform the correct processing of these data sets to obtain high precision values for the classification of liquid substances.

In 2020, Leon-Medina et al. [2] developed a methodology that seeks to improve the classification accuracy with an approach based on non-linear feature extraction of signals obtained with electronic tongue type sensor array devices. This methodology is composed of several stages: (1) Data unfolding, (2) Normalization, (3) Non-linear dimensionality reduction, (4) Classification by means of a supervised machine learning model and finally a (5) Cross validation [2]. The application of the methodology in each stage includes the execution of algorithms in the software Matlab®. These algorithms contain a series of parameters that must be configured. As a result of the application of the methodology, the value of the classification accuracy and the confusion matrix of the classification model used are obtained, together with their performance metrics.

Due to the number of stages and the different configuration options of the parameters in the algorithms, the need was generated to develop a tool that would facilitate the application of this methodology, guiding the user through the different stages and making the configuration of the algorithms more user-friendly. One of the main advantages of a graphical user interface (GUI) is that it makes an implemented system easy to use, understand and evaluate [4].

Section 2, describes two tests performed by the developed GUI, as well as the datasets used in each one and the operation of the GUI. Then, Section 3, illustrates the main findings obtained during the two tests applying the methodology of data processing through the GUI. Finally, Section 4 shows the main conclusions in data processing through the GUI.

2. Materials and Methods

The measurements of the responses of an electronic tongue system are discretized currents in time. In this way, a measurement is obtained at each instant of time for each of the electrodes that make up the electronic tongue device, obtaining a matrix of size $I \times K$ where I are the experimental tests and K are the time instants of the signal collected by each electrode. Due to the electronic tongue system has an array of sensors and taking J as the number of electrodes. A data unfolding procedure is executed to convert the three-dimensional matrix $I \times J \times K$, in a two-dimensional matrix $I \times (J \cdot K)$ [2]. Figure 1 shows an illustrative graph of the Data Unfolding process.

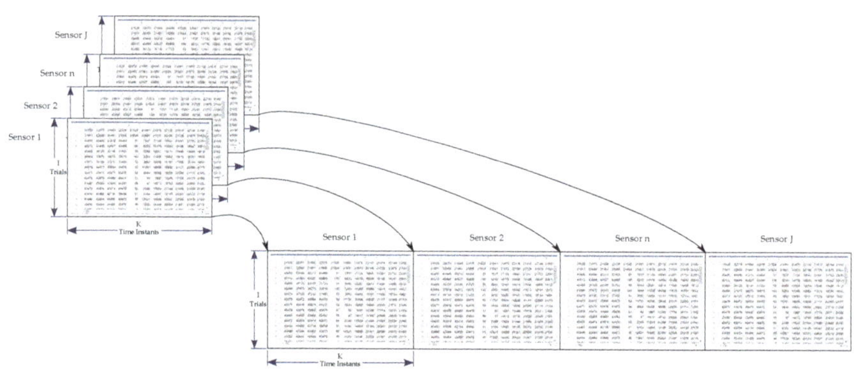

Figure 1. Data unfolding procedure.

In this work, two tests with the developed tool are performed using two different datasets. These tests are described below:

For the first test, a dataset obtained by means of a MLAPV electronic tongue developed by Liu et al. [5] is used. The electronic tongue consisted of a platinum pillar auxiliary sensor, an Ag/AgCl reference sensor, and six working electrodes made of different materials, gold, platinum, palladium, titanium, tungsten, and silver. In the experiment, the fourth titanium electrode was damaged, so it was not considered in the data analysis [5]. Seven liquids or aqueous matrices were used to collect the data from the ***first dataset***: (1) red wine, (2) Chinese liquor, (3) beer, (4) black tea, (5) oolong tea, (6) you maofeng and (7) you pu'er. Each one with three different concentrations (14%, 25% and 100%) of the original solution mixed with distilled water, to which three replications were made, that is, 9 samples for each liquid [2], for a total of 63 samples. With 2050 measurement points per sensor and

5 sensors in the electronic tongue, when performing the Unfolding procedure of the data (described above, see Figure 1), the dataset is composed of a matrix of size ***63*** × ***10,250***.

The second test uses a dataset obtained from the study by Zhang et al. [6]. This second dataset contains the data collected from an MLAPV electronic tongue with five working electrodes made of gold, silver, palladium, tungsten and silver. The auxiliary electrode used is platinum pillar and the reference electrode is Ag/AgCl [7]. For this study, 13 liquids or aqueous matrices (number of samples) were used: beer (19), red wine (8), white liqueur (6), black tea (9), tea Maofeng (9), pu'er tea (9), Oolong tea (9), coffee (9), milk (9), cola (6), vinegar (9), medicine (6), and salt (6), for a total of 114 samples [6]. Like the first dataset, in the ***second dataset*** there are 2050 measurement points per sensor and 5 sensors in the electronic tongue, when performing the Unfolding procedure of the data, the second dataset has a size of ***114*** × ***10,250***.

The developed GUI is an application made in Matlab® App Designer, it is made up of seven tabs. Only the first tab is enabled at the beginning of the GUI, as shown in Figure 2a). By means of the ***Browser*** button in the ***Data Selection*** section, the file containing the dataset previously ordered with the unfolding process is selected. Subsequently, the data is loaded in the GUI through the button ***Load***, after this, the size of the dataset is shown in the GUI, Figure 2b) illustrates this process.

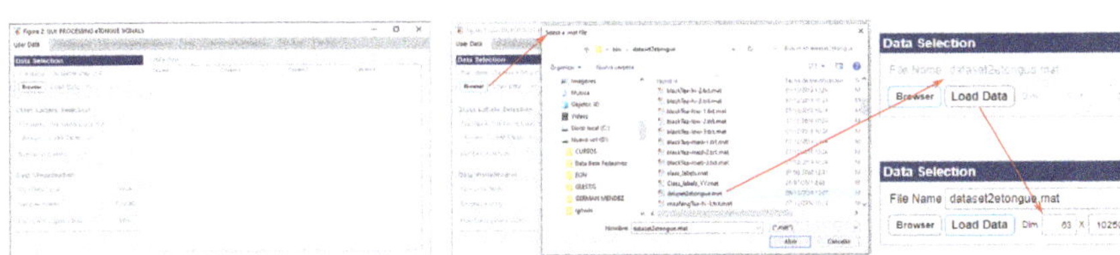

a) b)

Figure 2. (**a**) GUI Initial state. (**b**) dataset selection.

With the dataset loaded, the ***Browser*** button is enabled in the ***Class Labels Selection*** to select, in the same way as was done with the dataset, the file ***Class Labels***. Once this vector is loaded, the number of classes used can be viewed, see Figure 3a.

After selecting the data files, the ***Normalization*** tab is enabled, in which the method for data normalization can be selected, see Figure 3b. With normalized data, the ***Dimensionality Reduction*** tab is enabled where the Feature Extraction technique [8] to reduce the dimensionality of the data can be selected, additionally there is a ***Parameters*** **section** where it is possible to configure certain parameters depending on the selected dimensionality reduction technique, see Figure 3c. With the data in low dimensionality, the ***Plot*** tab is enabled for the selection and visualization of the variables in 2D and scatter plots, see Figure 3d. Simultaneously, the ***Classification/Validation*** tab is enabled where there are four classifiers, along with some parameters that can be configured depending on the selected classifier, see Figure 3e. Executing the classification stage, the ***Cross Validation*** section is enabled, which contains three validation techniques, see Figure 3f. At the end of the procedure, the ***Results*** tab is enabled, where the classification performance metrics [9] and the confusion matrix are shown, see Figure 3g. At the same time, the ***History*** tab is enabled, in which a summary of the different techniques and methods used in data processing is presented, see Figure 3h. Figure 3 shows the sequence of enabling the GUI tabs throughout the data processing in each of the stages.

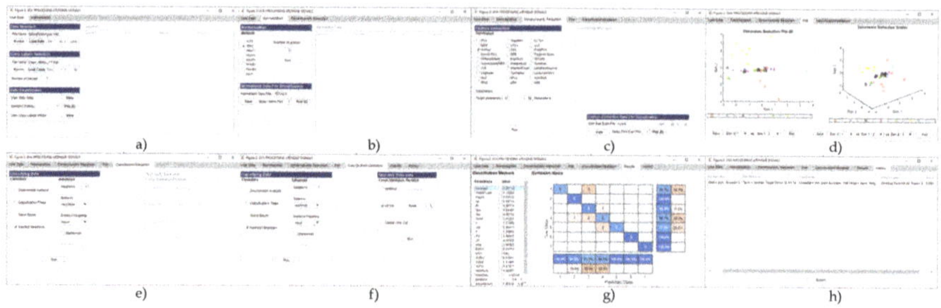

Figure 3. Sequence of enabling the stages in the developed GUI of the tool for classification of electronic tongue signals. (**a**) ***User Data Tab***: Selecting the dataset and vector from Class Labels; (**b**) ***Normalization Tab***: Data Normalization; (**c**) ***Dimensionality Reduction Tab***: Data dimensionality reduction; (**d**) ***Plot Tab***: 2D and scatter graphics display; (**e**) ***Classification Tab***: Classifier selection; (**f**) ***Validation Tab***: Selection of the cross-validation method; (**g**) ***Results Tab***: Visualization of the Confusion Matrix and metrics of the classification model; (**h**) ***History Tab***: Summary of tests carried out.

In relation to the plots int the GUI, the data loaded in the GUI, as well as the normalized data and data after the dimensionality reduction process can be visualized. A table or graph visualization can be obtained by each experiment in the corresponding tabs. In Figure 4, the original data are observed, in the Normalization and Dimensionality Reduction stage, the graphs are made for sample 7 in the same stages.

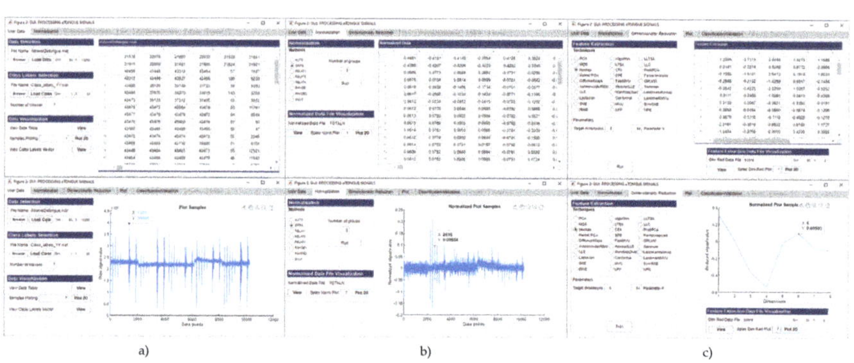

Figure 4. Viewing data as a sample chart or table. (**a**) Original data; (**b**) Standardized data; (**c**) Data after dimensionality reduction.

3. Results

Through the GUI, the following tests are performed with the datasets described above, see Section 2.

3.1. Comparison Plots 2D and Scatter

In the ***Plot*** tab the 2D and scatter graphs obtained after applying a dimensionality reduction technique are displayed. Figure 5 shows the graphs obtained with three different dimensionality reduction techniques applied to the first dataset. Additionally, new graphs generated by selecting different dimensions are observed, the label corresponds to each class of liquid in the dataset. Each graph can be saved in a file independently. The parameters used for each dimensionality reduction technique are described below:

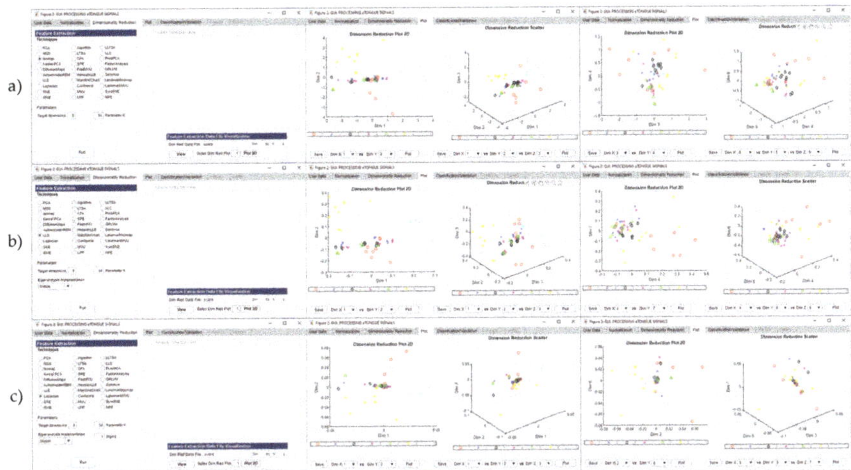

Figure 5. Data representation after dimensionality reduction. (**a**) Dimensionality reduction method = Isomap, Dim = 8, K = 54, Plot Dim1 2D = 1, 2 (Default), Plot Dim1 3D = 1, 2, 3 (Default), Plot Dim2 2D = 3, 4, Plot Dim2 3D = 4, 5, 6; (**b**) Dimensionality reduction method = Locally Linear Embedding (LLE), Dim = 8, K = 54, Plot Dim1 2D = 1, 2, Plot Dim1 3D = 1, 2, 3, Plot Dim2 2D = 4, 7, Plot Dim2 3D = 4, 6, 8; (**c**) Dimensionality reduction method = Laplacian Eigenmaps, Dim = 8, K = 54, Plot Dim1 2D = 1, 2, Plot Dim1 3D = 1, 2, 3, Plot Dim2 2D = 2, 8, Plot Dim2 3D = 3, 5, 7.

3.2. Classification Accuracy Behavior

Two tests are described below to observe the behavior of the classification accuracy. These tests are applied to the second dataset based on the developed methodology of [7]. First, the number of *k* neighbors of the *k*-NN Classifier is modified, varying its value from 1 to 16, but keeping the number of dimensions fixed at 8 in the PCA dimensionality reduction technique. In the second test, the number of PCA dimensions is varied from 3 to 16, but the number of neighbors is fixed equal to 2. In both tests, the group scaling method (GRPS) is used to normalize the data and 5-Fold Cross validation is performed. The results of the tests carried out are described below in Figure 6.

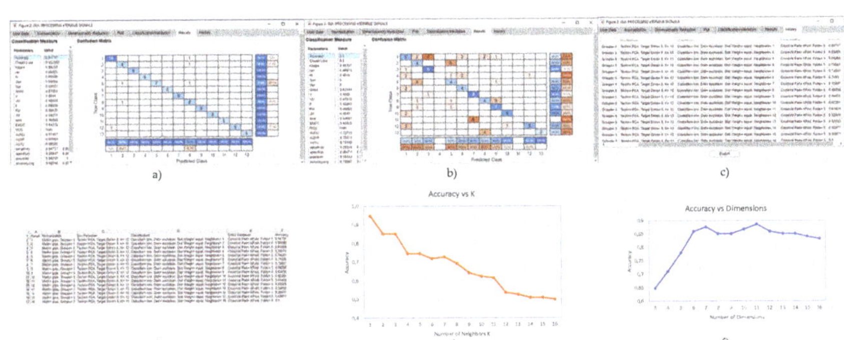

Figure 6. Confusion matrix results, and accuracy behavior varying the number of target dimensions and number of *k* neighbors. (**a**) Confusion matrix and performance metrics of the classification model for the Accuracy of 94.73% obtained in the first test with a parameter *k* = 2; (**b**) Confusion matrix and performance metrics of the classification model for the Accuracy of 50% obtained in the first test with a parameter ***k* = 16**; (**c**) Summary of the trials of the first trial displayed in the ***History tab*** of GUI; (**d**) Excel file exported from ***History tab*** for the first test; (**e**) Graph of Accuracy vs. Number of *k* Neighbors, obtained from the results in the first test. (**f**) Graph of Accuracy vs. Number of Dimensions, obtained from the results in the second test.

4. Conclusions

This work showed the development of a tool for the processing of data contained acquired by an electronic tongue type sensor array. First, the GUI design allows the user to be guided intuitively through the signal processing methodology by enabling the tabs, but at the same time it allows the user to choose the different techniques and methods, as well as the parameter configuration. Second, the GUI offers the visualization of the data, by means of tables or graphically, both the original data and those transformed in the Normalization and dimensionality reduction stages. Another advantage is the visualization of 2D and 3D scatter graphics, where the user can observe the distribution of the samples, according to the selected feature extraction technique, choosing between different combinations of dimensions. In the same way, this tool offers the visualization of the results in the confusion matrix and the performance classification metrics of the classification model, finally it provides a summary table of the tests carried out in such a way that the user can easily compare the results obtained.

Author Contributions: All authors contributed to the development of this work, specifically their contributions are as follow: conceptualization, D.A.T. and J.X.L.-M.; data organization and pre-processing, J.X.L.-M. and E.G.M.-L.; methodology, J.X.L.-M. and E.G.M.-L.; validation, J.X.L.-M. and D.A.T. All authors have read and agreed to the published version of the manuscript.

Funding: This research was funded by FONDO DE CIENCIA TECNOLOGÍA E INNOVACION FCTeI DEL SISTEMA GENERAL DE REGALÍAS SGR. The authors express their gratitude to the Administrative Department of Science, Technology and Innovation—Colciencias with the grant 779—"Convocatoria para la Formación de Capital Humano de Alto Nivel para el Departamento de Boyacá 2017" for sponsoring the research presented herein.

Informed Consent Statement: Not applicable.

Acknowledgments: Jersson X. Leon-Medina is grateful with Colciencias and Gobernación de Boyacá for his PhD fellowship. Jersson X. Leon-Medina thanks Miryam Rincón Joya from the Department of Physics of the National University of Colombia and Leydi Julieta Cardenas Flechas, who is currently a Ph.D. student, for their introduction to the electronic tongue sensor array field of research.

Conflicts of Interest: The authors declare no conflict of interest.

References

1. Leon-Medina, J.X.; Vejar, M.A.; Tibaduiza, D.A. Signal Processing and Pattern Recognition in Electronic Tongues: A Review. In *Pattern Recognition Applications in Engineering*; IGI Global: Hershey, PA, USA, 2020; pp. 84–108. [CrossRef]
2. Leon-Medina, J.X.; Anaya, M.; Pozo, F.; Tibaduiza, D. Nonlinear Feature Extraction Through Manifold Learning in an Electronic Tongue Classification Task. *Sensors* **2020**, *20*, 4834. [CrossRef] [PubMed]
3. Ayesha, S.; Hanif, M.K.; Talib, R. Overview and comparative study of dimensionality reduction techniques for high dimensional data. *Inf. Fusion* **2020**, *59*, 44–58. [CrossRef]
4. Djelouat, H.; Ali, A.A.S.; Amira, A.; Bensaali, F. An interactive software tool for gas identification. *J. Nat. Gas Sci. Eng.* **2018**, *55*, 612–624. [CrossRef]
5. Liu, T.; Chen, Y.; Li, D.; Wu, M. An active feature selection strategy for DWT in artificial taste. *J. Sens.* **2018**, *2018*, 9709505. [CrossRef]
6. Zhang, L.; Wang, X.; Huang, G.B.; Liu, T.; Tan, X. Taste recognition in E-tongue using local discriminant preservation projection. *IEEE Trans. Cybern.* **2018**, *49*, 947–960. [CrossRef] [PubMed]
7. Leon-Medina, J.X.; Cardenas-Flechas, L.J.; Tibaduiza, D.A. A data-driven methodology for the classification of different liquids in artificial taste recognition applications with a pulse voltammetric electronic tongue. *Int. J. Distrib. Sens. Netw.* **2019**, *15*. [CrossRef]
8. Van der Maaten, L. *An Introduction to Dimensionality Reduction Using Matlab*; Report MICC 07-07 Universiteit Maastricht; Faculty of Humanities & Sciences, MICC/IKAT: Maastricht, The Netherlands, 2007; Volume 1201, pp. 1–44.
9. Ballabio, D.; Grisoni, F.; Todeschini, R. Multivariate comparison of classification performance measures. *Chemom. Intell. Lab. Syst.* **2018**, *174*, 33–44. [CrossRef]

chemistry proceedings

Proceeding Paper

Electronic Nose for Bladder Cancer Detection †

Heena Tyagi [1,*], Emma Daulton [1], Ayman S. Bannaga [2,3], Ramesh P. Arasaradnam [2,3,4,5] and James A. Covington [1]

1 School of Engineering, University of Warwick, Coventry CV4 7AL, UK; e.daulton@warwick.ac.uk (E.D.); J.A.Covington@warwick.ac.uk (J.A.C.)
2 Department of Gastroenterology, University Hospital Coventry & Warwickshire, Coventry CV2 2DX, UK; Ayman.Bannaga@warwick.ac.uk (A.S.B.); R.Arasaradnam@warwick.ac.uk (R.P.A.)
3 Warwick Medical School, University of Warwick, Coventry CV4 7AL, UK
4 School of Health Sciences, Coventry University, Coventry CV1 5FB, UK
5 Leicester Cancer Research Centre, University of Leicester, Leicester LE1 7RH, UK
* Correspondence: Heena.Tyagi@warwick.ac.uk
† Presented at the 1st International Electronic Conference on Chemical Sensors and Analytical Chemistry, 1–15 July 2021, Available online: https://csac2021.sciforum.net/.

Abstract: This study outlines the use of an electronic nose as a method for the detection of VOCs as biomarkers of bladder cancer. Here, an AlphaMOS FOX 4000 electronic nose was used for the analysis of urine samples from 15 bladder cancer and 41 non-cancerous patients. The FOX 4000 consists of 18 MOS sensors that were used to differentiate the two groups. The results obtained were analysed using s MultiSens Analyzer and RStudio. The results showed a high separation with sensitivity and specificity of 0.93 and 0.88, respectively, using a Sparse Logistic Regression and 0.93 and 0.76 using a Random Forest classifier. We conclude that the electronic nose shows potential for discriminating bladder cancer from non-cancer subjects using urine samples.

Keywords: electronic nose; bladder cancer; AlphaMOS FOX 4000; VOCs

Citation: Tyagi, H.; Daulton, E.; Bannaga, A.S.; Arasaradnam, R.P.; Covington, J.A. Electronic Nose for Bladder Cancer Detection. *Chem. Proc.* **2021**, *5*, 22. https://doi.org/10.3390/CSAC2021-10438

Academic Editor: Nicole Jaffrezic-Renault

Published: 30 June 2021

1. Introduction

Bladder cancer (BC) is the eighth most common cancer worldwide. In the UK, there were 12,434 new cases and 6458 fatalities in 2020 [1]. Fortunately, the survival rate for bladder cancer remains good, with almost every three out of four people surviving the disease for one or more years [2]. Even with this high survival rate, there has been no significant improvement over the past ten years. The most common BC screening methods are cystoscopy, urine cytology, and urine tests, including bladder tumour antigen (BTA) test, nuclear matrix protein 22 (NMP22), urinary bladder cancer antigen (UBC), and fibrin degradation products (FDP) [3,4]. Unfortunately, none of these are effective enough for early diagnosis of BC and they are both expensive and invasive [5,6]. Therefore, there is a need for a more disease-specific, non-invasive, highly sensitive and low-cost screening test for BC.

The use of Volatile Organic Compounds (VOCs) has provided a new perspective for the early detection of cancer. The alterations in VOCs emitted from the body reflect the changes inside the body caused by this disease. VOCs can be measured from a range of different biological sources including urine [7], saliva [8], breath [9], faeces [10] and blood [11]. Measuring VOCs can be simple and non-invasive, mapping well onto the needs of a screening test [12]. Different studies that have previously been performed to analyse VOCs to diagnose BC, have mainly focused on urine and faeces [13,14]. The most common approach is to use GC-MS (Gas Chromatography-Mass Spectrometry). However, the analysis time is long (tens of minutes) and it is expensive, both in terms of equipment and running costs [15]. An alternative is to use an electronic nose (eNose), an instrument designed to mimic the human olfactory system. The eNose is widely used in the food and beverage industry [16], environment control, pharmaceutical companies [17] and biomedical applications [18]. Several previous studies have been undertaken for the detection of bladder cancer using the eNose. Van De Goor et al. used breath to distinguish head and neck, colon and bladder cancer using an electronic nose [19]. Another study conducted by Bernabei et al. was able to identify 100% of patients with urinary tract cancer (bladder and prostate cancer combined) from healthy controls [20]. A further study showed that bladder cancer was identified with the use of fluorescence urinary VOCs detection with a sensitivity of 84.2% and a specificity of 87.8% [21].

In our study, we aimed to identify and test the potential of urinary biomarkers to distinguish between BC and non-cancerous groups using an electronic nose. This is the first study conducted using an AlphaMOS FOX 4000 eNose to identify bladder cancer from non-cancerous samples using urinary VOCs.

2. Materials and Methods

2.1. Study Design

A total of 56 patients were recruited at University Hospital Coventry and Warwickshire NHS Trust, UK. This study was approved by Coventry and Warwickshire and North-East Yorkshire NHS Ethics Committees (Ref. 18717 and Ref. 260179). Out of 56 patients, 15 were confirmed with BC, and 41 non-cancerous (symptoms suggestive of but excluded after further investigations). The demographics for these groups are shown in Table 1.

Table 1. Demographic for the study.

Group	Bladder Cancer	Non-Cancerous
Number of samples	15	41
Mean Age (years)	70.0 (90–50)	62.5 (90–32)
Sex: Male/Female	12:3	24:12
Avg. BMI	24.4 (27.4–19.8)	30.9 (37.3–22.4)
Current Smoker (Number and % of patients)	1 (6.7%)	3 (8.3%)

The samples were collected and stored in standard sterile specimen containers and frozen for 2 h at −80 °C. The samples were then shipped to the University of Warwick for testing, where the samples were defrosted in a laboratory fridge at 4 °C and 3 mL of sample aliquoted into 10 mL glass vials [22].

2.2. AlphaMOS FOX 4000 (Toulouse, France)

The AlphaMOS Fox 4000 is an eNose that comprises 18 commercial metal oxide sensors (MOS) distributed in three temperature-controlled chambers. There are 6 p-type sensors and 12 n-type sensors. The output of the sensors is measured as resistance. The FOX 4000 is fitted with a CombiPAL HS-100 auto-sampler using a 2.5 mL gas syringe. In testing, the samples were placed in the autosampler, then were agitated, and heated to 40 °C for 10 min. The headspace was then injected into the eNose at a rate of 200 mL/min into a flow of 150 mL/min of zero air. Each sample was analysed for 180 s by all the 18 MOX sensors.

2.3. Statistical Analysis

The sensor's responses were extracted using AlphaSoft (AlphaMOS v12.36) and then analysed using a MultiSens Analyzer (JLM Innovation GmbH, Tübingen, Germany) and RStudio (Version 1.4.1106). AlphaSoft is a software product developed to control AlphaMOS instruments. The output generated by the program was processed and exported in ASCII format. These files were further analysed using a MultiSens Analyzer (JLM Innovation GmbH, Germany) for multivariate analysis. Due to the high dimensionality of the data, the maximum change in resistance was extracted per sensor and used as the input features for a PCA (Principal Component Analysis) and an LDA (Linear Discriminant Analysis). The feature matrix was also exported and further processed using a custom analysis pipeline created in RStudio. Here, 10-fold cross-validation was performed where the data was divided into 10 equally sized groups, with nine groups being used for model training and then applied to the 10th group as a test set. This was repeated 10 times until all the samples had been in a test group. SMOTE (Synthetic Minority Over-Sampling Technique) was performed on the data groups due to the high imbalance in the sample size for BC and the non-cancerous group. It generated syntenic balanced, which were then used to train the classifier [23]. This was undertaken inside the training fold so as not to affect the test result. Two classification models were applied to the data, specifically Random Forest (RF) and Sparse Logistic Regression (SLR), which we have successfully used before in similar studies [24]. From the resultant probabilities, statistical parameters were calculated, including Receiver Operator Characteristic (ROC) curve, area under the curve (AUC), sensitivity, specificity, positive predictive value (PPV), and negative predictive value (NPV).

3. Results

The typical output from the FOX 4000 eNose is shown in Figure 1, where each curve represents the response of a sensor to a BC urine sample. Here, the sensor response is defined as intensity, which is the change in resistance from the baseline divided by the baseline resistance.

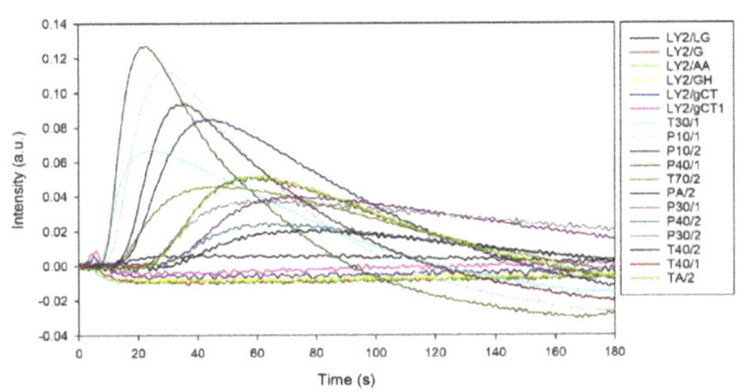

Figure 1. A typical output from AlphaMOS FOX 4000 to a BC urine sample.

The PCA results obtained from the MultiSens Analyzer are shown in Figure 2. The data shows that most of the sample variance can be plotted in the first principal component (85.3%). Furthermore, there is reasonable separation (though not perfect) between the BC and non-cancerous groups.

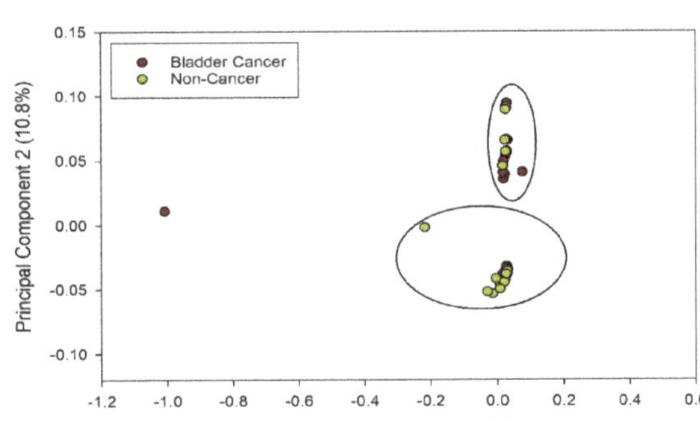

Figure 2. PCA output from BC and non-cancerous groups.

LDA was also performed on the data, as shown in Figure 3, to show the maximum potential separation between BC and Non-cancer samples.

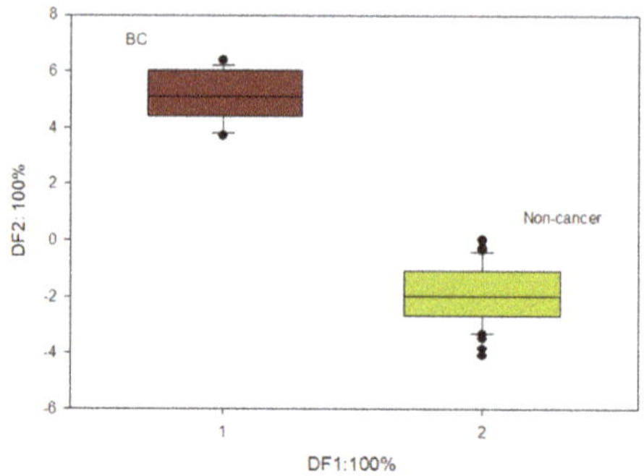

Figure 3. LDA output from BC and non-cancerous groups.

Finally, output statistical parameters were calculated, the results of which are shown in Table 2.

Table 2. Statistical output from FOX.

	Sparse Logistic Regression	Random Forest
AUC	0.92 (0.85–0.99)	0.86 (0.76–0.97)
Sensitivity	0.93 (0.68–0.99)	0.93 (0.68–0.99)
Specificity	0.88 (0.74–0.96)	0.76 (0.59–0.88)
PPV	0.74	0.58
NPV	0.97	0.97

The highest separation between the BC and non-cancerous group was obtained using Sparse Logistic Regression with an AUC (Area Under the Curve) of 0.92. The sensitivity and specificity obtained were 0.93 and 0.88, respectively. The high sensitivity and specificity signify that the SLR correctly predicted 36 out of 41 non-cancerous patients and was able to identify 14 BC patients out of 15.

The RF classifier was able to achieve a sensitivity of 0.93 with a specificity of 0.76. The AUC for this classifier was 0.86. With the RF classifier, the model was able to correctly identify 31 of the non-cancerous samples and 14 BC samples out of 15.

This shows that eNose can distinguish cancer samples from non-cancerous samples. The ROC curve for random forest classifier distinguishing BC and the non-cancerous group is shown in Figure 4.

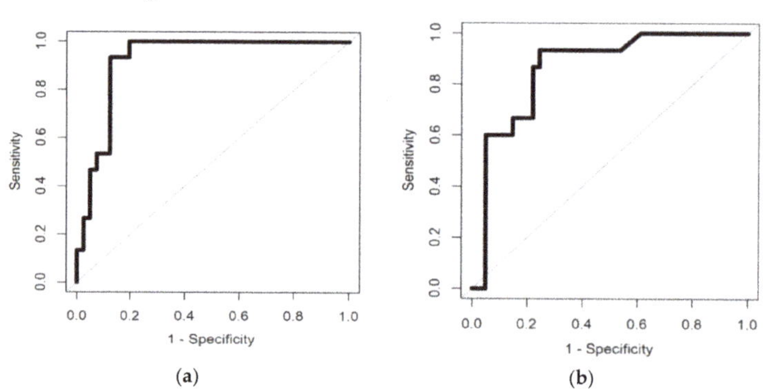

Figure 4. (**a**) illustrates ROC curve distinguishing BC from Non-cancerous group using Sparse Logistic Regression and (**b**) illustrates ROC curve distinguishing BC from Non-cancerous group using Random Forest classifier.

4. Discussion

In this paper, we have shown that the AlphaMOS FOX4000 Electronic nose was able to distinguish bladder cancer urine samples from non-cancerous samples based on their VOC profile. Our findings prove that eNose can be used to accurately separate these two groups. This is the first study to compare bladder cancer urine samples from non-cancerous samples using this eNose, which has the advantage of being fully automated allowing large numbers of samples to be tested easily.

In our study, we were able to separate the bladder cancer urine samples from the non-cancerous group with a high AUC of 0.92 and 0.86 using SLR and RF classifiers, respectively. For the classification of BC and non-cancerous groups using SLR, the sensitivity and the specificity obtained were 0.93 and 0.88, respectively. The threshold value for classification of the two groups was 0.13 and the p-value was <0.001. For the RF classifier, the sensitivity and specificity obtained were 0.93 and 0.76. The threshold value and p-value were 0.29 and <0.001, respectively.

We found that eNose was able to identify 14 BC samples out of 15, and 36 out of 41 non-cancerous samples using Sparse Logistic Regression classifier. However, our study is limited by the small number of samples, and we did not attempt to identify the specific VOC biomarkers involved. A previous study with a commercial eNose (Sensigent Cyranose 320) also showed high sensitivity and specificity [14], with comparable results to those found here. The sensitivity of eNose highly depends upon the material of the sensors and the environmental conditions, such as humidity and temperature [25]. A further limitation of our study was the lack of healthy controls for comparison. Further investigation is required to understand the specific chemicals associated with separating BC from non-cancer and to test samples from a larger patient group.

Author Contributions: Conceptualization, H.T., J.A.C. and R.P.A.; methodology, H.T. and E.D.; formal analysis, H.T.; investigation, H.T. and E.D.; resources, A.S.B., J.A.C. and R.P.A.; data curation, H.T.; writing—original draft preparation, H.T.; writing—review and editing, H.T., E.D. and J.A.C. visualization, H.T. and J.A.C.; supervision, J.A.C. and R.P.A. All authors have read and agreed to the published version of the manuscript.

Funding: This research received no external funding.

Institutional Review Board Statement: This study was approved by Coventry and Warwickshire and North-East Yorkshire NHS Ethics Committees (Ref. 18717 and Ref. 260179).

Informed Consent Statement: Informed consents were obtained from all subjects involved in the study.

Conflicts of Interest: The authors declare no conflict of interest.

References

1. GLOBOCAN. 2020 Estimated Number of Incident Cases, Bladder Cancer, All Ages. Available online: https://gco.iarc.fr (accessed on 9 June 2021).
2. Cancer Research UK Bladder Cancer Statistics. Available online: https://www.cancerresearchuk.org/health-professional/cancer-statistics/statistics-by-cancer-type/bladder-cancer#heading-Two (accessed on 9 June 2021).
3. DeGeorge, K.C.; Holt, H.R.; Hodges, S.C. Bladder Cancer: Diagnosis and Treatment. *Am. Fam. Physician* **2017**, *96*, 507–514.
4. Lotan, Y.; Svatek, R.S.; Malats, N. Screening for bladder cancer: A perspective. *World J. Urol.* **2008**, *26*, 13–18. [CrossRef]
5. Lotan, Y.; Roehrborn, C.G. Sensitivity and specificity of commonly available bladder tumor markers versus cytology: Results of a comprehensive literature review and meta-analyses. *Urology* **2003**, *61*, 109–118. [CrossRef]
6. Tilki, D.; Burger, M.; Dalbagni, G.; Grossman, H.B.; Hakenberg, O.W.; Palou, J.; Reich, O.; Rouprêt, M.; Shariat, S.F.; Zlotta, A.R. Urine Markers for Detection and Surveillance of Non–Muscle-Invasive Bladder Cancer. *Eur. Urol.* **2011**, *60*, 484–492. [CrossRef] [PubMed]
7. Silva, C.L.; Passos, M.; Câmara, J.S. Investigation of urinary volatile organic metabolites as potential cancer biomarkers by solid-phase microextraction in combination with gas chromatography-mass spectrometry. *Br. J. Cancer* **2011**, *105*, 1894–1904. [CrossRef]
8. Shigeyama, H.; Wang, T.; Ichinose, M.; Ansai, T.; Lee, S.-W. Identification of volatile metabolites in human saliva from patients with oral squamous cell carcinoma via zeolite-based thin-film microextraction coupled with GC–MS. *J. Chromatogr. B* **2019**, *1104*, 49–58. [CrossRef]

9. Boots, A.W.; van Berkel, J.J.B.N.; Dallinga, J.W.; Smolinska, A.; Wouters, E.F.; van Schooten, F.J. The versatile use of exhaled volatile organic compounds in human health and disease. *J. Breath Res.* **2012**, *6*, 27108. [CrossRef] [PubMed]
10. Garner, C.E.; Smith, S.; de Lacy Costello, B.; White, P.; Spencer, R.; Probert, C.S.J.; Ratcliffem, N.M. Volatile organic compounds from feces and their potential for diagnosis of gastrointestinal disease. *FASEB J.* **2007**, *21*, 1675–1688. [CrossRef]
11. Jia, C.; Yu, X.; Masiak, W. Blood/air distribution of volatile organic compounds (VOCs) in a nationally representative sample. *Sci. Total Environ.* **2012**, *419*, 225–232. [CrossRef]
12. Shirasu, M.; Touhara, K. The scent of disease: Volatile organic compounds of the human body related to disease and disorder. *J. Biochem.* **2011**, *150*, 257–266. [CrossRef] [PubMed]
13. Weber, C.M.; Cauchi, M.; Patel, M.; Bessant, C.; Turner, C.; Britton, L.E.; Willis, C.M. Evaluation of a gas sensor array and pattern recognition for the identification of bladder cancer from urine headspace. *Analyst* **2011**, *136*, 359–364. [CrossRef]
14. Heers, H.; Gut, J.M.; Hegele, A.; Hofmann, R.; Boeselt, T.; Hattesohl, A.; Koczulla, A.R. Non-invasive Detection of Bladder Tumors Through Volatile Organic Compounds: A Pilot Study with an Electronic Nose. *Anticancer Res.* **2018**, *38*, 833–837.
15. Zhu, S.; Corsetti, S.; Wang, Q.; Li, C.; Huang, Z.; Nabi, G. Optical sensory arrays for the detection of urinary bladder cancer-related volatile organic compounds. *J. Biophotonics* **2019**, *12*, e201800165. [CrossRef] [PubMed]
16. Ghasemi-Varnamkhasti, M.; Mohtasebi, S.S.; Rodriguez-Mendez, M.L.; Lozano, J.; Razavi, S.H.; Ahmadi, H. Potential application of electronic nose technology in brewery. *Trends Food Sci. Technol.* **2011**, *22*, 165–174. [CrossRef]
17. Wasilewski, T.; Migoń, D.; Gębicki, J.; Kamysz, W. Critical review of electronic nose and tongue instruments prospects in pharmaceutical analysis. *Anal. Chim. Acta* **2019**, *1077*, 14–29. [CrossRef]
18. Dragonieri, S.; Schot, R.; Mertens, B.J.A.; Le Cessie, S.; Gauw, S.A.; Spanevello, A.; Resta, O.; Willard, N.P.; Vink, T.J.; Rabe, K.F.; et al. An electronic nose in the discrimination of patients with asthma and controls. *J. Allergy Clin. Immunol.* **2007**, *120*, 856–862. [CrossRef]
19. Van De Goor, R.; Leunis, N.; Van Hooren, M.R.A.; Francisca, E.; Masclee, A.; Kremer, B.; Kross, K.W. Feasibility of electronic nose technology for discriminating between head and neck, bladder, and colon carcinomas. *Eur. Arch. Oto-Rhino-Laryngol.* **2017**, *274*, 1053–1060. [CrossRef]
20. Bernabei, M.; Pennazza, G.; Santonico, M.; Corsi, C.; Roscioni, C.; Paolesse, R.; Di Natale, C.; D'Amico, A. A preliminary study on the possibility to diagnose urinary tract cancers by an electronic nose. *Sens. Actuators B Chem.* **2008**, *131*, 1–4. [CrossRef]
21. Zhu, S.; Huang, Z.; Nabi, G. Fluorometric optical sensor arrays for the detection of urinary bladder cancer specific volatile organic compounds in the urine of patients with frank hematuria: A prospective case-control study. *Biomed. Opt. Express* **2020**, *11*, 1175–1185. [CrossRef] [PubMed]
22. Esfahani, S.; Wicaksono, A.; Mozdiak, E.; Arasaradnam, R.P.; Covington, J.A. Non-Invasive Diagnosis of Diabetes by Volatile Organic Compounds in Urine Using FAIMS and Fox4000 Electronic Nose. *Biosensors* **2018**, *8*, 121. [CrossRef]
23. Fancy, S.-A.; Rumpel, K. *GC-MS-Based Metabolomics BT-Biomarker Methods in Drug Discovery and Development*; Wang, F., Ed.; Humana Press: Totowa, NJ, USA, 2008; pp. 317–340. ISBN 978-1-59745-463-6.
24. Daulton, E.; Wicaksono, A.; Bechar, J.; Covington, J.A.; Hardwicke, J. The Detection of Wound Infection by Ion Mobility Chemical Analysis. *Biosensors* **2020**, *10*, 19. [CrossRef] [PubMed]
25. Wilson, A.D.; Baietto, M. Applications and Advances in Electronic-Nose Technologies. *Sensors* **2009**, *9*, 5099–5148. [CrossRef] [PubMed]

chemistry proceedings

MDPI

Abstract

Voltammetric Detection of Mercury Ions at Poly(azulene-EDTA)-like Screen Printed Modified Electrodes †

George-Octavian Buica [1,*], Georgiana-Luiza Tatu (Arnold) [1], Eleonora-Mihaela Ungureanu [1] and Gabriela Geanina Vasile [2]

1 Faculty of Applied Chemistry and Materials Science, University Politehnica of Bucharest, 1-7 Gheorghe, 060042 Bucharest, Romania; arnoldluiza1988@gmail.com (G.-L.T.); em_ungureanu2000@yahoo.com (E.-M.U.)
2 National Research and Development Institute for Industrial Ecology ECOIND Bucharest, 57-73 Drumul Podul Dambovitei Street, 060652 Bucharest, Romania; gabriela.vasile@incdecoind.ro
* Correspondence: buica_george@yahoo.com
† Presented at the 1st International Electronic Conference on Chemical Sensors and Analytical Chemistry, 1–15 July 2021; Available online: https://csac2021.sciforum.net/.

Abstract: In recent years, many applications have been developed for the detection of different toxic metals (As, Cd, Cu, Hg, Ni, Pb) in water samples. The classic analytical methods (ICP-MS, AAS with graphite furnace, ICP-EOS with ultrasonic nebulizer) not only require a longer analysis time (pretreatment of the sample and analysis), but also the costs involved are higher as a result of expensive equipment, costs associated with the method validation process and qualified staff. The use of modified electrodes for trace metals analysi from wastewater samples represents a modern approach which can provide accurate, fast results with selectivity and sensitivity. Thus, here we present the development of the previously obtained glassy carbon-modified electrodes based on poly(2,2′-(ethane-1,2-diylbis(2-(azulen-2-ylamino)-2-oxoethyl)azanediyl))diacetic acid, (polyL) in laboratory-scale studies. In order to analyze Hg(II) ion content from aqueous samples, an assembly system made of carbon screen-printed modified electrodes (SPEs) modified with polyL selective complexing polymeric films coupled with a portable potentiostat was used. The detection of Hg(II) ions was accomplished by chemical accumulation in an open circuit followed by anodic stripping using the differential pulse voltammetry technique. The calibration curve of the analytical method was situated in the range of 20 ppb to 150 ppb ($y = 0.0051x + 0.123$, $R^2 = 0.9951$), with a detection limit of 6 ppb. The precision value for the lower limit of the calibration curve was 20%, while for the upper limit, the value was 10.5%. The novelty of the method consists not only of the low cost of the analysis, but also of the possibility to provide real-time reliable information about the Hg(II) concentration in wastewater using a small and portable device.

Keywords: complexing polymer; modified electrode; voltammetric detection; mercury analysis

Citation: Buica, G.-O.; Tatu, G.-L.; Ungureanu, E.-M.; Vasile, G.G. Voltammetric Detection of Mercury Ions at Poly(azulene-EDTA)-like Screen Printed Modified Electrodes. *Chem. Proc.* **2021**, *5*, 23. https://doi.org/10.3390/CSAC2021-10630

Academic Editor: Nicole Jaffrezic-Renault

Published: 7 July 2021

Supplementary Materials: The following are available online at https://www.mdpi.com/article/10.3390/CSAC2021-10630/s1.

Author Contributions: G.-O.B. and G.G.V. designed the study and methodology, supervised, and wrote the main manuscript text. E.-M.U. and G.-L.T. performed the electrochemical experiments. All authors have read and agreed to the published version of the manuscript.

Funding: This research was funded by Romanian National Authority for Scientific Research, UEFISCDI, under grant PN-III-P2-2.1-PED-2019-0730, contract no. 293PED/2020.

Institutional Review Board Statement: Not applicable.

Informed Consent Statement: Not applicable.

Data Availability Statement: Not applicable.

Acknowledgments: The authors gratefully acknowledge the financial support of the Romanian National Authority for Scientific Research, UEFISCDI, under grant PN-III-P2-2.1-PED-2019-0730, contract no. 293PED/2020.

Conflicts of Interest: The authors declare no conflict of interest.

chemistry proceedings

MDPI

Abstract

Electrochemical Immunosensor for Simultaneous Determination of Emerging Autoimmune Disease Biomarkers in Human Serum †

Esther Sánchez-Tirado *, Sara Guerrero, Araceli González-Cortés, Lourdes Agüí, Paloma Yáñez-Sedeño and José Manuel Pingarrón

Department of Analytical Chemistry, Faculty of Chemistry, University Complutense of Madrid, Avda. Complutense s/n, 28040 Madrid, Spain; sguerr02@ucm.es (S.G.); aracelig@ucm.es (A.G.-C.); mlagui@ucm.es (L.A.); yseo@ucm.es (P.Y.-S.); pingarro@ucm.es (J.M.P.)

* Correspondence: esther.sanchez@ucm.es

† Presented at the 1st International Electronic Conference on Chemical Sensors and Analytical Chemistry, 1–15 July 2021; Available online: https://csac2021.sciforum.net/.

Abstract: Rheumatoid arthritis is an autoimmune disorder characterized by persistent erosive synovitis, systemic inflammation and the presence of autoantibodies, which play an important role in inducing inflammation and joint damage, releasing pro-inflammatory cytokines from monocytes and macrophages [1,2]. Likewise, neutrophil activating protein-2 (CXCL7) is a platelet-derived growth factor belonging to the CXC chemokine subfamily, which is expressed in serum, synovial fluid and synovial tissue of patients developing rheumatoid arthritis during the first twelve weeks, being useful to reflect local pathological changes [3]. Besides, matrix metalloproteinase-3 (MMP-3), which is induced by inflammatory cytokines such as interleukin-1 (IL-1) and tumor necrosis factor alpha (TNF-α) in rheumatoid synovium, degrades several extracellular matrix components of cartilage and plays central roles in rheumatoid joint destruction [4]. Therefore, monitoring serum CXCL7 and MMP-3 levels is useful for predicting the disease activity in rheumatoid arthritis. In this work, the construction and analytical performance of a dual electrochemical platform for the simultaneous determination of CXCL7 and MMP-3 is described. After the optimization of experimental variables involved in the preparation and implementation of the biosensor, the analytical usefulness of the developed configuration was demonstrated by its application to the determination of these biomarkers in serum samples from healthy individuals and patients with rheumatoid arthritis. To carry out the simultaneous determination of CXCL7 and MMP3 in human serum, just a fifty-fold sample dilution in PBS of pH 7.4 was required. In addition, the results obtained using the dual immunosensor were compared with those provided by the respective ELISA immunoassays, yielding no significant differences between the two methods. It is important to highlight that reagents consumption, four times smaller using the dual immunosensor than that required in the ELISA protocol, and an assay time of 2 h 50 min versus almost 5 h, counted in both cases after incubation of the capture antibody, are advantageous features of the dual immunosensor [5].

Keywords: rheumatoid arthritis; CXCL7; MMP-3; immunosensor; simultaneous determination; human serum samples

Citation: Sánchez-Tirado, E.; Guerrero, S.; González-Cortés, A.; Agüí, L.; Yáñez-Sedeño, P.; Pingarrón, J.M. Electrochemical Immunosensor for Simultaneous Determination of Emerging Autoimmune Disease Biomarkers in Human Serum. *Chem. Proc.* **2021**, *5*, 24. https://doi.org/10.3390/CSAC2021-10437

Academic Editor: Nicole Jaffrezic-Renault

Published: 30 June 2021

Supplementary Materials: The following are available online at https://www.mdpi.com/article/10.3390/CSAC2021-10437/s1.

Author Contributions: Conceptualization, P.Y.-S.; methodology, E.S.-T., S.G., A.G.-C. and L.A.; software, E.S.-T. and S.G.; validation, E.S.-T., S.G., A.G.-C. and L.A.; formal analysis, E.S.-T., S.G., A.G.-C. and L.A.; investigation, E.S.-T., S.G., A.G.-C. and L.A.; resources, A.G.-C., L.A. and P.Y.-S.; data curation, E.S.-T. and S.G.; validation, E.S.-T., S.G., A.G.-C. L.A., P.Y.-S.; writing—original draft preparation, E.S.-T., S.G., A.G.-C., L.A. and P.Y.-S.; writing—review and editing, P.Y.-S. and J.M.P.; visualization, E.S.-T., S.G., A.G.-C. and L.A.; resources, A.G.-C., L.A. and P.Y.-S.; supervision, A.G.-C., L.A., P.Y.-S.; funding acquisition, A.G.-C., P.Y.-S. All authors have read and agreed to the published version of the manuscript.

Funding: This research was funded by Spanish Ministerio de Ciencia, Innovación y Universidades, grant number RTI2018-096135-B-I00 and TRANSNANOAVANSENS-CM Program from the Comunidad de Madrid, grant number S2018/NMT-4349.

Institutional Review Board Statement: Not applicable.

Informed Consent Statement: Not applicable.

Data Availability Statement: Guerrero, S.; Sánchez-Tirado, E.; Agüí, L.; González-Cortés, A.; Yáñez-Sedeño, P.; Pingarrón, J.M. Simultaneous determina-tion of CXCL7 chemokine and MMP3 metalloproteinase as biomarkers for rheumatoid arthritis. *Talanta* **2021**, *234*, 122705.

Conflicts of Interest: The authors declare no conflict of interest.

References

1. Bos, W.H.; Wolbink, G.J.; Boers, M.; Tijhuis, G.J.; de Vries, N.; van der Horst-Bruinsma, I.E.; Tak, P.P.; van de Stadt, R.J.; van der Laken, C.J.; Dijkmans, B.A.C.; et al. Arthritis development in patients with arthralgia is strongly associated with anti-citrullinated protein antibody status: A prospective cohort study. *Ann. Rheum. Dis.* **2010**, *69*, 490–494. [CrossRef]
2. Scott, D.L.; Wolfe, F.; Huizinga, T.W.J. Rheumatoid arthritis. *Lancet* **2010**, *376*, 1094–1108. [CrossRef]
3. Guerrero, S.; Cadano, D.; Agüí, L.; Barderas, R.; Campuzano, S.; Yáñez-Sedeño, P.; Pingarrón, J.M. Click chemistry-assisted antibodies immobilization for immunosensing of CXCL7 chemokine in serum. *J. Electroanal. Chem.* **2019**, *837*, 246–253. [CrossRef]
4. Sugiyama, E. Role of matrix metalloproteinase-3 in joint destruction in rheumatoid arthritis. *Clin. Calcium* **2007**, *17*, 528–534. [PubMed]
5. Guerrero, S.; Sánchez-Tirado, E.; Agüí, L.; González-Cortés, A.; Yáñez-Sedeño, P.; Pingarrón, J.M. Simultaneous determination of CXCL7 chemokine and MMP3 metalloproteinase as biomarkers for rheumatoid arthritis. *Talanta* **2021**, *234*, 122705. [CrossRef] [PubMed]

chemistry proceedings

MDPI

Proceeding Paper

Advances in Visual Immunoassays for Sensitive Detection of Mycotoxins in Food—A Review †

Meijuan Liang [1,2,3,4,5], Qi Zhang [1,2,3,4,5,*] and Peiwu Li [1,2,3,4,5,*]

1 Oil Crops Research Institute, Chinese Academy of Agricultural Sciences, Wuhan 430062, China; liangmj@whu.edu.cn
2 Key Laboratory of Biology and Genetic Improvement of Oil Crops, Ministry of Agriculture, Wuhan 430062, China
3 Laboratory of Quality & Safety Risk Assessment for Oilseed Products (Wuhan), Ministry of Agriculture, Wuhan 430062, China
4 Key Laboratory of Detection for Mycotoxins, Ministry of Agriculture, Wuhan 430062, China
5 Quality Inspection & Test Center for Oilseed Products, Ministry of Agriculture, Wuhan 430062, China
* Correspondence: zhangqi01@caas.cn (Q.Z.); peiwuli@oilcrops.cn (P.L.)

† Presented at the 1st International Electronic Conference on Chemical Sensors and Analytical Chemistry (CSAC2021), 1–15 July 2021; Available online: https://sciforum.net/conference/CSAC2021.

Abstract: Mycotoxins are the toxic secondary metabolites naturally produced by fungi; their contamination in agricultural products and food severely threatens food safety and public health worldwide. The reliable, efficient, and sensitive quantification of mycotoxins in food has become increasingly challenging to tackle due to the complexity of food matrices and their low level. Visual detection has emerged as a popular trend toward miniaturization and simplification of mycotoxins assays yet is constrained with their limited sensitivity. This review mainly focuses on the various sensitive visual immunoassays for signal amplified detection of mycotoxins. These signal amplified immunoassays for the improved sensitivity of mycotoxins detection in food through nanomaterials for encapsulation enzyme, enzyme-mediated nanomaterials as the amplified signal readout, and nanozyme. Furthermore, the underlying principle and the advantages of visual immunoassays for mycotoxins have been proposed. And the challenges and perspectives have been proposed to develop improved efficient visual immunoassays for mycotoxins in food.

Keywords: mycotoxins; nanomaterials; catalysis; immunoassay; visualization

check for updates

Citation: Liang, M.; Zhang, Q.; Li, P. Advances in Visual Immunoassays for Sensitive Detection of Mycotoxins in Food—A Review. *Chem. Proc.* **2021**, 5, 25. https://doi.org/10.3390/CSAC2021-10443

Academic Editor: Huangxian Ju

Published: 30 June 2021

1. Introduction

Mycotoxins are toxic secondary metabolites secreted by fungi under suitable temperature and humidity pre- and/or post-harvest [1–3]. Mycotoxins can affect the quality and safety of agriculture products, the associated processed foodstuffs, feedstuff, and animals. Over 400 mycotoxins have recently been identified, the worldwide occurrence of mycotoxins involving aflatoxin (AF), ochratoxin (OT), zearalenone (ZEN), deoxynivalenol (DON), fumonisin (FB), and T-2 toxin [4,5]. It is well known that aflatoxin is the representative mycotoxins, including AFB_1, AFB_2, AFG_1, and AFG_2, which has been confirmed to be immunosuppressive, teratogenic, and mutagenic [6,7]. Meanwhile, AFB_1 could be metabolized into the toxic hydroxyl metabolite of AFM_1, which is widespread presence of milk and dairy products.

Additionally, ZEN with a strong estrogenic effect and OTA with neurotoxicity and hepatotoxicity could adversely affect animals and humans. To protect humans from exposure mycotoxins, strict standards of limiting mycotoxin levels in food and the associated products have been regulated in many countries worldwide [8]. The monitoring of mycotoxins has been recognized as a significant way to safeguard food safety. However, mycotoxins detection in food matrices is challenging due to their low levels and complex food matrices.

Accordingly, it is highly desirable to conduct the effective, reliable and sensitive analytical strategy for screening mycotoxins in food matrices.

Nowadays, many efforts have been made to detect mycotoxins in food, involving instrumental analysis [9–13] and immunoassays [14–16]. The instrumental analysis requires expensive, sophisticated instruments, a time-consuming sample preparation process, and well-trained staff, which is not suitable for rapid screening numerous samples, and precludes their wide application in resource-constrained regions [17]. Immunoassays have been extensively identified as promising specific recognition for quantifying mycotoxins thanks to their sensitivity, on-site, as well as high-throughput screening capability. The specific recognition interaction between antibody and antigen has generally favored highly selective and reliable monitoring of mycotoxins. Various signal transduction techniques have currently been utilized to conduct mycotoxins immunoassays, such as fluorescence [18–20], electrochemistry [21–24], chemiluminescence [25], and colorimetry [26–28]. Attractively, visual detection, a popular trend toward miniaturization and simplification analysis, is capable of directly observing the results by the naked eye without other sophisticated instruments [29–31].

Currently, various immunoassays involving enzyme-linked immunosorbent assay (ELISA) [32,33], lateral flow immunoassay (LFI) [34–37] have been demonstrated as an excellent platform for discrimination of mycotoxins. Among them, ELISA and LFI served as the representative visual immunoassay, have attracted continuous interest due to their advantages of simple, and on-sites for rapid screening mycotoxins. Yet, the sensitivity of these conventional visual detection methods require improvement to monitor trace amounts of mycotoxins in complex food matrices. Thus, numerous studies have currently been devoted to the construction of the visualized immunoassays for enhancing the sensitivity of mycotoxins detection via signal amplification.

Recently, the robust enzyme catalytic amplification has been confirmed to enhance the sensitivity of immunoassays. Particularly, elaborate enzymatic strategies for improving the limited enzyme amount and the catalytic activity have been engineered as efficient and sensitive immunoassays for high-performance sensing targeted analytes. The emerging nanomaterials with unique optical, electrical, magnetic, and catalytic properties provide new opportunities for improving enzymatic immunoassays [38–42]. More evidence has revealed that the integration of novel nanomaterials promoted sensitivity improvements on mycotoxins detection [43–45]. For instance, Au nanoparticles (AuNPs) functionalized with antibodies, effectively discriminating the immune complex and enzyme to catalytic reaction substrate, significantly elevated their analytical performance [46–48]. Accordingly, the combination of nanomaterials and enzymatic immunoassays provides a potent signal amplified platform for highly sensitive and specific rapidly screening of mycotoxins. Herein, we summarize the improvements on visual immunoassays of mycotoxins by integrating nanomaterials and enzymatic signal amplification. The improvements in sensitivity of mycotoxin in food were emphasized with the assistance of nanomaterials for encapsulation enzyme, enzyme-mediated nanomaterials as the amplified signal readout, and nanomaterials for enzyme-mimics. Challenges and outlook of mycotoxin detection have been proposed to develop improved and efficient visual immunoassays in food.

2. The Signal Amplified Strategies

Natural enzymes, as potent biocatalysts have been widely used in countless laboratories, medical and food safety fields thanks to their high catalytic activities, substrate specificity, good biocompatibility, and wide range of biocatalysis [49,50]. ELISA is a classical enzyme-based visual immunoassay, which mainly includes the sorbent substrate, immunorecognition and enzyme labels. The antigen or antibody serves as sorbent substrate to immobilize onto the supporting material, enzyme-labeled molecule then immobilized to sorbent [51]. The sensing principle of ELISA mainly relies on the specific immune reaction between antibodies and antigens. Generally, after precoating the antibody or antigen on the sorbent substrate through physical absorption, the antigen or antibody were captured via specific immuno-recognition, and further immobilized on the substrate. The enzyme-labeled antibody would bind to the antigens to form a bioconjugation. Significantly, the enzyme catalyzes the colorless chromogenic substrate to generate colorimetric output, and the resultant colorimetric signal is recorded by UV-vis spectrophotometer or microplate reader to quantify the analyte concentration [52,53]. The sensitivity of ELISA could be effectively enhanced by improving the absorbent substrate, the recognition element, enzyme label, or chromogenic reagent. Among them, natural enzymes represent robust signal amplification, which has been extensively utilized to develop the highly sensitive immunoassays for trace level mycotoxins because of the catalytically amplified signal.

The peroxidase activity of horseradish peroxidase (HRP) has been used in the traditional ELISA, where HRP served as signal amplification for catalysis H_2O_2 into hydroxyl radical (•OH) that can react with the colorless chromogenic substrate 3,3′,5,5′-tetramethylbenzidine (TMB), 2,2′-azino-bis-(3-ethylbenzthiazoline-6-sulfonic acid) (ABTS) or o-phenylenediamine (OPD) into blue TMB_{ox}, green $ABTS^{+\bullet}$, or yellow OPD_{ox} under acidic condition. The colorimetric signal intensity is associated with the anchored HRP-labeled antigen or antibody for catalysis chromogenic substrates [54]. Accordingly, the analytes can be quantified through a direct method or an enzyme-labeled secondary antibody. In the previous studies, HRP-labeled antibodies were the most commonly used in the traditional ELISA to realize the various mycotoxin detection in foods [55–59]. The aforementioned ELISA adopted enzyme-labeled secondary antibodies through chemical conjugation to generate a signal. Yet, the chemical conjugation of the enzyme might result in the loss of enzyme activity, low stability for reagents labeling, and decreased sensitivity and specificity of the ELISA [60]. More evidence was revealed that the fusion protein had been recognized as an immunological agent for mycotoxins detection since its good antigen binding and enzyme activity. A nanobody-alkaline phosphatase (ALP) fusion protein has been revealed to improve the sensitivity for FB_1 and OTA detection in argo-products [61–63].

Note that the enzyme-labeled antigen or antibody revealed the limited enzyme molecules. For instance, HRP-labeled conjugate always presented the limited HRP molecules with approximately 2–3 HRP per antibody [64], which remarkably weakened the enzymatic signal amplification and the sensitivity of immunoassays. Besides the limited enzyme molecules, the low economy of the conjugated enzyme might lead to an increase the production cost of the immunoassays [65,66]. Meanwhile, enzyme-label is susceptible to decreasing or even losing catalytic activity upon practical detection [67]. Thus, the efficient strategies of augmenting enzyme amounts contribute to amplifying the sensitivity of visual immunoassay. Various enzymatic signal amplification immunoassays using nanomaterials as a robust scaffold for enzyme immobilization, enzyme-mediated nanomaterials for amplified signal readout, and nanozyme as an alternative for natural enzyme have recently been used to improve the enzyme loading and catalytic activity.

2.1. Immobilized Natural Enzymes on Nanomaterials for Amplification

Increasing the enzyme amounts in the final antigen–antibody–enzyme complex facilitates the catalysis of the substrate and signal amplification in a single recognition reaction (Figure 1A). Attractively, nanomaterials can execute as excellent carriers for loading and immobilizing enzymes by virtue of their large surface area-to-volume ratio, high loading capacity, facile fabrication, ease of functionalization, and high chemical stability. The multienzymes and antibodies immobilized on the surface of a single nanomaterial to effectively amplify the detectable signal, and thus enhance the sensitivity [68]. The emerging nanomaterials of metal/metal oxides nanoparticles, silica nanoparticles [69], carbon nanomaterials, and metal-organic frameworks have been demonstrated as excellent carriers for immobilizing natural enzymes for sensitive analysis. For instance, Zhu et al. utilized botryoid-shaped Au/Ag nanoparticles (BSNPs) loading HRP–IgG to construct indirect competitive ELISA for amplified ochratoxin A (OTA) detection in four wheat samples. After precoating the OTA-OVA antigen, the analyte of OTA was introduced as a competing component, followed by the addition of an anti-OTA antibody. Thus HRP-IgG-BSNPs complex was used as the enzyme-labeled secondary antibody for catalysis colorless TMB into blue oxidized TMB with the assistance of H_2O_2. And the colorimetric intensity was recorded by a microplate reader to examine the OTA level. The high loading amount of HRP–IgG onto the BSNPs contributed to improved sensitivity of OTA with the IC_{50} of 0.05 ng/mL, which revealed a 30-fold improvement compared to the conventional ELISA [70].

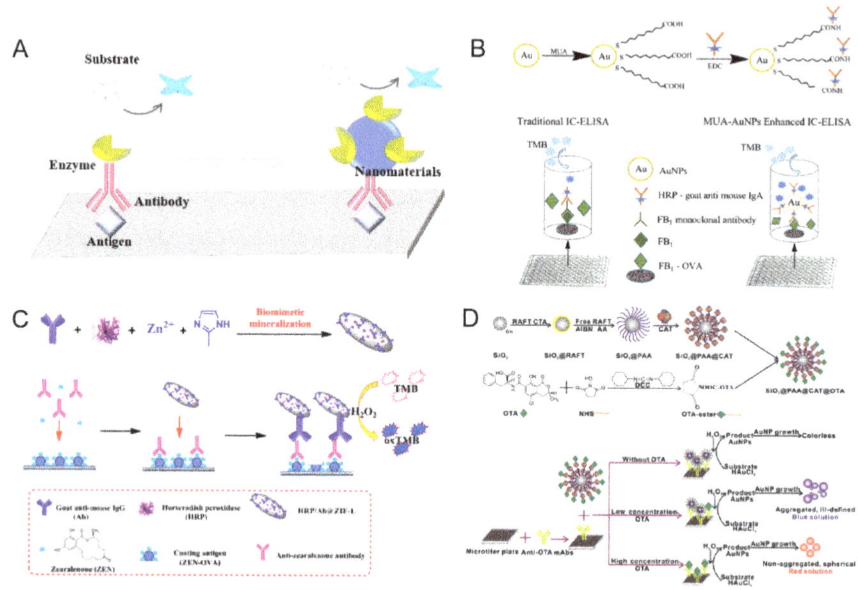

Figure 1. (**A**) The improved immunoassays using nanomaterials for immobilization natural enzymes. (**B**) AuNPs-HRP-goat anti-mouse IgA enhanced ELISA for FB_1. Reprinted from ref. [71]. Copyright 2018 Royal Society of Chemistry. (**C**) Zeolitic imidazolate framework-encapsulated HRP-based ELISA for ZEN. Reprinted from ref. [72]. Copyright 2021 Elsevier. (**D**) SiO_2 NPs carrying poly (acrylic acid)@CAT-based ELISA for OTA. Reprinted from ref. [73]. Copyright 2016 American Chemical Society.

Similarly, Li et al. [71] developed an indirect competitive ELISA for the total FB_1, FB_2, and FB_3 detection in maize samples based on AuNPs immobilized HRP-goat anti-mouse IgA. The enhanced sensitivity was approximately ten times compared to the conventional ELISA (Figure 1B). Liu et al. [72] developed metal-organic frameworks (MOFs)-loaded HRP and goat anti-mouse IgG for ZEN detection in argo-products. The LOD of this immunoassay achieved 0.5 ng/L for ZEN detection, which showed an approximately 126-fold enhancement relative to conventional HRP-based immunoassay (Figure 1C). Besides single nanomaterials, polymer-coated nanomaterials as enzyme containers have demonstrated to be the amplified strategies of conventional nanomaterials for further elevating the enzyme loading capacity of nanomaterials. Xiong's group presented that SiO_2 NPs carrying poly (acrylic acid) (PAA) brushes as a "CAT container" were used to amplify the sensitivity of OTA in various argo-products [73]. In this case, the SiO_2@PAA@CAT could generate a signal amplification for plasmonic ELISA by using catalase (CAT)-catalyzed the changed plasmonic signal readout of AuNPs. The LOD by naked eye and microplate reader was 10^{-18} and 5×10^{-20} g/mL, which was seven and eight orders of magnitude lower than that of CAT-based ELISA (10^{-11} g/mL by the naked eye) and HRP-based conventional ELISA (10^{-11} g/mL by the microplate reader) (Figure 1D).

2.2. Natural Enzyme-Mediated Nanomaterials for Amplified Signal Readout

In addition to the typical chromogenic substrate, natural enzyme-catalyzed products enable to regulate the color change of nanomaterials, especially for plasmonic property of AuNPs, achieving the visual detection of mycotoxins (Figure 2A). For instance, Xiong's group [74] developed a direct competitive ELISA through CAT-mediated AuNPs aggregation using HRP + H_2O_2 + tyramine system. In this case, phenol polymerization of tyramine by •OH from HRP-catalyzed H_2O_2 triggered AuNPs aggregation. The competitive antigen of OTA-labeled CAT was employed to catalyze H_2O_2 into H_2O and O_2. AuNPs presented monodisperse (red) without OTA, while the AuNPs aggregation (blue) was observed with OTA, and the extinction spectra of AuNPs were used as the signal recorder. The combined advantages of ultrahigh CAT catalytic activity and color change of AuNPs contributed to sensitively detecting OTA in corn samples. The IC_{50} and LOD (IC_{10}) of OTA were 84.75 and 17.8 pg/mL, which revealed a 2.9- and 2.7-fold enhancement compared with the conventional ELISA (Figure 2B).

Meanwhile, this group also utilized the glucose oxidase (GO_x) -catalyzed glucose into H_2O_2, which reduces Au^{3+} into Au^0 on the surface of Au seeds with an obvious color change for a direct competitive ELISA for FB_1 detection in maize samples. The IC_{50} was 1.86 ng/mL, approximately 13-fold lower than that of HRP-based conventional ELISA [75]. Apart from AuNPs, enzyme-assisted etching of Au nanorods (NRs) triggered visual detection of mycotoxins. HRP-assisted AuNRs-etching direct competitive ELISA was developed to sensitively detect AFB_1 in corn samples. The competitive antigen of AFB_1-labeled GO_x could catalyze glucose molecules into H_2O_2, and HRP simultaneously catalyze H_2O_2 to form •OH. The rod-like morphology AuNRs was chemically etched to spherical morphology by •OH, leading to visual signal output. The etching process of AuNRs efficiently occurred without AFB_1, yet the blocking of AuNRs etching was clearly presented in the presence of AFB_1. The decreased optical density and the apparent color change from bluish-green to red were collected by a microplate reader or the naked eye for qualitative AFB_1 detection. The method allowed sensitive determination of AFB_1 with IC_{50} of 22.3 pg/mL, which enhanced 32 times compared to the traditional ELISA [76].

Although these approaches achieved superior sensitivity, most of them rely on traditional single-signal readout mode. And these strategies might encounter the limitation of inaccuracy for mycotoxins evaluation, which was partly ascribed to external interferences, such as nonstandard test processes, different operators, or diverse surrounding environments [77–79]. Recent development in mycotoxins immunoassays enable the integration of visual and various signal transduction techniques into dual-signal strategies, and thus offering multi models for mycotoxins detection because of their self-calibration. Typically, by using the changed multiple color and LSPR shifts of Au nanobipyramids etched by •OH generated from HRP-catalyzed H_2O_2, and the changed photocurrent of CdS etched by the oxidized HRP. Wei et al. [80] developed an improved colorimetric and photoelectrometric immunoassay for ochratoxins (Figure 2C). The nanoliposomes as the vehicle for carrying more secondary antibodies and encapsulating HRP significantly amplified the detection signal, realizing the simultaneous detection of three ochratoxins (OTA, OTB, and OTC). The dual-modality immunoassay showed high sensitivity with LOD of 0.7 and 1.7 ng/L for photoelectrometric and colorimetric readouts, respectively. Attractively, the dual-modality response immunoassays showed a more accurate and reliable outcome compared with the single modality.

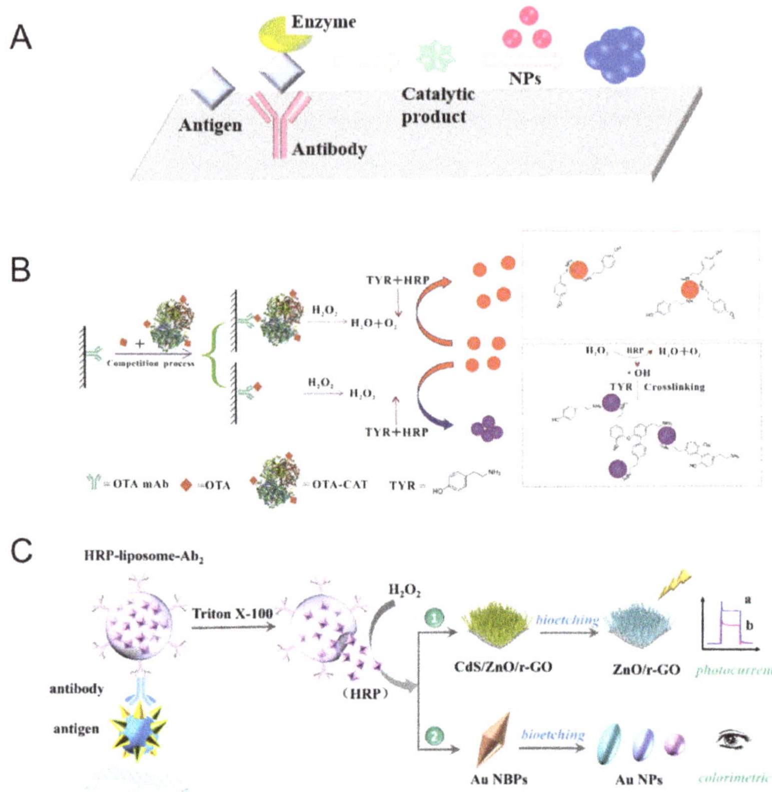

Figure 2. (**A**) The enzymes-catalyzed products-mediated nanomaterials for signal readout. (**B**) CAT-mediated AuNPs aggregation-based ELISA for OTA. Reprinted from ref [74]. Copyright 2018 Elsevier.(**C**) HRP-mediated Au nanobipyramids etching process-based immunoassay for ochratoxins. Reprinted from ref. [80]. Copyright 2019 American Chemical Society.

2.3. Nanozyme for Signal Amplification

Although natural enzymes are extensively used in various fields, their catalytic activities were still susceptible to the extreme environment, e.g., heat, pH, organic solvents, mechanical stress, heavy metal, etc. Meanwhile, they present many shortcomings, such as high expense, low recyclability, poor operational stability and limited practical applications, e.g., the preparation, reaction, and storage requirements [81–83]. Nanomaterials-based artificial enzymes (nanozyme) have been particularly attractive since the discovery of Fe_3O_4 NPs with peroxidase-like activity by Yan's group in 2007 [84]. Nanozymes are ideal candidates for alternative natural enzymes due to their high catalytic activity, tunable catalytic activity and types, multienzyme mimetic activity, high stability, low cost, durability and ease of functionalization [62]. Nowadays, various nanozymes have been served as catalytic labels for multi-category signal amplification in newly developed immunoassays. Numerous studies revealed that metal NPs (Au, Ag, Pt, Pd) [85,86], metal oxide NPs (Fe_3O_4, CeO_2, MnO_2, CuO) [87–92], carbon-based (graphene oxide, carbon nitride, carbon dots) [93–96], and MOFs-based nanomaterials [97–99] with peroxidase-, catalase-, oxidase-, superoxide dismutase-mimicking properties.

These nanozymes have been designed to amplify the sensing of mycotoxins (Figure 3A). For example, Xu et al. [100] developed an indirect competitive MOFs -linked immunosorbent assay for the high throughput and sensitive detection of AFB_1 in grain drinks.

Peroxidase-like activity of MOFs (MIL-88) was conjugated to a secondary antibody to substitute natural HRP-labeled secondary antibody. The MOFs-based immunoassay allowed to sensitively detect AFB_1 with the LOD of 0.009 ng/L with 20 times improvement compared to the conventional ELISA. The enhanced sensitivity might arise from their good dispersity, more active sites, and pores of MOFs-labeled antibodies promoted the catalytic reaction between MOFs-labeled antibody nanozyme and substrate. Significantly, the immunoassay could successfully decrease the occurrence of false positives and false negatives during the detection of AFB_1 (Figure 3B).

Furthermore, Zhu et al. [101] developed a competitive ELISA that was constructed to sensitively monitor OTA in millet samples through octahedral Cu_2O nanoparticles etching of Au nanobipyramids. Peroxidase-mimicking activity of Cu_2O could oxidize TMB in the presence of H_2O_2, and the yellow product TMB^{2+} could etch the Au nanobipyramids, triggering a significant longitudinal peak blue shift of local surface plasmon resonance. In this case, a dopamine-coated microplate was used to capture OTA antigens, and followed by the immunoreaction between OTA antibodies and the Cu_2O-labled secondary antibody. The growing concentration of OTA resulted in a decrease of Cu_2O-labled secondary antibody amount, further imposing adverse effects on the generation of catalytic product TMB^{2+} and the etching process of AuNRs (Figure 3C). The method allowed to sensitively detect OTA with LOD of 0.47 ng/L.

Apart from the single nanozyme for signal amplification, multienzyme-based cascade catalysis is another important signal transduction and amplification strategy. In the catalytic cascade system, the decreased diffusion path of intermediates between the enzymes enables the improvement of unstable intermediates, facilitating their efficiency and specificity [102–104]. Meanwhile, the single substrate can be converted into more signal molecule through the multienzyme-associated continuous catalysis reaction and contributes to the signal amplification [68,83,105]. Lai et al. [105] proposed a competitive cascade amplified immunoassay for AFB_1 detection in peanut samples by a combination of ascorbate oxidase (AO_x)/anti-AFB_1 antibody-labeled AuNPs and oxidase-mimics MnO_2 (Figure 3D). With the assistance of ascorbic acid (AA), a blue MnO_2-TMB system was converted into a colorless system because of the dissolution of MnO_2 into Mn^{2+}. Once introduced AO_x, the color change could be suppressed since AO_x catalysis AA to dehydroascorbic acid. The cascade signal amplification remarkably improved the sensitivity of AFB_1 with LOD of 6.5 pg/mL, which approximately enhanced 15-, 7-, and 38-fold compared to the existing commercialized AFB_1 kits (e.g., QuickingBiotech:100 ppt; Max Signals: 50 pg/mL; MyBioSource: 250 pg/mL). Similarly, Lai further developed a competitive immunoassay for sensitive screening AFB_1 in a peanut sample (LOD: 0.1 ng/mL), based on the just-in-time generation of an oxidase, mimics MnO_2 through the reaction $KMnO_4$ and Mn^{2+} with the assistance of AO_x [106].

Similar to ELISA, LFI is another important visual immunoassay for nanomaterials-labeled one-step immunochromatographic paper-based point-of-care tests. LFI is widely used in food safety owing to its low cost, speed, and ease of use [107–109]. The components of LFI mainly include a sample pad, a nitrocellulose (NC) membrane containing the test and control zones, conjugate and absorbent pads from cellulose, and a polyvinyl chloride backing card for assembling the components [110]. Once the sample solution is dropped onto the sample pad, it can migrate along the strips driven by capillary forces. Then, the sample dissolves the detection reagent in the conjugation pad, followed by flows along the strip within the porous membrane, where the analyte and the signal reporter were captured on the test line, thereby leading to the generation of a detectable signal. The sensing principle of LFI for analytes mainly includes the competitive and sandwich types. Generally, the competitive LFI is utilized to analyze mycotoxins due to their low-molecular weight. For the competitive LFI, the analyte competes with the same molecule, or the analyte blocks the capture agent attached on reporter tags in conjugation [111,112]. The resultant detectable signal intensity of the test line decreased upon the growing concentration of mycotoxins.

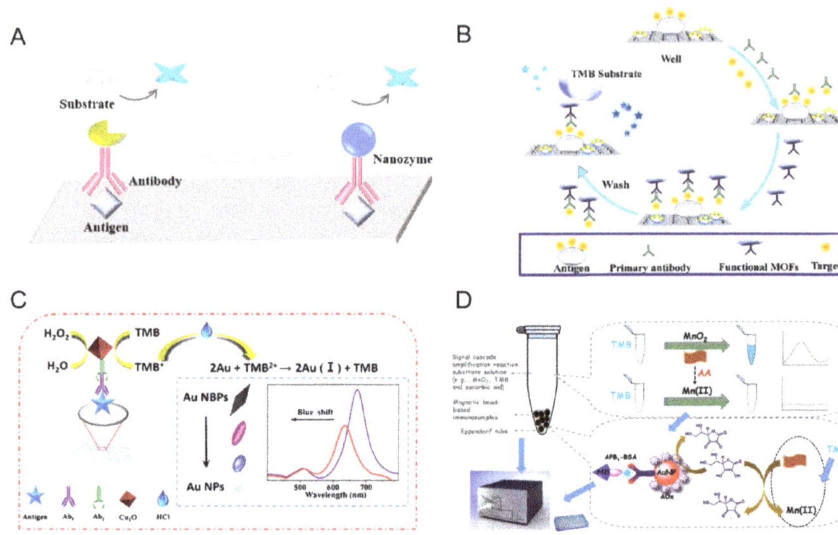

Figure 3. (**A**) Nanozyme-based immunoassays. (**B**) MOFs-linked immunosorbent assay for AFB_1 detection. Reprinted from ref. [100]. Copyright 2021 Elsevier.(**C**) Peroxidase-like activity of Cu_2O-based immunoassay for OTA detection. Reprinted from ref. [101]. Copyright 2021 Springer Nature. (**D**) MnO_2-AO_x cascade amplified immunoassay for AFB_1 detection. Reprinted from ref. [105]. Copyright 2017, Elsevier.

For colorimetric LFI, AuNPs are the common signal labeled material for visual output through non-covalent electrostatic adsorption of antibodies or antigens [113]. Au-nanomaterials-based LFI have been extensively developed for analysis multiplex mycotoxins including FB_1 [114], AFB_1 [115], OTA [116], ZEN [117] etc. In addition, natural enzymes also provide signals through conjugating to mycotoxin-protein and are executed as the signal transducer to achieve visual detection, such as HRP-labeled antibodies or /antigen for immunological recognition construction LFI [118,119]. Nowadays, numerous nanozymes have been used to label antibodies or antigens for rapid visual LFI. The evidence of Fe_3O_4 nanozyme for enhanced detection Ebola virus with 100 times enhancement compared to the conventional AuNPs-based LFI, revealing the signal amplification ability of nanozyme [120]. Various fascinating nanozyme, such as AuPt nanoflowers [121], Pt nanocatalyst [122], Pt-$Ni(OH)_2$ nanosheets [123], Prussian blue NPs (PBNPs) [124], have been used to construct LFI, and realized their widely application in food safety. For example, Tian et al. developed PBNPs as a marker signal LFI platform for OTA in soybeans samples. The new signal of PBNPs can be amplified via the TMB cascaded signal. The colorimetric signal of PBNPs accumulated on the test line through specific immune interactions, triggering the formation of a visible blue line. Meanwhile, the colorimetric signal could be further amplified via the peroxidase-mimic property of PBNPs. The resultant colorimetric images and grey intensity for OTA concentration were collected and analyzed by a smartphone and software Image J, respectively. This proposed LFI significantly improved the sensitivity of OTA with 2–3 orders of magnitude relative to commercial AuNPs-based LFI [125]. Although nanozyme have been extensively applied in food analysis, their poor substrate specificity, unclear mechanism, lack of standards and reference materials, and potential toxicity remained the major challenges for their further application.

3. Conclusions and Outlook

Mycotoxin contamination is a continuous global concern for food safety. Visual immunoassays remain simple, rapid, on-site detection of mycotoxins contamination as an alternative to traditional sophisticated techniques. The combination between conventional visual immunoassays and nanomaterials, novel visual immunoassays tend to be popular for mycotoxins by using the signal amplified strategies for tackling their inherent limited sensitivity. The representative immunoassays based on various nanomaterials could achieve the enhanced sensitive detection of mycotoxins using signal amplified strategies. Enzyme-immobilized onto nanomaterials, enzyme-mediated nanomaterials for amplified signal readout, nanozyme for amplifying the sensitivity of mycotoxins detection.

Although the aforementioned sensitive visual immunoassays for mycotoxins have revealed outstanding analytical performance and a fascinating prospect, many challenges still need to be tackled.

(1) The visual signal is obtained by the naked eye. Yet, the reliance on manual observation rather than instrumental measurement might cause large subjective uncertainty, as well as difficulty in quantitative data. The integration of digital technology [126] (e.g., machine vision) to simulate human visual ability and objective perception, the accurate and reliable results could be easily quantified, and thus reducing subjective errors in manual observations.

(2) Compared to the traditional immunoassays, the limited reproducibility and stability of nanomaterials-based immunoassays is the important obstacle for further application in food analysis due to their experimental and systemic factors. The standardization of nanomaterials preparation could effectively guarantee the reproducibility and stability of nanomaterials-based immunoassays.

(3) Most visual immunoassays are developed for single mycotoxin, while mycotoxins always co-occurred with the others in actual food samples. Thus, the simultaneous monitoring multi-mycotoxins by combing the multi-recognition elements in immunoassays facilitate to shorten the required time, save costs and alleviate the required labor.

(4) The integration of the visual analysis technology and multi-analysis technologies (e.g., magnetic, optical, and thermal properties, etc.), multi-signal immunoassays of mycotoxins contribute to minimum background signal and false-positive errors.

(5) Further exploiting the smart, automatic, miniaturized detector with the integration of smartphone, a portable and high-resolution device for the highly sensitive screening of mycotoxin contamination.

Author Contributions: M.L.: conceptualization, writing the original draft, and writing–review & editing. Q.Z.: conceptualization, project administration, and funding acquisition. P.L.: funding acquisition, project administration, and supervision. All authors have read and agreed to the published version of the manuscript.

Funding: This work was supported by the National Key R&D Program of China (2018YFC1602500), the Key Project of National Science Foundation of China (32030085), Agricultural Science and Technology Innovation Program of CAAS (CAAS-ZDRW202011), Natural Science Foundation of Hubei Province (2021CFB181).

Conflicts of Interest: The authors declare no conflict of interest.

References

1. Alshannaq, A.; Yu, J.-H. Occurrence, toxicity, and analysis of major mycotoxins in food. *Int. J. Environ. Res. Public Health* **2017**, *14*, 632. [CrossRef] [PubMed]
2. Sweeney, M.J.; Dobson, A.D.W. Mycotoxin production by *Aspergillus, Fusarium* and *Penicillium* species. *Int. J. Food Microbiol.* **1998**, *43*, 141–158. [CrossRef]
3. Chen, Y.; Chen, Q.; Han, M.; Zhou, J.; Gong, L.; Niu, Y.; Zhang, Y.; He, L.; Zhang, L. Development and optimization of a multiplex lateral flow immunoassay for the simultaneous determination of three mycotoxins in corn, rice and peanut. *Food Chem.* **2016**, *213*, 478–484. [CrossRef] [PubMed]
4. Luo, Y.; Liu, X.; Li, J. Updating techniques on controlling mycotoxins—A review. *Food Control* **2018**, *89*, 123–132. [CrossRef]
5. Yang, Y.; Li, G.; Wu, D.; Liu, J.; Li, X.; Luo, P.; Hu, N.; Wang, H.; Wu, Y. Recent advances on toxicity and determination methods of mycotoxins in foodstuffs. *Trends Food Sci. Technol.* **2020**, *96*, 233–252. [CrossRef]
6. Rossi, C.N.; Takabayashi, C.R.; Ono, M.A.; Saito, G.H.; Itano, E.N.; Kawamura, O.; Hirooka, E.Y.; Ono, E.Y.S. Immunoassay based on monoclonal antibody for aflatoxin detection in poultry feed. *Food Chem.* **2012**, *132*, 2211–2216. [CrossRef]

7. Zhou, S.; Xu, L.; Kuang, H.; Xiao, J.; Xu, C. Immunoassays for rapid mycotoxin detection: State of the art. *Analyst* **2020**, *145*, 7088–7102. [CrossRef]
8. Singh, J.; Mehta, A. Rapid and sensitive detection of mycotoxins by advanced and emerging analytical methods: A review. *Food Sci. Nutr.* **2020**, *8*, 2183–2204. [CrossRef] [PubMed]
9. Yu, L.; Ma, F.; Zhang, L.; Li, P. Determination of aflatoxin B-1 and B-2 in vegetable oils using Fe_3O_4/rGO magnetic solid phase extraction coupled with high-performance liquid chromatography fluorescence with post-column photochemical derivatization. *Toxins* **2019**, *11*, 621. [CrossRef]
10. Andrade, P.D.; Dantas, R.R.; Moura-Alves, T.L.d.S.d.; Caldas, E.D. Determination of multi-mycotoxins in cereals and of total fumonisins in maize products using isotope labeled internal standard and liquid chromatography/tandem mass spectrometry with positive ionization. *J. Chromatogr. A* **2017**, *1490*, 138–147. [CrossRef]
11. Zhang, Y.; Pei, F.; Fang, Y.; Li, P.; Zhao, Y.; Shen, F.; Zou, Y.; Hu, Q. Comparison of concentration and health risks of 9 Fusarium mycotoxins in commercial whole wheat flour and refined wheat flour by multi-IAC-HPLC. *Food Chem.* **2019**, *275*, 763–769. [CrossRef] [PubMed]
12. Medina, D.A.V.; Borsatto, J.V.B.; Maciel, E.V.S.; Lancas, F.M. Current role of modern chromatography and mass spectrometry in the analysis of mycotoxins in food. *TrAC Trends Anal. Chem.* **2021**, *135*, 116156. [CrossRef]
13. Alsharif, A.M.A.; Choo, Y.-M.; Tan, G.-H. Detection of five mycotoxins in different food matrices in the Malaysian market by using validated liquid chromatography electrospray ionization triple quadrupole mass spectrometry. *Toxins* **2019**, *11*, 196. [CrossRef]
14. Yan, T.T.; Zhang, Z.W.; Zhang, Q.; Tang, X.Q.; Wang, D.; Hu, X.F.; Zhang, W.; Chen, X.M.; Li, P.W. Simultaneous determination for A. flavus-metabolizing mycotoxins by time-resolved fluorescent microbead or gold-enabling test strip in agricultural products based on monoclonal antibodies. *Microchim. Acta* **2020**, *187*, 653. [CrossRef] [PubMed]
15. Yang, H.L.; Zhang, Q.; Liu, X.L.; Yang, Y.Y.; Yang, Y.; Liu, M.Y.; Li, P.W.; Zhou, Y. Antibody-biotin-streptavidin-horseradish peroxidase (HRP) sensor for rapid and ultra-sensitive detection of fumonisins. *Food Chem.* **2020**, *316*, 126356. [CrossRef]
16. Beloglazova, N.V.; Graniczkowska, K.; Ediage, E.N.; Averkieva, O.; De Saeger, S. Sensitive flow-through immunoassay for rapid multiplex determination of cereal-borne mycotoxins in feed and feed ingredients. *J. Agric. Food Chem.* **2017**, *65*, 7131–7137. [CrossRef]
17. He, Q.-H.; Xu, Y.; Wang, D.; Kang, M.; Huang, Z.-B.; Li, Y.-P. Simultaneous multiresidue determination of mycotoxins in cereal samples by polyvinylidene fluoride membrane based dot immunoassay. *Food Chem.* **2012**, *134*, 507–512. [CrossRef]
18. Li, M.; Wang, H.M.; Sun, J.D.; Ji, J.; Ye, Y.L.; Lu, X.; Zhang, Y.Z.; Sun, X.L. Rapid, on-site, and sensitive detection of aflatoxin M1 in milk products by using time-resolved fluorescence microsphere test strip. *Food Control* **2021**, *121*, 107616. [CrossRef]
19. Xu, Y.; Ma, B.; Chen, E.J.; Yu, X.P.; Ye, Z.H.; Sun, C.X.; Zhang, M.Z. Dual fluorescent immunochromatographic assay for simultaneous quantitative detection of citrinin and zearalenone in corn samples. *Food Chem.* **2021**, *336*, 127713. [CrossRef]
20. Li, H.; Wang, D.; Tang, X.; Zhang, W.; Zhang, Q.; Li, P. Time-resolved fluorescence immunochromatography assay (TRFICA) for aflatoxin: Aiming at increasing strip method sensitivity. *Front. Microbiol.* **2020**, *11*, 676. [CrossRef]
21. Kaminiaris, M.D.; Mavrikou, S.; Georgiadou, M.; Paivana, G.; Tsitsigiannis, D.I.; Kintzios, S. An impedance based electrochemical immunosensor for aflatoxin B-1 monitoring in pistachio matrices. *Chemosensors* **2020**, *8*, 121. [CrossRef]
22. Kudr, J.; Zhao, L.; Nguyen, E.P.; Arola, H.; Nevanen, T.K.; Adam, V.; Zitka, O.; Merkoci, A. Inkjet-printed electrochemically reduced graphene oxide microelectrode as a platform for HT-2 mycotoxin immunoenzymatic biosensing. *Biosens. Bioelectron.* **2020**, *156*, 112109. [CrossRef] [PubMed]
23. Kunene, K.; Weber, M.; Sabela, M.; Voiry, D.; Kanchi, S.; Bisetty, K.; Bechelany, M. Highly-efficient electrochemical label-free immunosensor for the detection of ochratoxin A in coffee samples. *Sens. Actuators B Chem.* **2020**, *305*, 127438. [CrossRef]
24. Hou, S.; Ma, Z.; Meng, H.; Xu, Y.; He, Q. Ultrasensitive and green electrochemical immunosensor for mycotoxin ochratoxin A based on phage displayed mimotope peptide. *Talanta* **2019**, *194*, 919–924. [CrossRef]
25. Quan, Y.; Zhang, Y.; Wang, S.; Lee, N.; Kennedy, I.R. A rapid and sensitive chemiluminescence enzyme-linked immunosorbent assay for the determination of fumonisin B1 in food samples. *Anal. Chim. Acta* **2006**, *580*, 1–8. [CrossRef] [PubMed]
26. Ren, X.; Lu, P.; Feng, R.; Zhang, T.; Zhang, Y.; Wu, D.; Wei, Q. An ITO-based point-of-care colorimetric immunosensor for ochratoxin A detection. *Talanta* **2018**, *188*, 593–599. [CrossRef]
27. Bazin, I.; Nabais, E.; Lopez-Ferber, M. Rapid visual tests: Fast and reliable detection of ochratoxin A. *Toxins* **2010**, *2*, 2230–2241. [CrossRef]
28. Garden, S.R.; Strachan, N.J.C. Novel colorimetric immunoassay for the detection of aflatoxin B1. *Anal. Chim. Acta* **2001**, *444*, 187–191. [CrossRef]
29. Yu, Z.; Cai, G.; Liu, X.; Tang, D. Pressure-based biosensor integrated with a flexible pressure sensor and an electrochromic device for visual detection. *Anal. Chem.* **2021**, *93*, 2916–2925. [CrossRef]
30. Wang, L.; He, K.; Wang, X.; Wang, Q.; Quan, H.; Wang, P.; Xu, X. Recent progress in visual methods for aflatoxin detection. *Crit. Rev. Food Sci. Nutr.* **2021**. [CrossRef]
31. Majdinasab, M.; Ben Aissa, S.; Marty, J.L. Advances in colorimetric strategies for mycotoxins detection: Toward rapid industrial monitoring. *Toxins* **2021**, *13*, 13. [CrossRef] [PubMed]
32. Liu, B.-H.; Chu, K.C.; Yu, F.-Y. Novel monoclonal antibody-based sensitive enzyme-linked immunosorbent assay and rapid immunochromatographic strip for detecting aflatoxin M1 in milk. *Food Control* **2016**, *66*, 1–7. [CrossRef]

33. Wang, Y.; Li, P.; Zhang, Q.; Hu, X.; Zhang, W. A toxin-free enzyme-linked immunosorbent assay for the analysis of aflatoxins based on a VHH surrogate standard. *Anal. Bioanal. Chem.* **2016**, *408*, 6019–6026. [CrossRef] [PubMed]
34. Li, Y.; Jin, G.; Liu, L.; Xiao, J.; Kuang, H. Fast determination of citreoviridin residues in rice using a monoclonal antibody-based immunochromatographic strip assay. *Food Agric. Immunol.* **2020**, *31*, 893–906. [CrossRef]
35. Badea, M.; Micheli, L.; Messia, M.C.; Candigliota, T.; Marconi, E.; Mottram, T.; Velasco-Garcia, M.; Moscone, D.; Palleschi, G. Aflatoxin M1 determination in raw milk using a flow-injection immunoassay system. *Anal. Chim. Acta* **2004**, *520*, 141–148. [CrossRef]
36. Qu, J.W.; Xie, H.J.; Zhang, S.Y.; Luo, P.J.; Guo, P.; Chen, X.X.; Ke, Y.B.; Zhuang, J.Y.; Zhou, F.M.; Jiang, W.X. Multiplex flow cytometric immunoassays for high-throughput screening of multiple mycotoxin residues in milk. *Food Anal. Methods* **2019**, *12*, 877–886. [CrossRef]
37. Li, Y.; Zhang, N.; Wang, H.; Zhao, Q. An immunoassay for ochratoxin a using tetramethylrhodamine-labeled ochratoxin a as a probe based on a binding-induced change in fluorescence intensity. *Analyst* **2020**, *145*, 651–655. [CrossRef]
38. Tian, Y.; Bu, T.; Zhang, M.; Sun, X.; Jia, P.; Wang, Q.; Liu, Y.; Bai, F.; Zhao, S.; Wang, L. Metal-polydopamine framework based lateral flow assay for high sensitive detection of tetracycline in food samples. *Food Chem.* **2021**, *339*, 127854. [CrossRef]
39. Tan, X.; Wang, X.; Zhang, L.; Liu, L.; Zheng, G.; Li, H.; Zhou, F. Stable and photothermally efficient antibody-covered $Cu_3(PO_4)_2$@polydopamine nanocomposites for sensitive and cost-effective immunoassays. *Anal. Chem.* **2019**, *91*, 8274–8279. [CrossRef]
40. Cheng, N.; Song, Y.; Zeinhom, M.M.A.; Chang, Y.-C.; Sheng, L.; Li, H.; Du, D.; Li, L.; Zhu, M.-J.; Luo, Y.; et al. Nanozyme-mediated dual immunoassay integrated with smartphone for use in simultaneous detection of pathogens. *ACS Appl. Mater. Interfaces* **2017**, *9*, 40671–40680. [CrossRef]
41. Yao, X.; Wang, Z.; Zhao, M.; Liu, S.; Su, L.; Dou, L.; Li, T.; Wang, J.; Zhang, D. Graphite-like carbon nitride-laden gold nanoparticles as signal amplification label for highly sensitive lateral flow immunoassay of 17β-estradiol. *Food Chem.* **2021**, *347*, 129001. [CrossRef]
42. Oh, S.; Kim, J.; Tran, V.T.; Lee, D.K.; Ahmed, S.R.; Hong, J.C.; Lee, J.; Park, E.Y.; Lee, J. Magnetic nanozyme-linked immunosorbent assay for ultrasensitive influenza a virus detection. *ACS Appl. Mater. Interfaces* **2018**, *10*, 12534–12543. [CrossRef]
43. Zhang, X.; Song, M.; Yu, X.; Wang, Z.; Ke, Y.; Jiang, H.; Li, J.; Shen, J.; Wen, K. Development of a new broad-specific monoclonal antibody with uniform affinity for aflatoxins and magnetic beads-based enzymatic immunoassay. *Food Control* **2017**, *79*, 309–316. [CrossRef]
44. Lai, W.; Wei, Q.; Zhuang, J.; Lu, M.; Tang, D. Fenton reaction-based colorimetric immunoassay for sensitive detection of brevetoxin B. *Biosens. Bioelectron.* **2016**, *80*, 249–256. [CrossRef]
45. Yan, C.; Wang, Q.; Yang, Q.; Wu, W. Recent advances in aflatoxins detection based on nanomaterials. *Nanomaterials* **2020**, *10*, 1626. [CrossRef] [PubMed]
46. Ambrosi, A.; Airo, F.; Merkoci, A. Enhanced gold nanoparticle based ELISA for a breast cancer biomarker. *Anal. Chem.* **2010**, *82*, 1151–1156. [CrossRef]
47. Zhou, Y.; Tian, X.L.; Li, Y.S.; Pan, F.G.; Zhang, Y.Y.; Zhang, J.H.; Yang, L.; Wang, X.R.; Ren, H.L.; Lu, S.Y.; et al. An enhanced ELISA based on modified colloidal gold nanoparticles for the detection of Pb(II). *Biosens. Bioelectron.* **2011**, *26*, 3700–3704. [CrossRef] [PubMed]
48. Zha, Y.H.; Zhou, Y. Functional nanomaterials based immunological detection of aflatoxin B-1: A review. *World Mycotoxin J.* **2020**, *13*, 151–162. [CrossRef]
49. Liang, M.; Yan, X. Nanozymes: From new concepts, mechanisms, and standards to applications. *Acc. Chem. Res.* **2019**, *52*, 2190–2200. [CrossRef]
50. Huang, Y.; Ren, J.; Qu, X. Nanozymes: Classification, catalytic mechanisms, activity regulation, and applications. *Chem. Rev.* **2019**, *119*, 4357–4412. [CrossRef]
51. Wu, L.; Li, G.; Xu, X.; Zhu, L.; Huang, R.; Chen, X. Application of nano-ELISA in food analysis: Recent advances and challenges. *TrAC Trends Anal. Chem.* **2019**, *113*, 140–156. [CrossRef]
52. Gao, Y.; Zhou, Y.; Chandrawati, R. Metal and metal oxide nanoparticles to enhance the performance of enzyme-linked immunosorbent assay (ELISA). *ACS Appl. Nano Mater.* **2020**, *3*, 1–21. [CrossRef]
53. Heussner, A.H.; Ausländer, S.; Dietrich, D.R. Development and characterization of a monoclonal antibody against ochratoxin b and its application in ELISA. *Toxins* **2010**, *2*, 1582–1594. [CrossRef]
54. Huang, X.; Liu, Y.; Yung, B.; Xiong, Y.; Chen, X. Nanotechnology-enhanced no-wash biosensors for in vitro diagnostics of cancer. *ACS Nano* **2017**, *11*, 5238–5292. [CrossRef] [PubMed]
55. Li, P.; Zhang, Q.; Zhang, W.; Zhang, J.; Chen, X.; Jiang, J.; Xie, L.; Zhang, D. Development of a class-specific monoclonal antibody-based ELISA for aflatoxins in peanut. *Food Chem.* **2009**, *115*, 313–317. [CrossRef]
56. He, T.; Wang, Y.; Li, P.; Zhang, Q.; Lei, J.; Zhang, Z.; Ding, X.; Zhou, H.; Zhang, W. Nanobody-based enzyme immunoassay for aflatoxin in agro-products with high tolerance to cosolvent methanol. *Anal. Chem.* **2014**, *86*, 8873–8880. [CrossRef]
57. Wu, Y.; Yu, J.; Li, F.; Li, J.; Shen, Z. A Calibration curve implanted enzyme-linked immunosorbent assay for simultaneously quantitative determination of multiplex mycotoxins in cereal samples, Soybean and Peanut. *Toxins* **2020**, *12*, 718. [CrossRef] [PubMed]

58. Guan, D.; Li, P.; Zhang, Q.; Zhang, W.; Zhang, D.; Jiang, J. An ultra-sensitive monoclonal antibody-based competitive enzyme immunoassay for aflatoxin M-1 in milk and infant milk products. *Food Chem.* **2011**, *125*, 1359–1364. [CrossRef]
59. Jiang, W.; Wang, Z.; Noelke, G.; Zhang, J.; Niu, L.; Shen, J. Simultaneous determination of aflatoxin b-1 and aflatoxin m-1 in food matrices by enzyme-linked immunosorbent assay. *Food Anal. Methods* **2013**, *6*, 767–774. [CrossRef]
60. Guesdon, J.-L. Immunoenzymatic techniques applied to the specific detection of nucleic acids: A review. *J. Immunol. Methods* **1992**, *150*, 33–49. [CrossRef]
61. Sun, Z.; Wang, X.; Chen, Q.; Yun, Y.; Tang, Z.; Liu, X. Nanobody-alkaline phosphatase fusion protein-based enzyme-linked immunosorbent assay for one-step detection of ochratoxin A in rice. *Sensors* **2018**, *18*, 4044. [CrossRef]
62. Tang, Z.; Wang, X.; Lv, J.; Hu, X.; Liu, X. One-step detection of ochratoxin A in cereal by dot immunoassay using a nanobody-alkaline phosphatase fusion protein. *Food Control* **2018**, *92*, 430–436. [CrossRef]
63. Shu, M.; Xu, Y.; Liu, X.; Li, Y.; He, Q.; Tu, Z.; Fu, J.; Gee, S.J.; Hammock, B.D. Anti-idiotypic nanobody-alkaline phosphatase fusion proteins: Development of a one-step competitive enzyme immunoassay for fumonisin B1 detection in cereal. *Anal. Chim. Acta* **2016**, *924*, 53–59. [CrossRef]
64. Yan, J.; Wang, J.; Zhao, M.P.; Chang, W.B. Determination of papaverine by biotin-avidin amplified ELISA. *Anal. Lett.* **2004**, *37*, 2977–2989. [CrossRef]
65. Zhang, X.; Wu, D.; Zhou, X.; Yu, Y.; Liu, J.; Hu, N.; Wang, H.; Li, G.; Wu, Y. Recent progress in the construction of nanozyme-based biosensors and their applications to food safety assay. *TrAC Trends Anal. Chem.* **2019**, *121*, 115668. [CrossRef]
66. Wang, Q.Q.; Wei, H.; Zhang, Z.Q.; Wang, E.K.; Dong, S.J. Nanozyme: An emerging alternative to natural enzyme for biosensing and immunoassay. *TrAC Trends Anal. Chem.* **2018**, *105*, 218–224. [CrossRef]
67. Niu, X.; Cheng, N.; Ruan, X.; Du, D.; Lin, Y. Review-nanozyme-based immunosensors and immunoassays: Recent developments and future trends. *J. Electrochem. Soc.* **2020**, *167*, 037508. [CrossRef]
68. Xiong, Y.; Leng, Y.; Li, X.; Huang, X.; Xiong, Y. Emerging strategies to enhance the sensitivity of competitive ELISA for detection of chemical contaminants in food samples. *TrAC Trends Anal. Chem.* **2020**, *126*, 115861. [CrossRef]
69. Lei, C.; Xu, C.; Nouwens, A.; Yu, C. Ultrasensitive ELISA(+) enhanced by dendritic mesoporous silica nanoparticles. *J. Mater. Chem. B* **2016**, *4*, 4975–4979. [CrossRef] [PubMed]
70. Zhu, Y.; Liu, C.-L.; Xie, Z.-J.; Liu, L.-Q.; Peng, C.-F.; Xue, F. Botryoid-shaped nanoparticles-enhanced ELISA for ochratoxin A. *Food Agric. Immunol.* **2017**, *28*, 299–309. [CrossRef]
71. Li, Z.; Sheng, W.; Liu, Q.; Li, S.; Shi, Y.; Zhang, Y.; Wang, S. Development of a gold nanoparticle enhanced enzyme linked immunosorbent assay based on monoclonal antibodies for the detection of fumonisin B-1, B-2, and B-3 in maize. *Anal. Methods* **2018**, *10*, 3506–3513. [CrossRef]
72. Liu, Z.; Wang, X.; Dong, F.; Li, Y.; Guo, Y.; Liu, X.; Xu, J.; Wu, X.; Zheng, Y. Ultrasensitive immunoassay for detection of zearalenone in agro-products using enzyme and antibody co-embedded zeolitic imidazolate framework as labels. *J. Hazard. Mater.* **2021**, *412*, 125276. [CrossRef] [PubMed]
73. Huang, X.; Chen, R.; Xu, H.; Lai, W.; Xiong, Y. Nanospherical brush as catalase container for enhancing the detection sensitivity of competitive plasmonic ELISA. *Anal. Chem.* **2016**, *88*, 1951–1958. [CrossRef]
74. Liang, Y.; Huang, X.; Chen, X.; Zhang, W.; Ping, G.; Xiong, Y. Plasmonic ELISA for naked-eye detection of ochratoxin A based on the tyramine-H_2O_2 amplification system. *Sens. Actuators B Chem.* **2018**, *259*, 162–169. [CrossRef]
75. Zhan, S.; Zheng, L.; Zhou, Y.; Wu, K.; Duan, H.; Huang, X.; Xiong, Y. A gold growth-based plasmonic ELISA for the sensitive detection of Fumonisin B1 in maize. *Toxins* **2019**, *11*, 323. [CrossRef] [PubMed]
76. Xiong, Y.; Pei, K.; Wu, Y.; Duan, H.; Lai, W.; Xiong, Y. Plasmonic ELISA based on enzyme-assisted etching of Au nanorods for the highly sensitive detection of aflatoxin B-1 in corn samples. *Sens. Actuators B Chem.* **2018**, *267*, 320–327. [CrossRef]
77. Wang, M.; Zhou, X.; Wang, S.; Xie, X.; Wang, Y.; Su, X. Fabrication of bioresource-derived porous carbon-supported iron as an efficient oxidase mimic for dual-channel biosensing. *Anal. Chem.* **2021**, *93*, 3130–3137. [CrossRef] [PubMed]
78. Huang, Y.; Ge, J.; Chen, H.; Wang, Z.; Han, J.; Xie, G.; Chen, S. Dual-signal readout aptasensor for electrochemical and colorimetric assay using a bifunctional Ni-Fe PBA probe. *Sens. Actuators B Chem.* **2021**, *327*, 128871. [CrossRef]
79. Chang, J.; Lv, W.; Li, Q.; Li, H.; Li, F. One-step synthesis of methylene blue-encapsulated zeolitic imidazolate framework for dual-signal fluorescent and homogeneous electrochemical biosensing. *Anal. Chem.* **2020**, *92*, 8959–8964. [CrossRef]
80. Wei, J.; Chen, H.; Chen, H.; Cui, Y.; Qileng, A.; Qin, W.; Liu, W.; Liu, Y. Multifunctional peroxidase-encapsulated nanoliposomes: Bioetching-induced photoelectrometric and colorimetric immunoassay for broad-spectrum detection of Ochratoxins. *ACS Appl. Mater. Interfaces* **2019**, *11*, 23832–23839. [CrossRef]
81. Payal, A.; Krishnamoorthy, S.; Elumalai, A.; Moses, J.A.; Anandharamakrishnan, C. A Review on recent developments and applications of nanozymes in food safety and quality analysis. *Food Anal. Methods* **2021**, *14*, 1537–1558. [CrossRef]
82. Jia, M.; Liao, X.; Fang, L.; Jia, B.; Liu, M.; Li, D.; Zhou, L.; Kong, W. Recent advances on immunosensors for mycotoxins in foods and other commodities. *TrAC Trends Anal. Chem.* **2021**, *136*, 116193. [CrossRef]
83. Xu, W.Q.; Jiao, L.; Wu, Y.; Hu, L.Y.; Gu, W.L.; Zhu, C.Z. Metal-organic frameworks enhance biomimetic cascade catalysis for biosensing. *Adv. Mater.* **2021**, *33*, 2005172. [CrossRef]
84. Gao, L.Z.; Zhuang, J.; Nie, L.; Zhang, J.B.; Zhang, Y.; Gu, N.; Wang, T.H.; Feng, J.; Yang, D.L.; Perrett, S.; et al. Intrinsic peroxidase-like activity of ferromagnetic nanoparticles. *Nat. Nanotechnol.* **2007**, *2*, 577–583. [CrossRef]

85. Zhi, L.-J.; Sun, A.-L. Platinum nanozyme-encapsulated poly(amidoamine) dendrimer for voltammetric immunoassay of pro-gastrin-releasing peptide. *Anal. Chim. Acta* **2020**, *1134*, 106–114. [CrossRef] [PubMed]
86. Yao, S.; Li, J.; Pang, B.; Wang, X.; Shi, Y.; Song, X.; Xu, K.; Wang, J.; Zhao, C. Colorimetric immunoassay for rapid detection of Staphylococcus aureus based on etching-enhanced peroxidase-like catalytic activity of gold nanoparticles. *Microchim. Acta* **2020**, *187*, 504. [CrossRef] [PubMed]
87. Li, J.; Cao, Y.; Hinman, S.S.; McKeating, K.S.; Guan, Y.; Hu, X.; Cheng, Q.; Yang, Z. Efficient label-free chemiluminescent immunosensor based on dual functional cupric oxide nanorods as peroxidase mimics. *Biosens. Bioelectron.* **2018**, *100*, 304–311. [CrossRef] [PubMed]
88. Lian, J.; Liu, P.; Jin, C.; Shi, Z.; Luo, X.; Liu, Q. Perylene diimide-functionalized CeO_2 nanocomposite as a peroxidase mimic for colorimetric determination of hydrogen peroxide and glutathione. *Microchim. Acta* **2019**, *186*, 332. [CrossRef]
89. Wu, J.; Yang, Q.; Li, Q.; Li, H.; Li, F. Two-dimensional MnO_2 nanozyme-mediated homogeneous electrochemical detection of organophosphate pesticides without the interference of H_2O_2 and color. *Anal. Chem.* **2021**, *93*, 4084–4091. [CrossRef]
90. Ge, J.; Yu, J.-H.; Yang, H.; Yang, D.; Cai, R. Human serum albumin templated MnO_2 nanosheets as an efficient biomimetic oxidase for biomolecule sensing. *J. Mater. Chem. B* **2020**, *8*, 11090–11095. [CrossRef] [PubMed]
91. Huang, X.; Xia, F.; Nan, Z. Fabrication of FeS_2/SiO_2 double mesoporous hollow spheres as an artificial peroxidase and rapid determination of H_2O_2 and glutathione. *ACS Appl. Mater. Interfaces* **2020**, *12*, 46539–46548. [CrossRef]
92. Chen, W.; Chen, J.; Liu, A.-L.; Wang, L.-M.; Li, G.-W.; Lin, X.-H. Peroxidase-like activity of cupric oxide nanoparticle. *ChemCatChem* **2011**, *3*, 1151–1154. [CrossRef]
93. Song, Y.; Qu, K.; Zhao, C.; Ren, J.; Qu, X. Graphene oxide: Intrinsic peroxidase catalytic activity and its application to glucose detection. *Adv. Mater.* **2010**, *22*, 2206–2210. [CrossRef]
94. Tian, J.; Liu, Q.; Asiri, A.M.; Qusti, A.H.; Al-Youbi, A.O.; Sun, X. Ultrathin graphitic carbon nitride nanosheets: A novel peroxidase mimetic, Fe doping-mediated catalytic performance enhancement and application to rapid, highly sensitive optical detection of glucose. *Nanoscale* **2013**, *5*, 11604–11609. [CrossRef] [PubMed]
95. Zhang, J.; Lu, X.; Tang, D.; Wu, S.; Hou, X.; Liu, J.; Wu, P. Phosphorescent carbon dots for highly efficient oxygen photosensitization and as photo-oxidative nanozymes. *ACS Appl. Mater. Interfaces* **2018**, *10*, 40808–40814. [CrossRef] [PubMed]
96. Shi, W.; Wang, Q.; Long, Y.; Cheng, Z.; Chen, S.; Zheng, H.; Huang, Y. Carbon nanodots as peroxidase mimetics and their applications to glucose detection. *Chem. Commun.* **2011**, *47*, 6695–6697. [CrossRef]
97. Xia, H.; Li, N.; Huang, W.; Song, Y.; Jiang, Y. Enzymatic cascade reactions mediated by highly efficient biomimetic quasi metal-organic frameworks. *ACS Appl. Mater. Interfaces* **2021**, *13*, 22240–22253. [CrossRef]
98. Yuan, A.; Lu, Y.; Zhang, X.; Chen, Q.; Huang, Y. Two-dimensional iron MOF nanosheet as a highly efficient nanozyme for glucose biosensing. *J. Mater. Chem. B* **2020**, *8*, 9295–9303. [CrossRef]
99. Guo, J.; Wu, S.; Wang, Y.; Zhao, M. A label-free fluorescence biosensor based on a bifunctional MIL-101(Fe) nanozyme for sensitive detection of choline and acetylcholine at nanomolar level. *Sens. Actuators B Chem.* **2020**, *312*, 128021. [CrossRef]
100. Xu, Z.; Long, L.-l.; Chen, Y.-q.; Chen, M.-L.; Cheng, Y.-H. A nanozyme-linked immunosorbent assay based on metal-organic frameworks (MOFs) for sensitive detection of aflatoxin B-1. *Food Chem.* **2021**, *338*, 128039. [CrossRef] [PubMed]
101. Zhu, H.; Liu, C.; Liu, X.; Quan, Z.; Liu, W.; Liu, Y. A multi-colorimetric immunosensor for visual detection of ochratoxin A by mimetic enzyme etching of gold nanobipyramids. *Microchim. Acta* **2021**, *188*, 62. [CrossRef] [PubMed]
102. Vázquez-González, M.; Wang, C.; Willner, I. Biocatalytic cascades operating on macromolecular scaffolds and in confined environments. *Nat. Catal.* **2020**, *3*, 256–273. [CrossRef]
103. Kuzmak, A.; Carmali, S.; von Lieres, E.; Russell, A.J.; Kondrat, S. Can enzyme proximity accelerate cascade reactions? *Sci. Rep.* **2019**, *9*, 455. [CrossRef]
104. Gao, Z.; Hou, L.; Xu, M.; Tang, D. Enhanced colorimetric immunoassay accompanying with enzyme cascade amplification strategy for ultrasensitive detection of low-abundance protein. *Sci. Rep.* **2014**, *4*, 3966. [CrossRef]
105. Lai, W.; Wei, Q.; Xu, M.; Zhuang, J.; Tang, D. Enzyme-controlled dissolution of MnO_2 nanoflakes with enzyme cascade amplification for colorimetric immunoassay. *Biosens. Bioelectron.* **2017**, *89*, 645–651. [CrossRef] [PubMed]
106. Lai, W.; Zeng, Q.; Tang, J.; Zhang, M.; Tang, D. A conventional chemical reaction for use in an unconventional assay: A colorimetric immunoassay for aflatoxin B-1 by using enzyme-responsive just-in-time generation of a MnO_2 based nanocatalyst. *Microchim. Acta* **2018**, *185*, 92. [CrossRef] [PubMed]
107. Liu, Y.; Zhan, L.; Qin, Z.; Sackrison, J.; Bischof, J.C. Ultrasensitive and highly specific lateral flow assays for point-of-care diagnosis. *ACS Nano* **2021**, *15*, 3593–3611. [CrossRef]
108. Bishop, J.D.; Hsieh, H.V.; Gasperino, D.J.; Weigl, B.H. Sensitivity enhancement in lateral flow assays: A systems perspective. *Lab Chip* **2019**, *19*, 2486. [CrossRef]
109. Yu, S.; He, L.; Yu, F.; Liu, L.; Qu, C.; Qu, L.; Liu, J.; Wu, Y.; Wu, Y. A lateral flow assay for simultaneous detection of Deoxynivalenol, Fumonisin B-1 and Aflatoxin B-1. *Toxicon* **2018**, *156*, 23–27. [CrossRef]
110. Bahadır, E.B.; Sezgintürk, M.K. Lateral flow assays: Principles, designs and labels. *TrAC Trends Anal. Chem.* **2016**, *82*, 286–306. [CrossRef]
111. Mahmoudi, T.; de la Guardia, M.; Shirdel, B.; Mokhtarzadeh, A.; Baradaran, B. Recent advancements in structural improvements of lateral flow assays towards point-of-care testing. *TrAC Trends Anal. Chem.* **2019**, *116*, 13–30. [CrossRef]

112. Xing, K.-Y.; Shan, S.; Liu, D.-F.; Lai, W.-H. Recent advances of lateral flow immunoassay for mycotoxins detection. *TrAC Trends Anal. Chem.* **2020**, *133*, 116087. [CrossRef]
113. Wu, L.; Wang, M.; Wei, D. Advances in gold nanoparticles for mycotoxin analysis. *Analyst* **2021**, *146*, 1793–1806. [CrossRef]
114. Ren, W.; Huang, Z.; Xu, Y.; Li, Y.; Ji, Y.; Su, B. Urchin-like gold nanoparticle-based immunochromatographic strip test for rapid detection of fumonisin B-1 in grains. *Anal. Bioanal. Chem.* **2015**, *407*, 7341–7348. [CrossRef]
115. Ji, Y.; Ren, M.; Li, Y.; Huang, Z.; Shu, M.; Yang, H.; Xiong, Y.; Xu, Y. Detection of aflatoxin B-1 with immunochromatographic test strips: Enhanced signal sensitivity using gold nanoflowers. *Talanta* **2015**, *142*, 206–212. [CrossRef] [PubMed]
116. Anfossi, L.; Di Nardo, F.; Giovannoli, C.; Passini, C.; Baggiani, C. Increased sensitivity of lateral flow immunoassay for ochratoxin A through silver enhancement. *Anal. Bioanal. Chem.* **2013**, *405*, 9859–9867. [CrossRef]
117. Xu, S.; Zhang, G.; Fang, B.; Xiong, Q.; Duan, H.; Lai, W. Lateral flow immunoassay based on polydopamine-coated gold nanoparticles for the sensitive detection of zearalenone in maize. *ACS Appl. Mater. Interfaces* **2019**, *11*, 31283–31290. [CrossRef]
118. Mirasoli, M.; Buragina, A.; Dolci, L.S.; Simoni, P.; Anfossi, L.; Giraudi, G.; Roda, A. Chemiluminescence-based biosensor for fumonisins quantitative detection in maize samples. *Biosens. Bioelectron.* **2012**, *32*, 283–287. [CrossRef] [PubMed]
119. dos Santos, G.P.; Correa, C.C.; Kubota, L.T. A simple, sensitive and reduced cost paper-based device with low quantity of chemicals for the early diagnosis of Plasmodium falciparum malaria using an enzyme-based colorimetric assay. *Sens. Actuators B Chem.* **2018**, *255*, 2113–2120. [CrossRef]
120. Duan, D.; Fan, K.; Zhang, D.; Tan, S.; Liang, M.; Liu, Y.; Zhang, J.; Zhang, P.; Liu, W.; Qiu, X.; et al. Nanozyme-strip for rapid local diagnosis of Ebola. *Biosens. Bioelectron.* **2015**, *74*, 134–141. [CrossRef] [PubMed]
121. Zhang, J.; Yu, Q.C.; Qiu, W.W.; Li, K.; Qian, L.S.; Zhang, X.J.; Liu, G.D. Gold-platinum nanoflowers as a label and as an enzyme mimic for use in highly sensitive lateral flow immunoassays: Application to detection of rabbit IgG. *Microchim. Acta* **2019**, *186*, 357. [CrossRef]
122. Loynachan, C.N.; Thomas, M.R.; Gray, E.R.; Richards, D.A.; Kim, J.; Miller, B.S.; Brookes, J.C.; Agarwal, S.; Chudasama, V.; McKendry, R.A.; et al. Platinum nanocatalyst amplification: Redefining the gold standard for lateral flow immunoassays with ultrabroad dynamic range. *ACS Nano* **2018**, *12*, 279–288. [CrossRef]
123. Cheng, N.; Shi, Q.; Zhu, C.; Li, S.; Lin, Y.; Du, D. Pt-Ni$(OH)_2$ nanosheets amplified two-way lateral flow immunoassays with smartphone readout for quantification of pesticides. *Biosens. Bioelectron.* **2019**, *142*, 111498. [CrossRef]
124. Liu, S.; Dou, L.; Yao, X.; Zhang, W.; Zhao, M.; Yin, X.; Sun, J.; Zhang, D.; Wang, J. Nanozyme amplification mediated on-demand multiplex lateral flow immunoassay with dual-readout and broadened detection range. *Biosens. Bioelectron.* **2020**, *169*, 112610. [CrossRef] [PubMed]
125. Tian, M.; Xie, W.; Zhang, T.; Liu, Y.; Lu, Z.; Li, C.M.; Liu, Y. A sensitive lateral flow immunochromatographic strip with prussian blue nanoparticles mediated signal generation and cascade amplification. *Sens. Actuators B Chem.* **2020**, *309*, 127728. [CrossRef]
126. Liu, H.; Li, Z.; Shen, R.; Li, Z.; Yang, Y.; Yuan, Q. Point-of-care pathogen testing using photonic crystals and machine vision for diagnosis of urinary tract infections. *Nano Lett.* **2021**, *21*, 2854–2860. [CrossRef] [PubMed]

chemistry proceedings

MDPI

Proceeding Paper

Reproductivity Study of Metal Oxide Gas Sensors Using Two Different Temperature Setups †

Giulia Zambotti [1,2,*] and Andrea Ponzoni [1,2]

[1] National Research Council, National Institute of Optics (CNR-INO), Unit of Brescia, Via Branze 45, 25123 Brescia, Italy; andrea.ponzoni@ino.cnr.it
[2] Department of Information Engineering, University of Brescia, Via Branze 38, 25123 Brescia, Italy
* Correspondence: giulia.zambotti@ino.cnr.it
† Presented at the 1st International Electronic Conference on Chemical Sensors and Analytical Chemistry, 1–15 July 2021; Available online: https://csac2021.sciforum.net/.

Abstract: The use of the electronic nose as a screening device is of great interest in various types of applications, including food quality control and environmental monitoring. It is an easy-to-use device and produces a much faster response than that obtained by classical chemical and microbiological techniques. The reproductivity of nominally identical electronic noses and sensors is critical. Four identical MOX sensors were compared using two different working methods, namely, the temperature modulation mode and isothermal mode. Each sensor was tested with two standard compounds, water and lactic acid, often identified in food matrices, which are potential applications of the electronic nose.

Keywords: sensor reproductivity; modulation of temperature; isothermal mode; electronic nose; MOX sensors

Citation: Zambotti, G.; Ponzoni, A. Reproductivity Study of Metal Oxide Gas Sensors Using Two Different Temperature Setups. *Chem. Proc.* **2021**, 5, 26. https://doi.org/10.3390/CSAC2021-10613

Academic Editor: Nicole Jaffrezic-Renault

Published: 6 July 2021

1. Introduction

Sensor reproductivity is an important issue to ensure the reliability of the final instrument, such as electronic noses, in which individual sensors are implemented [1]. The core of the electronic nose is made of an MOX sensor array [2], which may work in isothermal and/or temperature modulation mode. These two different types of working modes can affect the sensitivity of the sensor in respect to the gases.

For temperature modulation, a periodic signal is applied to the heater to periodically change the sensor temperature in order to activate and inactivate the oxidation–reduction reactions between the sensitive material and gases [3,4]. As a consequence, the sensor resistance changes periodically with time, and parameters describing this curve can be extrapolated and used as input (features) to the pattern recognition algorithm. These parameters play the same roles as those attributed to the responses of individual sensors that work at a constant temperature, i.e., a fixed voltage applied to the heater [5].

Electronic noses have been used for different types of applications, for example, in the environmental field [6], medicine [7], security and safety [8], and food control [9,10].

For commercial applications (medical, food, environmental), three different aspects are important to develop an effective electronic nose: the system must guarantee good performance in order to achieve sensitivity and specificity; the database and the pattern recognition software should work on different nominally identical electronic noses with minimal adaptation work; and sensors should be interchangeable with nominally identical ones in case of failure. For these reasons, in this work, the reproductivity of nominally identical sensors was tested comparing the features extrapolated with sensors working in the two modes: isothermal and temperature modulation.

2. Material and Methods

In this work, we used a JLM MOX STICK (JLMInnovation Gmbh) device (Figure 1) to perform experiments. This instrument includes sensor control software and the electronic part. The experiments were conducted using 4 commercial sensors (TGS 2620—Figaro Sensor) exposed to vapors from a pure solution of water and lactic acid.

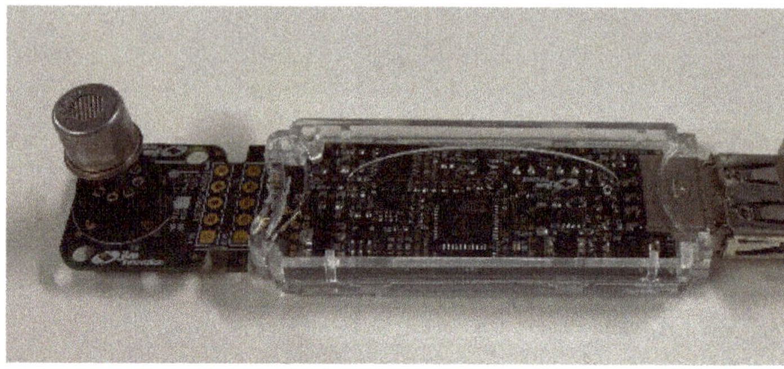

Figure 1. JLM MOX STICK device (JLMInnovation Gmbh) equipped with a commercial (Figaro TGS2620) sensor.

For the constant temperature mode, each sensor was tested with a constant voltage of 3.5 applied to the heater. For temperature modulation, the first part of the period lasted 10 s (4 volts) and the second part of the period lasted 10 s (3 volts). We chose these voltage and time values to have the same average temperature for both modes during a single thermal period.

Each sensor was turned on and exposed to ambient air (environmental temperature 21 °C) for one hour in order to stabilize it before performing the planned series of measurements.

Two vials were prepared using 10 mL of water and 10 mL of lactic acid. Both were sealed with parafilm and left in the room for one hour to create the headspace. A hole was then created in the parafilm, and the sensor was inserted. Measurement times were as follows: 10 min in contact with the compound vapors and 10 min in air to allow for baseline recovery. Each substance was replicated 3 times.

The following features were used for the elaboration on the temperature modulation dataset and to describe the resistance versus time curve. Using this method of measurement, it is possible to extrapolate several features to be used to analyze the data, unlike the isothermal method that allows the processing of a single parameter, R/R0. The value of R/R0 was calculated using the minimum resistance value reached by the gas in contact with the sensor (R), divided by the starting value of the sensor in air (R0). For the temperature modulation method, 2 significant features were extrapolated:

- Ratio-CH = ratio between the sensor resistance identified at the end of the cold half-period and at the beginning of the next hot half-period, i.e., Ratio-CH = R2/R3.
- DeltaR-C = change in sensor resistance measured during the cold half-period, i.e., DeltaR-C = R1 − R2.

All of this information is shown in Figure 2.

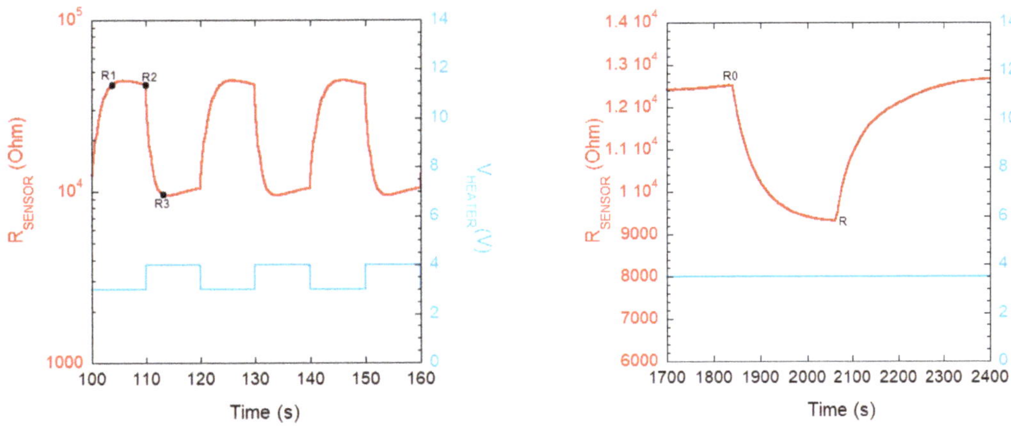

(a) (b)

Figure 2. (**a**) The light blue line represents the modulation of the heater voltage between 3 and 4 volts. The red curve is an example of sensor resistance measured during exposure to water vapors. DeltaR-C = R1 − R2; Ratio-CH = R2/R3. (**b**) The isothermal example shows the voltage applied to the heater with a light blue line, while the red curve shows an example of sensor resistance measured during exposure to water vapors (response calculated as R/R0).

3. Results and Discussion

Based on our previous experience of food quality control applications [9], we chose the most significant features (Ratio-CH and R/R0) to compare the results obtained from each individual sensor. The statistics of the recorded responses are shown in Figure 3.

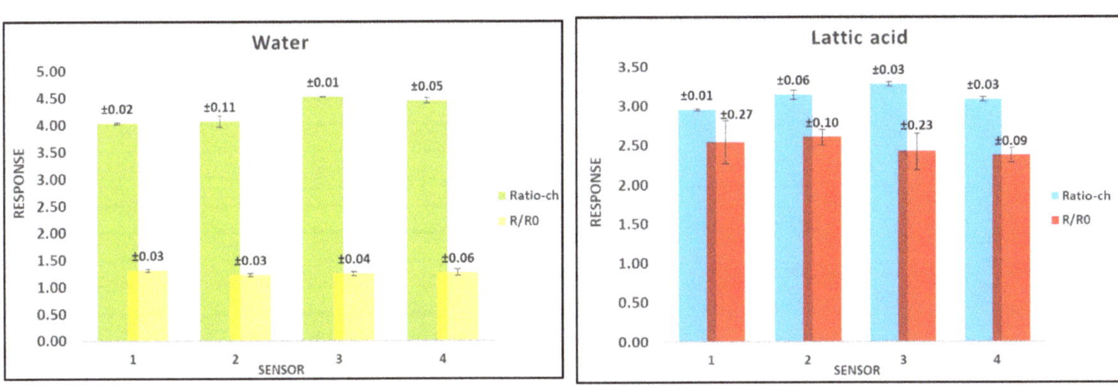

(a) (b)

Figure 3. (**a**) Comparison between Ratio-CH (green color) and R/R0 (yellow color) response to water vapors. (**b**) Response obtained to lactic acid vapors. Light blue rectangles relate to Ratio-CH, while red rectangles concern R/R0.

The value of each single response is the average of the single values obtained during the measurement session (replicated three times). The graphs also show the standard deviation corresponding to each result.

The first thing to note is that for both compounds analyzed and for both methods used, the sensors return a reproducible response. Evaluating, in detail, the response obtained by each method, we can see that the standard deviation (absolute value) for temperature modulation is always lower than that obtained using the isothermal method. This is further enhanced by normalizing the standard deviation to the average of the response intensity. Indeed, concerning water, the value changes from 0.5% to 2.6% for Ratio-CH, while for R/R0, it changes from 2.25% to 4.65%. It is possible to note the same situation in the measure of lactic acid (Ratio-CH 0.5–3.9%, R/R0 3.80–10.74%).

During the experience gained using the temperature modulation method, we realized that one of the features that introduce a large amount of information in the construction of the PCA (principal component analysis) plot is DeltaR-C, and therefore we decided to analyze it [7–9].

As can be seen in Figure 4, the reproductivity of the sensors towards the analyzed compounds is relatively poor in respect to what was observed concerning the Ratio-CH feature. There is an appreciable variance from sensor to sensor and within individual measurements. We find a standard deviation that ranges from 5.3% to 63% (water) and from 2.5% to 27.6% (lactic acid). However, in any case, it is able to return the most important information combined with other extracted features [7–9]. This means that the reproductivity of sensors depends on the given feature analyzed.

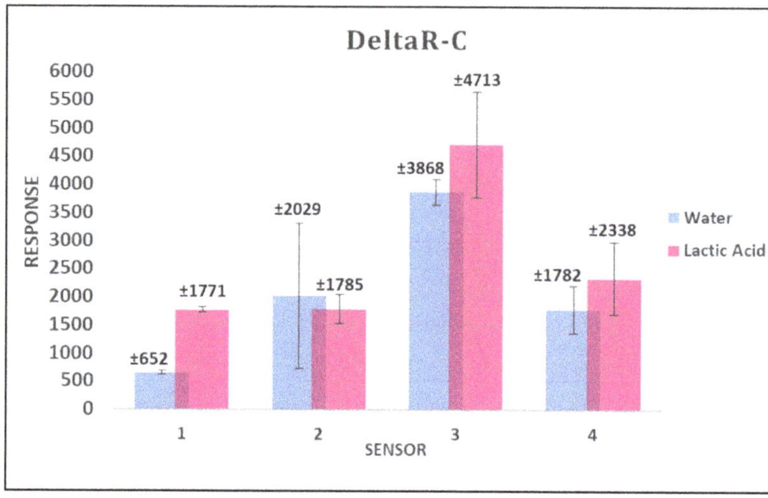

Figure 4. Sensor responses measured during exposure to water and lactic acid vapors.

4. Conclusions

In conclusion, two features were extrapolated and analyzed during experiments carried out using temperature modulation. Both these features were revealed to be useful during our old experiments dedicated to the detection of fish shelf life. The reproductivity of nominally identical sensors showed feature dependence, i.e., the feature Ratio-CH is better than the feature DeltaR-C. For a constant temperature, the normalized feature may be more or less repeatable than that extrapolated from the other mode depending on individual features, though the response extrapolated from this working mode benefits from the normalization to the reference air, while this does not occur for the temperature modulation mode.

Supplementary Materials: The following are available online at https://www.mdpi.com/article/10.3390/CSAC2021-10613/s1.

Author Contributions: Conceptualization, G.Z. and A.P.; methodology, G.Z. and A.P.; resources, G.Z. and A.P.; data curation, G.Z. and A.P.; writing—original draft preparation, G.Z. and A.P.; writing—review and editing, G.Z. and A.P. All authors have read and agreed to the published version of the manuscript.

Funding: This research was funded by Lombardia region and Fondazione Cariplo through the project EMPATIA@LECCO.

Institutional Review Board Statement: Not applicable.

Informed Consent Statement: Not applicable.

Data Availability Statement: The main data have been reported in figures (mean and std). Additional data are available from the corresponding author upon reasonable request.

Conflicts of Interest: The authors declare no conflict of interest.

References

1. Zonta, G.; Astolfi, M.; Casotti, D.; Cruciani, G.; Fabbri, B.; Gaiardo, A.; Gherardi, S.; Guidi, V.; Landini, N.; Valt, M.; et al. Reproducibility tests with zinc oxide thick-film sensors. *Ceram. Int.* **2020**, *46*, 6847–6855. [CrossRef]
2. Ponzoni, A.; Baratto, C.; Cattabiani, N.; Falasconi, M.; Galstyan, V.; Nunez-Carmona, E.; Rigoni, F.; Sberveglieri, V.; Zambotti, G.; Zappa, D. Metal Oxide Gas Sensors, a Survey of Selectivity Issues Addressed at the SENSOR Lab, Brescia (Italy). *Sensors* **2017**, *17*, 714. [CrossRef] [PubMed]
3. Krivetskiya, V.; Efitorov, A.; Arkhipenko, A.; Vladimirova, S.; Rumyantseva, M.; Dolenko, S.; Gaskov, A. Selective detection of individual gases and CO/H_2 mixture at low concentrations in air by single semiconductor metal oxide sensors working in dynamic temperature mode. *Sens. Actuators B Chem.* **2018**, *254*, 502–513. [CrossRef]
4. Hosseini-Golgoo, S.M.; Hossein-Babaei, F. Assessing the diagnostic information in the response patterns of a temperature-modulated tin oxide gas sensor. *Meas. Sci. Technol.* **2011**, *22*, 035201. [CrossRef]
5. Soprani, M.; Zambotti, G.; Ponzoni, A. Response Times of Metal-Oxide Chemiresistors: Comparison Between the Isothermal and Temperature Modulation Modes. *J. Nanosci. Nanotechnol.* **2021**, *21*, 2668–2674. [CrossRef] [PubMed]
6. Al Barakeh, Z.; Breuil, P.; Redon, N.; Pijolat, C.; Locoge, N.; Viricelle, J.-P. Development of a normalized multi-sensors system for low cost on-line atmospheric pollution detection. *Sens. Actuators B Chem.* **2017**, *241*, 1235–1243. [CrossRef]
7. Soprani, M.; Zambotti, G.; Gobbi, E.; Ponzoni, A. Application of a Micro-Machined Electronic Nose to Detect Escherichia Coli in Human Urine Samples. In Proceedings of the 8th GOSPEL Workshop, Gas Sensors Based on Semiconducting Metal Oxides: Basic Understanding & Application Fields, Ferrara, Italy, 20–21 June 2019. [CrossRef]
8. Vezzoli, M.; Ponzoni, A.; Pardo, M.; Falasconi, M.; Faglia, G.; Sberveglieri, G. Exploratory data analysis for industrial safety application. *Sens. Actuators B Chem.* **2008**, *131*, 100–109. [CrossRef]
9. Zambotti, G.; Soprani, M.; Gobbi, E.; Capuano, R.; Pasqualetti, V.; Di Natale, C.; Ponzoni, A. Early detection of fish degradation by electronic nose. In Proceedings of the IEEE International Symposium on Olfaction and Electronic Nose (ISOEN), Fukuoka, Japan, 26–29 May 2019. [CrossRef]
10. Gobbi, E.; Falasconi, M.; Zambotti, G.; Sberveglieri, V.; Pulvirenti, A.; Sberveglieri, G. Rapid diagnosis of Enterobacteriaceae in vegetable soups by a metal oxide sensor based electronic nose. *Sens. Actuators B Chem.* **2015**, *207*, 1104–1113. [CrossRef]

chemistry proceedings

MDPI

Proceeding Paper

First Principles Investigation of the Optoelectronic Properties of Molybdenum Dinitride for Optical Sensing Applications †

Amall A. Ramanathan

Department of Physics, The University of Jordan, Amman 11942, Jordan; amallahmad@gmail.com

† Presented at the 1st International Electronic Conference on Chemical Sensors and Analytical Chemistry, 1–15 July 2021; Available online: https://csac2021.sciforum.net/.

Abstract: The electronic and optical properties of the newly synthesized molybdenum dinitride (MoN_2) in the hypothetical 2H structure analogous to MoS_2 is investigated using the density functional theory (DFT) full potential linearized augmented plane wave (FP-LAPW) method and the modified Becke–Johnson (mBJ) approximation. The aim is to investigate the optoelectronic properties of this compound for potential optical sensing applications and compare with the capabilities of MoS_2 in this field. As compared to MoS_2, which is a semiconductor, MoN_2 is found to be a semi metal from the band structure plots. The dielectric function, optical conductivity and the optical constants, namely, the refractive index, the reflectivity, the extinction and absorption coefficients, are evaluated and compared with those of MoS_2 and discussed with reference to the sensing performance.

Keywords: layered materials; electronic structure; dielectric function; optical conductivity; optical constants; optical sensing

Citation: Ramanathan, A.A. First Principles Investigation of the Optoelectronic Properties of Molybdenum Dinitride for Optical Sensing Applications. *Chem. Proc.* **2021**, *5*, 27. https://doi.org/10.3390/CSAC2021-10429

Academic Editor: Elena Benito-Peña

Published: 30 June 2021

1. Introduction

The high potential of transition metal dichalcogenides (TMD) for electronic, sensing, photonic and thermoelectric device applications has been exploited this past decade, and especially MoS_2, a prototype TMD material, has shown a lot of promise [1–3]. It has been studied and characterized extensively for structural, electronic, optical and transport properties both in bulk and in the 2D limit [4–6]. Interest in TM nitrides has been rekindled because they exhibit a number of unique and advanced catalytic properties for photo and electrochemical catalysis [7,8]. There were no layered structures in any of these studies. Layered structures provide more flexibility in doping, the ease of going down to lower dimensions and materials design.

The search for layered nitrogen rich TM nitrides, particularly those of the MoS_2 type, led to the recent synthesis and discovery of 3R-MoN_2, which has the rhombohedral MoS_2 structure [9]. It was synthesized through a high P-T route of solid-state ion exchange and has shown great potential for applications in catalysis and hydrogenation. In addition, the very recent first principle study of MoN_2 monolayer by Zhang et al. [10] showed the 1H configuration to be the most stable among the structures considered in their study. Their study revealed the importance of 2D MoN_2 as a high-capacity electrode material for metal ion batteries. Further, the first principles study of Ramanathan and Khalifeh [11] has shown the 2H MoN_2 to be a promising thermoelectric material.

All the above interesting results for MoN_2 provide a strong motivation to study this compound. Considering that to date no optical characterization of MoN_2 has been performed, the present study is devoted to the determination of the electronic and optical properties from first principles and to look at the various possibilities for optical sensing applications of MoN_2. Since an optical sensor measures a physical property of light and, depending upon the sensor usage, converts it to a readable output, it is highly essential to characterize the optical properties of the new layered material MoN_2. The hypothetical

2H structure analogous to MoS_2 of MoN_2 is investigated using the DFT full potential linearized augmented plane wave (FP-LAPW) method and the mBJ approximation. In addition, the 2H MoS_2 optoelectronic properties are determined by the same method for the sake of completeness and comparison.

2. Calculation Details

The geometry of MoN_2 is optimized using the ABINIT software program [12,13] with the generalized gradient approximation (GGA) of Perdew, Burke and Ernzerhof (PBE) PAW (projector augmented wave) pseudopotentials [14]. All the structural calculations are performed with convergence criteria of less than 1×10^{-6} Ha for the self-consistent field (SCF) iterations and a threshold of less than 1 mRy/a.u. for the optimization of the geometries [15,16]. The fully relaxed MoS_2 lattice constant values are taken from our previous work [6].

The optimized structures and lattice constant values are then used with the WIEN2k [17] code to perform full-potential linearized-augmented plane wave (FP-LAPW) calculation employing GGA_PBE to obtain the ground state energy and electronic properties at a $20 \times 20 \times 4$ k-point grid. The optical properties are evaluated using denser grids of $40 \times 40 \times 5$ with the more accurate mBJ exchange correlation of Trans Blaha (TB-mBJ) [18].

3. Results and Discussion

3.1. Structural and Electronic

The 2H-MoN_2/MoS_2 unit cells have hexagonal symmetry and consist of two stacks of three atomic layers; each stack consists of a Mo atomic plane sandwiched between two N/S atomic planes respectively. The atoms are bonded covalently in plane and the stacks are held together by weak Van der Waals force.

The non-magnetic state is the ground state for both the layered compounds and the structural relaxation of the system with a complete relaxation of all the atoms which simultaneously gives us the equilibrium geometry. The lattice parameter a and c values of MoN_2 are 3.094 and 11.975 Å, respectively. The lattice parameter values of MoN_2 are much smaller than that of MoS_2 due to the shorter bond lengths of Mo–N as compared to Mo–S. The lattice parameters for MoS_2 a = 3.193 and c = 12.359 Å taken from previous work [6] using the LDA (local density approximation) are in good agreement with the experimental values [19] a_{exp} = 3.16 Å and c_{exp} = 12.29 Å and within 2.3 and 0.6%, respectively.

The equilibrium lattice constant values for MoN_2 and MoS_2 are used with the GGA-PBE WIEN2k code to extract the electronic band structures with a $20 \times 20 \times 4$ k-point grid.

We notice that there is a change in the electronic distribution of MoN_2 as compared to MoS_2 which is reflected in the band structure of MoN_2, as shown in Figure 1. The MoS_2 band structure is also shown on the right panel of Figure 1. The band structures illustrate the change in behavior of MoN_2 to a semi metal one from the semiconducting one of MoS_2. There is an overlap between the bottom of the conduction band and the top of the valence band in MoN_2. This semi-metal feature implies that there is a range of energies for which electrons and holes co-exist. The Mo 4d, S 3p and N 2p atomic orbitals play a decisive role in the band structure properties as shown by the total and partial density of states plots Figures S1 and S2 (a and b).

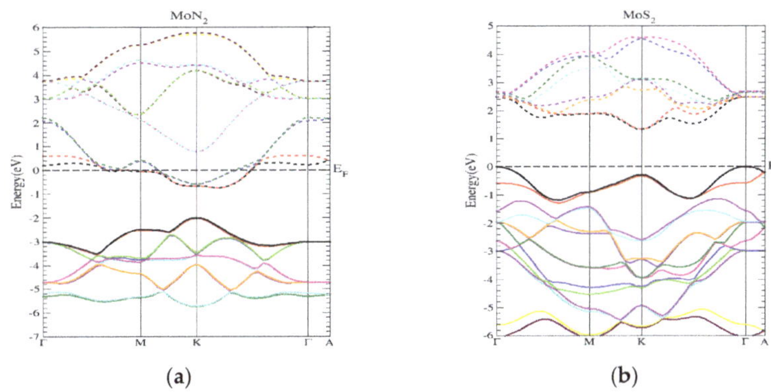

Figure 1. The band structures of (**a**) MoN_2 and (**b**) MoS_2.

3.2. Optical Properties

The TB-mBJ proves to be an excellent choice with a 40 × 40 × 5 grid for calculating the optical properties with a high degree of accuracy for MoN_2 and MoS_2. This section is devoted to the presentation and discussion of the results for the dielectric function and optical conductivity. In addition, the optical constants, namely, the refractive index, the reflectivity, the extinction and absorption coefficients, are obtained and interpreted.

The complex dielectric function ($\varepsilon = \varepsilon_1 + i\varepsilon_2$) is a function of the amount of light absorbed by the material. The imaginary part of dielectric function, $\varepsilon_2(\omega)$, which represents absorption behavior, can be calculated from the electronic band structure of solids. The real part of dielectric function, $\varepsilon_1(\omega)$, which represents the electronic polarization under incident light can be calculated according to Kramers–Kroing relation [20,21]. Figure 2 shows the real and imaginary plots for the dielectric function for MoN_2 and MoS_2 in the photon energy range of 0–14 eV. We see from the plots the anisotropy of the dielectric function. The general trend is the in-plane values are almost double that of the out-of-plane direction and the peaks for the zz direction are shifted more towards the right, towards higher photon energies. The dielectric plots show the high capability for absorption in the visible part of the spectrum for the in-plane direction and ultra-violet (UV) for the perpendicular direction, thereby effectively covering a wide range of energies.

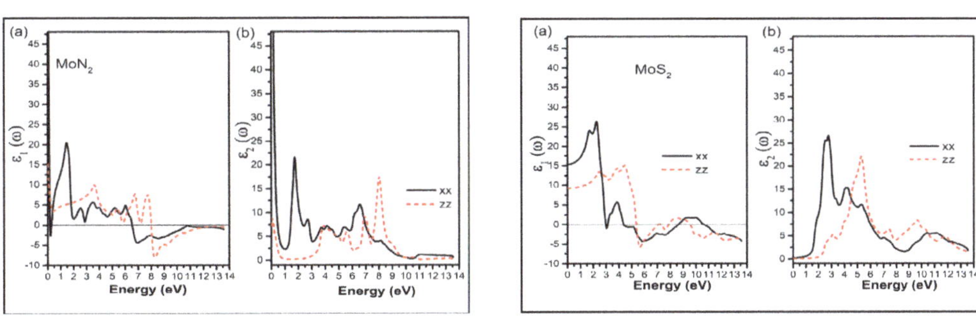

Figure 2. The dielectric function left panel MoN_2 and right panel MoS_2; (**a**) the real $\varepsilon_1(\omega)$ and (**b**) imaginary part $\varepsilon_2(\omega)$ for the in-plane (xx) and out of plane (zz) directions.

The complex index of refraction of the medium N is defined as $N = \sqrt{\varepsilon} = n + ik$, where n is the refractive index and k the extinction coefficient. These are depicted in Figure 3.

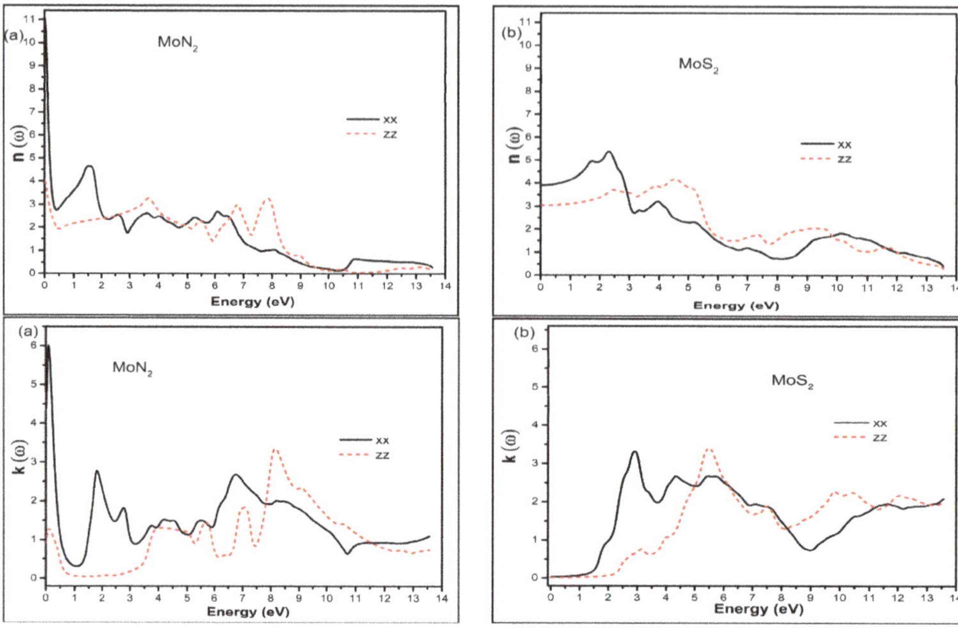

Figure 3. The refractive index top panel and extinction coefficient bottom panel for MoN_2 (**a**) and MoS_2 (**b**) in the xx in-plane and zz out of plane directions.

Once again, we see the anisotropy in the two directions for these optical constants. The amplitudes in the xx direction are larger and closer to the visible range for both the n (ω) and k (ω) compounds. We notice that MoN_2 has large static ($\omega = 0$) refractive index values of ~11 and 4 in the xx and zz directions, respectively. In contrast the corresponding values for MoS_2 are 4 and 3. The larger values of n (ω) imply higher electron density. In contrast to MoS_2, MoN_2 has peak extinction coefficient values at $\omega = 0$ of 4.4 and 0.9 in the xx and zz directions, respectively. Both MoN_2 and MoS_2 show low k (ω) values in the infrared and MoS_2 continues to have almost zero values up to 1.5 and 2.5 eV for the xx and zz directions, respectively. The first maxims of MoS_2 are at 3 eV of the spectra and the magnitude in the xx direction is almost six times larger than in the zz direction. The second k peak for the zz direction is in the UV region and slightly larger than the first xx peak. Since the extinction coefficient reflects the degree to which light is absorbed, we can see that the in-plane direction is most favorable for both compounds. However, in contrast to MoS_2, with MoN_2 we have strong absorption in infrared in addition to the visible region of the photon energy.

The conductivity and absorption coefficient graphs for MoN_2 and MoS_2 are shown in Figure 4. The graphs show for the xx in-plane direction maximum conductivity is in the UV region for MoN_2 whereas for MoS_2 it is in the visible part of the photon energy. The conductivity in the zz direction has peak positions in the UV region around 8 and 5 eV for MoN_2 and MoS_2, respectively.

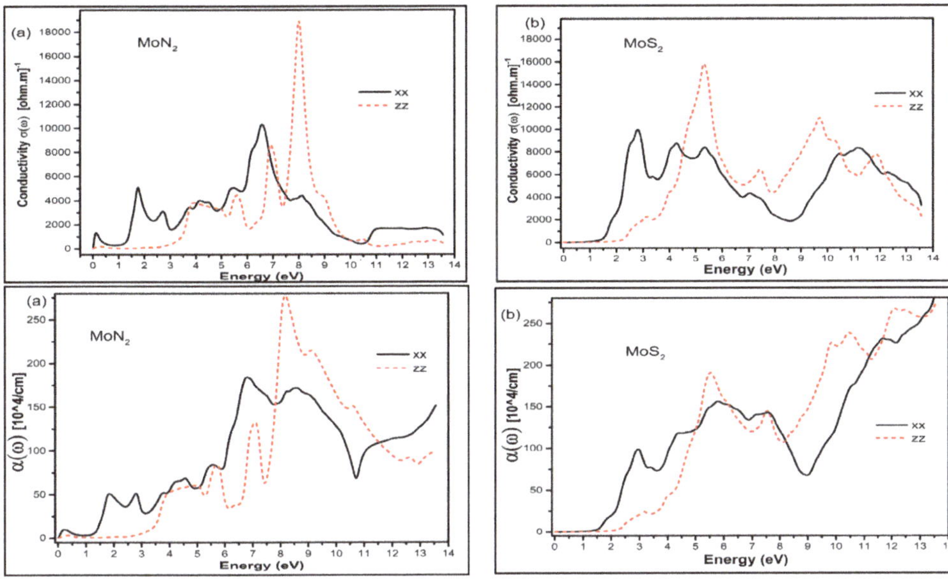

Figure 4. The optical conductivity top panel and absorption coefficient bottom panel for MoN_2 (**a**) and MoS_2 (**b**) in the xx in-plane and zz out of plane directions.

The absorption on the other hand shows the first peak in the visible and second broader peak with a much higher magnitude in the UV region for MoN_2; and a very broad peak of almost constant magnitude for MoS_2, covering the visible and the UV region in the in-plane xx direction. Beyond 10 eV, both MoN_2 and MoS_2 show a rise in the absorption coefficient. With respect to the zz direction, there are a set of small peaks beyond the visible and a sharp maximum value peak at around 9 eV of the UV region for MoN_2; whereas for MoS_2, the peaks are around 6, 10 and 12 eV in the UV region. These characteristics confirm the suitability of MoN_2 and MoS_2 for visible and UV sensing applications. The reflectance is depicted in Figure 5 and we observe large static reflectance value greater than 0.7 for MoN_2 that is double that of MoS_2 in the xx direction. In the zz direction, the values are 0.35 and 0.25 for MoN_2 and MoS_2 respectively. The graphs show that MoN_2 is a good infrared reflector, whereas MoS_2 reflects best just beyond the visible range.

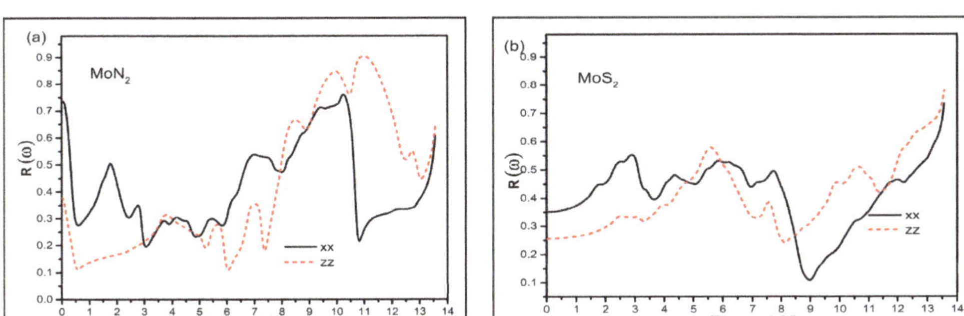

Figure 5. The reflectance R (ω) for MoN_2 (**a**) and MoS_2 (**b**) in the xx in-plane and zz out of plane directions in the photon energy range of 0–14 eV.

4. Conclusions

In conclusion, as opposed to MoS_2, MoN_2 is a semi metal as evinced from the band structure plots. Both the layered materials show anisotropy for all the optical properties with different magnitudes and peak positions, although the shapes of the graphs for the same property are similar in the two directions.

The large values of refractive index and good optical conductivity, absorption and reflectance results obtained reinstate the suitability of these materials for optical sensing applications. In comparison to MoS_2, MoN_2 shows suitability for infrared sensing applications in addition to visible and UV.

Supplementary Materials: The following are available online at https://www.mdpi.com/article/10.3390/CSAC2021-10429/s1.

Funding: No funding was received for this research work.

Institutional Review Board Statement: Not applicable.

Informed Consent Statement: Not applicable.

Data Availability Statement: Data can be provided upon reasonable request.

Conflicts of Interest: There are no conflict of interest to declare.

References

1. Onofrio, N.; Guzman, D.M.; Strachan, A. Novel doping alternatives for transition metal dichalcogenides from high-throughput DFT calculations. *J. Appl. Phys.* **2017**, *122*, 185102. [CrossRef]
2. Wang, Y.H.; Huang, K.J.; Wu, X. Recent advances in transition-metal dichalcogenides based electrochemical biosensors: A review. *Biosens. Bioelectron.* **2017**, *97*, 305. [CrossRef]
3. Vikraman, D.; Akbar, K.; Hussain, S.; Yoo, G.; Jang, J.-Y.; Chun, S.-H.; Jung, J.; Park, H.J. Direct synthesis of thickness-tunable MoS_2 quantum dot thin layers: Optical, structural and electrical properties and their application to hydrogen evolution. *Nano Energy* **2017**, *35*, 101. [CrossRef]
4. Ramanathan, A.A. Defect Functionalization of MoS_2 nanostructures as toxic gas sensors. *IOP Conf. Ser. Mater. Sci. Eng.* **2018**, *305*, 012001. [CrossRef]
5. Zhang, H.; Lu, S.B.; Zheng, J.; Du, J.; Wen, S.C.; Tang, D.Y.; Loh, K.P. Molybdenum disulfide (MoS_2) as a broadband saturable absorber for ultra-fast photonics. *Opt. Express* **2014**, *22*, 7249–7260. [CrossRef]
6. Ramanathan, A.A.; Khalifeh, J.M. Enhanced thermoelectric properties of suspended mono and bilayer of MoS_2 from first principles. *IEEE Trans. Nanotechnol.* **2018**, *17*, 974. [CrossRef]
7. Chen, W.-F.; Muckerman, J.T.; Fujita, E. Recent developments in transition metal carbides and nitrides as hydrogen evolution electrocatalysts. *Chem. Commun.* **2013**, *49*, 8896–8909. [CrossRef] [PubMed]
8. Tabata, M.; Maeda, K.; Higashi, M.; Lu, D.; Takata, T.; Abe, R.; Domen, K. Modified Ta_3N_5 powder as a photocatalyst for O_2 evolution in a two-step water splitting system with an iodate/iodide shuttle redox mediator under visible light. *Langmuir* **2010**, *26*, 9161–9165. [CrossRef] [PubMed]
9. Wang, S.M.; Ge, H.; Sun, S.L.; Zhang, J.Z.; Liu, F.M.; Wen, X.D.; Yu, X.H.; Wang, L.P.; Zhang, Y.; Xu, H.W.; et al. A New Molybdenum Nitride Catalyst with Rhombohedral MoS_2 Structure for Hydrogenation Applications. *J. Am. Chem. Soc.* **2015**, *137*, 4815. [CrossRef] [PubMed]
10. Zhang, X.; Yao, Y.; Yu, Z.; Wang, S.-S.; Guan, S.; Yang, H.Y.; Yang, S.A. Theoretical prediction of MoN_2 monolayer as a high capacity electrode material for metal ion batteries. *J. Mater. Chem. A* **2016**, *4*, 15224. [CrossRef]
11. Ramanathan, A.A.; Khalfeh, J.M. Thermoelectrics of $MoS_{2(1-x)}N_{2x}$ Compounds. *Phys. Sci. Biophys. J.* **2021**, *5*, 000167. [CrossRef]
12. Ramanathan, A.A. A DFT calculation of Nb and Ta (001) Surface Properties. *J. Mod. Phys.* **2013**, *4*, 432–437. [CrossRef]
13. Gonze, X.; Jollet, F.; Araujo, F.A.; Adams, D.; Amadon, B.; Applencourt, T.; Audouze, C.; Beuken, J.M.; Bieder, J.; Bokhanchuk, A. Recent developments in the ABINIT software package. *Comput. Phys. Commun.* **2016**, *205*, 106. [CrossRef]
14. Perdew, J.P.; Burke, K.; Ernzerhof, M. Generalized Gradient Approximation Made Simple. *Phys. Rev. Lett.* **1996**, *77*, 3865. [CrossRef]
15. Ramanathan, A.A.; Khalifeh, J.M. Substrate matters: Magnetic tuning of the Fe monolayer. *J. Magn. Magn. Mater.* **2017**, *426*, 450–453. [CrossRef]
16. Ramanathan, A.A.; Khalifeh, J.M.; Hamad, B.A. Evidence of surface magnetism in the V/Nb(0 0 1) system: A total energy pseudopotential calculation. *Surf. Sci.* **2008**, *602*, 607–613. [CrossRef]
17. Blaha, P.; Schwarz, K.; Tran, F.; Laskowski, R.; Madsen, G.K.H.; Marks, L.D. WIEN2k: An APW+lo program for calculating the properties of solids. *J. Chem. Phys.* **2020**, *152*, 074101. [CrossRef]
18. Tran, F.; Blaha, P. Accurate band gaps of semiconductors and insulators with a semilocal exchange-correlation potential. *Phys. Rev. Lett.* **2009**, *102*, 226401. [CrossRef] [PubMed]

19. Coehoorn, R.; Haas, C.; Dijkstra, J.; Flipse, C.J.F.; de Groot, R.A.; Wold, A. Electronic structure of $MoSe_2$, MoS_2, and WSe_2. I. Band-structure calculations and photoelectron spectroscopy. *Phys. Rev. B* **1987**, *35*, 6195. [CrossRef]
20. Hulthén, R. Kramers–Kronig relations generalized: On dispersion relations for finite frequency intervals. A spectrum-restoring filter. *J. Opt. Soc. Am.* **1982**, *72*, 794–803. [CrossRef]
21. Ramanathan, A.A.; Khalifeh, J.M. Electronic, magnetic and optical properties of $XScO_3$ (X = Mo, W) perovskites. *PeerJ Mater. Sci.* **2021**, *3*, e15. [CrossRef]

chemistry proceedings

MDPI

Proceeding Paper

Essential Oils as Possible Candidates to Be Included in Active Packaging Systems and the Use of Biosensors to Monitor the Quality of Foodstuff †

Anton Soria-Lopez [1], Maria Carpena [1], Bernabe Nuñez-Estevez [1,2], Paula Garcia-Oliveira [1,2], Nicolas Collazo [1], Paz Otero [1], Pascual Garcia-Perez [1], Hui Cao [1], Jianbo Xiao [1], Márcio Carocho [2], Lillian Barros [2], Jesus Simal-Gandara [1,2,*] and Miguel A. Prieto [1,2,*]

[1] Nutrition and Bromatology Group, Department of Analytical Chemistry and Food Science, Faculty of Science, Universidade de Vigo, E-32004 Ourense, Spain; anton.soria@uvigo.es (A.S.-L.); mcarpena@uvigo.es (M.C.); bernabe.nunez@uvigo.es (B.N.-E.); paula.garcia.oliveira@uvigo.es (P.G.-O.); nicolascollazojimenez@gmail.com (N.C.); paz.otero@uvigo.es (P.O.); pasgarcia@uvigo.es (P.G.-P.); hui.cao@uvigo.es (H.C.); jianboxiao@uvigo.es (J.X.)

[2] Centro de Investigação de Montanha (CIMO), Instituto Politécnico de Bragança, Campus de Santa Apolonia, 5300-253 Bragança, Portugal; mcarocho@ipb.pt (M.C.); lillian@ipb.pt (L.B.)

* Correspondence: jsimal@uvigo.es (J.S.-G.); mprieto@uvigo.es (M.A.P.)

† Presented at 1st International Electronic Conference on Chemical Sensors and Analytical Chemistry, 1–15 July 2021; Available online: https://csac2021.sciforum.net/.

Citation: Soria-Lopez, A.; Carpena, M.; Nuñez-Estevez, B.; Garcia-Oliveira, P.; Collazo, N.; Otero, P.; Garcia-Perez, P.; Cao, H.; Xiao, J.; Carocho, M.; et al. Essential Oils as Possible Candidates to Be Included in Active Packaging Systems and the Use of Biosensors to Monitor the Quality of Foodstuff. *Chem. Proc.* **2021**, *5*, 28. https://doi.org/10.3390/CSAC2021-10485

Academic Editor: Ye Zhou

Published: 30 June 2021

Abstract: Active packaging has gained interest in recent years. As well as protecting food from the environment, it can incorporate agents with specific properties to extend the shelf life of the food. As a requirement, it is essential that the active agent has a greater affinity for the food than for the packaging material and, in this sense, essential oils (EOs) are potential candidates to be included in this new packaging system. The use of EOs can add to food matrix antimicrobial and antioxidant properties, reduce the permeability of the packaging to water vapor and extend the shelf life of food products. However, their use has been limited because they can produce a strong flavor by interacting with other compounds present in the food matrix and modify the organoleptic characteristics. Although the nanoencapsulation of EOs can provide chemical stability and minimize the impact of the Eos on the organoleptic properties by decreasing their volatilization, some physical modifications have still been observed, such as plasticizing effects and color variations. In this sense, the quality of the food products and consumer safety can be increased by using sensors. This technology indicates when food products are degrading and informs us if specific packaging conditions have changed. This work focuses on highlighting the use of biosensors as a new methodology to detect undesirable changes in the food matrix in a short period of time and the use of nanotechnology to include EOs in active films of natural origin.

Keywords: active packaging; intelligent packaging; EOs; nanoencapsulation; biosensors

1. Introduction

There is a vast variety of foods that is sensitive to deterioration through the action of microorganisms and to the oxidation of lipids during storage. Packaging is used to protect foods against external and internal conditions, to ensure food safety and to avoid rapid deterioration caused by chemical and microbiological contamination. Furthermore, nowadays, consumers are more conscious about sustainability as a benefit of safe and healthy foods. As a result, the use of materials of natural origin (proteins, polysaccharides or lipids) as food packaging has gained attention in recent years. Simultaneously, two new technologies have emerged to protect foods and increase food shelf life, namely [1]: (i) intelligent packaging (IP) and (ii) active packaging (AP).

AP consists of the inclusion of chemical or bioactive compounds into the packaging system to ensure that the protective function of the packaging has a longer duration. This type of packaging interacts with the product, which can absorb or release components from the food [2,3]. Despite there being AP that contains different chemical additives, the use of bioactive compounds, such as essential oils (EOs), as new additives has gained attention. They are incorporated into films and coatings because of their important biological activities, such as their antioxidant and antimicrobial properties [4]. However, this technology presents some limitations to being applied at industrial level. For this reason, nanoencapsulation has emerged to improve food quality and reduce the limitations of AP in combination with active ingredients.

Regarding IP, its main objective is to control the conditions of packaged foods, such as the environment around them (i.e., storage conditions, food quality, sell by date, etc.). Biosensors are a type of sensor that belong to IP technology that have been widely employed at an industrial level in food processing during recent years. Specifically, electrochemical biosensors are the most used technology. However, they are still in the improvement phase of increasing their applications in packaging systems. These devices can detect undesirable changes or processes that can occur inside packaging systems, and transform them into a certain signal that can be easily analyzed [5,6].

This proceeding paper is focused on the use of EOs as possible natural additives or ingredients to be incorporated into the AP system, in the form of films or coatings, due to their antioxidant and microbial activities, and the use of biosensors as a possible tool to detect any undesirable changes inside of the packaging system. Finally, nanoencapsulation could be a suitable solution to improve food quality and safety even more.

2. Essential Oils in Active Packaging System

Active packaging (AP) is a novel method mainly utilized to prolong the shelf life of food products and to improve food quality and safety [3]. Many industries are interested in obtaining AP of natural origins that contains ingredients with bioactive compounds in order to avoid the use of chemical additives that can be harmful to human health, minimize the environmental impacts and ensure the acceptance of consumers. Essential oils (EOs) are one of the possible candidates for natural food additives.

EOs are volatile liquids of a lipid nature that can be obtained from plants. They are classified as GRAS (generally recognized as safe) food additives [7], thus their use has gained the attention of many researchers because of their antioxidant and antimicrobial activities. In addition, they can be used such as food preservatives or incorporated into edible films or coatings. Regarding food preservatives, their use is very limited due to their strong flavor and odor. As for edible films, there is a current trend in using materials such as polysaccharides, proteins or lipids as the edible films or coatings of packaging. EOs are used as additives or ingredients in edible emulsified films and coatings and can be incorporated into these edible matrices by several methods, emulsification being the most common. In fact, many studies are focused on using EOs as food additives in the packaging field to compete with the current packaging materials due to their substantial possibilities and adaptability [8–10].

2.1. Effect of the Incorporation of EOs in AP

The incorporation of EOs into the film matrix leads to a heterogeneous film structure featuring discontinuities, producing modifications to physical properties of the film such as tensile strength (TS), water vapor permeability (WVP), color, transparency and gloss [11]. Regarding TS, studies have shown different responses of TS when incorporating EOs into the film matrix [12–14]. In fact, the effect of the addition of EOs on the tensile properties of edible films depends on the specific interactions between the oil components and the polymer matrix. Concerning WVP, most studies have shown that the incorporation of EOs into the film matrix leads to an improvement of the water vapor barrier properties and a decrease in WVP [15,16]. Keeping this in mind, the hydrophobicity of EOs is a very fluctuating characteristic because it depends on various factors. Finally, color, transparency and gloss are influenced by the type and concentration of EOs [11]. So, the incorporation of EOs into the matrix leads to specific physical modifications in the packaging that can reduce the quality and safety of the food products.

On the other hand, EOs are known for the presence of chemical compounds with antioxidant and microbial properties, which can be applied to avoid oxidation and increase food quality [11]. The antioxidant activity of the EOs occurs through different mechanisms: acting as O_2 scavengers; producing a barrier against O_2.; and promoting a specific antioxidant action. In this sense, the incorporation of EOs can lead to the improvement of food quality and a reduction in food waste due to the oxidation [17]. However, their use is limited at an industrial level due to the possible migration of these compounds into the food product and, consequently, the modification of its organoleptic properties. Concerning microbial capacity, it depends on the characteristics of the EO and the type of microorganism. The antimicrobial action mainly inhibits the growth of food pathogens, thus ensuring protection against microbial deterioration [18]. Many industries are interested in obtaining packaging systems with antimicrobial properties, since they will promote a longer shelf life for food products and guarantee a better food quality. Table 1 shows some examples of AP where EOs with antimicrobial and antioxidant properties have been incorporated.

Table 1. Recent examples of active films containing EOs as the active agents, showing their main components and biological properties for packaged food products.

Film	EOs	Main Components of EOs	Biological Activity	Ref.
Gelatin	OLEO	Sabinene	Antimicrobial 2% OLEO (*B. subtilis*, *S. aureus*, *E. coli*, *P. aeruginosa*, *C. albicans*); Antioxidant DPPH 2% OLEO (52%)	[8]
Pectin	CEO	Cinnamaldehyde; L-linalool	Antimicrobial (*S. aureus*, *E. coli*, *L. monocytogenes*); Antioxidant: DPPH 1.5% CEO (64.73%)	[9]
Chitosan	PAEO	Caryophyllene; aromadendrene oxide; selinene	Antimicrobial (*S. aureus*, *S. typhimurium*, *K. pneumonia*, *P. aeruginosa*, *B. subtilis*)	[10]
Chitosan	ZEO	Thymol; γ-terpinene	Antioxidant DPPH (97.2%); Antimicrobial (*B. cereus*, *E. coli*, *P. aeruginosa*, *E. faecalis*, *S. aureus*, *A. flavus*)	[19]
Chitosan-GA	CEO	Cinnamaldehyde; L-linalool	Antioxidant: DPPH, maximum for 1:2 (Chitosan:GA)	[20]

Note: ZEO: *Zataria multiflora* essential oil; CEO: cinnamon essential oil; GA: gum arabic; OLEO: *Citrus sinensis* essential oil; OEO: oregano essential oil; PAEO: *Plectranthus amboinicus* essential oil; DPHH: 2,2-diphenyl-1-picrylhydrazyl.

2.2. *Nanoencapsulation*

Due to some disadvantages that EOs present when used as food additives (low solubility, high volatility, strong flavor, sensible to heat and light or the possibility of adversely affecting the organoleptic properties of food), many researchers began to focus their studies on the use of nanotechnologies in order to overcome these limitations and contribute to improving food preservation [21]. The nanoencapsulation technique consists of introducing an active agent (EO) into a polymer membrane with a diameter of 0.05–1 μm, known as a nanocapsule (Figure 1). This technique is used to protect the EOs against the previous limiting factors since this membrane acts as a barrier against the external environment, which preventions oxidation, masks unpleasant odors and taste and avoids the loss of the volatile substances of the EOs. In addition, nanoencapsulation allows the controlled release of the EOs from the capsule, meaning that the release of the active agents occurs at the ideal place and time. Likewise, the nanoencapsulation of EOs can improve their biological activities, since their bioavailability depends on surface/volume ratio and particle size [22,23]. Keeping this in mind, the lower the particle size, the higher the surface/volume ratio or stability during the incorporation into the matrix. Many studies that employed this technique using EOs have shown excellent results in the quality and shelf life of food products [24,25].

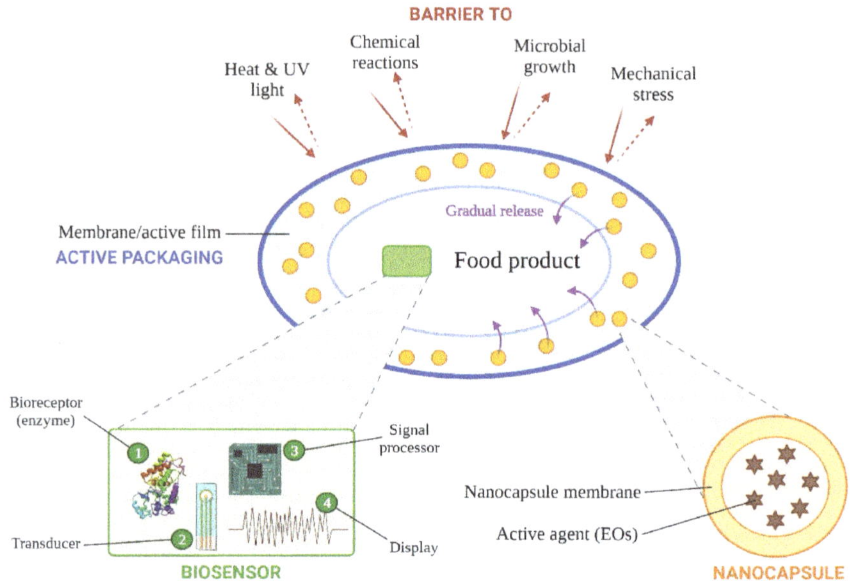

Figure 1. Biosensor structure, nanoencapsulation of EOs and functions of active films or coatings. Created with BioRender.com.

3. Biosensors

Besides AP, another technology has emerged in recent years, known as intelligent packaging (IP). IP is a packaging system that contains a certain device that provides information to the retailer or consumer about the state of the food product and its surrounding environment. In this sense, IP allows a constant communication about the state of the system with all steps of the supply chain, which is an important characteristic of this technology [3,26]. Keeping this in mind, this technology allows for the quick detection of unpleasant changes in the packaging system, an increase in food safety and the production of less food waste. Sensors, indicators and identification systems are the main components of IP, with sensors being the most common components and the ones that have received the most attention in recent years. All sensors contain: (i) a detection system, known as a receptor, which can detect specific analytes and transform its presence into an electric signal; (ii) a signal processor, known as a transducer, which is responsible for processing the generated signal; and (iii) an electronic system, which is responsible for displaying the measured properties. Depending on the type of analyte that they can detect, sensors can be chemical or biological, with the latter being the most promising technology to develop and improve IP systems. Biosensors are responsible for transforming biological responses into a processed signal, with enzymes, receptor proteins, antibodies and nucleic acids being the recognition elements (Figure 1) [5]. In fact, the use of enzymes as recognition elements is widely employed due to low production costs, lack of the need for additional instrumentation, small size and ease of use. Regarding the transducer group, biosensors can be optical, mass-based, calorimetric or electrochemical, with the latter being the most used and the one that has gained the most attention. Electrochemical biosensors consist of devices that measure the electrochemical signal that is proportional to the analyte concentration [27]. However, their current applications in IP are limited to certain conditions since the biosensor structure can present biological components that have harmful effects. Furthermore, important improvements are needed in the biosensor structures in order to avoid the pretreatment of food samples and include degradation markers in packaging systems [28]. Therefore, more studies are required to improve and reduce these limitations.

4. Conclusions

In recent decades, both AP and IP have emerged as technologies that protect foods and increase food shelf life. In addition, due to a large percentage of consumers being conscious of environmental sustainability, many industries have employed natural food additives or ingredients, such as Eos, to replace synthetic chemical additives and the use of natural materials (i.e., proteins, polysaccharides or lipids) to reduce major waste. However, there are some limitations concerning the use of EOs as active agents, such as their low solubility, high volatility, strong taste and flavor, sensibility to heat and light, changes in organoleptic properties and modifications of the physical properties of the films or coatings. Nanotechnology, specifically the nanoencapsulation of EOs, has gained attention during recent years and has been presented as a new alternative to improve the quality of food products as many studies suggest that they have major benefits. On the other hand, biosensors, specifically electrochemical biosensors, could be the most promising technology for IP systems. In fact, the combination of AP with nanocapsules containing EOs and biosensors could lead to important improvements in food safety, an extension of products' shelf life and higher protection against oxidation and food deterioration mediated by the action of microorganisms.

Author Contributions: Conceptualization, A.S.-L., M.C. (Maria Carpena), B.N.-E., P.G.-O. and N.C.; investigation, A.S.-L., M.C. (Maria Carpena), B.N.-E., P.G.-O., N.C., P.O., P.G.-P., H.C. and M.C. (Márcio Carocho); resources, J.X., L.B., J.S.-G. and M.A.P.; writing—original draft preparation, A.S.-L., Maria Carpena, B.N.-E., P.G.-O. and N.C.; writing—review and editing, P.O., P.G.-P., H.C. and M.C. (Márcio Carocho); visualization, J.X., L.B., J.S.-G. and M.A.P.; supervision, J.X., L.B., J.S.-G. and M.A.P.; project administration, J.X., L.B., J.S.-G. and M.A.P.; funding acquisition, J.X., L.B., J.S.-G. and M.A.P. All authors have read and agreed to the published version of the manuscript.

Funding: The JU receives support from the European Union's Horizon 2020 research and innovation program and the Bio Based Industries Consortium. The project, SYSTEMIC Knowledge hub on Nutrition and Food Security, has received funding from national research funding parties in Belgium (FWO), France (INRA), Germany (BLE), Italy (MIPAAF), Latvia (IZM), Norway (RCN), Portugal (FCT) and Spain (AEI) in a joint action of JPI HDHL, JPI-OCEANS and FACCE-JPI, launched in 2019 under the ERA-NET ERA-HDHL (n° 696295).

Institutional Review Board Statement: Not applicable.

Informed Consent Statement: Not applicable.

Data Availability Statement: Not applicable.

Acknowledgments: The research leading to these results was supported by: MICINN, for providing the Ramón y Cajal grant for M.A. Prieto (RYC-2017-22891); Xunta de Galicia, for providing the EXCELENCIA-ED431F 2020/12 program and the predoctoral grant for P. García-Oliveira (ED481A-2019/295); the University of Vigo, for providing the predoctoral grant for M. Carpena (Uvigo-00VI 131H 6410211); the EcoChestnut Project (Erasmus+ KA202) for supporting the work of B. Nuñez-Estevez; the Bio Based Industries Joint Undertaking (JU) under grant agreement No 888003 UP4HEALTH Project (H2020-BBI-JTI-2019) for supporting the work of P. Otero and P. Garcia-Perez; and the Ibero-American Program on Science and Technology (CYTED—AQUA-CIBUS, P317RT0003). The authors are grateful to the Foundation for Science and Technology (FCT, Portugal) for its financial support through the national funds FCT/MCTES to CIMO (UIDB/00690/2020), and to the national funding from FCT, P.I., through the institutional scientific employment program contract for L. Barros.

Conflicts of Interest: The authors declare no conflict of interest.

References

1. Yam, K.L.; Takhistov, P.T.; Miltz, J. Intelligent Packaging: Concepts and Applications. *Food Sci.* **2005**, *70*, R1–R10. [CrossRef]
2. Janjarasskul, T.; Suppakul, P. Active and intelligent packaging: The indication of quality and safety. *Crit. Rev. Food Sci. Nutr.* **2018**, *58*, 808–831. [CrossRef] [PubMed]
3. Firouz, M.S.; Mohi-Alden, K.; Omid, M. A critical review on intelligent and active packaging in the food industry: Research and development. *Food Res. Int.* **2021**, *141*, 110113. [CrossRef] [PubMed]
4. Sharma, S.; Barkauskaite, S.; Jaiswal, A.K.; Jaiswal, S. Essential oils as additives in active food packaging. *Food Chem.* **2021**, *343*, 128403. [CrossRef] [PubMed]
5. Fracchiolla, N.S.; Artuso, S.; Cortelezzi, A. Biosensors in clinical practice: Focus on oncohematology. *Sensors* **2013**, *13*, 6423–6447. [CrossRef] [PubMed]
6. Alocilja, E.C.; Radke, S.M. Market analysis of biosensors for food safety. *Biosens. Bioelectron.* **2003**, *18*, 841–846. [CrossRef]
7. Ruiz-Navajas, Y.; Viuda-Martos, M.; Sendra, E.; Perez-Alvarez, J.A.; Fernández-López, J. In vitro antibacterial and antioxidant properties of chitosan edible films incorporated with Thymus moroderi or Thymus piperella essential oils. *Food Control* **2013**, *30*, 386–392. [CrossRef]
8. Alparslan, Y.; Yapici, H.H.; Metin, C.; Baygar, T.; Günlü, A.; Baygar, T. Quality assessment of shrimps preserved with orange leaf essential oil incorporated gelatin. *LWT Food Sci. Technol.* **2016**, *72*, 457–466. [CrossRef]
9. Nisar, T.; Wang, Z.C.; Yang, X.; Tian, Y.; Iqbal, M.; Guo, Y. Characterization of citrus pectin films integrated with clove bud essential oil: Physical, thermal, barrier, antioxidant and antibacterial properties. *Int. J. Biol. Macromol.* **2018**, *106*, 670–680. [CrossRef]
10. Priya, N.V.; Vinitha, U.G.; Sundaram, M.M. Preparation of chitosan-based antimicrobial active food packaging film incorporated with Plectranthus amboinicus essential oil. *Biocatal. Agric. Biotechnol.* **2021**, *34*, 102021. [CrossRef]
11. Atarés, L.; Chiralt, A. Essential oils as additives in biodegradable films and coatings for active food packaging. *Trends Food Sci. Technol.* **2016**, *48*, 51–62. [CrossRef]
12. Pires, C.; Ramos, C.; Teixeira, B.; Batista, I.; Nunes, M.L.; Marques, A. Hake proteins edible films incorporated with essential oils: Physical, mechanical, antioxidant and antibacterial properties. *Food Hydrocoll.* **2013**, *30*, 224–231. [CrossRef]
13. Atarés, L.; Bonilla, J.; Chiralt, A. Characterization of sodium caseinate-based edible films incorporated with cinnamon or ginger essential oils. *J. Food Eng.* **2010**, *100*, 678–687. [CrossRef]
14. Zinoviadou, K.G.; Koutsoumanis, K.P.; Biliaderis, C.G. Physico-chemical properties of whey protein isolate films containing oregano oil and their antimicrobial action against spoilage flora of fresh beef. *Meat Sci.* **2009**, *82*, 338–345. [CrossRef] [PubMed]
15. Sánchez-González, L.; Vargas, M.; González-Martínez, C.; Chiralt, A.; Cháfer, M. Characterization of edible films based on hydroxypropylmethylcellulose and tea tree essential oil. *Food Hydrocoll.* **2009**, *23*, 2102–2109. [CrossRef]
16. Benavides, S.; Villalobos-Carvajal, R.; Reyes, J.E. Physical, mechanical and antibacterial properties of alginate film: Effect of the crosslinking degree and oregano essential oil concentration. *J. Food Eng.* **2012**, *110*, 232–239. [CrossRef]
17. Wang, Z.C.; Lu, Y.; Yan, Y.; Nisar, T.; Fang, Z.; Xia, N.; Guo, Y.; Chen, D.W. Effective inhibition and simplified detection of lipid oxidation in tilapia (Oreochromis niloticus) fillets during ice storage. *Aquaculture* **2019**, *511*, 634183. [CrossRef]
18. Ribeiro-Santos, R.; Andrade, M.; Sanches-Silva, A. Application of encapsulated essential oils as antimicrobial agents in food packaging. *Curr. Opin. Food Sci.* **2017**, *14*, 78–84. [CrossRef]
19. Carpena, M.; Nuñez-Estevez, B.; Soria-Lopez, A.; Garcia-Oliveira, P.; Prieto, M.A. Essential oils and their application on active packaging systems: A review. *Resources* **2021**, *10*, 7. [CrossRef]
20. Xu, T.; Gao, C.C.; Feng, X.; Yang, Y.; Shen, X.; Tang, X. Structure, physical and antioxidant properties of chitosan-gum arabic edible films incorporated with cinnamon essential oil. *Int. J. Biol. Macromol.* **2019**, *134*, 230–236. [CrossRef]
21. Shamaei, S.; Seiiedlou, S.S.; Aghbashlo, M.; Tsotsas, E.; Kharaghani, A. Microencapsulation of walnut oil by spray drying: Effects of wall material and drying conditions on physicochemical properties of microcapsules. *Innov. Food Sci. Emerg. Technol.* **2017**, *39*, 101–112. [CrossRef]
22. Pezzin, A.P.T. Microencapsulação: Inovação em diferentes áreas. *Heal. Environ. J.* **2015**, *7*, 12–20.
23. Martin, A.N.; Bustamante, P.; Chun, A.H.C. *Physical Pharmacy: Physical Chemical Principles in the Pharmaceutical Sciences*, 4th ed.; Lea & Febiger: Philadelphia, PA, USA, 1993.
24. Zohri, M.; Shafiee Alavidjeh, M.; Mirdamadi, S.S.; Behmadi, H.; Hossaini Nasr, S.M.; Eshghi Gonbaki, S.; Shafiee Ardestani, M.; Jabbari Arabzadeh, A. Nisin-Loaded Chitosan/Alginate Nanoparticles: A Hopeful Hybrid Biopreservative. *J. Food Saf.* **2013**, *33*, 40–49. [CrossRef]
25. Kamkar, A.; Molaee-Aghaee, E.; Khanjari, A.; Akhondzadeh-Basti, A.; Noudoost, B.; Shariatifar, N.; Alizadeh Sani, M.; Soleimani, M. Nanocomposite active packaging based on chitosan biopolymer loaded with nano-liposomal essential oil: Its characterizations and effects on microbial, and chemical properties of refrigerated chicken breast fillet. *Int. J. Food Microbiol.* **2021**, *342*, 109071. [CrossRef]
26. Dobrucka, R.; Cierpiszewski, R. Active and Intelligent Packaging Food-Research and Development-A Review. *Pol. J. Food Nutr. Sci.* **2014**, *64*, 7–15. [CrossRef]

27. Park, Y.W.; Kim, S.M.; Lee, J.Y.; Jang, W. Application of biosensors in smart packaging. *Mol. Cell. Toxicol.* **2015**, *11*, 277–285. [CrossRef]
28. Mustafa, F.; Andreescu, S. Chemical and biological sensors for food-quality monitoring and smart packaging. *Foods* **2018**, *7*, 168. [CrossRef]

chemistry proceedings

MDPI

Abstract

From Single Nanowires to Smart Systems: Different Ways to Assess Food Quality †

Matteo Tonezzer [1,*], Franco Biasioli [2] and Flavia Gasperi [3]

1 IMEM-CNR, Sede di Trento-FBK, Via alla Cascata 56/C, 38123 Povo, Trento, Italy
2 Research and Innovation Centre, Fondazione Edmund Mach, 38010 San Michele all'Adige, Trento, Italy; franco.biasioli@fmach.it
3 Research and Innovation Centre, Department of Food Quality and Nutrition, Fondazione Edmund Mach, Via Edmund Mach 1, 38010 San Michele all'Adige, Trento, Italy; flavia.gasperi@fmach.it
* Correspondence: matteo.tonezzer@cnr.it

† Presented at the 1st International Electronic Conference on Chemical Sensors and Analytical Chemistry, 1–15 July 2021. Available online: https://csac2021.sciforum.net.

Abstract: Recently, low-dimensional (1D, 2D) nanostructured materials have been attracting more and more interest as building blocks for innovative systems. Metal oxide nanowires are one of the most widely used materials for solid-state gas sensors, as they are simple to make, inexpensive, and sensitive to a wide range of gases and volatiles. Unfortunately, their broad sensitivity has a price to pay, which is very low selectivity. Fortunately, this flaw is not a problem for all applications. Where the boundary conditions are defined and "simple" (only the presence of a target gas is expected, without any interfering gases), a single traditional chemiresistor may be the best choice, while in cases where the variables are many, it is better to use an intelligent system. In this paper, we will show a resistive sensor based on a single SnO_2 nanowire which, working at three temperatures (200, 250, and 300 °C), is able to detect tens of ppb of ammonia (30 ppb at 300 °C). The limit of detection (LoD) was calculated as 3 N/S, where N is the standard deviation of the sensor signal in air and S is the sensor sensitivity. We will show that the performance of this nanosensor is excellent and can be used in various applications, including agri-food quality monitoring. We will demonstrate that the SnO_2 nanowire in a thermal gradient can act as a nano-electronic nose thanks to machine learning algorithms. The single nanowire-based sensor can estimate the total viable count with an error of 2.32% on mackerel fish samples stored at room temperature (25 °C) and in a fridge (4 °C). The integration of such a small (less than one square mm) and cheap device into the food supply chain would greatly reduce waste and the frequency of food poisoning.

Keywords: gas sensors; metal oxide; nanowire; electronic nose; machine learning

Citation: Tonezzer, M.; Biasioli, F.; Gasperi, F. From Single Nanowires to Smart Systems: Different Ways to Assess Food Quality. *Chem. Proc.* **2021**, *5*, 29. https://doi.org/10.3390/CSAC2021-10605

Academic Editor: Ye Zhou

Published: 5 July 2021

Supplementary Materials: The following are available online at https://www.mdpi.com/article/10.3390/CSAC2021-10605/s1.

Author Contributions: Conceptualization, M.T.; methodology, M.T.; software, M.T.; validation, M.T., F.B. and F.G.; formal analysis, M.T.; investigation, M.T.; resources, F.B., F.G.; data curation, M.T.; writing—original draft preparation, M.T.; writing—review and editing, M.T., F.B. and F.G.; visualization, M.T.; supervision, F.B. and F.G.; project administration, F.B. and F.G.; funding acquisition, F.B. and F.G. All authors have read and agreed to the published version of the manuscript.

Funding: This research received no external funding.

Institutional Review Board Statement: Not applicable.

Informed Consent Statement: Not applicable.

Data Availability Statement: The data presented in this study are openly available in Open Science Framework at doi:10.17605/OSF.IO/83SMW.

Conflicts of Interest: The authors declare no conflict of interest.

chemistry proceedings

Proceeding Paper

Silver Nanomaterials as Electron Mediators in a Bio-Electronic Tongue Dedicated to the Analysis of Milks. The Role of the Aspect Ratio of Nanoparticles vs. Nanowires †

Coral Salvo-Comino [1,2], Clara Perez-Gonzalez [1,3], Fernando Martin-Pedrosa [2,3], Cristina Garcia-Cabezon [2,3,*] and Maria Luz Rodriguez-Mendez [1,2,*]

[1] Group UVASENS, Escuela de Ingenierías Industriales, Universidad de Valladolid, Paseo del Cauce, 59, 47011 Valladolid, Spain; coralsalvocomino@gmail.com (C.S.-C.); claraperez.biotecnologia@gmail.com (C.P.-G.)
[2] BioecoUVA Research Institute, Universidad de Valladolid, 47011 Valladolid, Spain; fmp@eii.uva.es
[3] Department of Materials Science, Universidad de Valladolid, Paseo del Cauce, 59, 47011 Valladolid, Spain
* Correspondence: crigar@eii.uva.es (C.G.-C.); mluz@eii.uva.es (M.L.R.-M.)

† Presented at the 1st International Electronic Conference on Chemical Sensors and Analytical Chemistry, 1–15 July 2021; Available online: https://csac2021.sciforum.net/.

Abstract: The integration of silver nanomaterials as electron mediators in electrochemical biosensors can be crucial to improve the affinity with biomolecules and the electrochemical response. In this work, two voltammetric bioelectronics tongues (bioET) formed by biosensors based on the combination of enzymes with silver nanoparticles (AgNPs) (bioET-1) or silver nanowires (AgNWs) (bioET-2) have been developed and used to analyze milks. Each array was formed by four biosensors formed by enzymes (glucose oxidase, galactose oxidase, β-galactosidase and a blank), capable to detect compounds usually found in milks. Principal component analysis (PCA) has revealed the ability of both biosensor systems to discriminate between milk samples with different fat contents, but with some differences, attributed to the structure employed in the detection.

Keywords: silver nanowires; silver nanoparticles; electronic tongue; electrochemical biosensor; enzymes; milk

Citation: Salvo-Comino, C.; Perez-Gonzalez, C.; Martin-Pedrosa, F.; Garcia-Cabezon, C.; Rodriguez-Mendez, M.L. Silver Nanomaterials as Electron Mediators in a Bio-Electronic Tongue Dedicated to the Analysis of Milks. The Role of the Aspect Ratio of Nanoparticles vs. Nanowires. *Chem. Proc.* **2021**, *5*, 30. https://doi.org/10.3390/CSAC2021-10554

Academic Editor: Núria Serrano

Published: 1 July 2021

1. Introduction

Milk is an essential component of human diets, although its composition varies depending on the brand, storage period, animal origin and the components that make milk up. As a result, it is vital to assess the composition and quality of milk from the time it is obtained to the time it is consumed [1].

The implementation of biosensors in food industry has taken an essential role due to the fact that these devices are able to provide qualitative information with high specificity and selectivity with the advantages of being portable, miniaturizable, cheap, stable, fast and show effective on-line response [2,3].

The use of nanomaterials in electrochemical sensors and biosensors has attracted researchers' attention. However, using specialized sensors in the analysis of complicated matrices to determine specific parameters, is insufficient to generate useful data. The development of arrays of sensors may be the solution to avoid irrelevant information. For this reason, the implementation of electronic tongues (ET) may be the solution to determine important parameters and properties of complex samples and to be able of discriminate between them [4]. Metallic nanostructures are excellent sensing materials due to their high surface area, high aspect ratio, electrical conductivity and electrocatalytic characteristics, which provide good sensing properties for the detection of a wide range of analytes. Moreover, it has been demonstrated that their morphology is essential on the electrochemical response and their ability to improve the electron mobility.

Furthermore, due to their large specific surface area and high surface free energy, metallic nanomaterials may strongly adsorb biomolecules maintaining their bioactivity due to their biocompatibility [5].

This research aims to create a bioelectronic tongue (bioET) for milk analysis. Two electrochemical biosensor arrays in which AgNPs (bioET-1) or AgNWs (bioET-2) have been deposited as enzyme supports, were created and tested for their capacity to differentiate between milk samples with variable fat and nutritional content. For this purpose, unsupervised (PCA) multivariate classification methods were used to assess the bioETs' performance. Finally, the discrimination abilities of both sensor arrays were evaluated to determine which silver nanomaterial provided the best results.

2. Results and Discussion

2.1. Sensors Development

Metallic nanomaterials previously synthetized were deposited onto BDD substrates by a cast of 25 μL suspension of AgNPs or AgNWs. After drying, 25 μL of the correspondent enzyme (5 mg/mL) was drop-casted onto the modified electrode. Finally, 25 μL of Nafion® was drop-casted and let dry overnight at room temperature.

2.2. Milk Samples

The samples collection consisted of 6 groups of milks purchased in the supermarket, including milks with different fat content (skimmed, semi-skimmed and whole), and different nutritional content (classic and lactose-free).

2.3. Electrochemical Characterization

The electrochemical responses of the bioETs were studied by cyclic voltammetry in a 50% diluted milk 0.1 M KCl solution.

Figure 1 illustrates the electrochemical response of the developed sensor arrays with the bioETs for one type of milk (Classical Whole milk). As it can be observed, when the biological material is immobilized on the modified electrode surfaces, the increase in the intensity of the cathodic and anodic peaks is more pronounced in the presence of nanowires in the AgNWs bioET.

The results demonstrate that silver nanowires outperform silver nanoparticles in terms of electrocatalytic activity because the electrical conductivity and the electrocatalytic properties of the developed sensors have been enhanced when AgNWs are used as electron mediators. Furthermore, because AgNWs have a high surface-to-volume ratio, they have more surface active sites, allowing for improved enzyme immobilization [5].

As is well known, the ultimate electrocatalytic activity of electrode materials was impacted by surface shape, nanoparticle size and fabrication technique. For this reason, it can increase electrode surface area depending on the chosen morphology, but mass transport behavior of silver nanoparticles, for example, is difficult to define because nanoparticles are randomly spread or aggregated, whereas nanowires, increased in high proportion the electrical conductivity and mass transport in comparison with them [6].

Multivariate data analysis was used to evaluate the responses of the proposed bioETs in order to discriminate between different types of milks.

Data were evaluated using principal component analysis for this purpose (PCA).

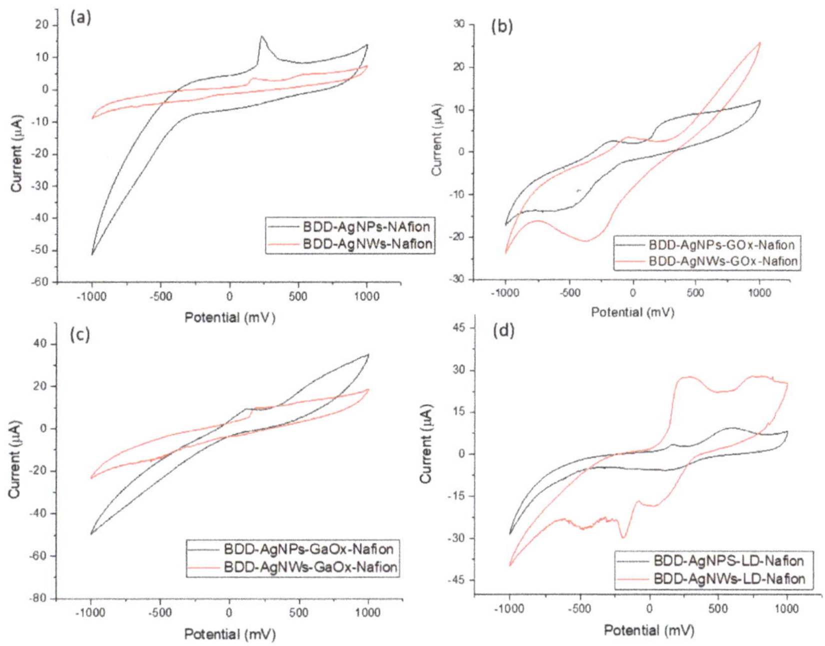

Figure 1. Electrochemical response of the sensors of AgNPs bioET (black) and AgNWs bioET (red) in a 50% diluted Classical Whole Milk in 0.1 M KCl solution, (**a**) sensors formed without enzyme, (**b**) biosensors with glucose oxidase immobilized, (**c**) biosensors with galactose oxidase immobilized and (**d**) biosensors with lactate dehydrogenase immobilized.

Figure 2 shows the scores plot of both bioETs. As Figure 2a shows the accumulated explained variance for the two first components of AgNWs bioET was distributed in 49% (PC-1) and 30% (PC2) and for the AgNPs biET (Figure 2b) was 49% (PC1) and 19% (PC2). As it can be shown in the figure, the scores plot of the array that uses nanoparticles displays certain overlapping between Classical Skimmed milks and Classical Whole, appearing in the same quadrant, it is obvious that the AgNPs bioET presents difficulties to effectively discriminate milks based on their fat content. However, in both cases a clear discrimination of milks according the nutritional contents was achieved. This result indicates that the AgNWs bioET provides a higher proportion of the explained variance for the same number of PCs.

These findings can be explained by the fact that immobilizing the enzyme on nanomaterials improves the enzyme's catalytic effectiveness greatly due to its operational stability. However, it has been demonstrated that the size, morphology and charge distribution of the nanomaterial can variate the effects on enzyme structure and corresponding activity [7]. When nanowires are used as support for enzyme immobilization, different reactivity and good enzyme immobilization is obtained through easier interactions between enzyme and material surface due to providing higher surface/volume ratio in comparison with nanoparticles, thus could be the reason why the discrimination capability of the bioET is higher when AgNWs are used as electrocatalytic material.

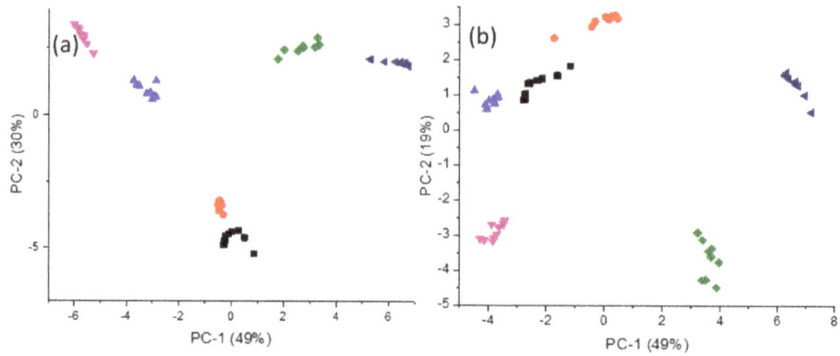

Figure 2. PCA score-plot analyzed using a 4 sensors array (**a**) AgNWs bioET or (**b**) AgNPs bioET, of the 6 milks of different fat content and nutritional characteristics: Classical Skimmed (black), Classical Skimmed Lactose free (red), Classical Whole (blue), Classical Whole Lactose free (pink), Classical Semi Skimmed (green) and Classical Semi Skimmed Lactose free (dark blue).

3. Conclusions

In this work, two bioelectronics tongues modified with AgNWs or AgNPs has been developed to discriminate between milks with different nutritional composition. The electrochemical responses based on cyclic voltammetry, of the two different bioETs have been crucial to evaluate the influence of the morphology of the conductive material. Furthermore, the statistical analysis based on PCA loading plots have demonstrated the high capability of the sensors arrays to discriminate between milks.

Finally, it can be concluded that the use of AgNWs could be a better choice because their excellent electrocatalytic properties.

Funding: This research was funded by Ministerio de Ciencia Innovación y Universidades-FEDERPlan Nacional (RTI2018-097990-B-100), Junta de Castilla y Leon-FEDER VA275P 18, Infraestructuras Red de Castilla y León (INFRARED) UVA01 and MINECO (BES-2016-077825).

Institutional Review Board Statement: Not applicable.

Informed Consent Statement: Not applicable.

Conflicts of Interest: The authors declare no conflict of interest.

References

1. Poghossian, A.; Geissler, H.; Schöning, M.J. Rapid methods and sensors for milk quality monitoring and spoilage detection. *Biosens. Bioelectron.* **2019**, *140*, 111272. [CrossRef] [PubMed]
2. Salvo-Comino, C.; García-Hernández, C.; García-Cabezón, C.; Rodríguez-Méndez, M.L. Discrimination of Milks with a Multisensor System Based on Layer-by-Layer Films. *Sensors* **2018**, *18*, 2716. [CrossRef] [PubMed]
3. Jiang, H.; Zhang, M.; Bhandari, B.; Adhikari, B. Application of electronic tongue for fresh foods quality evaluation: A review. *Food Rev. Int.* **2018**, *34*, 746–769. [CrossRef]
4. Manikandan, V.S.; Adhikari, B.; Chen, A. Nanomaterial based electrochemical sensors for the safety and quality control of food and beverages. *Analyst* **2018**, *143*, 4537–4554. [CrossRef] [PubMed]
5. Li, L.; Wang, P.; Shao, Q.; Huang, X. Metallic nanostructures with low dimensionality for electrochemical water splitting. *Chem. Soc. Rev.* **2020**, *49*, 3072–3106. [CrossRef] [PubMed]
6. Hovancová, J.; Šišoláková, I.; Oriňaková, R.; Oriňak, A. Nanomaterial-based electrochemical sensors for detection of glucose and insulin. *J. Solid State Electrochem.* **2017**, *21*, 2147–2166. [CrossRef]
7. Sharifi, M.; Sohrabi, M.J.; Hosseinali, S.H.; Kani, P.H.; Talaei, A.J.; Hasan, A.; Karim, A.Y.; Nanakali, N.M.Q.; Salihi, A.; Aziz, F.M.; et al. Enzyme immobilization onto the nanomaterials: Application in enzyme stability and prodrug-activated cancer therapy. *Int. J. Biol. Macromol.* **2020**, *143*, 665–676. [CrossRef] [PubMed]

chemistry proceedings

MDPI

Proceeding Paper

Development of a Bioelectronic Tongue Modified with Gold Nanoparticles for Dairy Analysis †

Clara Pérez-González [1,2], Coral Salvo-Comino [2,3], Fernando Martin-Pedrosa [1,2], Cristina García-Cabezón [1,2,*] and María Luz Rodríguez Méndez [2,3,*]

1 Material Science Departament, University of Valladolid, Escuela de Ingenierías Industriales Paseo del Cauce, 59, 47011 Valladolid, Spain; claraperez.biotecnologia@gmail.com (C.P.-G.); fmp@eii.uva.es (F.M.-P.)

2 Group UVaSens, University of Valladolid, Escuela de Ingenierías Industriales Paseo del Cauce, 59, 47011 Valladolid, Spain; coraldeugena@hotmail.com

3 Inorganic Chemistry Departament, University of Valladolid, Escuela de Ingenierías Industriales Paseo del Cauce, 59, 47011 Valladolid, Spain

* Correspondence: crigar@eii.uva.es (C.G.-C.); mluz@eii.uva.es (M.L.R.M.)

† Presented at the 1st International Electronic Conference on Chemical Sensors and Analytical Chemistry, 1–15 July 2021; Available online: https://csac2021.sciforum.net/.

Abstract: The objective of this work was to create an all-solid-state potentiometric bioelectronic tongue with an array of polymeric membrane-based sensors, which could then be used in the dairy sector. Membranes were modified with gold nanoparticles, and enzymes were covalently linked to the sensor's surface to create an array of sensors with greater sensitivity. The responses of the sensors modified with gold nanoparticles and covalently associated enzymes, showed higher sensitivities. Moreover, the developed bioelectronic tongue was able to perform the discrimination of milks with different nutritional characteristics by applying principal component analysis. In addition, the results obtained showed that by applying partial least squares analysis, the system could be used as a prediction system for different chemical parameters (such as acidity, proteins, lactose, etc.).

Keywords: electronic tongue; biosensor; potentiometric; dairy industry; nanoparticles

Citation: Pérez-González, C.; Salvo-Comino, C.; Martin-Pedrosa, F.; García-Cabezón, C.; Rodríguez Méndez, M.L. Development of a Bioelectronic Tongue Modified with Gold Nanoparticles for Dairy Analysis. *Chem. Proc.* **2021**, *5*, 31. https://doi.org/10.3390/CSAC2021-10553

Academic Editor: Núria Serrano

Published: 1 July 2021

1. Introduction

The concept of electronic tongues has expanded rapidly during recent years due to their large potential. E-tongues are based on sensor arrays with low selectivity and high cross-selectivity between multiple sensors [1]. Unlike other analytical methods, this type of device allows the acquisition of chemical information from different matrixes by applying suitable multivariate statistical, qualitative or quantitative data processing techniques, thereby targeting issues to overcome drawbacks, such as the requirement to pretreat samples, noise issues, and collinearity between variables [2].

E-tongues can implement a range of transduction principles, electrochemical sensors (potentiometric, amperometric, voltammetric, or impedimetric sensors) being the most common sensors applied in the development of e-tongues. Potentiometric sensors are based on the measurement of the differences in the interface potential created across a selective membrane. The interaction between the electrode and the solution determines this potential, which is related to the physicochemical properties of the solutions under inquiry [3].

The objective of this work was to design an all-solid-state potentiometric bioelectronic tongue (bio-ET) dedicated to the dairy sector, using an array of biosensors based on polymeric membranes operating in parallel. The membranes were modified with gold nanoparticles (AuNPs) to create an array of sensors with greater sensitivity [4]. Moreover, to further improve the sensor's selectivity, enzymes, including galactose oxidase, urease, and lactate dehydrogenase, were covalently attached to the PVC surface. The bio-ET has been applied to the analysis of milk samples with different nutritional contents.

2. Materials and Methods

The electronic tongue was constructed by combining sensors based on polymeric matrices that differ in composition. The polymeric membranes were made of PVC [poly(vinyl chloride)] mixed with an additive (oleyl alcohol), a plasticiser [(bis(1-butylpentyl) adipate, tris(2ethylhexyl)phosphate or 2-nitrophenyl-octylether] and gold nanoparticles creating an array of 27 sensors. Each of the polymeric mixtures were applied on solid conducting silver-epoxy supports. The bio-ET was composed of the sensor array, an Ag/AgCl reference electrode, and a data-collecting multiplexer.

The effect of gold nanoparticles and the bioelectronic tongue performance was tested by analysing six standard solutions at concentrations ranging from 1×10^{-3} to 1×10^{-1} mol/L. When applying the multivariate analysis technique, a set of sensors was selected for the construction of a bio-ET that was applied to twelve commercial milk samples. Potentiometric signals obtained from the array of sensors were processed using principal component analysis (PCA). Using partial least squares (PLS), correlations between the responses of the sensors and the chemical parameters were established.

3. Results

The results obtained in this research indicate that sensors with a higher percentage of gold nanoparticles in their composition showed higher sensitivities towards compounds found in milk. An example of the sensor behaviour is shown in Figure 1. This behaviour could be due to an increase in the sensitivity of the sensors towards specific ions on the interface of the polymeric membrane when AuNPs are included in the matrix composition.

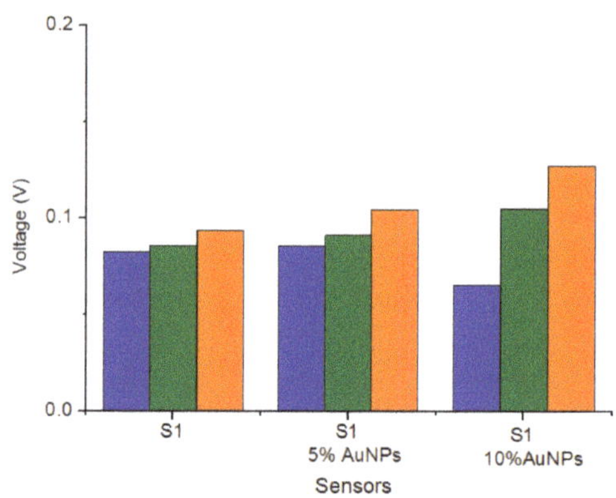

Figure 1. Each sensor's response to glucose at increasing concentrations (blue = 10^{-4} M, green = 10^{-3} M, and orange = 10^{-2} M) depending on the percentage of AuNPs applied to the membrane matrix.

Moreover, the sensor that combined gold nanoparticles with enzymes showed a greater ability to differentiate between increasing concentrations of products of interest found in milk, such as urea, lactic acid, galactose, etc., with deviations of voltages up to 65 mV between different samples.

Furthermore, using statistical analysis (PCA), the constructed bioelectronic tongue was able to classify milk with various nutritional features, resulting in four distinct groups that were also sorted according to the fat content of the samples (Figure 2).

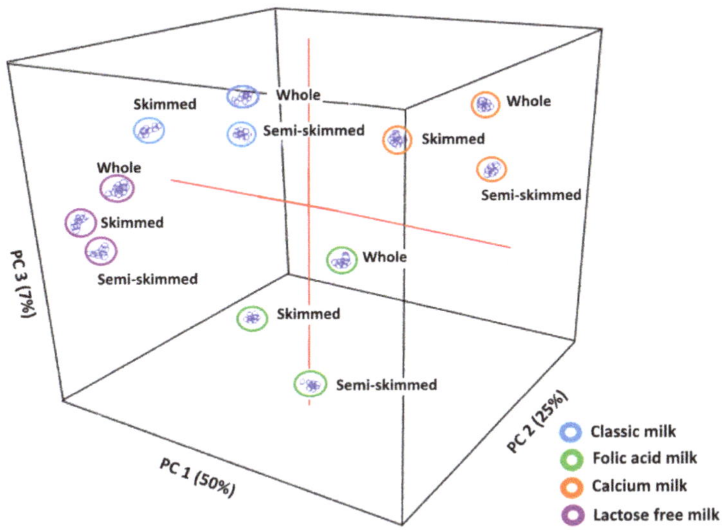

Figure 2. Classification of the milk samples studied according to the score diagram of the principal component analysis.

Additionally, the study's findings revealed that using partial least squares analysis (PLS), with regression coefficients above 0.85 for three variables in the physio-chemical parameters studied, the developed bioelectronic tongue could be used as a prediction system to determine the parameters, such as density, acidity, lactose, or fat content, of future milk samples (Table 1).

Table 1. Correlation parameters resulting from the regression of partial least squares analysis (PLS).

Parameters	$R2_C$	$RMSE_C$	$R2_P$	$RMSE_P$	Latent Variables
Acids	0.8683	0.2894	0.8483	0.2753	3
Density	0.8729	0.5172	0.8621	0.5736	3
%Proteins	0.8629	0.1027	0.8432	0.1251	3
%Fat	0.8512	0.4723	0.8404	0.4817	3
%Lactose	0.8895	0.0553	0.8813	0.0577	3

4. Conclusions

In this work, a bioelectronic tongue was developed and used to predict the chemical characteristics of milk. Sensors with higher concentrations of gold nanoparticles in their composition showed greater sensitivity towards the compounds of interest in milk (such as lactic acid, galactose, and urea). The system, using nine potentiometric sensors, could be successfully used to discriminate between milks, applying PCA based on their nutritional content. The bio-ET was successfully used to predict the acidity and density, in addition to the protein, lactose, and fat content of the milk, with low errors and high correlation coefficients for three factors. This device could be adapted for its implementation in the dairy industry.

Funding: MICINN-FEDER Plan Nacional (RTI2018-097990-B-100), Consejería de Educación Junta de Castilla y Leon- FEDER VA275P18 and «Infraestructuras Red de Castilla y León (INFRARED)» UVA01.

Institutional Review Board Statement: Not applicable.

Informed Consent Statement: Not applicable.

Conflicts of Interest: The authors declare no conflict of interest.

References

1. Ródriguez-Mendez, M.L.; De Saja, J.A.; González-Antón, R. Electronic Noses and Tongues in Wine Industry. *Front. Bioeng. Biotechnol.* **2016**, *4*, 81. [CrossRef] [PubMed]
2. Wasilewski, T.; Wojciech, K.; Jacek, G. Bioelectronic Tongue: Current Status and Perspectives. *Biosens. Bioelectron.* **2020**, *150*, 111923. [CrossRef] [PubMed]
3. Bakker, E.; Pretsch, E. Potentiometric Sensors for Trace-Level Analysis. *Trac Trends Anal. Chem.* **2005**, *24*, 199–207. [CrossRef] [PubMed]
4. Yin, T.A.; Qin, W. Applications of Nanomaterials in Potentiometric Sensors. *Trends Anal. Chem.* **2013**, *51*, 79–86. [CrossRef]

chemistry proceedings

MDPI

Abstract

Core Modulation of Porphyrins for Chemical Sensing †

Karolis Norvaiša [1,*] and Mathias Otto Senge [1,2]

1 School of Chemistry, Trinity Biomedical Sciences Institute, Trinity College Dublin, The University of Dublin, 152-160 Pearse Street, D08W9RT Dublin, Ireland; sengem@tcd.ie
2 Institute for Advanced Study (TUM-IAS), Technical University of Munich, Lichtenbergstrasse 2a, D-85748 Garching, Germany
* Correspondence: norvaiak@tcd.ie

† Presented at the 1st International Electronic Conference on Chemical Sensors and Analytical Chemistry, 1–15 July 2021; Available online: https://csac2021.sciforum.net/.

Abstract: The inner core system of metal-free ('free base') porphyrins has continually served as a ligand for various metal ions, but it was only recently studied in organocatalysis due its highly tunable basicity. Highly conjugated porphyrin systems offer spectrophotometric sensitivity toward geometrical and/or electronic changes and, thus, utilizing the porphyrin core for the selective detection of substrates in solution offers significant potential for a multitude of applications. However, solvation and dilution drastically affect weak interactions by dispersing the binding agent to its surroundings. Thus, the spectroscopic detection of N–H···X-type binding in porphyrin solutions is almost impossible without especially designing the binding pocket. Here, we present the first report on the spectroscopic detection of N–H···X-type interplay in porphyrins formed by weak interactions. Protonated 2,3,7,8,12,13,17,18-octaethyl-5,10,15,20-tetrakis(2-aminophenyl) porphyrin contains coordination sites for the selective binding of charge-bearing analytes, revealing characteristic spectroscopic responses. While electronic absorption spectroscopy proved to be a particularly useful tool for the detection of porphyrin–analyte interactions in the supramolecular complexes, X-ray crystallography helped to pinpoint the orientation, flexibility, and encapsulation of substrates in the corresponding atropisomers. This charge-assisted complexation of analytes in the anion-selective porphyrin inner core system is ideal for the study of atropisomers using high-resolution NMR, since it reduces the proton exchange rate, generating static proton signals. Therefore, we were able to characterize all four rotamers of the nonplanar 2,3,7,8,12,13,17,18-octaethyl-5,10,15,20-tetrakis(2-aminophenyl) porphyrin by performing 1D and 2D NMR spectroscopic analyses of host-guest systems consisting of benzenesulfonic acid (BSA) and each porphyrin atropisomer. Lastly, a detailed assignment of the symmetry operations that are unique to porphyrin atropisomers allowed us to accurately identify the rotamers using NMR techniques only. Overall, the N–H···X-type interplay in porphyrins formed by weak interactions that form restricted H-bonding complexes is shown to be the key to unravelling the atropisomeric enigma.

Keywords: porphyrins; sensing; atropisomers; NMR; UV–vis; nonplanar

Citation: Norvaiša, K.; Senge, M.O. Core Modulation of Porphyrins for Chemical Sensing. *Chem. Proc.* **2021**, *5*, 32. https://doi.org/10.3390/CSAC2021-10417

Academic Editor: Nicole Jaffrezic-Renault

Published: 30 June 2021

Supplementary Materials: The following are available online at https://www.mdpi.com/article/10.3390/CSAC2021-10417/s1.

chemistry proceedings

Proceeding Paper

Nanostructured Bismuth Electrodes for Non-Enzymatic Paracetamol Sensing: Development, Testing, and Computational Approach †

Mallikarjun Madagalam [1], Federica Catania [1], Mattia Bartoli [2,3,*], Alberto Tagliaferro [3,4] and Sandro Carrara [5]

1 Faculty of Science and Technology, Free University of Bolzano-Bozen, Piazza Università 1, 39100 Bolzano, Italy; Mallikarjun.Madagalam@unibz.it (M.M.); Federica.Catania@unibz.it (F.C.)
2 Center for Sustainable Future Technologies @POLITO, Istituto Italiano di Tecnologia, Via Livorno 60, 10144 Turin, Italy
3 National Consortium for Materials Science and Technology (INSTM), Via G. Giusti 9, 50121 Florence, Italy; alberto.tagliaferro@polito.it
4 Department of Applied Science and Technology, Politecnico di Torino, C.so Duca degli Abruzzi 24, 10129 Turin, Italy
5 Bio/CMOS Interfaces Group, ICLAB, EPFL, CH-2002 Neuchâtel, Switzerland; sandro.carrara@epfl.ch
* Correspondence: mattia.bartoli@iit.it
† Presented at the 1st International Electronic Conference on Chemical Sensors and Analytical Chemistry, 1–15 July 2021; Available online: https://csac2021.sciforum.net/.

Abstract: In this work, new Screen Printed Carbon-paste Electrodes (SPCEs) were developed through deposition of nanostructures of HO–$BiONO_3$ synthesized with or without surfactant additions. We performed a cyclic voltammetry study showing the improvement in performance of bismuth tailored electrodes for paracetamol sensing compared with bare SPCE. A computation study was also performed for investigating the interaction between paracetamol and bismuth species during the electron transfer process enlighten the preferential sites of interaction on the surface of modified SPEs.

Keywords: paracetamol; bismuth; SPCE

Citation: Madagalam, M.; Catania, F.; Bartoli, M.; Tagliaferro, A.; Carrara, S. Nanostructured Bismuth Electrodes for Non-Enzymatic Paracetamol Sensing: Development, Testing, and Computational Approach. *Chem. Proc.* **2021**, *5*, 33. https://doi.org/10.3390/CSAC2021-10427

Academic Editor: Nicole Jaffrezic-Renault

Published: 30 June 2021

1. Introduction

Paracetamol is among the most studied emerging pollutants [1] and one of the most used antipyretic drugs in the world [2]. Accordingly, the development of a new trustworthy paracetamol detection system is a key pillar in the sensors field for personalized therapy and for developing environmental monitoring. Among all available approaches, electrochemical techniques are the most useful considering production-cost, sensibility and reproducibility of sensing technologies [3]. The development of new electrode materials for paracetamol detection has attracted more interest in recent years. To improve the sensing performance, several studies used modified carbon electrodes tailored with several nanostructured materials [2,3]. Recently, Madagalam et al. [4] reported a new SPCE decorated with bismuth sub-nitrate as an effective electrochemical sensor for paracetamol detection. The choice of bismuth was driven by the great tuneability of bismuth species [5], together with its remarkable electrochemical performances [6,7].

Here, we move a step forward in the comprehension of the paracetamol detection by using bismuth tailored SPCEs through a computational study. Accordingly, we report a solid approach for the calculation of electron transfer rate with new structural information of transitional state geometry of paracetamol–bismuth sub-nitrate system.

2. Materials and Methods

2.1. Materials

Bismuth nitrate pentahydrate ($Bi(NO_3)_3 \cdot 5H_2O$), 1-butanol, and Polyethylene glycol (PEG) were used for material synthesis at Politecnico di Torino, Italy. Phosphate buffer saline (PBS), DI water, and paracetamol tablets made into powder were used for the preparation of material suspensions and electrolytic solution at EPFL, Neuchatel, Switzerland.

2.2. Methods

2.2.1. Material Synthesis and Modification of SPCEs

HO–BiONO$_3$ was synthesized According to Liu et al. [8] and by the modified synthesis reported by Madagalam et al. [4]. SPCEs were prepared according to the procedure described in [4]. A SPCEs consisting on a working electrode and a counter electrode in carbon, and a reference electrode in silver/silver-chloride is presented in this study.

2.2.2. Electrochemical Measurements

Cyclic Voltammetry measurements (CV) were performed under atmospheric conditions using AutoLab potentiostat and run accordingly with the procedure described in [4]. For an electrochemically reversible electron transfer process involving freely diffusing redox species, the Randles–Sevcik equation [9] describes the linear increase of the peak current i_p (A) with the square root of the scan rate ν (V/s):

$$i_p = 0.466nFAC\left(\frac{nF\nu D}{RT}\right)^{\frac{1}{2}}, \quad (1)$$

where n is the number of electrons transferred in the redox reaction, A (cm^2) is the electrode geometric surface area, D (cm^2/s) is the diffusion coefficient of the oxidized analyte, and C (mol/cm^3) is the bulk concentration of the analyte. The Randles–Sevcik equation (Equation (1)) is used to deduce the analyte behavior identifying the two scenarios when it is freely diffusing in solution or adsorbed on the surface of the electrode.

Kinetic parameters were calculated using the Laviron model [10], where the peak potentials are described as follows, with the cathodic peak potential being:

$$E_{pc} = E^0 - \left(\frac{RT}{\alpha nF}\right)ln\left[\frac{\alpha}{|m|}\right] \quad (2)$$

and the anodic peak potential being:

$$E_{pa} = E^0 + \left(\frac{RT}{(1-\alpha)nF}\right)ln\left[\frac{(1-\alpha)}{|m|}\right] \quad (3)$$

with m = (RT/F) (k/nν). R is the universal gas constant, n is the number of electrons involved in the redox reaction, T is the absolute temperature in Kelvin, F is the Faraday constant, ν is the scan rate, and E^0 is the surface standard potential. The kinetic rate constant k was calculated by substituting α value into the following equation:

$$\ln(k) = \alpha\ln(1-\alpha) + (1-\alpha)ln\alpha - ln\left(\frac{RT}{nF\nu}\right) - \alpha(1-\alpha)\frac{nF\Delta E_p}{RT} \quad (4)$$

2.2.3. Modelling

Computational simulations were run with HyperChem (HC) software using the following procedure.

A 3D structure is obtained in HC by the function *model build* for both paracetamol and Bi clusters. Then, the simulations started to evaluate the electronic properties and the geometry information required in the generalized Marcus model. Therefore, the potential energy must be minimized with respect to the system coordinated by means of geometry optimization. This is done by evaluating the potential energy as function of molecules coordinates according to a hybrid quantum/molecular mechanics approach (MM+) [11]. Then, a second simulation was run in line with the neglect of differential diatomic overlap approximation (PM3 basis set) [12]. Two different Bi clusters model the polymeric layer as shown in Figure 1.

Figure 1. $BiONO_3$ dimer on the left and $BiONO_3$ trimer on the right.

The dimer was assumed to be the minimum unit allowing the electron transfer (ET). Indeed, ET is depicted through two Bi–C "fake bonds" as shown in Figure 2 by supposing that each couple exchanges just one electron and the two charged particles are simultaneously acquired from paracetamol [1].

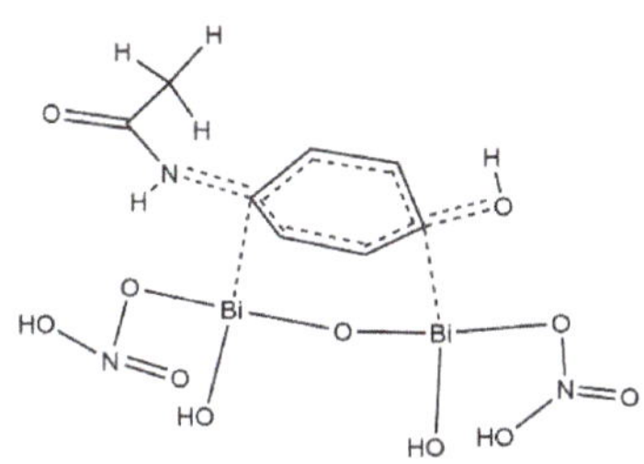

Figure 2. Transition state model: the double dashed lines on paracetamol defined the chemical bonds involved in the reaction while the single ones defined the Bi–C "fake bonds" through which the ET occurs.

The Figure 2 represents one of the hypothesized transition states (HTS) geometry. Indeed, the transitional state (TS) is identified as a unique crossing point in the potential energy surface (PES) defining the reaction coordinates of both reactants and products [13]. However, according to Tachiya et al. [14] the TS is not uniquely defined since it depends on the probability to achieve a certain value of the electrostatic potential distribution. For that reason, it is necessary to talk about "hypothetical" TS. Moreover, it is observed after each simulation that the interaction distance between each Bi–C couple is not equal:, the system is asymmetric, thus the average between the interaction distance is kept within the model according to the procedure reported in [14]. Afterwards, the simulations were repeated for the Bi trimer. The geometry optimization was performed two times in both cases by cutting the "fake bonds" before starting the second simulation. The solvation effect arising from the electrolyte wetting the electrode at the interface with the organic compound was considered as well. It was modelled by adding water molecules surrounding paracetamol and it was established that the ET would be ensured by a minimum amount of five molecules. Finally, the HTS in which the nitric functional groups in both dimer and trimer were replaced by hydroxyl group by supposing that the functionalized electrode undergone a cleaning pre-treatment in H_2SO_4 was simulated. The electron transfer rate constant in a CdS-Phenol system was analyzed to validate our model. According to Serpone et al. [15], we simulated the electrochemical redox process between a CdS powder and a phenol molecule in solution by defining a box containing water molecules, as shown in Figure 2. All of them were mirrored outside the box by reproducing the system in solution. The obtained ET rate constant was $k = 6.42{\cdot}10^{-5}\ s^{-1}$, in compliance with the literature constant ($k = 5.17{\cdot}10^{-5}\ s^{-1}$).

3. Results and Discussion

3.1. Electrochemical Measurements

SPCEs modified sensors were tested by Madagalam et al. [4] through CV and main data are summarized in Table 1.

The peaks detected during CV without and with paracetamol were distinguishable and confirmed that we observed higher peak currents when there was paracetamol in the electrolytic solution. PEG-coated HO–$BiONO_3$ sensor was found to have a potential shift of −75 mV, whereas HO–$BiONO_3$ sensor has a shift of −26 mV due to Nernst effect compared to the bare SPCE sensor. PEG-coated HO–$BiONO_3$ sensor was also found to have a higher oxidation peak current of 52.1 ± 3.0 μA. To study the behavior of the electrochemical system, CVs were run by varying the scan rate (50 mV/s–300 mV/s) while determining 1 mM paracetamol in 0.1 M PBS at pH 7. It was observed that the peak position changed by increasing the scan rate while the redox current increased linearly with the square root of the scan rate. From these observations and Equation (1), it was possible to give an indication that the electrochemical system was a freely diffusing quasi-reversible system. By plotting peak positions (E_{pa}, E_{pc}) versus ln(ν), a linear variation was observed and peak-to-peak separation (ΔE_p) increased linearly with ln(ν). According to Equations (2) and (3), it was found that n is approximately equal to 2 meaning that two electrons were participating in the redox reaction of paracetamol as reported in [1]. The α values are listed in Table 1 for different sensors together with the k values obtained at a scan rate of 100 mV/s for three different sensors. PEG-coated HO–$BiONO_3$ sensor had a higher rate constant of 42.0 ± 9.8 ms^{-1} with lower ΔE_p of 243 ± 10 mV compared to other sensors. This is a big advantage compared to the SPCE sensor since a higher rate constant gives rise to faster electrochemical reaction and lower ΔEp indicates a higher possibility of a reversible reaction (low resistance). Quantitative studies run by CVs measurements showed a higher sensitivity for PEG-coated HO–$BiONO_3$ compared with bare SPCE due to the nanostructure deposited on the surface [4].

Table 1. Main experimental outputs from electrochemical measurements of bare and tailored SPCEs.

Working Electrode	Peak Position (mV)	Peak Current (μA)	α	ΔEp (mV)	k (ms^{-1})	Sensitiviy (μA/mM)	Detection Limit (μM)
SPCE	459.9 ± 1.0	33.1 ± 1.5	0.44 ± 0.10	477 ± 45	0.60 ± 0.33	20.02 ± 0.23	2.34± 0.03
HO–$BiONO_3$	433.6 ± 6.1	42.0 ± 1.5	0.45 ± 0.04	392 ± 19	2.20 ± 0.82	34.88 ± 0.15	2.79 ± 0.01
PEG HO–$BiONO_3$	385 ± 30	52.1 ± 3.0	0.40 ± 0.04	243 ± 10	42.0 ± 9.8	43.50 ± 0.54	2.24 ± 0.03

3.2. Computational Evaluation

The electron transfer rate constant values coming from the computational simulations are presented in the following sections by underlying the affecting parameters on the transition state for both the HO–$BiONO_3$ dimer and trimer cases. A geometrical scheme as shown in Figure 3 acts as reference to describe the variations in the hypothetical transition state conformation depending on the functional groups bonded to the Bi active atoms. According to the assumption that paracetamol interacts with Bi cluster through its two carbon atoms, we can define:

(1) The symmetric axis passing through the organic compound C atoms that is a C_2 axis since the symmetry is maintained by a rotation of 180°
(2) The paracetamol plane *A* of symmetry on which the C ring lies that is a reflection plane
(3) A second plane *B* on which the active Bi atoms lie
(4) The dihedral angle α between these two planes
(5) The angle β defining the rotation of the paracetamol plane with respect to its symmetric axis.

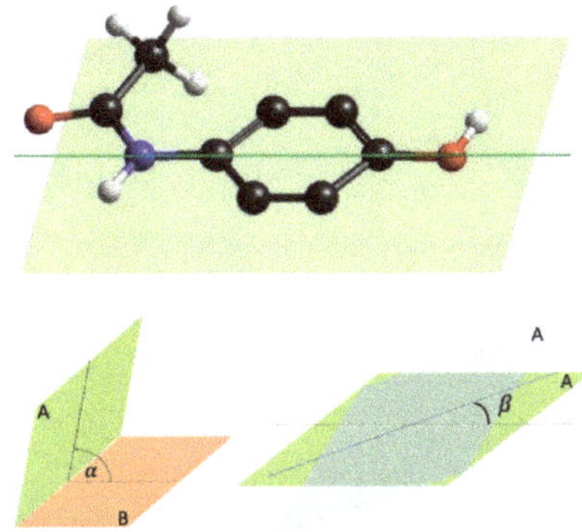

Figure 3. Paracetamol geometrical description on the left showing the symmetrical axis C_2 (right); dihedral angle is defined as the angle between the plane *A* (green) on which the paracetamol carbon ring lies and the plane *B* (orange) passing through the Bi atoms (center); description of paracetamol plane rotation of an angle β with respect to the symmetric axis (left).

3.2.1. HO–BiONO$_3$ Dimers

All the results are shown in Table 2. The different cases are labelled with the functional groups bonded to the two Bi atoms.

Table 2. Electron transfer rate constant and activation energy for Bi dimers.

Dimer	k (s^{-1})	Activation Energy (eV)
NO_3-NO_3	$1.2 \cdot 10^{-3}$	9.3
OH-OH	$2.6 \cdot 10^{-3}$	10.89
OH-NO_3	$6.9 \cdot 10^{-3}$	7.74
NO_3-OH	$2.78 \cdot 10^{-2}$	7.46

We firstly evaluated the characteristics of symmetric dimers with two nitric or hydroxylic groups as shown in Figure 4. In both cases, the electron transfer rate constant is of the order of magnitude of 10^{-3} s^{-1} a factor of 2 higher in the hydroxylic groups case. The activation energy difference is about 15%. This is reasonable, as the dihedral angle was slightly greater in NO_3-NO_3 than OH-OH by fixing the angle β, as can be seen in Figure 5. However, paracetamol approaching is hugely affected by the two functional groups. Indeed, the NO_3 group bonded with Bi (a) was above the Bi plane and moved closer to the paracetamol nitro groups by leading to a distortion of the Bi-O bond on that side. The surrounding water molecules improved that distortion and emphasized the steric hindrance. On the other hand, having four identical hydroxylic groups ensured a less steric effect since no strong repulsions between electron clouds can arise in that case.

A second comparison was made between the HTS associated to the two specular Bi-dimers having an OH group and a NO_3 alternatively bonded to each Bi atom. In this case, a larger dihedral angle resulted when NO_3 was bonded to the right Bi since the interaction with the paracetamol OH-group lead the nitric group position above the Bi plane. This also resulted in an ET rate constant equals to $6.9 \cdot 10^{-3}$ s^{-1} which is smaller than the NO_3-OH by a factor of 4. Regarding the activation energy, a difference of only 4% was obtained since the structures involved are the same in terms of atoms. From these simulations, we realized that having hydroxyl groups could enhance the sensing properties of the functionalized layer despite the steric effect arising from neighboring nitric groups.

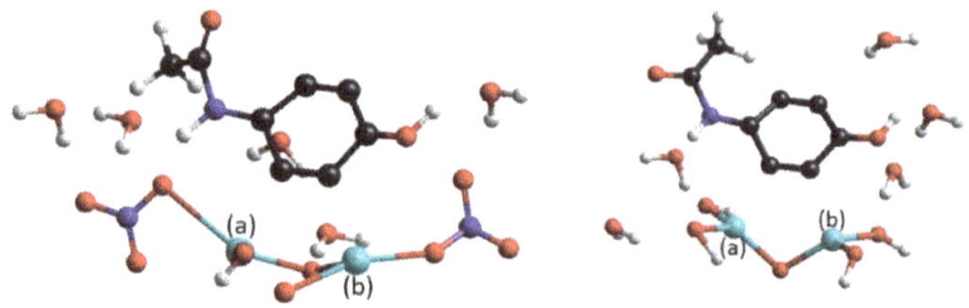

Figure 4. Molecular conformation of the HTS associated to the dimer case NO_3–NO_3 on the left and OH–OH on the right and two bismuth atoms (named **a** and **b** respectively).

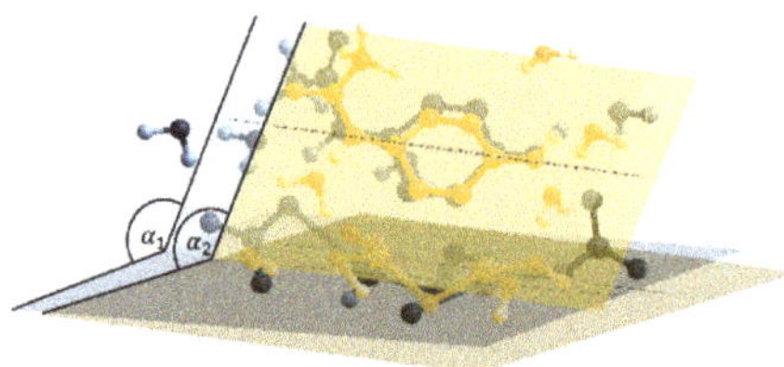

Figure 5. Overlapped molecular structure of HTS associated to the dimer NO3–NO3 (blue planes) and OH–OH (yellow planes): α angles comparison at fixed β.

3.2.2. HO–$BiONO_3$ Trimer

All the results are shown in Table 3. The different cases are labelled with the functional groups bonded to the three Bi atoms.

Table 3. k and activation energy for Bi trimers.

Trimer	k (s^{-1})	Activation Energy (eV)
NO_3-NO_3-NO_3	$7.08 \cdot 10^{-5}$	7.34
NO_3-OH-NO_3	$5.9 \cdot 10^{-2}$	5.75
NO_3-OH-OH	$8.46 \cdot 10^{-2}$	9.16
OH-OH-NO_3	$8.3 \cdot 10^{-2}$	9.87

When we simulated the HTS associated to the trimer cases, we introduced a further assumption by distinguishing among the reacting Bi atoms and the non-reacting ones since just two Bi atoms are supposed to interact with the paracetamol C atoms. Based on that, we observed that the steric effect was more evident in the NO_3-OH-NO_3 structure due to the orientation of the nitric group bonded with the non-reacting Bi atom towards the nitro group of the organic compound. On the contrary, no distortion occurs in the NO_3-NO_3-NO_3 because the functional groups bonded with reacting Bi lie in the same plane. However, lower activation energy of the transition state made the NO_3-OH-NO_3 more energetically favorable to the ET process and further confirmed that the presence of some OH groups should improve the efficiency of the ET process. Indeed, the ET rate constant was three orders of magnitude greater in NO_3-OH-NO_3 case than the NO_3-NO_3-NO_3 case.

Finally, the specular case for the trimer was simulated by replacing the two nitric groups with hydroxyl as shown in Figure 6. We observed that NO_3-OH-OH was strongly affected by the orientation of NO_3 above the Bi plane towards the paracetamol nitro group that is also above the Bi plane. Nevertheless, no distortion occurred, and it can be seen in Figure 7 that the dihedral angle was qualitatively the same as for the OH–OH–NO_3 case. These also let to ET values in the two specular trimers with a difference less than 2% and geometries with very similar the dihedral angles. Indeed, we obtained a value of 9.16 eV for NO_3–OH–OH and 9.87 eV for OH–OH–NO_3.

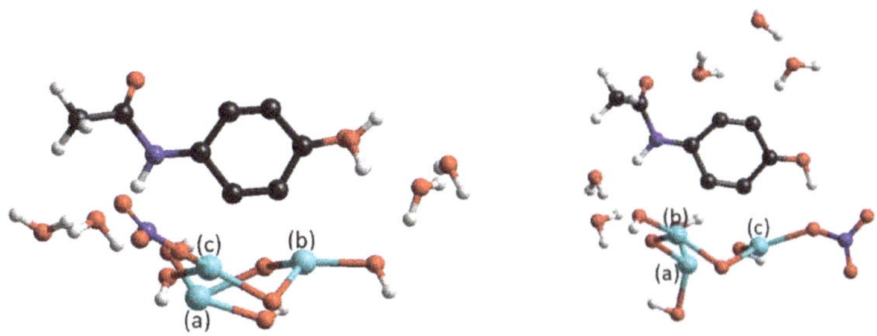

Figure 6. Molecular conformation of the HTS associated to the dimer case NO_3–OH–OH on the left and OH–OH–NO_3 on the right and three bismuth atoms (named **a**, **b** and **c** respectively).

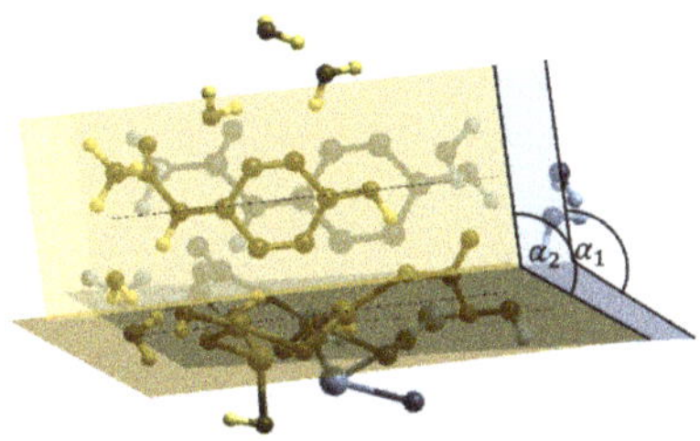

Figure 7. Overlapped molecular structure of HTS associated to the dimer OH–OH–NO_3 (blue planes) and NO_3–OH–OH (yellow planes): α angles comparison at fixed β.

4. Conclusions

As clearly shown from the data, the electrochemical sensing boost observed for HO–$BiONO_3$ tailored SPCEs is likely due to defective sites on HO–$BiONO_3$ particles sitting on the surface of the electrode. The hydroxyl functionalities played a relevant role in the paracetamol–bismuth interaction that defines the geometry of the transitional state as showed by the computational study. Accordingly, a more polar surface induced a better interaction with paracetamol.

This study has firstly described in a comprehensive way the effect of bismuth subnitrate chemistry in an electrochemical system considering both empirical and experimental data sets, leading to the foundation for a rational design of bismuth modified SPCEs.

Supplementary Materials: The following are available online at https://www.mdpi.com/article/10.3390/CSAC2021-10427/s1.

Author Contributions: Conceptualization, A.T., S.C. and M.B.; formal analysis, M.M. and F.C.; resources, A.T. and S.C.; data curation, M.M. and F.C.; writing—original draft preparation, M.M., F.C. and M.B.; writing—review and editing, M.M., F.C., M.B., A.T. and S.C.; supervision, M.B.; project administration, A.T. and S.C. All authors have read and agreed to the published version of the manuscript.

Funding: This research received no external funding.

Conflicts of Interest: The authors declare no conflict of interest.

References

1. Kang, X.; Wang, J.; Wu, H.; Liu, J.; Aksay, I.A.; Lin, Y. A graphene-based electrochemical sensor for sensitive detection of paracetamol. *Talanta* **2010**, *81*, 754–759. [CrossRef] [PubMed]
2. Li, J.; Liu, J.; Tan, G.; Jiang, J.; Peng, S.; Deng, M.; Qian, D.; Feng, Y.; Liu, Y. High-sensitivity paracetamol sensor based on Pd/graphene oxide nanocomposite as an enhanced electrochemical sensing platform. *Biosens. Bioelectron.* **2014**, *54*, 468–475. [CrossRef] [PubMed]

3. Li, M.; Wang, W.; Chen, Z.; Song, Z.; Luo, X. Electrochemical determination of paracetamol based on Au@ graphene core-shell nanoparticles doped conducting polymer PEDOT nanocomposite. *Sens. Actuators B Chem.* **2018**, *260*, 778–785. [CrossRef]
4. Madagalam, M.; Bartoli, M.; Tagliaferro, A.; Carrara, S. Bismuth-nanocomposites modified SPCEs for non-enzymatic electrochemical sensors. *IEEE Sens. J.* **2021**, *21*, 11155–11162. [CrossRef]
5. Bartoli, M.; Jagdale, P.; Tagliaferro, A. A Short Review on Biomedical Applications of Nanostructured Bismuth Oxide and Related Nanomaterials. *Materials* **2020**, *13*, 5234. [CrossRef] [PubMed]
6. Aliakbarinodehi, N.; Taurino, I.; Pravin, J.; Tagliaferro, A.; Piccinini, G.; De Micheli, G.; Carrara, S. Electrochemical nanostructured biosensors: Carbon nanotubes versus conductive and semi-conductive nanoparticles. *Chem. Pap.* **2015**, *69*, 134–142. [CrossRef]
7. Đurđić, S.; Vukojević, V.; Vlahović, F.; Ognjanović, M.; Švorc, L'.; Kalcher, K.; Mutić, J.; Stanković, D.M. Application of bismuth (III) oxide decorated graphene nanoribbons for enzymatic glucose biosensing. *J. Electroanal. Chem.* **2019**, *850*, 113400. [CrossRef]
8. Liu, G.-Q.; Zhong, H.; Li, X.-R.; Yang, K.; Jia, F.-F.; Cheng, Z.-P.; Zhang, L.-L.; Yin, J.-Z.; Guo, L.-P.; Qian, H.-Y. Research on nonenzymatic electrochemical sensor using HO-BiONO3 nanocomposites for glucose detection. *Sens. Actuators B Chem.* **2017**, *242*, 484–491. [CrossRef]
9. Bard, A.J.; Faulkner, L.R. *Electrochemical Methods Fundamentals and Applications*; Wiley: New York, NY, USA, 2001.
10. Laviron, E. General expression of the linear potential sweep voltammogram in the case of diffusionless electrochemical systems. *J. Electroanal. Chem. Interfacial Electrochem.* **1979**, *101*, 19–28. [CrossRef]
11. Brunk, E.; Rothlisberger, U. Mixed Quantum Mechanical/Molecular Mechanical Molecular Dynamics Simulations of Biological Systems in Ground and Electronically Excited States. *Chem. Rev.* **2015**, *115*, 6217–6263. [CrossRef] [PubMed]
12. Stewart, J.J.P. Optimization of parameters for semiempirical methods I. Method. *J. Comput. Chem.* **1989**, *10*, 209–220. [CrossRef]
13. Marcus, R.A. Chemical and electrochemical electron-transfer theory. *Annu. Rev. Phys. Chem.* **1964**, *15*, 155–196. [CrossRef]
14. Tachiya, M. Generalization of the Marcus equation for the electron-transfer rate. *J. Phys. Chem.* **1993**, *97*, 5911–5916. [CrossRef]
15. Serpone, N.; Maruthamuthu, P.; Pichat, P.; Pelizzetti, E.; Hidaka, H. Exploiting the interparticle electron transfer process in the photocatalysed oxidation of phenol, 2-chlorophenol and pentachlorophenol: Chemical evidence for electron and hole transfer between coupled semiconductors. *J. Photochem. Photobiol. A Chem.* **1995**, *85*, 247–255. [CrossRef]

chemistry proceedings

MDPI

Abstract

Plasmonic Hydrogel Nanocomposites with Combined Optical and Mechanical Properties for Biochemical Sensing †

Bruno Miranda [1,2,*], Rosalba Moretta [1], Selene De Martino [3], Principia Dardano [1], Ilaria Rea [1], Carlo Forestiere [2] and Luca De Stefano [1]

1 Institute of Applied Sciences and Intelligent Systems, Via P. Castellino 111, 80131 Naples, Italy; rosalba.moretta@na.isasi.cnr.it (R.M.); principia.dardano@na.isasi.cnr.it (P.D.); ilaria.rea@na.isasi.cnr.it (I.R.); luca.destefano@na.isasi.cnr.it (L.D.S.)

2 Department of Electrical Engineering and Information Technology, University of Naples "Federico II", Via Claudio 21, 80125 Naples, Italy; carlo.forestiere@unina.it

3 Materias s.r.l., Via Protopisani 50, 80131 Naples, Italy; selene.demartino@na.isasi.cnr.it

* Correspondence: bruno.miranda@na.isasi.cnr.it

† Presented at the 1st International Electronic Conference on Chemical Sensors and Analytical Chemistry, Online, 1–15 July 2021; Available online: https://csac2021.sciforum.net/.

Abstract: Localized surface plasmon resonance (LSPR) and metal-enhanced-fluorescence (MEF)-based optical biosensors exhibit unique properties compared to other sensing devices that can be exploited for the design point-of-care (POC) diagnostic tools [1]. Plasmonic devices exploit the capability of noble-metal nanoparticles of absorbing light at a well-defined wavelength. The increasing request for wearable, flexible and easy-to-use diagnostic tools has brought to the development of plasmonic nanocomposites, whose peculiar performances arise from the combination of the optical properties of plasmonic nanoparticles and mechanical properties of the polymeric matrix in which they are embedded [2,3]. An optical platform based on spherical gold nanoparticles (AuNPs) embedded in high molecular weight poly-(ethylene glycol) diacrylate (PEGDA) hydrogel is proposed. PEGDA hydrogel represents a biocompatible, flexible, transparent polymeric network to design wearable, 3D, plasmonic biosensors for the detection of targets with different molecular weights for the early diagnosis of disease. The swelling capability of PEGDA is directly correlated to the plasmonic decoupling of AuNPs embedded within the matrix. A study on the effect of swelling on the optical response of the PEGDA/AuNPs composites was investigated by using a biorecognition layer/target model system. Specifically, after the in situ chemical modification of the AuNPs within the hydrogel, the interaction biotin-streptavidin is monitored within the 3D hydrogel network. Additionally, metal-enhanced fluorescence is observed within the PEGDA/AuNPs nanocomposites, which can be exploited to achieve an ultra-low limit of detection. LSPR signal was monitored via transmission mode customized setup and MEF signal was detected via fluorescence and confocal microscopes. Label-free (LSPR-based) and fluorescence (MEF-based) signals of a high molecular weight target analyte were successfully monitored with relatively high resolutions and low limits of detection compared to the standard polymeric optical platforms available in the literature. The optimized platform could represent a highly reproducible and low-cost novel biosensor to be applied as a POC diagnostic tool in healthcare and food monitoring applications.

Keywords: optical biosensors; flexible hybrid materials; disease early diagnosis; nanofabrication techniques; nanocomposite materials; LSPR-based biosensors; metal-enhanced fluorescence

Citation: Miranda, B.; Moretta, R.; Martino, S.D.; Dardano, P.; Rea, I.; Forestiere, C.; Stefano, L.D. Plasmonic Hydrogel Nanocomposites with Combined Optical and Mechanical Properties for Biochemical Sensing. *Chem. Proc.* **2021**, *5*, 34. https://doi.org/10.3390/CSAC2021-10467

Academic Editor: Elisabetta Comini

Published: 30 June 2021

Supplementary Materials: The following are available online at https://www.mdpi.com/article/10.3390/CSAC2021-10467/s1.

Author Contributions: Conceptualization, L.D.S., I.R., and P.D.; methodology, B.M., C.F., S.D.M. and R.M.; validation, B.M. and C.F.; formal analysis, B.M.; investigation, B.M., and R.M.; resources, L.D.S.;

writing—original draft preparation, B.M., and R.M.; writing—review and editing, C.F., I.R., and P.D.; supervision, L.D.S. All authors have read and agreed to the published version of the manuscript.

Funding: This research received no external funding.

Institutional Review Board Statement: Not applicable.

Informed Consent Statement: Not applicable.

Data Availability Statement: The data that support the findings of this study are available from the corresponding author upon reasonable request.

Conflicts of Interest: The authors declare no conflict of interest.

References

1. Miranda, B.; Chu, K.-Y.; Maffettone, P.L.; Shen, A.Q.; Funari, R. Metal-Enhanced Fluorescence Immunosensor Based on Plasmonic Arrays of Gold Nanoislands on an Etched Glass Substrate. *ACS Appl. Nano Mater.* **2020**, *3*, 10470–10478. [CrossRef]
2. Miranda, B.; Rea, I.; Dardano, P.; De Stefano, L.; Forestiere, C. Recent Advances in the Fabrication and Functionalization of Flexible Optical Biosensors: Toward Smart Life-Sciences Applications. *Biosensors* **2021**, *11*, 107. [CrossRef] [PubMed]
3. Miranda, B.; Moretta, R.; De Martino, S.; Dardano, P.; Rea, I.; Forestiere, C.; De Stefano, L. A PEGDA hydrogel nanocomposite to improve gold nanoparticles stability for novel plasmonic sensing platforms. *J. Appl. Phys.* **2021**, *129*, 033101. [CrossRef]

chemistry proceedings

Proceeding Paper

Statistical Analysis for Selective Identifications of VOCs by Using Surface Functionalized MoS_2 Based Sensor Array †

Uttam Narendra Thakur *, Radha Bhardwaj and Arnab Hazra *

Department of Electrical & Electronics Engineering, Birla Institute of Technology and Science (BITS)-Pilani, Vidya Vihar, Pilani 333031, India; radikabhardwaj.rb@gmail.com
* Correspondence: unthakur08@gmail.com (U.N.T.); arnabhazra2013@gmail.com (A.H.)
† Presented at the 1st International Electronic Conference on Chemical Sensors and Analytical Chemistry, 1–15 July 2021; Available online: https://csac2021.sciforum.net/.

Abstract: Disease diagnosis through breath analysis has attracted significant attention in recent years due to its noninvasive nature, rapid testing ability, and applicability for patients of all ages. More than 1000 volatile organic components (VOCs) exist in human breath, but only selected VOCs are associated with specific diseases. Selective identification of those disease marker VOCs using an array of multiple sensors are highly desirable in the current scenario. The use of efficient sensors and the use of suitable classification algorithms is essential for the selective and reliable detection of those disease markers in complex breath. In the current study, we fabricated a noble metal (Au, Pd and Pt) nanoparticle-functionalized MoS_2 (Chalcogenides, Sigma Aldrich, St. Louis, MO, USA)-based sensor array for the selective identification of different VOCs. Four sensors, i.e., pure MoS_2, Au/MoS_2, Pd/MoS_2, and Pt/MoS_2 were tested under exposure to different VOCs, such as acetone, benzene, ethanol, xylene, 2-propenol, methanol and toluene, at 50 °C. Initially, principal component analysis (PCA) and linear discriminant analysis (LDA) were used to discriminate those seven VOCs. As compared to the PCA, LDA was able to discriminate well between the seven VOCs. Four different machine learning algorithms such as k-nearest neighbors (kNN), decision tree, random forest, and multinomial logistic regression were used to further identify those VOCs. The classification accuracy of those seven VOCs using KNN, decision tree, random forest, and multinomial logistic regression was 97.14%, 92.43%, 84.1%, and 98.97%, respectively. These results authenticated that multinomial logistic regression performed best between the four machine learning algorithms to discriminate and differentiate the multiple VOCs that generally exist in human breath.

Keywords: breath analysis; surface functionalized MoS_2; classification; discrimination

Citation: Thakur, U.N.; Bhardwaj, R.; Hazra, A. Statistical Analysis for Selective Identifications of VOCs by Using Surface Functionalized MoS_2 Based Sensor Array. *Chem. Proc.* **2021**, 5, 35. https://doi.org/10.3390/CSAC2021-10451

Academic Editor: Elisabetta Comini

Published: 30 June 2021

1. Introduction

In the field of medical diagnostic and health care systems, breath analysis has gained a lot of interest for the noninvasive detection of diseases and monitoring of health parameters [1,2]. More than 1000 volatile organic components (VOCs) are present in exhaled breath, but only some of them are considered disease markers [3,4]. In this context, the selective detection of the different VOCs using smart sensor systems has a high demand for efficient breath analysis. Selective detection can also be achieved using suitable pattern recognition algorithms on sensor signals. For the early detection of disease, the combination of a highly selective sensor and an effective machine learning algorithm is required. Diagnostics through breath is less time consuming compared to the clinical process and, at the same time, it is cost-efficient as does not require well-trained professionals and the sensors are less costly [5,6].

Chemiresistive sensors typically recognize a target VOC by changing their resistance depending upon the adsorption-desorption properties of the analyte to the detecting layer surface. An extensive variety of materials are used for VOC sensing, including thin metal films [7], metal oxides [8–10], polymers [11], etc. Accessible surface functionalization possibilities, high surface area to volume ratio, increased flexibility, high sensitivity, and tunable bandgap make two-dimensional molybdenum disulfide (MoS_2) an encouraging channel material to sense the VOC [12,13].

A pattern recognition algorithm also plays an essential role in the detection of VOC. A suitable classifier is required to achieve an effective classification rate in VOC sensing based on the sensor data. Different algorithms such as partial least squares discriminant analysis [14], canonical discriminant analysis [15], k-nearest neighbor [4,16], Discriminant function analysis [17], support vector machine [18], random forest [19], logistic regression [20], etc. were reported in the literature. In some of the reported literature, different types of neural network classifiers were used [21–24].

In the current study, we used principal component analysis (PCA) and linear discriminant analysis (LDA) to visualize the data in lesser dimensions compared to the original extent. Furthermore, four different supervised algorithms, k-nearest neighbor (kNN), decision tree, random forest, and multinomial logistic regression, were implemented to identify the best-suited algorithm based on their performance.

2. Material and Methods

2.1. Preparation of MoS_2 and Noble Metal Nanoparticles Solutions

All materials MoS_2 (Chalcogenides, Sigma Aldrich, St. Louis, MO, USA), gold (III) chloride ($AuCl_3$, 99%, Sigma Aldrich), palladium chloride ($PdCl_2$, 60%, Molychem, Mumbai, India), and chloroplatinic acid ($H_4PtCl_6xH_2O$, 40%, Molychem) were analytical grade and used without any further purification. The 0.2 wt% MoS_2 solution was prepared in deionized water and stirred for 1.5 h at room temperature to maintain homogeneity. Similarly, 0.1 MM aqueous solutions of noble metal nanoparticles (Au, Pd, Pt) were prepared by adding corresponding metal salts in deionized water with continuous stirring and dropwise diluted HCl was also added to obtain stable and uniform nanoparticles at room temperature.

Au, Pd and Pt nanoparticle-loaded MoS_2 samples were prepared using the spray coating technique. Firstly, MoS_2 solution was spray coated on washed SiO_2/Si substrate and dried at room temperature. In the final step, nanoparticle solutions were spray-coated on previously deposited MoS_2 and dried at room temperature.

Thermal annealing was performed for 4 h at 250 °C to provide crystallization and thermal stability in all 4 samples (MoS_2, Au-MOS_2, Pd-MoS_2, and Pt-MoS_2). The MoS_2 flakes were coated by very tiny Au NPs (12 nm), whereas the larger Pd NPs (54 nm) were deposited, maintaining a consistent spacing. Pt nanoparticles on the MoS_2 surface had the biggest size (77 nm) of the three.

2.2. Fabrication of Sensors

Au source and drain electrodes of 150 nm thickness were deposited on all four samples by using an electron beam evaporation unit. Sensors were then placed into a sensor holder, and further sensing performance was studied.

The sensor holder was placed in a glass sealed sensing chamber of 650 mL on a heating plate. The sensing performance of prepared sensors was examined by a static mode sensing setup where VOCs were injected using micro syringes (Hamilton micro syringe). The sensor was recovered by flowing 450 SCCM synthetic air by using a mass flow controller. The amount of injected VOC was calculated by using the formula: C (ppm) = 2.46 × (V1D/VM) × 103, where D (gm/mL), M (gm/mol), and V (L) represent density of the VOC, molecular weight of the VOC and volume of vaporization chamber, respectively [13,25,26]. Seven different VOCs, i.e., acetone, 2-propanol, benzene, ethanol, methanol, toluene, and xylene were tested during the study. Using a Keithley 6487 source meter, sensing performance was recorded applying 1 V constant bias. The sensitivity of the sensor was calculated by formula; Ra − Rv/Ra × 100 where Ra and Rv were the resistances of the sensor in the air and in target VOC.

To read the generated output of sensors stored in CSV file a python script was used. All the algorithms, analysis, and plotting were performed on Python 3.7 and Jupiter notebook as a platform.

3. Results and Discussion

3.1. VOC Sensing

As a reference ambient, synthetic air was used to perform the gas sensing measurements of the four different sensors: pure MoS_2, Au-MoS_2, Pd-MoS_2, and Pt-MoS_2. Figure 1 shows the change in resistance (MΩ) with respect to time at 50 °C. In the presence of VOCs, as the exposer time increased, the resistance offered by the sensor decreased. This decrease in resistance confirms that the sensor is an n-type property. In the presence of seven distinct VOCs, i.e., acetone, 2-propanol, benzene, ethanol, methanol, toluene, and xylene, four different sensors, i.e., pure MoS_2, Au-MoS_2, Pd-MoS_2, and Pt-MoS_2, were observed and stored for further processing of data.

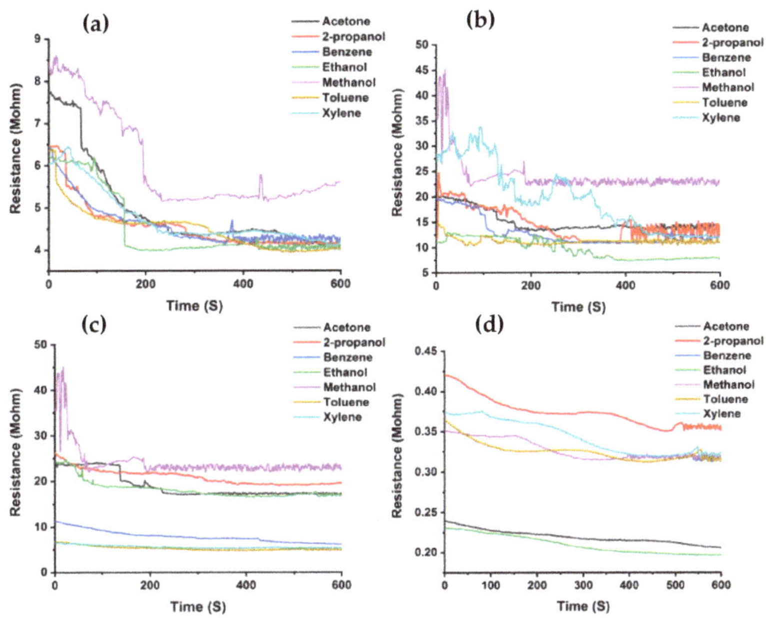

Figure 1. Change in resistance offered by sensors (**a**) MoS_2 (**b**) Au-MoS_2 (**c**) Pd-MoS_2 (**d**) Pt-MoS_2 with respect to time in presence of 7 VOCs.

3.2. Data Analysis

Figure 2 describes the influence of the volatile organic components (VOCs) on the outcomes of two-dimensionality reduction techniques: principal component analysis and linear discriminant analysis. The measurement parameters were kept constant during the experiment. The operating temperature was 50 °C, the response was taken up to 600 s. and the sample concentration was 100 ppm.

The response obtained by the four different sensors for seven different VOCs was used for principal component analysis (PCA). The three-dimensional plot between the first principal component (PC1), second principal component (PC2), and third principal component (PC3) is represented in Figure 2. As we have four independent variables (sensor responses), the maximum principal component obtained was four. Therefore, we have considered only the first three principal components contributing the most to the explained variance in this analysis. The total explained variance was 93.58%, in which PC1 contributed 52.52%, PC2 contributed 30.91%, and PC3 contributed 10.14%. All seven VOCs had their compact clusters, and they separated. Still, the separation between the cluster of acetone/2-propanol and benzene/toluene was quite small, which increased the possibility of misclassification.

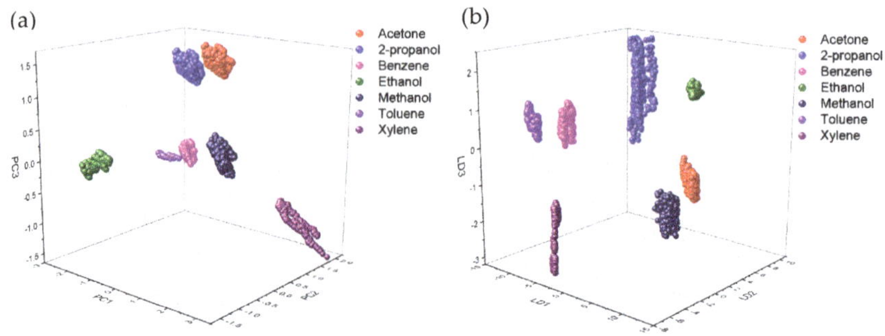

Figure 2. Scatter plot from the exposure of four sensors to seven VOCs in (**a**) PCA (**b**) LDA.

Taking account of the problem of discrimination between the different VOCs, a linear discrimination analysis was also performed. The same sensor response vector was used in the linear discriminant analysis (LDA). Figure 2b shows that the employment of the classifier allowed the discrimination of all seven VOCs. Thus, LDA is highly efficient for investigating the VOCs based on the sensor response. A three-dimensional plot is shown in Figure 2b, which clearly depicts the performance of the LDA on the raw data (sensor response vector). The different VOCs were densely clustered within their groups, and they were well separated from each other. So, there was a significantly lower probability of misclassification among the VOCs. The 2-propanol was slightly more stretched along the second linear discriminant function (LD2) axis, and xylene was along the third discriminant function (LD3). The three discriminant functions, LD1, LD2, and LD3 contributed 71.22%, 27.42% and 1.21%, respectively, the total resultant explained the variance for the classifier becoming 99.85%.

3.3. *VOC Identification*

The previously discussed LDA and PCA plot gives only the visual representation of the separation of VOCs based on the sensor response. The goal of the sensor setup is to design a generalized model based on the known data during the training phase and try to predict the class when an unknown data sample is encountered.

The supervised algorithm was performed in the current work to determine the VOCs; four different machine learning algorithms such as k-nearest neighbor (kNN), decision tree, random forest, and multinomial logistic regression were used to identify those seven VOCs. The normalized sensor response was fed to the algorithms, and the whole data set was divided into training testing data with 70% and 30%, respectively. The data set consisted of 4200 measurements of each sensor, with each class containing 600 data vectors and seven classes. So, 2940 vectors were used to train the model, and the remaining 1260 vectors were used to test the model. For identification of VOCs, above 84% was the classification accuracy for every classifier with an accuracy of 97.14%, 92.43%, 84.1%, 98.97% for kNN, decision tree, random forest, and multinomial logistic regression, respectively. A confusion matrix is used to calculate the classification accuracy, and the confusion matrix furnishes the observation into what components were mistakenly classified. Figure 3a shows the confusion matrix of kNN where 11 samples of toluene were classified as xylene and 10 samples of benzene were wrongly predicted as ethanol. Figure 3b is a representation of the confusion matrix obtained from the decision tree algorithm. The confusion matrix of the random forest and multinomial logistic regression are shown in Figure 3c,d, respectively. In multinomial logistic regression, only 12 benzene samples were identified as acetone, and one sample of xylene was identified as toluene.

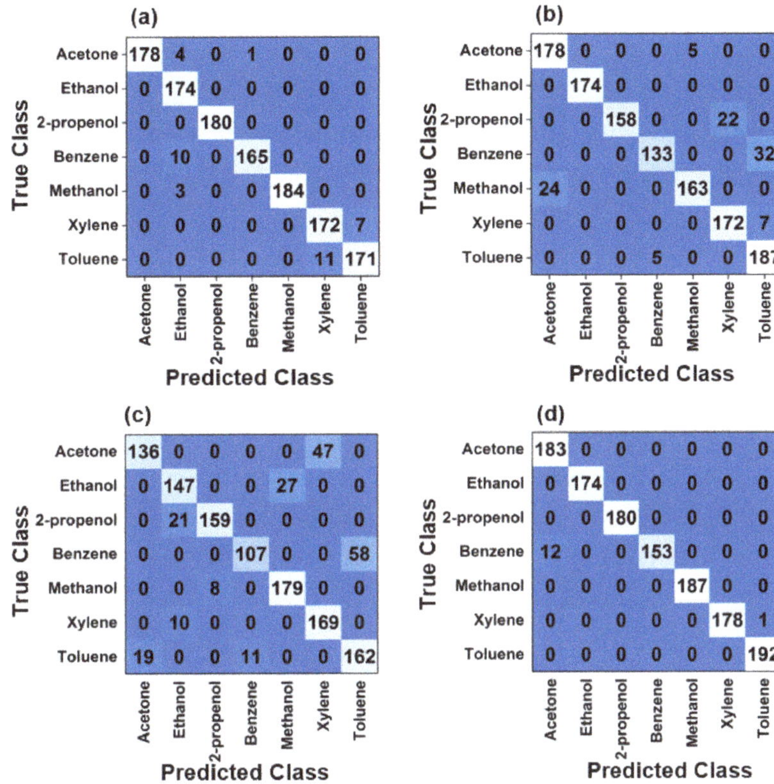

Figure 3. Confusion matrix of (**a**) k-nearest neighbor, (**b**) decision tree, (**c**) random forest, and (**d**) multinomial logistic regression.

4. Conclusions

The ability of a surface-functionalized MoS_2 sensor to distinguish between the various VOCs was appraised by PCA and LDA, in which LDA laid out the excellent separation between the VOCs. Further, to evaluate the effectiveness of the sensor output to identify the VOCs, four different machine learning-based (supervised) classification algorithms were implemented, and among them, the k-nearest neighbor and multinomial logistic regression were performed outstandingly with an accuracy of 97.14% and 98.97%, respectively. Thus, high selectivity and accuracy authenticate that the system discriminates and differentiates the multiple VOCs that generally exist in human breath.

Supplementary Materials: The following are available online at https://www.mdpi.com/article/10.3390/CSAC2021-10451/s1.

Author Contributions: U.N.T. is responsible for the experiment and analyzed data. R.B. is responsible for the research activity with fabrication of sensors and their characterization and conceived the experiment. A.H. designed the research. All authors have read and agreed to the published version of the manuscript.

Funding: This work was supported in part by SPARC grant (SPARC/2018-2019/P1394/SL), Ministry of Human resource development (MHRD), Govt. of India and Department of Biotechnology grant (Letter No. BT/PR28727/NNT/28/1569/ 2018).

Institutional Review Board Statement: Not applicable.

Informed Consent Statement: Not applicable.

Data Availability Statement: Not applicable.

Conflicts of Interest: The authors declare no conflict of interest.

References

1. Dragonieri, S.; Pennazza, G.; Carratu, P.; Resta, O. Electronic Nose Technology in Respiratory Diseases. *Lung* **2017**, *195*, 157–165. [CrossRef]
2. Sahatiya, P.; Kadu, A.; Gupta, H.; Gomathi, P.T.; Badhulika, S. Flexible, Disposable Cellulose-Paper-Based MoS_2/Cu_2S Hybrid for Wireless Environmental Monitoring and Multifunctional Sensing of Chemical Stimuli. *ACS Appl. Mater. Interfaces* **2018**, *10*, 9048–9059. [CrossRef]
3. Cikach, F.S.; Dweik, R.A. Cardiovascular Biomarkers in Exhaled Breath. *Prog. Cardiovasc. Dis.* **2012**, *55*, 34–43. [CrossRef]
4. Pereira, J.; Porto-Figueira, P.; Cavaco, C.; Taunk, K.; Rapole, S.; Dhakne, R.; Nagarajaram, H.; Câmara, J.S. Breath analysis as a potential and non-invasive frontier in disease diagnosis: An overview. *Metabolites* **2015**, *5*, 3–55. [CrossRef]
5. Arshak, K.; Moore, E.; Lyons, G.M.; Harris, J.; Clifford, S. A review of gas sensors employed in electronic nose applications. *Sens. Rev.* **2004**, *24*, 181–198. [CrossRef]
6. Tran, V.H.; Chan, H.P.; Thurston, M.; Jackson, P.; Lewis, C.; Yates, D.; Bell, G.; Thomas, P.S. Breath analysis of lung cancer patients using an electronic nose detection system. *IEEE Sens. J.* **2010**, *10*, 1514–1518. [CrossRef]
7. Pundt, A. Hydrogen in Nano-sized Metals. *Adv. Eng. Mater.* **2004**, *6*, 11–21. [CrossRef]
8. Hazra, A. Amplified Methanol Sensitivity in Reduced Graphene Oxide FET Using Appropriate Gate Electrostatic. *IEEE Trans. Electron. Devices* **2020**, *67*, 5111–5118. [CrossRef]
9. Gakhar, T.; Hazra, A. Oxygen vacancy modulation of titania nanotubes by cathodic polarization and chemical reduction routes for efficient detection of volatile organic compounds. *Nanoscale* **2020**, *12*, 9082–9093. [CrossRef]
10. Kanazawa, E.; Sakai, G.; Shimanoe, K.; Kanmura, Y.; Teraoka, Y.; Miura, N.; Yamazoe, N. Metal oxide semiconductor N_2O sensor for medical use. *Sens. Actuators B Chem.* **2001**, *77*, 72–77. [CrossRef]
11. Bai, H.; Shi, G. Gas Sensors Based on Conducting Polymers. *Sensors* **2007**, *7*, 267–307. [CrossRef]
12. Selamneni, V.; Raghavan, H.; Hazra, A.; Sahatiya, P. MoS 2/Paper Decorated with Metal Nanoparticles (Au, Pt, and Pd) Based Plasmonic-Enhanced Broadband (Visible-NIR) Flexible Photodetectors. *Adv. Mater. Interfaces* **2021**, *8*, 2001988. [CrossRef]
13. Bindra, P.; Hazra, A. Selective detection of organic vapors using TiO_2 nanotubes based single sensor at room temperature. *Sens. Actuators B Chem.* **2019**, *290*, 684–690. [CrossRef]
14. Di Natale, C.; Macagnano, A.; Martinelli, E.; Paolesse, R.; D'Arcangelo, G.; Roscioni, C.; Finazzi-Agrò, A.; D'Amico, A. Lung cancer identification by the analysis of breath by means of an array of non-selective gas sensors. *Biosens. Bioelectron.* **2003**, *18*, 1209–1218. [CrossRef]
15. Fens, N.; Roldaan, A.C.; Van Der Schee, M.P.; Boksem, R.J.; Zwinderman, A.H.; Bel, E.H.; Sterk, P.J. External validation of exhaled breath profiling using an electronic nose in the discrimination of asthma with fixed airways obstruction and chronic obstructive pulmonary disease. *Clin. Exp. Allergy* **2011**, *41*, 1371–1378. [CrossRef] [PubMed]
16. Haddi, Z.; Mabrouk, S.; Bougrini, M.; Tahri, K.; Sghaier, K.; Barhoumi, H.; El Bari, N.; Maaref, A.; Jaffrezic-Renault, N.; Bouchikhi, B. E-Nose and e-Tongue combination for improved recognition of fruit juice samples. *Food Chem.* **2014**, *150*, 246–253. [CrossRef] [PubMed]
17. Ghasemi-Varnamkhasti, M.; Mohammad-Razdari, A.; Yoosefian, S.H.; Izadi, Z.; Siadat, M. Aging discrimination of French cheese types based on the optimization of an electronic nose using multivariate computational approaches combined with response surface method (RSM). *Lwt* **2019**, *111*, 85–98. [CrossRef]
18. Saidi, T.; Zaim, O.; Moufid, M.; El Bari, N.; Ionescu, R.; Bouchikhi, B. Exhaled breath analysis using electronic nose and gas chromatography–mass spectrometry for non-invasive diagnosis of chronic kidney disease, diabetes mellitus and healthy subjects. *Sens. Actuators B Chem.* **2018**, *257*, 178–188. [CrossRef]
19. Nallon, E.C.; Schnee, V.P.; Bright, C.; Polcha, M.P.; Li, Q. Chemical Discrimination with an Unmodified Graphene Chemical Sensor. *ACS Sens.* **2016**, *1*, 26–31. [CrossRef]
20. Poli, D.; Carbognani, P.; Corradi, M.; Goldoni, M.; Acampa, O.; Balbi, B.; Bianchi, L.; Rusca, M.; Mutti, A. Exhaled volatile organic compounds in patients with non-small cell lung cancer: Cross sectional and nested short-term follow-up study. *Respir. Res.* **2005**, *6*, 71. [CrossRef]
21. Srivastava, A.K. Detection of volatile organic compounds (VOCs) using SnO_2 gas-sensor array and artificial neural network. *Sens. Actuators B Chem.* **2003**, *96*, 24–37. [CrossRef]
22. Fu, J.; Li, G.; Qin, Y.; Freeman, W.J. A pattern recognition method for electronic noses based on an olfactory neural network. *Sens. Actuators B Chem.* **2007**, *125*, 489–497. [CrossRef]
23. Dutta, R.; Morgan, D.; Baker, N.; Gardner, J.W.; Hines, E.L. Identification of Staphylococcus aureus infections in hospital environment: Electronic nose based approach. *Sens. Actuators B Chem.* **2005**, *109*, 355–362. [CrossRef]
24. Montuschi, P.; Santonico, M.; Mondino, C.; Pennazza, G.; Mantini, G.; Martinelli, E.; Capuano, R.; Ciabattoni, G.; Paolesse, R.; Di Natale, C.; et al. Diagnostic performance of an electronic nose, fractional exhaled nitric oxide, and lung function testing in asthma. *Chest* **2010**, *137*, 790–796. [CrossRef] [PubMed]

25. Bhardwaj, R.; Selamneni, V.; Thakur, U.N.; Sahatiya, P.; Hazra, A. Detection and discrimination of volatile organic compounds by noble metal nanoparticle functionalized MoS_2 coated biodegradable paper sensors. *New J. Chem.* **2020**, *44*, 16613–16625. [CrossRef]
26. Bindra, P.; Hazra, A. Electroless deposition of Pd/Pt nanoparticles on electrochemically grown TiO_2 nanotubes for ppb level sensing of ethanol at room temperature. *Analyst* **2021**, *146*, 1880–1891. [CrossRef] [PubMed]

chemistry proceedings

MDPI

Proceeding Paper

Evaluation of the Effect of Extracted Time Conditions on the Phenolic Content of Olive Pastes from *cv.* Arbequina and Discrimination Using a Lab-Made Potentiometric Electronic Tongue †

Ítala M. G. Marx [1,2,*], Nuno Rodrigues [1], Ana C. A. Veloso [3,4], José A. Pereira [1] and António M. Peres [1]

1 CIMO, Instituto Politécnico de Bragança, 5300-253 Bragança, Portugal; nunorodrigues@ipb.pt (N.R.); jpereira@ipb.pt (J.A.P.); peres@ipb.pt (A.M.P.)
2 LAQV/REQUIMTE, Faculdade de Farmácia da Universidade do Porto, 4050-313 Porto, Portugal
3 Instituto Politécnico de Coimbra, ISEC, DEQB, 3030-199 Coimbra, Portugal; anaveloso@isec.pt
4 CEB—Centre of Biological Engineering, Campus de Gualtar, University of Minho, 4710-057 Braga, Portugal
* Correspondence: itala.marx@ipb.pt

† Presented at the 1st International Electronic Conference on Chemical Sensors and Analytical Chemistry, 1–15 July 2021; Available online: https://csac2021.sciforum.net/.

Abstract: The present study investigated the effect of malaxation times (Mt) (0, 15, 30, 45 and 60 min), during the industrial extraction of *cv.* Arbequina oils at 25 °C on total phenolic content of olive pastes. Additionally, the possibility of applying a lab-made potentiometric electronic tongue (E-tongue), comprising 40 lipid/polymer sensor membranes with cross sensitivity, to discriminate the olive pastes according to the Mt, was evaluated. The results pointed out that the olive pastes' total phenolic contents significantly decreased (p-value < 0.001, one-way ANOVA) with the increase of the Mt (from 2.21 ± 0.02 to 1.99 ± 0.03 g gallic acid equivalents/kg olive paste), there being observed a linear decreasing trend (R-Pearson = −0.910). These findings may be tentatively attributed to the migration of the phenolic compounds from the olive pastes to the extracted oil and water phases, during the malaxation process. Finally, the E-tongue signals, acquired during the analysis of the olive pastes' methanolic extracts (methanol:water, 80:20 v/v), together with a linear discriminant analysis (LDA), coupled with a simulated annealing (SA) algorithm, allowed us to establish a successful classification model. The E-tongue-LDA-SA model, based on 11 selected non-redundant sensors, allowed us to correctly discriminate all the studied olive pastes according to the Mt (sensitivities of 100% for training and leave-one-out cross-validation). The satisfactory performance of the E-tongue could be tentatively explained by the known capability of lipid/polymeric sensor membranes to interact with phenolic compounds, through electrostatic interactions and/or hydrogen bonds, which total content depended on the Mt.

Keywords: electronic tongue; lipid sensor membranes; chemometrics; olive pastes; total phenolic content

check for updates

Citation: Marx, Í.M.G.; Rodrigues, N.; Veloso, A.C.A.; Pereira, J.A.; Peres, A.M. Evaluation of the Effect of Extracted Time Conditions on the Phenolic Content of Olive Pastes from *cv.* Arbequina and Discrimination Using a Lab-Made Potentiometric Electronic Tongue. *Chem. Proc.* **2021**, 5, 36. https://doi.org/10.3390/CSAC2021-10556

Academic Editor: Manel del Valle

Published: 1 July 2021

1. Introduction

The worldwide consumption of virgin olive oil (VOO) is associated with its appreciated sensory attributes as well as with the recognized health benefits, namely, the reduced risk of chronic diseases and increased longevity, mainly related to the unsaturated fatty acids and minor components like polyphenols [1]. One strategy to ensure the natural enrichment of olive oils in phenolic compounds is based on the optimization of the extraction conditions, namely, using different malaxation times and/or temperatures [2–4].

Several destructive and nondestructive analytical techniques (e.g., chromatography, electrochemical sensor devices and spectroscopy) have been applied to evaluate the olive oil physicochemical and quality characteristics, including the assessment of total and individual compositions in phenolics [5,6]. Nevertheless, most of the studies are focused on the olive oil evaluation after being extracted or during the storage period. In those studies, the proposed methodologies were not used as prognostic tools of olive oil quality, i.e., to predict the quality of the olive oil to be processed from measurements on the olive pastes collected during the olive oil extraction process. Actually, a small number of works have been published on the potential prediction of olive oil composition and quality before or during olive oil production [4,7,8]. In this context, this study aimed to evaluate the effect of malaxation time (Mt), during the industrial extraction of oils, on the total phenolic content (TPC) of *cv.* Arbequina pastes. Additionally, the use of a potentiometric lab-made electronic tongue (E-tongue) to estimate the TPC in olive pastes collected at different Mt, was also evaluated. This capability could allow establishing indirect correlations between the composition of olive pastes and the TPC of the *cv.* Arbequina oils industrially extracted. It is important to emphasize that E-tongues comprising lipid sensor membranes have been extensively used to determine the phenolic profile and the sensory sensations of olive oils [3,7], which versatility has been related to the low selectivity and cross-sensitivity of the sensors that mimic the behavior of the human biological gustatory receptors [9].

2. Materials and Methods

2.1. Olives and Olive Pastes Samples

Olives from the *cv.* Arbequina were harvested in mid-November 2017 from an orchard located in Trás-os-Montes region (northeast Portugal). Olive pastes were collected at 5 time-periods (0, 15, 30, 45 and 60 min) during the oil extraction at 25 °C, in an industrial olive mill (OLIMONTES, Macedo de Cavaleiros, Portugal. Five samples of olive pastes (~100 g) were collected from the malaxers during the extraction, totalizing 25 olive paste sub-samples (5 replicas × 5 time-periods). The TPC and the potentiometric profiles of the olive pastes were determined.

2.2. Olive Pastes

2.2.1. Analytical Extraction for TPC and Potentiometric Analysis

The methodology applied was previously described by Marx et al. [6]. The polar extract containing the phenolic compounds was collected to assess the TPC and to establish the potentiometric profiles.

2.2.2. TPC of Olive Paste Extracts

The TPC was determined following the methodology proposed by Singleton and Rossi [10] and previous described by Marx et al. [7]. Gallic acid was used as the external standard compound to establish the calibration curve ($R^2 > 0.999$), being the results expressed as g of Gallic acid equivalents (GAE) per kg of olive paste (g GAE/kg olive paste).

2.2.3. E-Tongue Apparatus and Potentiometric Analysis of Olive Paste Extracts

A lab-made potentiometric E-tongue, comprising two cylindrical arrays, was used. Each array contained 20 lipid polymeric cross-sensitive sensor membranes (1st array: S1:1 to S1:20; 2nd array: S2:1 to 2:20). The construction details, as well as the composition of the membranes were previously reported by Marx et al. [3]. The device was connected to an Agilent Data Acquisition unit (model 34970A), which was controlled by an Agilent BenchLink Data Logger software. For the olive pastes analysis, the TPC polar extract was used after a 1:5 (*v*/*v*) dilution in deionized water [7]. The diluted solution was analyzed with the E-tongue during 5 min to allow reaching a pseudo-equilibrium between the non-specific lipid polymeric membranes and the dissolved chemical compounds [7].

2.3. Statistical Analysis

The TPC of olive pastes were analyzed using the one-way ANOVA followed by the Tukey's post-hoc multi-comparison test. Linear discriminant analysis (LDA) was applied to evaluate the correct discrimination of the studied pastes based on the best subsets of

E-tongue sensors selected using the simulated annealing (SA) algorithm. The leave-one-out cross-validation (LOO-CV) variant was used to evaluate the predictive performance of the classification model and the repeated K-fold-CV. The quality of the results was assessed considering the sensitivity (i.e., the percentage of corrected classified samples). The statistical analysis was performed using the Sub-select and MASS packages of the open-source statistical program R (RStudio version Version 1.2.5033), at a 5% significance level, as previous detailed by Marx et al. [3,7].

3. Results and Discussion

3.1. TPC of Olive Pastes

The TPC of the olive pastes, collected at five time-periods (0, 15, 30, 45 and 60 min) during the industrial extraction of *cv.* Arbequina oils, were determined following the Folin-Ciocalteau spectrophotometric method and are shown in Table 1. According to the results, the TPC of the pastes linearly decreases with the Mt (*R*-Pearson = −0.910). However, until 30 min of malaxation, the observed decrease is not significant, there being observed a reduction of 0.46% between 15 and 30 min. On the other hand, after 30 min of malaxation, the reduction on the TPC of the studied pastes was more pronounced. Similar trends (negative correlation between the TPC of the olive paste and the Mt) have already been reported in the literature [4]. According to Trapani et al. [4], the decreasing trend was attributed to the enzymatic oxidation of phenolic compounds, probably due to the fact that during the malaxation process the olive paste was exposed to air. The knowledge of the TPC of olive pastes during malaxation could pave the way towards a real-time control of the impact of the Mt on the olive oils being extracted in order to promote the increase of the total phenolics in olive oils.

Table 1. Statistical analysis of TPC of olive pastes collected at five different malaxation times, during the industrial extraction of olive oil (average, standard deviation, *p*-value and *R*-Pearson).

Folin–Ciocalteau Spectrophotometric Method	Malaxation Time					*p*-Value [1]	*R*-Pearson
	0 min	15 min	30 min	45 min	60 min		
TPC (g GAE/kg olive paste)	2.21 ± 0.02 [A]	2.18 ± 0.02 [A]	2.17 ± 0.04 [A]	2.04 ± 0.03 [B]	2.00 ± 0.03 [B]	<0.0001	−0.910

[1] *p*-values for the one-way ANOVA. Different letters in the same row show statistically differences from the given mean ($p < 0.05$). n = 5.

3.2. Estimating TPC of Olive Pastes Based on the Potentiometric E-Tongue Analysis of Olive Paste Extracts

Among the 40 lipid polymeric sensors, it was possible to establish linear correlations (positive or negative, i.e., signal-on or signal-off) between the potentiometric signals recorded by the E-tongue sensor membranes and the decimal logarithm of TPC, for 75% of the sensors ($0.836 \leq R^2 \leq 0.998$). The sensors mean sensitivities varied from +3.7 to +376 mV/decade, or between −185 to −42 mV/decade. The linear correlations were obtained for 30 E-tongue sensors (1st array: S1:1, S1:2, S1:3, S4, S1:7, S1:8, S1:10, S1:12, S1:13, S1:15, S1:16, S1:17, S1:18, S1:19 and S1:20; 2nd array: S2:2, S2:3, S2:4, S2:5, S2:7, S2:8, S2:9, S2:10, S2:12, S2:13, S2:15, S2:17, S2:18, S2:19 and S2:20), and the mean TPC calculated by applying the referred linear correlations are shown in Table 2.

Table 2. TPC (mean ± standard deviation) of olive pastes estimated using the correlations established between the E-tongue signals and the decimal logarithm of TPC, for the five different malaxation times studied (minimum and maximum contents in brackets).

E-Tongue Analysis	Malaxation Time				
	0 min	15 min	30 min	45 min	60 min
Estimated TPC (g GAE/kg olive paste)	2.21 ± 0.03 (2.06–2.25)	2.13 ± 0.04 (2.03–2.22)	2.15 ± 0.03 (2.08–2.23)	2.05 ± 0.02 (1.99–2.09)	1.99 ± 0.01 (1.96–2.02)

The agreement between the experimental TPC (Table 1) and those estimated by the device (Table 2), pointed out that the E-tongue could be applied as a real-time analytical tool to estimate the TPC in olive pastes collected during the oil extraction, allowing establishing the best Mt of the olive pastes that would ensure the extraction of an olive oil rich in phenolic compounds.

Finally, the E-tongue signals acquired during the analysis of the olive pastes' methanolic extracts (methanol: water, 80:20 v/v), allowed the establishing of a successful classification LDA-SA model. The E-tongue-LDA-SA model (Figure 1), based on 11 selected non-redundant sensors, correctly discriminated all the studied olive pastes according to the Mt (sensitivities of 100% for training and LOO-CV) and 91 ± 12% for repeated K-fold-CV. The satisfactory performance of E-tongue could be tentatively attributed by the known capability of the lipid sensor membranes to interact with phenolic compounds, through electrostatic interactions and/or hydrogen bonds, which total content depended on the Mt [7].

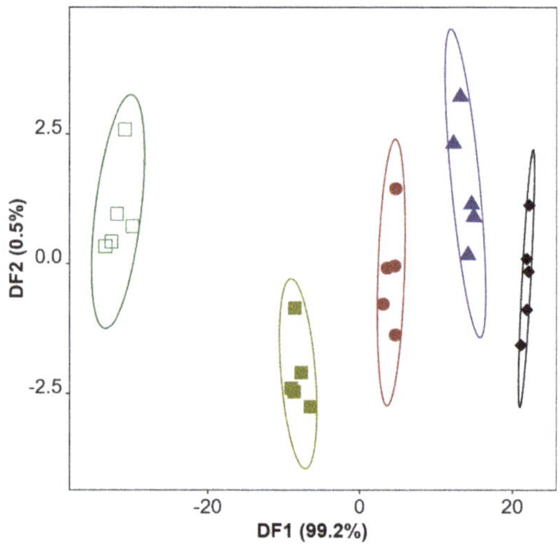

Figure 1. E-tongue-LDA-SA model performance regarding the supervising classification of *cv.* Arbequina olive pastes extracted at 0 min (□); 15 min (■); 30 min (●); 45 min (▲) and 60 min (♦) based on the potentiometric signals gathered by eleven lipid sensor membranes (1st array: S1:1, S1:8, S1:14, S1:17, S1:18, S1:20; 2nd array: S2:2, S2:3, S2:4, S2:5 and S2:18), selected using the SA algorithm from a set of 40 sensors.

4. Conclusions

The spectrophotometric evaluation of the olive pastes showed that until 30 min of malaxation, the TPC of the olive pastes were not significantly different. Oppositely, after 30 min of malaxation, the TPC of the pastes decreased, being the lowest contents determined for pastes after 60 min of malaxation. However, monitoring the TPC of olive pastes by spectrophotometry is a time-consuming task that requires several sample pre-treatments. Furthermore, this conventional spectrophotometric technique has some practical limitations, like the difficult regarding its implementation as an in-situ and *online* tool, besides being an invasive/destructive technique.

The present study showed that the potentiometric E-tongue analysis of extracts of olive pastes, collected during the industrial extraction of *cv.* Arbequina oils, coupled with chemometric tools, allowed estimating of the TPC. In addition, the E-tongue was capable to correctly discriminate all olive pastes studied according to the malaxation time.

Taking into account its portability, the lab-made E-tongue could be easily implemented in an industrial olive mill allowing estimating of the TPC of the olive pastes and, indirectly, establishing of the optimal malaxation time of the olive pastes to obtain a high-quality oil.

Supplementary Materials: The following are available online at https://www.mdpi.com/article/10.3390/CSAC2021-10556/s1.

Author Contributions: Conceptualization, Í.M.G.M., N.R., J.A.P. and A.M.P.; methodology, Í.M.G.M. and N.R.; software, A.M.P., and A.C.A.V.; validation, J.A.P. and A.M.P.; investigation, Í.M.G.M.; resources, Í.M.G.M. and N.R.; writing—original draft preparation, Í.M.G.M.; writing—review and editing, A.M.P. and J.A.P.; supervision, A.M.P.; funding acquisition, Í.M.G.M. and N.R. All authors have read and agreed to the published version of the manuscript.

Funding: Foundation for Science and Technology (FCT, Portugal) FCT/MCTES; CIMO (UIDB/00690/2020); CEB (UIDB/04469/2020); REQUIMTE-LAQV (UIDB/50006/2020); BioTecNorte operation (NORTE-01-0145-FEDER-000004) funded by the European Regional Development Fund under the scope of Norte2020—Programa Operacional Regional do Norte. Ph.D. research grant (SFRH/BD/137283/2018) provided by FCT.

Institutional Review Board Statement: Not applicable.

Informed Consent Statement: Not applicable.

Data Availability Statement: Not applicable.

Acknowledgments: The authors are grateful to the Foundation for Science and Technology (FCT, Portugal) for financial support by national funds FCT/MCTES to CIMO (UIDB/00690/2020), to CEB (UIDB/04469/2020), to REQUIMTE-LAQV (UIDB/50006/2020) and to BioTecNorte operation (NORTE-01-0145-FEDER-000004) funded by the European Regional Development Fund under the scope of Norte2020—Programa Operacional Regional do Norte. Ítala Marx acknowledges the Ph.D. research grant (SFRH/BD/137283/2018) provided by FCT. Nuno Rodrigues thanks the National funding by FCT-Foundation for Science and Technology, P.I., through the institutional scientific employment program contract.

Conflicts of Interest: The authors declare no conflict of interest.

References

1. Marković, A.K.; Torić, J.; Barbarić, M.; Brala, C.J. Hydroxytyrosol, tyrosol and derivatives and their potential effects on human health. *Molecules* **2019**, *24*, 2001. [CrossRef] [PubMed]
2. Guerrini, L.; Masella, P.; Angeloni, G.; Zanoni, B.; Breschi, C.; Calamai, L.; Parenti, A. The effect of an increase in paste temperature between malaxation and centrifugation on olive oil quality and yield: Preliminary results. *Ital. J. Food Sci.* **2019**, *31*, 451–458. [CrossRef]
3. Marx, Í.M.G.; Rodrigues, N.; Veloso, A.C.A.; Casal, S.; Pereira, J.A.; Peres, A.M. Effect of malaxation temperature on the physicochemical and sensory quality of *cv.* Cobrançosa olive oil and its evaluation using an electronic tongue. *LWT-Food Sci. Technol.* **2021**, *137*, 110426. [CrossRef]
4. Trapani, S.; Breschi, C.; Cecchi, L.; Guerrini, L.; Mulinacci, N.; Parenti, A.; Canutia, V.; Picchia, M.; Caruso, G.; Gucci, R. Indirect indice of oxidative damage to phenolic compounds for the implementation of olive paste malaxation optimization charts. *J. Food Eng.* **2017**, *207*, 24–34. [CrossRef]
5. Aleixandre-Tudó, J.L.; Castelló-Cogollos, L.; Aleixandre, J.L.; Aleixandre-Benavent, R. Bibliometric insights into the spectroscopy research field: A food science and technology case study. *Appl. Spectrosc. Rev.* **2020**, *55*, 873–906. [CrossRef]
6. Kalogiouri, N.P.; Aalizadeh, R.; Dasenaki, M.E.; Thomaidis, N.S. Application of High Resolution Mass Spectrometric methods coupled with chemometric techniques in olive oil authenticity studies—A review. *Anal. Chim. Acta* **2020**, *1134*, 150–173. [CrossRef]
7. Marx, Í.M.G.; Casal, S.; Rodrigues, N.; Veloso, A.C.A.; Pereira, J.A.; Peres, A.M. Estimating hydroxytyrosol-tyrosol derivatives amounts in *cv.* Cobrançosa olive oils based on the electronic tongue analysis of olive paste extracts. *LWT-Food Sci. Technol.* **2021**, *147*, 111542. [CrossRef]
8. Martínez-Gila, D.M.; Gámez-García, J.; Bellincontro, A.; Mencarelli, F.; Gómez-Ortega, J. Fast tool based on electronic nose to predict olive fruit quality after harvest. *Postharvest Biol. Technol.* **2020**, *160*, 111058. [CrossRef]
9. Wu, X.; Tahara, Y.; Yatabe, R.; Toko, K. Taste sensor: Electronic tongue with lipid membranes. *Anal. Sci.* **2020**, *36*, 147–159. [CrossRef] [PubMed]
10. Singleton, V.L.; Rossi, J.A. Colorimetry of total phenolics with phosphomolybdic-phosphotungstic acid reagents. *Am. J. Enol. Vitic.* **1965**, *16*, 144–158.

chemistry proceedings

MDPI

Abstract

Conductive Electrospun Nanofibers for Multifunctional Portable Devices †

Antonio Fotia [1], Patrizia Frontera [2], Lucio Bonaccorsi [2] and Angela Malara [2,*]

1 Department of Information Engineering, Infrastructures and Sustainable Energy, Mediterranea University of Reggio Calabria, Via Graziella Loc Feo di Vito, 89124 Reggio Calabria, Italy; antonio.fotia@unirc.it

2 Department of Civil, Energy, Environment and Material Engineering, Mediterranea University of Reggio Calabria, Via Graziella Loc Feo di Vito, 89124 Reggio Calabria, Italy; patrizia.frontera@unirc.it (P.F.); lucio.bonaccorsi@unirc.it (L.B.)

* Correspondence: angela.malara@unirc.it

† Presented at the 1st International Electronic Conference on Chemical Sensors and Analytical Chemistry, 1–15 July 2021; Available online: https://csac2021.sciforum.net/.

Abstract: The need to perform in situ sensing measurements lead to the development of innovative and smart field-portable devices. The advantages of such systems are remarkable since they are mainly battery-powered, lightweight and easy to carry and keep. Moreover, field-portable devices are easy to use and are able to give fast sensing responses. In the last few years, many efforts have been made in the development of new performing systems and the advantageous use of nanofibrous materials was assessed. To this purpose, the electrospinning has been recognized as the most powerful and facile technique for generating uniform nanofibers with controlled dimension and morphology. When conductive polymers are electrospun, very interesting electrical properties can be obtained along with the well-known ones that are typical of nanofibers. Among these polymers, polyaniline has been extensively used. In this work, an innovative hybrid material based on polyaniline/polyvinyl acetate/graphene oxide nanofibers was developed and tested as a sensor toward the detection of contaminants in aqueous media. Nanofibers, in the form of a compact mat, were deposited onto a support with suitable electrical contacts. Measurements were performed exploiting the excellent electrical properties of the realized nanofibers in both direct and alternating currents. When a direct current was used, the change in the nanofibers' resistance value was registered upon exposure to contaminated aqueous solutions and used to determine the presence or absence of contaminants, whereas when tests were performed with an alternating current, the quantitative determination of single species in contaminated solutions was also possible. In this way, by integrating the two different measurement methodologies, an opportunely designed multifunctional portable device will be developed for both qualitative and quantitative contaminants determinations.

Keywords: polyaniline; electrospinning; sensors; portable devices

Citation: Fotia, A.; Frontera, P.; Bonaccorsi, L.; Malara, A. Conductive Electrospun Nanofibers for Multifunctional Portable Devices. *Chem. Proc.* **2021**, *5*, 37. https://doi.org/10.3390/CSAC2021-10634

Academic Editor: Elisabetta Comini

Published: 7 July 2021

Supplementary Materials: The following are available online at https://www.mdpi.com/article/10.3390/CSAC2021-10634/s1.

Author Contributions: Conceptualization, P.F. and A.M.; methodology, A.F. and A.M.; formal analysis, P.F. and A.M.; investigation, A.F. and A.M.; data curation, A.F. and A.M.; writing—original draft preparation, A.F. and A.M.; writing—review and editing, P.F., L.B. and A.M.; supervision; P.F. and L.B. All authors have read and agreed to the published version of the manuscript.

Funding: This research received no external funding.

Institutional Review Board Statement: Not applicable.

Conflicts of Interest: The authors declare no conflict of interest.

chemistry proceedings

Proceeding Paper

Polysaccharide-Based Organic Frameworks with Embedded Nanoparticles: Advanced SPR Study on the Antiviral Activity of Gold Composites Derived from Glucuronoxylomannan †

Praskoviya Boltovets [1,*], Sergii Kravchenko [1], Oleksiy Kovalenko [2] and Borys Snopok [1,*]

1 Institute of Semiconductor Physics, National Academy of Sciences of Ukraine, Prospect Nauky, 41, 03039 Kyiv, Ukraine; 24kravchenko@gmail.com
2 Institute of Microbiology and Virology, National Academy of Sciences of Ukraine, Acad. Zabolotny Str, 154, 03680 Kyiv, Ukraine; udajko@ukr.net
* Correspondence: paraskeva2013@gmail.com (P.B.); snopok@isp.kiev.ua (B.S.)
† Presented at the 1st International Electronic Conference on Chemical Sensors and Analytical Chemistry, 1–15 July 2021; Available online: https://csac2021.sciforum.net/.

Abstract: The nanosized composites based on the natural polysaccharides and nanoparticles of noble metals are promising candidates for efficient antiviral drugs. However, the complexity of such objects, their diversity and novelty necessitate the development of new analytical methods for investigation of such supramolecular architectures. In this work, which was recently developed for SPR-based instrumentation, the concept of variative refraction (DViFA, density variations in fixed architectures) was used to elucidate the mechanism of the antiviral action of a polysaccharide with gold nanoparticles grown in it. The SPR data were confirmed by direct biological tests: the effect of the native polysaccharide glucuronoxylomannan (GXM) obtained from the fungus *Ganoderma adspersum* and gold nanocomposites thereon on the infection of *Datura stramonium* with tobacco mosaic virus (TMV) was investigated. Both drugs suppress the development of viral infections. However, if for high concentrations the characteristic activity of the composite is somewhat lower than for GXM, then with an increase in dilution, the effectiveness of the composite increases significantly, up to a twofold excess. It has been reasonably suggested that the mechanism of antiviral action is associated with the formation of clusters of viruses that are no longer capable of infecting cells.

Keywords: nanocomposites; variative refraction; surface plasmon resonance; antiviral activity

Citation: Boltovets, P.; Kravchenko, S.; Kovalenko, O.; Snopok, B. Polysaccharide-Based Organic Frameworks with Embedded Nanoparticles: Advanced SPR Study on the Antiviral Activity of Gold Composites Derived from Glucuronoxylomannan. *Chem. Proc.* **2021**, *5*, 38. https://doi.org/10.3390/CSAC2021-10475

Academic Editor: Elisabetta Comini

Published: 15 December 2021

1. Introduction

Nanosized composites combined organic compounds and inorganic nanoparticles extend our capabilities to form supramolecular architectures of advanced functionality. Complex macromolecules of biological origin, which can act as a matrix for the synthesis of such objects, are of particular interest. In this work, the polysaccharide glucuronoxylomannan with known immunomodulatory activity was chosen as such a matrix [1].

Earlier, we considered the potential of polysaccharide glucuronoxylomannan as an antiphytoviral agent. It was demonstrated that GXM isolated from *Tremella mesenterica* culture can suppress TMV infection in *Nicotiana tabacum* and *Datura stramonium* plants [2]. It was shown that GXM affects both the virus before infection and the processes that are proceeding immediately in the cell. In particular, polysaccharide can suppress virus reproduction and induce plant resistance to pathogens. It was suggested that GXM sterically block the virions, thereby suppressing its ability to infect [3]. In this regard, it is reasonable to evaluate the antiviral effectiveness of GXM nanocomposites with gold and well-known inhibitory agents, the effect of which is often associated with the activity of this metal ions in an aqueous medium [4]. It should be emphasized that the "ionic model" of the antiviral action of nanoparticles finds fewer and fewer supporters among researchers: the authors of many publications increasingly prefer various physical processes in the field of contact of nanoparticles with biological objects when describing one or another effect of nanoobjects on biological systems.

Indeed, the nanosized composites turned out to be a fundamentally new objects with unusual properties—in particular, their toxicity, which does not correlate with the properties of the material in the atomic-molecular state or in a solid. This is well illustrated by the example of the effect of nano-objects on viruses and bacteria, for which a large amount of experimental material has been collected. It is reasonable to conclude that the antiviral effects of nanoparticles are due to some physical phenomenon, rather than a chemical interaction of one type or another. Nanoscale analytes induce the change of the toxicity paradigm: physical effects come to the forefront, not the features of the chemical structure. This means that we need take into account not only the chemical composition of the object, but also its geometric characteristics, such as its size and shape [5].

The aim of this work was to study GXM nanocomposites with gold nanoparticles in order to elucidate their potential as antiviral drugs using the example of the TMV virions.

Since macromolecules of polysaccharides are complex branched structures, it is not possible to uniformly introduce pre-synthesized metal nanoparticles into these structures. However, taking into account the fact that polysaccharides contain a sufficiently large set of functional groups with the characteristics necessary for carrying out the corresponding redox reactions, in this work the synthesis of metal nanoparticles was carried out directly inside the polysaccharide matrix.

This polysaccharide can act both as a reducing agent and as a stabilizer due to the presence of the corresponding functional groups in the molecule. GXM consists of a linear backbone of $(1 \rightarrow 3)$-linked α-D-mannose with mainly xylose and glucuronic acid in the side chains [6]. Molecules of glucuronic acid contains carboxylic acid group, which gives acidic properties to GXM (acid polysaccharide). In order to clarify the peculiarities of the interaction of composites with viral particles at the molecular level, an analysis of their interactions with the tobacco mosaic virus was carried out using the SPR method. Antiviral activity of the complex was also tested in vivo at the *Datura stramonium* L.

2. Methods

Preparation of GXM. The principle of the method of obtaining preparations of β-glucan from the mycelium of the fungus *G. adspersum* was the same as in obtaining $\beta(1 \rightarrow 3)$-$\beta(1 \rightarrow 6)$-bound glucan ("ganoderan") from other species of *Ganoderma* sp. Namely: polysaccharide preparations were obtained from lyophilized mycelium of the fungus by sequential aqueous, alkaline and acid extraction [7]. First, to remove low molecular weight compounds to the dry mycelium of the fungus *G. adspersum*, thoroughly ground in a porcelain mortar, add 85% ethanol solution (1:5, v/v) and boil for 3 h. The alcohol extraction procedure was repeated three times. Each time the precipitate was separated by centrifugation (10,000× g, 15–20 min) and was used in further work. To obtain the aqueous fraction of β-glucan, water (1:5) was added to the mycelium purified from low molecular weight impurities and was boiled for 3 h. The procedure was repeated 5 times. The extracts were collected by centrifugation (10,000× g, 15–20 min) and combined. The resulting extract was dialyzed against running and distilled water and then evaporated to a minimum volume (1:5) on a rotary evaporator 1/5 by volume of a mixture of isoamyl alcohol and chloroform (1:10) was then added to the concentrate, the mixture was shaken vigorously for 10 min and then centrifuged (10,000× g, 20 min) to separate the aqueous and organic phases. Deproteinization of the extract was repeated once more, the organic extracts were removed, the aqueous ones were combined and dried in freeze-drying [8].

Synthesis of AuNPs. First, GXM water solution was prepared by dissolution of 3 mg GXM in 2.9 mL H_2O, and then aqueous solution of $HAuCl_4$ (0.1 mL, 30 mM) was added to it at violent stirring. The mixture was stirred during 1 min at room temperature and heated to 100 °C with boiling for 10 min. The AuNPs formation was confirmed by UV-Vis Spectroscopy and TEM.

Instrumentation and Measurements. The morphological, optical, and spectroscopic properties of AuNPs were examined using the following measurements. UV-vis spectra were acquired with Umico UV-Vis spectrophotometer (data not shown). TEM was performed at 100 kV using a JEOL-1011 (JEM, Japan) instrument (data not shown). Scanning spectrometer "BioHelper" (ISP NASU, Kiev, Ukraine) was used for SPR measurements with standard chips (50 nm Au/1.5 nm Cr/glass (n = 1.61)) and the protein A immobilization protocol described in detail in [9]. SPR measurements were carried out in a static mode without a sample flow in an open cell configuration (400 µL).

3. Results and Discussion

3.1. Molecular Level Analysis: Advanced SPR Study

Conventional SPR studies typically use "2D" interfacial architectures whose thickness is significantly smaller than the penetration depth of an exponentially decaying evanescent wave in a dielectric medium [10,11]. The use of SPR techniques for the investigation of "bigger" objects (e.g., intact virus particles so called virions, cells etc.) has some limitations, since their characteristic size exceeds tens of nanometers, which is required to ensure an adequacy of the quasi-linear approximation in SPR sensing.

In the conventional approach, the SPR response depends on the effective thickness of the analyte layer that is bound to the receptors on the surface—the density of both layers (receptor and analyte) is uniform—i.e., the variation of the SPR signal is due to the changes of the thickness. However, an SPR shift depends also on the change of the refractive index within the layer. Therefore, variations of the layer density can also affect the response value due to the variations of the refractive index inside the layer ("variative" refraction). One of the possible mechanisms is changing the packing of the objects of different sizes and shapes within interfacial architectures on the surface (e.g., a virus, an antibody, a small molecule etc.) If the thickness of the surface layer is fixed due to the constant form of the biggest interacting components (e.g., virion), the SPR shift is a single-valued function of the molecular assembly compactness. Exactly such an unconventional approach DViFA (density variations in fixed architectures) has been proposed by us early [12–14] for quantitative detection of native virus particles t (V) using SPR techniques.

Typical protocol includes the incubation of a virus-bearing material (V) with a definite concentration of a specific antibody (Ab) followed by the injection of statistically formed V-Ab complexes into SPR instrument with a sensitive surface covered by protein A. A shift of the minimum of SPR signal depended on the compactness of the V-Ab monolayer on the surface. To realize such an approach, we took advantage of (1) specificity of protein A from *Staphylococcus aureus* to the Fc-fragment of immunoglobulins as well as development of (2) statistically stable distribution of complexes with different numbers Abs per V specific for different Ab/V ratios.

To refine the DViFA approach, the investigation of TMV interaction with serum containing specific antibodies was performed, both with and without the presence of GXM-Au composites in the sample during the incubation phase. We investigated how the nanocomposite affects the interaction of the virus with specific antibodies. During the measurements, the concentration of TMV particles was constant (at constant concentration of 100 µg/mL), and the concentration of antibodies was changed by serial dilutions of serum in the range from 1:25 to 1:1600. The results obtained were compared with the data for a virus at the same concentration preincubated with GXM-Au complex (Figure 1):

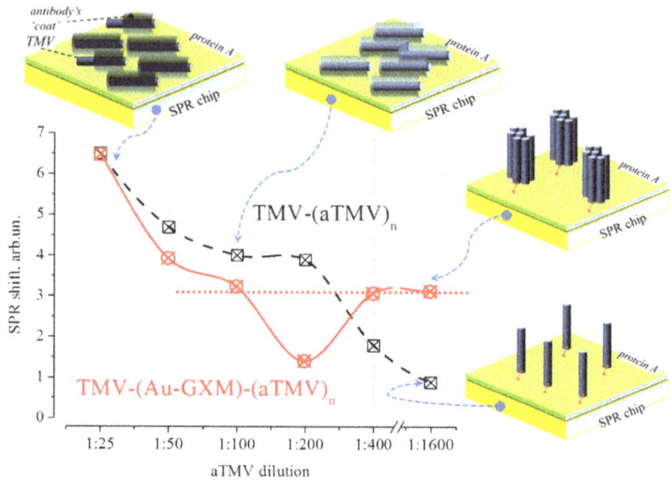

Figure 1. Comparison of the immobilization of TMV-aTMV complexes in the presence and in the absence of Au-GXM nanocomposites. The figures around the graph show schematic representations of the surface orientation of viral particles "on side" (at low dilutions) and "at end" (at large dilutions of antibodies). Individual virus particles or aggregates are attached to the surface by "bridging" antibodies, the Fc fragment of which is bound by protein A immobilized on the surface.

At dilution levels between 1:25 and 1:200, the antibodies fill the free surface of the V-Ab cylinders. Such complexes experience a mutual orienting effect on the surface modified by protein A and their configuration is typical for the classical "car parking" problem. The general structure of such an ensemble of viral particles on the surface remains practically unchanged up to a dilution of antibodies of about 1:200. The decrease in the SPR response in this range is due to a decrease in the number of antibodies associated with the virus, i.e., density of surface architecture. Some stabilization of the response in the range from 1:100 to 1:200 can be associated with two competing processes: on the one hand, a decrease in the number of surface-bound antibodies decreases the density, and on the other hand, it increases through some convergence of individual viral particles. In the presence of a GXM-Au composite, all the considered effects are less pronounced and are observed at low dilutions, since the GXM composite blocks some of the surface binding sites of Ab on the virion surface. At a dilution level of 1:200 and 1:100 for a virus without and in the presence of a nanocomposite, respectively, the amount of antibodies becomes insufficient to keep the virus "on side" on the surface and some of the viral particles become "at end", fixing pointwise on the surface with a chain of virus–antibody–protein A. In the absence of a GXM-Au composite, this process continues with further dilution (1:400–1:1600), leading to a decrease in the mass (number of attached virus particles) on the surface.

A radically different situation is observed in the case of the presence of a GXM gold nanocomposite: a decrease in the amount of antibodies at dilutions greater than 1:200 leads to an increase and subsequent stabilization of the response for dilutions up to 1:1600. In this case, the stabilization of the response occurs approximately at the same level, which corresponded to the densest structure of viral particles in the position "on the side". Since the diameter of the virus particle coated with antibodies is about 40 nm, and the length of the virus particle is 300 nm, then in the orientation "on side" and "at end" the surface ensemble of viral particles practically completely overlap the region of the highest concentration of the evanescent wave (less then 100 nm due to its exponential decay with distance from the surface). All this suggests that the presence of a polysaccharide nanocomposite stimulates the aggregation of virus particles into large clusters that are attached to the surface by only a few antibody molecules. Summarizing the obtained results, it can be argued that the nanocomposite stimulates the aggregation of viral particles into clusters, preventing their "independent" functioning.

3.2. Biological Experiments In Vivo

To verify the antiviral activity of the nanosized composite, the classical biological test has been performed. Aqueous solutions of Au-GXM (in the concentration range 1–500 μg/mL) were added to a suspension of TMV (10 μg/mL) and the mixture was incubated for 30 min at room temperature. Then, the left halves *Datura stramonium* L. were inoculated with mixture, whereas the right halves were infected with the virus at the same concentration without a composite.

The degree of viral infection suppression (in percentage terms) was calculated from the number of necrotic local lesions on the test and control leaves using the following expression [8]:

$$I = ((C - P)/C) \cdot 100\% \quad (1)$$

where I is the degree of viral inhibition in percent; C—local lesions number on the control half; P—number of local lesions on the test half. The results of the calculation of necrosis were subjected to statistical processing by parametric criteria, calculating their average number (M) and the ratio of these data in the experiment and control, as well as the average error of the ratio (m). On the graph, the data of statistical evaluation of the results were expressed as M ± m (Figure 2).

Figure 2. Percentage of necrosis per leaf with native glucan (GXM) and Au-glucan composite (Au-GXM). Insert: *Datura stramonium* L. leaves infected with TMV in a mixture with native glucan GXM (labeled "Gl_{10}") and GXM-Au nanocomposite (labeled "Gl+Au").

The in-vivo experiment showed that both native GXM and GXM-Au composites inhibit the development of viral infection: this is manifested in a much smaller number of necrosis on the experimental halves of the leaves of the studied plants (see Insert in Figure 2).

In order to compare the effectiveness of potent antiviral drugs, a concentration dependence of the percentage of necrosis on the experimental leaf halves to the necrosis on the control leaf halves on the drug concentration was constructed (i.e., the lower the value, the more effectively the virus is suppressed). For the preparation of native GXM, a clear linear (exponential in linear coordinates) dependence of the degree of viral infection suppression on the concentration was demonstrated. The difference in the concentration range of 0.5 mg/mL–0.001 mg/mL is c.a. 50%. It was shown that at high concentrations (0.5 mg/mL) the activity of the GXM is higher than the activity of the composite. However, the effectiveness of the GXM-Au composite is significantly less dependent on the concentration of the drug. This leads to the fact that the antiviral activity of the composite is much higher at low concentrations. In particular, for a concentration of 0.001 mg/mL, the degree of viral infection suppression was more than twice better than that for Au composites in respect to native glucan.

The results of molecular analysis and in vivo studies suggest that polysaccharide matrices with embedded gold nanoparticles have a stronger antiviral effect at low concentrations in comparison with natural polysaccharides. The mechanism of this action is due to the fact that the metal composite induces the aggregation of viral particles into clusters incapable of subsequent infection of plant cells.

Supplementary Materials: The following are available online at https://www.mdpi.com/article/10.3390/CSAC2021-10475/s1.

Author Contributions: Conceptualization, B.S.; methodology, P.B. and S.K.; software, B.S.; validation, O.K. and P.B.; formal analysis, B.S.; investigation, S.K.; resources, O.K.; data curation, O.K.; writing—original draft preparation, P.B.; writing—review and editing, P.B. and B.S.; visualization, S.K.; supervision, B.S.; project administration, P.B.; funding acquisition, B.S. All authors have read and agreed to the published version of the manuscript.

Funding: This research was funded by the National Academy of Sciences and Ministry of Education and Science of Ukraine.

Institutional Review Board Statement: Not applicable.

Informed Consent Statement: Not applicable.

Conflicts of Interest: The authors declare no conflict of interest.

References

1. Chlubnovhá, I.; Sylla, B.; Nugier-Chauvin, C.; Daniellou, R.; Legentil, L.; Kralová, B.; Ferrières, V. Natural glycans and glycoconjugates as immunomodulating agents. *Nat. Prod. Rep.* **2011**, *28*, 937–952. [CrossRef] [PubMed]
2. Boltovets, P.M.; Kravchenko, S.O.; Kovalenko, O.G.; Snopok, B.A. Mushroom derived glycane as capping and reducing agent for pH-dependent growth of gold nanoparticles. *FEBS Open Bio* **2018**, *8*, 472.
3. Podgorsky, V.S.; Kovalenko, A.G.; Boltovets, P.N.; Snopok, B.A.; Polishchuk, E.N. Complex formation of glucuronoxylomannan *Tremella mesenterica* Ritz. Fr. with tobacco mosaic virus such as one of the mechanisms polysaccharide's antiviral activity. *Rep. NAS Ukr.* **2013**, *12*, 157–165.
4. Yu, Q.; Li, J.; Zhang, Y.; Wang, Y.; Liu, L.; Li, M. Inhibition of gold nanoparticles (AuNPs) on pathogenic biofilm formation and invasion to host cells. *Sci. Rep.* **2016**, *6*, 26667. [CrossRef] [PubMed]
5. Snopok, B.A.; Snopok, O.B. Nanoscale–Specific Analytics: How to Push the Analytic Excellence in Express Analysis of CBRN. In *Advanced Nanomaterials for Detection of CBRN. NATO Science for Peace and Security Series A: Chemistry and Biology*; Bonča, J., Kruchinin, S., Eds.; Springer: Dordrecht, The Netherlands, 2020. [CrossRef]
6. Vinogradov, E.; Petersen, B.; Duubs, J.O.; Wasser, S. The structure of glucuroxylomannan produced by culinary-medicinal yellow brain mushroom (*Tremella mesenterica* Ritz.: Fr., Heterobasidomycetes) grown as on cell biomass in submerged culture. *Carbohydr. Res.* **2004**, *339*, 1483–1489. [CrossRef] [PubMed]
7. Musaki, A.; Sone, Y.; Yoshida, M.; Takeuchi, K. Beta-Glucan. U.S. Patent US4769363A, 6 September 1988.
8. Kovalenko, O.G.; Polishchuk, E.N.; Wasser, S.P. Glycans of higher basisdiomycetes mushroom ganoderma adspersum (Schulzer) donk: Isolation and antyphytoviral activity. *Biotechnology* **2010**, *3*, 83–91. (In Ukrainian)
9. Snopok, B.; Yurchenko, M.; Szekely, L.; Klein, G.; Kasuba, E. SPR based immuno-capture approach for in vitro analysis of protein complex formation: Mapping of MRS18-2 binding site on retinoblastoma protein. *Anal. Bioanal. Chem.* **2006**, *386*, 2063–2073. [CrossRef] [PubMed]
10. Snopok, B. Biosensing under Surface Plasmon Resonance Conditions, Chapter 19. In *21st Century Nanoscience—A Handbook*; Sattler, K., Ed.; CRC Press & Taylor and Francis Group: Boca Raton, FL, USA, 2020. [CrossRef]
11. Snopok, B.A. Theory and Practical Application of Surface Plasmon Resonance for Analytical Purposes. *Theor. Exp. Chem.* **2012**, *48*, 283–306. [CrossRef]
12. Boltovets, P.M.; Polischuk, O.M.; Kovalenko, O.G.; Snopok, B.A. A simple SPR-based method for the quantification of the effect of potential virus inhibitors. *Analyst* **2013**, *138*, 480–486. [CrossRef] [PubMed]
13. Boltovets, P.M.; Snopok, B.A.; Boyko, V.R.; Shevchenko, T.P.; Dyachenko, N.S.; Shirshov, Y.M. Detection of plant viruses using a surface plasmon resonance via complexing with specific antibodies. *J. Virol. Methods* **2004**, *121*, 101–106. [CrossRef] [PubMed]
14. Boltovets, P.M.; Boyko, V.R.; Kostikov, I.Y.; Dyachenko, N.S.; Snopok, B.A.; Shirshov, Y.M. Simple method for plant virus detection: Effect of antibody immobilization technique. *J. Virol. Methods* **2002**, *105*, 141–146. [CrossRef]

chemistry proceedings

MDPI

Proceeding Paper

Metal–Peptide Complexes—A Novel Class of Molecular Receptors for Electrochemical Phosphate Sensing †

Aleksandra Tobolska [1,2,*], Nina E. Wezynfeld [1], Urszula E. Wawrzyniak [1], Wojciech Bal [3] and Wojciech Wróblewski [1]

1 Chair of Medical Biotechnology, Faculty of Chemistry, Warsaw University of Technology, Noakowskiego 3, 00-664 Warsaw, Poland; nwezynfeld@ch.pw.edu.pl (N.E.W.); uwawrzyniak@ch.pw.edu.pl (U.E.W.); wuwu@ch.pw.edu.pl (W.W.)

2 Faculty of Chemistry, University of Warsaw, Pasteura 1, 02-093 Warsaw, Poland

3 Institute of Biochemistry and Biophysics, Polish Academy of Sciences, Pawińskiego 5a, 02-106 Warsaw, Poland; wbal@ibb.waw.pl

* Correspondence: atobolska@ch.pw.edu.pl

† Presented at the 1st International Electronic Conference on Chemical Sensors and Analytical Chemistry, 1–15 July 2021; Available online: https://csac2021.sciforum.net/.

Abstract: Amyloid-β (Aβ) peptides are crucial in the pathology of Alzheimer's disease. On the other hand, their metal complexes possess distinctive coordination properties that could be of great importance in the selective recognition of (bio)analytes, such as anions. Here, we report a novel group of molecular receptors for phosphate anions recognition: metal–peptide complexes of Aβ peptides, which combine features of synthetic inorganic ligands and naturally occurring binding proteins. The influence of the change in the metal ion center on the coordination and redox properties of binary Cu(II)/Ni(II)-Aβ complexes, as well as the affinity of these complexes towards phosphate species, were analyzed. This approach offers the possibility of fine-tuning the receptor affinity for desired applications.

Keywords: metal–peptide complexes; voltammetry; molecular receptors

Citation: Tobolska, A.; Wezynfeld, N.E.; Wawrzyniak, U.E.; Bal, W.; Wróblewski, W. Metal–Peptide Complexes—A Novel Class of Molecular Receptors for Electrochemical Phosphate Sensing. *Chem. Proc.* **2021**, *5*, 39. https://doi.org/10.3390/CSAC2021-10449

Academic Editor: Nicole Jaffrezic-Renault

Published: 30 June 2021

1. Introduction

The determination of phosphate anions in body fluids provides information about various disorders such as hyperparathyroidism or vitamin D deficiency [1]. Therefore, the monitoring of phosphate levels is of interest for human health. Chemical sensors are an ideal alternative to classic analytical methods, but their construction requires the synthesis of appropriate receptors, selectively binding to the analyte.

Amyloid β peptides (Aβ) related to Alzheimer's disease are well known for their neurotoxic properties [2]. However, metal complexes, with their N-terminally truncated analogs, have unique coordination properties that could be employed in the design of potential receptors for biorelevant anionic species [3,4]. The $A\beta_{5\text{-}9}$ (Arg-His-Asp-Ser-Gly-NH_2) peptide possesses a His-2 binding motif, and thus forms stable complexes with transition metal ions such as Cu(II) or Ni(II). At pH 7.4, the Cu(II) or Ni(II) ion is bound by three nitrogen atoms (3N) from the His residue, the N-terminal amine, and the peptide backbone amide [5]. The resulting chelates also exhibit a labile coordination site, enabling ternary interactions. Hence, metal–peptide complexes offer the possibility of fine-tuning their affinity for desired applications by altering the amino acid sequence and the metal ion center.

The present work explores and compares the coordination and redox properties of Cu(II) and Ni(II) complexes of the $A\beta_{5\text{-}9}$ peptide, followed by their ability to interact with biologically relevant phosphate anions and nucleotides.

2. Materials and Methods

2.1. Chemicals and Reagents

All chemicals were purchased from Merck and Sigma Aldrich. All solutions were prepared daily with deionized water (18 MΩ·cm). In order to avoid Cu(II)/Ni(II) contamination, the glassware was raised with 6 M HNO_3 followed by deionized water. AMP and ATP stock solutions were adjusted to pH 7.0–7.4 and kept on ice during measurements to prevent nucleotides hydrolysis.

2.2. Peptide Synthesis

Synthesis of the $A\beta_{5\text{-}9}$ peptide was performed according to the Fmoc/tBu strategy [6] on a Prelude™ peptide synthesizer (Protein Technologies, Inc., Tucson, AZ, USA). The crude was purified by HPLC with the UV detection (Waters, Milford, MA, USA) at 220 nm. The purity of the peptide was verified by ESI-MS (Waters, Milford, MA, USA).

2.3. Voltammetry

Electrochemical measurements (CV, DPV) were performed using the CHI 1030 potentiostat (CH Instrument, Austin, TX, USA) in a three-electrode arrangement with a GCE (BASi, 3 mm diameter) as a working electrode, an Ag/AgCl electrode (Mineral, Warsaw, Poland) as a reference, and a platinum wire as an auxiliary electrode. The GCE was sequentially polished with the alumina powder (1.0 μm and 0.3 μm) on a Buehler polishing cloth. Then, the working electrode was sonicated for 1 min and rinsed thoroughly with deionized water. All voltametric experiments were carried out in 100 mM KNO_3 at pH 7.4 under argon. The pH was adjusted with small aliquots of concentrated KOH or HNO_3 solutions. The peptide-to-metal(II) ratio was 1.0:0.9.

3. Results and Discussion

The electrochemical response for the $A\beta_{5\text{-}9}$ metal complexes recorded at pH 7.4 is depicted in Figure 1. The binary Cu(II)-$A\beta_{5\text{-}9}$ complex enabled both the reduction (Figure 1A, blue line) and oxidation of Cu(II) ions (Figure 1B, blue line). The exchange of the metal center complex to Ni(II) caused significant differences in the redox behavior, with a decrease in metal center oxidation by 188 mV compared to Cu(II) complexes (Figure 1B, green line). We did not observe any signals associated with Ni(II) reduction (Figure 1A, green line).

Distinct electrochemical properties are likely caused by the differences in geometry and the stabilities of the $A\beta_{5\text{-}9}$ complexes. Under studied conditions, Cu(II) complex is square-planar, while Ni(II) complexes are mostly octahedral. Additionally, the conditional stability constant of the Cu(II)-$A\beta_{5\text{-}9}$ (5.8×10^{12} M^{-1}) is about five orders magnitude higher than for Ni(II)-$A\beta_{5\text{-}9}$ (1.7×10^{6} M^{-1}) [3,5].

Considering the application of the metal complexes of $A\beta_{5\text{-}9}$ as a recognition element, we studied their response to selected anionic species. Observed changes in the oxidation potentials of the metal center upon the addition of 10 mM phosphates were similar (~150–160 mV) for Cu(II) and Ni(II) complexes of $A\beta_{5\text{-}9}$. Comparable sensitivity of chelates is probably related to a similar Lewis acidity of the metal centers. Furthermore, both complexes exhibit a good selectivity towards phosphates in the presence of chlorides and sulfates (Table 1). Aside from phosphates, only acetates, among other tested analytes, interacted with the metal–peptide complexes, causing changes in redox activity. Nevertheless, the voltametric signals for acetates occurred at different potentials than for phosphates.

Since the intercellular level of organic phosphates can be 20 times higher than inorganic phosphates, we decided to investigate the affinities of the studied metal–peptide complexes for selected nucleotides (AMP, ATP). Similar to phosphate anions, the presence of nucleotides shifted the oxidation peak to less positive values. However, in contrast to Cu(II)-$A\beta_{5\text{-}9}$, the signal of the metal center oxidation for Ni(II)-$A\beta_{5\text{-}9}$ occurs at different potentials for mono- and triphosphates (see Table 1). We suggest that this is due to the ability of the octahedral nickel complex to interact with more than one phosphate group of ATP as a result of the chelate effect.

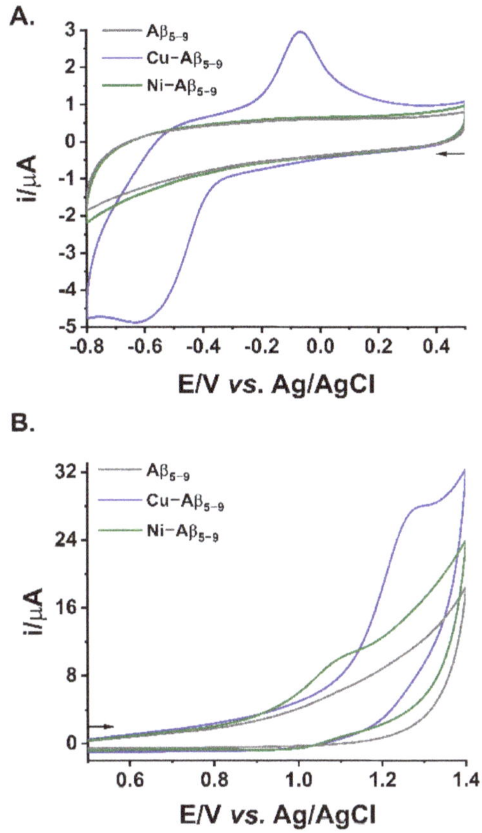

Figure 1. CV curves registered for 0.5 mM $A\beta_{5\text{-}9}$ in absence (grey line) and in the presence of 0.45 mM Cu(II) (blue line), or 0.45 mM Ni(II) (green line) scanned towards negative (**A**) and positive (**B**) potentials and recorded in 100 mM KNO_3 at pH 7.4, scan rate v = 0.1 V/s. Analyzed based on data published in the literature [3–5].

Table 1. Comparison of the affinity towards selected anions and nucleotides of the Cu(II)-$A\beta_{5\text{-}9}$ and Ni(II)-$A\beta_{5\text{-}9}$ complexes. $\Delta E_{M(II)/M(III)}$ is the difference of the potential values of Cu(II) or Ni(II) oxidation of the respective ternary system and the binary complex. Calculated based on results published previously [3,4].

Anion	$\Delta E_{Cu(II)/Cu(III)}$ (mV)	$\Delta E_{Ni(II)/Ni(III)}$ (mV)
Cl^-	−2	−8
SO_4^{2-}	−8	−21
CH_3COO^-	−84	−68
$H_2PO_4^-/HPO_4^{2-}$	−150	−156
AMP	−136	−126
ATP	−152	−230

4. Conclusions

Metal–peptide complexes of peptides possessing the His-2 motif ensure there are labile coordination sites, enabling ternary interactions with phosphate anions and nucleotides. Such interactions lead to a strong electrochemical response, which could be valuable for designing a promising class of peptide-based molecular receptors with potential applications as recognition elements in electrochemical biosensors and in vitro clinical diagnostics.Our research proved that the change in the metallic center of the A$\beta_{5\text{-}9}$ complex significantly influences its coordination properties and redox activity. Nevertheless, altering the metal center from Cu(II) to Ni(II) does not change the sensitivity of the complex toward phosphate anions.

Supplementary Materials: The following are available online at https://www.mdpi.com/article/10.3390/CSAC2021-10449/s1.

Author Contributions: Conceptualization, W.B., N.E.W. and U.E.W.; methodology, W.W., U.E.W. and N.E.W.; formal analysis, A.T., U.E.W. and N.E.W.; investigation, A.T.; resources, W.W. and W.B.; data curation, A.T., U.E.W. and N.E.W.; writing—original draft preparation, A.T.; writing—review and editing, A.T., N.E.W. and W.W.; visualization, A.T.; supervision, N.E.W., W.W. and W.B.; project administration, N.E.W. and A.T.; funding acquisition, N.E.W. and A.T. All authors have read and agreed to the published version of the manuscript.

Funding: This work was financially supported by the Warsaw University of Technology under the program Excellence Initiative, Research University (ID-UB), BIOTECHMED-1 project no. PSP 504/04496/1020/45.010407 (N.E.W.) and implemented as a part of the Operational Program Knowledge Education Development 2014–2020 (Project No POWR.03.02.00-00-I007/16-00) co-financed by the European Social Fund (A.T.).

Institutional Review Board Statement: Not applicable.

Informed Consent Statement: Not applicable.

Data Availability Statement: The data presented in this study are available on request from the corresponding author.

Acknowledgments: We gratefully acknowledge the assistance of Dawid Płonka (Institute of Biochemistry and Biophysics, Polish Academy of Sciences) in peptide synthesis.

Conflicts of Interest: The authors declare no conflict of interest. The funders had no role in the design of the study; in the collection, analyses, or interpretation of data; in the writing of the manuscript, or in the decision to publish the results.

References

1. Ramakrishnam Raju, M.V.; Harris, S.M.; Pierre, V.C. Design and applications of metal-based molecular receptors and probes for inorganic phosphate. *Chem. Soc. Rev.* **2020**, *49*, 1090–1108. [CrossRef] [PubMed]
2. Rauk, A. Why is the amyloid beta peptide of Alzheimer's disease neurotoxic? *Dalton Trans.* **2008**, 1273–1282. [CrossRef] [PubMed]
3. Tobolska, A.; Wezynfeld, N.E.; Wawrzyniak, U.E.; Bal, W.; Wróblewski, W. Tuning Receptor Properties of Metal-Amyloid Beta Complexes. Studies on the Interaction between Ni(II)-A$\beta_{5\text{-}9}$ and Phosphates/Nucleotides. *Inorg. Chem.* **2021**. [CrossRef] [PubMed]
4. Tobolska, A.; Wezynfeld, N.E.; Wawrzyniak, U.E.; Bal, W.; Wróblewski, W. Copper(II) complex of N-truncated amyloid-β peptide bearing a His-2 motif as a potential receptor for phosphate anions. *Dalton Trans.* **2021**, *50*, 2726–2730. [CrossRef] [PubMed]
5. Wezynfeld, N.E.; Tobolska, A.; Mital, M.; Wawrzyniak, U.E.; Wiloch, M.Z.; Płonka, D.; Bossak-Ahmad, K.; Wroblewski, W.; Bal, W. A$\beta_{5\text{-}x}$ Peptides: N-Terminal Truncation Yields Tunable Cu(II) Complexes. *Inorg. Chem.* **2020**, *59*, 14000–14011. [CrossRef] [PubMed]
6. Chan, W.C.; White, P.D. Fmoc Solid Phase Peptide Synthesis: A Practical Approach. In *Fmoc Solid Phase Peptide Synthesis: A Practical Approach*; Chan, W.C., White, P.D., Eds.; Oxford University Press: New York, NY, USA, 2000; pp. 41–76.

chemistry proceedings

MDPI

Proceeding Paper

Evaluation of Olive Oil Quality Grade Using a Portable Battery-Operated Sensor System †

Marco Grossi [1,*], Enrico Valli [2], Alessandra Bendini [2], Tullia Gallina Toschi [2] and Bruno Riccò [1]

1 Department of Electrical Energy and Information Engineering "Guglielmo Marconi" (DEI), University of Bologna, 40126 Bologna, Italy; bruno.ricco@unibo.it

2 Department of Agricultural and Food Sciences (DISTAL), University of Bologna, 47521 Cesena, Italy; enrico.valli4@unibo.it (E.V.); alessandra.bendini@unibo.it (A.B.); tullia.gallinatoschi@unibo.it (T.G.T.)

* Correspondence: marco.grossi8@unibo.it; Tel.: +39-051-2093038

† Presented at the 1st International Electronic Conference on Chemical Sensors and Analytical Chemistry, 1–15 July 2021; Available online: https://csac2021.sciforum.net/.

Abstract: Olive oil quality is normally assessed by chemical analysis as well as sensory analysis to detect the presence of organoleptic defects. Two of the most important parameters that define the quality of olive oil are the free acidity and the peroxide index. These chemical parameters are usually determinated by manual titration procedures that must be carried out in a laboratory by trained personnel. In this paper, a portable sensor system to evaluate the quality grade of olive oil is presented. The system is battery operated and characterized by small dimensions, a light weight and quick measurement response. The working principle is based on the measurement of the electrical conductance of an emulsion between a hydro-alcoholic solution and the olive oil sample. Tests have been carried out on a set of 17 olive oil samples. The results have shown how for fresh olive oil samples, the olive oil's free acidity can be estimated from the electrical conductance of the emulsion. In the case of oxidized olive oil, the measured electrical conductance is also the function of the oxidation level, and a conductance threshold can be set to discriminate between extra virgin olive oils and lower-quality grade oils. The proposed system can be a low-cost alternative to standard laboratory analysis to evaluate the quality grade of olive oil.

Keywords: olive oil; free acidity; peroxide index; electrochemical sensors; portable systems; electrical conductance; in situ measurements

Citation: Grossi, M.; Valli, E.; Bendini, A.; Toschi, T.G.; Riccò, B. Evaluation of Olive Oil Quality Grade Using a Portable Battery-Operated Sensor System. *Chem. Proc.* **2021**, *5*, 40. https://doi.org/10.3390/CSAC2021-10614

Academic Editor: Tobias Placke

Published: 6 July 2021

1. Introduction

Olive oil is a vegetable lipid highly appreciated for its beneficial effects on human health [1]. Olive oil quality is normally assessed by chemical analysis as well as sensory analysis to detect the presence of organoleptic defects. Two of the most important parameters that define the quality of olive oil are free acidity, defined as the amount of fatty acids no longer linked to their parent triglyceride molecules, which is affected by the quality of the olives used to produce the oil as well as the production process, and the peroxide index, expressed as milliequivalents of active oxigen for a kg of oil, which is an indicator of the oil primary oxidation and is affected by the storage conditions [2]. The official techniques for measuring these chemical parameters are manual titration procedures that must be carried out in a laboratory by trained personnel [3].

In the case of small industrial environments, such as olive oil mills and small packaging centers, which cannot afford an internal laboratory for quality analysis, the olive oil samples to be tested must be shipped to an external laboratory, and this results in high costs for the analysis and long response times. Thus, the development of simple and quick techniques for the analysis of quality grade of olive oil is important to allow in situ measurements directly in the industrial environment. A substantial research has been carried out in recent years towards the development of portable and low-cost sensor systems for the quality analysis of food products. Some examples include a portable biosensor system for bacterial concentration detection in raw milk [4], a system for the detection of chicken meat freshness [5], a low-cost handheld system for rapid non-destructive testing of fruit firmness [6], a system for the characterization of ice-cream properties with electrical impedance [7], a system for the determination of solid fat content in fats and oils [8] and an optical system for the assessment of lycopene content in tomatoes [9]. Many portable sensor systems presented in the literature are designed using a microcontroller as the core device of the system as well as commercial electronic chips to realize the analog measurement system and the communication system with an external PC. More recently, substantial research has been carried out for the development of smartphone-based sensor systems, since modern mobile phones integrate powerful processors for data analysis, a rich sensor set (camera, accelerometer, gyroscope, light sensor, etc.) as well as peripherals for wireless and wired communication [10].

In this paper, a battery-operated portable sensor system for the quality analysis of olive oil is presented. The system working principle is based on the measurement of the electrical characteristics of an emulsion between a hydro-alcoholic solution and the olive oil sample [11]. Tests on a set of 17 olive oils have shown how the system can discriminate between extra virgin olive oils (EVOOs) lower-quality grade olive oils and thus represents a low-cost and accurate alternative to standard laboratory analysis for in situ olive oil quality assessment in a real production environment.

2. Materials and Methods

A portable sensor system to evaluate the quality grade of olive oil samples has been designed and built. The system, shown in Figure 1a, is characterized by small size (11 × 15 × 5 cm), light weight (350 g), quick measurement response (30 s) and can be powered by a USB port or batteries (3 AAA alkaline batteries).

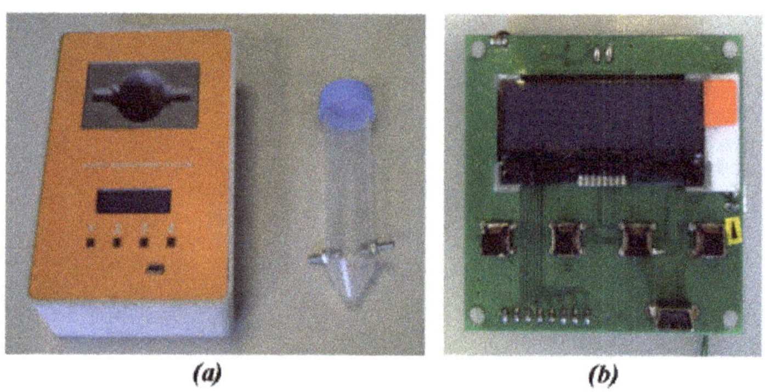

Figure 1. Photograph of the portable sensor system for olive oil quality grade detection (**a**) and designed electronic board (**b**).

The system working principle is based on electrical impedance spectroscopy (EIS) measurements [12] on an emulsion between a hydro-alcoholic solution and the olive oil tested. The emulsion electrical properties were measured using a 50 mL Falcon vial modified with a couple of cap-shaped stainless-steel electrodes (hereafter the sensor). In more detail, the following steps were carried out:

- The reagent (15 mL), a hydro-alcoholic solution of 60% ethanol and 40% distilled water, was added to the sensor vial.
- The reagent electrical conductance was measured using the portable system to check if it was suitable for the measurement (i.e., it was not degraded).
- The olive oil being tested (1 mL) was added to the sensor vial.
- The sensor vial was vigorously stirred for about 15 s to create the emulsion.

- The emulsion electrical conductance and the environmental temperature were measured using the portable system, and these values were used to estimate the olive oil quality grade.

The system's primary function is the measurement of the olive oil free acidity. In fact, in presence of the hydro-alcoholic solution, the free fatty acid molecules dissociate and generate ions that contribute to the increase in the emulsion electrical conductance which, once compensated for by variations of the environmental temperature, can be used to estimate the olive oil's free acidity. In the case of fresh olive oil samples, which are characterized by low values of peroxide index, a very good correlation exists between the emulsion electrical conductance and the oil free acidity measured with the standard titration technique. However, when olive oil storage conditions are not adequate, that is, if the oil is exposed to heath or light, oxidation takes place, and this results in the generation of non-volatile compounds (such as aldehydes, ketones and hydrocarbons) that also contribute to the increase in the emulsion electrical conductance. Thus, by setting a threshold for the emulsion electrical conductance, olive oil top quality grade (EVOO) can be distinguished from lower-quality grades, virgin olive oils (VOOs) and lampante olive oils (LOOs).

All electrical measurements, data processing and data filing are carried were using the electronic board that is shown in Figure 1b. The electronic board integrates a LCD screen to display the measurement results, four buttons for user interaction and a USB port that can be used to power the sensor system as well as to transfer the measured data to a PC for further data analysis. The core device of the electronic board is a microcontroller produced by ST Microelectronics (STM32L152RCT6A) that is responsible for the generation of the test signal, the signals acquisitions, the signals processing and the control of all the electronic components of the board. Different commercial electronic chips are integrated on the electronic board to design the analog circuits for the measurement of the emulsion electrical conductance G_m. A schematic diagram of the measurement circuit is shown in Figure 2. A sinewave test signal $V_{IN}(t)$ is generated with the 12-bit DAC integrated in the microcontroller and applied to the sensor vial electrodes. The current through the electrodes is converted to a voltage signal $V_{OUT}(t)$ using a trans-impedance amplifier. The sinewave voltage signals $V_{IN}(t)$ and $V_{OUT}(t)$ are acquired with the 12-bit ADC integrated in the microcontroller, processed to calculate the sinewave parameters and the emulsion electrical conductance G_m. A temperature sensor (MCP9700A) is integrated in the electronic board to make measurements of environmental temperature to compensate variations of G_m with temperature.

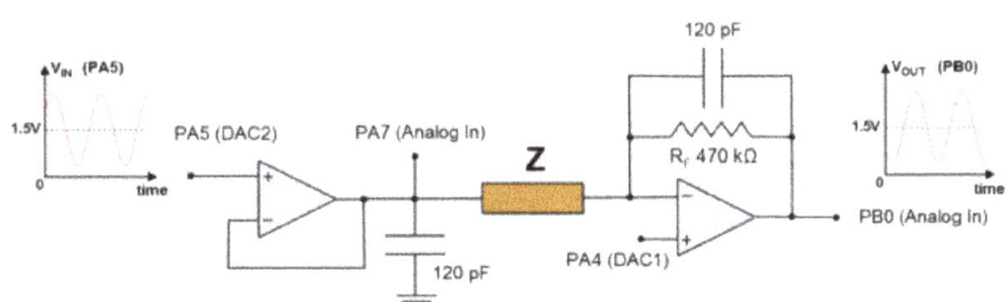

Figure 2. Schematic of the circuit for the measurement of the emulsion electrical conductance [13].

The estimation of the olive oil free acidity is carried out with the following steps:

- The emulsion electrical conductance at a generic temperature $G_{m,T}$ and the value of temperature T are measured.
- A compensation model, implemented in the microcontroller, calculates the emulsion electrical conductance at 23.5 °C $G_{m,23.5°C}$ from the measured values of $G_{m,T}$ and T.
- The olive oil free acidity is estimated from the calculated $G_{m,23.5°C}$ using a calibration curve equation stored in the microcontroller memory.

3. Results and Discussion

The portable sensor system was used to evaluate the quality grade of a set of 17 olive oil samples (11 fresh olive oil samples characterized by a peroxide index < 20 and 6 oxidized olive oil samples characterized by a peroxide index > 20). All samples were tested with the portable sensor system, the quality parameters (free acidity and peroxide index) were determined using the reference manual titration techniques, and the oil quality category was defined as suggested by the EU Reg. 2019-1604.

3.1. Analysis of Fresh Olive Oil Samples

The subset of 11 olive oil samples characterized by a peroxide index < 20 was tested with the portable sensor system. In Figure 3 the emulsion electrical conductance at 23.5 °C ($G_{m,23.5\,°C}$) is plotted vs. the free acidity determined by the reference titration technique. A correlation exists between $G_{m,23.5\,°C}$ and the olive oil free acidity. The best-fit curve that correlates the two parameters is defined by the following equation:

$$FA = \left(\frac{G_{m,23.5°C} - 0.6856}{2.6662}\right)^2 \quad (1)$$

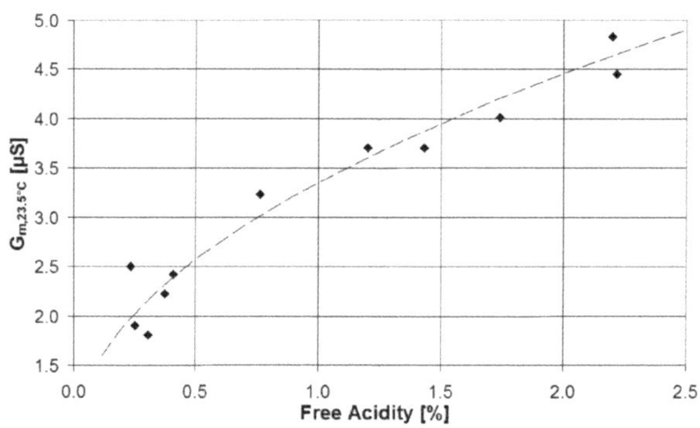

Figure 3. Scatter plot of the emulsion electrical conductance measured at 23.5 °C vs. free acidity for the subset of olive oil samples featuring a peroxide index <20 meq O_2/kg oil.

Equation (1) was used to estimate the free acidity for all the tested olive oil samples. The values of the estimated free acidity as well as the free acidity determined with the reference titration technique are reported in Table 1. As can be seen the free acidity estimated with the portable sensor system is very close to the value obtained with the reference titration technique, and the error in the estimated free acidity is never higher than 0.23%.

Table 1. Estimated values of the free acidity for the subset of olive oil samples featuring a peroxide index <20 meq O_2/kg oil.

Sample Number	Free Acidity (%)	Estimated Free Acidity (%)
1	0.25	0.21
2	0.31	0.18
3	0.41	0.42
4	0.38	0.33
5	0.76	0.91
6	1.20	1.28
7	2.20	2.42
8	0.24	0.46
9	1.74	1.56
10	1.43	1.28
11	2.22	1.99

3.2. Analysis of the Full Set of Olive Oil Samples

The full set of 17 olive oil samples (6 EVOOs, 3 VOOs and 8 LOOs) was tested with the portable sensor system. In the case of oxidized samples, characterized by a peroxide index > 20, the presence of non-volatile compounds contributes to the increase in the emulsion electrical conductance. This was verified by performing the following experiment: a sample, with a free acidity of 0.42% and peroxide index of 11.07 meq O_2/kg of oil, was subjected to UV radiation for a total time of 3 weeks, and the emulsion electrical conductance at 23.5 °C was measured at time intervals of 1 week. The results, presented in Table 2, confirm that even after 1 week there was a substantial increase in the emulsion electrical conductance due to the products of oil oxidation.

Table 2. Emulsion electrical conductance at 23.5 °C as function of UV stress time.

UV Stress Time	$G_{m,23.5\,°C}$ (μS)
No UV stress	2.26
1 week UV stress	5.09
2 weeks UV stress	5.22
3 weeks UV stress	5.12

The results for the full set of samples is presented in Figure 4, where each sample is represented by a circle of different color depending on the quality grade (EVOO, VOO and LOO), while the circle diameter represents the emulsion electrical conductance at 23.5 °C. In general, samples of lower-quality grades were characterized by higher values of the circle diameter. The results show that by setting a suitable threshold value for the emulsion electrical conductance at 23.5 °C ($G_{m,23.5°C,TH}$ = 2.7 μS), EVOOs can be distinguished from lower-quality oils (VOOs and LOOs) with good accuracy. In this case, all 11 samples of lower-quality grades (3 VOOs and 8 LOOs) were correctly classified. In the case of EVOOs, five samples were correctly classified, and the only misclassified sample featured a free acidity value (0.76%) that was close to the threshold between EVOO and VOO (0.8%).

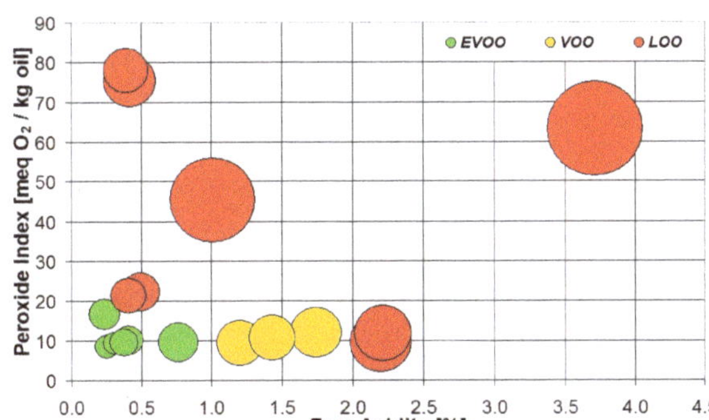

Figure 4. Emulsion electrical conductance at 23.5 °C as function of free acidity and peroxide index for the full set of olive oil samples.

4. Conclusions

A portable battery-operated sensor system for the evaluation of olive oil quality grade has been presented. The system is characterized by its small size, light weight and quick measurement response. It can be used for the in situ evaluation of olive oil quality grade in small industrial environments that cannot afford an internal laboratory.

The system working principle is based on the measurement of the electrical conductance of an emulsion between a hydro-alcoholic solution and the olive oil sample. The emulsion electrical conductance is mainly affected by the free acidity as well as the oxidation level of the sample. Tests on 17 olive oil samples demonstrated how EVOO samples can be differentiated from lower quality oils with good accuracy.

Supplementary Materials: The following are available online at https://www.mdpi.com/article/10.3390/CSAC2021-10614/s1.

Author Contributions: Conceptualization, M.G.; methodology, M.G. and A.B.; software, M.G.; validation, M.G. and A.B.; investigation, M.G., E.V. and A.B.; data curation, M.G. and A.B.; writing—original draft preparation, M.G.; writing—review and editing, M.G., E.V. and A.B.; supervision, T.G.T. and B.R.; project administration, T.G.T.; funding acquisition, T.G.T. All authors have read and agreed to the published version of the manuscript.

Funding: This work was developed in the context of the project OLEUM "Advanced solutions for assuring authenticity and quality of olive oil at global scale" funded by the European Commission within the Horizon 2020 Programme (GA no. 635690).

Institutional Review Board Statement: Not applicable.

Informed Consent Statement: Not applicable.

Data Availability Statement: Not applicable.

Conflicts of Interest: The authors declare no conflict of interest.

References

1. Bendini, A.; Cerretani, L.; Carrasco-Pancorbo, A.; Gomez-Caravaca, A.M.; SeguraCarretero, A.; Fernandez-Gutierrez, A.; Lercker, G. Phenolic molecules in virgin olive oils: A survey of their sensory properties, health effects, antioxidant activity and analytical methods. An overview of the last decade. *Molecules* **2007**, *12*, 1679–1719. [CrossRef]
2. Tena, N.; Wang, S.C.; Aparicio-Ruiz, R.; García-Gonzaléz, D.L.; Aparicio, R. Indepth assessment of analytical methods for olive oil purity, safety, and quality characterization. *J. Agric. Food Chem.* **2015**, *63*, 4509–4526. [CrossRef]
3. European Commission. Commission Regulation (EEC) No 2568/91 of 11 July 1991 on the characteristics of olive oil and olive-residue oil and on the relevant methods of analysis. *Off. J. Eur. Union* **1991**, *248*, 1–83.
4. Grossi, M.; Lanzoni, M.; Pompei, A.; Lazzarini, R.; Matteuzzi, D.; Riccò, B. A portable biosensor system for bacterial concentration measurements in cow's raw milk. In Proceedings of the 4th IEEE International Workshop on Advances in Sensors and Interfaces, Savelletri di Fasano, Italy, 28–29 June 2011; pp. 132–136.
5. Magwili, G.V.; Cruz, F.R.G.; De Pedro, R.A.C.; Evangelista, R.L.C.; Icaro, K.P.G.; Villarosa, K.A. Non-invasive Moisture Content Prediction and Characterization of Chicken Meat Freshness by Bioelectrical Impedance Spectroscopy. In Proceedings of the 11th International Conference on Humanoid, Nanotechnology, Information Technology, Communication and Control, Environment, and Management (HNICEM), Laoag, Philippines, 29 November–1 December 2019; pp. 1–5.
6. Das, A.J.; Wahi, A.; Kothari, I.; Raskar, R. Ultra-portable, wireless smartphone spectrometer for rapid, non-destructive testing of fruit ripeness. *Sci. Rep.* **2016**, *6*, 32504. [CrossRef] [PubMed]
7. Grossi, M.; Lanzoni, M.; Lazzarini, R.; Riccò, B. Automatic ice-cream characterization by impedance measurements for optimal machine setting. *Measurement* **2012**, *45*, 1747–1754. [CrossRef]
8. Grossi, M.; Valli, E.; Glicerina, V.T.; Rocculi, P.; Gallina Toschi, T.; Riccò, B. Practical determination of solid fat content in fats and oils by single-wavelength near-infrared analysis. *IEEE Trans. Instrum. Meas.* **2020**, *69*, 585–592. [CrossRef]
9. Mignani, A.G.; Ciaccheri, L.; Mencaglia, A.A.; Tuccio, L.; Agati, G. Application of a LED-based reflectance sensor for the assessing in situ the licopene content of tomatoes (*Lycopersicon esculentum* Mill.). *Sens. Agric. Food Qual. Saf.* **2015**, *9488*, 948806. [CrossRef]
10. Grossi, M. A sensor-centric survey on the development of smartphone measurement and sensing systems. *Measurement* **2019**, *135*, 572–592. [CrossRef]
11. Grossi, M.; Di Lecce, G.; Gallina Toschi, T.; Riccò, B. Fast and accurate determination of olive oil acidity by electrochemical impedance spectroscopy. *IEEE Sens. J.* **2014**, *14*, 2947–2954. [CrossRef]
12. Barsoukov, E.; Macdonald, J.R. *Impedance Spectroscopy: Theory, Experiment, and Applications*, 2nd ed.; John Wiley & Sons: New York, NY, USA, 2005.
13. Grossi, M.; Palagano, R.; Bendini, A.; Riccò, B.; Servili, M.; García-González, D.L.; Gallina Toschi, T. Design and in-house validation of a portable system for the determination of free acidity in virgin olive oil. *Food Control* **2019**, *104*, 208–216. [CrossRef]

chemistry proceedings

Abstract

Optical Characterization of Acetone-Sensitive Thin Films of poly(vinyl alcohol)-g-poly(methyl acrylate) †

Katerina Lazarova [1,*], Silvia Bozhilova [2], Sijka Ivanova [2], Darinka Christova [2] and Tsvetanka Babeva [1,*]

1 Institute of Optical Materials and Technologies "Acad. J. Malinowski", Bulgarian Academy of Sciences, Akad. G. Bonchev Str., bl. 109, 1113 Sofia, Bulgaria
2 Institute of Polymers, Bulgarian Academy of Sciences, Akad. G. Bonchev Str., bl. 103-A, 1113 Sofia, Bulgaria; s.bozhilova@polymer.bas.bg (S.B.); sivanova@polymer.bas.bg (S.I.); dchristo@polymer.bas.bg (D.C.)
* Correspondence: klazarova@iomt.bas.bg (K.L.); babeva@iomt.bas.bg (T.B.); Tel.: +359-02-979-3521 (K.L.)
† Presented at the 1st International Electronic Conference on Chemical Sensors and Analytical Chemistry, 1–15 July 2021; Available online: https://csac2021.sciforum.net/.

Citation: Lazarova, K.; Bozhilova, S.; Ivanova, S.; Christova, D.; Babeva, T. Optical Characterization of Acetone-Sensitive Thin Films of poly(vinyl alcohol)-g-poly(methyl acrylate). *Chem. Proc.* **2021**, *5*, 41. https://doi.org/10.3390/CSAC2021-10416

Published: 30 June 2021

Organic solvents are widely used as reaction media and/or for the separation and purification of synthetic products in chemical and pharmaceutical industries. Many of those solvents, among them being acetone, are considered to be harmful to human health. The detection of vapors of such volatile solvents present in the air can be achieved using multiple devices and materials [1], but the method of optical detection has a few important advantages, such as room temperature detection without the need for electrical power supply and easy detection when it is based on color/reflectance change. To achieve this, acetone-sensitive copolymers were designed by grafting poly(methyl acrylate) side chains onto poly(vinyl alcohol) precursors. Copolymer aqueous dispersions were used for thin-film deposition on silicon substrates by applying the spin-coating method. Optical properties of the film-refractive index, n, and extinction coefficient, k, as well as thickness, d, were determined from measured reflectance spectra, R, by using the two-stage nonlinear curve fitting method [2]. To evaluate the sensing properties of the films, they were placed in a quartz cell, and the atmosphere inside was constantly changed from air to argon to acetone using a homemade bubbler system (Figure 1).

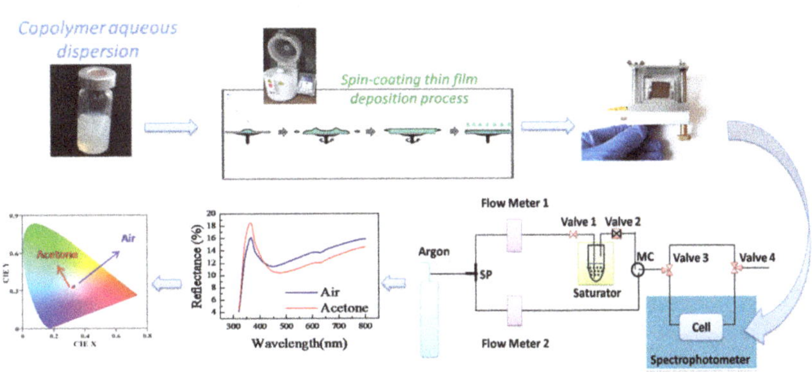

Figure 1. Scheme of the detection of acetone vapor process.

Reflectance spectra were measured before and during exposure to acetone vapors and were used to calculate optical constants and the thickness of the films in the presence of acetone vapors. When exposed to the vapors, the copolymer side chains swelled due to the absorption of acetone, and as a result, the film thickness increased while its refractive index decreased. This led to a shift of reflectance spectrum toward longer wavelengths and a subsequent change of the color of the film (Figure 1). The calculated sensitivity of polymer thin films was about $1.2 \times 10^{-4}\%$ per 1 ppm but could be further increased by two approaches. Firstly, the thickness of the polymer films could be optimized in order for the optical response to be maximized. Secondly, different multilayers structures could be designed using polymer films as building blocks.

According to the selectivity experiments, which are in progress, the initial results are very promising: the optical response of the films exposed to relative humidity of up to 80% RH is more than 10 times less as compared to acetone vapors response.

In conclusion, thin films of poly(vinyl alcohol)-g-poly(methyl acrylate) were successfully deposited using the spin-coating method on silicon substrates. A reaction toward acetone vapors and a very weak humidity response were demonstrated by measuring reflectance changes. Optimization approaches for sensitivity enhancement were discussed.

Supplementary Materials: The following are available online at https://www.mdpi.com/article/10.3390/CSAC2021-10416/s1.

Funding: This research received no external funding.

Acknowledgments: S. Bozhilova acknowledges the National Scientific Program for young scientists and postdoctoral fellows, funded by the Bulgarian Ministry of Education and Science (MES) with DCM 577/17.08.2018.

Conflicts of Interest: The authors declare no conflict of interest.

References

1. Malik, R.; Tomer, V.K.; Mishra, Y.K.; Lin, L. Functional gas sensing nanomaterials: A panoramic view. *Appl. Phys. Rev.* **2020**, *7*, 021301. [CrossRef]
2. Lazarova, K.; Vasileva, M.; Marinov, G.; Babeva, T. Optical characterization of sol-gel derived Nb_2O_5 thin films. *Opt. Laser Technol.* **2014**, *58*, 114–118. [CrossRef]

chemistry proceedings

MDPI

Proceeding Paper

Characterization and Comparison of the Relative Humidity Response of Hydromorphic Polymers in Long-Period Fiber Grating Structures †

Bernardo Dias [1,2,*], João P. Mendes [1,2,3], José M. M. M. de Almeida [1,4] and Luís C. C. Coelho [1,2,*]

1 INESC TEC–Institute for Systems and Computer Engineering, Technology and Science, Faculty of Sciences, University of Porto, 4169-007 Porto, Portugal; joao.p.mendes@inesctec.pt (J.P.M.); jmmma@utad.pt (J.M.M.M.d.A.)
2 Department of Physics and Astronomy, Faculty of Sciences, University of Porto, 4169-007 Porto, Portugal
3 Chemistry Research Unit, Faculty of Sciences, University of Porto, 4169-007 Porto, Portugal
4 Department of Physics, School of Science and Technology, University of Trás-os-Montes e Alto Douro, 5001-801 Vila Real, Portugal
* Correspondence: bernardo.s.dias@inesctec.pt (B.D.); luis.c.coelho@inesctec.pt (L.C.C.C.)
† Presented at the 1st International Electronic Conference on Chemical Sensors and Analytical Chemistry, 1–15 July 2021; Available online: https://csac2021.sciforum.net/.

Abstract: Relative humidity monitorization is of extreme importance on scientific and industrial applications, and fiber optics-based sensors may provide solutions where other types of sensors have limitations. In this work, fiber optics' sensors were fabricated by combining Long-Period Fiber Gratings with three different humidity-responding polymers, namely Poly(vinyl alcohol), Poly(ethylene glycol) and Hydromed™ D4. The performance of the multiple sensors was experimentally tested and crossed with numerical simulations, which provide a comparison with the expected response given the optical properties of the materials.

Keywords: relative humidity sensors; long-period fiber gratings; hydromorphic polymers; optical sensors

Citation: Dias, B.; Mendes, J.P.; de Almeida, J.M.M.M.; Coelho, L.C.C. Characterization and Comparison of the Relative Humidity Response of Hydromorphic Polymers in Long-Period Fiber Grating Structures. *Chem. Proc.* **2021**, *5*, 42. https://doi.org/10.3390/CSAC2021-10461

Academic Editor: Nicole Jaffrezic-Renault

Published: 30 June 2021

1. Introduction

The real time monitoring of relative humidity on scientific and industrial applications is of extreme importance and many types of sensors have been developed, mostly based on capacitive or resistive structures which may display some flaws, such as not being immune to electromagnetic radiation, and not fit to extreme and harsh environments (such as underwater applications). The usage of hydromorphic polymers in optical fiber structures is a thoroughly explored field of research with many publications associated [1–4], some of which include the usage of Poly(vinyl alcohol) (PVA), Poly(ethylene glycol) (PEG) or the combination of the two [1]. The polymers display a refractive index (RI) that decreases with the absorption of water molecules, also displaying considerable swelling effects. These changes can be tracked by analyzing the spectral characteristics of specific optical structures such as long-period fiber gratings (LPFG) [5,6]. LPFGs consist of a periodic modulation to the RI of the core of the fiber, resulting in coupling of light from the fundamental core mode to the co-propagating cladding modes, which creates rejection bands in the transmission spectrum at specific wavelengths with high sensitivity to the surrounding RI. By coating LPFGs with the aforementioned materials, this refractometer can be used to measure relative humidity.

Preliminary results regarding LPFGs fabricated in single mode optical fibers and coated with three different humidity-responding polymers PVA, PEG and a Hydrogel (Hydromed™ D4) are presented. In the case of Hydrogel, no such work seems to have been published previously. When placed in an environment with varying humidity, the polymers used as coating of the LPFG will both swell and change their RI, resulting in changes in the coupling conditions of the light to the cladding modes. Besides the experimental work, in order to have a better understanding of the effect of the coatings in the LPFG, simulations based on coupled-mode theory [7] were implemented, which allow understanding of the range of action of the LPFGs and optimization of their performance.

2. Materials and Methods

In order to test the capabilities of the polymers, several solutions of different concentrations were made, and thin layers were deposited around LPFGs previously fabricated in single mode fibers (Corning SMF28e) with the electric arc technique following the procedure described in [6]. This allowed the calibration of the wavelength shift and optical power shift induced by the coating process, meaning that for a given LPFG it is possible to choose the concentration of the coating solution that gives both visibility of the rejection band and high sensitivity to the RH variation. At the same time, given that the RH response of the sensor is a result of changes in RI and the thickness of the coating layer, testing LPFGs coated with solutions of different concentrations will also allow exploration of different responses caused by this variation of parameters.

2.1. Polymers Used for Humidity Sensing

Three different polymers were chosen and compared, a hydrogel (HydroMed™ D4), PVA (polyvinyl alchohol) and PEG (polyethylene glycol).

PVA is a water-soluble polymer, which has been studied extensively in the past as a coating for fiber optics-based humidity sensors [1–4]. It is a hydrophilic polymer that absorbs and desorbs water, with variations in the RI between 1.49 to 1.45 (at 1310 nm) according to [2] or between 1.49 to 1.34 (at 1550 nm) according to [4].

PEG is a polymer derived from petroleum containing ether linkages in its main chain, thus being referred to as a polyether. While there are some studies using this polymer for humidity sensing [8], only one application of this polymer to LPFG was found [1], which relied on a combination of PVA and PEG. The RI of PEG is lower than that of PVA, varying from 1.455 to 1.413 (n_D) [8].

HydroMed™ D is a series of ether-based hydrophilic urethanes fabricated by AdvanSource Biomaterials [9]. These materials were designed to respond by expanding when in contact with water molecules, with each product having different responses to different water contents. The polymer is distributed in granular form, which can be dissolved in many different solvents, of which ethanol was chosen.

2.2. Fabrication of the Coating Solutions and Coating Process

The solutions required different procedures in order to be dissolved. Generally, a solvent with low boiling point such as ethanol is preferred because it will lead to a faster evaporation and consequently the coating process will be faster. Nevertheless, given the low solubilities of PVA and PEG in ethanol, deionized water was used instead, which requires a longer time to fully evaporate and thus deposit the coating. In the case of hydrogel, the solvent was ethanol. Solutions with three different concentrations were produced, namely 10, 7.5 and 5% *wt*/*wt*, prepared by dissolving the high purity granules with the solvent, in the corresponding ratio. The PVA solution was obtained by adding PVA to water in a concentration of 10% *wt*/*wt*, and stirring for 3 h at 60 °C. The same procedure was applied to obtain the 5 and 7.5% concentration solutions and in the case of the PEG Scheme 50, 75 and 100% *wt*/*wt*.

The fabricated LPFGs were coated with all the polymers by stretching and dipping horizontally with a small angle between the longitudinal axes of the fiber and a V groove filled with the polymer. The fiber was then left at room temperature to ensure solvent evaporation.

This coating process was done for each sensor, thus justifying the need for different concentrations of the solutions, which have different viscosities and thus will create thicker coatings the more concentrated the solution is. In addition, calibration of the appropriate concentration of the solution to a given LPFG, in order to ensure that the wavelength and optical power shifts do not render the LPFG outside certain desirable parameters (such as minimum wavelength in the interval 1500 to 1600 nm), was performed.

2.3. Humidity Measurements

In order to calibrate the fabricated sensors, an experimental setup was devised in which the environment's humidity could be controlled and measured (Figure 1).

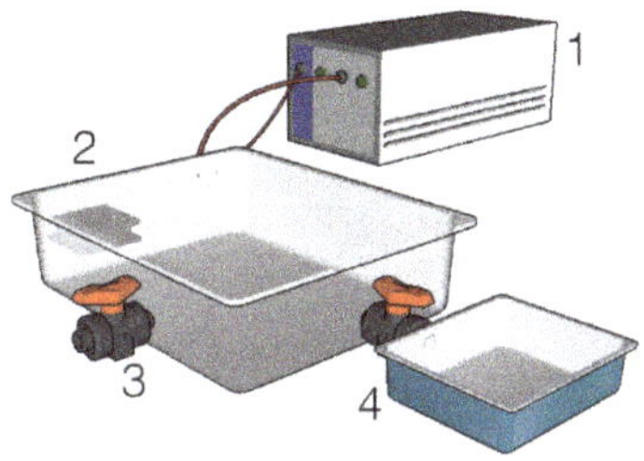

Figure 1. Experimental setup created to measure changes in LPFG spectra in varying values of relative humidity. 1—Interrogation unit; 2—Humidity Chamber; 3—Exterior valve; 4—Humidity valve.

This chamber was made from a container with two valves that connected to the exterior (where the RH was around 50%), and a side container with water connected by a valve to the main one. A small fan was also placed inside the container in order to promote faster diffusion of the water molecules in the air. This setup allowed the variation of internal humidity in varying rates, depending on the fan speed.

The humidity chamber allowed for the insertion of two LPFGs at once, one with the humidity sensor and another for thermal compensation. The two fibers were placed in a stand with weights in their extremities, guaranteeing that the sensor was fully stretched at all times. The fiber was connected to an interrogation unit (HBM Fibersensing FS22 Braggmeter) on the outside, which recorded the spectra at all times. Also inside the container was a humidity and temperature sensor (DHT22), which has a typical accuracy of $\pm$2%RH and $\pm$0.5 °C and a working range of 0%RH to 100%RH and -40 °C to 80 °C. This sensor was connected to an Arduino Uno unit which displayed the humidity and temperature values every 12 s, allowing to record every LPFG sensor spectra along with the respective humidity and temperature value.

2.4. Experimental Procedure

In order to provide a preliminary characterization of the sensors, the LPFG spectra were taken in descending RH values. Firstly, the valve to the side container with water was opened and the fan was turned on for 15 min. This ensured that the chamber attained the maximum possible values of humidity, which was recorded at 99%RH. After this process, the fan was turned off and after a minute of stabilization, the valve to the external environment was opened, and the measurements initiated. Due to the opened valve, the relative humidity decreased by 1% every 2 to 3 min, which provided an ideal stable decrease in humidity for measurements. The spectra were recorded for every decrease in 1%RH. The data were plotted in real time to determine the spectral evolution and the working range of the fabricated sensor.

3. Results

3.1. Humidity Measurements

As described in the experimental procedure, for each fabricated sensor its response to RH variations was tested in its working range for both descending and ascending variations. In this section, the results obtained by direct comparison from LPFGs coated with the same polymers at different concentrations are summarized, in order to choose one that has an optimal performance.

3.1.1. PVA Coated LPFGs

Figure 2 shows the results obtained from LPFGs coated with thin films of PVA from solutions with different concentrations.

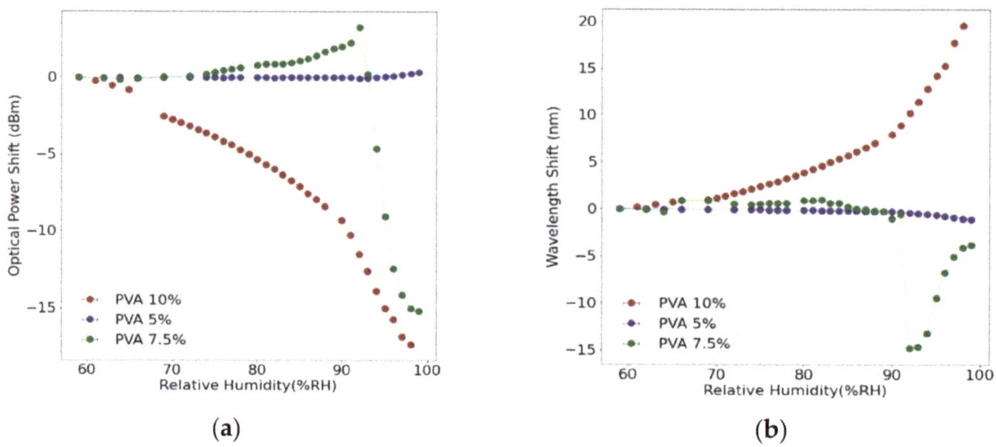

Figure 2. Variation of the spectra of the three PVA coated LPFG sensors and their measured response: (**a**) Optical Power Shift response; (**b**) Wavelength Shift Response.

The sensor coated with a 5% *wt*/*wt* solution displays very low response to RH variations, meaning that the obtained layer is too thin and making the sensor not suitable. On the other hand, the 7.5 and 10% sensors display very different behaviors, with the 10% *wt*/*wt* one displaying the typical LPFG wavelength shift curve due to the external RI being near the cladding index ($n_{clad} = 1.44$), but the 7.5% sensor displays a transition characteristic from guided to leaky modes, meaning that in this case the external RI transitioned from $n < n_{clad}$ to $n > n_{clad}$, which disagrees with the RI values reported in [2] (in which the PVA's RI as a value of 1.45 at 100%RH). The discontinuity seen in Figure 2b shows that at 91%RH the layer of PVA matches the cladding index, meaning that the fiber is suddenly thicker, and the mode has a discontinuous transition. This process also renders the 7.5% sensor not suitable. On the other hand, the 10% *wt*/*wt* sensor shows very good response with high sensitivity especially in the region above 90%RH, but overall good performance in the tested range.

Considering the response of the LPFG coated with 10% *wt*/*wt* PVA solution, a comparison with the 3-layer coupled-mode theory simulations of LPFGs [7] was established, in order to check if the response of the sensor matches the values of [2] or [4]. Good agreement is seen between the simulations considering the RI variations of 1.345 to 1.43 reported in [4] and the experimental data (Figure 3), showing also that in this case the coating layer can be approximated to an infinite medium, because the penetration depth of the evanescent field of the cladding mode is inferior to the PVA coating thickness. The simulations show that the 10% *wt*/*wt* is working on the high sensitivity zone of the LPFG, justifying the excellent performance of the sensor seen in Figure 2. Given the excellent performance of this sensor, the RH cycle was repeated for increasing RH values and a second LPFG was fabricated, in order to test the reproducibility of this structure. The results shown in Figure 3 show that excellent reproducibility was observed.

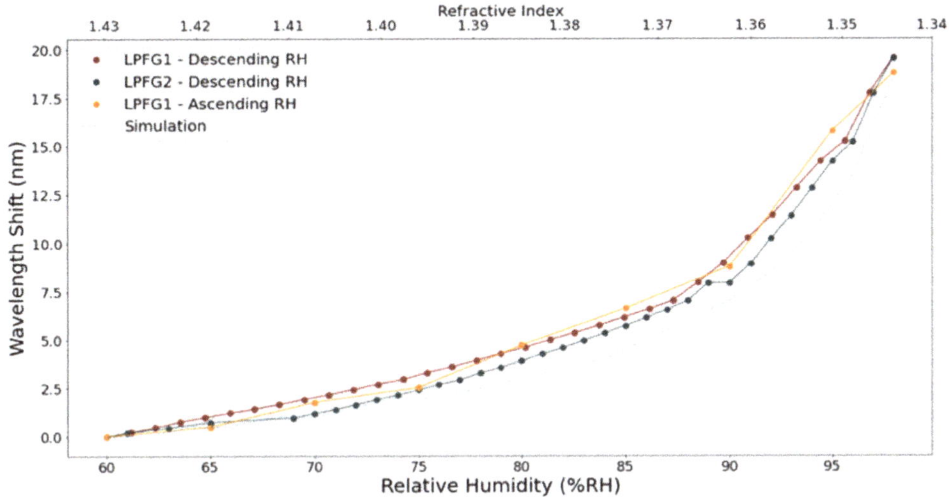

Figure 3. Comparison between the experimental wavelength shift of two 10% *wt*/*wt* PVA LPFG and the simulations, as function the external refractive index and relative humidity.

3.1.2. PEG-Coated LPFGs

Figure 4 shows the optical power and wavelength shifts in varying humidity for the cases of the PEG-coated LPFGs. In this case, the transition associated with the external medium having $n > n_{clad}$ is seen in all cases, and at lower RH values than PVA (80 to 87%RH), as can be seen in Figure 4b. This result was to be expected, given that the RI of PEG is lower than that of PVA, ranging from around 1.44 to 1.413 in the range of 60 to 99%RH (in the Sodium line) [8], meaning that the transition will happen at lower RH values. This renders the sensors not suitable in wavelength shift response, even though their optical power shift response (Figure 3a) could be used, in the 80 to 95%RH in the case of the 75% *wt*/*wt* coated LPFG.

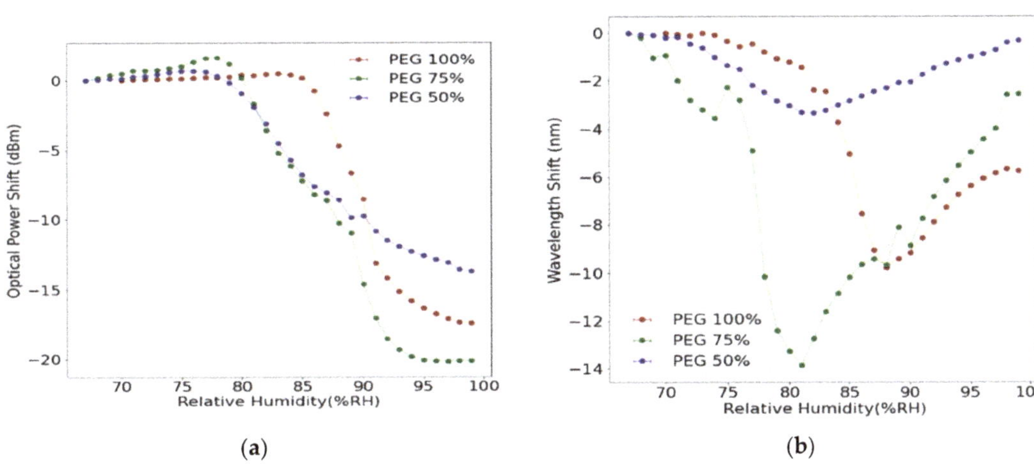

Figure 4. Variation of the spectra of the three PEG-coated LPFG sensors and their measured response: (**a**) Optical Power Shift response; (**b**) Wavelength Shift Response.

3.1.3. Hydrogel Coated LPFGs

Figure 5 shows the optical power and wavelength shifts in varying humidity for the cases of the Hydrogel coated LPFGs. The responses in both wavelength and optical power shift seem similar to the case of the PEG coated LPFGs, with the transition of $n > n_{clad}$ to $n < n_{clad}$ happening in the range of 83% to 92%RH, which makes the wavelength shifts an unsuitable figure of merit for relative humidity monitoring. On the other hand, the optical power shift presents a suitable curve that enables the use of the structures as sensors, even though the sensitivity is considerably less than both the PVA and PEG cases.

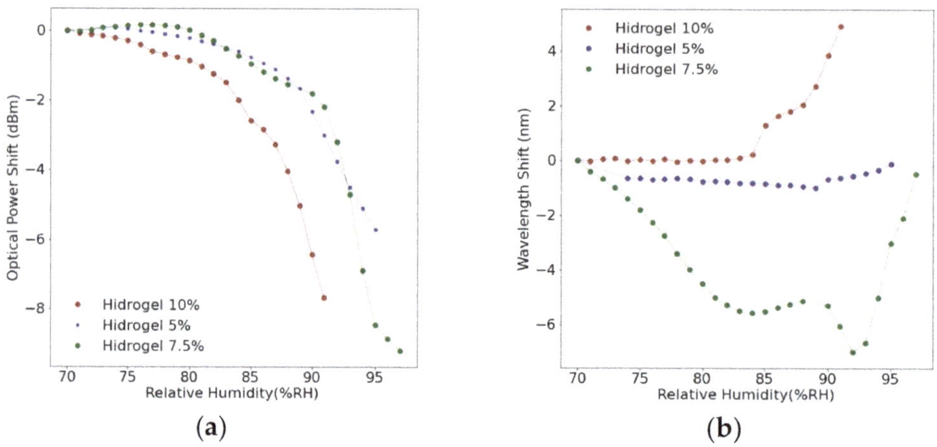

Figure 5. Variation of the spectra of the three Hydrogel coated LPFG sensors and their measured response: (**a**) Optical Power Shift response; (**b**) Wavelength Shift Response.

4. Conclusions

The response curves of LPFGs coated with three different humidity responding polymers (PVA, PEG and a Hydrogel) were obtained, with humidities varying from 60%RH (or 70%RH, depending on the polymer) to 100%RH. Of all the fabricated sensors, the 10% *wt/wt.* PVA coated LPFG displays the best properties for relative humidity sensing, which were verified when comparing to numerical LPFG simulations.

Due to the variation of the optical properties of the polymers with varying humidity, namely the fact that the polymer's RI becomes larger than the cladding RI at low RH values, a transition from guided to leaky modes is seen, displaying a non-linear wavelength shift response, which renders it as an unsuitable figure of merit for RH sensing for most sensors. This conclusion shows that further work is needed to quantify the response of the three polymers to humidity variations, in order to optimize the sensors' performance and make them commercially viable. By fabricating new materials, which could possibly result from mixing the ones here studied, the desired optical properties (refractive index slightly below the cladding RI for all RH values and linear response) may be attained, creating a highly sensitive, viable solution for industrial and scientific application.

Supplementary Materials: The following are available online at https://www.mdpi.com/article/10.3390/CSAC2021-10461/s1.

Author Contributions: Conceptualization, B.D., J.M.M.M.d.A., L.C.C.C.; methodology, B.D., J.P.M., J.M.M.M.d.A., L.C.C.C.; formal analysis, B.D., J.M.M.M.d.A., L.C.C.C.; investigation, B.D., J.P.M., J.M.M.M.d.A., L.C.C.C.; writing—original draft preparation, B.D.; writing—review and editing, B.D., J.P.M., J.M.M.M.d.A., L.C.C.C.; supervision, J.M.M.M.d.A., L.C.C.C.; project administration, J.M.M.M.d.A., L.C.C.C. All authors have read and agreed to the published version of the manuscript.

Funding: This work has received funding from the project "SolSensors—Development of Advanced Fiber Optic Sensors for Monitoring the Durability of Concrete Structures", with reference POCI-01-0145-FEDER-031220, supported by COMPETE 2020 and the Lisbon Regional Operational Program in its FEDER component, and by the budget of FCT Foundation for Science and Technology, I.P. Luis Coelho acknowledges the support from FCT research contract grant CEECIND/00471/2017.

Conflicts of Interest: The authors declare no conflict of interest. The funders had no role in the design of the study; in the collection, analyses, or interpretation of data; in the writing of the manuscript, or in the decision to publish the results.

References

1. Wang, Y.; Liu, Y.; Zou, F.; Jiang, C.; Mou, C.; Wang, T. Humidity Sensor Based on a Long-Period Fiber Grating Coated with Polymer Composite Film. *Sensors* **2019**, *19*, 2263. [CrossRef]
2. Gastón, A.; Pérez, F.; Sevilla, J. Optical Fiber Relative-Humidity Sensor with Polyvinyl Alcohol Film. *Appl. Opt.* **2004**, *43*, 4127–4132. [CrossRef]
3. Dong, X.; Li, T.; Liu, Y.; Li, Y.; Zhao, C.-L.; Chan, C.C. Polyvinyl Alcohol-Coated Hybrid Fiber Grating for Relative Humidity Sensing. *J. Biomed. Opt.* **2011**, *16*, 077001. [CrossRef]
4. Miao, Y.; Liu, B.; Zhang, H.; Li, Y.; Zhou, H.; Sun, H.; Zhang, W.; Zhao, Q. Relative Humidity Sensor Based on Tilted Fiber Bragg Grating With Polyvinyl Alcohol Coating. *IEEE Photonics Technol. Lett.* **2009**, *21*, 441–443. [CrossRef]
5. Vengsarkar, A.M.; Lemaire, P.J.; Judkins, J.B.; Bhatia, V.; Erdogan, T.; Sipe, J.E. Long-Period Fiber Gratings as Band-Rejection Filters. *J. Lightwave Technol.* **1996**, *14*, 58–65. [CrossRef]
6. Rego, G. Arc-Induced Long Period Fiber Gratings. *J. Sens.* **2016**, *2016*, e3598634. [CrossRef]
7. Erdogan, T. Cladding-Mode Resonances in Short- and Long-Period Fiber Grating Filters. *J. Opt. Soc. Am. A JOSAA* **1997**, *14*, 1760–1773. [CrossRef]
8. Acikgoz, S.; Bilen, B.; Demir, M.M.; Menceloglu, Y.Z.; Skarlatos, Y.; Aktas, G.; Inci, M.N. Use of Polyethylene Glycol Coatings for Optical Fibre Humidity Sensing. *Opt. Rev.* **2008**, *15*, 84. [CrossRef]
9. Hydromed™ Product Datasheet. Available online: http://www.advbiomaterials.com/products/hydrophilic/HydroMed.pdf (accessed on 1 June 2021).

chemistry proceedings

MDPI

Proceeding Paper

Identification, Quantification, and Method Validation of Anthocyanins †

Paula Garcia-Oliveira [1,2], Antia G. Pereira [1,2], Maria Fraga-Corral [1,2], Catarina Lourenço-Lopes [1,2], Franklin Chamorro [1], Aurora Silva [1,3], Pascual Garcia-Perez [1], Fatima Barroso [3], Lillian Barros [2], Isabel C. F. R. Ferreira [2], Jesus Simal-Gandara [1,*] and Miguel A. Prieto [1,2,*]

1 Nutrition and Bromatology Group, Faculty of Food Science and Technology, Universidade de Vigo, E32004 Ourense, Spain; paula.garcia.oliveira@uvigo.es (P.G.-O.); antia.gonzalez.pereira@uvigo.es (A.G.P.); maria.fraga.corral@hotmail.es (M.F.-C.); clopes@uvigo.es (C.L.-L.); franklin.noel.chamorro@uvigo.es (F.C.); mass@isep.ipp.pt (A.S.); pasgarcia@uvigo.es (P.G.-P.)

2 Centro de Investigação de Montanha (CIMO), Instituto Politécnico de Bragança, Campus de Santa Apolónia, 5300-253 Bragança, Portugal; lillian@ipb.pt (L.B.); iferreira@ipb.pt (I.C.F.R.F.)

3 REQUIMTE/LAQV, Instituto Superior de Engenharia do Porto, Instituto Politécnico do Porto, Rua Dr António Bernardino de Almeida 431, 4200-072 Porto, Portugal; mfb@isep.ipp.pt

* Correspondence: jsimal@uvigo.es (J.S.-G.); mprieto@uvigo.es (M.A.P.)

† Presented at the 1st International Electronic Conference on Chemical Sensors and Analytical Chemistry, 1–15 July 2021; Available online: https://csac2021.sciforum.net/.

Abstract: Nowadays, anthocyanins have gained scientific and industrial attention due to their biological activities and coloring properties. In this regard, anthocyanins have been proposed for use in the development of new nutraceutical foods to replace synthetic additives as well as to be value-added ingredients. The aim of this study was to evaluate current data on identification and quantification techniques and the validation process of such methods. Our results showed that anthocyanins have been identified by different methods, including nuclear magnetic resonance and chromatography-based techniques. Although problems have been described in this validation, most of the reports showed positive results on the validation parameters, suggesting that the current analytical technology offers a satisfactory identification and quantification of anthocyanins.

Keywords: anthocyanins; plant; extraction; validation

Citation: Garcia-Oliveira, P.; Pereira, A.G.; Fraga-Corral, M.; Lourenço-Lopes, C.; Chamorro, F.; Silva, A.; Garcia-Perez, P.; Barroso, F.; Barros, L.; Ferreira, I.C.F.R.; et al. Identification, Quantification, and Method Validation of Anthocyanins. *Chem. Proc.* **2021**, *5*, 43. https://doi.org/10.3390/CSAC2021-10680

Academic Editor: Manel del Valle

Published: 14 July 2021

1. Introduction

Anthocyanins are soluble glycosides linked by an O-glucosidic bond between an aglycone and a sugar molecule. It is a group belonging to the flavonoids that are found naturally in various plant sources, including fruits such as berries or grapes and flowers such as hibiscus, forming part of the secondary metabolites of plants [1]. Therefore, the extraction of anthocyanins is usually carried out from plant matrices. In the last few years, these compounds have attracted great interest due to their diverse beneficial properties. Specifically, anthocyanins present a high antioxidant capacity, which is attributed to the presence of phenolic hydroxyl groups in their chemical structure [2] (Figure 1). Furthermore, several studies have reported that a daily intake of this compound has a preventive and protective effect against cardiovascular diseases, diabetes, and heart disease [3–5]. These compounds also have coloring properties, covering ranges from salmon pink to red and from violet to dark blue. Thus, they are considered as an interesting source of natural colorants. To date, more than 20 structures are known, among which are orange pelargonidin, orange red cyanidin and peonidin, bluish-red delphinidin, and bluish-red malvidin and petunidin (Figure 1).

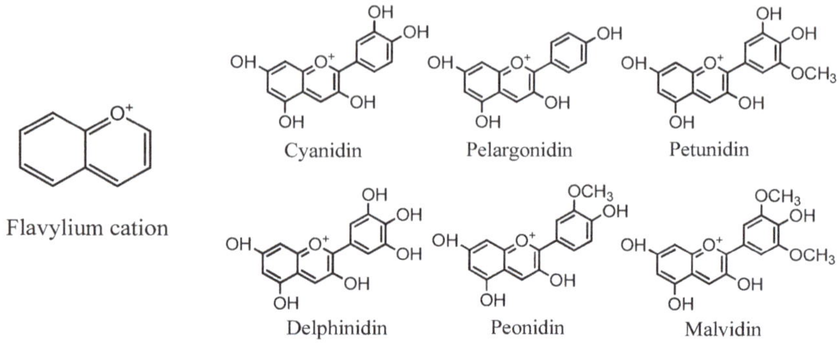

Figure 1. (**Left**) Base group of anthocyanins' structure. (**Right**) Most common anthocyanins in the nature.

Natural dyes have several advantages, such as non-toxicity and eco-friendliness, as their acquisition has a low environmental impact. Therefore, different industries have a great interest in identifying new and economically viable sources of anthocyanins to use them as new functional ingredients and/or colorants in food [6]. As the range of application of these compounds increases, it is necessary to design efficient extraction methods with better yields and to develop suitable analytical methods for the identification and quantification of anthocyanins. Therefore, the main objective of the study was to evaluate the current data on the identification and quantification techniques used and the validation process of these methods.

2. Identification and Quantification Techniques

The search for new ingredients of natural origin and the study of their bioactivities has been going on for some decades due to the growing demand for natural products with beneficial health properties, including the group of anthocyanins. Various techniques have been used for the identification and characterization of anthocyanins in different matrices, including mass spectrometry (MS), nuclear magnetic resonance (NMR), and high-performance liquid chromatography (HPLC). In Table 1, several examples of studies employing these techniques have been collected.

Table 1. Examples of studies employing different methods to identify and quantify anthocyanins.

Identification Technique	Source	Compounds	Ref.
MS; EI-MS; FAB-MS	Black rice, orchids, bilberries	DEL, CYA, PET, and MAL derivatives	[7–10]
NMR	Maqui berries, grapes, sumac, black currant, blue flowers, sweet potato, chokeberry	DEL, CYA, PET, MAL, and PEO derivatives	[11–15]
HPLC; HPLC-DAD; HPLC-MS/MS; HPLC-ESI/MS	Blueberries, hibiscus, red cabbage, cranberry, strawberry, grapefruits, grape skin, *Euterpe oleracea*	DEL, CYA, PET, PEO, and MAL derivatives	[6,16–22]

Abbreviation: delphinidin: DEL; cyanidin: CYA; petunidin: PET; malvidin: MAL; peonidin: PEO.

Regarding the application of MS-based technologies, several studies have obtained suitable results. For example, a work employed electron impact (EI)-MS to evaluate the degradation products of cyanidin-3-*O*-glucoside when the compound was subjected to oxidation radicals. Up to six derived products were detected, and two new molecules, 2-(3,4-dihydroxyphenyl)-4,6-dihydroxybenzofuran-3-carboxylic acid) and 2-*O*-(3,4-dihydroxybenzoyl)-2,4 glucose esters, 6-trihydroxyphenylacetic acid glucose ester, were identified. When analyzing samples of black rice stored for long periods, these two new compounds were identified, being useful to distinguish fresh rice from that stored for a long time [7]. Similarly, another EI-MS study investigated the degradation products of anthocyanin glycosides when they were exposed to the microflora of pig intestine. According to the results, anthocyanin glycosides underwent significant changes, suggesting that the antioxidant or anticancer activities observed for anthocyanin glycosides are due to these degradation products [9]. Fast atom bombardment (FAB)-MS has been also used in the field of anthocyanins to evaluate different structural modifications. For instance, the antioxidant effect of several anthocyanin fractions isolated from bilberry extracts against pyridinium bisretinoid A2E (a prooxidant compound) was determined by this technology. According to the results, the majority of the anthocyanins were delphinidin 3-galactoside, cyanidin 3-galactoside, delphinidin 3-glucoside, and cyanidin 3-glucoside [10]. In orchid flowers, FAB-MS successfully identified eight pigments, of which two were new structures: 3-*O*-[6-*O*-(malonyl)-β-glucopyranoside]-7,3′-di-[6-*O*-(trans-synapoyl)-β-glucopyranoside] and its demalonyl derivative [8].

Regarding NMR techniques, they have been used to identify anthocyanins and study their exact structural characteristics, to establish their mechanisms of action, which can lead to a better application of these compounds as functional ingredients [12]. For example, the structure of two major acylated peonidin derivatives obtained from sweet potato were studied using NMR [14]. Another study evaluated the anthocyanins composition of chokeberry showing the presence of the structure of cyanidin galactoside and cyanidin arabinoside along the second stage of the fruit ripening [15].

Finally, high-performance liquid chromatography (HPLC) is the most widely used method for the identification and quantification of anthocyanins. In general, the analytical parameters used in the literature show very uniform conditions for the identification of anthocyanins. The most used column is C_{18}, while mobile phase composition mainly corresponds to water, acetonitrile, and methanol with acid modifiers, such as formic acid. The acid presence in mobile phases ensures that anthocyanin compounds are going to be mobilized in their cationic flavylium form, which has been described as possessing its highest absorbance, around 520 nm [23]. During the HPLC analysis of anthocyanins, as well as other compounds, the retention times and peak areas can be strongly influenced by the column temperature, mobile phase composition, or the complexity of the matrix in which they are embedded. The detection of anthocyanins is often performed by diode array detectors (DAD), mass-spectrometry detectors (both MS or MS/MS) which are, most of the time, coupled to an electrospray ionization source (ESI) [6,16]. These methodologies have been shown to provide satisfactory results for the identification and quantification of anthocyanins. Nevertheless, the use of ultra-high pressure liquid chromatography (UPLC) provided better resolution, shorter elution times, and lower consumption of mobile phases than conventional HPLC methodologies. UPLC also presents a high performance in the efficiency of the peaks of identification [23]. The anthocyanin profile of diverse vegetal samples has been evaluated by HPLC-DAD. For example, in grape, glucoside derivates of delphinidin, cyanidin, pelargonidin, peonidin, petunidin, and malvidin were identified [20]. Similarly, in grape skin samples, petunidin-3-*O*-glucoside and malvidin-3-*O*-glucoside were the major compounds [21]. HPLC-DAD can be also coupled to MS, which provides a more accurate identification, since mass information is considered in the analysis and data processing. HPLC-DAD-MS has been employed with different matrixes, such as strawberries, where cyanidin-3-*O*-glucoside, pelargonidin 3-glucoside, pelargonidin-3-*O*-rutinoside (tentative), pelargonidin-3-*O*-succinyl-glucoside, or pelargonidin-3-*O*-arabinoside were identified [22]. Regarding HPLC coupled to MS, this approach has been employed to analyze the anthocyanin composition of different samples. For example, cyanidin-3-*O*-glucoside, cyanidin-3-*O*-rutinoside, and perlagonidin-3-*O*-glucoside have been identified in *Euterpe edulis* extracts [24]. In strawberry, glucoside derivates of cyanidin, delphinidin, pelargonidin, and malvidin were identified. In muscadine grapes, 3,5-di-*O*-glucoside of cyanidin, delphinidin, and petunidin were identified as the major anthocyanins [25].

3. Validation of Methods

There are different approaches to develop a validation plan, depending on the type of technique used, the field of application of the method, and the type of samples analyzed. The scientific literature shows many examples of the development and optimization of methods for anthocyanins detection. However, when these techniques are validated, just few of them indicated the guideline used to perform this complex process.

3.1. Selectivity

To achieve a selective method, analytes are first isolated from another family of analytes or matrix interferences. Pre-treatment of anthocyanin samples includes the use of different techniques such as ultrasound or microwave-assisted extraction (UAE or MAE) or the use of solid phase extraction (SPE) cartridges. When methods for detecting anthocyanins were validated, no selectivity/specificity issues were found; the most selective instrument was ultra-high-performance liquid chromatography (UPLC) coupled to a photodiode array detector (PAD) or to mass spectrometry (UPLC-MS) against spectroscopic ones, according to the literature [6,26].

When analyzing anthocyanins together with non-anthocyanin compounds under similar conditions, resolution issues have been reported. In general, the most common option when using HPLC techniques is the selection of C_{18} columns and the modification of the mobile phase's acidity, by increasing the percentage of acid or by changing the type of acid [27]. However, other authors have also increased the resolution peak between anthocyanins and non-anthocyanin compounds by performing two different injections using C_{18} columns with different conditions [28]. The last option described is the use of a fluorinated C_{18} column, which has been demonstrated to provide better results in terms of peak separation, symmetry, and short analysis time [26].

3.2. Linearity, Limit of Detection (LOD), and Quantification (LOQ)

For anthocyanins, it has been scientifically demonstrated that they can be detected with calibration curves with ranges from 0.01 to 800 μg/mL, using different techniques. Validation studies in which calibration curves have been carried out with concentrations within these ranges have shown high linearity with an $R^2 \geq 0.99$. For example, Grace et al. (2019) validated an LC-MS method. All calibration curves showed good linearity in the range of 0.04–40 μg/mL, with a regression coefficient (r^2) ≥ 0.99 [6]. Fibigr et al. (2017) developed an UHPL-UV method, also achieving similar results [26]. There are few exceptions with low anthocyanin concentrations [6,26,29] or in some specific cases. For example, the quantification of malvidin-3-*O*-glucoside by a spectrophotodensitometry method required a polynomial adjustment instead of a linear one [30].

As mentioned before, anthocyanins have been extensively analyzed using different techniques such as HPLC-DAD, UPLC-DAD, UPLC-UV, HPLC-MS, or capillary zone electrophoresis (CZE), among others, but most of the validation studies have been performed using chromatographic techniques. According to the literature, the ranges of the limit of detection and limit of quantification that have been reported were 0.01–3.7 and 0.03–8.5 μg/mL (ppm), respectively, when using chromatographic-based methods, coupled to diverse detectors [6,26,31]. In general, chromatographic methods achieved good performance results as observed in the studies of Grace et al. (2019) or Fibigr et al. (2017). In the first one, the limits of detection and quantification were 0.06–0.40 μg/mL and 0.12–1.20 μg/mL, respectively [6]. On the second example, quite similar results were reported, with limits of detection and quantification being 0.11–0.14 μg/mL and 0.36–0.47 μg/mL, respectively [26].

3.3. Accuracy and Precision

In most of the validated methods carried out for anthocyanins, both accuracy and precision are determined by adding known amounts of standard solutions to the samples or by using commercial standards. In general, the results of validation studies for accuracy were very good and the relative standard deviation for repeatability, and intermediate precision ranged from <1% to <10%, meaning that the methods are acceptable for possible routine use. Most of the validation studies employ chromatographic methods [6,26,29]. To cite an example, an LC-MS method has been recently validated. The results showed a good precision for all analytes tested; the relative standard deviation in intra and inter-day was less than 10% in both cases, while the reproducibility of all analytes was under 5% [6]. Thus, this method allowed characterizing simultaneously and selectively different anthocyanins and non-anthocyanidins, being a promising method for the analysis of these compounds.

3.4. Stability and Robustness

Regarding stability, several factors affect the stability of anthocyanins, such as their chemical structure, pH, light, or storage temperature and time, among others [23,32]. Minimal variations have been shown in studies of anthocyanin stability when these compounds are stored at low temperatures for long periods of time. However, rapid degradation processes were observed when matrixes or purified anthocyanins were stored at room temperature, as reported in the study of Gras et al., who employed UHPLC-PDA to evaluate anthocyanins from black carrot. In the study, significant losses of 8 to 14% were observed when standards (9 μg/L) were stored at room temperature for 24 h, showing that standard solution should be evaluated as soon as possible to avoid inaccurate results [23].

Regarding robustness, to our knowledge, few validation studies have evaluated these parameters. Nevertheless, studies in which it has been carried out displayed no significant differences in the total amount of anthocyanins extracted when the method was exposed to experimental variations such as changes in the pH, temperature, source, age and concentration of samples, variable standards, or solvents. This consistency in the results indicated that validated methods are robust and may be applied for the routine detection of anthocyanins [32,33]. For example, in the study of Canuto et al. (2016), the robustness of the developed reverse phase LC method was estimated introducing small variations in the mobile phase pH, column temperature, and mobile phase flow rate. According to their results, no significant results were detected in the determination of Cyn-3-glu and Plg-3-glu, demonstrating the robustness of the method [32].

4. Conclusions

There is a wide variety of identification and detection methods that have been demonstrated to be efficient for the analysis of anthocyanins. However, there is a lack of papers in which different methodologies are compared by using the same kind of samples. Therefore, it makes it difficult to conclude which method is more advantageous over others. On the one hand, the literature reviewed suggest that LC-MS represents a quick and efficient technique. However, it may present difficulties regarding the complexity of the sample matrix, which can cause ion suppression. Another issue is the determination of the best anthocyanin source. The variability of the experimental conditions, in terms of extraction protocols, analyzing parameters, and quantification, hinders the selection of the most appropriate matrix for obtaining the most efficient recovery of anthocyanins. Therefore, the standardization of extraction and analytical protocols may be critical to permit the real comparison of these experimental results.

Author Contributions: Conceptualization, F.B., L.B., I.C.F.R.F., J.S.-G. and M.A.P.; investigation, P.G.-O., A.G.P., C.L.-L., F.C. and A.S.; writing—original draft preparation, P.G.-O., A.G.P., C.L.-L., F.C. and A.S.; writing—review and editing, M.F.-C. and P.G.-P.; supervision, F.B., L.B., I.C.F.R.F., J.S.-G. and M.A.P. All authors have read and agreed to the published version of the manuscript.

Funding: The JU receives support from the European Union's Horizon 2020 research and innovation program and the Bio-Based Industries Consortium. The project SYSTEMIC Knowledge hub on Nutrition and Food Security has received funding from national research funding parties in Belgium (FWO), France (INRA), Germany (BLE), Italy (MIPAAF), Latvia (IZM), Norway (RCN), Portugal (FCT), and Spain (AEI) in a joint action of JPI HDHL, JPI-OCEANS and FACCE-JPI launched in 2019 under the ERA-NET ERA-HDHL (n° 696295).

Acknowledgments: The research leading to these results was supported by MICINN supporting the Ramón y Cajal grant for M.A. Prieto (RYC-2017-22891); by Xunta de Galicia for supporting the program EXCELENCIA-ED431F 2020/12, the post-doctoral grant of M. Fraga-Corral (ED481B-2019/096), the pre-doctoral grants of P. Garcia-Oliveira (ED481A-2019/295) and A. González Pereira (ED481A-2019/0228), the program BENEFICIOS DO CONSUMO DAS ESPECIES TINTORERA-(CO-0019-2021) that supports the work of F. Chamorro. The authors are grateful to the Ibero-American Program on Science and Technology (CYTED—AQUA-CIBUS, P317RT0003), to the Bio-Based Industries Joint Undertaking (JU) under grant agreement No 888003 UP4HEALTH Project (H2020-BBI-JTI-2019) that supports the work of P. Garcia-Perez and C. Lourenço-Lopes. The authors would like to thank the EU and FCT for funding through the project PTDC/OCE-ETA/30240/2017—SilverBrain—From sea to brain: Green neuroprotective extracts for nanoencapsulation and functional food production (POCI-01-0145-FEDER-030240). The authors are grateful to the Foundation for Science and Technology (FCT, Portugal) for financial support through national funds FCT/MCTES to CIMO (UIDB/00690/2020); and to the national funding by FCT, P.I., through the institutional scientific employment program-contract for L. Barros contract.

Conflicts of Interest: The authors declare no conflict of interest.

References

1. Cavalcanti, R.N.; Santos, D.T.; Meireles, M.A.A. Non-thermal stabilization mechanisms of anthocyanins in model and food systems-An overview. *Food Res. Int.* **2011**, *44*, 499–509. [CrossRef]
2. Ge, Q.; Ma, X. Composition and antioxidant activity of anthocyanins isolated from Yunnan edible rose (*An ning*). *Food Sci. Hum. Wellness* **2013**, *2*, 68–74. [CrossRef]
3. Sasaki, R.; Nishimura, N.; Hoshino, H.; Isa, Y.; Kadowaki, M.; Ichi, T.; Tanaka, A.; Nishiumi, S.; Fukuda, I.; Ashida, H.; et al. Cyanidin 3-glucoside ameliorates hyperglycemia and insulin sensitivity due to downregulation of retinol binding protein 4 expression in diabetic mice. *Biochem. Pharmacol.* **2007**, *74*, 1619–1627. [CrossRef]
4. Tsuda, T.; Horio, F.; Uchida, K.; Aoki, H.; Osawa, T. Dietary cyanidin 3-O-beta-D-glucoside-rich purple corn color prevents obesity and ameliorates hyperglycemia in mice. *J. Nutr.* **2003**, *133*, 2125–2130. [CrossRef] [PubMed]
5. Xu, J.W.; Ikeda, K.; Yamori, Y. Upregulation of endothelial nitric oxide synthase by cyanidin-3-glucoside, a typical anthocyanin pigment. *Hypertension* **2004**, *44*, 217–222. [CrossRef]
6. Grace, M.H.; Xiong, J.; Esposito, D.; Ehlenfeldt, M.; Lila, M.A. Simultaneous LC-MS quantification of anthocyanins and non-anthocyanin phenolics from blueberries with widely divergent profiles and biological activities. *Food Chem.* **2019**, *277*, 336–346. [CrossRef]
7. Kamiya, H.; Yanase, E.; Nakatsuka, S. Novel oxidation products of cyanidin 3-*O*-glucoside with 2,2′-azobis-(2,4-dimethyl)valeronitrile and evaluation of anthocyanin content and its oxidation in black rice. *Food Chem.* **2014**, *155*, 221–226. [CrossRef]
8. Tatsuzawa, F.; Saito, N.; Seki, H.; Yokoi, M.; Yukawa, T.; Shinoda, K.; Honda, T. Acylated anthocyanins in the flowers of Vanda (Orchidaceae). *Biochem. Syst. Ecol. Biochem. SYST ECOL* **2004**, *32*, 651–664. [CrossRef]
9. Keppler, K.; Humpf, H.U. Metabolism of anthocyanins and their phenolic degradation products by the intestinal microflora. *Bioorgan. Med. Chem.* **2005**, *13*, 5195–5205. [CrossRef] [PubMed]
10. Jang, Y.P.; Zhou, J.; Nakanishi, K.; Sparrow, J.R. Anthocyanins protect against A2E photooxidation and membrane permeabilization in retinal pigment epithelial cells. *Photochem. Photobiol.* **2005**, *81*, 529–536. [CrossRef] [PubMed]
11. Acevedo De la Cruz, A.; Hilbert, G.; Rivière, C.; Mengin, V.; Ollat, N.; Bordenave, L.; Decroocq, S.; Delaunay, J.C.; Delrot, S.; Mérillon, J.M.; et al. Anthocyanin identification and composition of wild *Vitis* spp. accessions by using LC-MS and LC-NMR. *Anal. Chim. Acta* **2012**, *732*, 145–152. [CrossRef] [PubMed]
12. Kirby, C.W.; Wu, T.; Tsao, R.; McCallum, J.L. Isolation and structural characterization of unusual pyranoanthocyanins and related anthocyanins from Staghorn sumac (*Rhus typhina* L.) via UPLC-ESI-MS, 1H, 13C, and 2D NMR spectroscopy. *Phytochemistry* **2013**, *94*, 284–293. [CrossRef]
13. Brauch, J.E.; Reuter, L.; Conrad, J.; Vogel, H.; Schweiggert, R.M.; Carle, R. Characterization of anthocyanins in novel Chilean maqui berry clones by HPLC–DAD–ESI/MSn and NMR-spectroscopy. *J. Food Compos. Anal.* **2017**, *58*, 16–22. [CrossRef]
14. Zhang, J.L.; Luo, C.L.; Zhou, Q.; Zhang, Z.C. Isolation and identification of two major acylated anthocyanins from purple sweet potato (*Ipomoea batatas* L. cultivar Eshu No. 8) by UPLC-QTOF-MS/MS and NMR. *Int. J. Food Sci. Technol.* **2018**, *53*, 1932–1941. [CrossRef]
15. Zielińska, A.; Siudem, P.; Paradowska, K.; Gralec, M.; Kaźmierski, S.; Wawer, I. *Aronia melanocarpa* fruits as a rich dietary source of chlorogenic acids and anthocyanins: 1H-NMR, HPLC-DAD, and chemometric studies. *Molecules* **2020**, *25*, 3234. [CrossRef]
16. Müller-Maatsch, J.; Gurtner, K.; Carle, R.; Björn Steingass, C. Investigation into the removal of glucosinolates and volatiles from anthocyanin-rich extracts of red cabbage. *Food Chem.* **2019**, *278*, 406–414. [CrossRef]
17. Grajeda-Iglesias, C.; Salas, E.; Barouh, N.; Baréa, B.; Figueroa-Espinoza, M.C. Lipophilization and MS characterization of the main anthocyanins purified from hibiscus flowers. *Food Chem.* **2017**, *230*, 189–194. [CrossRef]

18. Wang, Y.; Johnson-Cicalese, J.; Singh, A.P.; Vorsa, N. Characterization and quantification of flavonoids and organic acids over fruit development in American cranberry (Vaccinium macrocarpon) cultivars using HPLC and APCI-MS/MS. *Plant Sci.* **2017**. [CrossRef]
19. Karaaslan, N.M.; Yaman, M. Anthocyanin profile of strawberry fruit as affected by extraction conditions. *Int. J. Food Prop.* **2018**, *20*, S2313–S2322. [CrossRef]
20. Hohnová, B.; Šťavíková, L.; Karásek, P. Determination of anthocyanins in red grape skin by pressurised fluid extraction and HPLC. *Czech J. Food Sci.* **2008**, *26*, 39–41. [CrossRef]
21. Benmeziane, F.; Cadot, Y.; Djamai, R.; Djermoun, L. Determination of major anthocyanin pigments and flavonols in red grape skin of some table grape varieties (vitis vinifera SP.) by high-performance liquid chromatography-photodiode array detection (HPLC-DAD). *OENO One* **2016**, *50*, 125–135. [CrossRef]
22. Lopes-Da-Silva, F.; De Pascual-Teresa, S.; Rivas-Gonzalo, J.; Santos-Buelga, C. Identification of anthocyanin pigments in strawberry (cv Camarosa) by LC using DAD and ESI-MS detection. *Eur. Food Res. Technol.* **2002**, *214*, 248–253. [CrossRef]
23. Gras, C.C.; Carle, R.; Schweiggert, R.M. Determination of anthocyanins from black carrots by UHPLC-PDA after ultrasound-assisted extraction. *J. Food Compos. Anal.* **2015**, *44*, 170–177. [CrossRef]
24. Vieira, G.S.; Marques, A.S.F.; Machado, M.T.C.; Silva, V.M.; Hubinger, M.D. Determination of anthocyanins and non-anthocyanin polyphenols by ultra performance liquid chromatography/electrospray ionization mass spectrometry (UPLC/ESI–MS) in jussara (*Euterpe edulis*) extracts. *J. Food Sci. Technol.* **2017**, *54*, 2135–2144. [CrossRef]
25. Huang, Z.; Wang, B.; Williams, P.; Pace, R.D. Identification of anthocyanins in muscadine grapes with HPLC-ESI-MS. *LWT Food Sci. Technol.* **2009**, *42*, 819–824. [CrossRef]
26. Fibigr, J.; Šatínský, D.; Solich, P. A UHPLC method for the rapid separation and quantification of anthocyanins in acai berry and dry blueberry extracts. *J. Pharm. Biomed. Anal.* **2017**, *143*, 204–213. [CrossRef] [PubMed]
27. Chen, S.; Xiang, Y.; Deng, J.; Liu, Y.; Li, S. Simultaneous Analysis of Anthocyanin and Non-Anthocyanin Flavonoid in Various Tissues of Different Lotus (Nelumbo) Cultivars by HPLC-DAD-ESI-MS n. *PLoS ONE* **2013**, *8*, e62291. [CrossRef]
28. Gonçalves, G.A.; Soares, A.A.; Correa, R.C.G.; Barros, L.; Haminiuk, C.W.I.; Peralta, R.M.; Ferreira, I.C.F.R.; Bracht, A. Merlot grape pomace hydroalcoholic extract improves the oxidative and inflammatory states of rats with adjuvant-induced arthritis. *J. Funct. Foods* **2017**, *33*, 408–418. [CrossRef]
29. Li, X.; Hilgers, M.; Cunningham, M.; Chen, Z.; Trzoss, M.; Zhang, J.; Kohnen, L.; Lam, T.; Creighton, C.; Gc, K.; et al. Structure-based design of new DHFR-based antibacterial agents: 7-aryl-2,4-diaminoquinazolines. *Bioorgan. Med. Chem. Lett.* **2011**, *21*, 5171–5176. [CrossRef]
30. Made, I.; Wirasuta, I.M.A.G.; Kadek, N.; Suryani, A.; Made, P.; Armitasari, N.; Mirah, P.; Dewi, K. Validation Assay of Total Flavonoids Content in Ipomoea batatas L., as Rutin Equivalent, by Using Thin Layer Spectrophotodensitometry. *Int. J. Pharm. Sci. Rev. Res.* **2019**, *53*, 30–33.
31. Bordonaba, J.G.; Crespo, P.; Terry, L.A. A new acetonitrile-free mobile phase for HPLC-DAD determination of individual anthocyanins in blackcurrant and strawberry fruits: A comparison and validation study. *Food Chem.* **2011**, *129*, 1265–1273. [CrossRef]
32. Canuto, G.A.B.; Oliveira, D.R.; Da Conceição, L.S.M.; Farah, J.P.S.; Tavares, M.F.M. Development and validation of a liquid chromatography method for anthocyanins in strawberry (*Fragaria* spp.) and complementary studies on stability, kinetics and antioxidant power. *Food Chem.* **2016**, *192*, 566–574. [CrossRef]
33. Brown, P.N.; Shipley, P.R. Determination of anthocyanins in cranberry fruit and cranberry fruit products by high-performance liquid chromatography with ultraviolet detection: Single-laboratory validation. *J. AOAC Int.* **2011**, *94*, 459–466. [CrossRef]

chemistry proceedings

MDPI

Proceeding Paper

Portable Electrochemical Detection of Illicit Drugs in Smuggled Samples: Towards More Secure Borders †

Marc Parrilla [1,2], Amorn Slosse [3], Robin Van Echelpoel [1,2], Noelia Felipe Montiel [1,2], Filip Van Durme [3] and Karolien De Wael [1,2,*]

1 A-Sense Lab, Department of Bioscience Engineering, University of Antwerp, Groenenborgerlaan 171, 2020 Antwerp, Belgium; marc.parrillapons@uantwerpen.be (M.P.); robin.vanechelpoel@uantwerpen.be (R.V.E.); noelia.felipemontiel@uantwerpen.be (N.F.M.)

2 NANOlab Center of Excellence, University of Antwerp, Groenenborgerlaan 171, 2020 Antwerp, Belgium

3 Drugs and Toxicology Department, National Institute for Criminalistics and Criminology (NICC), Vilvoordsesteenweg 100, 1120 Brussels, Belgium; amorn.slosse@just.fgov.be (A.S.); filip.vandurme@just.fgov.be (F.V.D.)

* Correspondence: karolien.dewael@uantwerpen.be; Tel.: +32-32653335

† Presented at the 1st International Electronic Conference on Chemical Sensors and Analytical Chemistry, 1–15 July 2021; Available online: https://csac2021.sciforum.net/.

Abstract: Illicit drug consumption is posing critical concerns in our society causing health issues, crime-related activities and the disruption of the border trade. The smuggling of illicit drugs urges the development of new tools for rapid on-site identification in cargos. Current methods used by law enforcement offices rely on presumptive color tests and portable spectroscopic techniques. However, these methods sometimes exhibit inaccurate results due to commonly used cutting agents or because the drugs are smuggled (hidden or mixed) in colored samples. Interestingly, electrochemical sensors can deal with these specific problems. Herein, it is presented an electrochemical device that uses low-cost screen-printed electrodes for the electrochemical detection of illicit drugs by square-wave voltammetry (SWV) profiling. A library of electrochemical profiles is built upon pure and mixtures of illicit drugs with common cutting agents. This library allows the design of a tailor-made script that shows the identification of each drug through a user-friendly interface. Finally, the results obtained from the analysis of different samples from confiscated cargos at an end-user laboratory present a promising alternative to current methods offering low-cost and rapid testing in the field.

Keywords: electrochemical sensors; square-wave voltammetry; screen-printed electrodes; illicit drugs; forensics; portable device

Citation: Parrilla, M.; Slosse, A.; Van Echelpoel, R.; Montiel, N.F.; Van Durme, F.; De Wael, K. Portable Electrochemical Detection of Illicit Drugs in Smuggled Samples: Towards More Secure Borders. *Chem. Proc.* **2021**, *5*, 44. https://doi.org/10.3390/CSAC2021-10612

Academic Editor: Nicole Jaffrezic-Renault

Published: 6 July 2021

1. Introduction

The consumption of drugs of abuse is causing critical issues in our society due to health issues, crime-related activities and the disruption of the border trade [1]. These illicit drugs can enter the illegal market through external borders (e.g., natural drugs) or by internal production (e.g., synthetic drugs). The smuggling of illicit drugs such as cocaine and heroin in Europe urges the development of new tools for rapid on-site identification in cargos. Besides, the production of synthetic drugs increases internal trafficking, thus demanding simple and straightforward devices to detect illicit drugs in the field. Current methods used by law enforcement offices rely on presumptive color tests [2] and portable spectroscopic techniques (e.g., near-infrared [3] and Raman spectroscopy [4]). However, these methods sometimes exhibit inaccurate results due to commonly used cutting agents or because the drugs are colored. Besides, drug traffickers are generating innovative ways to overcome traditional detection methods such as mixing with conventional goods (e.g., tcharcoal, food) or adding colorants or other substances to avoid the on-site determination by current methods. Therefore, new devices that can overcome the current problems are necessary to cope with the determination of smuggled illicit drugs in common goods.

The devices for on-site analysis must be portable, low-cost and user-friendly in order to be implemented and used by law enforcement officers [5]. Electrochemical sensors can provide the aforementioned features, and importantly, they can deal with current challenges, providing more reliable results in comparison to commercially available devices [6,7]. In this direction, portable and wearable electrochemical sensors have been designed for the detection of illicit drugs in different configurations [8], including glove-based sensors [9]. The electrochemical approach is based on the characteristic electrochemical profile (EP) of each compound that reveals the electroactive moieties of the target compound. Following this strategy, cocaine [10], ketamine [11], heroin [12] and synthetic cathinones [13] have been detected by using low-cost screen-printed electrodes (SPEs). Amphetamine is a special case as it is not electroactive in the potential window of commercial carbon SPEs. Therefore, an in situ derivatization by employing 1,2-Naphthoquinone-4-sulphonic acid sodium salt allows for its electrochemical detection [14]. Overall, the most used illicit drugs can be determined by electrochemical methods at certain conditions.

Herein, we present an electrochemical device that uses low-cost SPEs for the electrochemical detection of illicit drugs by square-wave voltammetry (SWV) profiling. A pH strategy based on the profiling of each illicit drug using specific buffers allows detection of the most-encountered illicit drugs (i.e., cocaine, MDMA, heroin and amphetamine). Hence, the electrochemical interrogation of the illicit drugs exhibits the oxidation of the electroactive moieties in each drug at a certain potential, with the exception of amphetamine that uses an in situ derivatization to unravel its oxidation peak. A library of electrochemical profiles is built upon pure and mixtures of illicit drugs with common cutting agents. This library allows the design of a tailor-made script that shows the identification of each drug through a user-friendly interface. Finally, the results obtained from the analysis of different samples from confiscated cargos at different end-users sites present a promising alternative to current methods. Overall, the fast analysis of samples with a portable electrochemical device exhibits a straightforward on-site detection aiming to facilitate the tasks of law enforcement agents in the field, thus providing a more secure border management and a safer society.

2. Methods

2.1. Materials

Standards of D,L-amphetamine · HCl, methamphetamine · HCl, 3,4-methylenedioxym ethamphetamine · HCl (MDMA), cocaine · HCl and heroin · HCl, were purchased from Chiron AS, Trondheim, Norway. Standards of paracetamol, caffeine and creatine were provided by National Institute for Criminalistics and Criminology (NICC, Brussels, Belgium). Confiscated samples of amphetamine, MDMA, cocaine and heroin were also provided by the NICC. Analytical grade salts of potassium chloride, potassium phosphate, sodium borate, sodium bicarbonate, sodium acetate and potassium hydroxide were purchased from Sigma-Aldrich (Overijse, Belgium). 1,2-Naphthoquinone-4-sulphonic acid sodium salt (NQS) (>98 %) was purchased from Tokyo Chemical Industry Co., LTD., Tokyo, Japan.

2.2. Methods

Square wave voltammograms and cyclic voltammograms were recorded using a MultiPalmSens4 or EmStat Pico potentiostats (PalmSens, Houten, The Netherlands) with PSTrace/MultiTrace. Disposable ItalSens SPEs (PalmSens, Houten, The Netherlands), containing a graphite working electrode (Ø = 3 mm), a carbon counter electrode and a (pseudo) silver reference electrode were used for all measurements. The SWV parameters that were used: potential range of 0.0–1.4 V, frequency 10 Hz, 25 mV amplitude and 5 mV step potential. All the square wave voltammograms are background corrected using the PSTrace software.

Electrochemical tests were performed in 20 mM buffer solutions with 100 mM KCl by applying 60 μL of the solution onto the SPE. Phosphate buffer, acetate buffer and hydrogen carbonate buffer were used for the detection of cocaine and heroin, MDMA and amphetamine, respectively. Preanodized SPEs for heroin detection were performed by applying 1.5 V for 60 s in PBS solution at pH 7 by drop casting 60 μL on the SPE [12].

The composition of the confiscated samples was previously analyzed in the forensic laboratory at NICC with gas chromatography-mass spectrometry (GC-MS) to subsequently validate the electrochemical approach. Besides, the confiscated samples were also analyzed by a handheld Raman spectrometer (Bruker Bravo, Ettingen, Germany).

3. Results and Discussion

3.1. Electrochemical Profiling of Illicit Drugs

The electrochemical profiling is based on the interrogation by SWV of the target molecules which exhibit an oxidation process at a certain potential. Hence, cocaine oxidation peak might differ from heroin oxidation peak, showing the possibility to identify the target by the specific peak potential [15]. Fortunately, illicit drugs contain moieties which are electroactive at the potential window of carbon SPEs. These moieties are usually secondary or tertiary amines that allow its oxidation, thus showing a peak signal during the electrochemical scan by SWV. However, some illicit drugs share some moieties which can exhibit some overlap in the oxidation potential. In previous works, our research group has optimized the detection of illicit drugs by exploring certain conditions during the SWVs. For example, the anodic pretreatment to unravel the phenolic group oxidation of the 6-monoacetylmorphine (6-MAM) (a byproduct of heroin degradation at pH 12) [12], the cathodic pretreatment to avoid the suppressing effect of some adulterants [13] or the use of different pH (as some moieties are not oxidizable in certain pH at SPE) [11]. Besides, some illicit drugs such as amphetamine need a derivatization step to allow its electrochemical profiling employing low-cost carbon SPEs. In this case, a simple mixing step with NQS launches a chemical reaction to a product that is electroactive at the SPE [14]. An oxidation peak was observed at 1.15 V due to NQS oxidation at the carbon SPE. Since this occurs outside the potential window for illicit drugs (from 0.6 to 1.03 V) in pH 10 it will not affect their identification.

3.2. Generating the Library of Electrochemical Profiles

Figure 1 shows the SWVs of pure illicit drugs (i.e., cocaine, heroin, MDMA and amphetamine) at pH 5, pH 10, pH 12 and pH 12 using anodic pretreated SPEs to record the specific electrochemical profiles of the molecules at certain conditions. Figure 1A,B shows the differences between anodic pretreated SPEs. Although similar profiles were obtained, a clear peak separation occurred in the MDMA signal. The pH dependence on the oxidation of illicit drugs is dramatically shown in the analysis at pH 5, where only heroin and MDMA exhibited electroactivity (Figure 1C). This fact assists in the proper identification of the unknown sample by a simple dual pH test. Finally, the pH assessment clearly exhibits that amphetamine is not electroactive, and only after the derivatization step (Figure 1D), an oxidation peak appears.

A similar electrochemical approach was performed employing the most encountered cutting agents (i.e., paracetamol, levamisole, lidocaine, caffeine, phenacetin, benzocaine, procaine and lactose [12,14,16]) (Figure 2A–D). Particularly interesting is the effect of the anodic pretreatment on the paracetamol signal exhibiting a sharper oxidation peak (Figure 2B). This permits the proper identification of the 6-MAM peak, thus avoiding peak overlap [12]. Besides, the effect of pH on the electrochemical signal showed a pH dependence as the oxidation peak shifts towards higher potentials at acidic pHs (Figure 2C). As pH 10 with NQS is targeted for the detection of amphetamine, only common cutting agents encountered in amphetamine real samples are explored (Figure 2D). Considering the profiling of the cutting agents, most of the peak potentials do not fall in the same position as the illicit drugs, thus allowing for a suitable identification in real samples.

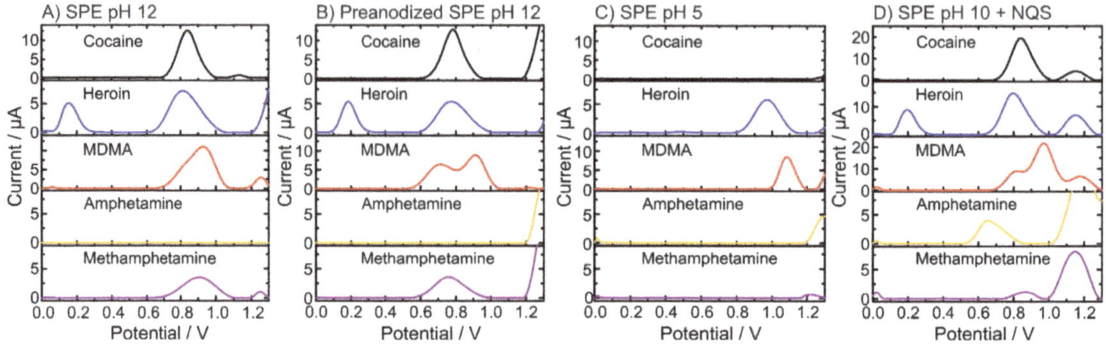

Figure 1. Electrochemical profiles of illicit drugs (0.5 mM) obtained by square-wave voltammetry (SWV) using SPE at different pH: (**A**) pH 12; (**B**) pH 12 using preanodized SPE; (**C**) pH 5; and (**D**) pH 10 including the derivatizing agent NQS.

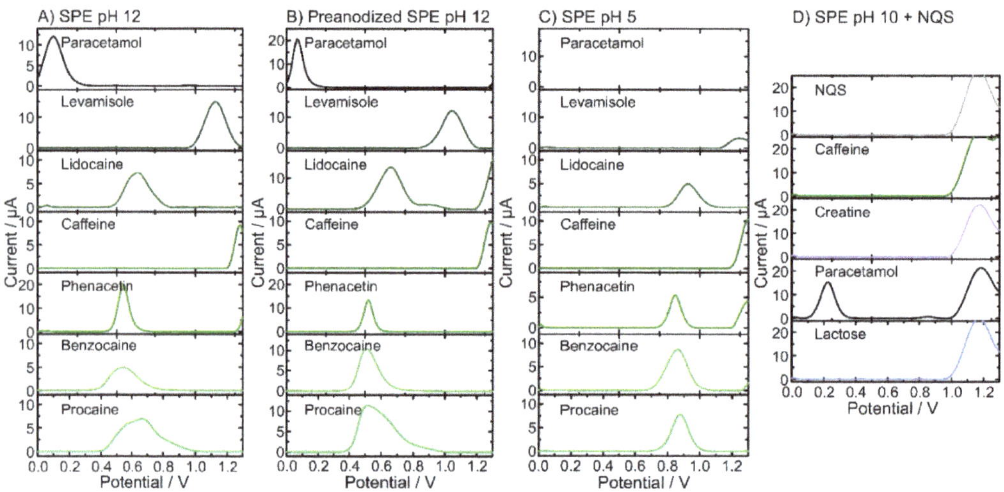

Figure 2. Electrochemical profiles of common cutting agents (0.5 mM) obtained by square-wave voltammetry (SWV) using SPE at different pH: (**A**) pH 12; (**B**) pH 12 using preanodized SPE; (**C**) pH 5; and (**D**) pH 10 including the derivatizing agent NQS.

After building the library of electrochemical profiles with several conditions, a custom-made script (Matlab R2018b, MathWorks, Natick, MA, USA) is integrated. This script enhances the peak separation and facilitates the identification of the compounds in the sample. In brief, the script removes the background signal and applies a top-hat filter that provides an enhanced separation of overlapped peaks which permits a successful identification of the substances based on the peak potential of each drug. Therefore, the peak potential of each drug and cutting agent is introduced in the script to properly identify the drug and display it through a user-friendly interface.

3.3. Testing the Portable Electrochemical Device with Confiscated Samples

The electrochemical device consists of a miniaturized potentiostat with Bluetooth connectivity, a disposable SPE, a sampling container, a disposable spatula and a disposable pipette (Figure 3A). The sampling procedure consists of collecting the powder (either powder, liquid, crystal or impregnated material) with the disposable spatula into a tube containing 15 mL of the suitable buffer (Figure 2B). After shaking thoroughly, a drop of the solution is deposited on the SPE with the disposable pipette (Figure 2C). Subsequently, the operation is started on the user-friendly interface launching the electrochemical method, subsequent data treatment and results display (Figure 2D). For the analysis of confiscated samples, the strategies employing pH 12, preanodized SPE in pH 12, pH 5 and pH 10 with NQS were employed for cocaine, heroin, MDMA and amphetamine, respectively.

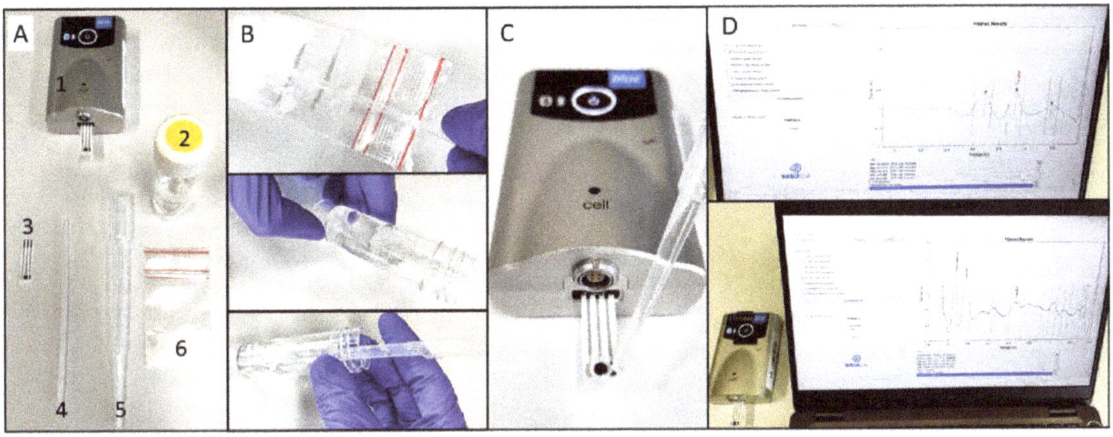

Figure 3. On-site detection of illicit drugs with the portable electrochemical device. (**A**) Elements of the electrochemical device (1—potentiostat, 2—buffer container, 3—SPE, 4—disposable spatula, 5—disposable pipette, 6—confiscated sample); (**B**) Sampling procedure; (**C**) deposition of the solution on the setup ready for the electrochemical interrogation and (**D**) user-friendly interface showing the results of the analysis.

The reliability of the electrochemical device was evaluated in 40 confiscated samples provided by NICC (Table 1). After the analysis, the 40 samples were all positive for the corresponding illicit drug using the described sampling method in comparison to the standard methods (GC-MS). Besides, a portable Raman spectrometer was also used as a commonly used method in border settings. The electrochemical reader and portable Raman spectrometer exhibited an accuracy of 100% and 50%, respectively, calculated employing (observed detection by the method/actual detection by the GC-MS) $\times$ 100. Therefore, the electrochemical device outperformed the Raman device, particularly in heroin and amphetamine detection. It is worth mentioning that the low performance of the Raman device could be attributed to the colored nature of the samples, thus exhibiting one of the flaws of current methods. Overall, the electrochemical device is positioned as a reliable alternative for its use in the field due to its affordability, reliability and user-friendliness.

Table 1. Results of the analysis by the analytical methods and composition of the confiscated samples.

Sample	GC-MS	Electrochemical Reader	Portable Raman
1	Cocaine	Cocaine	Cocaine mixture
2	Cocaine, caffeine, levamisole	Cocaine	Cocaine mixture
3	Cocaine, levamisole	Cocaine	Cocaine mixture
4	Cocaine, lidocaine, levamisole, phenacetin	Cocaine, lidocaine	Lidocaine
5	Cocaine, phenacetin, levamisole	Cocaine, phenacetin	Cocaine mixture
6	Cocaine, levamisole	Cocaine	Cocaine mixture
7	Cocaine base	Cocaine	Benzyl benzoate
8	Cocaine, levamisole	Cocaine, levamisole	Cocaine mixture
9	Cocaine, caffeine	Cocaine	Cocaine mixture
10	Cocaine, lidocaine, levamisole, caffeine, phenacetin	Cocaine, lidocaine	Lidocaine

Table 1. *Cont.*

Sample	GC-MS	Electrochemical Reader	Portable Raman
11	Heroin, caffeine, 6-MAM, Noscapine, Papaverine	Heroin	Normorphine
12	Heroin, caffeine, 6-MAM, Morphine, Noscapine, Papaverine	Heroin	Cetyltrimethylammonium chloride
13	Heroin, caffeine, paracetamol, 6-MAM, Noscapine, Papaverine	Heroin, paracetamol	Cetyltrimethylammonium chloride
14	Heroin, caffeine, Dextromethorphan, 6-MAM, Noscapine	Heroin	Unknown
15	Heroin, 6-MAM, Noscapine, Papaverine	Heroin	Cetyltrimethylammonium chloride
16	Heroin, paracetamol, 6-MAM, morphine, caffeine, codeine, noscapine, papaverine	Heroin, paracetamol	Cetyltrimethylammonium chloride
17	Heroin, paracetamol, 6-MAM, caffeine, noscapine, papaverine	Heroin, paracetamol	Cetyltrimethylammonium chloride
18	Heroin, paracetamol, 6-MAM, caffeine, noscapine, papaverine	Heroin	Unknown
19	Heroin, paracetamol, 6-MAM, caffeine, noscapine, papaverine	Heroin, paracetamol	Cetyltrimethylammonium chloride
20	Heroin, paracetamol, 6-MAM, caffeine, noscapine, papaverine	Heroin, paracetamol	Cetyltrimethylammonium chloride
21	MDMA	MDMA	MDMA tablet
22	MDMA	MDMA	MDMA crystals
23	MDMA	MDMA	MDMA tablet
24	MDMA, levamisole	MDMA	MDMA crystals
25	MDMA	MDMA	MDMA tablet
26	MDMA	MDMA	MDMA tablet
27	MDMA	MDMA	MDMA crystals
28	MDMA	MDMA	MDMA crystals
29	MDMA	MDMA	MDMA crystals
30	MDMA	MDMA	MDMA crystals
31	Amphetamine	Amphetamine	Paraform
32	Amphetamine	Amphetamine	Norephedrine HCl
33	Amphetamine	Amphetamine	1_(*S*)-2-diphenylmethylpyrrolidine HCl_(*S*)-desoxy-D2PM
34	Amphetamine	Amphetamine	1_(*S*)-2-diphenylmethylpyrrolidine HCl_(*S*)-desoxy-D2PM
35	Amphetamine, caffeine	Amphetamine	Amphetamine
36	Amphetamine, caffeine	Amphetamine	Unknown
37	Amphetamine	Amphetamine	1-phenyl-1-propanol
38	Amphetamine	Amphetamine	1-phenyl-1-propanol
39	Amphetamine	Amphetamine	1-phenyl-1-propanol
40	Amphetamine, caffeine	Amphetamine	Amphetamine

4. Conclusions

In this work, the analysis of confiscated samples from illicit drugs is presented by the use of a portable electrochemical device. First, the construction of a library from several electrochemical profiles of standards of illicit drugs and common cutting agents at different conditions by SPE is performed. After the selection of the suitable conditions and the integration of the peak potentials of each target into a tailor-made script, the electrochemical device is ready for on-site analysis. The examination of 40 confiscated samples with the electrochemical device and a portable Raman spectrometer showed an outstanding performance of the electrochemical device in front of the Raman device according to the GC-MS identification. Overall, the electrochemical device based on SPE is presented as a promising alternative to current rapid and on-site methods for the detection of illicit drugs at border and coast controls.

Supplementary Materials: The following are available online at https://www.mdpi.com/article/10.3390/CSAC2021-10612/s1.

Author Contributions: Conceptualization, M.P.; methodology, M.P., A.S. and N.F.M.; software, R.V.E.; validation, M.P., A.S. and F.V.D.; M.P.; investigation, M.P.; resources, K.D.W. and F.V.D.; data curation, M.P. and A.S.; writing—original draft preparation, M.P.; writing—review and editing, M.P.; visualization, M.P.; supervision, K.D.W.; project administration, K.D.W.; funding acquisition, K.D.W. All authors have read and agreed to the published version of the manuscript.

Funding: This research was funded by European Union's Horizon 2020 research and innovation program, grant number 833787, Bordersens.

Conflicts of Interest: The authors declare no conflict of interest. The funders had no role in the design of the study; in the collection, analyses, or interpretation of data; in the writing of the manuscript, or in the decision to publish the results.

References

1. European Monitoring Centre for Drugs and Drug Addiction. *European Drug Report: Trends and Developments*; European Monitoring Centre for Drugs and Drug Addiction: Lisbon, Portugal, 2020.
2. Philp, M.; Fu, S. A review of chemical 'spot' tests: A presumptive illicit drug identification technique. *Drug Test. Anal.* **2018**, *10*, 95–108. [CrossRef] [PubMed]
3. Kranenburg, R.F.; Verduin, J.; Weesepoel, Y.; Alewijn, M.; Heerschop, M.; Koomen, G.; Keizers, P.; Bakker, F.; Wallace, F.; van Esch, A.; et al. Rapid and robust on-scene detection of cocaine in street samples using a handheld near-infrared spectrometer and machine learning algorithms. *Drug Test. Anal.* **2020**, *12*, 1404–1418. [CrossRef] [PubMed]
4. Kranenburg, R.F.; Verduin, J.; Ridder, R.; Weesepoel, Y.; Alewijn, M.; Heerschop, M.; Keizers, P.H.J.; Esch, A.; Asten, A.C. Performance Evaluation of Handheld Raman Spectroscopy for Cocaine Detection in Forensic Case Samples. *Drug Test. Anal.* **2020**, *13*, 1054–1067. [CrossRef] [PubMed]
5. de Araujo, W.R.; Cardoso, T.M.G.; da Rocha, R.G.; Santana, M.H.P.; Muñoz, R.A.A.; Richter, E.M.; Paixão, T.R.L.C.; Coltro, W.K.T. Portable analytical platforms for forensic chemistry: A review. *Anal. Chim. Acta* **2018**, *1034*, 1–21. [CrossRef] [PubMed]
6. Florea, A.; de Jong, M.; De Wael, K. Electrochemical strategies for the detection of forensic drugs. *Curr. Opin. Electrochem.* **2018**, *11*, 34–40. [CrossRef]
7. De Rycke, E.; Stove, C.; Dubruel, P.; De Saeger, S.; Beloglazova, N. Recent developments in electrochemical detection of illicit drugs in diverse matrices. *Biosens. Bioelectron.* **2020**, *169*, 112579. [CrossRef] [PubMed]
8. Teymourian, H.; Parrilla, M.; Sempionatto, J.R.; Montiel, N.F.; Barfidokht, A.; Van Echelpoel, R.; De Wael, K.; Wang, J. Wearable Electrochemical Sensors for the Monitoring and Screening of Drugs. *ACS Sens.* **2020**, *5*, 2679–2700. [CrossRef] [PubMed]
9. de Jong, M.; Sleegers, N.; Kim, J.; Van Durme, F.; Samyn, N.; Wang, J.; De Wael, K. Electrochemical fingerprint of street samples for fast on-site screening of cocaine in seized drug powders. *Chem. Sci.* **2016**, *7*, 2364–2370. [CrossRef] [PubMed]
10. De Jong, M.; Florea, A.; Eliaerts, J.; Van Durme, F.; Samyn, N.; De Wael, K. Tackling Poor Specificity of Cocaine Color Tests by Electrochemical Strategies. *Anal. Chem.* **2018**, *90*, 6811–6819. [CrossRef] [PubMed]
11. Schram, J.; Parrilla, M.; Sleegers, N.; Samyn, N.; Bijvoets, S.M.; Heerschop, M.W.J.; van Nuijs, A.L.N.; De Wael, K. Identifying Electrochemical Fingerprints of Ketamine with Voltammetry and Liquid Chromatography–Mass Spectrometry for Its Detection in Seized Samples. *Anal. Chem.* **2020**, *92*, 13485–13492. [CrossRef] [PubMed]
12. Felipe Montiel, N.; Parrilla, M.; Beltrán, V.; Nuyts, G.; Van Durme, F.; De Wael, K. The opportunity of 6-monoacetylmorphine to selectively detect heroin at preanodized screen printed electrodes. *Talanta* **2021**, *226*, 122005. [CrossRef] [PubMed]
13. Schram, J.; Parrilla, M.; Sleegers, N.; Van Durme, F.; Van Den, J.; van Nuijs, A.L.N.; De Wael, K. Electrochemical profiling and LC-MS characterization of synthetic cathinones: From methodology to detection in forensic samples. *Drug Test. Anal.* **2021**, *13*, 1282–1294. [CrossRef] [PubMed]
14. Parrilla, M.; Montiel, N.F.; Van Durme, F.; Wael, K. De Derivatization of amphetamine to allow its electrochemical detection in illicit drug seizures. *Sens. Actuators B Chem.* **2021**, *337*, 129819. [CrossRef]
15. Moro, G.; Barich, H.; Driesen, K.; Felipe Montiel, N.; Neven, L.; Mendonça, C.D.; Shanmugam, S.T.; Daems, E.; Wael, K. De Unlocking the full power of electrochemical fingerprinting for on-site sensing applications. *Anal. Bioanal. Chem.* **2020**, *412*, 5955–5968. [CrossRef] [PubMed]
16. Broséus, J.; Gentile, N.; Esseiva, P. The cutting of cocaine and heroin: A critical review. *Forensic Sci. Int.* **2016**, *262*, 73–83. [CrossRef] [PubMed]

chemistry proceedings

MDPI

Abstract

The Inhibition Study of Cytochrome bd Oxidase Using the Enzyme-Based Electrochemical Sensor †

Iryna Makarchuk [1,*], Anton Nikolaev [1], Alexander Thesseling [2], Lisa Dejon [3], Daniel Lamberty [3], Laura Stief [3], Thorsten Friedrich [2], Petra Hellwig [1], Hamid R. Nasiri [4] and Frederic Melin [1]

1 Laboratoire de Bioélectrochimie et Spectroscopie, UMR 7140, Chimie de la Matière Complexe, Université de Strasbourg-CNRS, 67000 Strasbourg, France; teterev3000@mail.ru (A.N.); hellwig@unistra.fr (P.H.); fmelin@unistra.fr (F.M.)

2 Institut für Biochemie, Albert-Ludwigs-Universität, 79104 Freiburg, Germany; thesseling@bio.chemie.uni-freiburg.de (A.T.); friedrich@bio.chemie.uni-freiburg.de (T.F.)

3 Fachbereich 11 Organische Chemie, Universität des Saarlandes, 66123 Saarbrücken, Germany; lisa.dejon@boehringer-ingelheim.com (L.D.) daniellamberty@hotmail.com (D.L.); laura.stief@uni-saarland.de (L.S.)

4 Institute of Microbiology, University of Hohenheim, 70599 Stuttgart, Germany; Nasiri@nmr.uni-frankfurt.de

* Correspondence: makarchuk.iryna@etu.unistra.fr

† Presented at the 1st International Electronic Conference on Chemical Sensors and Analytical Chemistry, 1–15 July 2021; Available online: https://csac2021.sciforum.net/.

Abstract: Membrane proteins that participate in multiple vital functions of every living organism such as transport, signaling and respiration, provide 80 to 90% of the relevant targets for the pharmaceutical industries. The family of cytochrome bd oxidase enzymes is of great interest for the development of future antibiotics as they are found only in the respiratory chain of the prokaryotes and they are believed to be involved in bacterial adaptability mechanisms. They catalyze the reduction of molecular oxygen in water and oxidation of quinols and contribute to the proton motive force required for ATP synthesis. Due to their hydrophobic nature, membrane proteins are more difficult to handle than soluble proteins. Protein film voltammetry is a very convenient technique, because it allows for working at a very low concentration and for optimizing the electrode surface to the nature of the enzyme. Here, we have developed a biosensor for the study of terminal oxidases based on their immobilization on gold nanoparticles modified with a self-assembled monolayer of thiols. The stability of the protein films can be optimized by varying the nature of thiols and amount of lipids. This enzyme-based electrochemical sensor was successfully used for the inhibition screening of a target-focused library of 34 compounds which belong to the families of quinones, naphthoquinones, phenols, quinolones, coumarins and flavonoids against cytochrome bd oxidase. Moreover, the developed device was applied for the study of the catalytic reaction of the enzyme with small gaseous signaling molecules.

Keywords: membrane proteins; bioelectrochemistry; inhibition

Citation: Makarchuk, I.; Nikolaev, A.; Thesseling, A.; Dejon, L.; Lamberty, D.; Stief, L.; Friedrich, T.; Hellwig, P.; Nasiri, H.R.; Melin, F. The Inhibition Study of Cytochrome bd Oxidase Using the Enzyme-Based Electrochemical Sensor. *Chem. Proc.* **2021**, *5*, 45. https://doi.org/10.3390/CSAC2021-10555

Academic Editor: Nicole Jaffrezic-Renault

Published: 1 July 2021

Supplementary Materials: The poster presentation is available online at https://www.mdpi.com/article/10.3390/CSAC2021-10555.

Author Contributions: F.M., H.R.N. and P.H. designed research and analyzed data. I.M. and A.N. performed the electrochemical experiments and analyzed data. I.M. wrote the abstract and prepared the poster. T.F. and P.H. raised funding. A.T. and T.F. provided the protein samples. L.D., D.L., L.S., A.S. and H.R.N. provided the inhibitors. All authors reviewed and edited the abstract and poster.

Institutional Review Board Statement: Not applicable.

Informed Consent Statement: Not applicable.

Data Availability Statement: The data presented in this study are available in [I. Makarchuk, A. Nikolaev, A. Thesseling, et al, Identification and optimization of quinolone-based inhibitors against cytochrome bd oxidase using an electrochemical assay, Electrochimica Acta 381 (2021) 138293, https://doi.org/10.1016/j.electacta.2021.138293] and in the Supplementary Materials here.

Conflicts of Interest: The authors declare no conflict of interest.

chemistry proceedings

MDPI

Proceeding Paper

Electrochemical Measurement System for Chlorides in Drinking and Wastewater †

Diana A. Toriz-Gutiérrez [1], Humberto Ramírez-Gasca [2], Luis E. Cárdenas-Galindo [1] and Eloisa Gallegos-Arellano [2,*]

1 Química Industrial, Universidad Tecnológica de Salamanca, Av. Universidad Tecnológica 200, Salamanca Gto. C.P. 36766, Mexico; dtoriz@utsalamanca.edu.mx (D.A.T.-G.); lcardenas@utsalamanca.edu.mx (L.E.C.-G.)

2 Mecatrónica, Universidad Tecnológica de Salamanca, Av. Universidad Tecnológica 200, Salamanca Gto. C.P. 36766, Mexico; hramirez@utsalamanca.edu.mx

* Correspondence: egallegos@utsalamanca.edu.mx; Tel.: +52-464-64-73861 (ext. 120)

† Presented at the 1st International Electronic Conference on Chemical Sensors and Analytical Chemistry, 1–15 July 2021; Available online: https://csac2021.sciforum.net/.

Abstract: This paper presents a system for the measurement of chlorides in drinking and wastewater, based on an electrochemical process using a selective electrode as a transducer, which was developed by this group. The measurement for the concentration is carried out by introducing the implemented electrode (considered as reference) in the sample that will be analyzed; then a current is passed producing a potential difference in the system. Different aqueous solutions of sodium chloride (NaCl) were used, ranging between 35 and 3546 µg of chloride ions (Cl^-). As a data acquisition and monitoring system for the analysis, an ATmega 328P microcontroller was used as the main capture element for subsequent interpretation through graphics. The experimental results show that it was possible to detect a potential difference in the electrochemical measurement system that corresponded to 35 µg of chloride ions (Cl^-), making clear the detection process and the selectivity of chloride ions. It is important to mention that with this measurement system and the applied methodology, results are obtained in real time using a small sample volume and without generate ng extra liquid waste, compared to the application of the traditional analytical titrimetric method. Finally, this chloride measurement system is inexpensive and can be used in drinking and wastewater measurements.

Keywords: measurement system; data acquisition; electrodes; electrochemical process; ion selectivity; chloride ions

Citation: Toriz-Gutiérrez, D.A.; Ramírez-Gasca, H.; Cárdenas-Galindo, L.E.; Gallegos-Arellano, E. Electrochemical Measurement System for Chlorides in Drinking and Wastewater. *Chem. Proc.* **2021**, *5*, 46. https://doi.org/10.3390/chemproc2021005046

Academic Editor: Nicole Jaffrezic-Renault

Published: 17 December 2021

1. Introduction

Measurement by electrometric methods continues to be useful in the analytical identification processes of ion-selective electrodes (ISEs) of potentiometric sensors [1,2] because only small dimensions are involved and measurements can be made requiring little sample volume with respect to the activity of the ions in solution [3,4]. With the increasing demand for environmentally friendly technologies and in situ measurements, these designs facilitate quick readings and detection of low concentrations [1,2], as observed when obtaining measurements of concentrations of 35 µg/L in NaCl solutions [5].

Due to the diverse range of applications and reduced operating times, there is a need for measurement electrodes that are selective and able to detect a specific ion of a species that interacts with it using potential difference [4,6]. Several electrodes have been developed for selective ion determination which involve use of an ionophore to generate exchange in a solution enabling determination of the influence of the exchange ion percentage [7–10].

In this study, a chloride measurement system for drinking and wastewater is proposed based on an electrochemical process using a selective electrode as a transducer which was developed by this group. The ionophore that is proposed is an $AgNO_3$ solution with an extra modification that relates to the use of a metal that interacts with a natural organic membrane, modifying the attraction of the ion, which enhances the selectivity to the desired ion (Cl^-) [3,11], in addition to the copper reference electrode [9].

2. Materials and Methods

The experimental arrangement of the electrochemical measurement system, shown in Figure 1, is made up of two essential blocks: the electrochemical half-cell and the data acquisition system.

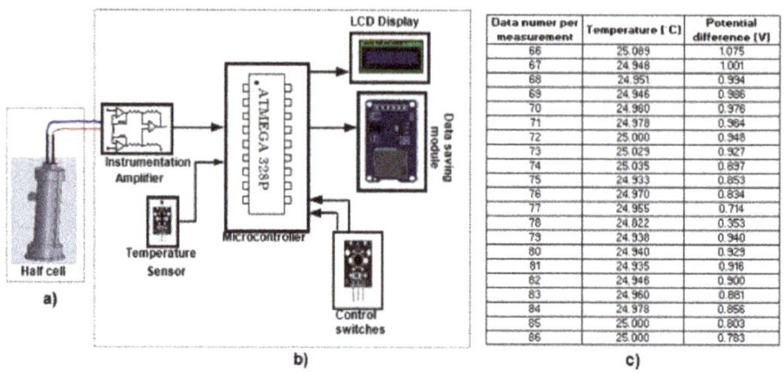

Data numer per measurement	Temperature [°C]	Potential difference [V]
66	25.089	1.075
67	24.948	1.001
68	24.951	0.994
69	24.946	0.986
70	24.960	0.976
71	24.978	0.964
72	25.000	0.948
73	25.029	0.927
74	25.035	0.897
75	24.933	0.853
76	24.970	0.834
77	24.955	0.714
78	24.822	0.353
79	24.938	0.940
80	24.940	0.929
81	24.935	0.916
82	24.946	0.900
83	24.960	0.881
84	24.978	0.856
85	25.000	0.803
86	25.000	0.783

Figure 1. Experimental arrangement of the electrochemical measurement system. (**a**) Electrochemical half-cell; (**b**) Data acquisition system; (**c**) Temperature and potential difference data obtained directly from the measuring device.

For the half-cell (Figure 1a), a 3D design was produced and printed to ensure the electrode separation was constant and to reduce movements in the experimental arrangement. The electrochemical half-cell was constructed with an active noble metal electrode (X) and a copper reference electrode, which are submerged in different solutions, such as $AgNO_3$ and NaCl, respectively, according to Creus (2011) [12] and previous research by Toriz, Loredo and Marcial [13].

The measurement of chloride ions (Cl^-) was achieved through reaction in the half-cell taking as analytes NaCl solutions at different concentrations, in which an ionic migration was generated through an organic (natural) membrane. The organic membrane and the inert metallic material (X) allow the potential difference to remain active during the measurement.

The coefficients of total ionic activity are related to the individual coefficients of ionic activity, detecting free energy in the system that connects the thermodynamics of the reaction with the values of the electromotive force (*emf*) [10] and that can only be applied in reversible reactions. This can be represented by Equation (1).

$$\varepsilon^{\circ} = \frac{-\Delta F^{\circ}}{n\mathfrak{F}} \tag{1}$$

where $\mathcal{E}^{\circ}$ is the potential difference, ΔF° is the total free energy of the system, n is the number of moles used in the reaction, and $\mathfrak{F}$ = Farad (free energy value = 96,500 Joules: approximate value = 23,052 cal).

The electromotive force (*emf*) depends on the concentration of the cation according to the Nernst equation [14]. According to the literature, the relationship between the signs ΔF° and $\mathcal{E}^{\circ}$ correspond directly to a spontaneous reaction (Maron and Prutton) [9].

Referring to a standard *emf* when the activity is a unit, the fundamental equation can be related to the thermodynamic equation and the *emf* values. The values of $\mathcal{E}^{\circ}$ can be tabulated for each reaction of a half-cell and added algebraically to obtain the value of $\mathcal{E}^{\circ}$.

To calculate the total free energy of the system, the relationship with thermodynamics is considered, calculating the *emf* from enthalpy and entropy tables for the reference half-cell [9,10], where T = 298 K, applying the relationship described in Equation (2):

$$\Delta F^{\circ} = \Delta H - T\Delta S \quad (2)$$

The calculation of the total ΔF° corresponds to the sum of the cathodes of the system; the reaction of each cell is stated as in Equation (3):

$$\Delta F^{\circ} = \mu_{Ag^{\circ}} + \mu_{Cu^{\circ}} + \mu_{Cl^{-}} - \mu_{AgCl} - \mu_{Ni} \quad (3)$$

Substituting the values for the half-cell reaction of the system (total ΔF°), Equation (4) is obtained:

$$\Delta F^{\circ} = 1000(0 + 23.6 + (-40.02) - (-30.36) + 7.2) - 298(17.67 - 23.6 + 13.17 - 22.9 - 7.2) \quad (4)$$

Solving Equation (4), the results presented in Equation (5) are obtained:

$$\Delta F^{\circ} = 21140 + 6812.28 = 27952.28 \quad (5)$$

Considering n = 1 and substituting the values in the equation, the standard potential of the system will be defined by Equation (6):

$$\varepsilon^{\circ} = \frac{27952.28}{23052} = 1.2126\ volt \quad (6)$$

The value obtained from Equation (6) represents the *emf* of the ionic activity with respect to the constant concentration of the ionic activity of the cell. It provides the voltage difference with respect to the element. This theoretical value was obtained experimentally using the measurements presented in Figure 1c. The relationship of this value and the concentration was used to establish the scale.

The data acquisition system is mainly comprised of an ATMEGA328P microcontroller that multiplexes two analog channels to store the temperature data and the potential difference of the half-cell in the microSD memory module for subsequent graphic analysis. The data from the potential difference of the electrodes are acquired through an instrumentation amplifier with unity gain to reduce the effects of electrical noise that may exist in the system (Figure 1b). The system also has an LCD module that prints the values of the analog channels sampled in the microcontroller (temperature and potential difference of the half-cell) every 5 s to verify changes in real time. When acquiring measurements with this system, the data id was saved in an Excel file where the first column shows the temperature and the second column the potential difference (Figure 1c).

3. Results and Discussion

The data stored in the microSD memory shown in Figure 1b were treated by a moving average filter with an N length window, which allows smoothing of the signal obtained from the potential difference measured in the half-cell.

The most significant data captured at different concentrations are shown in Figure 2, in which it is observed that the signals reached a stable state in the potential difference of each, which can be used for their subsequent characterization. The stabilization of the voltage occurs approximately in the same period, after the ionic migration of the oxidation-reduction reaction (transient response).

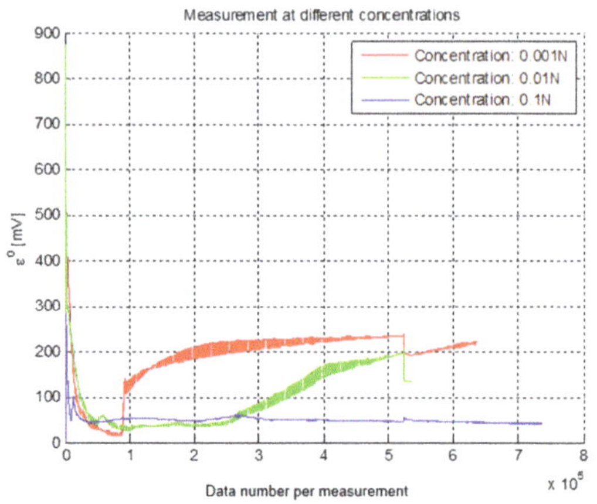

Figure 2. Graph of *emf* measurement at different concentrations of NaCl.

The data show the behavior of the concentration and the *emf* obtained for each concentration [5,15]. A scale was established to observe the change in electric potential as a function of the ionic activity in the solution [16].

Figure 3 shows that the values of the half-cell ($\mathcal{E}^{\circ}$) are proportional to the logarithm of the ion concentration in the solution, in addition to presenting a Nernstian slope, as shown in Equation 7, which presents an r^2 coefficient of 0.9953 and a correlation of 0.99.

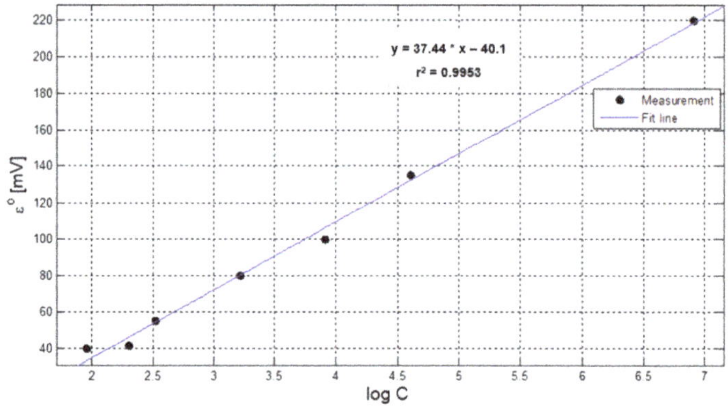

Figure 3. Graph of half-cell values ($\mathcal{E}^{\circ}$) as a function of the logarithm of the concentration.

It is important to mention that the value of the half-cell ($\mathcal{E}^{\circ}$ = 1.2126 V) corresponds to the standard electrode data and confirms a reversible reaction and the ion selectivity; while the value of the half-cell l ($\mathcal{E}^{\circ}$ = −40.47) represents the value of the ion that will be detected in the analyte.

$$\varepsilon^{\circ} = -37.54\ln(C) - 40.47 \quad (7)$$

It is of note that it was possible to optimize the resolution of the analog-digital converter of the ATMEGA 328P microcontroller so that it can detect potential differences of 1.05 mV, in addition to increasing the number of samples per second that are processed through this internal microcontroller module.

4. Conclusions

In this study, it was possible to implement a novel chloride ion measurement system using a selective membrane (natural) and a support structure to assemble a half-cell based on a design made in a 3D printer. The half-cell obtained $\mathcal{E}^\circ = 1.2126$ V; this data corresponds to the value reported for a standard electrode, ion-selective and for a reversible reaction. The indicated value of the ion-selective in the half cell ($\mathcal{E}^\circ = -40.47$), represents chloride ions (Cl^-).

The behavior of ionic activity at different concentrations shows a logarithmic trend and a typical Nernstian response and is verified with the coefficient (r^2 of 0.9953) of the fitted curve.

Based on the measurements made, the results obtained and the functionality of the assembled device, it was determined that the electrode can be used as a potential chloride ion meter for wastewater and drinking water.

Author Contributions: Conceptualization, D.A.T.-G., H.R.-G., L.E.C.-G. and E.G.-A.; methodology, D.A.T.-G., H.R.-G., L.E.C.-G. and E.G.-A.; software, D.A.T.-G., H.R.-G., L.E.C.-G. and E.G.-A.; validation, D.A.T.-G., H.R.-G., L.E.C.-G. and E.G.-A.; formal analysis, D.A.T.-G., H.R.-G., L.E.C.-G. and E.G.-A.; investigation, D.A.T.-G., H.R.-G., L.E.C.-G. and E.G.-A.; resources, D.A.T.-G., H.R.-G., L.E.C.-G. and E.G.-A.; data curation, D.A.T.-G., H.R.-G., L.E.C.-G. and E.G.-A.; writing—original draft preparation, D.A.T.-G., H.R.-G., L.E.C.-G. and E.G.-A.; writing—review and editing, D.A.T.-G., H.R.-G., L.E.C.-G. and E.G.-A.; visualization, D.A.T.-G., H.R.-G., L.E.C.-G. and E.G.-A.; supervision, D.A.T.-G., H.R.-G., L.E.C.-G. and E.G.-A.; project administration, D.A.T.-G., H.R.-G., L.E.C.-G. and E.G.-A.; funding acquisition, D.A.T.-G., H.R.-G., L.E.C.-G. and E.G.-A. All authors have read and agreed to the published version of the manuscript.

Funding: This research received no external funding.

Institutional Review Board Statement: Not applicable.

Informed Consent Statement: Not applicable.

Data Availability Statement: Not applicable.

Acknowledgments: We want to thank Nicolás Flores Sámano and for help in the design and 3D printing, Roxana Robles García and María Guadalupe Martínez Muñoz for their help in the laboratory work and experimental tests, and finally Diana Priscila Vicencio Toriz for the English revision.

Conflicts of Interest: The authors declare no conflict of interest. The funders had no role in the design of the study; in the collection, analyses, or interpretation of data; in the writing of the manuscript; or in the decision to publish the results.

References

1. Arada-Pérez, M.; Yazdani-Pedram, M.; Pérez-Saavedra, J. Sensores electroquímicos basados en sales cuaternarias de amonio. *Rev. Cuba. Química* **2008**, *20*, 31–38.
2. Xiao, K.P.; Bühlmann, P.; Nishizawa, S.; Amemiya, S.; Umezawa, U. A chloride ion-selective solvent polymeric membrane electrode based on a hydrogen bond forming ionophore. *Anal. Chem.* **1997**, *69*, 1038–1044. [CrossRef]
3. Meyerhoff, M.E.; Opdycke, W.N. Ion-Selective Electrodes. *Adv. Clin. Chem.* **1986**, *25*, 1–47. [CrossRef] [PubMed]
4. Pérez-Olmos, R.; Rios, A.; Fernández, J.R.; Lapa, R.A.S.; Lima, J.L.F.C. Construction and evaluation of ion selective electrodes for nitrate with a summing operational amplifier. *Appl. Tob. Análisis. Talanta.* **2001**, *53*, 741–748. [CrossRef]
5. Midgley, D. Limits of detection of ion-selective electrodes: Theory and practice. *Trans. Inst. Meas. Control* **1987**, *9*, 25–36. [CrossRef]
6. Gómez-Biedma, S.; Soria, M. Análisis electroquímico. *Rev. Diagn. Biol.* **2002**, *51*, 18–27.
7. Ewing, G.W. *Métodos Instrumentales de Análisis Químico*; Mc Graw Hill: New York, NY, USA, 1978; pp. 282–369. ISBN 968-6046-10-0.
8. Levine, I.N. *Fisicoquímica*, 5th ed.; Mc Graw Hill Interamérica: Aravaca Madrid, Spain, 2004; pp. 515–571. ISBN 84-481-3787-6.
9. Maron, S.H.; Prutton, C.F. *Fundamentos de Fisicoquímica*; Limusa: Johannesburg, South Africa, 2012; pp. 479–553. ISBN 968-18-0164-4.
10. Eggers, D.F.; Gregory, N.W.; Halsey, G.D.; Rabinovitch, B.S. *Fisicoquímica*; Limusa Wiley: Mexico City, México, 1967; ISBN 0471064998.
11. Arada-Pérez, M.A.; Pérez-Marín, L.; Calvo-Quintana, J.; Alonso-Chamarro, J.; Tacoronte-Morales, J. Evaluación de un electrodo selectivo a nitrato de membrana líquida con cloruro de trioctilmetilamonio sobre un soporte conductor. *Rev. Mex. De Ing. Química* **2002**, *1*, 23–28.
12. Creus, S.A. *Instrumentación Industrial*, 8th ed.; Alfaomega Editores: Ciudad de México, México, 2011; pp. 370–376. ISBN 978-607-707-042-9.

13. Toriz, G.D.A.; Loredo, A.R.; Marcial, M.D. Elaboración de Electrodo para medición de Cloruros en agua. *Acad. J.* **2013**, *5*, 3544–3549.
14. Somoza-Chuay, J.A.; Pavoni-Oliver, S.; Eirez-Izquierdo, J.E.; Bistel-Esquivel, R.A. Análisis de respuestas potenciométricas de electrodos de Vidrio/ITO y Vidrio/ITO/PANI-LS en la medición de pH. *Ing. Electrónica Automática Y Comun.* **2019**, *40*, 1–9.
15. Ben-Rayana, M.C.; Burnett, R.W.; Covington, A.K.; D'Orazio, P.; Fogh-Andersen, N.; Jacobs, E.; Kataky, R.; Külpmann, W.R.; Kuwa, K.; Larsson, L.; et al. Recommendation for measuring and reporting chloride by ISEs in undiluted serum, plasma or blood. *Clin. Chem. Lab. Med.* **2006**, *44*, 346–352. [CrossRef]
16. de Vera, G.; Climenta, M.A.; Antón, C.; Hidalgo, A.; Andrade, C. Determination of the selectivity coefficient of a chloride ion selective electrode in alkaline media simulating the cement paste pore solution. *J. Electroanal. Chem.* **2010**, *639*, 43–49. [CrossRef]

chemistry proceedings

MDPI

Proceeding Paper

Poly(bromocresol purple)-Based Voltammetric Sensor for the Simultaneous Quantification of Ferulic Acid and Vanillin †

Anastasiya Zhupanova * and Guzel Ziyatdinova

Analytical Chemistry Department, Kazan Federal University, Kremleyevskaya, 18, 420008 Kazan, Russia; Ziyatdinovag@mail.ru

* Correspondence: Zhupanova.Nastya@mail.ru

† Presented at the 1st International Electronic Conference on Chemical Sensors and Analytical Chemistry, 1–15 July 2021; Available online: https://csac2021.sciforum.net/.

Abstract: Natural phenolic antioxidants are extensively studied compounds due to their positive health effect and wide distribution in human diets. The simultaneous occurrence in samples requires selective methods for their determination. Electrochemical sensor based on the polyaminobenzene sulfonic acid functionalized single-walled carbon nanotubes (f-SWCNT) and electropolymerized bromocresol purple has been developed for the simultaneous quantification of ferulic acid and vanillin. The electrode has been characterized by scanning electron microscopy (SEM) and electrochemical methods, and the effectivity of the developed modifier has been confirmed. Thus, the novel sensitive voltammetric sensor is simple to fabricate, reliable, cost-effective, and can be applied for foodstuff screening.

Keywords: electrochemical sensors; carbon nanomaterials; electropolymerization; dyes; natural phenolics; antioxidants

check for updates

Citation: Zhupanova, A.; Ziyatdinova, G. Poly(bromocresol purple)-Based Voltammetric Sensor for the Simultaneous Quantification of Ferulic Acid and Vanillin. *Chem. Proc.* **2021**, *5*, 47. https://doi.org/10.3390/CSAC2021-10441

Academic Editor: Ye Zhou

Published: 30 June 2021

1. Introduction

Natural phenolic antioxidants are extensively studied compounds in modern electroanalysis due to their positive health effect and wide distribution in the human diet [1]. Simultaneous occurrence in samples requires selective methods for their determination. Among a wide range of natural phenolics, vanillin and its biological precursor ferulic acid [2] are of practical interest. High-performance liquid [3,4] and thin-layer chromatography [5] are usually applied for this purpose. Both phenolics under consideration are electrochemically active, which makes it possible to use electrochemical methods for their quantification. Although various types of electrochemical sensors have been developed for the simultaneous quantification of natural phenolics of different classes [6–8], ferulic acid and vanillin are not considered as analytes.

Thus, the current work is focused on the development of an electrochemical sensor based on a poly(bromocresol purple)-modified electrode for the simultaneous quantification of ferulic acid and vanillin.

2. Materials and Methods

Bromocresol purple (90% purity), 99% vanillin from Sigma-Aldrich (Steinheim, Germany), and 99% ferulic acid from Aldrich (Steinheim, Germany) were used. Their standard 10 mM solutions were prepared in ethanol (rectificate). The exact dilution was used for the preparation of less concentrated solutions.

Polyaminobenzene sulfonic acid functionalized single-walled carbon nanotubes (f-SWCNT) ($d \times l$ is 1.1 nm × 0.5–1.0 μm) were purchased from Sigma-Aldrich (Steinheim, Germany). A homogeneous 1.0 mg mL^{-1} suspension of f-SWCNT was obtained by ultrasonic dispersion for 30 min in dimethylformamide.

All reagents had chemical-grade purity. Double distilled water was used for the measurements. The experiments were carried out at laboratory temperature (25 ± 2 °C).

Electrochemical measurements were carried out on the potentiostat/galvanostat Autolab PGSTAT 302N with FRA 32M module (Eco Chemie B.V., Utrecht, The Netherlands) and NOVA 1.10.1.9 software. The 10 mL glassy electrochemical cell with a working glassy carbon electrode (GCE) with a 7.07 mm^2 geometric surface area (CH Instruments, Inc., Bee Cave, TX, USA) or modified electrode, a silver-silver chloride saturated KCl reference electrode, and a platinum wire as the counter electrode was used.

An "Expert-001" pH meter (Econix-Expert Ltd., Moscow, Russian Federation) equipped with the glassy electrode was applied for pH measurements.

Scanning electron microscopy (SEM) was carried out on the high-resolution field emission scanning electron microscope MerlinTM (Carl Zeiss, Oberkochen, Germany) at the accelerating voltage of 5 kV and emission current of 300 pA.

3. Results and Discussion

3.1. Characterization of the Electrodes

Bromocresol purple forms a nonconducting film which is confirmed by the disappearance of the oxidation peak with an increase of the cycles number and which is typical for the electropolymerization of phenolics [9].

The conditions of bromocresol purple potentiodynamic electropolymerization (monomer concentration, number of scans, supporting electrolyte pH, electrolysis parameters) have been optimized in order to find the best voltammetric response of the co-existed ferulic acid and vanillin. The peak potential separation of 170 mV on the polymer-based electrode is not affected by electropolymerization conditions, while oxidation currents change significantly. The best response has been obtained for the poly(bromocresol purple) obtained by 10-fold potential cycling from 0.1 to 1.2 V with a scan rate of 100 mV s^{-1} from the 25 μM monomer solution in 0.1 M phosphate buffer pH 7.0.

The electrodes have been characterized by SEM (Figure 1). The data obtained confirm the successful immobilization of the nanomaterial on the electrode surface.

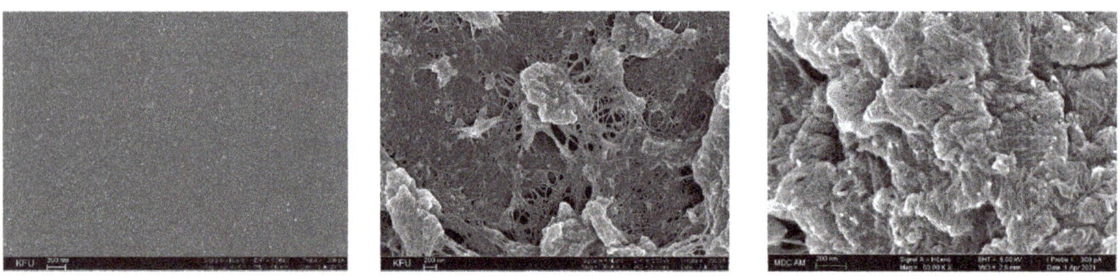

(a) (b) (c)

Figure 1. SEM images for (**a**) CGE; (**b**) CGE/f-SWCNT; and (**c**) CGE/f-SWCNT/Poly(bromocresol purple).

The effective surface area of the electrodes has been evaluated using 1.0 mM $[Fe(CN)_6]^{4-}$ as a redox probe under conditions of cyclic voltammetry and chronoamperometry (for GCE). The Cottrell equation for GCE and Randles–Sevcik equation for the cyclic voltammetry data have been applied. A 5.1-fold increase of the effective surface area for the polymer-modified electrode vs. bare GCE has been obtained, which leads to the ferulic acid and vanillin oxidation current increasing. Electrochemical impedance spectroscopy (EIS) has been performed in the presence of $[Fe(CN)_6]^{4-/3-}$ as a redox probe at 0.23 V. EIS data show a 7.2-fold lower charge transfer resistance for the poly(bromocresol purple)-based sensor in comparison to GCE, which means an increase of the electron transfer rate.

3.2. Simultaneous Quantification of Natural Phenolic Antioxidants

The well-resolved oxidation peaks of the ferulic acid and vanillin at 0.732 and 0.903 V, respectively, with a potential separation of 170 mV have been obtained on the created sensor (Figure 2).

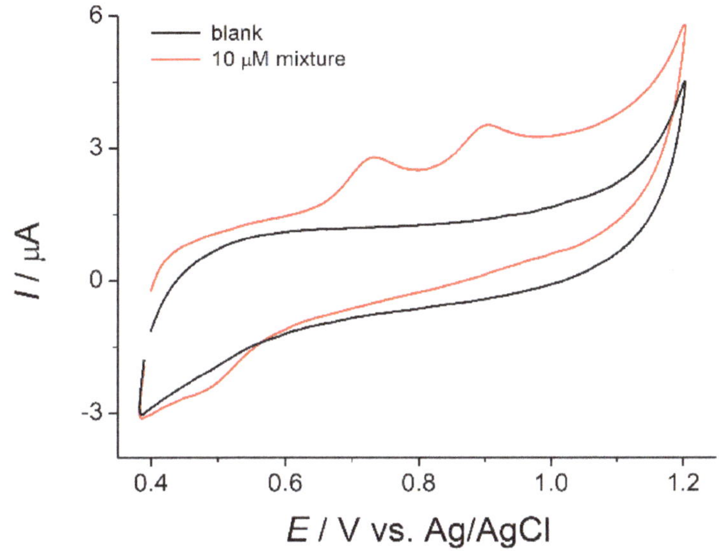

Figure 2. Cyclic voltammogram of 10 μM mixture of ferulic acid and vanillin on the poly(bromocresol purple)–based electrode in Britton–Robinson buffer pH 2.0.

The analytes' electrooxidation parameters have been studied. Both phenolics are oxidized under a diffusion control and irreversible with the participation of two electrons and protons (Figure 3).

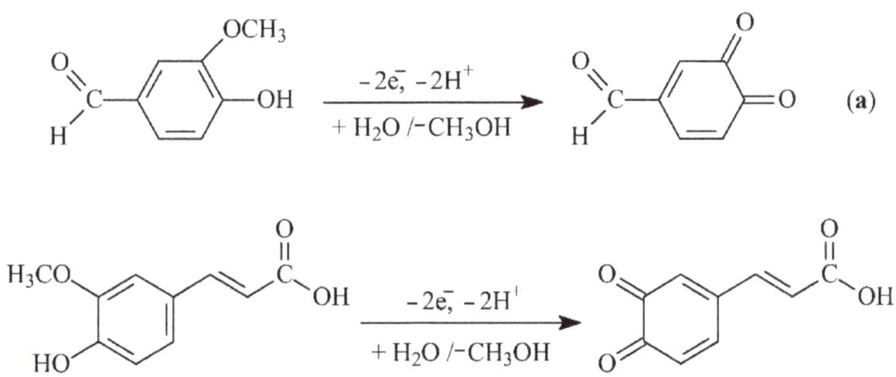

Figure 3. Electrooxidation scheme of (**a**) vanillin and (**b**) ferulic acid.

The developed sensor has been operated under conditions of differential pulse voltammetry. The pulse parameters have been optimized, and it has been found that a modulation amplitude of 75 mV and modulation time of 25 ms provide the best response of the target analytes. The sensor allows a direct simultaneous quantification of ferulic acid and vanillin in the ranges of 0.1–5.0 and 5.0–25 μM for both analytes (Figure 4) with detection limits of 72 and 64 nM, respectively. The accuracy of determination has been tested on the model mixtures of ferulic acid and vanillin. The relative standard deviation and recovery values obtained confirm the absence of a random error and the precision of the developed sensor.

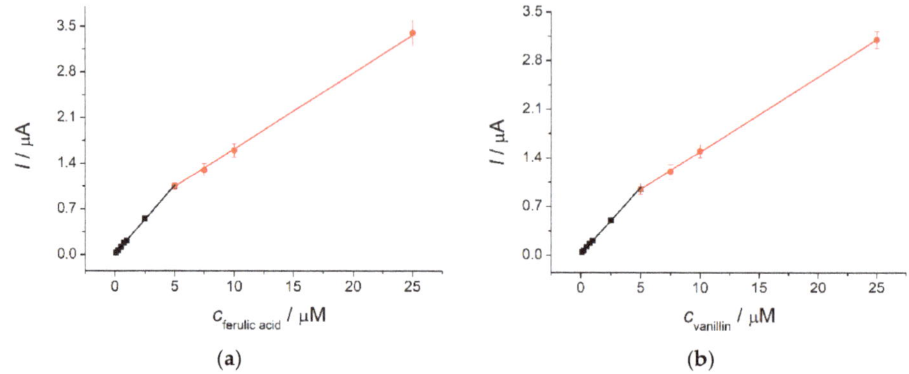

Figure 4. Calibration plots for (**a**) ferulic acid and (**b**) vanillin based on the differential pulse voltammetry data on the poly(bromocresol purple)-based electrode in Britton—Robinson buffer pH 2.0.

Thus, the novel sensitive voltammetric sensor is simple to fabricate, reliable, cost-effective, and can be applied for the foodstuff screening.

Funding: This research received no external funding.

Institutional Review Board Statement: Not applicable.

Informed Consent Statement: Not applicable.

Data Availability Statement: The data presented in this study are available on request from the corresponding author.

Conflicts of Interest: The authors declare no conflict of interest.

References

1. Bhuyan, D.J.; Basu, A. Phenolic Compounds potential health Benefits and toxicity. In *Utilisation of Bioactive Compounds from Agricultural and Food Waste*, 2nd ed.; Vuong, Q.V., Ed.; CRC Press: Boca Raton, FL, USA, 2017; Chapter 2; pp. 27–59.
2. Gallage, N.J.; Hansen, E.H.; Kannangara, R.; Olsen, C.E.; Motawia, M.S.; Jørgensen, K.; Holme, I.; Hebelstrup, K.; Grisoni, M.; Møller, B.L. Vanillin formation from ferulic acid in *Vanilla planifolia* is catalysed by a single enzyme. *Nat. Commun.* **2014**, *5*, 4037. [CrossRef] [PubMed]
3. Ranadive, A.S. Vanillin and related flavor compounds in vanilla extracts made from beans of various global origins. *J. Agric. Food Chem.* **1992**, *40*, 1922–1924. [CrossRef]
4. Sinha, A.K.; Verma, S.C.; Sharma, U.K. Development and validation of an RP-HPLC method for quantitative determination of vanillin and related phenolic compounds in *Vanilla planifolia*. *J. Sep. Sci.* **2007**, *30*, 15–20. [CrossRef] [PubMed]
5. Hingse, S.S.; Digole, S.B.; Annapure, U.S. Method development for simultaneous detection of ferulic acid and vanillin using high-performancethin layer chromatography. *J. Anal. Sci. Technol.* **2014**, *5*, 21. [CrossRef]
6. Ziyatdinova, G.K.; Guss, E.V.; Morozova, E.V.; Budnikov, H.C. An electrode based on electropolymerized sunset yellow for the simultaneous voltammetric determination of chlorogenic and ferulic acids. *J. Anal. Chem.* **2021**, *76*, 371–380. [CrossRef]
7. Zhupanova, A.; Guss, E.; Ziyatdinova, G.; Budnikov, H. Simultaneous voltammetric determination of flavanones using an electrode based on functionalized single-walled carbon nanotubes and polyaluminon. *J. Anal. Lett.* **2020**, *53*, 2170–2189. [CrossRef]
8. Guss, E.V.; Ziyatdinova, G.K.; Zhupanova, A.S.; Budnikov, H.C. Voltammetric determination of quercetin and rutin in their simultaneous presence on an electrode modified with polythymolphthalein. *J. Anal. Chem.* **2020**, *75*, 526–535. [CrossRef]
9. Yang, G.-J.; Qu, X.-L.; Zhu, A.-P.; Wang, C.-Y.; Qu, Q.-S.; Hu, X.-Y. Characterization, growth mechanism and application of network poly(bromophenol blue). *J. Electroanal. Chem.* **2007**, *604*, 48–56. [CrossRef]

chemistry proceedings

MDPI

Proceeding Paper

Label-Free Anti-Human IgG Biosensor Based on Chemical Modification of a Long Period Fiber Grating Surface †

João P. Mendes [1,2,3,*], Luís C. C. Coelho [2,4], Viviana P. Pereira [1], Manuel A. Azenha [1,3], Pedro A. S. Jorge [2,4] and Carlos M. Pereira [1,3]

1 Chemistry Research Unit of University of Porto, Faculty of Sciences, University of Porto, 4169-007 Porto, Portugal; viviana.p.pereira@gmail.com (V.P.P.); mazenha@fc.up.pt (M.A.A.); cmpereir@fc.up.pt (C.M.P.)
2 INESC TEC—Institute for Systems and Computer Engineering, Technology and Science, Faculty of Sciences, University of Porto, 4169-007 Porto, Portugal; lcoelho@inesctec.pt (L.C.C.C.); pedro.jorge@fc.up.pt (P.A.S.J.)
3 Chemistry and Biochemistry Department, Faculty of Sciences, University of Porto, 4169-007 Porto, Portugal
4 Department of Physics and Astronomy, Faculty of Sciences, University of Porto, 4169-007 Porto, Portugal
* Correspondence: joao.p.mendes@inesctec.pt

† Presented at the 1st International Electronic Conference on Chemical Sensors and Analytical Chemistry, 1–5 July 2021; Available online: https://csac2021.sciforum.net/.

Abstract: This work introduces a method specially developed to produce a biorecognition element based on modified Stöber silica nanoparticles by the covalent immobilization of the human IgG. The sensing structure is based on long period fiber gratings (LPFG), specially developed to allow the interaction of the electromagnetic wave with the target analytes through its evanescent field. The surface was modified by the immobilization of the IgG-modified nanoparticles serving has recognition elements for specific target molecules. The resulting configuration was tested in the presence of anti-human IgG, recording the refractometric response of the modified LPFG in contact with different amounts of analyte. The selectivity of the sensor was also assessed.

Keywords: optical fiber; long period gratings; evanescent field; chemical immobilization; biosensor

Citation: Mendes, J.P.; Coelho, L.C.C.; Pereira, V.P.; Azenha, M.A.; Jorge, P.A.S.; Pereira, C.M. Label-Free Anti-Human IgG Biosensor Based on Chemical Modification of a Long Period Fiber Grating Surface. *Chem. Proc.* **2021**, *5*, 48. https://doi.org/10.3390/CSAC2021-10454

Academic Editor: Huangxian Ju

Published: 30 June 2021

1. Introduction

Biosensors are powerful allies for food safety, drug discovery, environmental monitoring and clinical diagnosis [1–3]. The sensing methodology comprises a bioreceptor, a recognition element and a transducer whose properties changes upon analyte binding [4]. Typically, the bioreceptor is immobilized on the surface of the transducer and the binding event can be, e.g., mechanically, electrically or optically transduced [5,6] producing an increase in mass, a change in electrical resistivity or changes in the refractive index at the surface of the used material allowing to be measured. In recent years, optical biosensors are an active field of research worldwide, presenting rapid progress [7–9]. In this perspective, optical biosensors based on refractometric sensing schemes have been developed with great successes in the last few decades [7]. Moreover, optical fibers (OF) based on evanescent wave sensing are an excellent platform to develop high-stability and high-sensitive optical biosensors [10]. The quantitative and/or qualitative measurements result from the interaction of the biorecognition element with the evanescent field of light at the fiber surface. Its good biocompatibility makes them appropriate for biochemical functionalization, creating very sensitive structures targeting viruses, drugs and proteins [11]. In recent years, several authors have reported transduction scheme's using optical fibers for optical biosensing. Lobry et al. demonstrated a plasmon-assisted tilted fiber Bragg gratings (TFBGs) based biosensor for non-enzymatic D-glucose using polydopamine-immobilized concanavalin A [12]. More recently, Liyanage et al. developed a label-free sensitive tapered optical fiber plasmonic biosensor targeting microRNAs. The sensing platform comprises different types of gold nanoparticles immobilized on the surface of the fiber to enhance the evanescent mode, followed by self-assembled ssDNA probes [13]. Using a long-period fiber grating (LPFG) platform, Liu et al. demonstrated the use of a LPFG coated with graphene oxide (GO) nanosheets using the changes in resonant intensity to measure different concentrations of hemoglobin adsorbed by the GO layer [14]; Chiavaioli et al. reported a D-shaped single mode optical fiber (SMF) nanocoated with different metals for IgG/anti-IgG assays, reaching limit of detections (LOD) around to the femtomolar values [15]; and Dey et al. presented a sensitivity-enhanced LPFG near the turning point by etching the fiber with hydrofluoric acid to detect anti-mouse IgG [16]. In a similar approach to this work, Liu et al. demonstrated an LPFG-coated with silica nanoparticles modified with gold nanoparticles (AuNPs). The gold surface was modified with anti-IgM receptors and the sensing platform was tested to understand its suitability for the detection of human IgM antibodies [17].

In this work, a method based on silica nanoparticles (prepared based on Stöber [18] method), immobilized in the surface of commercial SMF28 OF, serving as recognition elements for specific target molecules is presented. The nanoparticles surface was functionalized by the introduction of an aminosilane (APTMS) followed by the covalent immobilization of the immunoglobulin G from human serum (human-IgG). The antibody was activated by the EDC/NHS protocol to allow the interaction of the amine exposed groups, located on the surface of the silica nanoparticles, with the activated carboxyl acid groups of the human-IgG molecules. The resulting template was immobilized onto the surface of an OF by electrostatic interactions between the negative charges of the fiber surface and the positively charged amine groups located in the IgG molecules. The sensing structure is based on LPFGs, specially developed to allow the interaction of the electromagnetic wave with the target analytes through its evanescent field. The refractometric system comprises a Braggmetter unit (HBK, FiberSensing, Darmstadt, Germany) working in a wavelength range from 1500 to 1600 nm and a reference LPFG to correct possible false interactions. The resulting configuration was tested in the presence of anti-human IgG, recording the refractometric response of the modified LPFG in contact with different amounts of analyte.

2. Materials and Methods

2.1. Chemical Reagents

Silica nanoparticles were prepared following the Stöber method [19], using tetraethyl orthosilicate (TEOS; Sigma-Aldrich, St. Louis, MO, USA; ≥98%) and ammonium hydroxide solution (NH_4OH; Sigma-Aldrich, 28% m/m) as reagents. The functionalization of the nanoparticles surface was attained using the following reagents: (3-aminopropyl) trimethoxysilane solution (Sigma-Aldrich, St. Louis, MO, USA; 97%), anhydrous toluene (Sigma-Aldrich, St. Louis, MO, USA; 99.8%), phosphate buffered saline (PBS; pH 7.4, tablets, Sigma-Aldrich, St. Louis, MO, USA), 2-(*N*-morpholino)ethanesulfonic acid (MES; Sigma-Aldrich, St. Louis, MO, USA), *N*-(3-dimethylaminopropyl)-*N*′-ethylcarbodiimide hydrochloride (EDC; Sigma-Aldrich, St. Louis, MO, USA; ≥99%), *N*-hydroxysuccinimide (NHS; Sigma-Aldrich, St. Louis, MI, USA; ≥98%) and immunoglobulin G from human serum (human-IgG; Sigma-Aldrich, St. Louis, MO, USA; ≥95%). For the LPFG surface cleaning were used sodium hydroxide anhydrous (NaOH; Sigma-Aldrich, St. Louis, MO, USA; ≥98%) and hydrochloric acid (HCl; Sigma-Aldrich, St. Louis, MO, USA; 37%). For surface activation were used sulfuric acid (H_2SO_4; Sigma-Aldrich, St. Louis, MO, USA; 95%–98%) and hydrogen peroxide solution (H_2O_2; Sigma-Aldrich, St. Louis, MO, USA; 30%) to perform piranha solution. For the affinity and selectivity assays were used the human IgG, the human serum albumin (HAS; Sigma-Aldrich, St. Louis, MO, USA; ≥98%) and the anti-human IgG (Fab specific; antibody produced in goat, Sigma-Aldrich). Ultra-pure water (type II-analytical grade, <1 $\mu S \cdot cm^{-1}$) and ethanol (Labchem, Zelienople, PA, USA; 96%) were also used.

2.2. Synthesis of the SiO_2 Nanoparticles

The SiO_2 nanoparticles were synthesized according with the Stöber method. Briefly, in a proper container the ethanol, the ultra-pure water and the TEOS, were mixed in that order.

The mixture was sonicated for 20 min and was added, under stir, the ammonia hydroxide and the final mixture was stirred for 24 h at room temperature. The resulted solution was centrifuged at 6000 rpm for 10 min and the beads were redispersed/centrifuged (five times) in deionized water and acetone. The resulted beads were dried at 40 °C for 12 h in the oven. The average size of the nanoparticles was determined by W130i Dynamic Light Scattering (DLS, AvidNano, Wycombe, UK), showing an average diameter ranging from 300–400 nm and were evaluated by attenuated total reflectance (FTIR-ATR, Bruker, Billerica, MA, USA).

2.3. Immobilization of the Biorecognition Molecule onto the SiO_2 Surface

The dry beads were incubated in freshly prepared APTMS solution 2% (v/v) in anhydrous Toluene for 24 h at room temperature in a closed container, using 10 mg/mL of beads concentration. After incubation, the nanoparticles were centrifuged at 6000 rpm for 10 min and redispersed/centrifuged for five times in acetone and ethanol. From this step resulted amino-functionalized silica nanoparticles that were verified by ATR-IR. The beads were redispersed in the PBS solution in a concentration of 5 mg/mL. A solution of the biorecognition element in MES buffer (pH 5.5) and the activation of the template was prepared by adding EDC (10× molar excess) to NHS (10× molar excess) and incubating for 30 min at room temperature. The previous prepared beads solution was added to the template solution and the pH was adjusted to 7.4–8.0. The incubation carried out for 4 h at room temperature without stirring (just swirled the mixture every half-hour). The resulted modified nanoparticles were centrifuged at 6000 rpm for 10 min and redispersed/centrifuged in deionized water. The nanoparticles were assessed by FTIR-ATR.

2.4. Working Principle of the Evanescent Wave Based Sensors, Long-Period Fiber Grating Fabrication and Surface Modification

In this work, a long-period fiber grating was microfabricated on the optical fiber surface. This grating works as a wavelength selective filter, displaying a spectrum with several resonances resulting from the combination of the mode of the core and the different cladding modes [20]. Figure 1 shows a description of a LPFG on an optical fiber and the resultant transduced optical signal from the interaction between the recognition molecule and the target.

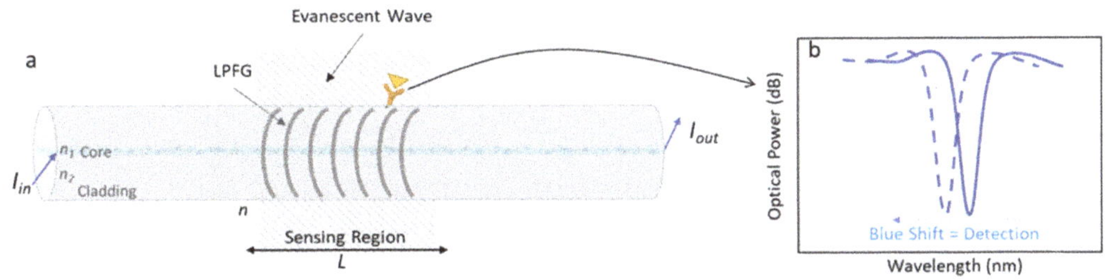

Figure 1. (**a**) Schematic figure of a long period fiber grating; and (**b**) the optical signal resultant from the biorecognition.

Moreover, the LPFGs were fabricated by the induced electric-arc technique following the protocol published by Rego (2016) [20] by creating a modulation in the propagating mode refractive index which, in this case, is achieved point-by-point through electric arc discharges with a current of 9 mA and a duration of 1 s along 30 to 50 mm with a period of 415 µm. Afterwards, the LPFG was chemically modified by the immobilization of the biorecognition molecule on the fiber surface. The sensitive section of the optical fiber was cleaned with a 2 M NaOH solution for 10 min followed by immersion in a 0.5 M HCl solution for 2 h. After washing with deionized water, the sensitive surface was activated with piranha solution (3:1 v/v) for 1 h at 60 °C. Finally, the LPFG was washed with deionized water and kept in the oven for 10 min to completely dry and was cooled with pure nitrogen. The process is schematically presented in Figure 2.

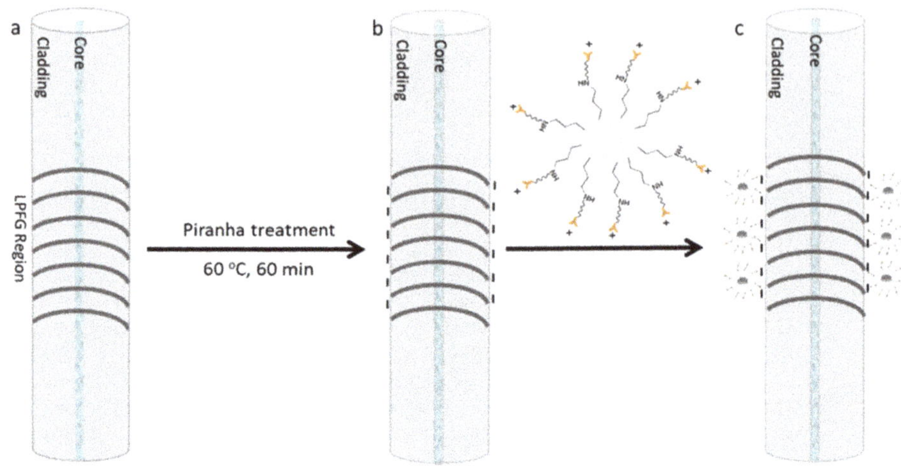

Figure 2. Schematic figure of the LPFG surface chemical modification: (**a**) Bare LPFG; (**b**) activated surface with negative electrical charges; and (**c**) modified surface by the template immobilization.

2.5. *Affinity and Selectivity Assays*

The modified LPFGs were tested in the presence of the anti-human IgG in different concentrations ranging from 1.5×10^{-2} to 9 µg/mL to attest the affinity of the sensing platform. To verify the selectivity of the modified optical fiber, the sensing scheme was exposed to a 9 µg/mL of human IgG, HSA and anti-human IgG solutions, in the same experimental conditions. In order to obtain the most trustable values, was used a bare LPFG as a reference signal. All data will be presented as the differential between sensing LPFG and reference LPFG.

3. Results and Discussion

3.1. *SiO_2 Nanoparticles Bare, SiO_2-NH_2 and SiO_2-NH_2-IgG FTIR-ATR Spectra*

To confirm the introduction of the new functional groups after each step of the SiO_2 nanoparticles surface (nanoSiO_2) modification, were made FTIR-ATR analysis. Figure 3 shows the spectra of the bare SiO_2 (nanoSiO_2), the aminated SiO_2 (nanoSiO_2-NH_2) and the obtained nanospheres after the incubation of the IgG (nanoSiO_2-NH_2-IgG). In the main figure, the absorption peaks at 3000–3500 cm^{-1} are related to the stretching -OH bands, the absorption peaks at 1000–1150 cm^{-1} are assigned to the Si-O-Si asymmetric stretching bands and the peaks at 800–950 cm^{-1} are appointed to the asymmetric bending of Si-OH. In the Figure 3a, the asymmetric deformation vibration of the -NH_2 at around 1550 cm^{-1} is displayed, suggesting that the amino groups were successfully fixed in the silica nanoparticle surface. Figure 3b show a peak at around 2900 cm^{-1} that are attributed to the presence of methyl groups of the APTMS structure. Finally, Figure 3c shows the carboxylate peak at 1650 cm^{-1}, assigned to the presence of the IgG molecule. This evaluation is similar to the evaluation made by Feifel and his co-worker when the authors proved the possibility to create electro-active cytochrome C multilayers by using carboxyl-modified SiO_2 nanoparticles [21]. Moreover, Hernandez-Leon et al. also showed parallel spectra when the authors modified a core-shell SiO_2 nanobeads for capture low molecular weight proteins and peptides [22].

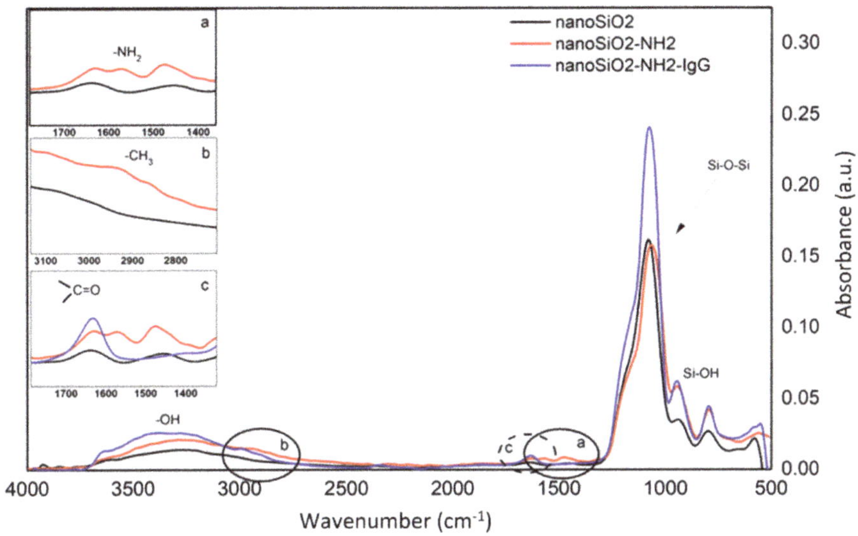

Figure 3. Obtained absorbance spectra from FTIR-ATR analysis of bare SiO_2 nanoparticles (black line), after amino-functionalized SiO_2 surface (inset graphs **a** and **b**; red line), and after IgG immobilization (inset graph **c**; blue line).

3.2. Affinity and Slectivity Assays

The SiO_2/IgG-modified LPFG probe was tested in the presence of different concentrations of anti-human IgG to attest the affinity of the sensing platform. The sensing LPFG (sensLPFG) was placed in an experimental chamber as well as the reference LPFG (refLPFG). Both gratings were exposed to a freshly prepared standard target solutions ranging from 1.5×10^{-2} to 9 μg/mL in PBS. After 10 min of exposure time, the LPFGs were washed three times with fresh PBS and three times with deionized water. All data were obtained measuring the LPFGs in deionized water at 22 °C. Figure 4a show the experimental data of the wavelength shift (sensLPFG–refLPFG) versus the anti-human IgG. Data are reported as a mean value with standard deviation (n = 3). Other similar works were described recently, such as the interferometric optical fiber biosensor for IgG/anti-IgG immunosensing presented by Wang et al., reporting a limit of detection around of 50 ng/mL [23]. Another approach was demonstrated by Han et al. that combined a Bragg acoustic reflector with an Au electrode and an aluminum nitride piezoelectric thin film, to develop a biosensor for anti-human IgG detection by immobilization of the human IgG antibody onto the modified Au electrode. The sensing platform was able to detect anti-human IgG concentrations smaller than 0.4 mg/mL [24]. The sensing platform presented in this work is able to detect the referenced target below to 0.1 μg/mL. Additionally, the data resulting from the linear fitting of Figure 4a, displaying a sensibility (S) (i.e., the slope) about $|S| = 59$ pm/(μg/mL).

To validate the specificity of the sensing platform, the same protocol was followed in the presence of human IgG antibody, the HSA protein and, finally, the anti-human IgG. Figure 4b show the resulting data after 10 min of incubation time for each target in 9 μg/mL (in PBS). The results are reported as a mean value with standard deviation (n = 3). These results showed the specificity of the built sensing platform to the proposed target, revealing a very relevant wavelength shift when exposed to it. By the other side, the shifts showed by the LPFG in the presence of the other targets are not relevant.

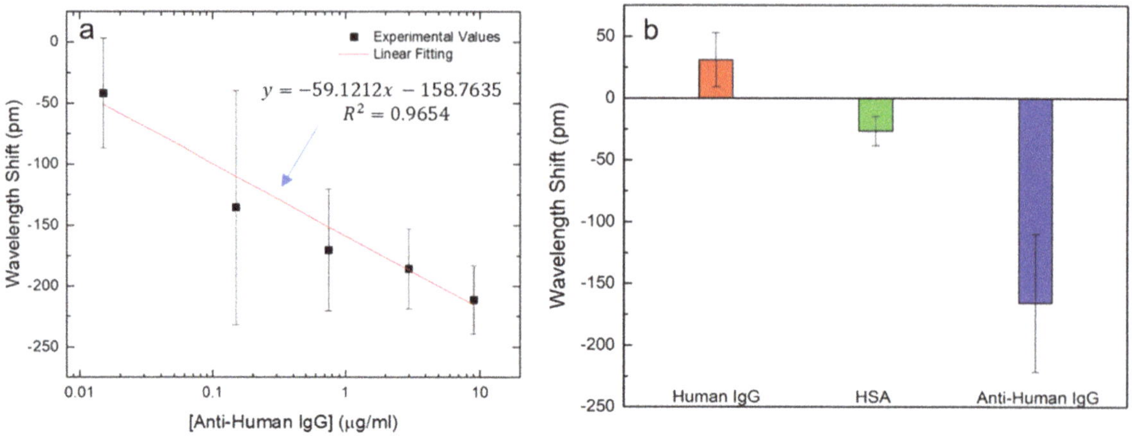

Figure 4. (**a**) Resonance shift vs. the anti-human IgG concentration (1.5 × 10^{-2} to 9 μg/mL). Data is reported as a mean value (n = 3) with standard deviation; (**b**) Resonance shift obtained by 10 min incubation in 9 μg/mL in human IgG, HSA protein and anti-human IgG. Data is reported as a mean value (n = 3) with standard deviation.

4. Conclusions

In this work, a sensing platform for the detection of the anti-human IgG antigen was developed by chemical modification of long period fiber grating surface. The sensing methodology is based on refractometric changes due to the interactions between the biorecognition molecule and the target. The surface of the optical fiber was changed by immobilization of IgG- modified silica nanoparticles. The FTIR-ATR spectra proved that the biorecognition molecule was successfully attached onto the SiO_2 nanoparticles surface and specificity assays demonstrated the selectivity of the method. The use of IgG-modified nanoparticles can bring some advantages, increasing the number of receptors available to interact with the target.

The low-cost and easy-to-use optical sensor reported here can detect anti-human IgG concentrations below 0.1 μg/mL by promoting specific antibody/antigen interactions. In the next step, we aim to imprint molecularly the analogue synthetic molecule of this template. The goal is to produce highly sensitive and selective molecularly imprinted polymers using the template of this work, combining them with highly sensitive optical platforms.

Supplementary Materials: The following supporting information can be downloaded at https://www.mdpi.com/article/10.3390/CSAC2021-10454/s1.

Author Contributions: Conceptualization, J.P.M. and L.C.C.C.; software, L.C.C.C.; formal analysis, J.P.M.; investigation, J.P.M. and V.P.P.; data curation, J.P.M.; writing—original draft preparation, J.P.M.; writing—review and editing, L.C.C.C., P.A.S.J. and C.M.P.; supervision, M.A.A., P.A.S.J. and C.M.P. All authors have read and agreed to the published version of the manuscript.

Funding: This work was financed by National Funds through the Portuguese funding agency, FCT—Fundação para a Ciência e a Tecnologia, within projects "UIDB/50014/2020" and "UIDB/00081/2020 (CIQUP)"and ANI through project "FAMEST"—POCI-01-0247-FEDER-024529 under program P2020 | COMPETE.

Acknowledgments: João Mendes would like to thank FCT for the PhD research grant SFRH/BD/130674/2017 and Luís Coelho acknowledges the support from FCT research contract grant CEECIND/00471/2017.

Conflicts of Interest: The authors declare no conflict of interest.

References

1. Tertis, M.; Hosu, O.; Florea, A.; Cristea, C. Biosensors for Clinical Samples: Consideration and Approaches. In *Immunodiagnostic Technologies from Laboratory to Point-Of-Care Testing*; Suman, P., Chandra, P., Eds.; Springer: Singapore, 2021.
2. Chocarro-Ruiz, B.; Fernandez-Gavela, A.; Herranz, S.; Lechuga, L.M. Nanophotonic label-free biosensors for environmental monitoring. *Curr. Opin. Biotechnol.* **2017**, *45*, 175–183. [CrossRef]
3. Balbinot, S.; Srivastav, A.M.; Vidic, J.; Abdulhalim, I.; Manzano, M. Plasmonic biosensors for food control. *Trends Food Sci. Technol.* **2021**, *111*, 128–140. [CrossRef]
4. Peltomaa, R.; Glahn-Martínez, B.; Benito-Peña, E.; Moreno-Bondi, M.C. Optical Biosensors for Label-Free Detection of Small Molecules. *Sensors* **2018**, *18*, 4126. [CrossRef]
5. Wollenberger, U. Chapter 2 Third generation biosensors—Integrating recognition and transduction in electrochemical sensors. In *Comprehensive Analytical Chemistry*; Elsevier: Amsterdam, The Netherlands, 2005; pp. 65–130.
6. Ong, J.J.; Pollard, T.D.; Goyanes, A.; Gaisford, S.; Elbadawi, M.; Basit, A.W. Optical biosensors—Illuminating the path to personalized drug dosing. *Biosens. Bioelectron.* **2021**, *188*, 113331. [CrossRef]
7. Chen, Y.; Liu, J.; Yang, Z.; Wilkinson, J.S.; Zhou, X. Optical biosensors based on refractometric sensing schemes: A review. *Biosens. Bioelectron.* **2019**, *144*, 111693. [CrossRef]
8. Chen, C.; Wang, J. Optical biosensors: An exhaustive and comprehensive review. *Analyst* **2020**, *145*, 1605–1628. [CrossRef]
9. Khansili, N.; Rattu, G.; Krishna, P.M. Label-free optical biosensors for food and biological sensor applications. *Sens. Actuators B Chem.* **2018**, *265*, 35–49. [CrossRef]
10. Zheng, Y.; Zhu, Z.-W.; Xiao, W.; Deng, Q.-X. Review of fiber optic sensors in geotechnical health monitoring. *Opt. Fiber Technol.* **2020**, *54*, 102127. [CrossRef]
11. Monfared, Y.E. Overview of Recent Advances in the Design of Plasmonic Fiber-Optic Biosensors. *Biosensors* **2020**, *10*, 77. [CrossRef]
12. Lobry, M.; Lahem, D.; Loyez, M.; Debliquy, M.; Chah, K.; David, M.; Caucheteur, C. Non-enzymatic D-glucose plasmonic optical fiber grating biosensor. *Biosens. Bioelectron.* **2019**, *142*, 111506. [CrossRef] [PubMed]
13. Liyanage, T.; Lai, M.; Slaughter, G. Label-free tapered optical fiber plasmonic biosensor. *Anal. Chim. Acta* **2021**, *1169*, 338629. [CrossRef] [PubMed]
14. Liu, C.; Xu, B.; Zhou, L.; Sun, Z.; Mao, H.; Zhao, J.; Zhang, L.; Chen, X. Graphene oxide functionalized long period fiber grating for highly sensitive hemoglobin detection. *Sens. Actuators B Chem.* **2018**, *261*, 91–96. [CrossRef]
15. Chiavaioli, F.; Zubiate, P.; Del Villar, I.; Zamarreño, C.R.; Giannetti, A.; Tombelli, S.; Trono, C.; Arregui, F.J.; Matias, I.R.; Baldini, F. Femtomolar Detection by Nanocoated Fiber Label-Free Biosensors. *ACS Sens.* **2018**, *3*, 936–943. [CrossRef]
16. Dey, T.; Tombeli, S.; Biswas, P.; Giannetti, A.; Basumallick, N.; Baldini, F.; Bandyopadhyay, S.; Trono, C. Realization of enhanced evanescent field long period fiber grating near turn around point for label-free immunosensing. In Proceedings of the 1st International Electronic Conference on Biosensors, Basel, Siwtzerland, 2–17 November 2020.
17. Liu, L.; Marques, L.; Correia, R.; Morgan, S.P.; Lee, S.-W.; Tighe, P.; Fairclough, L.; Korposh, S. Highly sensitive label-free antibody detection using a long period fibre grating sensor. *Sens. Actuators B Chem.* **2018**, *271*, 24–32. [CrossRef]
18. Stöber, W.; Fink, A.; Bohn, E. Controlled growth of monodisperse silica spheres in the micron size range. *J. Colloid Interface Sci.* **1968**, *26*, 62–69. [CrossRef]
19. Tadanaga, K.; Morita, K.; Mori, K.; Tatsumisago, M. Synthesis of monodispersed silica nanoparticles with high concentration by the Stöber process. *J. Sol-Gel Sci. Technol.* **2013**, *68*, 341–345. [CrossRef]
20. Rego, G. Arc-Induced Long Period Fiber Gratings. *J. Sens.* **2016**, *2016*, 3598634. [CrossRef]
21. Feifel, S.C.; Lisdat, F. Silica nanoparticles for the layer-by-layer assembly of fully electro-active cytochrome c multilayers. *J. Nanobiotechnol.* **2011**, *9*, 59. [CrossRef]
22. Hernandez-Leon, S.G.; Sarabia-Sainz, J.A.-I.; Montfort, G.R.-C.; Guzman-Partida, A.M.; Robles-Burgueño, M.D.R.; Vazquez-Moreno, L. Novel Synthesis of Core-Shell Silica Nanoparticles for the Capture of Low Molecular Weight Proteins and Peptides. *Molecules* **2017**, *22*, 1712. [CrossRef]
23. Wang, B.-T.; Wang, Q. An interferometric optical fiber biosensor with high sensitivity for IgG/anti-IgG immunosensing. *Opt. Commun.* **2018**, *426*, 388–394. [CrossRef]
24. Han, C.; Wang, X.; Zhao, Q.; Teng, L.; Zhang, S.; Lv, H.; Liu, J.; Ma, H.; Wang, Y. Solidly mounted resonator sensor for biomolecule detections. *RSC Adv.* **2019**, *9*, 21323–21328. [CrossRef]

chemistry proceedings

MDPI

Proceeding Paper

Precipitation of Iron Oxide in Hydrogel with Superparamagnetic and Stimuli-Responsive Properties †

Alice Mieting [1,*], Sitao Wang [1], Mia Schliephake [2], Daniela Franke [1], Margarita Guenther [1], Stefan Odenbach [2] and Gerald Gerlach [1]

1 Faculty of Electrical Engineering and Information Technology, Institute of Solid State Electronics, Dresden University of Technology, 01069 Dresden, Germany; sitao.wang@tu-dresden.de (S.W.); daniela.franke@tu-dresden.de (D.F.); margarita.guenther@tu-dresden.de (M.G.); gerald.gerlach@tu-dresden.de (G.G.)

2 Faculty of Mechanical Engineering, Institute of Mechatronic Mechanical Engineering, Dresden University of Technology, 01069 Dresden, Germany; mia.schliephake@tu-dresden.de (M.S.); stefan.odenbach@tu-dresden.de (S.O.)

* Correspondence: alice.mieting@tu-dresden.de

† Presented at the 1st International Electronic Conference on Chemical Sensors and Analytical Chemistry, 1–15 July 2021; Available online: https://sciforum.net/conference/CSAC2021.

Abstract: In this work, we present a template-based preparation of iron oxide-containing hydrogels (ferrogels) with ionic sensitive and superparamagnetic properties. The influence of the cross-linked template polyacrylamide and the concentration of the iron salts and sodium hydroxide on the precipitation of the iron oxide particles is investigated with respect to the stability of the ferrogels. Scanning electron microscope images show cubic particles, which can be semiquantitatively classified in three groups of particle size with respect to the dilution level. Magnetic hysteresis curves reveal a sigmoidal shape without remanence and coercivity for all samples. The higher cross-linked ferrogels, in comparison with the lower cross-linked ferrogels, possess a steady-state degree of swelling in ultrapure water and a stimuli-sensitive deswelling over a wide range of varying ionic strengths. Thus, they are suitable candidates for applications in sensing and microfluidics.

Keywords: stimuli-responsive hydrogel; superparamagnetic; iron oxide; coprecipitation; ferrogel

check for updates

Citation: Mieting, A.; Wang, S.; Schliephake, M.; Franke, D.; Guenther, M.; Odenbach, S.; Gerlach, G. Precipitation of Iron Oxide in Hydrogel with Superparamagnetic and Stimuli-Responsive Properties. *Chem. Proc.* **2021**, *5*, 49. https://doi.org/10.3390/chemproc2021005049

Academic Editor: Nicole Jaffrezic-Renault

Published: 17 December 2021

1. Introduction

Hydrogels are cross-linked, usually hydrophilic polymers, which are suitable candidates for applications in sensor [1] and actuator technology [2] due to their stimuli-sensitive swelling and viscoelastic properties. The incorporation of iron oxide particles into hydrogels results in novel composite materials with enhanced chemical and physical properties. Various approaches have demonstrated the sensitive and adsorptive properties of iron oxide particles with respect to heavy metal ions [3–5], pH [6], and biomolecules [7,8], as well as photocatalytic activity [9,10].

The aim of this work is the investigation of how the mechanical stability of the template structure and the concentration of the synthesis solutions influence the properties of in situ precipitated iron oxide particles. Understanding the structure–property relations of such novel composite materials relates to ongoing topics of scientific applications and research in engineering [11] and biomedicine [12] as well as in the treatment of contaminated water [13].

In this study, the wet chemical precipitation of iron oxides from iron salts with sodium hydroxide in the stoichiometric ratio of magnetite ($Fe^{II}(Fe^{III})_2O_4$) is investigated in two various cross-linked hydrogels: a higher cross-linked hydrogel, which has already been used in piezoresistive sensors [14], named sensor hydrogel/ferrogel, and a lower cross-linked hydrogel, which is used in actuator setups [15], named actuator hydrogel/ferrogel

(Figure 1A). It is expected that in template-based precipitation the differently crosslinked hydrogels will affect the particle shape and size and in turn the swelling properties as well as the magnetic properties of the resulting ferrogels. On the other hand, different dilution levels of iron salts and sodium hydroxide will be used to investigate a suitable synthesis concentration with respect to the sensitivity and stability of the ferrogels (Figure 1B).

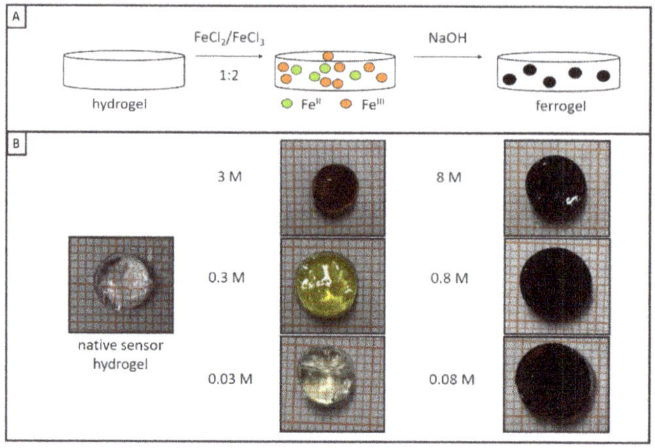

Figure 1. Scheme of the precipitation of iron oxide in hydrogel using iron chloride and sodium hydroxide solutions (**A**). Overview of the transformation of a sensor hydrogel to a ferrogel by soaking the samples in iron salt solution with orange-yellow coloring and forming black-brown-colored ferrogels in sodium hydroxide at different dilution levels (**B**).

2. Materials and Methods

2.1. Synthesis of Iron Oxide in Hydrogels

The monomer acrylamide (AAm), crosslinker *N,N′*-methylene-bis-acrylamide (BIS), ammonium peroxodisulfate (APS), ferric chloride hexahydrate ($FeCl_3 \cdot 6H_2O$), ferrous chloride tetrahydrate ($FeCl_2 \cdot 4H_2O$), and sodium chloride (NaCl) were purchased from Sigma-Aldrich, Saint Louis, MO, USA. *N,N,N′,N′*-tetramethylethylenediamine (TEMED) was purchased from Carl Roth, Karlsruher, Germany.

2.1.1. Sensor Hydrogels

For the synthesis of sensor hydrogel, 1.6 M (8 mmol, 0.5686 g) AAm and 1.5 mol% (0.12 mmol, 0.0185 g) BIS were dissolved in 4.156 mL ultrapure water and placed in an ice bath to cool. Polymerization was initiated by adding 300 µL of 0.072 M APS solution (0.022 mmol APS) and 2.1 mol% (0.168 mmol, 25.4 µL) TEMED. The cooled solution was placed in glass tubes with a diameter of about 6 mm, sealed, and left overnight at room temperature for polymerization. The polymerized polyacrylamide (PAAm) hydrogels were removed from the glass tubes and washed in ultrapure water for 5 days. Discs of about 2 mm thickness were cut from each of the cylindrical samples for in situ precipitation of the iron oxides in hydrogel.

2.1.2. Actuator Hydrogels

The actuator hydrogel was synthesized and handled in the same procedure as the sensor hydrogel, but with the following composition: 2.8 M (14 mmol, 1.0 g) AAm and 0.03 mol% (4.2 µmol, 0.7 mg) BIS were dissolved in 3.8 mL ultrapure water. Polymerization was initiated by adding 300 µL of 0.15 M APS solution (0.045 mmol APS) and 0.48 mol% (0.1 mmol, 10.2 µL) TEMED.

2.1.3. Coprecipitation of Iron Oxide in Hydrogels

In order to prepare for precipitation, the sliced native hydrogel discs were rinsed with nitrogen in ultrapure water free of oxygen.

The first step was to disperse the iron chloride solution into the hydrogel. Thus, 12 mL of a 3 M mixture of iron (III) chloride hexahydrate (24 mmol, 6.4870 g) and iron (II) chloride tetrahydrate (12 mmol, 2.3860 g) was prepared in a 2:1 molar ratio. After that, a 1:10 dilution and a 1:100 dilution, each in 10 mL, were made from the 3 M iron salt solution. Each sensor and actuator hydrogel was placed in 5 mL of the appropriate iron chloride concentration for 24 h.

For the precipitation of iron oxide, 12 mL of an 8 M sodium hydroxide solution (96 mmol, 3.8397 g) was prepared, and a 1:10 dilution and a 1:100 dilution, each in 10 mL, were made from it. The iron salt-soaked hydrogels were transferred to 5 mL of the appropriate concentration of sodium hydroxide and left overnight. Finally, the ferrogels were washed until the pH of the water was neutral.

In general, the precipitation was performed under nitrogen atmosphere, and all solutions and the ultrapure water were used in a degassed condition. However, the following experiments and investigations were carried out to characterize the hydrogels under ambient conditions so that oxidation of magnetite (Fe_3O_4) to maghemite (γ-Fe_2O_3) is suggested.

2.2. Characterization Methods

2.2.1. Scanning Electron Microscopy (SEM)

A piece of each sample was air-dried and sputtered with a 5 nm thick gold layer for subsequent secondary electron imaging of the embedded iron oxide particles with a SEM (Zeiss Supra 40VP; Schottky emitter) at a fixed stage width and 7 keV.

2.2.2. Vibrating Sample Magnetometer (VSM)

The magnetization of the air-dried samples was measured on a Lake Shore VSM 7407 in the magnetic field range of $\pm$17.5 kOe at room temperature.

2.2.3. Swelling Experiments

In order to measure the impact of the changed environmental conditions on the hydrogels, the respective masses of the samples were weighed before and after addition of the stimulus. Without stimulus, the sample was in ultrapure water and had a mass m_0. The mass m_i of the sample with stimulus was determined after 24 h or after the corresponding long-term point. The degree of swelling was calculated as follows:

$$Q = \frac{m_i - m_0}{m_0} \times 100\%. \quad (1)$$

3. Results and Discussion

3.1. Morphological Properties

The scanning electron microscope images in Figure 2 of the sensor and actuator ferrogels of each dilution level show cubic particles between 50 and 300 nm and, in some cases, 1 µm in size. Sensor and actuator ferrogels of undiluted concentrations (Figure 2a,b) and the actuator ferrogel of dilution level 1:10 (Figure 2e) additionally exhibit particles smaller than 20 nm distributed in the sample in a lawnlike manner and cannot be resolved with the currently used equipment. Due to the nonplanar arrangement of the particles, quantitative evaluation was not performed.

Figure 2. Representative SEM images of sensor ferrogels (**a–c**) and actuator ferrogels (**d–f**) of the different dilution levels. Scale bar: 1 µm.

3.2. Magnetic Properties

Magnetization curves with an S-shaped profile without hysteresis gaining magnetic saturation characterize superparamagnetic behavior [16,17]. For all prepared ferrogels, the magnetization curves show a sigmoidal curve shape without coercivity and without remanence (Figure 3).

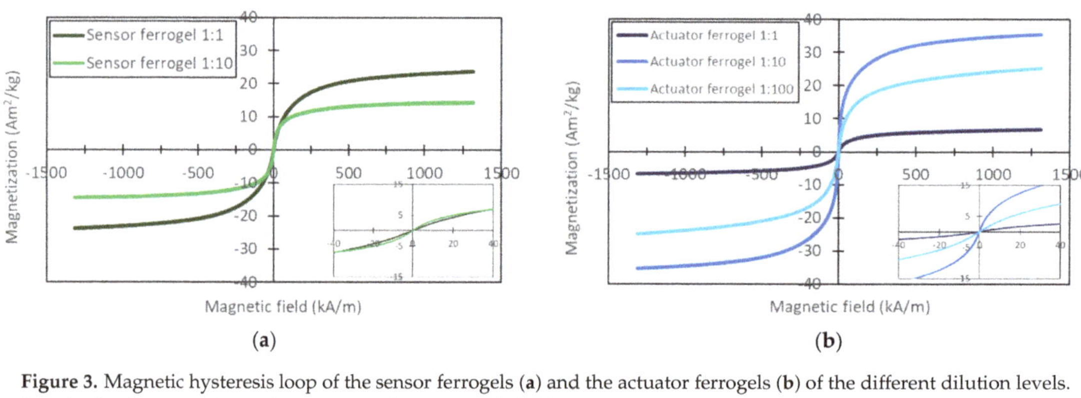

Figure 3. Magnetic hysteresis loop of the sensor ferrogels (**a**) and the actuator ferrogels (**b**) of the different dilution levels. Inset in the diagrams shows the curves at a low magnetic field.

The magnetization curve for the sensor ferrogel of dilution level 1:100 is not plotted in Figure 3a because the mass of this sample could not be determined due to sample instability in the dried state, making a quantitative comparison of the curves impossible. Except for the actuator ferrogel of dilution level 1:1, the saturation magnetization in both ferrogel types decreases with the dilution level, reflecting the lower content of magnetic particles in the ferrogel.

3.3. Reversibility of Ferrogel Swelling

Figure 4 depicts the swelling degrees of the ferrogels and native hydrogels alternating in ultrapure water and 1 M NaCl. Both ferrogels show a deswelling of around 80% in solution with increased ionic strength. It can be concluded that in aqueous solution, the ferrogels are in the swollen state due to electrostatic repulsions of charged surface groups of the iron oxides. An increased ionic strength in the solution leads to electrostatic neutralization of the ionized groups of the iron oxide with the dissolved ions.

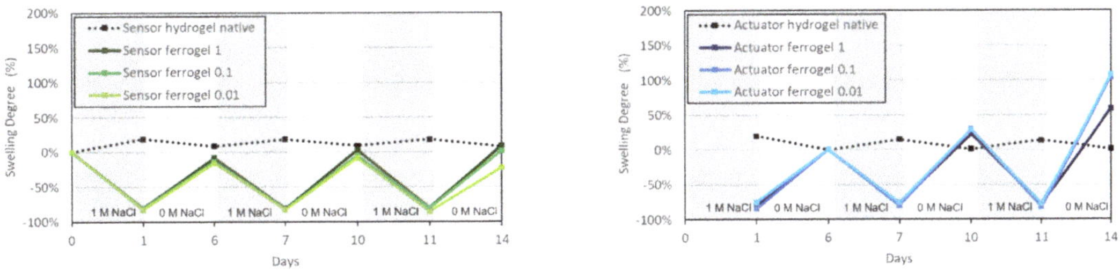

(a) (b)

Figure 4. Swelling cycles showing the repeatability of the deswelling in 1 M NaCl solution and swelling in ultrapure water (0 M NaCl) of the sensor ferrogels (**a**) and actuator ferrogels (**b**).

Deswelling of the ferrogel occurs due to a reduction in electrostatic repulsions within the composite material comparable to the swelling behavior of pH or ion-sensitive hydrogels with fixed ionic groups in the polymer network [18]. In contrast, the native hydrogels show minor swelling under increased ionic strength due to osmotically induced swelling mechanisms.

The sensor ferrogels obtain their initial steady-state degree of swelling in water after three water/NaCl cycles (Figure 4a).

Due to the increased volume in water after synthesis, it was not possible to remove the actuator ferrogels from the sample containers so that detecting the mass could just start in the deswollen state. Furthermore, the actuator ferrogels show an increase in swelling degree during the second cycle in water and reach a swelling degree up to 50–100% in the third water cycle (Figure 4b).

3.4. Sensitivity of the Sensor Ferrogel to Ionic Strength

Figure 5 depicts the swelling levels of the sensor ferrogels over a wide range of varying ionic strengths from nM to 5 M.

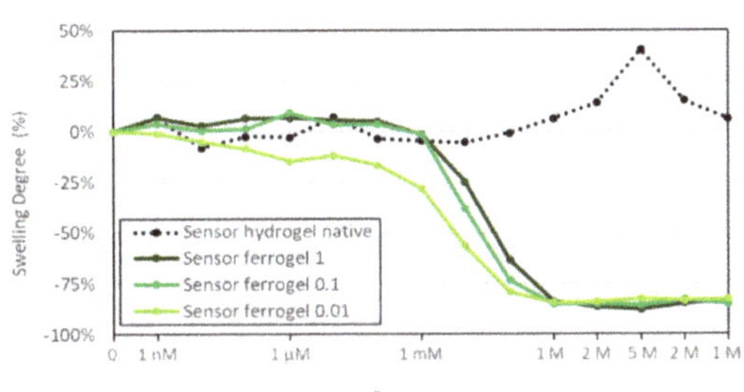

Figure 5. Swelling sensitivity of sensor ferrogels in NaCl solutions versus ionic strength.

A decay of the swelling curves between 1 mM and 1 M NaCl shows the concentration range of the deswelling that applies to all three dilution levels. Interesting for sensory applications is the flattened curve starting at 1 nM to 1 mM of the 1:100 sample. For this ferrogel, there seems to be an optimal balance between the charge density of ionized iron oxide particles under changing ionic strength and the mechanical stability of the hydrogel, so it can be used as a stimuli-responsive ferrogel in sensory applications.

4. Conclusions

The use of different cross-linked hydrogels as templates for the wet chemical precipitation of iron oxide producing fairly homogeneous cubic-shaped particles with superparamagnetic characteristic curves independent of the applied concentration of iron salts and base was presented. The ion-sensitive swelling properties of the sensor ferrogels and the reversibility of their swelling make them suitable candidates for applications in piezoresistive sensors. Their magnetic properties allow applications under magnetic field control in microfluidics and medicine. Due to their strong swelling in water, the lower cross-linked actuator ferrogels could be used as adsorption materials for the remediation of contaminated water.

Supplementary Materials: The following are available online at https://www.mdpi.com/article/10.3390/chemproc2021005049/s1.

Author Contributions: Conceptualization, A.M., D.F. and M.G.; methodology, A.M., S.W. and M.S.; writing—original draft preparation, A.M.; writing—review and editing, D.F., M.S., S.O. and G.G.; supervision, D.F. and M.G.; project administration, G.G.; funding acquisition, G.G. All authors have read and agreed to the published version of the manuscript.

Funding: This research was funded by the German Research Foundation (DFG) in the course of the Research Training Group "Hydrogel-Based Microsystems" (RTG 1865).

Institutional Review Board Statement: Not applicable.

Informed Consent Statement: Not applicable.

Data Availability Statement: Not applicable.

Acknowledgments: We thank the Institute of Electronic Packaging Technology (German abbreviation IAVT) at TU Dresden for the access to their scanning electron microscope (SEM).

Conflicts of Interest: The authors declare no conflict of interest.

References

1. Guenther, M.; Gerlach, G. *Hydrogels for Chemical Sensors*; Springer: Berlin/Heidelberg, Germany, 2009. [CrossRef]
2. Hines, L.; Petersen, K.; Lum, G.Z.; Sitti, M. Soft Actuators for Small-Scale Robotics. In *Advanced Materials*; Wiley-VCH: Hoboken, NJ, USA, 2017; p. 1603483. [CrossRef]
3. Bobik, M.; Korus, I.; Dudek, L. The Effect of Magnetite Nanoparticles Synthesis Conditions on Their Ability to Separate Heavy Metal Ions. *Arch. Environ. Prot.* **2017**, *43*, 3–9. [CrossRef]
4. Darezereshki, E.; Khodadadi Darban, A.; Abdollahy, M.; Jamshidi-Zanjani, A. Influence of Heavy Metals on the Adsorption of Arsenate by Magnetite Nanoparticles: Kinetics and Thermodynamic. *Environ. Nanotechnol. Monit. Manag.* **2018**, *10*, 51–62. [CrossRef]
5. Hayashi, K.; Matsuyama, T.; Ida, J. A Simple Magnetite Nanoparticle Immobilized Thermoresponsive Polymer Synthesis for Heavy Metal Ion Recovery. *Powder Technol.* **2019**, *355*, 183–190. [CrossRef]
6. Sivudu, K.S.; Rhee, K.Y. Preparation and Characterization of PH-Responsive Hydrogel Magnetite Nanocomposite. *Colloids Surf. A Physicochem. Eng. Asp.* **2009**, *349*, 29–34. [CrossRef]
7. Zhao, W.; Odelius, K.; Edlund, U.; Zhao, C.; Albertsson, A.C. In Situ Synthesis of Magnetic Field-Responsive Hemicellulose Hydrogels for Drug Delivery. *Biomacromolecules* **2015**, *16*, 2522–2528. [CrossRef] [PubMed]
8. Liu, T.Y.; Hu, S.H.; Liu, K.H.; Liu, D.M.; Chen, S.Y. Preparation and Characterization of Smart Magnetic Hydrogels and Its Use for Drug Release. *J. Magn. Magn. Mater.* **2006**, *304*, 397–399. [CrossRef]
9. Kim, J.H.; Jang, J.W.; Jo, Y.H.; Abdi, F.F.; Lee, Y.H.; Van De Krol, R.; Lee, J.S. Hetero-Type Dual Photoanodes for Unbiased Solar Water Splitting with Extended Light Harvesting. *Nat. Commun.* **2016**, *7*, 13380. [CrossRef] [PubMed]
10. Tilley, S.D.; Cornuz, M.; Sivula, K.; Grätzel, M. Light-Inducedwater Splitting with Hematite: Improved Nanostructure and Iridium Oxide Catalysis. *Angew. Chem. Int. Ed.* **2010**, *49*, 6405–6408. [CrossRef] [PubMed]
11. Liu, X.; Liu, J.; Lin, S.; Zhao, X. Hydrogel Machines. *Mater. Today* **2020**, *36*, 102–124. [CrossRef]
12. Krishnan, K.M. Biomedical Nanomagnetics: A Spin through Possibilities in Imaging, Diagnostics, and Therapy. *IEEE Trans. Magn.* **2010**, *46*, 2523–2558. [CrossRef]
13. Muya, F.N.; Sunday, C.E.; Baker, P.; Iwuoha, E. Environmental Remediation of Heavy Metal Ions from Aqueous Solution through Hydrogel Adsorption: A Critical Review. *Water Sci. Technol.* **2016**, *73*, 983–992. [CrossRef] [PubMed]
14. Erfkamp, J.; Guenther, M.; Gerlach, G. Hydrogel-Based Sensors for Ethanol Detection in Alcoholic Beverages. *Sensors* **2019**, *19*, 1199. [CrossRef] [PubMed]

15. Keplinger, C.; Sun, J.Y.; Foo, C.C.; Rothemund, P.; Whitesides, G.M.; Suo, Z. Stretchable, Transparent, Ionic Conductors. *Science* **2013**, *341*, 984–987. [CrossRef] [PubMed]
16. Unni, M.; Rinaldi, C. Chapter 4: Magnetic characterization of iron oxide nanoparticles for biomedical applications. In *Biomedical Nanotechnology*; Humana Press: New York, NY, USA, 2017. [CrossRef]
17. Teja, A.; Koh, P.Y. Synthesis, Properties, and Applications of Magnetic Iron Oxide Nanoparticles. *Prog. Cryst. Growth Charact. Mater.* **2009**, *55*, 22–45. [CrossRef]
18. Li, W.; Zhao, H.; Teasdale, P.R.; John, R.; Zhang, S. Synthesis and Characterisation of a Polyacrylamide-Polyacrylic Acid Copolymer Hydrogel for Environmental Analysis of Cu and Cd. *React. Funct. Polym.* **2002**, *52*, 31–41. [CrossRef]

chemistry proceedings

MDPI

Proceeding Paper

Critical Variables Influencing the Ultrasound-Assisted Extraction of Bioactive Compounds—A Review †

Anxo Carreira-Casais [1], Maria Carpena [1], Antia G. Pereira [1,2], Franklin Chamorro [1], Anton Soria-Lopez [1], Pascual Garcia Perez [1], Paz Otero [1], Hui Cao [1], Jianbo Xiao [1], Jesus Simal-Gandara [1,*] and Miguel A. Prieto [1,2,*]

[1] Nutrition and Bromatology Group, Department of Analytical and Food Chemistry, Faculty of Science, University of Vigo, Ourense Campus, E-32004 Ourense, Spain; anxocc@uvigo.es (A.C.-C.); mcarpena@uvigo.es (M.C.); antia.gonzalez.pereira@uvigo.es (A.G.P.); franklin.noel.chamorro@uvigo.es (F.C.); anton.soria@uvigo.es (A.S.-L.); pasgarcia@uvigo.es (P.G.P.); paz.otero@uvigo.es (P.O.); hui.cao@uvigo.es (H.C.); jianbo.xiao@uvigo.es (J.X.)

[2] Centro de Investigaçao de Montanha (CIMO), Instituto Politécnico de Bragança, Campus de Santa Apolonia 10, 5300-253 Bragança, Portugal

* Correspondence: jsimal@uvigo.es (J.S.-G.); mprieto@uvigo.es (M.A.P.)

† Presented at the 1st International Electronic Conference on Chemical Sensors and Analytical Chemistry, 1–15 July 2021; Available online: https://csac2021.sciforum.net/.

Abstract: Ultrasound-assisted extraction (UAE) is a novel methodology, belonging to the so-called "Green Chemistry", which has gained interest in recent years due to the potential to recover bioactive compounds, especially those from plant matrices. It is widely recognized that the extraction of molecules by UAE gives rise to higher or similar yields than those obtained by traditional extraction methods. UAE has certain advantages inherent to Green Chemistry extraction methods, such as short extraction time and low solvent consumption. The aim of this review is to critically present the different variables and parameters that can be modified in UAE, such as ultrasound power, time, temperature, solvent, and solid to solvent ratio that influence yield and extraction performance.

Keywords: ultrasound-assisted extraction; critical variables; power; temperature; time; solvent

Citation: Carreira-Casais, A.; Carpena, M.; Pereira, A.G.; Chamorro, F.; Soria-Lopez, A.; Perez, P.G.; Otero, P.; Cao, H.; Xiao, J.; Simal-Gandara, J.; et al. Critical Variables Influencing the Ultrasound-Assisted Extraction of Bioactive Compounds—A Review. *Chem. Proc.* **2021**, *5*, 50. https://doi.org/10.3390/CSAC2021-10562

Academic Editor: Huangxian Ju

Published: 1 July 2021

1. Introduction

Ultrasound-assisted extraction (UAE) is a technique that belongs to the group of novel extraction methods, together with microwave assisted extraction (MAE), enzyme assisted extraction (EAE) or high-pressure assisted extraction (HPAE) [1,2]. UAE promotes the extraction of compounds of interest, lowering the consumption of resources, such as solvent and energy, whereas achieving remarkably higher extraction yields [3,4]. In addition, UAE is a multipurpose method that lends itself to be combined with other extraction methods, both conventional and novel [5]. UAE has been applied to obtain extracts rich in bioactive compounds, such as phenolic compounds, pigments, polysaccharides, and amino acids, among others from plant matrices [1,3,6,7].

This methodology is based on the principle of cavitation, which leads to cell collapse of the matrix and allows the release of their inner substances. Several variables are relevant for the performance of UAE, including the solid–liquid ratio, the type of solvents used, the extraction time and the ultrasound power applied. Besides, ultrasound power and extraction time are closely linked to a fifth important factor, which is temperature. In practical terms, a correct optimization of these variables is essential to obtain a correct performance, resulting in a maximal extraction yield. In addition, temperature can affect the integrity of the bioactive compounds, since most of them are thermolabile. Considering that high ultrasound power linked to long extraction periods may lead to sample damage, temperature control is essential for a correct design of the cooling reactor and the optimization of UAE extraction protocols. Keeping all this in mind, this critical review is focused on the influence of all the variables that affect UAE, to analyze the critical factors involved in the optimization of this technique. In addition, response surface methodology (RSM) can be a representative tool to generate meta-models. RSM allows to analyze and optimize several variables at the same time and minimize the loss of matrices [8,9].

2. Variables Influencing Ultrasound-Assisted Extraction

To obtain good extraction yields, an optimization of the variables that influence the process is required. Among the variables that affect UAE, there are three types of parameters, as follows: physical, medium-dependent, and matrix-dependent parameters. Regarding the first, physical parameters are related to the ultrasonic waves applied during UAE and the equipment used. In this sense, those attributed to ultrasonic waves are ultrasound power, frequency, and ultrasound intensity (UI), whereas those related with ultrasound equipment are extraction time, and shape and size of the ultrasonic reactor. Medium-dependent parameters are related with the space in which ultrasound waves are transmitted from the emitting source to the matrix. Solvent properties, temperature and the presence of gases are examples of medium-dependent parameters. Finally, matrix-dependent parameters are those that have a significant influence in the extraction of target compounds and considerably affect the effectiveness of the extraction. Type of matrix, structure, pre-treatment, particle size, or solid–liquid ratio are examples of those parameters [3,10]. Therefore, a correct design of the process, optimization of the variables, and appropriate equipment is needed to obtain extraction yields comparable to those obtained by the so-called traditional methods [11].

Regarding the ultrasound power, the use of high values usually improves extraction yields due to the generation of strong shear forces, so it is considered as one of the critical parameters to be optimized. Furthermore, higher ultrasound power reduces the time of extraction. For example, a study showed good results of the extraction of β-d glucans at a high extraction power (590 W), in only 58 min [12]. However, the use of high ultrasound power without control can overheat the reactor producing degradation of labile compounds and solvent evaporation [13]. In addition, the higher ultrasound power, the higher UI, so that when UI reaches the maximum value can produce liquid agitation and the consequent loss of ultrasound wave and the reduction of cavitation efficiency [14].

Regarding the extraction time, UAE allows to obtain good extraction yields with relatively short processing times (maximum 60 min) since longer times may cause undesirable changes in the extracted compounds. In this sense, optimized time commonly ranges between 20–60 min, minimizing the energy consumption and reducing the compounds' exposure to the process [15]. For example, one study shows that 7.25 min are enough to extract pigments from annatto seeds [16], while 37 min are needed to extract betacyanin and betaxanthin in bougainvillea flowers [17]. In the case of amino acid extraction, shorter extraction time was needed (6 min) [18,19], whereas other authors were able to extract polysaccharides from purple glutinous rice bran (*Oryza sativa*) with an extraction time of 20 min at 70 °C [20] (Table 1).

The type or polarity of the solvent used is closely linked to the nature of the compounds to be extracted. In addition, due to current concern for the environment, eco-friendly solvents are preferred. Predominantly, an aqueous medium is generally chosen for the extraction of polar compounds used in food matrices, while in the case of other organic compounds, ethanol, and methanol are usually employed. However, despite the use of methanol tends to obtain better extraction yields, ethanol is preferably chosen because of its lower toxicity [10]. For example, distilled water and ethanol are usually used for pigments extraction, while water is the most common solvent for the extraction of polysaccharides and amino acids [12,20,21] (Table 1). Furthermore, European Directive 2010/59/EU lists the solvents that can be used for the extraction of compounds from foodstuffs, as well as their uses and limitations [22]. In addition to the suitable solubility of the compounds of interest, it is also important to consider the vapor pressure, the surface tension, and the viscosity of the solvent, since those may affect cavitation and the extraction yield [3]. The solid-to-solvent ratio used for each compound does not follow a certain pattern. It depends especially on the type of solvent and the matrix used. Table 1 shows that for the extraction of phenolic compounds, the most used solvent is ethanol and the solid-to-solvent ratio varies between 0.025 g/mL [23] and 0.1 g/mL [24], whereas ratios vary between from 0.058 g/mL for the extraction of betacyanin and betaxanthin in bougainvillea flowers [17] to 0.14 g/mL for the extraction of a natural pigment from annatto seeds [16].

Finally, high extraction temperatures not only affect the extraction yield but could also have negative effects due to the possible degradation of thermolabile compounds [14]. For this reason, the cooling system must allow the extraction of compounds avoiding the overheating of the medium by controlling the temperature of the system. For example, it is possible to extract phenolic compounds with temperatures up to 75 °C, β-d glucans at 81 °C or amino acids at 70 °C with optimal yields [12,18,25]. The increase in the temperature caused by the ultrasound probe itself is fundamentally produced when high ultrasound power is applied. The temperature increase produces a decrease in both viscosity and surface tension and induces an increase in the vapor pressure. Thus, too high temperatures can be harmful for the propagation of ultrasounds through the medium [13]. For these reasons, the optimization of the extraction temperature must be focused on both protecting the structure and function of the target components and improving the extractive properties of the solvent. Generally, the temperature does not exceed 80 °C, and it commonly works around 50 °C (Table 1).

Table 1. Optimized extraction conditions for phenolic acids, pigments, polysaccharides, and amino acids from different vegetal matrices by UAE.

Compounds	Matrix	Solvent Type and Ratio	Optimized Parameters	Ref.
TFC	Blueberry pomace	EtOH/water 50%/50%; 0.05 g/mL	ET: 60 min; UPA: 64 W; T: 40 °C	[24]
TFC	*Zea mays* waste	EtOH/water 70%/30% 0.09 g/mL	ET: 40 min; UPA: 50 W; T: ND	[23]
Stilbenes	Grape canes	EtOH/water 60%/40% 0.025 g/mL	ET: 10 min; UPA: 200 W; T: 75 °C	[25]
Anthocyanins	Grape skins	EtOH 60% acidified pH = 3; 0.033 g/mL	ET: 28 min; UPA: 400 W; T: 50 °C	[26]
Natural pigment	Annatto seeds	Dw, previously treated with chloroform/ratio solid solvent 0.14 g/mL	ET: 7.25 min; UPA: 200 W; T: 72.7 °C. (Duty cycle of 0.8 s)	[16]
Betacyanin and betaxanthin	*Bougainvillea glabra* flowers	Distilled water 100% 0.058 g/mL	ET: 37 min; UPA: 88 W; T: 55 °C	[17]
Natural yellow pigment	*Physalis pubescens* L.	EtOH 75% 0.083 g/mL	ET: 14 min; UPA: 180 W; T: ND; ultrasonic interval time of 10.55 s	[27]
B-d-glucan	*Ganoderma lucidum*	Distilled water 0.00004 g of fiber/mL	ET: 58 min; UPA: 590 W; T: 81 °C	[12]
Polysaccharides	*Oryza sativa* L.	Distilled water 0.05 g/mL	ET: 20 min; UPA: 150 W; T: 70 °C	[20]
PSMP	Perilla seed meal	Distilled water 0.038 g/mL	ET: 52 min; UPA: 229 W; T: 43 °C	[21]
Amino acids	Grapes	Distilled water 0.1 g/mL to 0.05 g/mL not significant differences	ET: 6 min; UPA: 140 W; T: 70 °C	[19]
Amino acids	*Apocynum venetum*	Distilled water 0.00047 g/mL	ET: 32 min; UPA: 187 W; T: NI	[18]

Abbreviations: TFC: total flavonoid content. UPA: ultrasonic power amplitude. T: temperature; ET: extraction time; PSMP: perilla seed meal polysaccharides; NI: not included; Dw: dry weight.

3. Conclusions

UAE is a useful method for obtaining different compounds of interest from plant matrices, since remarkably higher extraction yields are obtained with short extraction times. This extraction method belongs to "Green Chemistry" because it allows to decrease the consumption of resources, such as solvent and energy. However, it is still necessary to optimize the more relevant variables that influence the effectiveness of UAE, such as ultrasound power, extraction time and temperature, type of solvent, and solid-to-solvent ratio. Different studies have shown extraction yields of different bioactive compounds, such as phenolic compounds, polysaccharides, pigments, and amino acids, using UAE with short extraction times (maximum 60 min), medium ultrasound power (between 200–500 W), temperatures around 50 °C (maximum 80 °C), and environmentally friendly solvents (distilled water and ethanol). Therefore, UAE can be appropriately applied to obtain bioactive compounds through an efficient and eco-friendly process, considering and optimizing the different critical variables that affect the process.

Supplementary Materials: The following are available online at https://www.mdpi.com/article/10.3390/CSAC2021-10562/s1.

Author Contributions: Conceptualization, A.C.-C., A.G.P., M.C., P.O.; F.C., and A.S.-L.; methodology, P.O., and P.G.P.; validation, P.O.; formal analysis, P.O., and P.G.P.; writing—original draft preparation, A.C.-C., P.O., and M.C.; writing—review and editing, A.C.-C., A.S.-L., M.C., P.O., and M.A.P.; visualization, M.C., and A.G.P.; supervision, P.O., H.C., J.X., J.S.-G., and M.A.P.; project administration, M.A.P., and J.S.-G. All authors have read and agreed to the published version of the manuscript.

Funding: The JU receives support from the European Union's Horizon 2020 research and innovation program and the Bio Based Industries Consortium. The project SYSTEMIC Knowledge hub on Nutrition and Food Security has received funding from national research funding parties in Belgium (FWO), France (INRA), Germany (BLE), Italy (MIPAAF), Latvia (IZM), Norway (RCN), Portugal (FCT), and Spain (AEI) in a joint action of JPI HDHL, JPI-OCEANS and FACCE-JPI, launched in 2019 under the ERA-NET ERA-HDHL (n° 696295).

Institutional Review Board Statement: Not applicable.

Informed Consent Statement: Not applicable.

Data Availability Statement: Data sharing not applicable.

Acknowledgments: The research leading to these results was supported (1) by MICINN supporting the Ramón y Cajal grant for M.A. Prieto (RYC-2017-22891) and the FPU grants for A. Carreira-Casais (FPU2016/06135) and A. Soria-Lopez (FPU2020/06140), (2) by Xunta de Galicia for supporting the predoctoral grants of A.G. Pereira (ED481A-2019/0228) and M. Carpena (ED481A 2021/313), and (3) by the program BENEFICIOS DO CONSUMO DAS ESPECIES TINTORERA-(CO-0019-2021) that supports the work of F. Chamorro. The authors are grateful to the Ibero-American Program on Science and Technology (CYTED—AQUA-CIBUS, P317RT0003), and to the Bio Based Industries Joint Undertaking (JU) under grant agreement No 888003 UP4HEALTH Project (H2020-BBI-JTI-2019) that supports the work of P. Otero and P. Garcia-Perez.

Conflicts of Interest: The authors declare no conflict of interest.

References

1. Vinatoru, M.; Mason, T.J.; Calinescu, I. Ultrasonically Assisted Extraction (UAE) and Microwave Assisted Extraction (MAE) of Functional Compounds from Plant Materials. *TrAC Trends Anal. Chem.* **2017**, *97*, 159–178. [CrossRef]
2. Dias, A.L.B.; Arroio Sergio, C.S.; Santos, P.; Barbero, G.F.; Rezende, C.A.; Martínez, J. Ultrasound-Assisted Extraction of Bioactive Compounds from Dedo de Moça Pepper (*Capsicum Baccatum* L.): Effects on the Vegetable Matrix and Mathematical Modeling. *J. Food Eng.* **2017**, *198*, 36–44. [CrossRef]
3. Chemat, F.; Rombaut, N.; Sicaire, A.G.; Meullemiestre, A.; Fabiano-Tixier, A.S.; Abert-Vian, M. Ultrasound Assisted Extraction of Food and Natural Products. Mechanisms, Techniques, Combinations, Protocols and Applications. A Review. *Ultrason. Sonochem.* **2017**, *34*, 540–560. [CrossRef] [PubMed]
4. Chemat, F.; Zill-E-Huma; Khan, F.M. Applications of Ultrasound in Food Technology: Processing, Preservation and Extraction. *Ultrason. Sonochem.* **2011**, *18*, 813–835. [CrossRef]

5. Luque-García, J.L.; Luque De Castro, M.D. Ultrasound-Assisted Soxhlet Extraction: An Expeditive Approach for Solid Sample Treatment—Application to the Extraction of Total Fat from Oleaginous Seeds. *J. Chromatogr. A* **2004**, *1034*, 237–242. [CrossRef]
6. Saleh, I.A.; Vinatoru, M.; Mason, T.J.; Abdel-Azim, N.S.; Aboutabl, E.A.; Hammouda, F.M. A Possible General Mechanism for Ultrasound-Assisted Extraction (UAE) Suggested from the Results of UAE of Chlorogenic Acid from *Cynara scolymus* L. (Artichoke) Leaves. *Ultrason. Sonochem.* **2016**, *31*, 330–336. [CrossRef] [PubMed]
7. Wen, C.; Zhang, J.; Zhang, H.; Dzah, C.S.; Zandile, M.; Duan, Y.; Ma, H.; Luo, X. Advances in Ultrasound Assisted Extraction of Bioactive Compounds from Cash Crops—A Review. *Ultrason. Sonochem.* **2018**, *48*, 538–549. [CrossRef]
8. Chakraborty, S.; Uppaluri, R.; Das, C. Optimization of Ultrasound-Assisted Extraction (UAE) Process for the Recovery of Bioactive Compounds from Bitter Gourd Using Response Surface Methodology (RSM). *Food Bioprod. Process.* **2020**, *120*, 114–122. [CrossRef]
9. Jang, S.; Lee, A.Y.; Lee, A.R.; Choi, G.; Kim, H.K. Optimization of Ultrasound-Assisted Extraction of Glycyrrhizic Acid from Licorice Using Response Surface Methodology. *Integr. Med. Res.* **2017**, *6*, 388–394. [CrossRef]
10. Predescu, N.C.; Papuc, C.; Nicorescu, V.; Gajaila, I.; Goran, G.V.; Petcu, C.D.; Stefan, G. The Influence of Solid-to-Solvent Ratio and Extraction Method on Total Phenolic Content, Flavonoid Content and Antioxidant Properties of Some Ethanolic Plant Extracts. *Rev. Chim.* **2016**, *67*, 1922–1927.
11. Ghassempour, A.; Heydari, R.; Talebpour, Z.; Fakhari, A.R.; Rassouli, A.; Davies, N.; Aboul-Enein, H.Y. Study of New Extraction Methods for Separation of Anthocyanins from Red Grape Skins: Analysis by HPLC and LC-MS/MS. *J. Liq. Chromatogr. Relat. Technol.* **2008**, *31*, 2686–2703. [CrossRef]
12. Alzorqi, I.; Singh, A.; Manickam, S.; Al-Qrimli, H.F. Optimization of Ultrasound Assisted Extraction (UAE) of β- d -Glucan Polysaccharides from Ganoderma Lucidum for Prospective Scale-Up. *Resour. Technol.* **2017**, *3*, 46–54. [CrossRef]
13. Chuang, S.S.; Tamis-LeMonda, C.S. *Sonocrystallization of Fats*; Springer: New York, NY, USA, 2013; pp. 1–66. [CrossRef]
14. Carreira-Casais, A.; Otero, P.; Garcia-Perez, P.; Garcia-Oliveira, P.; Pereira, A.G.; Carpena, M.; Soria-Lopez, A.; Simal-Gandara, J.; Prieto, M.A. Benefits and Drawbacks of Ultrasound-Assisted Extraction for the Recovery of Bioactive Compounds from Marine Algae. *Int. J. Environ. Res. Public Health* **2021**, *18*, 9153. [CrossRef] [PubMed]
15. Aguilar-Hernández, G.; De Los Ángeles Vivar-Vera, M.; De Lourdes García-Magaña, M.; González-Silva, N.; Pérez-Larios, A.; Montalvo-González, E. Ultrasound-Assisted Extraction of Total Acetogenins from the Soursop Fruit by Response Surface Methodology. *Molecules* **2020**, *25*, 1139. [CrossRef]
16. Yolmeh, M.; Habibi Najafi, M.B.; Farhoosh, R. Optimisation of Ultrasound-Assisted Extraction of Natural Pigment from Annatto Seeds by Response Surface Methodology (RSM). *Food Chem.* **2014**, *155*, 319–324. [CrossRef] [PubMed]
17. Maran, J.P.; Priya, B.; Nivetha, C.V. Optimization of Ultrasound-Assisted Extraction of Natural Pigments from Bougainvillea Glabra Flowers. *Ind. Crops Prod.* **2015**, *63*, 182–189. [CrossRef]
18. Jin, Y.; Wang, C.Y.; Hu, W.; Huang, Y.; Xu, M.L.; Wang, H.; Kong, X.; Chen, Y.; Dong, T.T.; Qin, Q.; et al. An Optimization of Ultra-Sonication-Assisted Extraction from Flowers of Apocynum Venetum in Targeting to Amount of Free Amino Acids Determined by UPLC-MS/MS. *Food Qual. Saf.* **2019**, *3*, 52–60. [CrossRef]
19. Carrera, C.; Ruiz-Rodríguez, A.; Palma, M.; Barroso, C.G. Ultrasound-Assisted Extraction of Amino Acids from Grapes. *Ultrason. Sonochem.* **2015**, *22*, 499–505. [CrossRef]
20. Surin, S.; You, S.G.; Seesuriyachan, P.; Muangrat, R.; Wangtueai, S.; Jambrak, A.R.; Phongthai, S.; Jantanasakulwong, K.; Chaiyaso, T.; Phimolsiripol, Y. Optimization of Ultrasonic-Assisted Extraction of Polysaccharides from Purple Glutinous Rice Bran (*Oryza sativa* L.) and Their Antioxidant Activities. *Sci. Rep.* **2020**, *10*, 10410. [CrossRef]
21. Zhang, H.; Li, H.; Zhang, Z.; Hou, T. Optimization of Ultrasound-Assisted Extraction of Polysaccharides from Perilla Seed Meal by Response Surface Methodology: Characterization and in Vitro Antioxidant Activities. *J. Food Sci.* **2021**, *86*, 306–318. [CrossRef]
22. Ministerio de Sanidad BOE. Real Decreto 1101/2011, de 22 de Julio, Por El Que Se Aprueba La Lista Positiva de Los Disolventes de Extracción Que Se Pueden Utilizar En La Fabricación de Productos Alimenticios y de Sus Ingredientes. Real Decreto 1101/2011. 2011. Available online: https://www.boe.es/eli/es/rd/2011/07/22/1101 (accessed on 10 October 2021).
23. Aourabi, S.; Sfaira, M.; Mahjoubi, F. Optimization of Ultrasound-Assisted Extraction of Polyphenol Content from *Zea mays* Hairs (Waste). *Sci. World J.* **2020**, *2020*, 1–10. [CrossRef] [PubMed]
24. Bamba, B.S.B.; Shi, J.; Tranchant, C.C.; Xue, S.J.; Forney, C.F.; Lim, L.T. Influence of Extraction Conditions on Ultrasound-Assisted Recovery of Bioactive Phenolics from Blueberry Pomace and Their Antioxidant Activity. *Molecules* **2018**, *23*, 1685. [CrossRef] [PubMed]
25. Piñeiro, Z.; Marrufo-Curtido, A.; Serrano, M.J.; Palma, M. Ultrasound-Assisted Extraction of Stilbenes from Grape Canes. *Molecules* **2016**, *21*, 784. [CrossRef] [PubMed]
26. Tan, J.; Li, Q.; Xue, H.; Tang, J. Ultrasound-Assisted Enzymatic Extraction of Anthocyanins from Grape Skins: Optimization, Identification, and Antitumor Activity. *J. Food Sci.* **2020**, *85*, 3731–3744. [CrossRef] [PubMed]
27. Wang, H.; Shi, L.; Yang, X.; Hong, R.; Li, L. Ultrasonic-Assisted Extraction of Natural Yellow Pigment from *Physalis pubescens* L. and Its Antioxidant Activities. *J. Chem.* **2018**, *2018*, 1–6. [CrossRef]

chemistry proceedings

MDPI

Proceeding Paper

Plants of the Family Asteraceae: Evaluation of Biological Properties and Identification of Phenolic Compounds †

Marta Barral-Martinez [1], Paula Garcia-Oliveira [1,2], Bernabe Nuñez-Estevez [1,2], Aurora Silva [1,3], Tiane C. Finimundy [2], Ricardo Calhelha [2], Marija Nenadic [4], Marina Sokovic [4], Fatima Barroso [3], Jesus Simal-Gandara [1], Isabel C. F. R. Ferreira [2], Lillian Barros [2,*] and Miguel A. Prieto [1,2,*]

1 Nutrition and Bromatology Group, Faculty of Food Science and Technology, Universidade de Vigo, E32004 Ourense, Spain; marta.barral@uvigo.es (M.B.-M.); paula.garcia.oliveira@uvigo.es (P.G.-O.); bernabe.nunez@uvigo.es (B.N.-E.); mass@isep.ipp.pt (A.S.); jsimal@uvigo.es (J.S.-G.)

2 Centro de Investigação de Montanha (CIMO), Instituto Politécnico de Bragança, Campus de Santa Apolonia, 5300-253 Bragança, Portugal; tiane@ipb.pt (T.C.F.); rcalhelha@ipb.pt (R.C.); iferreira@ipb.pt (I.C.F.R.F.)

3 REQUIMTE/LAQV, Instituto Superior de Engenharia do Porto, Rua Dr António Bernardino de Almeida 431, 4200-072 Porto, Portugal; mfb@isep.ipp.pt

4 Institute for Biological Research "Siniša Stanković", University of Belgrade, Bulevar despota Stefana 142, 11060 Belgrade, Serbia; marija.nenadic@bio.bg.ac.rs (M.N.); mris@ibiss.bg.ac.rs (M.S.)

* Correspondence: lillian@ipb.pt (L.B.); mprieto@uvigo.es (M.A.P.)

† Presented at the 1st International Electronic Conference on Chemical Sensors and Analytical Chemistry, 1–15 July; Available online: https://csac2021.sciforum.net/.

Abstract: The present study focused on the biological analysis of five plants: *Achillea millefolium*, *Arnica montana*, *Calendula officinalis*, *Chamaemelum nobile* and *Taraxacum officinale*. The results indicated that *A. montana* extracts showed the highest content of phenolic compounds. Regarding the biological properties, *A. millefolium* had outstanding antioxidant activity, while *C. officinalis* had the highest rate of antimicrobial and antifungal activity. The anti-inflammatory and cytotoxic activities reflected that *C. nobile* showed the highest effect. In enzyme assays, *C. nobile* and *C. officinalis* extracts showed the highest inhibitory effects on acetylcholinesterase and butyrylcholinesterase enzymes. Overall, this study provides scientific evidence for the evaluation of the potential of medicinal plant extracts for the development of new products.

Keywords: medicinal plants; beneficial effects; biological properties; phenolic compounds

check for updates

Citation: Barral-Martinez, M.; Garcia-Oliveira, P.; Nuñez-Estevez, B.; Silva, A.; Finimundy, T.C.; Calhelha, R.; Nenadic, M.; Sokovic, M.; Barroso, F.; Simal-Gandara, J.; et al. Plants of the Family Asteraceae: Evaluation of Biological Properties and Identification of Phenolic Compounds. *Chem. Proc.* **2021**, *5*, 51. https://doi.org/10.3390/CSAC2021-10486

Academic Editor: Manel del Valle

Published: 30 June 2021

Publisher's Note: MDPI stays neutral with regard to jurisdictional claims in published maps and institutional affiliations.

1. Introduction

Currently, medicinal plants have great relevance due to their reported beneficial health properties. Many studies reflect that their biological properties, such as antioxidant, antitumor and antimicrobial activities, are related to different bioactive compounds, including phenolic compounds. Although some of their mechanisms of action are unknown, in many cases it has been shown that various natural phenolic compounds are related to bioactive properties, and this has aroused the interest of the scientific community [1]. Several medicinal plants are still used for therapeutic purposes, employed in different formulas (decoctions, infusions, ointments, etc.) but, in general, their use has been reduced. However, these plants can be re-valorized for the recovery of bioactive compounds with applications in the food, cosmetic and pharmaceutical industries [2]. In particular, plants from Asteraceae family are promising candidates, due to their beneficial properties and bioactive compounds.

The present study focused on five medicinal plants from the Asteraceae family, namely, *Achillea millefolium* L., *Arnica montana* L., *Calendula officinalis* L., *Chamaemelum nobile* L. and *Taraxacum officinale* (L.) Weber ex F. H. Wigg., all belonging to the Asteraceae family. These plants have been widely used in traditional medicine for the treatment of various disorders, but their use has been reduced. *A. millefolium*, *T. officinale* and *C. officinalis* are the most studied of these plants, and some of their terpenoids, flavonoids, phenolic acids and carotenoids have been described as bioactive compounds [3,4]. The bioactive compounds of *C. nobile* have not been studied in depth, although this plant is well is known to be especially beneficial for digestive health. Bioactive compounds identified so far include terpenoids, flavonoids, coumarins and other compounds such as esters of angelic and tyglic acids, among others [5]. However, the main bioactive compounds in *A. montana* have been demonstrated to be the so-called sesquiterpene lactones, which are related to its anti-inflammatory effects [6]. On this basis, the study focused on the determination of phenolic compounds and the evaluation of the biological properties of these plants, to deepen knowledge about the bioactive compounds and evaluate their possible use in future bio-based applications.

2. Materials and Methods

2.1. Sample Extraction

The samples were acquired in 2020 from Soria Natural and Pinisan and were received at room temperature, dried and crushed to facilitate and improve the efficiency of the extraction processes. Then, the samples were sieved with a sieve (pore size < 2 mm). The samples were extracted by solid–liquid extraction. A 5 g sample of each species was extracted with 100 mL of methanol–water (60:40 *v*/*v*). Extraction was carried out at 45 °C for 3 h. Then, the extracts were freeze-dried using Telstar LyoAlfa 15 equipment to obtain dry extracts that were used in the subsequent analyses.

2.2. Determination of Phenolic Compounds

The identification of phenolic compounds was carried out using a Dionex UltiMate 3000 UPLC system (Thermo Scientific, San Jose, CA, USA) [7]. The determination was performed using a diode array detector (DAD) and mass spectrometry (MS) (LTQ XL mass spectrometer, Thermo Finnigan, San Jose, CA, USA) working in negative mode. Data acquisition was carried out with an Xcalibur® data system (Thermo Finnigan, San Jose, CA, USA). The phenolic compounds were identified according to their chromatographic characteristics, by their retention, absorption spectra and mass characteristics in comparison to the obtained standard compounds and the literature. For quantitative analysis, calibration curves were prepared with appropriate standards. The results were expressed in mg per g of dry extract. Analyses were performed in triplicate.

2.3. Determination of the Main Biological Properties

2.3.1. Assessment of Antioxidant Activity

To evaluate the antioxidant activity, the lipid peroxidation inhibition in porcine (*Sus scrofa*) brain homogenates was analyzed, evaluating the decrease in thiobarbituric acid reactive substances (TBARS), as previously described in Pineda et al. [8]. Brain tissue was homogenized in Tris-HCl buffer (20 mM, pH 7.4) and then centrifuged at 3000 *g* for 10 min. An aliquot of the supernatant was incubated with the extracts at different concentrations in the presence of FeSO4 (10 mM) and ascorbic acid (0.1 mM) for 1h at 37 °C. Trichloroacetic acid (28%) and thiobarbituric acid (2%) were added to stop the reaction at 80 °C, and stirred for 20 min. After centrifugation, the color intensity of the malondialdehyde complex in the supernatant was measured via its absorbance at 532 nm. Using the dose–response values of the results obtained, a parameter that summarized the potential antioxidant effect of each sample was obtained, i.e., the concentration necessary to produce 50% of the antioxidant response (EC_{50}) [7].

2.3.2. Assessment of Antimicrobial Activity

The dried extracts were dissolved in distilled water (10 mg/mL) and the procedure described by Soković et al. [9] was followed. The activity was studied against threeGram-negative bacteria: *Escherichia coli*, *Salmonella typhimurium* and *Enterobacter cloacae* and three Gram-positive bacteria: *Bacillus cereus*, *Listeria monocytogenes* and methicillin-resistant

Staphylococcus aureus (MRSA). For antifungal assays, six micromycetes were tested: *Aspergillus fumigatus* (human isolate), *Aspergillus niger* (ATCC 6275), *Aspergillus versicolor* (ATCC11730), *Penicillium funiculosum* (ATCC 36839), *Trichoderma viride* (IAM 5061) and *Penicillium verrucosum var. cyclopium* (food isolate). The minimum inhibitory concentration (MIC), minimum bactericidal concentration (MBC) and minimum fungicidal concentration were determined.

2.3.3. Assessment of Anti-Inflammatory Properties

The dried extracts were dissolved in distilled water (8 mg/mL) and serial dilutions (1–8 mg/mL) were prepared and tested using a RAW 264.7 murine macrophage cell line. Lipopolysaccharide was used to stimulate inflammation and the production of nitric oxide was measured as described previously [10]. The results obtained were expressed as EC_{50} values (μg/mL) and dexamethasone was used as a positive control.

2.3.4. Cytotoxic Properties

Cytotoxicity was assessed using four tumor cell lines: AGS (human gastric adenocarcinoma cell line), CaCo (Caucasian colon adenocarcinoma), MCF-7 (break adenocarcinoma cell line), NCI- H460 (lung cancer). The Vero cell line was used as a control. Cytotoxic activity was measured using the sulforhodamine B assay [11]. The results obtained were expressed as GI_{50} values, i.e., the concentration of extract that inhibited 50% of net cell growth, and ellipticin was used as a positive control.

2.3.5. Enzymatic Activity

A previously developed colorimetric method was used [12]. This consists of detecting the inhibition of acetylcholinesterase (AChE) and butyrylcholinesterase (BuChE) activity via the increase in yellow coloring due to the production of thiocholine. These two enzymes have been reported to be involved in neurological disorders. In addition, inhibition of AChe has been recognized as a possible avenue for the symptomatic treatment of Alzheimer's disease [13]. The assay was carried out using three buffers: A with 50 mM Tris–HCl, pH 8; B with 50 mM Tris–HCl, pH 8, 0.1% BSA and C with 50 mM Tris–HCl, pH 8, 0.1 M NaCl and 0.02 M $MgCl_2$. The inhibitory capacity of the extracts was tested at concentrations of 1 and 2 mg/mL.

3. Results and Discussion

The phenolic profile of the selected plants showed great variability, both in quantity and in the identified phenolic compounds (Table 1; Figure 1). The plant with the highest content of phenolic compounds was *A. montana*, with a concentration of 119 mg/mL, where the most representative compound was 5-*O*-caffeolyquinic acid. The extracts of *C. nobile* presented a total phenolic content of 100 mg/mL and, in this case, the major compound was the flavonoid luteolin-*O*-pentosylhexoside. *A. millefolium* extracts achieved a total phenolic content of 81 mg/mL, and the most representative compound was 3-*O*-caffeoylquinic acid. *T. officinale* extracts had a phenolic content of 18 mg/mL and were rich in 3-*O*-caffeoylquinic acid. Finally, *C. officinalis* extracts had the lowest phenolic content, at 14.1 mg/mL, with 3-*O*-caffeoylquinic acid as the major compound. These results coincide with those of other studies, e.g., in the case of the plant *A. montana*, where 5-*O*-caffeoylquinic acid has also previously been reported as the main phenolic acid in the ethanolic extract of this plant [14]. However, it should be noted that the content of phenolic acids can be influenced by external factors such as the type of solvent used in the extraction process, and it is also related to the growing conditions of the plants, as mentioned in previous studies on *A. montana* and other species belonging to the Asteraceae family [15].

Table 1. Total phenolic content and main phenolic compounds identified (mg/mL).

Plants	Main Phenolic Compounds	Quantification	TPC
Achillea millefolium	3-*O*-Caffeoylquinic acid	18.85 ± 0.03	81
Arnica montana	5-*O*-Caffeolyquinic acid	23.9 ± 0.3	119
Calendula officinalis	3-*O*-Caffeoylquinic acid	9.8 ± 0.2	14.1
Chamaemelum nobile	Luteolin-*O*-pentosylhexoside	49.6 ± 0.5	100
Taraxacum officinale	3-*O*-Caffeoylquinic acid	6.74 ± 0.4	18

TPC: total phenolic compounds.

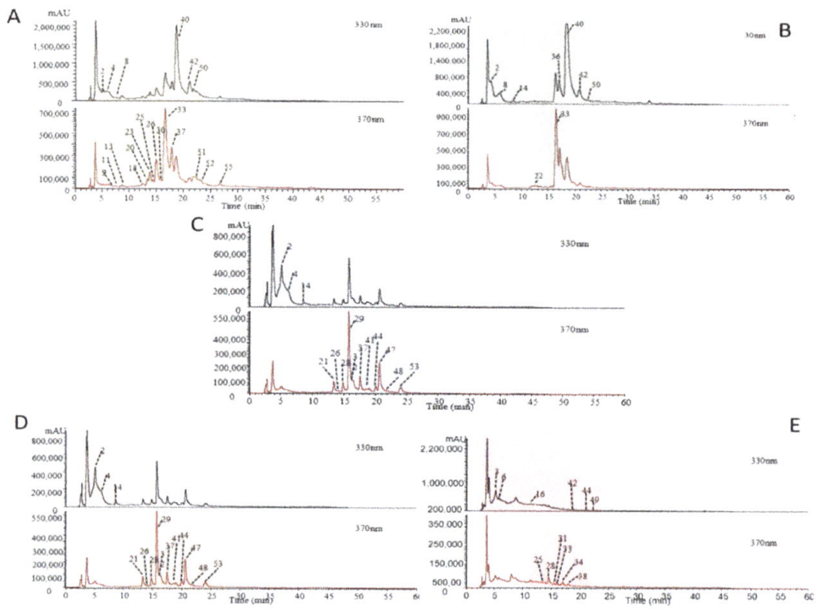

Figure 1. Representative chromatogram of the phenolic compounds identified: (**A**) *A. millefolium*; (**B**) *A. montana*; (**C**) *C. officinalis*; (**D**) *C. nobile*; (**E**) *T. officinale*.

Regarding antioxidant activity, the extracts of *A. millefolium* showed exceptional activity, with an EC_{50} value of 0.013 mg/mL. The extracts of *A. montana*, *C. nobile* and *C. officinalis* showed similar EC_{50} values (0.2, 0.2 and 0.25 mg/mL, respectively). Finally, the extracts of *T. officinale* showed the lowest antioxidant activity with an EC_{50} of 0.035 mg/mL. These results are presented in Figure 2. In previous studies, this assay has been employed to evaluate the antioxidant activity of *A. millefolium*, *C. officinalis* and *C. nobile*, reporting significant results. To the best of our knowledge, no study has used the TBARS assay to evaluate *A. montana* and *T. officinale*, but their antioxidant properties have been corroborated by various studies showing positive results [16–18].

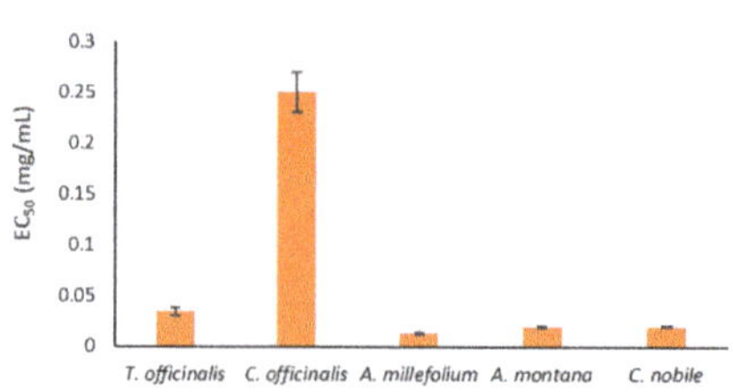

Figure 2. Antioxidant activity in traditional plants of the family Asteraceae: *A. millefolium*, *A. montana*, *C. officinalis*, *C. nobile* and *T. officinale*.

Regarding antimicrobial activity, all the plant extracts displayed significant antimicrobial effects, with *C. officinalis* being the most remarkable. This plant presented MIC values ranging from 0.25 to 0.5 mg/mL for all the tested bacteria and fungi. MBC and MFC values ranged between 0.5 and 1 mg/mL. The most susceptible bacteria were the Gram-positive species, while *T. viride* was the most susceptible fungus. *T. officinale* also showed relevant antibacterial potential, while *C. nobile* was also effective against fungi species. The antimicrobial potential of these species has been previously confirmed. Focusing on *C. officinalis*, a study reported that petal extracts of this plant showed comparable antibacterial effects against Gram-positive and Gram-negative bacteria using the disk diffusion method [19]. The results found in the literature are very similar to those obtained experimentally and therefore corroborate the hypothesis that the *C. officinalis* plant could be used as a possible source of antimicrobial compounds.

According to the results (Figure 3), *C. nobile* extracts showed the greatest effects in both assays, with EC_{50} values of 15.21 µg/mL for anti-inflammatory activity and GI_{50} values between 54 and 10.3 µg/mL in the case of cytotoxic activity. *A. millefolium* also showed significant results, with an EI_{50} of 30 µg/mL for anti-inflammatory activity and GI_{50} values ranging between 42 and 125 µg/mL. Considering the *C. nobile* results, the anti-inflammatory and the cytotoxic properties of this plant have been reported previously, showing positive results [20,21], and therefore this plant could be a promising source of anti-inflammatory and cytotoxic extracts.

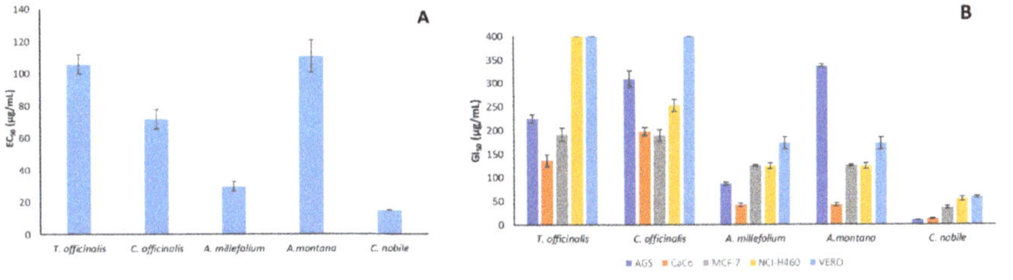

Figure 3. (**A**) Anti-inflammatory and (**B**) cytotoxic activity of *A. millefolium*, *A. montana*, *C. officinalis*, *C. nobile* and *T. officinale*.

Finally, *C. nobile* showed the highest inhibitory effects on AChE activity for the extract concentrations tested (1 and 2 mg/mL), causing an inhibition of >35% and >60%, respectively. In the case of the BuChE enzyme, *C. officinalis* caused an inhibition of >50% in both concentrations tested. *C. nobile* also showed a remarkable inhibitory effect against this enzyme, with an inhibition of >40% at 2 mg/mL and >20% at 1 mg/mL. To the best of our knowledge, no previous studies have evaluated the enzymatic activity of the selected plants.

4. Conclusions

All the plants studied had diverse phenolic compositions and biological activities. Regarding phenolic compounds, *A. montana* extracts showed the highest content. Regarding bioactivities, *A. millefolium* showed high antioxidant activity and *C. officinalis* showed the best antimicrobial and antifungal activities. In the case of anti-inflammatory and cytotoxic activities, *C. nobile* extracts achieved the best results. Finally in the enzyme assays, both *C. nobile* and *C. officinalis* extracts showed the highest inhibitory effects. Therefore, this study provides scientific evidence of the potential of medicinal plants as a source of extracts and bioactive compounds that may be considered for the development of new products.

Supplementary Materials: The poster presentation is available online at: https://www.mdpi.com/article/10.3390/CSAC2021-10486/s1.

Author Contributions: Conceptualization, M.B.-M., P.G.-O., L.B. and M.A.P.; methodology, M.B.-M., B.N.-E., A.S., F.B., M.N., M.S., T.C.F., R.C. and I.C.F.R.F.; software, M.B.-M., B.N.-E. and A.S.;

validation, F.B., I.C.F.R.F., M.A.P. and L.B.; formal analysis, A.S., F.B., M.N., M.S., T.C.F., R.C. and I.C.F.R.F.; investigation, M.B.-M., P.G.-O., A.S. and F.B.; writing—original draft preparation, M.B.-M. and P.G.-O.; writing—review and editing, M.B.-M. and P.G.-O.; visualization, P.G.-O., T.C.F. and L.B.; supervision, M.A.P., L.B. and J.S.-G. All authors have read and agreed to the published version of the manuscript.

Funding: The JU receives support from the European Union's Horizon 2020 research and innovation program and the Bio Based Industries Consortium. The project SYSTEMIC Knowledge hub on Nutrition and Food Security has received funding from national research funding parties in Belgium (FWO), France (INRA), Germany (BLE), Italy (MIPAAF), Latvia (IZM), Norway (RCN), Portugal (FCT) and Spain (AEI) in a joint action of JPI HDHL, JPI-OCEANS and FACCE-JPI launched in 2019 under the ERA-NET ERA-HDHL (No. 696295).

Acknowledgments: The research leading to these results was supported by MICINN by supporting the Ramón y Cajal grant for M.A. Prieto (RYC-2017-22891); by Xunta de Galicia by supporting the program EXCELENCIA-ED431F 2020/12 and the pre-doctoral grant of P. García-Oliveira (ED481A-43 2019/295); by the EcoChestnut Project (Erasmus+ KA202) that supports the work of B. Nuñez-Estevez; by the program Grupos de Referencia Competitiva (GRUPO AA1-GRC 2018) that supports the work M. Barral-Martínez; by the Bio Based Industries Joint Undertaking (JU) under grant agreement No 888003; by the UP4HEALTH Project (H2020-BBI-JTI-2019) that supports the work of P. Garcia-Perez and by the Ibero-American Program on Science and Technology (CYTED—AQUA-CIBUS, P317RT0003). The authors are grateful to the Foundation for Science and Technology (FCT, Portugal) for financial support through national funds FCT/MCTES to CIMO (UIDB/00690/2020); and to the national funding by FCT, P.I., through the institutional scientific employment program contract for L. Barros.

Conflicts of Interest: The authors declare no conflict of interest.

References

1. Shabab, S.; Gholamnezhad, Z.; Mahmoudabady, M. Protective effects of medicinal plant against diabetes induced cardiac disorder: A review. *J. Ethnopharmacol.* **2021**, *265*, 113328. [CrossRef]
2. Garcia-Oliveira, P.; Fraga-Corral, M.; Pereira, A.G.; Lourenço-Lopes, C.; Jimenez-Lopez, C.; Prieto, M.A.; Simal-Gandara, J. Scientific basis for the industrialization of traditionally used plants of the Rosaceae family. *Food Chem.* **2020**, *330*, 127197. [CrossRef] [PubMed]
3. Nicolaus, C.; Junghanns, S.; Hartmann, A.; Murillo, R.; Ganzera, M.; Merfort, I. *In vitro* studies to evaluate the wound healing properties of *Calendula officinalis* extracts. *J. Ethnopharmacol.* **2017**, *196*, 94–103. [CrossRef]
4. Lass, C.; Vocanson, M.; Wagner, S.; Schempp, C.M.; Nicolas, J.F.; Merfort, I.; Martin, S.F. Anti-inflammatory and immune-regulatory mechanisms prevent contact hypersensitivity to *Arnica montana* L. *Exp. Dermatol.* **2008**, *17*, 849–857. [CrossRef]
5. Calvo, M.I.; Cavero, R.Y. Medicinal plants used for neurological and mental disorders in Navarra and their validation from official sources. *J. Ethnopharmacol.* **2015**, *169*, 263–268. [CrossRef] [PubMed]
6. Vidic, D.; Ćavar Zeljković, S.; Dizdar, M.; Maksimović, M. Essential oil composition and antioxidant activity of four Asteraceae species from Bosnia. *J. Essent. Oil Res.* **2016**, *28*, 445–457. [CrossRef]
7. Bessada, S.M.F.; Barreira, J.C.M.; Barros, L.; Ferreira, I.C.F.R.; Oliveira, M.B.P.P. Phenolic profile and antioxidant activity of *Coleostephus myconis* (L.) Rchb.f.: An underexploited and highly disseminated species. *Ind. Crops Prod.* **2016**, *89*, 45–51. [CrossRef]
8. Pinela, J.; Barros, L.; Dueñas, M.; Carvalho, A.M.; Santos-Buelga, C.; Ferreira, I.C.F.R. Antioxidant activity, ascorbic acid, phenolic compounds and sugars of wild and commercial Tuberaria lignosa samples: Effects of drying and oral preparation methods. *Food Chem.* **2012**, *135*, 1028–1035. [CrossRef]
9. Soković, M.; Glamočlija, J.; Marin, P.D.; Brkić, D.; Van Griensven, L.J.L.D. Antibacterial effects of the essential oils of commonly consumed medicinal herbs using an in vitro model. *Molecules* **2010**, *15*, 7532–7546. [CrossRef]
10. Svobodova, B.; Barros, L.; Sopik, T.; Calhelha, R.C.; Heleno, S.; Alves, M.J.; Walcott, S.; Kuban, V.; Ferreira, I.C.F.R. Non-edible parts of *Solanum stramoniifolium* Jacq.—A new potent source of bioactive extracts rich in phenolic compounds for functional foods. *Food Funct.* **2013**, *8*, 2013–2021. [CrossRef]
11. Abreu, R.M.V.; Ferreira, I.C.F.R.; Calhelha, R.C.; Lima, R.T.; Vasconcelos, M.H.; Adega, F.; Chaves, R.; Queiroz, M.J.R.P. Anti-hepatocellular carcinoma activity using human HepG2 cells and hepatotoxicity of 6-substituted methyl 3-aminothieno[3,2-b]pyridine-2- carboxylate derivatives: *In vitro* evaluation, cell cycle analysis and QSAR studies. *Eur. J. Med. Chem.* **2011**, *46*, 5800–5806. [CrossRef] [PubMed]
12. Gawlik-Dziki, U.; Świeca, M.; Sugier, D. Seeds of Arnica montana and Arnica chamissonis as a potential source of natural antioxidants. *Herba Pol.* **2009**, *55*, 60–71.
13. Thu, K.; Phyu Myint Professor, P.; Phyu Phyu Myint Professor, C.; Phyu Myint, P. Pharmacological activities of *Cuscuta reflexa* (Shwe-nwe) stem and *Taraxacum officinale* Weber ex F.H. Wigg. (Dai-Si) leaf extracts. *J. Med. Plants Stud.* **2019**, *7*, 109–112.

14. Craciunescu, O.; Constantin, D.; Gaspar, A.; Toma, L.; Utoiu, E.; Moldovan, L. Evaluation of antioxidant and cytoprotective activities of *Arnica montana* L. and *Artemisia absinthium* L. ethanolic extracts. *Chem. Cent. J.* **2012**, *6*, 97. [CrossRef]
15. Abou Baker, D.H. *Achillea millefolium* L. ethyl acetate fraction induces apoptosis and cell cycle arrest in human cervical cancer (HeLa) cells. *Ann. Agric. Sci.* **2020**, *65*, 42–48. [CrossRef]
16. Gaspar, A.; Craciunescu, O.; Trif, M.; Moisei, M.; Moldovan, L. Antioxidant and anti-inflammatory properties of active compounds from *Arnica montana* L. *Rom. Biotechnol. Lett.* **2014**, *19*, 9353–9365.
17. Kriplani, P.; Guarve, K.; Baghael, U.S. *Arnica montana* L.—A plant of healing. *J. Pharm. Pharmacol.* **2017**, *69*, 925–945. [CrossRef]
18. Díaz, K.; Espinoza, L.; Madrid, A.; Pizarro, L.; Chamy, R. Isolation and Identification of Compounds from Bioactive Extracts of *Taraxacum officinale* Weber ex F. H. Wigg. (Dandelion) as a Potential Source of Antibacterial Agents. *Evid.-Based Complement. Altern. Med.* **2018**, 1–8. [CrossRef]
19. Efstratiou, E.; Hussain, A.I.; Nigam, P.S.; Moore, J.E.; Ayub, M.A.; Rao, J.R. Antimicrobial activity of *Calendula officinalis* petal extracts against fungi, as well as Gram-negative and Gram-positive clinical pathogens. *Complement. Ther. Clin. Pract.* **2012**, *18*, 173–176. [CrossRef]
20. Guimarães, R.; Barros, L.; Dueñas, M.; Calhelha, R.C.; Carvalho, A.M.; Santos-Buelga, C.; Queiroz, M.J.R.P.; Ferreira, I.C.F.R. Nutrients, phytochemicals and bioactivity of wild Roman chamomile: A comparison between the herb and its preparations. *Food Chem.* **2013**, *136*, 718–725. [CrossRef] [PubMed]
21. Zhao, J.; Khan, S.I.; Wang, M.; Vasquez, Y.; Yang, M.H.; Avula, B.; Wang, Y.; Avonto, C.; Smillie, T.J.; Khan, I.A. Octulosonic Acid Derivatives from Roman Chamomile (*Chamaemelum nobile*) with Activities against Inflammation and Metabolic Disorder. *J. Nat. Prod.* **2014**, *77*, 509–515. [CrossRef] [PubMed]

chemistry proceedings

MDPI

Proceeding Paper

Characterization of Unpleasant Odors in Poultry Houses Using Metal Oxide Semiconductor-Based Gas Sensor Arrays and Pattern Recognition Methods †

Mohammed Moufid [1,2], Carlo Tiebe [3], Nezha El Bari [2], Matthias Bartholmai [3] and Benachir Bouchikhi [1,*]

1 Sensor Electronic & Instrumentation Group, Department of Physics, Faculty of Sciences, Moulay Ismaïl University of Meknes, Zitoune, Meknes B.P. 11201, Morocco; mohammed.moufid.2@gmail.com
2 Biosensors and Nanotechnology Group, Department of Biology, Faculty of Sciences, Moulay Ismaïl University of Meknes, Zitoune, Meknes B.P. 11201, Morocco; n_elbari@hotmail.com
3 Bundesanstalt für Materialforschung und-prüfung (BAM), 8.1 Sensors, Measurement and Testing Methods, 12203 Berlin, Germany; Carlo.tiebe@bam.de (C.T.); matthias.bartholmai@bam.de (M.B.)
* Correspondence: benachir.bouchikhi@gmail.com; Tel.: +212-5355-38870; Fax: +212-5355-36808
† Presented at the 1st International Electronic Conference on Chemical Sensors and Analytical Chemistry, 1–15 July 2021; Available online: https://csac2021.sciforum.net/.

Abstract: In this study, the ability of an electronic nose developed to analyze and monitor odor emissions from three poultry farms located in Meknes (Morocco) and Berlin (Germany) was evaluated. Indeed, the potentiality of the electronic nose (e-nose) to differentiate the concentration fractions of hydrogen sulfide, ammonia, and ethanol was investigated. Furthermore, the impact change of relative humidity values (from 15% to 67%) on the responses of the gas sensors was reported and revealed that the effect remained less than 0.6%. Furthermore, the relevant results confirmed that the developed e-nose system was able to perfectly classify and monitor the odorous air of poultry farms.

Keywords: poultry odorous air monitoring; electronic nose; gas sensors; pattern recognition methods

Citation: Moufid, M.; Tiebe, C.; El Bari, N.; Bartholmai, M.; Bouchikhi, B. Characterization of Unpleasant Odors in Poultry Houses Using Metal Oxide Semiconductor-Based Gas Sensor Arrays and Pattern Recognition Methods. *Chem. Proc.* **2021**, *5*, 52. https://doi.org/10.3390/CSAC2021-10481

Academic Editor: Elisabetta Comini

Published: 30 June 2021

1. Introduction

Unpleasant odors are an inherent part of poultry production. They come from the wastes and emissions of animals. In addition, poultry farms in close proximity to the population generate volatile organic compounds (VOCs) in the air, resulting in a foul odor similar to that of rotten eggs and waste [1]. Moreover, odor nuisance from poultry farms has raised serious concerns about the quality of human life worldwide. Unpleasant smelling air affects the mental and physical health of the population and causes anger [2]. Similarly, the health of hens and farm workers can be threatened by the malodorous chemical compounds (such as hydrogen sulfide and ammonia) emanating from poultry farms [3]. Therefore, appropriate methods and techniques are needed to characterize and monitor odorous air samples, thereby determining the impacts of smelly air on chickens, humans, and the agricultural environment.

Although olfactometric techniques are the most widely used to analyze odorous air based on the perception of a group of human sniffers, these techniques are very expensive and do not provide information on chemical composition [4]. Analytical methods are widely used for the quantitative analysis of odorous air samples to identify unknown organic compounds and their concentration [5]. However, they require a qualified operator, and are expensive, time-consuming, and non-portable. The problems associated with the use of conventional odor air analysis methods could be replaced by the application of faster and cheaper e-nose technology. Electronic noses are devices equipped with an array of gas sensors combined with pattern recognition methods that provide a specific signature of the analyte [6,7]. The last few decades have seen a significant increase in interest in chemical sensors, which is reflected in the growing number of papers and conferences on this topic. For this purpose, these instruments are used in various applications for routine, rapid, and inexpensive assessment related to the environmental sector [8,9]. Similarly, in recent years, odor emissions from livestock farms have received increased attention due to their large number resulting in the production of high concentrations of hydrogen sulfide and ammonia [10,11]. Electronic noses are also applied in other fields, including biomedical [12,13], pharmaceutical [14], food [15], and security [16].

In this work, the ability of an electronic nose to discriminate malodorous VOCs from three poultry farm sites was investigated. In parallel, the monitoring of malodorous air emissions from a poultry farm as a function of the time and date of collection was carried out. In addition, the effect of relative humidity on the response of the gas sensors was checked. The sensitivity to hydrogen sulfide, ammonia, and ethanol was tested. Pattern recognition methods such as principal components analysis (PCA), discriminant function analysis (DFA), and support vector machines (SVMs) were used for processing the data from gas sensor responses.

2. Materials and Methods

2.1. Odorous Air Samples Collection

Odorous air samples were collected using 2L Tedlar bags in three poultry sheds located in the Faculty of Sciences of Meknes (FSM), as well as in the agglomeration of Meknes (Morocco) and in Berlin (Germany). In parallel, to verify the ability of the e-nose to monitor odorous air samples from a poultry farm, odorous air samples were collected from the poultry farm of Meknes at different times and on three different days, with an interval of two days of one week. In total, 126 odorous air samples were collected (14 samples from each time collection performed at 09:00 a.m., 12:00 p.m., and 18:00 p.m.).

2.2. Gas Sensor System

The developed electronic nose system consists of six MQ-type chemical sensors (MQ-3, MQ-4, MQ-5, MQ-8, MQ-9, and MQ-135) from Winsen Electronics Technology Co., Ltd. (Zhengzhou, China) (Figure 1). The sensor chamber also contains a relative humidity sensor (HIH 4000) and a temperature sensor (LM35) to monitor the environmental conditions during measurements. All the sensors were installed in a Teflon chamber with a volume of 270 cm^3. The collected odorous air samples were transferred to the sensor chamber using a Tedlar bag and a micro air pump. The sensor responses were acquired by NI-USB 6212 data acquisition from National Instruments (Austin, TX, USA). It allows signal conversion and preparation for further analysis by changing the analog signal produced by the sensors into its discrete digital representation.

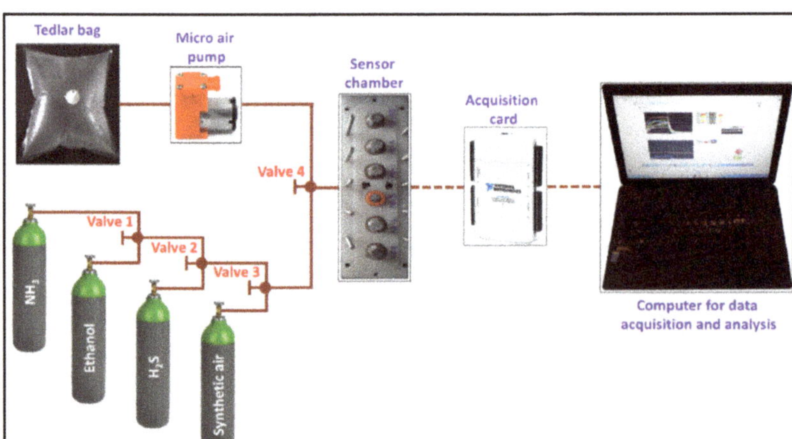

Figure 1. Electronic nose system developed for odorous air analysis and monitoring.

2.3. Sensing Measurements

During analysis, the samples were pumped for 5 min into the sensor chamber at a flow rate of 250 mL/min. The data were acquired every second. After each measurement, synthetic air was injected into the sensor chamber for 5 min to clean the surface of the sensors to return to their baselines.

The concentration fractions of hydrogen sulfide (6 ppm), ammonia (7 ppm), and ethanol (3 ppm) were adjusted with the Gas Mixing System (GMS) of BAM, Berlin, Germany.

The temperature measured inside the sensor chamber during the measurements was approximately 27.5 ± 3.1 °C.

2.4. Data Analysis

2.4.1. Features Extraction

In this study, three features were extracted from the response of each gas sensor:

- $\Delta G = (G_S - G_0)$: The difference in conductance between G_S and G_0 is the average value of the conductance in the last and first 60 s, respectively.
- A: The area under the curve of the sensor conductance between the first and the last minute of sample measurement. This area was calculated using the trapeze method.
- dG/dt: The slope of the sensor response, determined dynamically in a range of 60 to 540 s.

Eighteen variables were defined the developed system (6 sensors × 3 extracted features).

2.4.2. Pattern Recognition Methods

The extracted features were treated by using pattern recognition methods (PCA, DFA, and SVMs) to estimate the performance of the e-nose to classify and monitor odorous air samples from a poultry farm.

The aim of PCA is the multidimensionality reduction of a dataset by finding new orthogonal directions (principal components) which contain the maximum information. The main advantage of PCA is the ability to present the results on two- or three-dimensional graphs. Groups of points can be visualized on the plots, which makes it possible to assess the contribution of the sample to a particular group. The PCA algorithm generates linear combinations of principal components [17].

DFA is a linear method, but it differs from PCA in that it utilizes the cluster information that was given during the training (supervised method), while the PCA does not care about the relationship of the data points with the specified clusters. The DFA method helps to have the best discrimination by reducing distances between samples (variance within classes) and maximizing distances between clusters (variance between classes) [18].

SVMs objective is to increase the quantity called margin to distinguish clusters with a specific hyperplane. The margin is a distance calculated between the nearest points contained in different groups. SVMs have two techniques. The first is to consider one vs. one or one vs. all and the other is to use all the data in a single formulation. This work applied a 2nd degree polynomial kernel. Indeed, a leave-one-out cross-validation technique was used to determine the prediction accuracy. The second-order of a radial basis function (polynomial) kernel was employed to project the training data to a space that maximizes the margin hyper plane [19].

3. Results and Discussion

3.1. Sensor Calibration through Relative Humidity Variation

Since humidity has a great impact on the electrical conductivity of resistive gas sensors, it is necessary to calibrate them using different relative humidity values. Figure 2 represents the normalized conductance of the sensor arrays at three different relative humidity values. We can see from this figure that when the relative humidity values increase from 15% to 67%, the variation of the conductance does not exceed 0.6% for all the sensors. In conclusion, it can be noticed that the calibration of the gas sensors at the considered relative humidity values led to a slight difference in the responses of the gas sensor arrays.

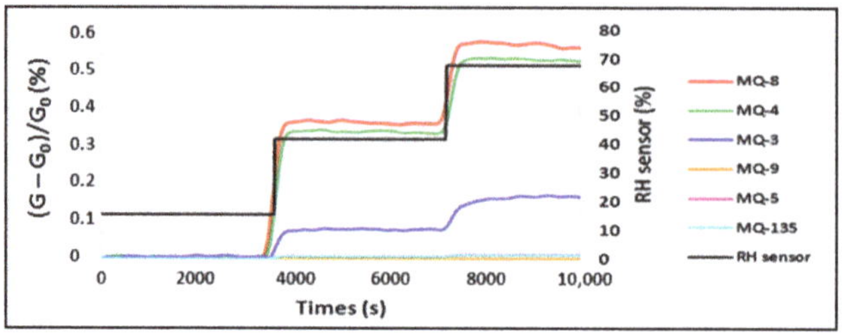

Figure 2. Conductance changes of sensor arrays and relative humidity in function of measurement time by adjusting RH values from 15% to 67%.

3.2. Sensing Behavior of Gas Sensor Arrays to Hydrogen Sulfide, Ammonia, and Ethanol at Room Temperature

Ammonia and hydrogen sulfide are harmful gases generated during the bacterial decomposition of livestock manure [20]. Therefore, it is necessary to test the sensitivity of gas sensors using ammonia and hydrogen sulfide gases. Figure 3 shows a plot of the time dependence of the difference in conductance ($G - G_0$) when the sensors are exposed to hydrogen sulfide (Figure 3a), ammonia (Figure 3b), and ethanol (Figure 3c) at fixed concentration fractions of 6, 7, and 3 ppm, respectively. It can be seen from this figure that all the gas sensors were sensitive to the three tested synthetic gases, except for the MQ-8 sensor. Furthermore, each gas sensor had a different response to the three gases studied, which means that the electronic nose was able to differentiate between these gases.

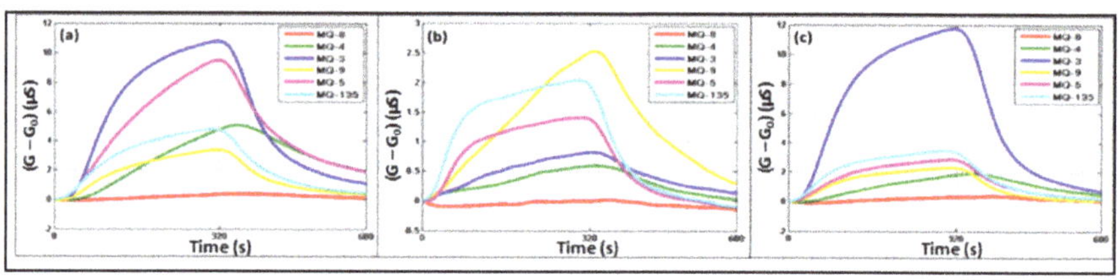

Figure 3. Conductance changes of gas sensor arrays in the presence of (**a**) hydrogen sulfide, (**b**) ammonia, and (**c**) ethanol with concentration fractions of 6, 7, and 3 ppm, respectively.

3.3. Classification Results of the Odorous Air Samples Collected from Poultry Sheds and Clean Air

3.3.1. Radar Plots

Figure 4 shows the results of the radar plot corresponding to the odorous air samples collected from three poultry sheds and synthetic air as the control. In this figure, it can be seen that the odorous air pattern (fingerprints) differed from one site to another. Indeed, the smallest one corresponds to the control.

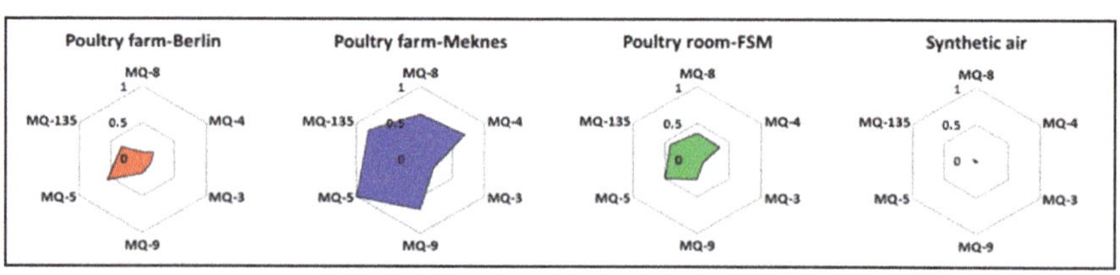

Figure 4. Radar plots of poultry odorous air samples and synthetic air (control) expressed as the difference in conductance ($\Delta G = (G - G_0)$) extracted from gas sensor responses.

3.3.2. PCA Classification

The PCA method was applied to the database gathered from gas sensor responses upon exposure to odorous air samples collected from the three poultry farms and synthetic air as a control. Figure 5 represents the projections of the experimental results onto a two-dimensional (2D) graph. In this figure, it can be seen that all the groups of odorous air samples are clearly separated, with no overlap with synthetic air (control). In fact, the first two principal components account for 94% of the data variance.

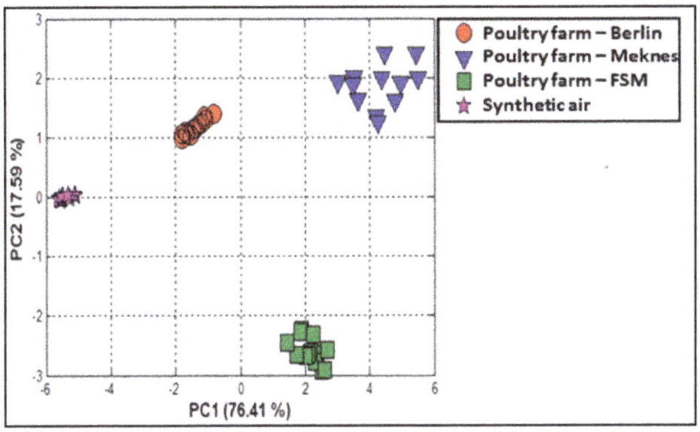

Figure 5. PCA plot performed on poultry odorous air samples and synthetic air by using the features (ΔG, area, and slope) extracted from gas sensor responses.

3.4. Classification Results Depending on the Time and Date of Samples Collection

3.4.1. DFA Classification

DFA was applied to the database gathered from the sensors' responses to verify the ability of the developed e-nose to monitor odorous air samples collected at different dates and times in a poultry farm in Meknes. Figure 6 shows the DFA plot with 89% of the data variance explained by the first two discriminant functions (DFs). It can be seen from this figure that all the clusters are separated from each other. Furthermore, DF1 discriminates odorous air samples based on the date of collection from the poultry shed, while DF2 separates them based on the time of collection. The DFA results prove that the e-nose system was capable of clearly discriminating odorous air samples from a poultry farm according to the date and time of collection.

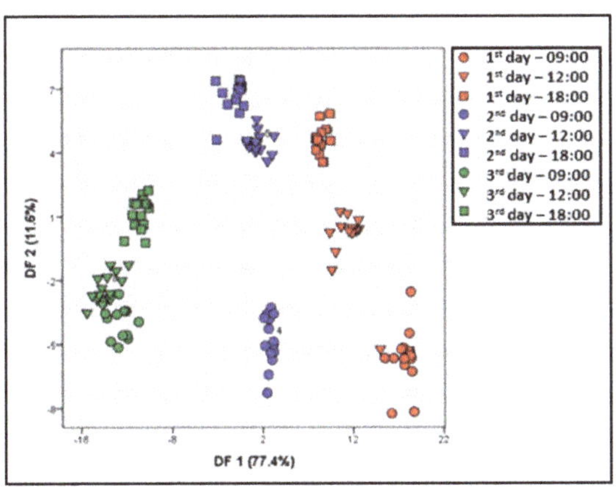

Figure 6. DFA plot performed on odorous air samples collected at different times and dates on a poultry farm (Meknes) using the features (ΔG, area, and slope) extracted from gas sensor array responses.

3.4.2. SVM Classification

An SVM is a supervised learning method. It was applied to the same dataset as DFA to verify the ability of the developed e-nose to monitor odorous air samples in a poultry farm from Meknes city. Table 1 shows the SVM confusion matrix for odorous air samples recognition. In this study, only two misclassified samples were observed in the data matrix. Therefore, a 98.41% accuracy in the recognition of the odorous air samples was achieved. This outcome was in good agreement with the obtained DFA results. This finding confirms that the e-nose system was able to monitor odorous air samples from a poultry farm.

Table 1. SVM classification results of odorous air samples collected at different times and dates on a poultry farm from Meknes using the features (ΔG, area, and slope) extracted from gas sensor responses (total score: 98.41%).

	1st Day–09:00	1st Day–12:00	1st Day–18:00	2nd Day–09:00	2nd Day–12:00	2nd Day–18:00	3rd Day–09:00	3rd Day–12:00	3rd Day–18:00
1st day–09:00	13	1							
1st day–12:00	1	13							
1st day–18:00			14						
2nd day–09:00				14					
2nd day–12:00					14				
2nd day–18:00						14			
3rd day–09:00							14		
3rd day–12:00								14	
3rd day–18:00									14

4. Conclusions

The present study demonstrated that the low-cost, portable, and easy-to-use e-nose system was able to distinguish odorous air samples from poultry sheds based on the sampling site, and also depending on the date and time of collection. The effect of relative humidity on gas sensor responses was also investigated and showed that when relative humidity increased from 15% to 67%, there was a slight difference in sensor responses that did not exceed 0.6%. Similarly, the sensitivity of the sensor array to hydrogen sulfide, ammonia, and ethanol was tested and showed that all gas sensors are sensitive to these three synthetic gases, with the exception of the MQ-8 sensor. Radar plots revealed a significant change in odorous air sample patterns based on the sampling sites. In addition, PCA showed that the e-nose system was able to clearly distinguish odorous air samples from three sites of poultry farms without any overlap with the unpolluted air samples (synthetic air) with 94% of the data variance. In order to monitor odorous air samples from a poultry farm according to their date and time of collection, the database was also processed by DFA and SVM. These two pattern recognition methods show a clear discrimination between the studied samples with a success rate of 89% and 98.41%, respectively. We can conclude that the developed e-nose system can be effectively used as a fast, easy-to-use, and inexpensive tool for the analysis and monitoring of odorous air samples from poultry farms.

Author Contributions: Investigation, Formal analysis, Methodology, software, Experimentation, Original Draft Preparation: M.M.; design of experiment for gas mixing system to investigate the sensor system in the laboratory, scientific exchange of the time-series data and coordination of the tasks in BAM: C.T.; Review & Editing: N.E.B. and M.B.; Conceptualization, Visualization, Supervision, Validation, Resources, Funding acquisition, Review & Editing: B.B. All authors have read and agreed to the published version of the manuscript.

Funding: PMARS N°2015-87 Moroccan–German scientific research cooperation.

Institutional Review Board Statement: Not applicable.

Informed Consent Statement: Not applicable.

Acknowledgments: We would like to thank Moulay Ismaïl University of Meknes for financial support of the project "Research support". This work is funded by the federal Ministry of Education and Research (Germany) and Ministry of Higher Education, Scientific Research and Executive Training (Morocco) under the Framework program of Moroccan–German scientific research cooperation, project under grant agreement n° PMARS N°2015-87. Additionally, we have to thank Michael Hofann and Jihed Ben Majed of BAM, Berlin, Germany for his help and advice.

Conflicts of Interest: The authors declare no conflict of interest.

References

1. Abdullah, A.H.; Shakaff, A.Y.; Adom, A.H.; Zakaria, A.; Saad, F.S.; Kamarudin, L.M. Poultry farm malodor monitoring using portable electronic nose system. *Chem. Eng.* **2012**, *30*, 55–60.
2. Horton, R.A.; Wing, S.; Marshall, S.W.; Brownley, K.A. Malodor as a trigger of stress and negative mood in neighbors of industrial hog operations. *Am. J. Public. Health* **2009**, *99*, 610–615. [CrossRef] [PubMed]
3. McCubbin, D.R.; Apelberg, B.J.; Roe, S.; Divita, F. Livestock Ammonia Management and Particulate-Related Health Benefits. *Environ. Sci. Technol.* **2002**, *36*, 1141–1146. [CrossRef] [PubMed]
4. Pan, L.; Yang, S.X. A new intelligent electronic nose system for measuring and analysing livestock and poultry farm odors. *Environ. Monit. Assess.* **2007**, *135*, 399–408. [CrossRef] [PubMed]
5. Murphy, K.R.; Wenig, P.; Parcsi, G.; Skov, T.; Stuetz, R.M. Characterizing odorous emissions using new software for identifying peaks in chemometric models of gas chromatography-mass spectrometry datasets. *Chemom. Intell. Lab. Syst.* **2012**, *118*, 41–50. [CrossRef]
6. Deisingh, A.K. *Sensors for Chemical and Biological Applications*; Taylor and Francis: Boca Raton, FL, USA, 2010.
7. Gardner, J.W.; Bartlett, P.N. A brief history of electronic noses. *Sens. Actuators B Chem.* **1994**, *18*, 210–211. [CrossRef]
8. Gębicki, J.; Dymerski, T.; Namieśnik, J. Monitoring of odour nuisance from landfill using electronic nose. *Chem. Eng. Trans.* **2014**, *40*, 85–90.
9. Li, H.; Luo, D.; Sun, Y.; GholamHosseini, H. Classification and Identification of Industrial Gases Based on Electronic Nose Technology. *Sensors* **2019**, *19*, 5033. [CrossRef] [PubMed]
10. Akter, S.; Cortus, E.L. Comparison of Hydrogen Sulfide Concentrations and Odor Annoyance Frequency Predictions Downwind from Livestock Facilities. *Atmosphere* **2020**, *11*, 249. [CrossRef]
11. Wang, Y.-C.; Han, M.-F.; Jia, T.-P.; Hu, X.-R.; Zhu, H.-Q.; Tong, Z.; Lin, Y.-T.; Wang, C.; Liu, D.-Z.; Peng, Y.-Z.; et al. Emissions, measurement, and control of odor in livestock farms: A review. *Sci. Total Environ.* **2021**, *776*, 145735. [CrossRef] [PubMed]
12. Saidi, T.; Moufid, M.; Zaim, O.; El Bari, N.; Bouchikhi, B. Voltammetric electronic tongue combined with chemometric techniques for direct identification of creatinine level in human urine. *Measurement* **2018**, *115*, 178–184. [CrossRef]
13. Diouf, A.; Bouchikhi, B.; El Bari, N. A nonenzymatic electrochemical glucose sensor based on molecularly imprinted polymer and its application in measuring saliva glucose. *Mater. Sci. Eng. C* **2019**, *98*, 1196–1209. [CrossRef] [PubMed]
14. Wasilewski, T.; Migoń, D.; Gębicki, J.; Kamysz, W. Critical review of electronic nose and tongue instruments prospects in pharmaceutical analysis. *Anal. Chim. Acta* **2019**, *1077*, 14–29. [CrossRef] [PubMed]
15. Tahri, K.; Duarte, A.A.; Carvalho, G.; Ribeiro, P.A.; Da Silva, M.G.; Mendes, D.; El Bari, N.; Raposo, M.; Bouchikhi, B. Distinguishment, identification and aroma compound quantification of Portuguese olive oils based on physicochemical attributes, HS-GC/MS analysis and voltammetric electronic tongue. *J. Sci. Food Agric.* **2018**, *98*, 681–690. [CrossRef] [PubMed]
16. Bonfanti, M.E. From sniffer dogs to emerging sniffer devices for airport security: An opportunity to rethink privacy implications? *Sci. Eng. Ethics* **2014**, *20*, 791–807. [CrossRef] [PubMed]
17. Wold, S.; Esbensen, K.; Geladi, P. Principal component analysis. *Chemom. Intell. Lab. Syst.* **1987**, *2*, 37–52. [CrossRef]
18. Brown, M.T.; Tinsley, H.E. Discriminant analysis. *J. Leis. Res.* **1983**, *15*, 290–310. [CrossRef]
19. Ivanciuc, O. Applications of Support Vector Machines in Chemistry. *Rev. Comput. Chem.* **2007**, *23*, 291–400. [CrossRef]
20. Park, J.; Kang, T.; Heo, Y.; Lee, K.; Kim, K.; Lee, K.; Yoon, C. Evaluation of short-term exposure levels on ammonia and hydrogen sulfide during manure-handling processes at livestock farms. *Saf. Health Work* **2020**, *11*, 109–117. [CrossRef] [PubMed]

chemistry proceedings

MDPI

Abstract

Europium-Doped Ceria Nanocrystals as Nanozyme Fluorescent Probes for Biosensing †

Ali Othman, Akhtar Hayat and Silvana Andreescu *

Department of Chemistry and Biomolecular Science, Clarkson University, Potsdam, NY 13699-5810, USA; aothman@clarkson.edu (A.O.); akhtarhayat@cuilahore.edu.pk (A.H.)

* Correspondence: eandrees@clarkson.edu

† Presented at the 1st International Electronic Conference on Chemical Sensors and Analytical Chemistry, 1–15 July 2021; Available online: https://csac2021.sciforum.net/.

Abstract: Molecular nanoprobes with intrinsic enzyme-like activity represent a new wave of technology for rapid and sensitive detection of molecular targets. This work reports synthesis and characterization of novel and well-dispersed europium-doped ceria nanocrystals (EuCe NCs) with self-integrated catalytic and fluorescence sensing functions. The NCs have an average size of ~5 nm and exhibit bright and stable fluorescence for more than 6 months in aqueous media. Their dual cooperative function as both a catalyst and fluorescent probe was explored to develop a universally applicable fluorescence-based biosensing method to monitor enzyme reactions and quantitatively measure clinically relevant molecules. Sensing capabilities are demonstrated for detection of H_2O_2, glucose/glucose oxidase, lactate/lactate oxidase, phosphatase activity, and the catecholamine neurotransmitter, dopamine. Results indicate that EuCe NCs not only provide high enzyme-mimetic activity, but also impart direct fluorescence sensing ability enabling all-in-one recognition, catalytic amplification, and the detection of biomolecular targets. The EuCe nanozyme offers a stable alternative to the more complex systems based on the combined use of natural enzymes and fluorescent dyes. The high stability and fluorescence detection capabilities demonstrate that EuCe NCs have the potential to be used as a generic platform in chemical and biological sensing and bioimaging applications.

Keywords: ceria nanocrystals; europium doping; nanozyme; fluorescent probe; bioanalytical applications

Citation: Othman, A.; Hayat, A.; Andreescu, S. Europium-Doped Ceria Nanocrystals as Nanozyme Fluorescent Probes for Biosensing. *Chem. Proc.* **2021**, *5*, 53. https://doi.org/10.3390/CSAC2021-10549

Academic Editor: Huangxian Ju

Published: 1 July 2021

Supplementary Materials: The following supporting information can be downloaded at: https://www.mdpi.com/article/10.3390/CSAC2021-10549/s1.

chemistry proceedings

MDPI

Proceeding Paper

Electrode Modified with Tin(IV) Oxide Nanoparticles and Surfactants as Sensitive Sensor for Hesperidin †

Elvira Yakupova * and Guzel Ziyatdinova

Analytical Chemistry Department, Kazan Federal University, Kremleyevskaya, 18, 420008 Kazan, Russia; Ziyatdinovag@mail.ru

* Correspondence: elviraeakupova96@mail.ru

† Presented at the 1st International Electronic Conference on Chemical Sensors and Analytical Chemistry, 1–15 July 2021; Available online: https://csac2021.sciforum.net/.

Abstract: Tin(IV) oxide nanoparticles in combination with surfactants were used as a sensitive layer in a sensor for hesperidin. The effect of the surfactant's nature and concentration on the hesperidin response was evaluated. The best parameters were registered in the case of 500 μM cetylpyridinium bromide (CPB) as a dispersive agent. The SEM and electrochemical data confirmed the increase in sensor surface effective area and electron transfer rate. The sensor gave a linear response to hesperidin in the ranges of 0.10–10 and 10–75 μM with a detection limit of 77 nM. The approach was successfully tested on orange juices and validated using ultra-HPLC.

Keywords: electrochemical sensors; metal oxide nanoparticles; surfactants; flavonoids; food analysis

Citation: Yakupova, E.; Ziyatdinova, G. Electrode Modified with Tin(IV) Oxide Nanoparticles and Surfactants as Sensitive Sensor for Hesperidin. *Chem. Proc.* **2021**, *5*, 54. https://doi.org/10.3390/CSAC2021-10615

Academic Editor: Ye Zhou

Published: 6 July 2021

1. Introduction

Hesperidin (Figure 1) is the major flavonoid of *Citrus* L. fruits [1], with a wide spectrum of biological activity that results in its application in medicine [2]. However, it can show a pro-oxidant effect in high concentrations that is typical for phenolic antioxidants [3]. Therefore, simple, sensitive, and selective methods for hesperidin determination are required.

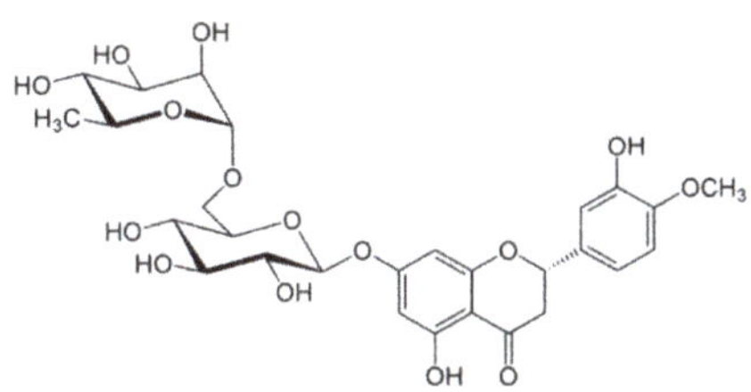

Figure 1. Hesperidin structure.

Electrochemical sensors can be successfully used for this purpose due to the ability of hesperidin to be oxidized at the electrode surface. The advantages of electrochemical methods, such as their simplicity, portability, and cost-efficiency in combination with reliability, sensitivity, and sufficient selectivity make them an attractive tool for practical applications. Nevertheless, hesperidin is almost disregarded as an analyte in electroanalysis compared to other natural flavonoids. Thus, the development of electrochemical sensors for hesperidin quantification is of interest from scientific and practical points of view.

The hanging drop mercury electrode has been used for hesperidin quantification [4,5]. Significant interference effects from a wide range of inorganic and organic compounds of different classes, as well as the toxicity of mercury, make these methods inapplicable in laboratory practice. Boron-doped diamond [6] and pencil graphite [7] electrodes allow the limitations mentioned above to be partially overcome, although the selectivity of the hesperidin response is still insufficient. At the present time, chemically modified electrodes are applied in hesperidin analysis [8–15] since they provide higher sensitivity and selectivity of response. The analytical characteristics of hesperidin reported for the existing electrochemical sensors are presented in Table 1. Nevertheless, the selectivity of the sensor's response is often not considered or is insufficient, and there is limited applicability to real samples. The narrow linear dynamic range of the sensor response in some cases also complicates the analysis of real samples. Therefore, further development of electrochemical sensors for hesperidin that are free of these limitations is required.

Table 1. Figures of merit of electrochemical sensors for hesperidin.

Transducer	Modifier (from Inner to Outer Layer)	Detection Mode	Linear Range/μM	Limit of Detection/nM	Ref.
BDDE [1]	–	AdSSWV [2]	4.19–115	1200	[6]
PGE [3]	–	DPV [4] AdADPV [5]	0.1–12 0.05–1.0	85.8 19.0	[7]
BPPGE [6]	Multi-walled carbon nanotubes	AdSSWV	0.02–0.4 and 0.4–30	7.3	[8]
GCE [7]	Single-walled carbon nanotubes/Electrochemically reduced graphene oxide	LSV [8]	0.05–3.0	20	[9]
CPE [9]	Mesoporous SiO_2 nanoparticles	AdADPV	0.5–25	250	[10]
GCE	Reduced graphene oxide/Gold nanoparticles	A [10]	0.050–8.0	8.2	[11]
GCE	Ultrafine activated carbon/Gold nanoparticles/Poly-*o*-aminothiophenol based molecularly imprinted polymer	DPV	0.080–30	45	[12]
GCE	Polyaminobenzene sulfonic acid functionalized single-walled carbon nanotubes/Polyaluminon	DPV	0.10–2.5 and 2.5–25	29	[13]
CPE	Nano-graphene-platelet/Brilliant green composite	DPV	0.1–7.0 and 7.0–100.0	50	[14]
PGE	Electrochemically reduced graphene oxide/Poly(2,6-pyridinedicarboxylic acid)/dsDNA	DPV	0.82–82	240	[15]

[1] Boron-doped diamond electrode. [2] Adsorptive stripping square-wave voltammetry. [3] Pencil graphite electrode. [4] Differential pulse voltammetry. [5] Adsorptive anodic differential pulse voltammetry. [6] Basal-plane pyrolytic graphite electrode. [7] Glassy carbon electrode. [8] Linear sweep voltammetry. [9] Carbon paste electrode. [10] Amperometry.

Electrochemically inert metal oxide nanoparticles (CeO_2, TiO_2, ZnO, SnO_2, Fe_3O_4, etc.) are prospective nanomaterials widely used for voltammetric sensor creation [16–22]. The combination of metal oxide nanoparticles with surfactants as dispersive agents has led to significant improvement in the voltammetric response of phenolic antioxidants [16–19], caused by stabilization of the nanoparticle dispersions on the one hand and, on the other hand, preconcentration of the analyte at the sensitive layer of the sensor surface via electrostatic or hydrophobic interactions. Another important aspect to be taken into account is the increase in the sensor conductivity due to the presence of surfactants since the metal oxide nanoparticles mentioned above are semiconductors. This type of electrode surface modifier has been successfully applied in the electroanalysis of natural phenolic antioxidants, particularly eugenol [16], thymol [17], quercetin and rutin [18], vanillin [19], and gallic acid [20,21], and in the simultaneous detection of synapic and syringic acids and rutin [22]. The sensors show high sensitivity and selectivity for the response and are easy to prepare, which is an advantage over other modified electrodes. No electrochemical sensors based on metal oxide nanoparticles for hesperidin determination have been reported to date, although it is of practical interest.

The current study was focused on the creation and application of a novel voltammetric sensor for hesperidin based on a glassy carbon electrode (GCE) modified with tin(IV) oxide nanoparticles and surfactants. Attention was paid to the evaluation of the effect of the surfactant's nature and concentration on the hesperidin voltammetric response. The electrodes under investigation were characterized by scanning electron microscopy (SEM) and electrochemical methods. The analytical aspects of hesperidin detection are discussed.

2. Materials and Methods

Hesperidin (94% purity) from Sigma (Steinheim, Germany) was used as a standard. A 0.40 mM stock solution was prepared in methanol (cp grade). Naringin (95% purity), 99% ascorbic and 98% caffeic acids, 95% quercetin trihydrate, and 85% morin hydrate from Sigma (Steinheim, Germany), 95% chlorogenic acid from Aldrich (Steinheim, Germany), and 97% rutin trihydrate from Alfa Aesar (Heysham, UK) were used in the interference test. For these, 10 mM stock solutions in methanol were prepared in 5.0 mL flasks. Less-concentrated solutions were obtained by exact dilution.

Tin(IV) oxide nanoparticles ($D < 100$ nm) were purchased from Aldrich (Steinheim, Germany). Their 1 mg mL^{-1} dispersions in water and the surfactants were prepared by sonication for 10 min in a WiseClean WUC-A03H ultrasonic bath (DAIHAN Scientific Co., Ltd., Wonju-si, Korea). Cetylpyridinium bromide (CPB) (98% purity), 97% *N*-lauroylsarcosine sodium salt (LSS), and Triton X-100 from Aldrich (Steinheim, Germany), 99% cetyltrimethylammonium bromide (CTAB) and Brij® 35 from Acros Organics (Geel, Belgium), sodium dodecylsulfate (SDS) (Ph. Eur.) from Panreac (Barcelona, Spain), and cetyltriphenylphosphonium bromide (CTPPB) synthesized in the Department of Organoelement Compounds Chemistry of Kazan Federal University were used as dispersive agents. Their 1.0 mM solutions were prepared in distilled water.

Other reagents were chemical grade purity. Double-distilled water was used for the measurements. The experiments were carried out at laboratory temperature (25 ± 2 °C).

Voltammetric measurements were carried out on an Autolab PGSTAT12 potentiostat/galvanostat (Eco Chemie B.V., Utrecht, The Netherlands) with GPES software, version 4.9.005. Electrochemical impedance spectroscopy was performed on an Autolab PGSTAT302N potentiostat/galvanostat with a FRA32M module (Eco Chemie B.V., Utrecht, Netherlands) and NOVA 1.10.1.9 software. A 10 mL glassy electrochemical cell with working GCE with a 7.07 mm^2 geometric surface area (BASi® Inc., West Lafayette, IN, USA) or a modified electrode, a silver–silver-chloride-saturated KCl reference electrode, and a platinum wire as the counter electrode was used.

An Expert-001 pH meter (Econix-Expert Ltd., Moscow, Russian Federation) equipped with the glassy electrode was used for pH measurements.

SEM was carried out on a Merlin™ high-resolution field-emission scanning electron microscope (Carl Zeiss, Oberkochen, Germany) at an accelerating voltage of 5 kV and an emission current of 300 pA.

3. Results and Discussion

3.1. Voltammetric Characteristics of Hesperidin on Modified Electrodes

The voltammetric behavior of hesperidin on bare GCE and the modified electrodes was studied in 0.1 M phosphate buffer at pH 7.0. Hesperidin is irreversibly oxidized in two steps. The second step is less pronounced. Therefore, the first oxidation peak was used (Table 2). Modification of the electrode surface with tin(IV) oxide nanoparticles provided an insignificant increase in the hesperidin oxidation currents. Furthermore, these values were still insufficient for sensitive hesperidin quantification. The use of surfactants provided stabilization of the nanoparticle dispersions and preconcentration of the hesperidin on the electrode surface via hydrophobic interaction, leading to an increase in the oxidation currents for all the surfactants under investigation. The effect of surfactant concentration in the range of 10–500 μM on the hesperidin response was evaluated. The oxidation potentials were cathodically shifted. The oxidation currents were statistically significantly increased. Higher oxidation currents were obtained in the case of cationic surfactants. The best hesperidin response was registered on the sensor based on tin(IV) oxide nanoparticles dispersed in 500 μM cetylpyridinium bromide. This agrees well with literature data for a cerium-dioxide-nanoparticles-based sensor for eugenol [16] and a tin-dioxide-nanoparticles-based sensor for vanillin [19].

Table 2. Voltammetric characteristics of hesperidin on GCE and modified electrodes ($n = 5$; $p = 0.95$).

Electrode	$c_{surfactant}$/mM	E_{ox1}/V	I_{ox1}/μA
GCE	0	0.623	0.100 ± 0.004
SnO_2-H_2O/GCE	0	0.603	0.120 ± 0.003
SnO_2-SDS/GCE	0.1	0.573	0.300 ± 0.006
SnO_2-LSS/GCE	0.1	0.553	0.254 ± 0.005
SnO_2-Triton X100/GCE	0.1	0.573	0.333 ± 0.008
SnO_2-Brij® 35/GCE	0.1	0.573	0.171 ± 0.004
SnO_2-CPB/GCE	0.1	0.583	0.440 ± 0.009
SnO_2-CTAB/GCE	0.1	0.563	0.271 ± 0.006
SnO_2-CTPPB/GCE	0.1	0.593	0.323 ± 0.008

3.2. Electrodes Characterization via SEM and Electrochemical Methods

SEM shows the presence of spherical and rhomboid structures and their aggregates of size 30–200 nm for SnO_2-H_2O/GCE, in contrast to the relatively smooth surface of GCE (Figure 2a,b). The application of CPB as dispersive agent provides more uniform coverage consisting of spherical particles of size 20–40 nm forming a porous surface leading to an increase in the electrode surface area (Figure 2c).

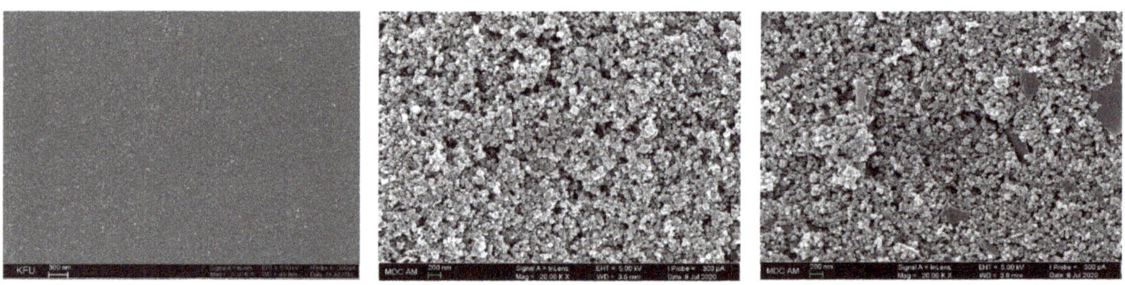

(**a**) (**b**) (**c**)

Figure 2. SEM characterization of bare GCE (**a**), SnO_2-H_2O/GCE (**b**), and SnO_2-CPB/GCE (**c**).

The electroactive surface area of the modified electrode is significantly increased compared with bare GCE (34.7 ± 0.3 mm^2 for SnO_2-CPB/GCE and 8.9 ± 0.3 mm^2 for GCE), as confirmed by cyclic voltammetry (for SnO_2-CPB/GCE) and chronoamperometry (for GCE and SnO_2-H_2O/GCE) using $[Fe(CN)_6]^{4-}$ ions as a standard. These data explain the increase in hesperidin oxidation currents on the modified electrode. Electrochemical impedance spectroscopy was performed in the presence of $[Fe(CN)_6]^{4-/3-}$ as a redox probe at 0.23 V. Fitting of the impedance spectra using the Randles equivalent circuit showed 554-fold less charge transfer resistance for the modified electrode in comparison to GCE, indicating a dramatic increase in the electron transfer rate. The constant phase element value for SnO_2-CPB/GCE was 4.7-fold higher than for the GCE due to the porous structure of the modified electrode and the increase in the surface total charge due to the presence of positively charged CPB. Thus, the developed sensor can be considered as a candidate for analytical applications.

3.3. Analytical Characterization of the Sensor

Hesperidin quantification using the developed sensor was performed in adsorptive differential pulse mode since surface-controlled electro-oxidation has been proved. The highest oxidation currents for hesperidin were obtained in 0.1 M phosphate buffer at pH 7.0. The variation of preconcentration time at the open circuit potential showed the highest oxidation currents for 120 s of accumulation. The evaluation of the effect of the pulse parameters showed that the best response was registered at a pulse amplitude of 100 mV and a pulse time of 50 ms.

The sensor gave a linear response to hesperidin in the ranges of 0.10–10 and 10–75 μM (Figure 3) with a detection limit of 77 nM. The calibration-plot parameters are presented in Table 3.

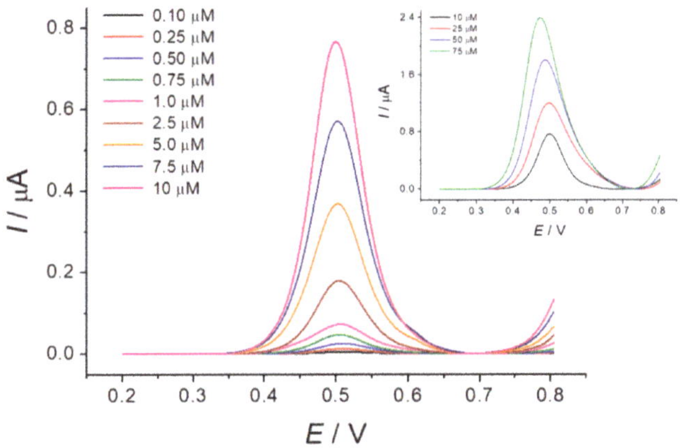

Figure 3. Baseline-corrected differential pulse voltammograms of hesperidin on SnO_2-CPB/GCE in 0.1 M phosphate buffer at pH 7.0, with t_{acc} = 120 s, ΔE_{pulse} = 100 mV, t_{pulse} = 50 ms, υ = 10 mV s^{-1}.

Table 3. Hesperidin calibration-plot parameters (I = a + bc(M)).

Linear Dynamic Range/μM	A ± SD/μA	(b ± SD) × 10^{-3}/μA M^{-1}	R^2
0.10–10	−0.009 ± 0.002	78.0 ± 0.4	0.9998
10–75	0.52 ± 0.01	25.3 ± 0.2	0.9997

The analytical characteristics obtained were significantly better [10,15] or comparable to other sensors based on the modified electrodes [12–14]. However, the sensor developed is simpler, relatively cheaper, and less tedious to prepare. The accuracy of the hesperidin determination was tested for the model solutions using the added–found method (Table 4). The recovery of 98.4–100% confirmed the high accuracy of the developed sensor. The relative standard deviation was less than 3.5%, indicating the absence of random errors of quantification and the high reproducibility of the sensor response since surface renewal was performed after each measurement.

Table 4. Quantification of hesperidin in model solutions (n = 5; p = 0.95).

Added/μg	Found/μg	RSD/%	R/%
0.244	0.24 ± 0.01	3.5	98.4
1.83	1.82 ± 0.05	2.1	99.5
12.2	12.2 ± 0.1	0.92	100.0
24.4	24.4 ± 0.4	1.3	100.0
183	183 ± 2	0.41	100.0

The sensor selectivity in the presence of a 1000-fold excess of inorganic ions (K^+, Mg^{2+}, Ca^{2+}, NO_3^-, Cl^-, and SO_4^{2-}), glucose, rhamnose, and sucrose, as well as a 1000-fold excess of ascorbic acid was demonstrated. Another important advantage was the high selectivity to hesperidin in the presence of other flavonoids and phenolic acids. A 10-fold excess of naringin, quercetin, rutin, morin, and caffeic and chlorogenic acids, despite the fact that they are electroactive, did not result in an interference effect in the hesperidin response.

3.4. Application to Real Samples

The sensor's applicability to the analysis of real samples was successfully tested on orange juices. The following sample preparation was applied before the measurements: 6 mL of juice was mixed with 6 mL of methanol, sonicated for 15 min, and filtered through 0.45 μm pore size nylon membrane filters [23].

There is a well-defined oxidation peak of hesperidin on the differential pulse voltammograms of orange juices (commercial and fresh) that was confirmed by the standard addition method (Figure 4). Recovery values of 99–100% indicated the absence of matrix effects in the determination.

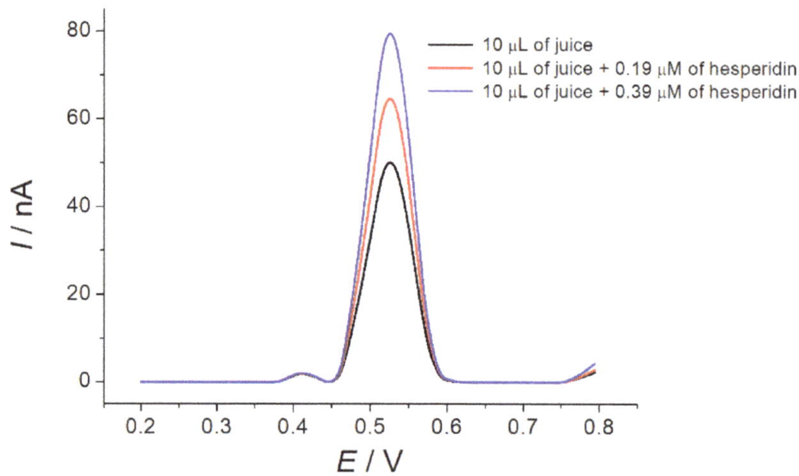

Figure 4. Typical baseline-corrected differential pulse voltammograms of orange juice at the SnO_2-CPB/GCE in 0.1 M phosphate buffer at pH 7.0, with t_{acc} = 120 s, ΔE_{pulse} = 100 mV, t_{pulse} = 50 ms, υ = 10 mV s^{-1}.

The results of the hesperidin quantification in orange juices using the developed sensor are presented in Figure 5. Validation with independent ultra-HPLC with mass-spectrometric detection results was performed (Figure 5). The relative standard deviation for both methods did not exceed 2%, proving the absence of random errors. The *t*-test values (0.290–1.08) were less than the critical value of 2.45, confirming the absence of systematic errors in the determination. Similarly, the *F*-test results (1.17–2.57) were less than critical value of 6.59, indicating the uniform precision of the methods used.

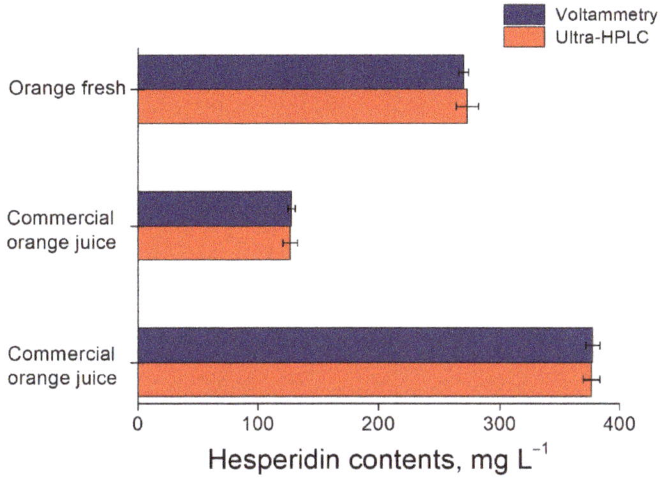

Figure 5. Quantification of hesperidin in orange juices using electrochemical sensor and ultra-HPLC.

4. Conclusions

A sensor based on tin(IV) oxide nanoparticles and CPB provide an improvement in the voltammetric and analytical characteristics of hesperidin. The surfactant provides stabilization of the nanomaterial dispersion and the accumulation of analyte on the sensor surface. The novel voltammetric sensor is highly sensitive, selective, and reliable and can be recommended for the preliminary screening of citrus juices as an alternative to chromatography.

Author Contributions: Conceptualization, G.Z.; methodology, G.Z. and E.Y.; investigation, E.Y.; writing—original draft preparation, G.Z.; writing—review and editing, G.Z.; visualization, G.Z. and E.Y. All authors have read and agreed to the published version of the manuscript.

Funding: This research received no external funding.

Institutional Review Board Statement: Not applicable.

Informed Consent Statement: Not applicable.

Data Availability Statement: The data presented in this study are available on request from the corresponding author.

Acknowledgments: The authors thank Irina Galkina (Department of High Molecular and Organoelement Compounds, Kazan Federal University) for the synthesis and granting of CTPPB, Rustam Davletshin (Department of High Molecular and Organoelement Compounds, Kazan Federal University) for the chromatographic measurements, and Aleksei Rogov (Laboratory of Scanning Electron Microscopy, Interdisciplinary Center for Analytical Microscopy, Kazan Federal University) for the SEM measurements.

Conflicts of Interest: The authors declare no conflict of interest.

References

1. Khan, M.K.; Zill-E-Huma; Dangles, O. A comprehensive review on flavanones, the major citrus polyphenols. *J. Food Compos. Anal.* **2014**, *33*, 85–104. [CrossRef]
2. Li, C.; Schluesener, H. Health-promoting effects of the citrus flavanone hesperidin. *Crit. Rev. Food Sci. Nutr.* **2017**, *57*, 613–631. [CrossRef]
3. Ziyatdinova, G.; Budnikov, H. Natural phenolic antioxidants in bioanalytical chemistry: State of the art and prospects of development. *Russ. Chem. Rev.* **2015**, *84*, 194–224. [CrossRef]
4. Obendorf, D.; Reichart, E. Determination of hesperidin by catholic stripping voltammetry in orange juice and helopyrin, a phytopharmaceutical preparation. *Electroanalysis* **1995**, *7*, 1075–1081. [CrossRef]

5. Temerk, Y.M.; Ibrahim, M.S.; Kotb, M. Square-wave cathodic adsorptive stripping voltammetric determination of 3-hydroxyflavone, morin and hesperidin in bulk form and biological fluids in absence and presence of Cu(II). *J. Braz. Chem. Soc.* **2011**, *22*, 2056–2064. [CrossRef]
6. Yiğit, A.; Yardım, Y.; Şentürk, Z. Square-wave adsorptive stripping voltammetric determination of hesperidin using a boron-doped diamond electrode. *J. Anal. Chem.* **2020**, *75*, 653–661. [CrossRef]
7. David, I.G.; Numan, N.; Buleandră, M.; Popa, D.-E.; Lițescu, S.C.; Riga, S.; Ciobanu, A.M. Rapid voltammetric screening method for the assessment of bioflavonoid content using the disposable bare pencil graphite electrode. *Chemosensors* **2021**, *9*, 323. [CrossRef]
8. Sims, M.J.; Li, Q.; Kachoosangi, R.T.; Wildgoose, G.G.; Compton, R.G. Using multiwalled carbon nanotube modified electrodes for the adsorptive striping voltammetric determination of hesperidin. *Electrochim. Acta* **2009**, *54*, 5030–5034. [CrossRef]
9. Wu, J.; Wang, L.; Wang, Q.; Zou, L.; Ye, B. The novel voltammetric method for determination of hesperetin based on a sensitive electrochemical sensor. *Talanta* **2016**, *150*, 61–70. [CrossRef] [PubMed]
10. Sun, D.; Wang, F.; Wu, K.; Chen, J.; Zhou, Y. Electrochemical determination of hesperidin using mesoporous SiO_2 modified electrode. *Microchim. Acta* **2009**, *167*, 35–39. [CrossRef]
11. Gao, Y.; Wu, X.; Wang, H.; Lu, W.; Guo, M. Highly sensitive detection of hesperidin using AuNPs/rGO modified glassy carbon electrode. *Analyst* **2018**, *143*, 297–303. [CrossRef]
12. Sun, B.; Hou, X.; Li, D.; Gou, Y.; Hu, F.; Li, W.; Shi, X. Electrochemical sensing and high selective detection of hesperidin with molecularly imprinted polymer based on ultrafine activated carbon. *J. Electrochem. Soc.* **2019**, *166*, B1644–B1652. [CrossRef]
13. Zhupanova, A.; Guss, E.; Ziyatdinova, G.; Budnikov, H. Simultaneous voltammetric determination of flavanones using an electrode based on functionalized single-walled carbon nanotubes and polyaluminon. *Anal. Lett.* **2020**, *53*, 2170–2189. [CrossRef]
14. Manasa, G.; Mascarenhas, R.J.; Bhakta, A.K.; Mekhalif, Z. Nano-graphene-platelet/Brilliant-green composite coated carbon paste electrode interface for electrocatalytic oxidation of flavanone hesperidin. *Microchem. J.* **2021**, *160 B*, 105768. [CrossRef]
15. Tığ, G.A.; Bolat, E.Ö.; Zeybek, B.; Pekyardımcı, Ş. Hesperidin-dsDNA interaction based on electrochemically reduced graphene oxide and poly-(2,6-pyridinedicarboxylic acid) modified glassy carbon electrode. *Hacet. J. Biol. Chem.* **2016**, *44*, 487–497.
16. Ziyatdinova, G.; Ziganshina, E.; Romashkina, S.; Budnikov, H. Highly sensitive amperometric sensor for eugenol quantification based on CeO_2 nanoparticles and surfactants. *Electroanalysis* **2017**, *29*, 1197–1204. [CrossRef]
17. Ziyatdinova, G.; Ziganshina, E.; Nguyen Cong, P.; Budnikov, H. Voltammetric determination of thymol in oregano using CeO_2-modified electrode in Brij® 35 micellar medium. *Food Anal. Meth.* **2017**, *10*, 129–136. [CrossRef]
18. Ziyatdinova, G.K.; Zakharova, S.P.; Ziganshina, E.R.; Budnikov, H.C. Voltammetric determination of flavonoids in medicinal plant materials using electrodes modified by cerium dioxide nanoparticles and surfactants. *J. Anal. Chem.* **2019**, *74*, 816–824. [CrossRef]
19. Ziyatdinova, G.K.; Antonova, T.S.; Mubarakova, L.R.; Budnikov, H.C. An amperometric sensor based on tin dioxide and cetylpyridinium bromide nanoparticles for the determination of vanillin. *J. Anal. Chem.* **2018**, *73*, 801–808. [CrossRef]
20. Chikere, C.; Faisal, N.H.; Lin, P.K.T.; Fernandez, C. Zinc oxide nanoparticles modified-carbon paste electrode used for the electrochemical determination of gallic acid. *J. Phys. Conf. Ser.* **2019**, *1310*, 012008. [CrossRef]
21. Tashkhourian, J.; Nami Ana, S.F.; Hashemnia, S.; Hormozi-Nezhad, M.R. Construction of a modified carbon paste electrode based on TiO_2 nanoparticles for the determination of gallic acid. *J. Solid State Electrochem.* **2013**, *17*, 157–165. [CrossRef]
22. Pwavodi, P.C.; Ozyurt, V.H.; Asir, S.; Ozsoz, M. Electrochemical sensor for determination of various phenolic compounds in wine samples using Fe_3O_4 nanoparticles modified carbon paste electrode. *Micromachines* **2021**, *12*, 312. [CrossRef] [PubMed]
23. Feng, X.; Zhang, Q.; Cong, P.; Zhu, Z. Simultaneous determination of flavonoids in different citrus fruit juices and beverages by high-performance liquid chromatography and analysis of their chromatographic profiles by chemometrics. *Anal. Methods* **2012**, *4*, 3748–3753. [CrossRef]

chemistry proceedings

MDPI

Proceeding Paper

Optical Biosensor for the Detection of Hydrogen Peroxide in Milk †

Helena Vasconcelos [1,2,*], Ana Matias [2], Pedro Jorge [2], Cristina Saraiva [1], João Mendes [2], João Araújo [2], Bernardo Dias [2], Paulo Santos [2], José M. M. M. Almeida [2,3] and Luís C. C. Coelho [2]

1 School of Agrarian and Veterinary Sciences, University of Trás-os-Montes e Alto Douro, 5001-801 Vila Real, Portugal; crisarai@utad.pt

2 INESC TEC-Institute for Systems and Computer Engineering, Technology and Science and Faculty of Sciences, University of Porto, 4169-007 Porto, Portugal; ana.b.teixeira@inesctec.pt (A.M.); pedro.jorge@inesctec.pt (P.J.); joaomendes.quimica@gmail.com (J.M.); up201703919@edu.fc.up.pt (J.A.); up201103504@edu.fc.up.pt (B.D.); paulo.s.santos@inesctec.pt (P.S.); jmmma@utad.pt (J.M.M.M.A.); lcoelho@inesctec.pt (L.C.C.C.)

3 Department of Physics, School of Science and Technology, University of Trás-os-Montes e Alto Douro, 5001-801 Vila Real, Portugal

* Correspondence: helenavasconcelos88@gmail.com

† Presented at the 1st International Electronic Conference on Chemical Sensors and Analytical Chemistry, 1–15 July 2021; Available online: https://csac2021.sciforum.net/.

Abstract: Over the years, the food industry's concern to provide safe food that does not cause harm or illness to consumers has increased. The growing demand for the detection of compounds that can contaminate food is increasingly important. Hydrogen peroxide is frequently used as a substance to control the growth of microorganisms in milk, thus increasing its shelf life. Here, a strategy is presented for the detection of hydrogen peroxide as a milk adulterant, using a single shot membrane sensor. The lowest concentration measured with this technique was 0.002% *w*/*w* of H_2O_2 in semi-fat milk.

Keywords: chemiluminescence; hydrogen peroxide; optical sensor; food safety; food fraud; quality assessment

check for updates

Citation: Vasconcelos, H.; Matias, A.; Jorge, P.; Saraiva, C.; Mendes, J.; Araújo, J.; Dias, B.; Santos, P.; Almeida, J.M.M.M.; Coelho, L.C.C. Optical Biosensor for the Detection of Hydrogen Peroxide in Milk. *Chem. Proc.* **2021**, *5*, 55. https://doi.org/10.3390/CSAC2021-10466

Academic Editors: Huangxian Ju and Elena Benito Peña

Published: 30 June 2021

1. Introduction

Milk is one of the most complete foods for humans, containing nutrients including carbohydrates, proteins, fats, minerals, and vitamins [1].

Owing to its rich composition, milk becomes a substrate for the growth of undesirable microorganisms that can easily deteriorate the product. To prevent this from happening, prohibited substances are fraudulently added [2]. Hydrogen peroxide (H_2O_2), hypochlorite, formaldehyde, potassium dichromate, and salicylic acid are examples of substances used as adulterants that need monitoring and quality control as they are toxic to humans [3].

In the case of H_2O_2, it is widely used in the dairy industry as an antimicrobial agent, thus helping to preserve raw milk in the absence of refrigeration [4]. Despite its conventional use, when added to milk, H_2O_2 can cause a decrease in the nutritional value of the food due to the destruction of vitamins A and E, which generate reactive and cytotoxic oxygen species, including hydroxyl radicals, that can initiate oxidation and damage nucleic acids, lipids, and proteins. Consequently, when ingested, milk can lead to negative effects on the health of the population, especially in immunocompromised people [2,4].

In the USA, hydrogen peroxide is used in cheese production in concentrations up to 0.05% *w*/*w*, however, in other countries, its addition is prohibited due to its toxic effects. A peroxide concentration > 0.1% *w*/*w* has been proven to induce cancer in the duodenum of mice and present short-term genotoxicity [3].

Here, a study is presented for the detection and quantification of H_2O_2 using a chemiluminescence technique. A small, low-cost hydroxyethyl cellulose sensitive membrane combined with a high-sensitive photodetector is used to measure H_2O_2 concentrations in semi-fat milk samples.

2. Materials and Methods

The sensing methodology is based on the detection of a luminescence signal from the chemical reaction within a solid membrane produced with hydroxyethyl cellulose (HEC, Sigma Aldrich, Taufkirchen, Germany), luminol, sodium phosphate, cobalt (II) chloride hexahydrate, sodium lauryl sulphate (SLS), and ethylenediaminetetraacetic acid (EDTA).

The procedure established by Omanovic-Miklicanin [5] was refined to establish experimental protocols. For the determination of H_2O_2 in very low concentrations, the sensor sensitivity should be as high as possible. Therefore, the systematic optimization of the membrane was necessary. Only one constituent was varied at a time, keeping the remaining constituents unchanged. After membrane optimization, the final concentrations of these constituents were set to luminol (0.2 mg), sodium phosphate (8.6 mg), SLS (60 μL, 34.36 mmol/L), cobalt hydroxide (100 μL, 5.0 mmol/L), EDTA (2 μL, 20 μmol/L), and HEC (150 mg) was added to 10 mL of Milli-Q® water.

The membrane solution was placed on a magnetic stirrer for 30 min. Individual 3D printed cups were used, and 1000 μL of membrane solution was added and dried for 4 h (T = 70 °C). After drying, the membranes were stored in a desiccator under a vacuum. For the measurement procedure, the membrane was placed directly onto the membrane holder on top of the detector. The light emission was measured by adding 500 μL of the sample solution.

For straight and rapid spectrophotometric H_2O_2 detection, a detection module was built containing a highly sensitive detection system with a photodiode (model S8746-01 Hamamatsu Japan), a dedicated amplification system with variable gain, and an embed controlling unit. The sensitive optoelectronic system was isolated inside of a custom-made 3D printed case allowing the easy replacement of the sensing membrane and allowing the sample pipetting, preventing the detection of the ambient light. This module was powered with a low noise power source, and the data was acquired and analyzed with a user-friendly graphical interface (GUI) and a raspberry pi (Figure 1).

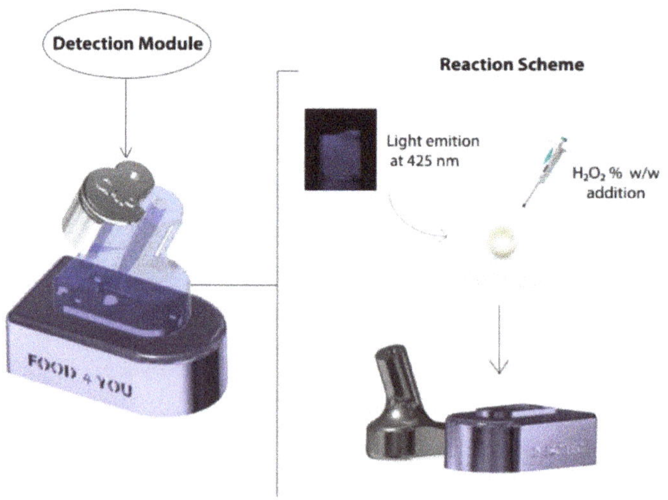

Figure 1. Schematic diagram of the analyte detection.

3. Results and Discussion

Semi-fat milk samples were adulterated with H_2O_2 concentrations from 0.001% *w*/*w* to 0.006% *w*/*w* by diluting a standard 30% *w*/*w* solution of H_2O_2. The variation of the chemiluminescent intensity is presented in Figure 2 for all samples, together with the time integral of the decaying chemiluminescent signal for each H_2O_2 concentration.

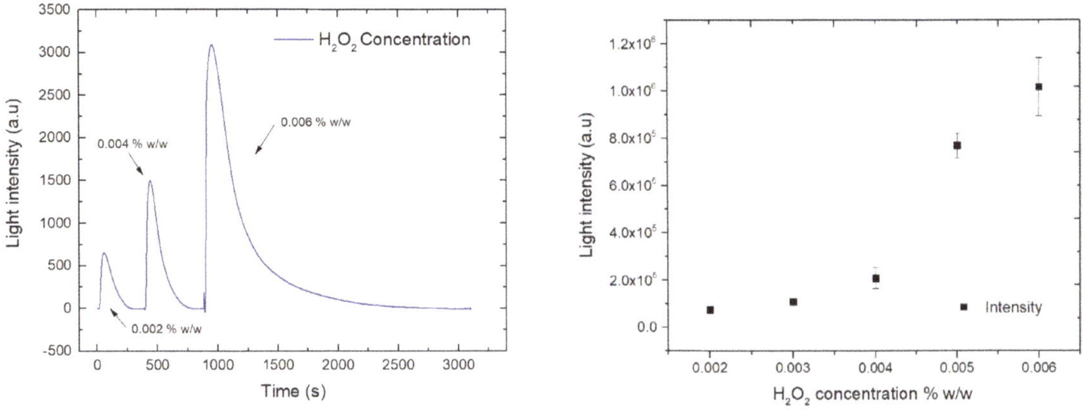

Figure 2. (**a**) Variation of the intensity of the light emission for the concentrations of 0.002, 0.004, and 0.006% *w*/*w* as a function of time; (**b**) Time integral of the decay time for each H_2O_2 concentration.

Taking into consideration that 0.05 % *w*/*w* of H_2O_2 is the defined limit for the FDA in milk for cheese production [6], the developed sensor would be suitable for determinations of H_2O_2 as a fraud controller in milk samples within the legal limits of different countries. Moreover, to achieve a more practical approach to the commonly time-consuming sample preparation methods, the pre-treatment step was successfully eliminated. In fact, the optimized sensor requires minimal solvent use and waste production. When compared with other methods available for the determination of H_2O_2 presence in milk, this portable biosensor is an easy and reliable method that ensures the required sensitivity while offering a low time of analysis and no need for additional laboratory equipment.

The methodology developed and optimized demonstrates that it is possible to detect very low concentrations of H_2O_2 (down to 0.001 % *w*/*w* in an aqueous system). As the H_2O_2 concentration increased, the intensity of the emitted light and the reaction time increased. Low limits of detection were achieved, thus indicating the applicability of this assay to real samples exhibiting the required sensitivity for the analytical determination of H_2O_2 in biological samples such as milk.

In this work, the reaction of H_2O_2 and luminol catalyzed by cobalt hydroxide was used to detect H_2O_2 in milk; however, another spectrophotometric method was described by Lima et al. [2] for the detection of H_2O_2 in milk, using the reaction between hydrogen peroxide and guaiacol, catalyzed by peroxidase, producing a red product, where a low detection limit was obtained.

4. Conclusions

The proposed sensor provided to be a rapid, cost-effective, and environmentally friendly approach for the determination of hydrogen peroxide as a milk adulterant. This optimized and validated method has a very good linearity range when the sample is in its liquid state, where concentrations of H_2O_2 as low as 0.001% *w*/*w* can be detected with good repeatability. As a practical application for this methodology under controlled conditions, an adulterated milk sample was analyzed. Concentrations of H_2O_2 of 0.002% *w*/*w* to 0.006% were detected, and the method was calibrated for semi-fat milk, proving that the limit of detection and linearity range of the proposed method are suitable for the analysis of milk samples in loco, which can add value to the food fraud department. Moreover, the reagents required are commonly used in analytical laboratories, are inexpensive, and can be consumed in low amounts (500 µL), thus resulting in negligible and nontoxic waste generation. In addition to the mentioned advantageous features, the proposed method validation is comparable to those found in the literature.

Supplementary Materials: The supporting information can be downloaded at: https://www.mdpi.com/article/10.3390/CSAC2021-10466/s1.

Author Contributions: Writing—review and editing, H.V., A.M., J.M., J.A., B.D. and P.S.; supervision, L.C.C.C., P.J., C.S. and J.M.M.M.A. All authors have read and agreed to the published version of the manuscript.

Funding: This research was funded by National Funds through the Portuguese funding agency, FCT—Fundação para a Ciência e a Tecnologia, within project UIDB/50014/2020. Helena Vasconcelos acknowledges the support from FCT grant SFRH/BD/120064/2016 and Luís Coelho acknowledges the support from FCT research contract grant CEECIND/00471/2017.

Institutional Review Board Statement: Not applicable.

Informed Consent Statement: Not applicable.

Acknowledgments: This work was financed by National Funds through the Portuguese funding agency, FCT—Fundação para a Ciência e a Tecnologia, within project UIDB/50014/2020. Helena Vasconcelos acknowledges the support from FCT grant SFRH/BD/120064/2016 and Luís Coelho acknowledges the support from FCT research contract grant CEECIND/00471/2017.

Conflicts of Interest: The authors declare no conflict of interest.

References

1. Handford, C.E.; Campbell, K.; Elliott, C.T. Impacts of Milk Fraud on Food Safety and Nutrition with Special Emphasis on Developing Countries. *Compr. Rev. Food Sci. Food Saf.* **2016**, *15*, 130–142. [CrossRef] [PubMed]
2. Lima, L.S.; Rossini, E.L.; Pezza, L.; Pezza, H.R. Bioactive Paper Platform for Detection of Hydrogen Peroxide in Milk. *Spectrochim. Acta—Part A Mol. Biomol. Spectrosc.* **2020**, *227*, 117774. [CrossRef] [PubMed]
3. Lima, M.J.A.; Sasaki, M.K.; Marinho, O.R.; Freitas, T.A.; Faria, R.C.; Reis, B.F.; Rocha, F.R.P. Spot Test for Fast Determination of Hydrogen Peroxide as a Milk Adulterant by Smartphone-Based Digital Image Colorimetry. *Microchem. J.* **2020**, *157*, 105042. [CrossRef]
4. Robinson, B.R.; D'Amico, D.J. Hydrogen Peroxide Treatments for the Control of Listeria Monocytogenes on High-Moisture Soft Cheese. *Int. Dairy J.* **2021**, *114*, 104931. [CrossRef]
5. Omanovic-Miklicanin, E.; Valzacchi, S. Development of New Chemiluminescence Biosensors for Determination of Biogenic Amines in Meat. *Food Chem.* **2017**, *235*, 98–103. [CrossRef] [PubMed]
6. CFR—Code of Federal Regulations Title 21. Available online: https://www.accessdata.fda.gov/scripts/cdrh/cfdocs/cfcfr/cfrsearch.cfm (accessed on 6 June 2021).

chemistry proceedings

Proceeding Paper

Locally Linear Embedding as Nonlinear Feature Extraction to Discriminate Liquids with a Cyclic Voltammetric Electronic Tongue †

Jersson X. Leon-Medina [1,*], Maribel Anaya [2] and Diego A. Tibaduiza [3]

[1] Departamento de Ingeniería Mecánica y Mecatrónica, Universidad Nacional de Colombia, Cra 45 No. 26-85, Bogota 111321, Colombia
[2] MEM (Modelling-Electronics and Monitoring Research Group), Faculty of Electronics Engineering, Universidad Santo Tomás, Bogota 110231, Colombia; maribelanaya@usantotomas.edu.co
[3] Departamento de Ingeniería Eléctrica y Electrónica, Universidad Nacional de Colombia, Cra 45 No. 26-85, Bogota 111321, Colombia; dtibaduizab@unal.edu.co
* Correspondence: jxleonm@unal.edu.co
† Presented at the 1st International Electronic Conference on Chemical Sensors and Analytical Chemistry, 1–15 July 2021; Available online: https://csac2021.sciforum.net/.

Abstract: Electronic tongues are devices used in the analysis of aqueous matrices for classification or quantification tasks. These systems are composed of several sensors of different materials, a data acquisition unit, and a pattern recognition system. Voltammetric sensors have been used in electronic tongues using the cyclic voltammetry method. By using this method, each sensor yields a voltammogram that relates the response in current to the change in voltage applied to the working electrode. A great amount of data is obtained in the experimental procedure which allows handling the analysis as a pattern recognition application; however, the development of efficient machine-learning-based methodologies is still an open research interest topic. As a contribution, this work presents a novel data processing methodology to classify signals acquired by a cyclic voltammetric electronic tongue. This methodology is composed of several stages such as data normalization through group scaling method and a nonlinear feature extraction step with locally linear embedding (LLE) technique. The reduced-size feature vector input to a *k*-Nearest Neighbors (*k*-NN) supervised classifier algorithm. A leave-one-out cross-validation (LOOCV) procedure is performed to obtain the final classification accuracy. The methodology is validated with a data set of five different juices as liquid substances.Two screen-printed electrodes voltametric sensors were used in the electronic tongue. Specifically the materials of their working electrodes were platinum and graphite. The results reached an 80% classification accuracy after applying the developed methodology.

Keywords: electronic tongue; locally linear embedding; cyclic voltammetry; *k*-Nearest Neighbors; classification; machine learning

Citation: Leon-Medina, J.X.; Anaya, M.; Tibaduiza, D.A. Locally Linear Embedding as Nonlinear Feature Extraction to Discriminate Liquids with a Cyclic Voltammetric Electronic Tongue. *Chem. Proc.* **2021**, *5*, 56. https://doi.org/10.3390/CSAC2021-10426

Academic Editors: Huangxian Ju and Manel del Valle

Published: 30 June 2021

1. Introduction

Discriminating between different types of liquid substance is a daily task in the food industry. This procedure can be used to preserve the flavor of a product, identify adulterations, confirm the presence of a specific liquid, among others [1]. Generally, the analysis of liquid food products is carried out using a panel of previously trained experts [2] who allow tasting and identifying a specific flavor. This through the training of the human sense of taste. However, over time this ability may be deteriorated and human reliability may be a risk factor for the process. Another method used in the analysis of liquids is high-performance liquid chromatography (HPLC) [3], but this type of analysis is expensive and must be performed in laboratories with specialized equipment. As an alternative to the two mentioned methods, the electronic tongue sensor array has emerged because its advantages such as portability, reliability and low price [4]. Inspired by the human sense of taste and the behavior of taste buds, electronic tongues use an array of non-selective sensors to capture signals from a specific liquid. An electronic tongue uses sensors of different materials and subsequently a sensor data fusion analysis based on pattern recognition algorithms to perform classification tasks of different liquids.

One of the applications of electronic tongue is the discrimination of different fruit juices. For example, in 2011 Dias et al. [5] developed a potentiometric electronic tongue using linear discriminant analysis (LDA) to differentiate four beverage groups including juices of orange, pineapple, mango and peach. In other work, eleven fruit juice varieties were correctly classified by a potentiometric electronic tongue using a Fuzzy ARTMAP neural network [6]. Several sensors composed the electronic tongue sensor arrays; thus, the datasets acquired are very large in size. To deal with this inconvenience, in 2012, Kiranmayee et al. [7] developed a method based on segmentation of the voltammetric signal with the objective to reduce the size of the signal maintaining meaningful information to discriminate the analyzed classes. The developed method was satisfactorily applied to an eight-juices dataset, reducing the data size by 78.94%. A common problem observed in the previous works is that the signals acquired with the electronic tongue have a high dimensionality. This work presents a novel methodology to correctly classify the signals acquired from a cyclic voltammetric electronic tongue.

In this work, the cyclic voltammetry technique was used to perform experiments on five different juices; two screen-printed electrodes (SPE) voltametric sensors were used. The working electrode materials were platinum and graphite. The amount of data captured when performing cyclic voltammetry experiments is high; therefore, these data have high dimensionality. This work uses the Locally Linear Embedding (LLE) [8] method to perform a dimensionality reduction of the original data. This dimensionality reduction serves as feature extraction method that is used as input of a *k*-Nearest Neighbor (*k*-NN) [9] classifier used as supervised machine learning method. In order to classify the five different juices a Leave-One-Out cross validation procedure is executed due to the small quantity of samples in the dataset, along to prevent over-fitting [10]. The results show a correct classification procedure of the juices evidenced with a high classification accuracy. The remainder of this papers is as follows: Section 2 describes the materials and methods including the experimental setup and the cyclic voltammetry tests performed. Following, Section 3 presents the data processing results including data unfolding, data scaling, dimensionality reduction, classification, and cross validation. Finally, the Section 4 outlines the main conclusions of this work.

2. Materials and Methods

2.1. Experimental Setup for the Acquisition of the Juice Dataset

The methodology developed in this work is used to classify 5 different classes of juices. This dataset of juices was obtained by conducting experiments on 5 different juices from a company located in the city of Tunja in the department of Boyacá-Colombia. Cyclic voltammetry tests were performed on each one of the 5 juices. For each juice, five experiments were performed, as shown in Table 1.

Table 1. Description of the type of juice in the dataset registered with the EVAL-AD5940ELCZ potentiostat.

ID	Juice	Number of Samples
1	BICHES FRUITS	5
2	GREEN APPLE	5
3	RED FRUITS	5
4	PASSION FRUIT	5
5	ORANGE	5

Experiments were performed on the different juices using the EVAL-AD5940ELCZ [11] electrochemical evaluation board from Analog Devices. This board is commanded by the evaluation board EVAL-ADICUP3029, which is an Arduino- and PMOD-compatible development board that includes Bluetooth and WiFi connectivity [12]. The EVAL-ADICUP3029 board uses the ADuCM3029 ultra low power Arm Cortex-M3 processor as the main device. The ADuCM3029 is an integrated mixed-signal microcontroller system for processing, control, and connectivity. The integration of the EVAL -AD5940ELCZ and EVAL-ADICUP3029 boards is used as potentiostat equipment. This system provide only 1 channel in such a way that a cyclic voltammogram was obtained at a time, In the experimentation the sensor had to be changed to perform each cyclic voltammetry experiment. This electronic tongue used two screen-printed electrode voltammetric sensors from the BVT technologies company [13]. Specifically, the types of these two sensors were: AC1.W2.R2 DW = 1 and AC1.W4.R2 DW = 1. These type of sensors uses the same material for their working and auxiliary electrodes. The first sensor used as working and auxiliary electrode platinum and the second sensor used graphite. Silver covered by AgCl was used as reference electrode in both sensors. The hardware used to obtain the data set of five juices is depicted in Figure 1 left.

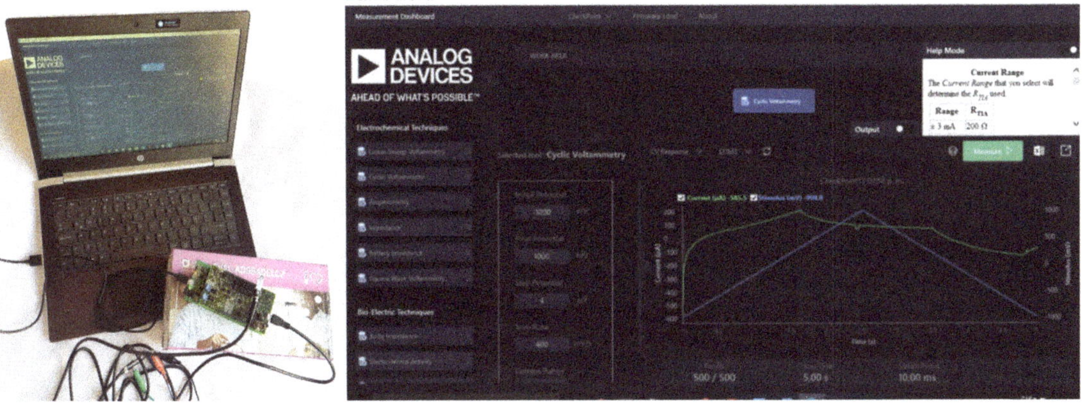

Figure 1. (**Left**) Integrated electronic tongue system: Computer with Sensor Pal software, USB cable to the EVAL-AD5940ELCZ and EVAL-ADICUP3029 attached boards, colored crocodile cable, sensor connection cable and SPE Sensor. (**Right**) Cyclic voltammetry test in Sensor Pal software from Analog Devices, blue line represents the excitation ramp signal, while the green line shows the response signal.

2.2. Cyclic Voltammetry Tests to Obtain the Juice Data Set

The Sensor Pal command software from Analog Devices was used to perform the cyclic voltammetry tests. The parameters used in the development of these experiments are shown in Table 2. The ramp-type drive signal shown in blue in Figure 1 right has a total duration of 4 s. The data points of each voltammogram is equal to 500, since there is a period of 8 ms for each sample. The scan rate used is equal to 500 mV/s. Results shown by the green line in the unfolding voltammogram present data current in the ordinate axis in the order of μA. According to Table 1, five measurements were taken per analyte. In this sense the two sensors are referred to one measure in Table 1. Thus, in total, five measures × 2 sensors = 10 voltammograms were acquired by each juice.

Figures 2 left and right show the cyclic voltammograms for two different juices with both the platinum and graphite sensors in the electronic tongue system by using the boards EVAL-AD5940ELCZ and EVAL-ADICUP3029 as potentiostat. In particular, Figure 2 left depicts the cyclic voltammograms obtained for green apple juice showing that the voltammogram obtained by the graphite sensor reaches higher positive current values than the platinum sensor. In contrast, Figure 2 right shows the cyclic voltammograms for an experiment in red fruit juice, the magnitude of the current obtained by the graphite sensor is clearly lower than with the platinum sensor.

Table 2. Parameters used in cyclic voltammetry tests to obtain the juice data set.

Parameters	Value	
Initial potential	−1000	mV
Final potential	1000	mV
Potential step	4	μV
Scan rate	500	mV/s
Current range	±450	μA
Calibration resistance	12,000	Ω
Load resistance	100	Ω

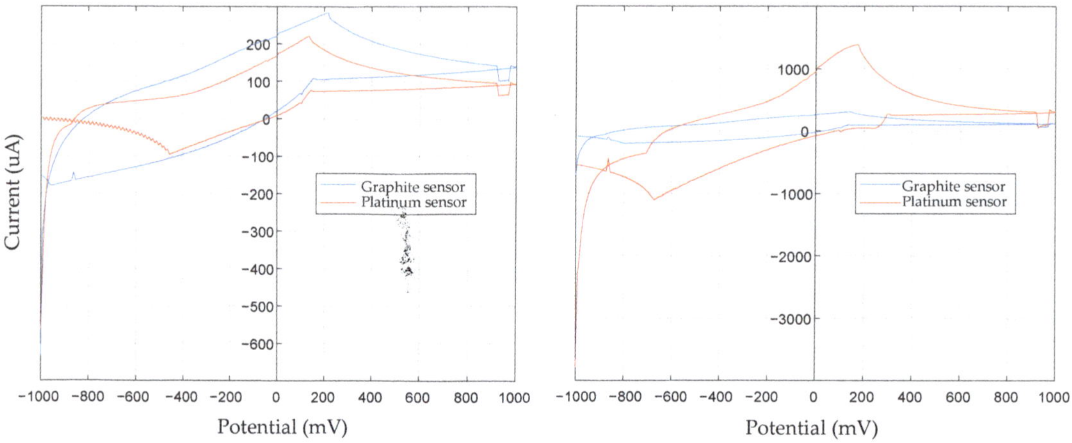

Figure 2. (**Left**) Cyclic voltammograms obtained by platinum and graphite sensors in experiment # 2 on juice # 2 (green apple). (**Right**) Cyclic voltammograms obtained by platinum and graphite sensors in experiment # 5 on juice # 3 (red fruits).

2.3. Dimensionality Reduction

Due to the high dimensionality of the data obtained when performing cyclic voltammetry experiments and how the data are unfolded creating a two-dimensional matrix, it is necessary to carry out a dimensionality reduction process. There are different methods of dimensionality reduction, these can be classified as linear or non-linear. Within the linear methods is the principal component analysis (PCA) [9]. In this method, the greatest amount of variance of the data is represented in a low-dimensional linear space. The data normalization affects the result of the embedding performed by different dimensionality reduction methods.

However, the data obtained by the electronic tongue can form a highly nonlinear manifold. To deal with this issue, different nonlinear dimensionality reduction methods have been developed [14]. These methods are based on the construction of a neighborhood graph and the idea that nearby points in the high-dimensional space can preserve this property in a low-dimensional space. One of the parameters that must be tuned in the dimensionality reduction process is the target dimension d. These target dimensions define the sized of the reduced feature matrix. Specifically, d defines the number of columns that the reduced feature matrix will have. The Locally Linear Embedding (LLE) method solely preserves manifold local properties.

3. Data Processing Results

3.1. Data Unfolding

The unfolding of the cyclic voltammogram data obtained by each sensor is carried out according to the group scaling method [15]. For each experiment carried out, the unfolding of the two sensors is performed, obtaining a signal of 1000 data points. Figure 3 left shows an unfolded signal by juice number 3 (red fruits). In this case, the ordinates correspond to current measurements in μA and the abscissa to data points. Since 25 juice samples were considered in total, the matrix size X is equal to 25×1000.

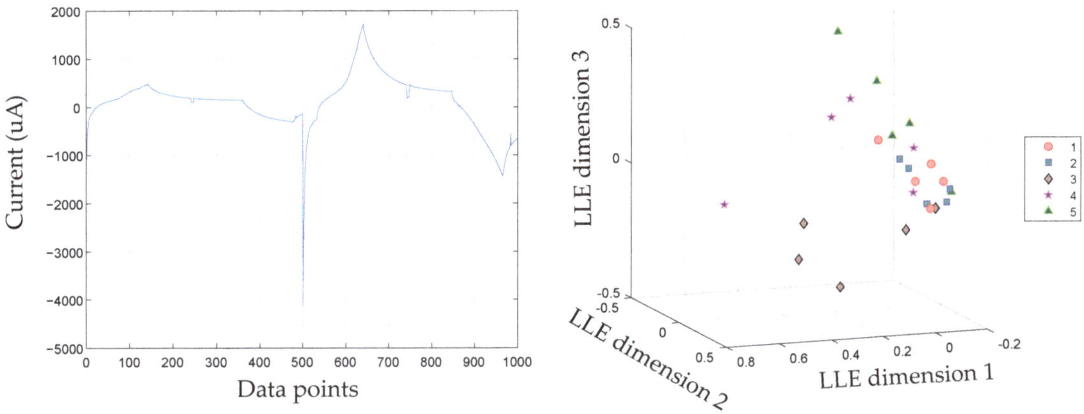

Figure 3. (**Left**) Example of unfolded signal of the cyclic voltammograms obtained by the platinum and graphite sensors. (**Right**) Three-dimensional scatter plot for the first three dimensions obtained using the LLE method on the juice data set.

3.2. Dimensionality Reduction Results

The next step in the juice recognition methodology using a cyclic voltammetry electronic tongue is to reduce the dimensionality of the data. In this case, the Locally Linear Embedding (LLE) algorithm was used, which allows to carry out the feature extraction process. The results of the first 3 dimensions after applying the LLE algorithm to the juice dataset are illustrated in the scatter diagram of Figure 3 right. Classes 3 and 5 are the ones better separated (according to the 3D-view shown in Figure 3 right). Thus, the use of a machine learning classifier algorithm is necessary. In this case, the classifying algorithm was k-Nearest Neighbors.

In order to compare the behavior of different methods [15] to perform the dimensionality reduction stage the PCA, Laplacian Eigenmaps, Isomap and t-distributed stochastic neighbor embedding (t-SNE) were selected to determine their behavior in terms of classification accuracy. In addition, a parameter tuning is performed for each manifold learning dimensionality reduction algorithm used. In this case there are 3 algorithms that in common need to create a neighborhood graph, which has the parameter k, on the other hand the algorithm t-SNE needs the calibration of its perplexity parameter p. Figure 4 shows the behavior in the classification accuracy with respect to the variation of each of these parameters. The used range for the parameters was from 4 to 24 since for the neighborhood graphs the minimum value of k= 4 and the maximum of 24 because there are 25 total samples in the data set. This same range was used for the perplexity p value. As can be seen in Figure 4, the LLE method is the one that achieves the highest accuracy values. Particularly when k = 22 the LLE method reaches 80% of classification accuracy.

3.3. Classification and Cross Validation

The LLE algorithm needs the definition of the destination dimension, to find this parameter, a study of the change of the destination dimension d vs. the classification accuracy obtained by the algorithm k-NN with $k = 1$ was carried out and Euclidean distance was considered. The cross-validation process executed was leaving one out (LOOCV) due to the small number of samples in the juice data set.

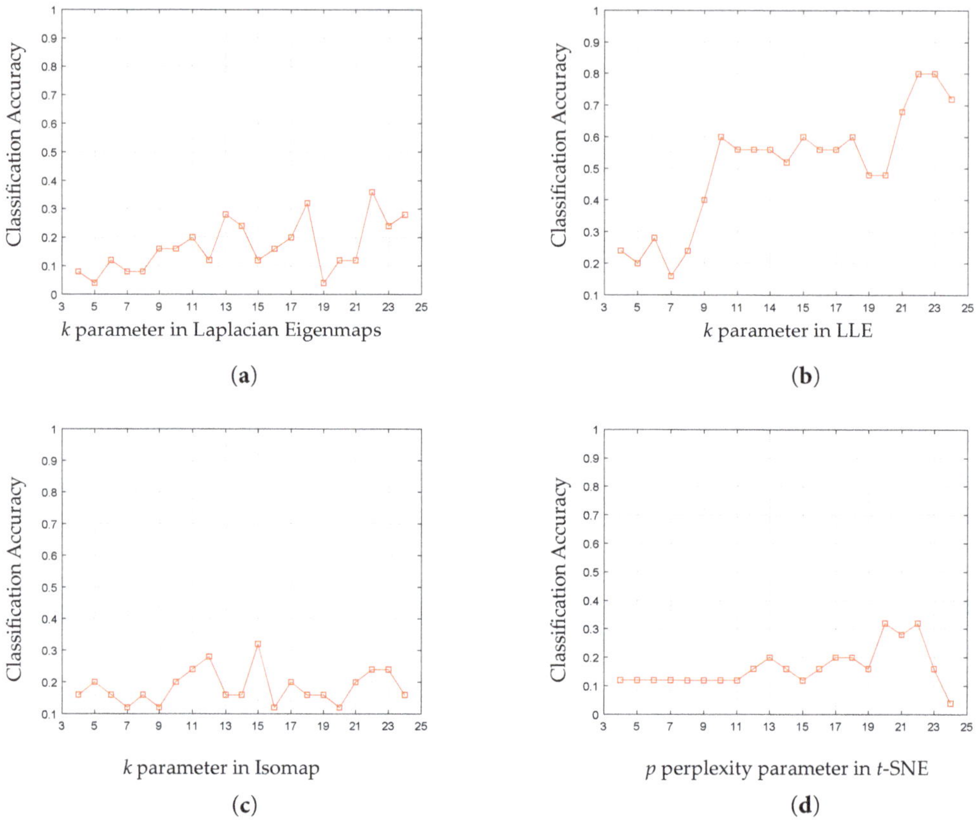

Figure 4. Behavior of the classification accuracy obtained by each manifold learning algorithm used in the data set of commercial fruit-based products. The target dimension was set at $d = 9$ in all cases. (**a**) Laplacian Eigenmaps, (**b**) LLE, (**c**) Isomap y (**d**) t-SNE.

Influence of Target Dimensions Variation

Since the number of dimensions at the input of the k-NN classifier algorithm can vary, a study was carried out to determine the best classification accuracy for each of the dimensionality reduction algorithms studied. In Table 3, it can be seen how the LLE method is the one with the best performance in terms of classification accuracy, reaching an accuracy value of 80% with 9 dimensions. As it can be seen in Table 3 the accuracy behavior tends to increase as d is increased up to a maximum of $d = 9$ for a classification accuracy of 80%. After the dimension $d = 9$ accuracy tends to decrease, in this sense the optimum size selection was defined as $d = 9$. Therefore, the feature matrix size at the input of the k-NN classifier is equal to 25×9.

Figure 5 shows the results of the confusion matrix for the mentioned accuracy of 80%. In this case, class 2 was correctly classified, there was 1 error for classes 1,3 and 4; finally, the class that was classified worst was class 5 with two errors. Overall, of the 25 total samples, 20 were classified well and five were classified badly.

Table 3. Classification accuracy behavior of each dimensionality reduction method when varying the number of dimensions at the input of the *k*-NN classifier algorithm for the classification of 5 different commercial fruit-based products.

Dimension	PCA	Laplacian	LLE	Isomap	*t*-SNE
2	0.1200	0.2000	0.1200	0.1600	0.1600
3	0.0800	0.3600	0.1600	0.0800	0.0400
4	0.1200	0.3200	0.2400	0.1200	0.2000
5	0.1600	0.2000	0.2400	0.1200	0.3200
6	0.1600	0.2000	0.6000	0.2400	0.2000
7	0.1600	0.2000	0.4800	0.2400	0.0800
8	0.1600	0.2000	0.7600	0.2400	0.1600
9	0.1600	0.2400	**0.8000**	0.1600	0.0800
10	0.1600	0.2000	0.6800	0.2000	0.1600
11	0.1600	0.2000	0.6400	0.1600	0.2800
12	0.1600	0.2400	0.6400	0.1600	0.2800
13	0.1600	0.3600	0.6800	0.1600	0.1600
14	0.1600	0.3600	0.7200	0.1600	0.2400
15	0.1600	0.3600	0.6400	0.1600	0.2400
16	0.1600	0.3600	0.6400	0.1600	0.1600
17	0.1600	0.3600	0.6400	0.1600	0.2400

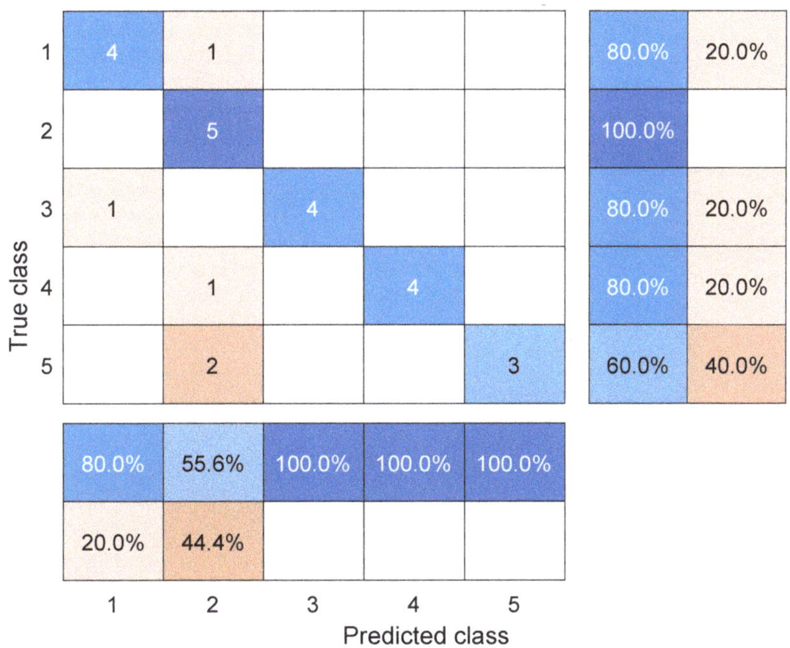

Figure 5. Confusion matrix for the juice data set obtained using 9 target dimensions of the LLE algorithm. The classification accuracy is equal to 80%.

4. Conclusions

This work presented a computational framework for processing the signals obtained by a cyclic voltammetry electronic tongue sensor array. The classification accuracy obtained by the developed methodology in a dataset of five different juices showed the advantages of apply this methodology as classification method. It processes the raw complete voltammograms obtained by each working electrode and unfolded them to create a two dimensional matrix. This matrix was normalized applying the group scaling method. Then, the locally linear embedding method is used as a nonlinear feature extraction approach to obtain the feature matrix at the input of a k-NN classifier. As future work, the developed methodology will be applied for classify other kind of substances and other approaches related to semi-supervised classification will be tested.

Author Contributions: All authors contributed to the development of this work, specifically their contributions are as follows: conceptualization, D.A.T. and J.X.L.-M.; data organization and pre-processing, J.X.L.-M. and M.A.; methodology, J.X.L.-M. and M.A.; validation, J.X.L.-M. and D.A.T. All authors have read and agreed to the published version of the manuscript.

Funding: This research was funded by FONDO DE CIENCIA TECNOLOGÍA E INNOVACION FCTeI DEL SISTEMA GENERAL DE REGALÍAS SGR. The authors express their gratitude to the Administrative Department of Science, Technology and Innovation—Colciencias with the grant 779—"Convocatoria para la Formación de Capital Humano de Alto Nivel para el Departamento de Boyacá 2017" for sponsoring the research presented herein.

Institutional Review Board Statement: Not applicable.

Informed Consent Statement: Not applicable.

Acknowledgments: J.X.L.-M. is grateful with Colciencias and Gobernación de Boyacá. J.X.L.-M. thanks Miryam Rincón Joya from the Department of Physics of the National University of Colombia and Leydi Julieta Cardenas Flechas, for their introduction to the electronic tongue sensor array field of research.

Conflicts of Interest: The authors declare no conflict of interest.

References

1. Podrażka, M.; Bączyńska, E.; Kundys, M.; Jeleń, P.S.; Witkowska Nery, E. Electronic tongue—A tool for all tastes? *Biosensors* **2018**, *8*, 3. [CrossRef] [PubMed]
2. Ross, C.F. Considerations of the use of the electronic tongue in sensory science. *Curr. Opin. Food Sci.* **2021**, *40*, 87–93. [CrossRef]
3. Zabadaj, M.; Ufnalska, I.; Chreptowicz, K.; Mierzejewska, J.; Wróblewski, W.; Ciosek-Skibińska, P. Performance of hybrid electronic tongue and HPLC coupled with chemometric analysis for the monitoring of yeast biotransformation. *Chemom. Intell. Lab. Syst.* **2017**, *167*, 69–77. [CrossRef]
4. Qiu, S.; Wang, J.; Gao, L. Qualification and quantisation of processed strawberry juice based on electronic nose and tongue. *LWT-Food Sci. Technol.* **2015**, *60*, 115–123. [CrossRef]
5. Dias, L.; Peres, A.M.; Barcelos, T.P.; Morais, J.S.; Machado, A. Semi-quantitative and quantitative analysis of soft drinks using an electronic tongue. *Sens. Actuator B Chem.* **2011**, *154*, 111–118. [CrossRef]
6. Haddi, Z.; Mabrouk, S.; Bougrini, M.; Tahri, K.; Sghaier, K.; Barhoumi, H.; El Bari, N.; Maaref, A.; Jaffrezic-Renault, N.; Bouchikhi, B. E-Nose and e-Tongue combination for improved recognition of fruit juice samples. *Food Chem.* **2014**, *150*, 246–253. [CrossRef] [PubMed]
7. Kiranmayee, A.; Panchariya, P.; Sharma, A. New data reduction algorithm for voltammetric signals of electronic tongue for discrimination of liquids. *Sens. Actuators A Phys.* **2012**, *187*, 154–161. [CrossRef]
8. Roweis, S.T.; Saul, L.K. Nonlinear dimensionality reduction by locally linear embedding. *Science* **2000**, *290*, 2323–2326. [CrossRef] [PubMed]
9. Leon-Medina, J.X.; Cardenas-Flechas, L.J.; Tibaduiza, D.A. A data-driven methodology for the classification of different liquids in artificial taste recognition applications with a pulse voltammetric electronic tongue. *Int. J. Distrib. Sens. Netw.* **2019**, *15*, 1550147719881601. [CrossRef]
10. Liu, T.; Chen, Y.; Li, D.; Wu, M. An active feature selection strategy for DWT in artificial taste. *J. Sens.* **2018**, *2018*, 9709505. [CrossRef]
11. Analog Devices (2021) EVAL-AD5940ELCZ Electrochemical Evaluation Board. Available online: https://www.analog.com/en/design-center/evaluation-hardware-and-software/evaluation-boards-kits/eval-ad5940elcz.html#eb-overview (accessed on 20 May 2021).
12. Analog Devices (2021) EVAL-ADICUP3029. Available online: https://www.analog.com/en/design-center/evaluation-hardware-and-software/evaluation-boards-kits/eval-adicup3029.html#eb-overview (accessed on 20 May 2021).
13. BVT Technologies (2021) ELECTROCHEMICAL SENSOR Type: AC1.W*.R* (*). Available online: https://bvt.cz/wp-content/uploads/2020/08/01-AC1.pdf (accessed on 20 May 2021).

14. Van der Maaten, L. *An Introduction to Dimensionality Reduction Using Matlab*; Report MICC 07-07; Universiteit Maastricht: Maastricht, The Netherlands, 2007; p. 62.
15. Leon-Medina, J.X.; Anaya, M.; Pozo, F.; Tibaduiza, D. Nonlinear Feature Extraction Through Manifold Learning in an Electronic Tongue Classification Task. *Sensors* **2020**, *20*, 4834. [CrossRef] [PubMed]

chemistry proceedings

MDPI

Proceeding Paper

Applied Voltage Effect in Lbl Sensors While Detecting 17α-Ethinylestradiol in Water Samples †

Paulo M. Zagalo [1,*], Cátia Magro [2], Paulo A. Ribeiro [1] and Maria Raposo [1,*]

[1] Laboratory of Instrumentation, Biomedical Engineering and Radiation Physics (LIBPhys-UNL), Department of Physics, NOVA School of Science and Technology, NOVA University Lisbon, 2829-516 Caparica, Portugal; pfr@fct.unl.pt

[2] Department of Physics, NOVA School of Science and Technology, NOVA University Lisbon, 2829-516 Caparica, Portugal; c.magro@fct.unl.pt

* Correspondence: p.zagalo@campus.fct.unl.pt (P.M.Z.); mfr@fct.unl.pt (M.R.)

† Presented at the 1st International Electronic Conference on Chemical Sensors and Analytical Chemistry, 1–15 July 2021; Available online: https://csac2021.sciforum.net/.

Abstract: The effect of the applied voltage on impedance spectra, measured on sensors based on solid supports with interdigitated electrodes (IDE) that are either covered or not with a layer-by-layer film prepared with polyethylenimine and poly (sodium 4-styrenesulfonate), was analyzed to detect 17α-ethinylestradiol(EE2) in mineral water and tap water. The results show that the sensor response is strongly affected by the applied voltage, the presence of film, and the water matrix, meaning that electrochemical reactions develop near the IDE. However, for low values of applied voltage, the sensor response is reproducible with negligible electrochemical reactions, allowing us to conclude that 25 mV is the appropriate voltage.

Keywords: voltage effect; impedance; sensor; layer-by-layer; interdigitate electrodes; electrochemical reactions

check for updates

Citation: Zagalo, P.M.; Magro, C.; Ribeiro, P.A.; Raposo, M. Applied Voltage Effect in Lbl Sensors While Detecting 17α-Ethinylestradiol in Water Samples. *Chem. Proc.* **2021**, *5*, 57. https://doi.org/10.3390/CSAC2021-10460

Academic Editor: Ye Zhou

Published: 30 June 2021

Publisher's Note: MDPI stays neutral with regard to jurisdictional claims in published maps and institutional affiliations.

1. Introduction

With the advent of more modern and advanced technologies, it has become possible to detect chemical compounds at very low concentrations. These compounds were previously hidden and invisible to former methods of analysis in water bodies such as ponds, lakes, rivers, underground waters, muds, or even wastewaters [1,2]. Considering that not only human lives, but also a vast majority of Earth's fauna and flora, rely deeply on the central and invaluable role that fresh and clean water plays, it is undoubtedly and alarmingly necessary to strive and succeed in finding novel ways to detect, monitor and conceivably remove these substances from water bodies. Some of these emerging contaminants are included in the category of pharmaceuticals and personal care products (PPCP), such as triclosan (5-chloro-2-(2,4-dichlorophenoxy) phenol) (TCS) which is a well-known and commonly used compound in toothpaste, shampoos and lotions due to its bacteriostatic and antimicrobial properties [3]. Amongst these PPCPs, there is a group of substances designated as endocrine disrupting compounds (EDC) and it is within this collection of composites that 17α-ethinylestradiol (EE2) is inserted [4].

More specifically, EE2 is a synthetic hormone that is commonly used in the manufacturing of women's oral birth control pills, a widespread and rather mainstream contraceptive. This compound finds its way into wastewater through regular bodily excretions and, even though this water enters wastewater treatment plants where it is treated and cleaned, it is not possible to achieve a 100% removal rate of EE2 from water [5,6]. The treated water that still contains low concentrations of EE2 is discharged into large bodies of water such as rivers and seas where it will subsequently impact the life cycles of both fauna and flora [7–9].

Given that some PPCP and EDC effects on the environment are already known or are the object of several studies, a certain degree of precautions has been taken. In the case of EE2, these include regulations such as the European Union Decision 840 of June 5th 2018 and the Regulation of synthetic estrogen 17α-ethinylestradiol in water bodies in Europe, the United States, and Brazil, and also other measures and guidelines which aim to limit and/or ban its usage [10–12].

This work is but a part of a major venture that aims to develop and put into practice cheaper, easier to fabricate and more user-friendly sensor devices capable of detecting and monitoring EE2 in different water bodies, while still retaining or improving the efficiency and detection limits of already existing devices of this nature. The combination of a range of sensors which would then work as a whole, such as an electronic tongue system (ET), to better detect harmful molecules in water bodies is not only a viable option but also a fascinating one due to its possibilities and versatility [13]. To that end, the use of sensorial units as interdigitated electrodes (IDE) coupled with an impedimetric system has been demonstrated as being a useful and rather simple method to study, analyze and infer the properties and variations of a multitude of medium samples, from environmental to biomedical ones [14,15].

IDE's versatility of being deployable in a myriad of applications without the added labor and expenses of modifying their core geometry and/or configuration has catapulted these devices into wide use in a vast array of fields and areas, also allowing for the possibility of lab-on-chip systems based on IDE. These types of sensors present further features and advantages that make them coveted and chosen by researchers worldwide in their works, such as the advent of more powerful and precise technological advances that paved the way for the production of increasingly smaller electrodes, at highly competitive costs due to processes as mass-fabrication [16–18].

In the present work, by combining interdigitated gold electrodes with thin-film techniques, for instance, layer-by-layer (LbL), and impedance spectroscopy the aforementioned goal of developing sensors to detect and monitor EE2 in different water bodies is made possible. In particular, the effect of different voltage levels applied to the IDE sensor devices and how it would impact the detection capabilities of EE2 in mineral water (MW) and tap water (TW) was analyzed. Although a myriad of studies and works have been conducted and performed on the subject of IDEs, none (to the best of the authors' knowledge) have+ analyzed the influence of the variation of voltage levels on IDE sensors and its impact on the sensors' performance and sensitivity when detecting 17α-ethinylestradiol in different water matrices.

2. Materials and Methods

In the course of the experimental work conducted for this study, gold IDE (200 μm/200 μm) deposited onto ceramic substrates were used as sensor devices to detect 17α-ethinylestradiol (EE2) in two different water matrices with distinct complexity levels. The water samples used were a commercial Portuguese mineral water (MW) (pH = 5.7 ± 0.3) and tap water (TW) (pH = 6.8 ± 0.1). These matrices were chosen to observe the way various water complexities interact and impact these electrical measurements performed as well as the detection of the hormone. Regarding the IDE, for each of the matrices two distinct types of sensors were prepared: naked (uncoated) IDE and thin film IDE. The thin films deposited onto the IDE were prepared through the technique of Layer-by-Layer (LbL), where the alternate deposition of polyelectrolytes (positive and negative) builds up to the formation of thin film bilayers [19]. The positive polyelectrolyte used was polyethylenimine (PEI) and its negatively charged counterpart was poly(sodium 4-styrenesulfonate) (PSS), prepared using aqueous solutions with a 10^{-2} M concentration of both polyelectrolytes. Through this technique, thin films of PEI/PSS with 5 bilayers were produced ($[PEI/PSS]_5$).

Solutions of both MW and TW were spiked with a concentration of 10^{-12} M of EE2. Subsequently, both types of sensors were immersed in these solutions and electrical measurements were conducted whilst applying varying AC voltages to the IDE sensors: 25 mV,

50 mV, 100 mV, 200 mV, 300 mV, 400 mV, 500 mV, 600 mV, 700 mV, 800 mV, 900 mV, and 1000 mV. The aforementioned electrical measurements were performed using a Solartron 1260 Impedance Analyzer, with a frequency range of [1 − 1M] Hz.

Both polyelectrolytes (PEI and PSS) and EE2 standards were purchased from Sigma-Aldrich (Darmstadt, Germany).

3. Results

Figures 1 and 2 show the loss tangent, real, and imaginary spectra for the uncoated sensors and those coated with a $(PEI/PSS)_5$ LbL film, respectively, while the sensors were immersed in aqueous MW and TW solutions, spiked with a fixed concentration of 10^{-12} M of EE2. The voltage applied on the IDE sensors was then sequentially altered from an initial value of 25 mV up to a maximum value of 1000 mV.

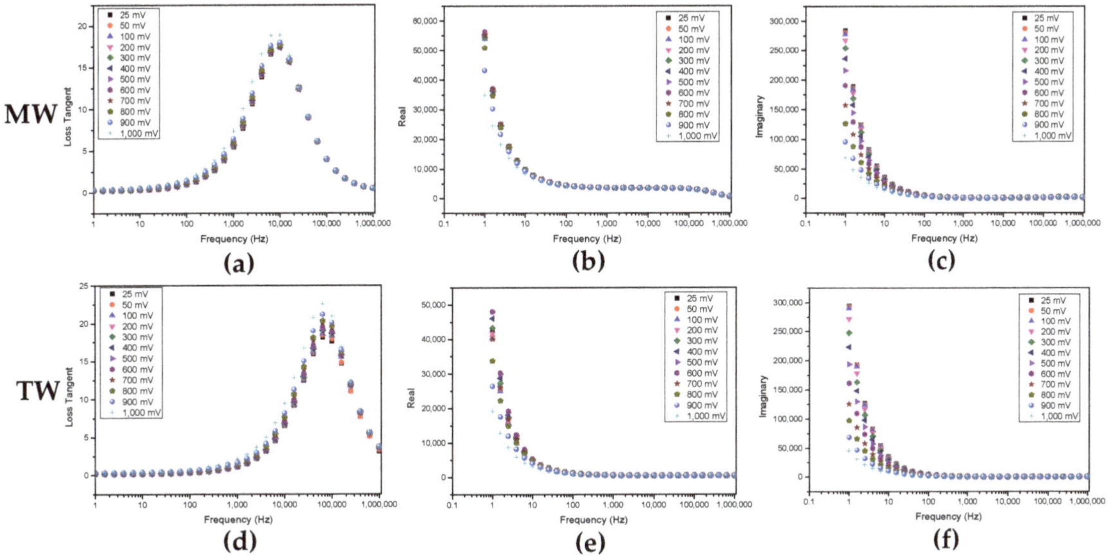

Figure 1. Comparison of derived information from impedance experimental data as functions of frequency for uncoated sensors after immersion in solutions spiked with EE2 in MW and TW: (**a**,**d**) Loss tangent; (**b**,**e**) Real; (**c**,**f**) Imaginary.

Through an analysis of the plots shown in Figures 1 and 2, the roles that different types of water play become apparent. In both figures, when the transition from mineral water to tap water is performed, the degree of distinction and separation between the various voltage values applied to the sensors greatly increases, while also displaying a higher tendency to achieve an ordered sequence between voltages. Although this behavior can be observed in all spectra above, it is particularly prominent in the loss tangent spectra.

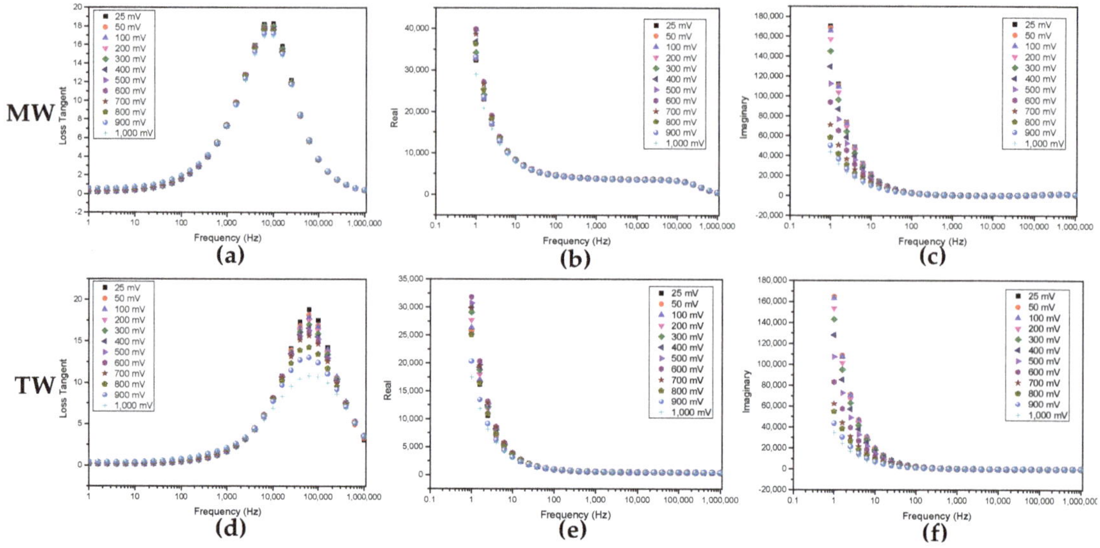

Figure 2. Comparison of derived information from impedance experimental data as functions of frequency for (PEI/PSS)$_5$ sensors after immersion in solutions spiked with EE2 in MW and TW: (**a**,**d**) Loss tangent; (**b**,**e**) Real; (**c**,**f**) Imaginary.

4. Discussion

By delving deeper into the analysis of Figures 1 and 2, in the spectra of loss tangent for TW (Figures 1d and 2d), it becomes possible to infer that the increase in conductivity of the water medium translates itself in a right shift of the curves towards higher frequencies, which in turn underlines an increase in signal intensity for that range of frequency. Building upon what was stated in the previous section, there is an evident improvement regarding the distinction and separation of the different voltage values, not only when transitioning from MW to a medium richer and densely filled with electric charges as is the case for TW, but also by adding the (PEI/PSS)$_5$ thin film coating onto the surface of the IDE sensors. There is one more interesting behavior to note, which is the inversion of the voltage ordered sequence observable in Figure 1a,d from lowest (25 mV) to highest (1000 mV) to the one displayed in Figure 2a,d which transitions from highest to lowest. This behavior change is most likely due to the addition of the thin film coating, given that it introduces other electrochemical reactions on the IDE surface.

To the best of the authors' knowledge, this type of study has not yet been conducted, with regard to an analysis of how LbL thin films behave when subjected to different applied voltages, while detecting PPCP or EDC such as in this case with EE2. However, the need to perform this experiment arose from the lack of information regarding this subject while conducting experimental works within our research group that are directly related to both this kind of sensor, (LbL films on IDE) and emerging compounds such as triclosan and EE2 [20–23]. This type of film ([PEI/PSS]$_5$) was chosen since it is viable, as demonstrated in previous works conducted where several types of thin film combinations were studied to ascertain which one (or ones) would be the most suitable to use in experiments that aim to detect (and possibly monitor) noxious compounds in water bodies [14]. These films have been found to be the most promising among all of the thin films that were analyzed, particularly as the complexity and pH of the water matrices increased. The reason being that these factors have a deep impact on this thin film's stability, namely on the sulfonate group present within the chemical composition and the structure of the outer layer (PSS), which achieves more stable behaviour when in the presence of pH of approximately 7 and above [23,24].

Figure 3 presents curves that illustrate the behavior of the maximum loss tangent and the applied voltage for each IDE upon the collection of data through impedance spectroscopy.

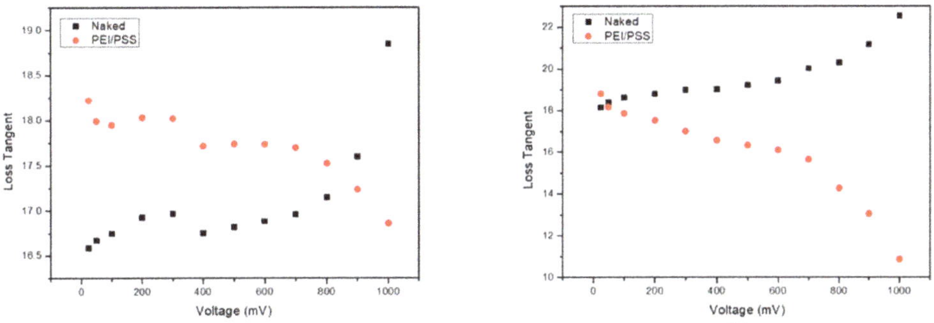

Figure 3. Plots of loss tangent values at the peak vs voltage for both types of IDE used (naked and $[PEI/PSS]_5$) for: (**a**) mineral water; (**b**) tap water.

From these plots, one can surmise, for both MW and TW, that by increasing the voltage that is applied to each IDE device, a decrease in the polarization of the thin film sensors occurs, which results in these sensor devices exhibiting a poorer signal response to the external electrical stimuli. It was also possible to observe that the higher the voltage, the less reproducible the sensors tend to be. This effect could be due to a combination of factors, namely the sensors becoming more prone to external noise at higher voltages and possible structural damage on either the IDE Au layer or the $(PEI/PSS)_5$ thin film itself, or of both simultaneously. These results are in accordance with the results observed by Magro et al. [14] which demonstrated the IDE are damaged by electrical measurements, inducing irreproducibility in the electrical measurements, which emphasizes the necessity of the presence of a thin film covering the IDE and the use of low voltage to measure the impedance spectra.

5. Conclusions

This work set out to study and understand how the effect of varying applied voltage to IDE sensors influences their overall responses while detecting EE2 in complex water matrices. For this purpose, two types of sensor devices consisting of ceramic solid supports coated with Au IDE were used, namely, naked sensors (no thin film) and $(PEI/PSS)_5$ thin film sensors. From the impedance spectra, it was possible to observe that by increasing the complexity of the water matrix, in this case from MW to TW, the sensors exhibit, while detecting EE2, a strong dependence with the applied voltage. Furthermore, by depositing a thin film of (PEI/PSS) with 5 bilayers, the abovementioned sensors' spectra was also found to change with the increase of the applied voltage with an opposite behavior of that achieved with the uncoated electrodes. This points to the increase in chemical reactions with an increase in the electric field between the electrodes. These reactions differ depending on whether the electrodes are covered by the thin film or not. Therefore, one can conclude that the sensor impedance response is strongly affected by the voltage applied and by the water matrix, meaning that electrochemical reactions are developed near the IDE electrodes. However, for low values of applied voltage the sensor is reproducible, and the electrochemical reactions are negligible. Moreover, the chemical reactions are dependent on the presence and type of the LbL film deposited on the electrodes. Finally, one can conclude that, to avoid chemical reactions, the applied voltage should be reduced to 25 mV when one can use this technique to characterize this kind of sensor. However, further studies and analyses should be conducted to investigate this subject further, while also striving to optimize the experimental system used.

Author Contributions: Conceptualization, P.M.Z., M.R.; methodology, P.M.Z.; software, P.M.Z., C.M.; validation, P.M.Z.; formal analysis, P.M.Z., C.M., M.R.; literature review, P.M.Z., M.R.; resources, M.R., P.A.R.; data curation, P.M.Z., M.R.; writing—original draft preparation, P.M.Z.; writing—review and editing, P.M.Z., M.R.; visualization, P.M.Z.; supervision, P.A.R., M.R.; project administration, P.A.R., M.R.; funding acquisition, P.A.R., M.R. All authors have read and agreed to the published version of the manuscript.

Funding: Fundação para a Ciência e a Tecnologia (FCT), Portugal, through research project grants PTDC/FIS-285 NAN/0909/2014, UID/FIS/00068/2019 and the Bilateral Project entitled "Deteção de Estrogénio- um 286 Contaminante Emergente—em Corpos Hídricos" within the scope of "Cooperação Transnacional_FCT 287 (Portugal)-CAPES (Brazil) 2018".

Institutional Review Board Statement: Not applicable.

Informed Consent Statement: Not applicable.

Acknowledgments: The authors acknowledge the Fundação para a Ciência e a Tecnologia (FCT), Portugal, through research project grants PTDC/FIS-NAN/0909/2014, UID/FIS/00068/2019 and the Bilateral Project entitled "Deteção de Estrogénio- um Contaminante Emergente—em Corpos Hídricos" within the scope of "Cooperação Transnacional_FCT (Portugal)-CAPES (Brazil) 2018". The authors also acknowledge the research center Laboratory for Instrumentation, Biomedical Engineering and Radiation Physics (LIBPhys)—FCT/UNL. PM Zagalo acknowledges his fellowship PD/BD/142767/2018 from RABBIT Doctoral Programme.

Conflicts of Interest: The authors declare no conflict of interest.

References

1. Houtman, C.J. Emerging contaminants in surface waters and their relevance for the production of drinking water in Europe. *J. Integr. Environ. Sci.* **2010**, *7*, 271–295. [CrossRef]
2. Wintgens, T.; Salehi, F.; Hochstrat, R.; Melin, T. Emerging contaminants and treatment options in water recycling for indirect potable use. *Water Sci. Technol.* **2008**, *57*, 99–107. [CrossRef] [PubMed]
3. Marques, I.; Magalhães-Mota, G.; Pires, F.; Sério, S.; Ribeiro, P.A.; Raposo, M. Detection of traces of triclosan in water. *Appl. Surf. Sci.* **2017**, *421*, 142–147. [CrossRef]
4. Water Quality Association. PPCP & EDC Fact Sheet, (Ccl Iii). 2014. Available online: https://www.wqa.org/Portals/0/Technical/Technical%20Fact%20Sheets/2014_Ps-PCPs-EDCs.pdf (accessed on 1 March 2020).
5. Johnson, A.C.; Williams, R.J. A model to estimate influent and effluent concentrations of estradiol, estrone, and ethinylestradiol at sewage treatment works. *Environ. Sci. Technol.* **2004**, *38*, 3649–3658. [CrossRef] [PubMed]
6. Thomas, K.V.; Hurst, M.R.; Matthiessen, P.; Waldock, M.J. Characterization of estrogenic compounds in water samples collected from United Kingdom estuaries. *Environ. Toxicol. Chem.* **2001**, *20*, 2165–2170. [CrossRef] [PubMed]
7. Heberer, T. Tracking persistent pharmaceutical residues from municipal sewage to drinking water. *J. Hydrol.* **2002**, *266*, 175–189. [CrossRef]
8. Alistair, B.A. *Boxall. New and Emerging Water Pollutants Arising from Agriculture*; OECD: York, UK, 2012.
9. Adeel, M.; Song, X.; Wang, Y.; Francis, D.; Yang, Y. Environmental impact of estrogens on human, animal and plant life: A critical review. *Environ. Int.* **2017**, *99*, 107–119. [CrossRef] [PubMed]
10. Vella, K. Commission Implementing Decision (EU) 2018/840 of 5 June 2018. *Off. J. Eur. Union* **2018**, *141*, 9–12.
11. da Cunha, D.L.; Camargo da Silva, S.M.; Bila, D.M.; da Mota Oliveira, J.L.; de Novaes Sarcinelli, P.; Larentis, A.L. Regulation of the synthetic estrogen 17α-ethinylestradiol in water bodies in Europe, the United States, and Brazil. *Cad. De Saúde Pública* **2016**, *32*, e00056715. [CrossRef]
12. Nagpal, N.; Meays, C. *Water Quality Guidelines for Pharmaceutically-Active-Technical Appendix*; Ministry of Environment of British Columbia Province: Victoria, BC, Canada, 2009; pp. 1–86.
13. Magro, C.; Mateus, E.; Raposo, M.; Ribeiro, A.B. Overview of electronic tongue sensing in environmental aqueous matrices: Potential for monitoring emerging organic contaminants. *Environ. Rev.* **2019**, *27*, 202–214. [CrossRef]
14. Magro, C.; Zagalo, P.; Pereira-da-Silva, J.; Pires Mateus, E.; Branco Ribeiro, A.; Ribeiro, P.; Raposo, M. Polyelectrolyte Based Sensors as Key to Achieve Quantitative Electronic Tongues: Detection of Triclosan on Aqueous Environmental Matrices. *Nanomaterials* **2020**, *10*, 640. [CrossRef] [PubMed]
15. Ibrahim, M.; Claudel, J.; Kourtiche, D.; Nadi, M. Geometric parameters optimization of planar interdigitated electrodes for bioimpedance spectroscopy. *J. Electr. Bioimpedance* **2013**, *4*, 13–22. [CrossRef]

16. Mukhopadhyay, S.C. Sensing and Instrumentation for a Low Cost Intelligent Sensing System. In Proceedings of the SICE-ICASE International Joint Conference, Busan, Korea, 18–21 October 2006; pp. 1075–1080.
17. Mukhopadhyay, S.C.; Gooneratne, C.P.; Demidenko, S.; Sen Gupta, G. Low cost sensing system for dairy products quality monitoring. In Proceedings of the 2005 International Instrumentation and Measurement Technology Conference, Ottawa, ON, Canada, 16–19 May 2005; pp. 244–249.
18. Van Gerwen, P.; Laureyn, W.; Laureys, W.; Huyberechts, G.; Op De Beeck, M.; Baert, K.; Suls, F.; Sansen, W.; Jacobs, P.; Hermans, L.; et al. Nanoscaled interdigitated electrode arrays for biochemical sensors. *Sens. Actuators B* **1998**, *49*, 73–80. [CrossRef]
19. MacKay, S.; Hermansen, P.; Wishart, D.; Chen, J. Simulations of interdigitated electrode interactions with gold nanoparticles for impedance-based biosensing applications. *Sensors* **2015**, *15*, 22192–22208. [CrossRef] [PubMed]
20. Pereira-da-Silva, J.; Zagalo, P.M.; Magro, C.; Pinto, I.; Ribeiro, P.A.; Raposo, M. Detection and Adsorption of Triclosan on Sensors Based on PAH/PVS Thin-Films. In Proceedings of the 2019 IEEE 6th Portuguese Meeting on Bioengineering (ENBENG), Caparica, Portugal, 6–8 September 2019; IEEE: Piscataway, NJ, USA, 2019; pp. 1–4.
21. Pereira-da-Silva, J.; Zagalo, P.; Magalhães-Mota, G.; Ribeiro, P.; Raposo, M. Adsorption of Triclosan on Sensors Based on PAH/PAZO Thin-Films: The Effect of pH. In *Computational Intelligence Methods for Bioinformatics and Biostatistics*; Raposo, M., Ribeiro, P., Sério, S., Staiano, A., Ciaramella, A., Eds.; Springer: Cham, Switzerland, 2020; pp. 319–325.
22. Zagalo, P.M.; Magro, C.; Pereira-da-Silva, J.; Bouchikhi, B.; Bari, N.E.; Ribeiro, P.A.; Raposo, M. Detection of Triclosan in Tuned Solutions by pH and Ionic Strength Using PAH/PAZO Thin Films. *Proceedings* **2019**, *15*, 25. [CrossRef]
23. Zagalo, P.M.; Ribeiro, P.A.; Raposo, M. Detecting Traces of 17α-Ethinylestradiol in Complex Water Matrices. *Sensors* **2020**, *20*, 7324. [CrossRef] [PubMed]
24. Smith, R.N.; McCormick, M.; Barrett, C.J.; Reven, L.; Spiess, H.W. NMR studies of PAH/PSS polyelectrolyte multilayers adsorbed onto silica. *Macromolecules* **2004**, *37*, 4830–4838. [CrossRef]

chemistry proceedings

MDPI

Proceeding Paper

Smart Sensing for Antibiotic Monitoring in Mineral and Surface Water: Development of an Electronic Tongue Device †

Cátia Magro [1,*], Tiago Moura [1], Paulo A. Ribeiro [2], Maria Raposo [2] and Susana Sério [2,*]

[1] Department of Physics, NOVA School of Science and Technology, NOVA University Lisbon, 2829-516 Caparica, Portugal; ta.moura@campus.fct.unl.pt

[2] Laboratory of Instrumentation, Biomedical Engineering and Radiation Physics (LIBPhys-UNL), Department of Physics, NOVA School of Science and Technology, NOVA University Lisbon, 2829-516 Caparica, Portugal; pfr@fct.unl.pt (P.A.R.); mfr@fct.unl.pt (M.R.)

* Correspondence: c.magro@fct.unl.pt (C.M.); susana.serio@fct.unl.pt (S.S.)

† Presented at the 1st International Electronic Conference on Chemical Sensors and Analytical Chemistry, 1–15 July 2021; Available online: https://csac2021.sciforum.net/.

Abstract: Sensors are considered the future monitoring tools, since, compared to traditional sampling and analysis techniques, they provide fast response on the output data in a timely, continuous, safe, and cost-effective fashion. Antibiotics are important pharmaceuticals with a large variety of applications. However, the overconsumption of these drugs is under the spotlight, since traces of antibiotics are being found in aquatic ecosystems and may lead to the development of antibiotic resistance. Thus, in this work, sensors consisting of ceramic or glass BK7 solid supports with interdigitated gold electrodes coated with five bilayers of polyethyleneimine (PEI) and poly(sodium 4-styrenesulfonate) (PSS) thin films were developed and able to distinguish clarithromycin concentrations between 10^{-15} M and 10^{-5} M in mineral and surface water matrices. In mineral water, the ceramic support sensors have shown high reproducibility, whereas glass support sensors are not reproducible for this matrix. For the surface water matrix, both types of sensors proved to be reproducible. The surface water's principal component analysis, obtained for an electronic tongue composed of the aforementioned sensors, demonstrated the concept's ability to discriminate between different concentrations of the target compound, although no significant pattern of trend was achieved.

Keywords: environmental monitoring; antibiotics; clarithromycin; electronic tongue; impedance spectroscopy

Citation: Magro, C.; Moura, T.; Ribeiro, P.A.; Raposo, M.; Sério, S. Smart Sensing for Antibiotic Monitoring in Mineral and Surface Water: Development of an Electronic Tongue Device. *Chem. Proc.* **2021**, *5*, 58. https://doi.org/10.3390/CSAC2021-10606

Academic Editor: Huangxian Ju

Published: 5 July 2021

1. Introduction

The consumption of antibiotics has grown substantially since their discovery, and they represent an important class of pharmaceuticals employed for treating bacterial infections by killing bacteria or preventing them from spreading. Antimicrobial drugs are widely utilized in human and veterinary medicine as well as in agriculture [1]. Although these chemicals allow us to live longer and have healthier lives, their overconsumption poses a great threat, mainly regarding the development of antibiotic resistance [2].

Clarithromycin is a macrolide, a class of antibiotic drugs produced by multiple *streptomyces* strains, mostly effective against Gram-positive bacteria [3]. It is paramount to understand the occurrence and fate of macrolides in the environment, since traces of these drugs are frequently detected in surface and ground waters. Accordingly, the European Commission included clarithromycin, together with other macrolides, in the 2nd Watch List of Emerging Water Pollutants. This surface water watch list was developed with the purpose of obtaining high-quality monitoring data regarding several potential water pollutants in order to establish their environmental and health risks, thus emphasizing the importance of developing efficient techniques to detect and quantify such pollutants.

Analytical methods based on liquid chromatography (LC) and mass spectrometry (MS) provide low detection limits but require intensive sample preparation, expensive equipment, and experienced operators and are not designed for in situ analysis [4,5]. Sensors are considered the future of monitoring tools and present many advantages compared to traditional techniques [6]. Sensor devices are low cost, simple to operate, and can be used for continuous, fast, and reliable in situ monitoring [7–9]. Even though sensors are not as selective as LC/MS methods, they have proved to be able to identify multiple analytes simultaneously [6]. Moreover, sensors have the ability to work as smart devices that may be incorporated in monitoring systems with real-time data transmission.

The electronic tongue (e-tongue) refers to a device that consists of an array of non-specific chemical and/or physical sensors that display cross-sensitivity to target compounds in a liquid matrix [10]. E-tongue devices may rely on potentiometry, voltammetry, or impedance spectroscopy as transducing methods [11].

The present work aims to explore the potential of the e-tongue concept to monitor different clarithromycin concentrations in two environmental aqueous matrices with incremental complexity. Relying on the discussion of reproducibility, an array of sensors coated with PEI/PSS thin films, which previously showed more mechanical stability [12], was studied as a potential smart device to monitor clarithromycin in mineral and surface water.

2. Materials and Methods

Clarithromycin (Sigma-Aldrich, Steinheim, Germany) solutions with concentrations between 10^{-15} M and 10^{-5} M were prepared by sequential dilutions of a mother solution with a concentration of 10^{-4} M. All dilutions, as well as the mother solution, were prepared with experimental matrix/MeOH (9:1) solutions. The experimental matrices utilized to prepare the solutions were a commercial Portuguese mineral water (MW) and a surface water (SW) collected from Tagus River at Porto Brandão, Caparica, Portugal. Lastly, solutions without clarithromycin, containing only experimental matrix/MeOH (9:1), were prepared for the MW and SW matrices to be used as the blank standard (0 M).

The sensors used in this work consist of interdigitated gold electrodes (IDE) deposited on ceramic and glass BK7 solid supports purchased from DropSens (Oviedo, Asturias, Spain). The dimensions of the ceramic support are 22.8 mm × 7.6 mm × 1 mm; the width of each "finger" and the spacing between "fingers" are both 200 µm. The glass support's dimensions are 22.8 mm × 7.6 mm × 0.7 mm, and the width of each "finger", like the spacing between "fingers", is 10 µm. The sensor devices were coated with thin films of polyethyleneimine (PEI) and poly(sodium 4-styrenesulfonate) (PSS) produced by the layer-by-layer (LbL) technique [13]. The substrates were alternately immersed in positively and negatively charged polyelectrolyte solutions with concentrations of 10^{-2} M for 30 s. After the adsorption of each layer, the substrates were immersed in water in order to remove any polyelectrolyte molecules that were not completely adsorbed. At the end of the deposition of each bilayer, the substrates were dried with nitrogen gas stream (99% purity, Air Liquide, Algés, Portugal). The thin films of PEI/PSS were prepared with 5 bilayers, $(PEI/PSS)_5$. Prior to the deposition of thin films, all sensors were cleaned with ethanol and ultra-pure water. Thereafter, the substrates were dried with compressed nitrogen gas (99% purity, Air Liquide, Algés, Portugal).

The electrical analysis was achieved by impedance spectroscopy measurements of the sensor devices when immersed in aqueous matrices spiked with a sequence of increasing concentrations of clarithromycin, from 0 to 10^{-5} M. The impedance spectra were obtained with a Solartron 1260 Impedance Analyzer (Solartron Analytical, AMETEK Scientific Instruments, Berwyn, PA, USA) in the frequency range of 1 Hz to 1 MHz by applying an alternate voltage with an amplitude of 25 mV. Each measurement was performed at room temperature (≈23 °C).

Principal component analysis (PCA) was performed, with respect to the normalized (Z-score normalization: $\frac{value - \mu}{\vartheta}$, μ and ϑ being the mean value and the standard deviation of the samples, respectively) impedance spectroscopy data, to reduce the size of the data and to obtain a new space of orthogonal components in order to detect and explain different concentration patterns. The clarithromycin detection in the target matrices was further evaluated by an array of sensors, composed of all of the produced thin films, using the e-tongue concept.

3. Results and Discussion

3.1. Sensor Reproducibility

To draw conclusions on the reproducibility of the $(PEI/PSS)_5$ sensors, each solution of clarithromycin was analyzed with two identical sensors produced under the same conditions. The average of the two sensors' data was calculated as a function of the frequency for each matrix and type of sensor (glass or ceramic support). The standard deviation was used as a measure of the uncertainty. In Figure 1 is depicted the average of the impedance magnitude measured by (a) two ceramic support and (b) two glass support sensors when immersed in the MW matrix. Figure 1a reveals that the ceramic support sensors coated with $(PEI/PSS)_5$ are reproducible when monitoring clarithromycin in MW, since the values of the standard deviation are small (two orders of magnitude smaller than the average), meaning that the results of both sensors are similar. Furthermore, Figure 1b shows that there is a significant discrepancy between the measurements of the two glass support sensors, which results in larger standard deviation values. These results can be explained, since the spacing between the IDE "fingers" is smaller in the glass support sensors, and, thus, the electric field generated between them, when an AC voltage signal is applied, has a larger magnitude. Additionally, the MW matrix has a lower conductivity, containing ion species in lower concentrations and affecting, therefore, the interactions between the matrix and the sensor. Consequently, at lower frequency values (<10 Hz), measurements of the glass sensors immersed in MW are compromised (see Figure 1b, region highlighted by a red dashed rectangle).

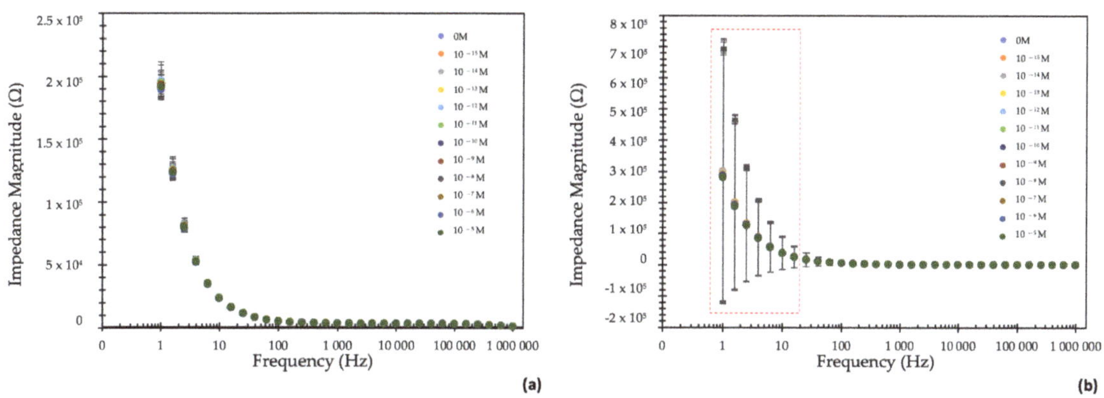

Figure 1. Reproducibility of the impedance magnitude measured by (**a**) two ceramic support and (**b**) two glass support sensors in a MW matrix.

Regarding the analysis of clarithromycin solutions prepared with SW, Figure 2a,b shows the average of the capacitance measured by two ceramic support sensors and two glass support sensors, respectively.

Figure 2a reveals that there is not a considerable difference between the results of both ceramic sensors. As stated before, this conclusion can be drawn from the low values of the standard deviation (10^{-10}–10^{-8} F). Figure 2b provides similar conclusions; thus, both types of sensors show good reproducibility when monitoring SW matrices. However, for the SW matrix, the glass support sensors coated with $(PEI/PSS)_5$ thin films present a more noticeable sensitivity in discriminating clarithromycin concentrations in the frequency range of 1–100 Hz.

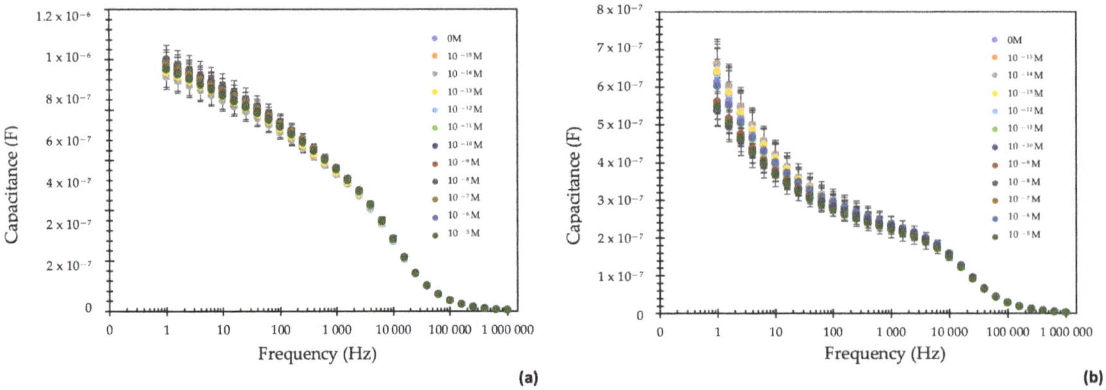

Figure 2. Reproducibility of the capacitance measured by (**a**) two ceramic support and (**b**) two glass support sensors in a SW matrix.

On the other hand, the sensors with a ceramic support showed superior reproducibility for both experimental matrices. Furthermore, the ceramic sensors produced more reliable results in the frequency region of 1–100 Hz for MW. For SW, both types of sensors achieved better reproducibility in the measurements from 1 Hz to 100 Hz.

3.2. Principal Component Analysis: E-Tongue

The results of the principal component analysis applied to the e-tongue concept will be presented and discussed only for SW. The e-tongue concept was not applied for MW, since the glass support sensors were not reproducible, as evidenced by the analysis of the impedance electrical characterization (Figure 1b). Thus, Figure 3 displays the PCA plots obtained for an array of sensors composed of two ceramic support sensors, for MW (Figure 3a), and two ceramic support sensors combined with two glass support sensors for SW (Figure 3b). In both target matrices, the array of sensors was coated with $(PEI/PSS)_5$ thin films.

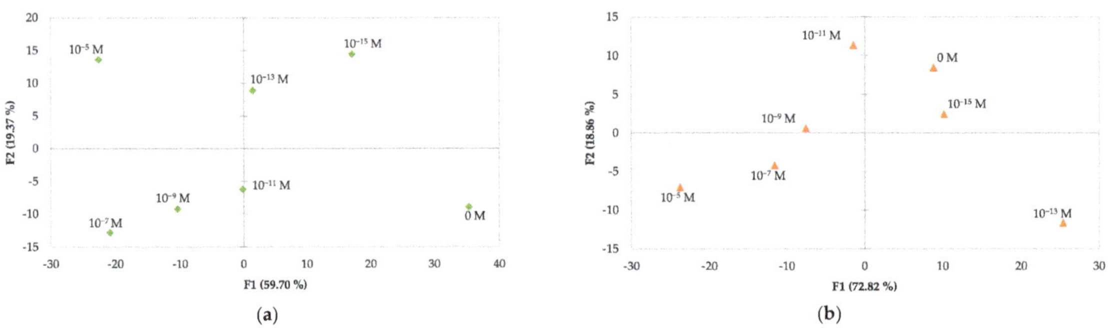

Figure 3. PCA plots for clarithromycin concentrations (0–10^{-5} M) discriminated by an e-tongue for (**a**) MW and (**b**) SW.

In Figure 3a, the first two principal components account for 91.68% of the total variance. The PCA plot reveals that the ceramic support sensors provide the ability to discriminate between different clarithromycin concentrations and the blank solution. There can also be observed a pattern in the concentration decay, with 10^{-13} M as an outlier. Figure 3b shows the PCA plot for an array of sensors identical to the one discussed above but immersed in SW solutions. In this case, the first two principal components accounted for 79.07% of the total variance. The plot reveals that the e-tongue concept provides the ability to discriminate between non-doped and doped SW solutions. Additionally, the sensors were able to distinguish the different clarithromycin concentrations.

4. Conclusions

Sensors composed of ceramic or glass BK7 solid supports, with interdigitated gold electrodes, were coated with five bilayers of PEI/PSS thin films produced by the LbL technique. An electronic tongue consisting of an array of these sensors was shown to provide the ability to distinguish between clarithromycin concentrations in the range of 10^{-15} M to 10^{-5} M in surface water.

The electrical analysis of the samples was performed with impedance spectroscopy by immersing the sensors in the water samples with different clarithromycin concentrations. An average of the measurements obtained with two identical sensors and the associated standard deviation were used to study the reproducibility of the sensors. In the MW matrix, the ceramic sensors showed reproducibility. The opposite can be said for the glass support sensors, which for lower frequencies struggle to identify the target compound. In the SW matrix, both types of sensors, ceramic or glass support, were proven to be highly reproducible.

Results of the principal component analysis of the impedance data did not show a clear pattern or trend but was able to distinguish between doped and non-doped solutions, both for MW and SW matrices. To achieve better results, the e-tongue concept requires a wider variety of thin films deposited on the sensors, such as, for example, metal oxides or carbon-based thin films.

Author Contributions: Conceptualization, C.M., T.M., M.R. and S.S.; methodology, T.M.; software, T.M.; validation, C.M., M.R. and S.S.; formal analysis, C.M, M.R. and S.S.; investigation, T.M. and C.M.; resources, P.A.R., M.R. and S.S.; data curation, T.M.; writing—original draft preparation, C.M., T.M., M.R. and S.S.; writing—review and editing, C.M., M.R. and S.S.; supervision, C.M., M.R. and S.S.; project administration, M.R.; funding acquisition, P.A.R., M.R. and S.S. All authors have read and agreed to the published version of the manuscript.

Funding: The research leading to these results has received funding from the Portuguese funding agency FCT—Fundação para a Ciência e a Tecnologia—within projects PTDC/FIS-NAN/0909/2014, UID/FIS/04559/2020 to LIBPhys-UNL from the FCT/MCTES/PIDDAC and the Bilateral Project entitled "Deteção de Estrogénio- um Contaminante Emergente em Corpos Hídricos" within the scope of "Cooperação Transnacional_FCT (Portugal)-CAPES (Brazil) 2018".

Institutional Review Board Statement: Not applicable.

Informed Consent Statement: Not applicable.

Acknowledgments: C. Magro acknowledges NOVA.ID of NOVA-FCT for her post-doc fellowship.

Conflicts of Interest: The authors declare no conflict of interest.

References

1. Silbergeld, E.K.; Graham, J.; Price, L.B. Industrial food animal production, antimicrobial resistance, and human health. *Annu. Rev. Public Health* **2008**, *29*, 151–169. [CrossRef] [PubMed]
2. Zimdahl, R.L. Antibiotics. In *Six Chemicals That Changed Agriculture*; Elsevier: Amsterdam, The Netherlands, 2015; pp. 165–182.
3. Hagren, V.; Peippo, P.; Lövgren, T. Detecting and controlling veterinary drug residues in poultry. In *Food Safety Control in the Poultry Industry*; Mead, G.C., Ed.; Woodhead Publishing Limited: Southston, UK, 2005; pp. 44–82.
4. Zhang, Y.; Zhu, Y.; Zeng, Z.; Zeng, G.; Xiao, R.; Wang, Y.; Hu, Y.; Tang, L.; Feng, C. Sensors for the environmental pollutant detection: Are we already there? *Coord. Chem. Rev.* **2021**, *431*, 213681. [CrossRef]
5. Rawtani, D.; Khatri, N.; Tyagi, S.; Pandey, G. Nanotechnology-based recent approaches for sensing and remediation of pesticides. *J. Environ. Manag.* **2018**, *206*, 749–762. [CrossRef] [PubMed]

6. Magro, C.; Mateus, E.P.; Fátima Raposo, M.; Ribeiro, A.B. Emerging Contaminants in Wastewater: Sensor Potential for Monitoring Electroremediation Systems. In *Electrokinetic Remediation for Environmental Security and Sustainability*; Wiley: Hoboken, NJ, USA, 2021; pp. 413–432.
7. Ciosek, P.; Wróblewski, W. Sensor arrays for liquid sensing- electronic tongue systems. *Analyst* **2007**, *132*, 963–978. [CrossRef] [PubMed]
8. Kruss, S.; Hilmer, A.J.; Zhang, J.; Reuel, N.F.; Mu, B.; Strano, M.S. Carbon nanotubes as optical biomedical sensors. *Adv. Drug Deliv. Rev.* **2013**, *65*, 1933–1950. [CrossRef]
9. Barsan, M.M.; Ghica, M.E.; Brett, C.M.A. Electrochemical sensors and biosensors based on redox polymer/carbon nanotube modified electrodes: A review. *Anal. Chim. Acta* **2015**, *881*, 1–23. [CrossRef] [PubMed]
10. Vlasov, Y.; Legin, A. Non-selective chemical sensors in analytical chemistry: From "electronic nose" to "electronic tongue". *Fresenius J. Anal. Chem.* **1998**, *361*, 255–260. [CrossRef]
11. Latha, R.S.; Lakshmi, P.K. Electronic tongue: An analytical gustatory tool. *J. Adv. Pharm. Technol. Res.* **2012**, *3*, 3–8.
12. Magro, C.; Zagalo, P.; Pereira-da-Silva, J.; Pires Mateus, E.; Branco Ribeiro, A.; Ribeiro, P.; Raposo, M. Polyelectrolyte Based Sensors as Key to Achieve Quantitative Electronic Tongues: Detection of Triclosan on Aqueous Environmental Matrices. *Nanomaterials* **2020**, *10*, 640. [CrossRef] [PubMed]
13. Oliveira, O.N.; Raposo, M.; Dhanabalan, A. Langmuir-blodgett and self-assembled polymeric films. In *Handbook of Surfaces and Interfaces of Materials*; Nalwa, H.S., Ed.; Elsevier: Burlington, VT, USA, 2001; pp. 1–63, ISBN 978-0-12-513910-6.

chemistry proceedings

MDPI

Abstract

Antimony Tin Oxide—Prussian Blue Screen-Printed Electrodes for Electrochemical Sensing of Potassium Ions †

Cecilia Lete [1], Mariana Marin [1], Francisco Javier del Campo [2,3,*], Ioana Diaconu [4] and Stelian Lupu [4,*]

1 Institute of Physical Chemistry "Ilie Murgulescu" of the Romanian Academy, 202 Splaiul Independentei, 060021 Bucharest, Romania; clete@icf.ro (C.L.); mmaria@icf.ro (M.M.)
2 BCMaterials, Basque Center for Materials, Applications and Nanostructures, UPV/EHU Science Park, 48940 Leioa, Spain
3 Ikerbasque Basque Foundation for Science, Plaza Euskadi 5, 48009 Bilbao, Spain
4 Department of Analytical Chemistry and Environmental Engineering, Faculty of Applied Chemistry and Materials Science, University Politehnica of Bucharest 1-7 Gh. Polizu Street, 011061 Bucharest, Romania; ioana.diaconu@upb.ro
* Correspondence: javier.delcampo@bcmaterials.net (F.J.d.C.); stelian.lupu@upb.ro (S.L.)
† Presented at the 1st International Electronic Conference on Chemical Sensors and Analytical Chemistry, 1–15 July 2021; Available online: https://csac2021.sciforum.net/.

Abstract: In this work, the characterization and the electro-analytical applications of antimony tin oxide (ATO)–Prussian blue (PB) screen printed electrodes (SPE) are presented. The ATO conducting particles have been used recently in the development of screen-printed electrodes due to their excellent spectroelectrochemical properties. PB is a transition metal hexacyanoferrate with high electrocatalytic properties towards various biologically active compounds like hydrogen peroxide, besides its outstanding electrochromic properties. A combination of ATO and PB ingredients into a screen-printing paste provided a versatile and cost-effective way in the development of novel electrode materials for electrochemical sensing. The ATO-PB electrode material displayed good electrochemical properties demonstrated by means of cyclic voltammetry and electrochemical impedance measurements. In addition, the PB provided a high selectivity towards potassium ions in solution due to its zeolitic structures and excellent redox behavior. The cyclic voltammetric responses recorded at the ATO-PB-SPE device in the presence of potassium ions revealed a linear dependence of the cathodic peak current and cathodic peak potential of the Prussian blue/Everitt's salt redox system on the potassium concentrations ranging from 0.1 to 10 mM. This finding could be exploited in the development of an electrochemical sensor for electro-inactive chemical species. The potential application of the ATO-PB electrode in the electrochemical sensing of electro-active species like caffeic acid was also studied. An increase of the anodic peak current of the PB/ES redox wave in the presence of caffeic acid was observed. These results point out to the potential analytical applications of the ATO-PB electrode in the sensing of both electro-active and electro-inactive species.

Keywords: antimony tin oxide; Prussian blue; screen-printed electrodes; cyclic voltammetry; electrochemical impedance spectroscopy; electrochromic material; electrochemical sensing

Citation: Lete, C.; Marin, M.; del Campo, F.J.; Diaconu, I.; Lupu, S. Antimony Tin Oxide—Prussian Blue Screen-Printed Electrodes for Electrochemical Sensing of Potassium Ions. *Chem. Proc.* **2021**, *5*, 59. https://doi.org/10.3390/CSAC2021-10639

Academic Editor: Huangxian Ju

Published: 1 July 2021

Supplementary Materials: The following supporting information can be downloaded at: https://www.mdpi.com/article/10.3390/CSAC2021-10639/s1.

Funding: This research received no external funding.

Institutional Review Board Statement: Not applicable.

Informed Consent Statement: Not applicable.

Data Availability Statement: The data are available in the PP presentantion.

Conflicts of Interest: The authors declare no conflict of interest.

chemistry proceedings

MDPI

Proceeding Paper

Towards Low Temperature VOCs Chemoresistors: Graphene Oxide Versus Porphyrin-Based Materials †

Eleonora Pargoletti [1,2], Francesca Tessore [1,2], Gabriele Di Carlo [1,2], Gian Luca Chiarello [1] and Giuseppe Cappelletti [1,2,*]

1 Dipartimento di Chimica, Università degli Studi di Milano, Via Golgi 19, 20133 Milano, Italy; eleonora.pargoletti@unimi.it (E.P.); francesca.tessore@unimi.it (F.T.); gabriele.dicarlo@unimi.it (G.D.C.); gianluca.chiarello@unimi.it (G.L.C.)
2 Consorzio Interuniversitario Nazionale per la Scienza e Tecnologia dei Materiali (INSTM), Via Giusti 9, 50121 Florence, Italy
* Correspondence: giuseppe.cappelletti@unimi.it; Tel.: +39-0250314228

† Presented at the 1st International Electronic Conference on Chemical Sensors and Analytical Chemistry, 1–15 July 2021; Available online: https://csac2021.sciforum.net/.

Abstract: The sensing of gas molecules is of fundamental importance for environmental monitoring, the control of chemical processes, and non-invasive medical diagnostics based on breath analysis in humans. Herein, the synthesis of hybrid materials (SnO_2/graphene oxide-GO and SnO_2/porphyrins composites) with ad hoc properties led to chemoresistors able to reduce the acetone sensing temperature, guaranteeing acceptable LOD values. As such, boosted potentialities, especially in terms of tuned selectivity and low water interference, may be obtained.

Keywords: VOC chemoresistors; hybrid materials; low-T sensing

Citation: Pargoletti, E.; Tessore, F.; Carlo, G.D.; Chiarello, G.L.; Cappelletti, G. Towards Low Temperature VOCs Chemoresistors: Graphene Oxide Versus Porphyrin-Based Materials. *Chem. Proc.* **2021**, *5*, 60. https://doi.org/10.3390/CSAC2021-10418

Academic Editor: Elisabetta Comini

Published: 30 June 2021

1. Introduction

Volatile Organic Compounds (VOCs) are a huge class of molecules emitted from a large variety of both biogenic and anthropogenic sources [1]. They are considered as a critical factor for air pollution, and they give rise to serious problems for both the environment and human health [1,2] due to their easy diffusivity, volatility, and toxicity even at low concentrations [3]. Moreover, some VOCs, present in the human's breath can be considered as biomarkers of specific illnesses, as they are strictly correlated to several metabolic processes. Among them, acetone can be considered a biomarker for type I diabetes as its concentration in breath varies from 300 to 900 ppb in healthy people to more than 1800 ppb for diabetics [4].

For all these reasons, the monitoring of these compounds has become mandatory. A promising solution for the detection and quantification of VOCs consists in the implementation of chemoresistive gas sensors, based on the electrical resistance variation of the sensing material in the presence of target molecules. The key to facing this challenge is the development of miniaturized chemical sensors, capable of selective sensing a few ppb of VOCs, giving stable and reproducible responses in the presence of high concentrations of interfering species, such as water vapor and other gases [5].

Notably, n-type semiconductor metal oxide (MOS such as SnO_2, WO_3, ZnO, and TiO_2) devices have already been used quite extensively for several applications. They are compact, low-cost, easy to produce and use, and able to detect a wide variety of gaseous species [6,7]. Although these features make such kind of sensors a convenient alternative to the traditional and most sophisticated analytical techniques (e.g., mass spectrometry and gas chromatography), there are also some drawbacks. Specifically, they can only operate at high temperatures (200–400 °C), showing short lifetime and low selectivity, so that it is difficult to selectively analyze multiple species in complex matrices [3].

To overcome these problems, the coupling of metal oxides with other matrices (such as carbonaceous- or porphyrin-based materials) [8,9] seems to be the key factor to create nanocomposites capable of sensing at low temperatures (<100 °C), simultaneously achieving good selectivity and sensitivity (ppb level) towards a target compound.

Therefore, the present work is aimed at evaluating and comparing the sensing performances of SnO_2 coupled with different porphyrins and graphene oxide (GO) towards acetone molecules, at mild temperatures (150 °C and 75 °C) under UV light, in a fixed SnO_2/matrix weight ratio.

2. Materials and Methods

2.1. Hybrid Synthesis, Electrodes Preparation and Sensing Tests

SnO_2 nanoparticles were chosen to be grown onto graphene oxide (GO) material by following a very easy hydrothermal method, already reported in our previous works [6–8]. According to earlier studies, we adopted 32:1 salt precursor-to-GO weight ratio since it resulted as the optimal one in terms of sensing performances at low operating temperatures [3].

Three different Zn^{II} porphyrins were synthesized according to previous papers [10–12], namely ZnTPP, ZnTPP(F_{20}), and ZnTPP(F_{20}CN); their relative chemical structures together with the UV-Vis spectra in CH_2Cl_2 solution are reported in Figure 1. The UV-Vis patterns are the ones typically observed for porphyrins metal complexes [13], with an intense ($\varepsilon \approx 10^5$ M^{-1} cm^{-1}) Soret or B band at 400–450 nm and two weaker ($\varepsilon \approx 10^3$–10^4 M^{-1} cm^{-1}) Q bands at 500–600 nm (see the inset in Figure 1).

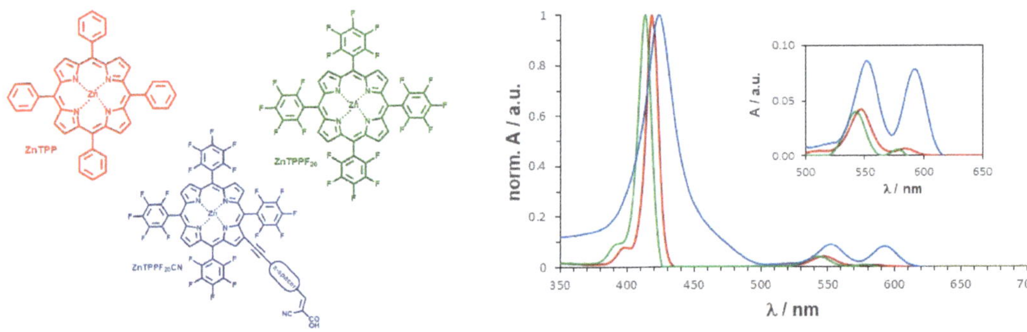

Figure 1. Investigated Zn^{II} porphyrins and their UV-Vis spectra.

Then, the SnO_2-porphyrin sensors were obtained by depositing onto Pt-based interdigitated electrodes (IDEs), through a hot-spraying method [6–8], a first layer of porphyrin (0.6 mg mL^{-1} in EtOH) followed by a thin film of SnO_2 (2.5 mg mL^{-1} in EtOH). The mass ratio between SnO_2 and porphyrin (verified by a microbalance) is 32:1 as in the case of the SnO_2/GO composite.

Sensing measurements towards acetone at 150 °C and 75 °C under UV light (Jelosil HG500 lamp; effective irradiation power: 30 mW cm^{-2}) were performed, adopting the chamber, already described elsewhere [14]. The tests have been carried up to 150 °C since the porphyrin complexes degrade at higher temperatures [15]. The sensor response is reported as: $(R_{air}/R_{acetone}) - 1$, where R_{air} is the film resistance in air and $R_{acetone}$ is the film resistance at a given concentration of the acetone gas. We also computed the sensor response (t_{res}) and the recovery times (t_{rec}).

2.2. Powders and Porphyrins Characterizations

SnO_2 and SnO_2/GO samples were characterized by specific surface area measurements (Micromeritics Tristar II 3020, (Norcross, GA, USA), X-ray Powder Diffraction (XRPD, Philips PW 3710, Texas City, TX, USA) analyses and Diffuse Reflectance Spectroscopy (DRS,

Shimadzu UV-2600, Kyoto, Japan) to evaluate powders optical band gaps (E_g) by means of Kubelka-Munk equation [3,8].

The goodness of the as-synthesized powders has been verified through ^{1}H- and ^{19}F-NMR spectroscopy in $CDCl_3$. The NMR spectra are fully in agreement with the recent literature [10–12].

3. Results and Discussion

Hybrid sensing materials, such as SnO_2-porphyrins and SnO_2-graphene oxide composites, arouse interest thanks to the possible complementary features between the two components, showing cooperative and synergistic behavior [6,9].

In the case of SnO_2/GO hybrid, the formation of nano-heterojunctions with a three-dimensional SnO_2 network has been verified by a combination of physical and chemical characterizations. Specifically, Table 1 reports the specific surface area (S_{BET}), together with total pores volume ($V_{tot.\ pores}$), the average domain size by X-ray diffraction analysis ($<d^{XRPD}>$) and optical band gap (E_g) of pure SnO_2 and GO, together with the composite (SnO_2/GO) sample. Moreover, HRTEM, XPS, Raman and responsivity analyses (already reported in our previous works [3,8]) corroborate that a nano-heterojunction occurs when tin oxide particles are grown onto GO sheets, allowing an intimate contact between the semiconductor and the graphene oxide matrix. This fact leads to a signal intensity of three times higher with respect to that of the pure SnO_2 (Figure 2a) in the case of 20 ppm of acetone at 150 °C under UV light. Notably, the SnO_2/GO was able to reach a LOD of 100 ppb of acetone thanks to the synergistic effect between n-type MOS and p-type GO [3,6]. Furthermore, the response (t_{res} around 60 s) and recovery (t_{rec} around 90 s) times seem to be comparable with those of pure SnO_2.

Table 1. Surface area (S_{BET}), total pore volume ($V_{tot.\ pores}$), crystallite domain size by XRD analysis ($<d^{XRD}>$) and optical band gap (E_g, by Kubelka-Munk extrapolation).

Sample	S_{BET} ($m^2\ g^{-1}$)	$V_{tot.\ pores}$ ($cm^3\ g^{-1}$)	$<d^{XRD}>$ (nm)	E_g (eV)
GO	30	0.020	11	–
SnO_2	67	0.210	15	3.6
SnO_2/GO	55	0.133	8	3.4

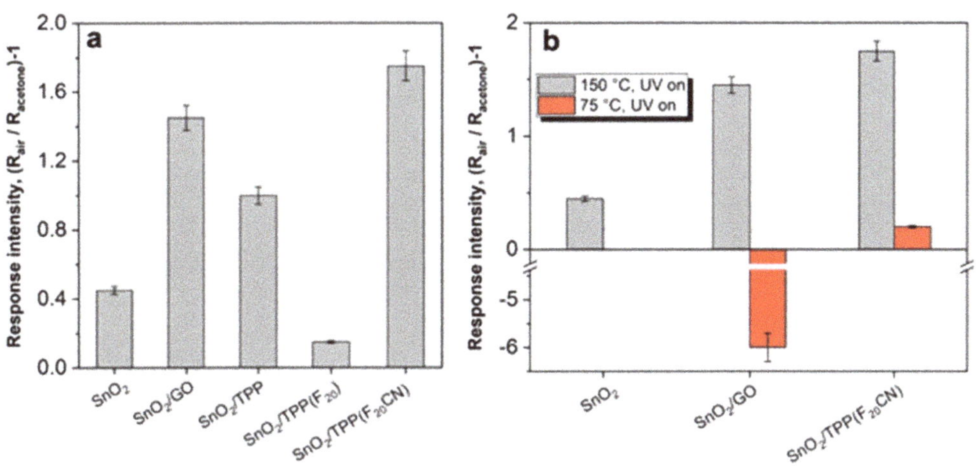

Figure 2. Sensor response intensities for both pure SnO_2 and hybrid materials towards 20 ppm of acetone under UV light at (**a**) 150 °C and (**b**) 75 °C.

Then, the sensors obtained overlapping a SnO_2 and porphyrin layers were tested. Specifically, to evaluate the effect of the porphyrin matrix on the sensing properties of SnO_2, the responses of SnO_2/ZnTPP and SnO_2/ZnTPP(F_{20}) were compared to that of pristine SnO_2 as shown in Figure 2a. The combination of SnO_2 with the ZnTPP porphyrin matrix has undoubtedly a beneficial effect on the sensing performance, as reported in the recent literature [9,15,16]. Li et al. observed that light has a beneficial influence in the gas sensing by ZnO nanorods functionalized with porphyrins [16]. They asserted that light activates a charge transfer from the porphyrin to the ZnO, simultaneously creating a depletion of electrons, which favors the charge transfer from the donor-absorbed species.

Moreover, the main interfering species in the gas sensing process is humidity, especially at low T: Chen et al. [17] observed that the moisture can adsorb on the semiconductor oxide surface, interacting with the acetone molecules and leading to a dramatic change in the final sensor behavior. Indeed, a fluorine modified porphyrin, named as SnO_2/ZnTPP(F_{20}), was synthesized and tested, since fluorine atoms may confer hydrophobic character leading to a possible reduction of the water interference. Unfortunately, no positive results were obtained and the signal intensity halved compared to that of the pure SnO_2 powders, probably due to the strong electron density attractor capability of F-groups [10–12]. Notably, the sensor response of SnO_2/ZnTPP(F_{20}CN) at 150 °C (Figure 2a) is the most intense one among the tested hybrid materials, since this Zn^{II} porphyrin carries a cyano-acrylic moiety able to bind SnO_2 and to impart a proper directionality to charge-injection [11,12]. Moreover, the CN group acts as a buffer towards the strong electron acceptor behavior of F atoms, guaranteeing concomitantly the desired hydrophobicity to prevent the water interference. All the porphyrin-based sensors reached LOD values of 200 ppb at 150 °C, notwithstanding an increase in the response times of around 25–30% and the recovery times significantly longer (around 200 s).

Finally, further tests were carried out at 75 °C (Figure 2b): while no acetone response (20 ppm) was appreciable in the case of pure SnO_2, a reversed behavior in conductance with SnO_2/GO sample occurred. This phenomenon is reported to be typical of metal oxide semiconductors, operating at low temperatures due to a greater amount of adsorbed oxygen species and moisture [17]. Instead, under these conditions SnO_2/ZnTPP(F_{20}CN) produces a positive response, even if the LOD is 600 ppb, corroborating the synergistic effect between the fluorine species and the anchor group.

We believe that these findings can provide guidelines for the engineering of miniaturized chemoresistive sensors for low-temperature detection of acetone molecule. The excellent performances of the SnO_2-GO nano-heterojunctions and especially of SnO_2/ZnTPP(F_{20}CN) composite can pave the way for the development of tunable low-cost materials for the fabrication of optoelectronic devices for various applications.

Author Contributions: Conceptualization, E.P., F.T. and G.C.; methodology, E.P., G.L.C. and G.D.C.; validation, E.P.; formal analysis, E.P.; investigation, E.P.; data curation, E.P. and G.D.C.; writing—original draft preparation, G.D.C.; writing—review and editing, E.P., F.T. and G.D.C.; supervision, G.D.C. All authors have read and agreed to the published version of the manuscript.

Funding: This work received financial support from the Università degli Studi di Milano through the "PSR2019 Azione A" projects.

Institutional Review Board Statement: Not applicable.

Informed Consent Statement: Not applicable.

Data Availability Statement: Data are contained within the article.

Conflicts of Interest: The authors declare no conflict of interest.

References

1. Guo, Y.; Wen, M.; Li, G.; An, T. Recent advances in VOC elimination by catalytic oxidation technology onto various nanoparticles catalysts: A critical review. *Appl. Catal. B Environ.* **2021**, *281*, 119447. [CrossRef]
2. Ren, F.; Gao, L.; Yuan, Y.; Zhang, Y.; Alqrni, A.; Al-Dossary, O.M.; Xu, J. Enhanced BTEX gas-sensing performance of CuO/SnO_2 composite. *Sens. Actuators B Chem.* **2016**, *223*, 914–920. [CrossRef]
3. Pargoletti, E.; Hossain, U.H.; Di Bernardo, I.; Chen, H.; Tran-Phu, T.; Chiarello, G.L.; Lipton-Duffin, J.; Pifferi, V.; Tricoli, A.; Cappelletti, G. Engineering of SnO_2–Graphene Oxide Nanoheterojunctions for Selective Room-Temperature Chemical Sensing and Optoelectronic Devices. *ACS Appl. Mater. Interfaces* **2020**, *12*, 39549–39560. [CrossRef] [PubMed]
4. Güntner, A.T.; Abegg, S.; Königstein, K.; Gerber, P.A.; Schmidt-Trucksäss, A.; Pratsinis, S.E. Breath Sensors for Health Monitoring. *ACS Sens.* **2019**, *4*, 268–280. [CrossRef] [PubMed]
5. Tricoli, A.; Nasiri, N.; De, S. Wearable and Miniaturized Sensor Technologies for Personalized and Preventive Medicine. *Adv. Funct. Mater.* **2017**, *27*, 1605271. [CrossRef]
6. Pargoletti, E.; Hossain, U.H.; Di Bernardo, I.; Chen, H.; Tran-Phu, T.; Lipton-Duffin, J.; Cappelletti, G.; Tricoli, A. Room-temperature photodetectors and VOC sensors based on graphene oxide-ZnO nano-heterojunctions. *Nanoscale* **2019**, *11*, 22932–22945. [CrossRef] [PubMed]
7. Pargoletti, E.; Verga, S.; Chiarello, G.L.; Longhi, M.; Cerrato, G.; Giordana, A.; Cappelletti, G. Exploring $Sn_xTi_{1-x}O_2$ Solid Solutions Grown onto Graphene Oxide (GO) as Selective Toluene Gas Sensors. *Nanomaterials* **2020**, *10*, 761. [CrossRef]
8. Pargoletti, E.; Tricoli, A.; Pifferi, V.; Orsini, S.; Longhi, M.; Guglielmi, V.; Cerrato, G.; Falciola, L.; Derudi, M.; Cappelletti, G. An electrochemical outlook upon the gaseous ethanol sensing by graphene oxide-SnO_2 hybrid materials. *Appl. Surf. Sci.* **2019**, *483*, 1081–1089. [CrossRef]
9. Ekrami, M.; Magna, G.; Emam-djomeh, Z.; Yarmand, M.S.; Paolesse, R.; Di Natale, C. Porphyrin-Functionalized Zinc Oxide Nanostructures for Sensor Applications. *Sensors* **2018**, *18*, 2279. [CrossRef] [PubMed]
10. Di Carlo, G.; Orbelli Biroli, A.; Pizzotti, M.; Tessore, F.; Trifiletti, V.; Ruffo, R.; Abbotto, A.; Amat, A.; De Angelis, F.; Mussini, P.R. Tetraryl ZnII porphyrinates substituted at β-pyrrolic positions as sensitizers in dye-sensitized solar cells: A comparison with meso-substituted push-pull ZnII porphyrinates. *Chem. Eur. J.* **2013**, *19*, 10723–10740. [CrossRef] [PubMed]
11. Orbelli Biroli, A.; Tessore, F.; Di Carlo, G.; Pizzotti, M.; Benazzi, E.; Gentile, F.; Berardi, S.; Bignozzi, C.A.; Argazzi, R.; Natali, M.; et al. Fluorinated Zn^{II} porphyrins for dye-sensitized aqueous photoelectrosynthetic cells. *ACS Appl. Mater. Interfaces* **2019**, *11*, 32895–32908. [CrossRef] [PubMed]
12. Berardi, S.; Caramori, S.; Benazzi, E.; Zabini, N.; Niorettini, A.; Orbelli Biroli, A.; Pizzotti, M.; Tessore, F.; Di Carlo, G. Electronic properties of electron-deficient Zn(II) porphyrins for HBr splitting. *Appl. Sci.* **2019**, *9*, 2739. [CrossRef]
13. Di Carlo, G.; Orbelli Biroli, A.; Tessore, F.; Pizzotti, M.; Mussini, P.R.; Amat, A.; De Angelis, F.; Abbotto, A.; Trifiletti, V.; Ruffo, R. Physicochemical investigation of the panchromatic effect on β-substituted Zn^{II} porphyrinates for DSScs: The role of the π bridge between a dithienylethylene unit and the porphyrinic ring. *J. Phys. Chem. C* **2014**, *118*, 7307–7320. [CrossRef]
14. Americo, S.; Pargoletti, E.; Soave, R.; Cargnoni, F.; Trioni, M.I.; Chiarello, G.L.; Cerrato, G.; Cappelletti, G. Unveiling the acetone sensing mechanism by WO_3 chemiresistors through a joint theory-experiment approach. *Electrochim. Acta* **2021**, *371*, 137611. [CrossRef]
15. Anjali, K.; Jayaraj, L.K.N.; Ayyamperumal, C. Zinc-tetraphenylporphyrin grafted on functionalised mesoporous SBA-15: Synthesis and utilisation for nitroaldol condensation. *J. Porous Mater.* **2020**, *27*, 1191–1201. [CrossRef]
16. Li, X.; Cheng, Y.; Kang, S.; Mu, J. Preparation and enhanced visible light-driven catalytic activity of ZnO microrods sensitized by porphyrin heteroaggregate. *Appl. Surf. Sci.* **2010**, *256*, 6705–6709. [CrossRef]
17. Chen, H.; Bo, R.; Shrestha, A.; Xin, B.; Nasiri, N.; Zhou, J.; Di Bernardo, I.; Dodd, A.; Saunders, M.; Lipton-Duffin, J.; et al. NiO–ZnO Nanoheterojunction Networks for Room-Temperature Volatile Organic Compounds Sensing. *Adv. Opt. Mater.* **2018**, *6*, 1800677. [CrossRef]

chemistry proceedings

MDPI

Proceeding Paper

Field Nitrogen Dioxide and Ozone Monitoring Using Electrochemical Sensors with Partial Least Squares Regression †

Rachid Laref *, Etienne Losson, Alexandre Sava and Maryam Siadat

Laboratoire de Conception, Optimisation et Modélisation des Systèmes, LCOMS EA 7306 Université de Loraine, 57000 Metz, France; etienne.losson@univ-lorraine.fr (E.L.); alexandre.sava@univ-lorraine.fr (A.S.); maryam.siadat@univ-lorraine.fr (M.S.)

* Correspondance: rachid.laref@cnrs-imn.fr

† Presented at the 1st International Electronic Conference on Chemical Sensors and Analytical Chemistry, 1–15 July 2021; Available online: https://csac2021.sciforum.net/.

Abstract: Low-cost gas sensors detect pollutants gas at the parts-per-billion level and may be installed in small devices to densify air quality monitoring networks for the spread analysis of pollutants around an emissive source. However, these sensors suffer from several issues such as the impact of environmental factors and cross-interfering gases. For instance, the ozone (O_3) electrochemical sensor senses nitrogen dioxide (NO_2) and O_3 simultaneously without discrimination. Alphasense proposes the use of a pair of sensors; the first one, NO2-B43F, is equipped with a filter dedicated to measure NO_2. The second one, OX-B431, is sensitive to both NO_2 and O_3. Thus, O_3 concentration can be obtained by subtracting the concentration of NO_2 from the sum of the two concentrations. This technique is not practical and requires calibrating each sensor individually, leading to biased concentration estimation. In this paper, we propose Partial Least Square regression (PLS) to build a calibration model including both sensors' responses and also temperature and humidity variations. The results obtained from data collected in the field for two months show that PLS regression provides better gas concentration estimation in terms of accuracy than calibrating each sensor individually.

Keywords: partial least square regression; gas sensors; electrochemical sensors; air pollution monitoring

Citation: Laref, R.; Losson, E.; Sava, A.; Siadat, M. Field Nitrogen Dioxide and Ozone Monitoring Using Electrochemical Sensors with Partial Least Squares Regression. *Chem. Proc.* **2021**, *5*, 61. https://doi.org/10.3390/CSAC2021-10622

Academic Editor: Huangxian Ju

Published: 6 July 2021

1. Introduction

Urban air pollution is a major preoccupation [1]. Government organizations encourage research on low-cost gas sensors to improve their performances in order to complement the actual air pollution monitoring networks by providing better spatiotemporal resolution of the pollutants' spread [2]. Today, low-cost sensors such as electrochemical sensors can sense most pollutant gases at the magnitude of parts per billion (ppb) [3]. However, several limitations inhibit systems based on these sensors to reach high performance similar to the regular instruments [4]. Among these limitations is the influence of environmental factors—essentially, the temperature and humidity and the interfering gases present in the ambient air, particularly in the case of measuring O_3 and NO_2 [5]. The existing commercial electrochemical sensors for measuring O_3 respond simultaneously to O_3 and NO_2, without discrimination, because NO_2 and O_3 are reducible at similar potentials on carbon or gold electrodes [6]. Therefore, the responses of these sensors are proportional to the combined concentration of O_3 and NO_2. This nonselectivity of sensors becomes an obstacle for air monitoring applications where NO_2 and O_3 are present simultaneously with the same order of concentration magnitude. In this paper, we evaluate electrochemical sensors for O_3 and NO_2 for in field and in real application conditions. We propose to calibrate simultaneously the two sensors using Partial Least Square regression (PLS) while considering also the temperature and humidity variations. The remainder of this paper is organized as follows: we present first the experimental set up and data collection, then, the calibration procedure with results, and finally, a conclusion.

2. Experiment Set Up and Data Collection

In this work, we focused on measuring the concentration of O_3 and NO_2 in ambient air because they are the principal pollutant gases in French cities that are still exceeding the limited values defined by the European directives [7]. Therefore, we designed a device constituted of two electrochemical sensors provided by Alphasense LTD—NO2-B41F and OX-B431—dedicated to measure NO_2 and oxidizing gases, respectively. The device also contains the sensors' conditioning circuits, the gas exposure chamber, and the data acquisition unit. The conditioning circuits consist of potentiostat circuits that allow amplifying and converting the sensor electrode currents to voltages. The device is placed inside the air monitoring station managed by ATMO Grand Est agency. This station is located beside a highway, crossing Metz city, France. Our device works in dynamic mode for air sampling; thus, a pump and a mass flow controller are placed on the exposure chamber exit to generate a constant and continuous airflow by suction (Figure 1). We set the airflow rate to 500 mL/min, in order to obtain the same airflow rate as the ATMO Grand Est O_3 and NO_2 analyzers. The collected data represent the voltages of the sensor responses with a data sampling frequency of 200 Hz. Sensor responses are then averaged over a period of 10 s and recorded on a computer using Matlab software. Finally, recorded data are averaged again every 15 min in order to comply with the reference data provided by ATMO Grand Est. Data are collected continuously from 22 February 2019 to 14 April 2019.

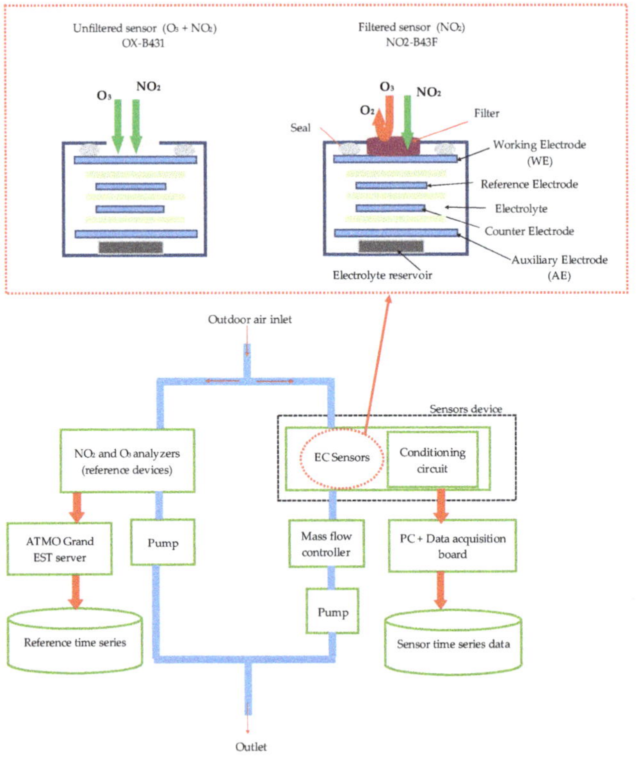

Figure 1. Experiment setup diagram and the schematic of filtered and unfiltered electrochemical sensors.

3. Sensors Calibration

To quantify O_3 and NO_2 concentrations, the manufacturer Alphasense recommends the use of a pair of electrochemical sensors: a model OX-B431 that responds to both gases (O_3 + NO_2) and a model NO2-B43F that responds only to NO_2. The NO2-B43F sensor is equipped with a manganese dioxide filter, which catalyzes O_3 into oxygen, thus preventing the sensor from responding to O_3 present in the environment (Figure 1). To determine the O_3 concentration, the contribution of NO_2 to the response of the OX-B431 sensor must be removed. Therefore, we first need to calculate the NO_2 concentration with the NO2-B43F sensor and then subtract it from the concentration provided by the OX-B431 sensor. The calibration procedure of this pair of sensors is performed as follows:

- Calibrate the NO2-B43F sensor for measuring NO_2 according to Equation (1):

$$[NO_2] = (WE_{NO2\text{-}B43F} - AE_{NO2\text{-}B43F})\, \alpha_1 + \alpha_2, \tag{1}$$

  where $[NO_2]$ is the concentration of the NO_2; $WE_{NO2\text{-}B43F}$ and $AE_{NO2\text{-}B43F}$ are signals of the working and auxiliary electrodes of NO2-B43F sensor, respectively; α_1 and α_2 are regression coefficients that can be determined by a simple linear regression.
- Calibrate the OX-B431 sensor to measure the mixture ($NO_2 + O_3$) according to Equation (2):

$$[NO_2 + O_3] = (WE_{OX\text{-}B431}) - AE_{OX\text{-}B431})\, b_1 + b_2, \tag{2}$$

  where $WE_{OX\text{-}B431}$ and $AE_{OX\text{-}B431}$ are signals of the working and auxiliary electrodes of OX-B431 sensor, respectively; b_1 and b_2 are regression coefficients determined by a simple linear regression.

The concentration of O_3 will be the difference between the concentration obtained by the OX-B431 sensor and the concentration obtained by the NO2-B43F sensor:

$$[O_3] = [NO_2 + O_3] - [NO_2], \tag{3}$$

The concentrations of both O_3 and NO_2 are typically 5 to 120 $\mu g/m^3$ at the roadside, so intelligent data analysis is required to differentiate each gas concentration. Our proposition is to combine both sensors' signals in the same equation plus the temperature and humidity variations:

$$[NO_2] = c_0 + c_1\, WE_{NO2\text{-}B43F} + c_2\, AE_{NO2\text{-}B43F} + c_3\, WE_{OX\text{-}B431} + c_4\, AE_{OX\text{-}B431} + c_5\, T + c_6\, H, \tag{4}$$

$$[O_3] = d_0 + d_1\, WE_{NO2\text{-}B43F} + d_2\, AE_{NO2\text{-}B43F} + d_3\, WE_{OX\text{-}B431} + d_4\, AE_{OX\text{-}B431} + d_5\, T + d_6\, H, \tag{5}$$

where c_0, $c_1 \ldots .c_6$ and d_0, $d_1 \ldots .d_6$ are regression coefficients determined by using PLS [8]; T and H are the temperature and humidity, respectively.

The comparison between calibration of each sensor individually and the combination of the two sensors' signals with temperature and humidity variation using PLS regression shows that the concentration estimation is better in the case of using PLS regression than the case of calibrating each sensor individually. Figure 2 illustrates that in the case of using PLS regression, the root-mean-square errors (RMSE) are 4.71 $\mu g/m^3$ and 6.89 $\mu g/m^3$ for NO_2 and O_3, respectively, whereas in the case of using each sensor individually, the RMSE values were 6.34 $\mu g/m^3$ and 8.76 $\mu g/m^3$, respectively. We note also that the estimation of NO_2 is better than the estimation of O_3 in both calibration cases. The reason behind this is that NO_2 estimation depends essentially on one sensor, whereas the estimation of O_3 depends on both sensors.

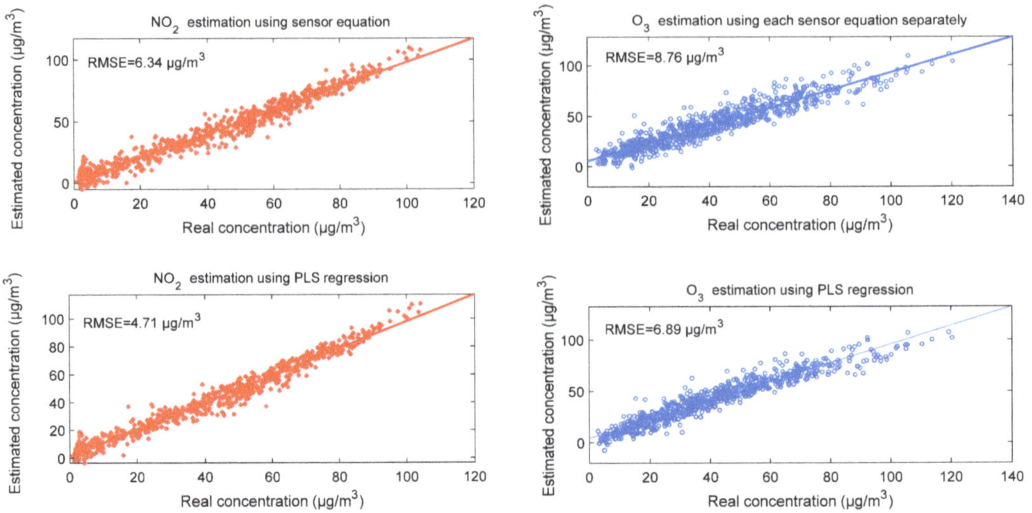

Figure 2. Correlation between reference concentration and estimated concentration of NO_2 and O_3.

4. Conclusions

In this work, electrochemical sensors calibration is proposed using PLS regression. First, we deployed a device to collect data in real outdoor conditions; then, we proposed multiple linear regression to estimate simultaneously nitrogen dioxide and ozone concentration. We found that the use of a pair with PLS regression is better than calibrating each sensor individually, as the RMSE reduced from 8.76 $\mu g/m^3$ to 6.89 $\mu g/m^3$ for ozone concentration estimation.

Author Contributions: The work presented here was carried out in collaboration between all authors. Conceptualization, R.L.; methodology, R.L.; validation, E.L., A.S., and M.S.; writing—original draft, R.L.; writing—review and editing, E.L., A.S., and M.S. All authors have read and agreed to the published version of the manuscript.

Funding: This research received no external funding.

Institutional Review Board Statement: Not applicable.

Informed Consent Statement: Not applicable.

Data Availability Statement: Not applicable.

Acknowledgments: We would like to thank ATMO GRAND EST agency for their support, including access to monitoring stations and data from their analyzers. The authors would also to thank Damien Durant from ATMO GRAND EST, head of metrological unit for his assistance and his helpful advice.

Conflicts of Interest: The authors declare no conflict of interest.

References

1. Helmut, M. Air pollution in cities. *Atmos. Environ.* **1999**, *33*, 4029–4037. [CrossRef]
2. Castell Balaguer, N.; Dauge, F.R.; Schneider, P.; Vogt, M.; Lerner, U.; Fishbain, B.; Broday, D.; Bartonova, A. Can commercial low-cost sensor platforms contribute to air quality monitoring and exposure estimates? *Environ. Int.* **2017**, *99*, 293–302. [CrossRef] [PubMed]
3. Wei, P.; Ning, Z.; Ye, S.; Sun, L.; Yang, F.; Wong, K.C.; Westerdahl, D.; Louie, P.K. Impact Analysis of Temperature and Humidity Conditions on Electrochemical Sensor Response in Ambient Air Quality Monitoring. *Sensors* **2018**, *18*, 59. [CrossRef] [PubMed]
4. Maag, B.; Zhou, Z.; Thiele, L.B. A Survey on Sensor Calibration in Air Pollution Monitoring Deployments. *IEEE Internet Things J.* **2018**, *5*, 4857–4870. [CrossRef]

5. Laref, R.; Losson, E.; Sava, A.; Siadat, M. Field Evaluation of Low Cost Sensors Array for Air Pollution Monitoring. In Proceedings of the 10th IEEE International Conference on Intelligent Data Acquisition and Advanced Computing Systems: Technology and Applications, Metz, France, 18–21 September 2019; pp. 849–853. [CrossRef]
6. Hossain, M.; Saffell, J.; Baron, R. Differentiating NO_2 and O_3 at Low Cost Air Quality Amperometric Gas Sensors. *ACS Sens.* **2016**, *1*, 1291–1294. [CrossRef]
7. Le Moullec, A. Bilan de la qualité de l'air extérieur en France en 2017. 2018, p. 36. Available online: https://www.statistiques.developpement-durable.gouv.fr/sites/default/files/2018-10/datalab-45-bilan-qualite-air-exterieur-france-2017-octobre2018.pdf (accessed on 5 July 2021).
8. Jurs, P.C.; Bakken, G.A.; McClelland, H.E. Computational methods for the analysis of chemical sensor array data from volatile analytes. *Chem. Rev.* **2000**, *100*, 2649–2678. [CrossRef] [PubMed]

chemistry proceedings

MDPI

Proceeding Paper

Tropomyosin Analysis in Foods Using an Electrochemical Immunosensing Approach †

Ricarda Torre [1], Maria Freitas [1], Estefanía Costa-Rama [2], Henri P. A. Nouws [1,*] and Cristina Delerue-Matos [1]

[1] REQUIMTE/LAQV, Instituto Superior de Engenharia do Porto, Instituto Politécnico do Porto, Dr. António Bernardino de Almeida 431, 4200-072 Porto, Portugal; rdvdt@isep.ipp.pt (R.T.); maria.freitas@graq.isep.ipp.pt (M.F.); cmm@isep.ipp.pt (C.D.-M.)

[2] Departamento de Química Física y Analitica, Universidad de Oviedo, Av. Julián Clavería 8, 33006 Oviedo, Spain; costaestefania@uniovi.es

* Correspondence: han@isep.ipp.pt

† Presented at the 1st International Electronic Conference on Chemical Sensors and Analytical Chemistry, 1–15 July 2021; Available online: https://csac2021.sciforum.net/.

Abstract: A screen-printed carbon electrode was used as the transducer for the development of an electrochemical immunosensor for the determination of tropomyosin (a major shrimp allergen) in food samples. Monoclonal and polyclonal antibodies were used in a sandwich-type immunoassay. The analytical signal was electrochemically obtained using an alkaline phosphatase-labelled secondary antibody and a 3-indoxyl phosphate/silver nitrate substrate. The total assay time was 2 h 50 min and allowed the quantification of tropomyosin between 2.5 and 20 ng mL^{-1}, with a limit of detection of 1.7 ng mL^{-1} The immunosensor was successfully applied to the analysis of commercial food products.

Keywords: seafood allergy; tropomyosin; shrimp; food allergy; screen-printed electrodes; electrochemical biosensor

check for updates

Citation: Torre, R.; Freitas, M.; Costa-Rama, E.; Nouws, H.P.A.; Delerue-Matos, C. Tropomyosin Analysis in Foods Using an Electrochemical Immunosensing Approach. *Chem. Proc.* **2021**, *5*, 62. https://doi.org/10.3390/CSAC2021-10471

Academic Editor: Ye Zhou

Published: 30 June 2021

1. Introduction

Over the past few years, food allergies have increasingly been regarded as a significant worldwide public health problem. Among shellfish allergies, shrimp is the predominant crustacean causing over 80% of allergic reactions that can result in severe hypersensitivity such as urticaria and asthma, and it is a major cause of anaphylaxis [1,2].

Tropomyosin (TPM), a major common allergenic protein found in seafood, is relatively resistant to peptic acidic digestion, which causes a continuous effect of the protein on the immune system. To protect the consumer from harmful allergens and potentially life-threatening reactions, food manufacturers are required to label and highlight shellfish-allergenic ingredients on food packages [3].

Currently, multiple technical approaches have been developed to identify the presence of shrimp tropomyosin in food, including enzyme-linked immunosorbent assays (ELISA), DNA detection, polymerase chain reaction (PCR), microarray and qualitative/semi-quantitative lateral flow assays. Although ELISA is the most commonly used method for TPM detection and quantification, it presents some disadvantages such as the long and tedious steps in the analysis procedure, long analysis times and high costs [4,5]. An alternative way to determine TPM in foods is through the use of electrochemical immunosensors. These sensors provide highly selective, sensitive, fast and cheap analysis and are suitable for in situ applications. Therefore, in this work, a simple voltametric immunosensor for the determination of TPM in commercial food products was developed. The immunoassay was based on a sandwich-type assay using screen-printed carbon electrodes (SPCE) as transducers. Monoclonal and polyclonal antibodies were used to capture and detect TPM. To obtain the analytical signal, an alkaline phosphatase-labelled secondary antibody and

3-indoxyl phosphate/silver nitrate (enzymatic substrate) were employed; the enzymatically deposited silver was analyzed by linear sweep voltammetry [6–8].

The applicability of the immunosensor was assessed by analyzing different food samples.

2. Materials and Methods

2.1. Instrumentation

Linear sweep voltametric analyses were performed using an Autolab PGSTAT204 potentiostat/galvanostat from Methrohm Autolab. Disposable screen-printed carbon electrodes (DRP-110) with a carbon working electrode (WE, d = 4 mm), a carbon counter electrode and a silver pseudoreference electrode were purchased from Methrohm DropSens.

2.2. Reagents and Solutions

Tris(hydroxymethyl)aminomethane (Tris, ≥99.8%), magnesium nitrate hexahydrate ($Mg(NO_3)_2$, 99%), nitric acid (HNO_3, ≥65%), 3-indoxyl phosphate (3-IP, ≥98%), silver nitrate ($AgNO_3$, ≥99.9995%), β-casein from bovine milk (≥98%), and bovine serum albumin (BSA) were obtained from Sigma-Aldrich.

Mouse IgG_1 monoclonal antibody (capture antibody, C-Ab), purified natural shrimp tropomyosin standard (antigen) and rabbit polyclonal antiserum shrimp tropomyosin (detection antibody, D-Ab) were purchased from Indoor Biotechnologies. An alkaline phosphatase goat anti-rabbit IgG antibody (AP-Ab) was supplied by Invitrogen. Throughout the work, ultra-pure water (resistivity = 18.2 MΩ cm), obtained from a Millipore (Simplicity 185) water purification system, was used. Working solutions of BSA, the antibodies and the antigen were prepared in 0.1 M Tris-HNO_3 pH 7.4 buffer (Buffer 1, B1). A second buffer (B2, 0.1 M Tris-HNO_3 pH 9.8 containing $Mg(NO_3)_2$ (2×10^{-2} M)) was used to prepare the solution containing 3-IP (1×10^{-3} M) and $AgNO_3$ (4×10^{-4} M).

2.3. Sample Preparation

Shrimp, shrimp sauce and crab and chicken paste were used to evaluate the immunosensor's applicability to food analysis. Samples were prepared as follows: (a) 1 g of sample was mixed with 10 mL of Tris-HNO_3 (pH 8.2, 1% NaCl) at 60 °C during 15 min in a water bath; (b) the resulting suspension was then centrifuged at 2500 rpm for 20 min and (c) the supernatant was divided in aliquots and stored at −20 °C until use.

2.4. Immunosensor Assay and Electrochemical Measurements

The representative scheme of the immunosensor assay and detection strategy is presented in Figure 1. The WE of the SPCE was coated with C-Ab (10 µL, 20 µg mL^{-1}) and left to incubate overnight at 4 °C. After rinsing the sensor with buffer B1, surface blocking was carried out using 40 µL of a 2-% (m/V) BSA solution during 30 min. After this, the sensor was washed with buffer B1 and incubated with 40 µL of a previously mixed (10 min before use) solution containing the antigen, the detection antibody (1:2000) and BSA (1% (m/V)) during 60 min. After rinsing with buffer B1, 40 µL of an AP-Ab solution (1:40,000) was placed on the sensor for 60 min. The sensor was then rinsed with buffer B2, and the enzymatic reaction was carried out by depositing 40 µL of a mixed solution containing 3-IP and silver nitrate on the SPCE for 20 min. LSV was used to record the analytical signal (potential range: −0.03 V to +0.4 V; scan rate: 50 mV/s). All analyses were performed in triplicate and carried out at room temperature (20 ± 1 °C).

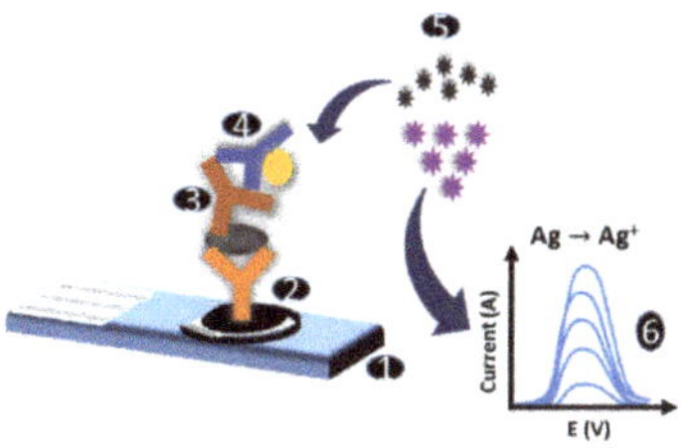

Figure 1. Schematic representation of the developed immunoassay. (1) Screen-printed carbon electrode; (2) C-Ab immobilization; (3) addition of a mixture containing standard/sample and D-Ab; (4) addition of AP-Ab; (5) addition of the enzymatic substrate (3-IP) and silver ions; and (6) voltametric detection of Ag^0.

3. Results and Discussion

3.1. Optimization Studies

The immunosensing strategy was based on a sandwich-type assay performed on bare SPCEs as transducers. In the first phase of the immunosensor development, two different surface blockers were tested: β-casein (2% (m/V)) and BSA (2% (m/V)). As can be observed in Figure 2, when BSA was used, the highest peak current intensity (i_p) and signal-to-blank ratio (S/B) was obtained.

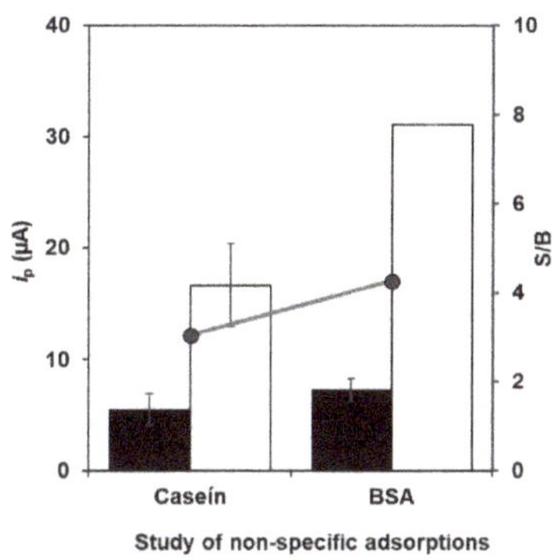

Figure 2. Peak current intensities (i_p) obtained for the study of the surface blocker (casein and BSA, both at 2% (m/V)). Black bars: blank assay. White bars: TPM (10 ng mL^{-1}). Results are presented as average ± standard deviation (n = 3). Experimental conditions: C-Ab—10 μg mL^{-1}; D-Ab—1:250 dilution; AP-Ab—1:20,000 dilution; 3-IP—1.0×10^{-3} M; and $AgNO_3$—4.0×10^{-4} M.

In order to select the optimum concentrations of both the capture and detection antibodies, a standard solution of tropomyosin (10 ng mL^{-1}) was used. First, for fixed dilutions of D-Ab (1:250) and AP-Ab (1:20,000), different C-Ab concentrations of between 2.5 and 20 μg mL^{-1} were tested. The obtained results reveal that a concentration of 20 μg mL^{-1} resulted in the highest peak current intensity and S/B ratio. After this and maintaining the AP-Ab dilution at 1:20,000, different D-Ab dilutions (between 1:250 and 1:12,000) were tested. The selected dilution was 1:2000 because the highest i_p and lowest blank signal were obtained. After selecting the C-Ab concentration (20 μg mL^{-1}) and D-Ab dilution (1:2000), different assay formats were studied in order to reduce the number of incubation steps and, subsequently, the assay time. Different steps were combined and the most adequate combination, the previous mixing of the antigen with the D-Ab, led to a 60-min reduction in the assay time. The next studies were performed to select the optimum AP-Ab dilution by testing dilutions of between 1:10,000 and 1:40,000. A 1:40,000 dilution was selected because a low blank signal and the highest S/B ratio were observed. After this, the AP-Ab incubation time was studied between 15 and 60 min, with the best results obtained for the 60 min incubation time. A summary of the optimization studies is indicated in Table 1.

Table 1. Optimization of the different experimental variables involved in the construction of the immunosensor for TPM analysis.

Variable	Studied Range	Selected Value
(C-Ab), μg mL^{-1}	2.5–20	20
(D-Ab), dilution	1:250–1:12,000	1:2000
(AP-Ab), dilution	1:10,000–1:40,000	1:40,000
AP-Ab incubation time, min	15–60	60

3.2. Analytical Performance

To establish the performance characteristics of the immunosensor, standard solutions with different TPM concentrations (2.5–50 ng mL^{-1}) were analyzed. A linear relationship was observed between 2.5 and 20 ng mL^{-1} (i_p (μA) = 0.787 (tropomyosin) (ng mL^{-1}) + 5.45, r = 0.990, n = 5). Examples of voltammograms in the linear range (Figure 3a) and the calibration plot (Figure 3b) are shown in Figure 3. The limit of detection (LOD) was calculated as three times the standard deviation of the blank divided by the slope and the value obtained was 1.7 ng mL^{-1}. The limit of quantification (LOQ) was calculated as 10 times the standard deviation of the blank divided by the slope, obtaining a concentration of 5.7 ng mL^{-1}. The coefficient of variation of the method was <9%.

3.3. Selectivity and Interference Studies

The selectivity of the sensor towards TPM was evaluated by analyzing other allergens such as Ara h 1 (peanut allergen, 250 ng mL^{-1}), Cyp C 1 (fish allergen, 20 ng mL^{-1}) and Ovalbumin (GAL d 2, chicken egg allergen, 1% (m/V)). Examples of the obtained voltammograms are shown in Figure 3c. Besides these allergens, histamine (6.8 mg mL^{-1}), a biogenic amine and the most important fish freshness indicator, was also included in this study. The signal for all these compounds was similar to the blank signal, confirming the selectivity of the proposed sensor. Besides this, TPM was mixed with each of the compounds to evaluate their interference in the analysis. The obtained signals were nearly the same as the one obtained for a 10-ng mL^{-1} TPM solution, which indicates that the other allergens and histamine did not significantly interfere in the analysis.

3.4. Applicability to Food Analysis

The feasibility of the sensor for the determination of TPM in commercial food samples was tested. Shrimp, shrimp sauce and crab paste were analyzed, obtaining TPM concentrations of 80.42 ± 2.7 μg g^{-1}, 170.4 ± 1.80 ng g^{-1} and 21.6 ± 4.13 ng g^{-1}, respectively. The developed immunosensor was also used to detect the presence of TPM in chicken paste. As expected, this sample gave a negative result (no significant difference when compared with the blank signal), so the TPM concentration was below the sensor's LOD. Examples of the obtained voltammograms are shown in Figure 3d.

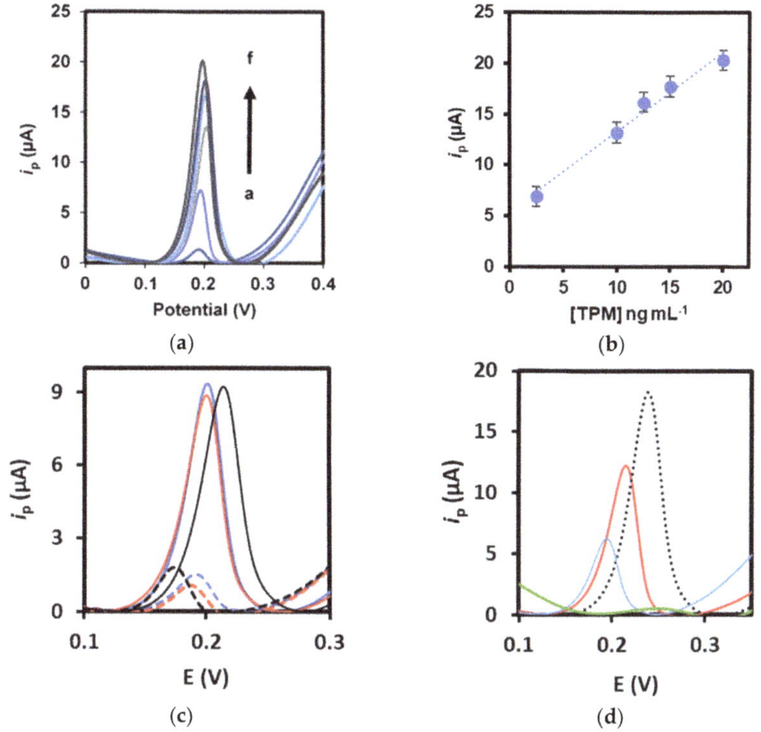

Figure 3. (**a**) Examples of voltammograms in the linear range (a—blank; b—2.5 ng mL^{-1}; c—10 ng mL^{-1}; d—12.5 ng mL^{-1}; e—15 ng mL^{-1}; and f—20 ng mL^{-1}). (**b**) Calibration plot. (**c**) Examples of voltammograms obtained in the selectivity and interference studies: TPM (10 mg L^{-1}, blue line, control) combined with Cyp C 1 (20 ng mL^{-1}, red line) and Ovalbumin (1% (m/V), black line) and blank (0 ng mL^{-1}, blue dashed line, control) with the addition of Cyp c 1 (200 ng mL^{-1}, red dashed line) and Ovalbumin (1% (m/V), black dashed line). (**d**) Examples of voltammograms obtained in the analysis of food samples (shrimp sauce—black dashed line; shrimp—red line; crab paste—blue line; and chicken paste—green line). Experimental conditions: C-Ab—20 μg mL^{-1}; BSA—2% (m/V); mixture of standard TPM solutions with D-Ab—1:2000; AP-Ab—1:40,000; 3-IP—1×10^{-3} M; and $AgNO_3$—4×10^{-4} M.

4. Conclusions

The current trends in analytical chemistry are focused on the development of simple and in situ analysis devices to ensure food safety. In this work, a simple immunosensor for tropomyosin analysis was developed. This immunoassay only takes 2 h 50 min, and it requires 40 μL of sample to perform the analysis. The sensor can determine tropomyosin in a concentration range between 2.5 and 20 ng mL^{-1} and a limit of detection of 1.7 ng mL^{-1} was achieved. The developed methodology fulfills the requirements of (bio)sensor construction such as small size and the use of low amounts of reagents and samples. Moreover, it allows the possibility of decentralized analysis, which could be useful for the control of tropomyosin, avoiding cases of food allergy.

Author Contributions: Conceptualization, M.F. and H.P.A.N.; methodology, M.F. and H.P.A.N.; validation, all authors; formal analysis, M.F. and R.T.; investigation, M.F. and R.T.; resources, H.P.A.N. and C.D.-M.; data curation, M.F. and R.T.; writing—original draft preparation, R.T.; writing—review and editing, M.F. and H.P.A.N.; visualization, R.T.; supervision, M.F., H.P.A.N., E.C.-R. and C.D.-M.;

project administration, H.P.A.N.; funding acquisition, H.P.A.N. and C.D.-M. All authors have read and agreed to the published version of the manuscript.

Funding: This research was funded by the European Union (FEDER funds through COMPETE POCI-01-0145-FEDER-030735) and National Funds (Fundação para a Ciência e a Tecnologia) through the project PTDC/QUI-QAN/30735/2017—TracAllerSens—Electrochemical sensors for the detection and quantification of trace amounts of allergens in food products.

Institutional Review Board Statement: Not applicable.

Informed Consent Statement: Not applicable.

Data Availability Statement: Not applicable.

Acknowledgments: This work was also supported by Portuguese national funds (FCT/MCTES, Fundação para a Ciência e a Tecnologia and Ministério da Ciência, Tecnologia e Ensino Superior) through the project UIDB/50006/2020.

Conflicts of Interest: The authors declare no conflict of interest.

References

1. Jiang, D.; Ji, J.; Sun, X.; Zhang, Y.; Zhang, G.; Tang, L. Mast cell-based electrochemical biosensor for quantification of the major shrimp allergen Pen a 1 (tropomyosin). *Biosens. Bioelectron.* **2013**, *50*, 150–156. [CrossRef]
2. Angulo-Ibáñez, A.; Eletxigerra, U.; Lasheras, X.; Campuzano, S.; Merino, S. Electrochemical tropomyosin allergen immunosensor for complex food matrix analysis. *Anal. Chim. Acta* **2019**, *1079*, 94–102. [CrossRef]
3. Wang, Y.; Rao, Z.; Zhou, J.; Zheng, L.; Fu, L. A chiral assembly of gold nanoparticle trimer-based biosensors for ultrasensitive detection of the major allergen tropomyosin in shellfish. *Biosens. Bioelectron.* **2019**, *132*, 84–89. [CrossRef]
4. Mohamad, A.; Rizwan, M.; Keasberry, N.A.; Nguyen, A.S.; Lam, T.D.; Ahmed, M.U. Gold-microrods/Pd-nanoparticles/polyaniline-nanocomposite-interface as a peroxidase-mimic for sensitive detection of tropomyosin. *Biosens. Bioelectron.* **2020**, *155*, 112108. [CrossRef]
5. Chinnappan, R.; Rahamn, A.A.; AlZabn, R.; Kamath, S.; Lopata, A.L.; Abu-Salah, K.M.; Zourob, M. Aptameric biosensor for the sensitive detection of major shrimp allergen, tropomyosin. *Food Chem.* **2020**, *314*, 126133. [CrossRef] [PubMed]
6. Fanjul-Bolado, P.; Hernández-Santos, D.; González-García, M.B.; Costa-García, A. Alkaline Phosphatase-Catalyzed Silver Deposition for Electrochemical Detection. *Anal. Chem.* **2007**, *79*, 5272–5277. [CrossRef] [PubMed]
7. Neves, M.M.P.S.; González-García, M.B.; Nouws, H.P.A.; Costa-García, A. An electrochemical deamidated gliadin antibody immunosensor for celiac disease clinical diagnosis. *Analyst* **2013**, *138*, 1956–1958. [CrossRef] [PubMed]
8. Neves, M.M.P.S.; González-García, M.B.; Nouws, H.P.A.; Costa-García, A. Celiac disease detection using a transglutaminase electrochemical immunosensor fabricated on nanohybrid screen-printed carbon electrodes. *Biosens. Bioelectron.* **2012**, *31*, 95–100. [CrossRef] [PubMed]

chemistry proceedings

Proceeding Paper

Simultaneous Sensing of Codeine and Diclofenac in Water Samples Using an Electrochemical Bi-MIP Sensor and a Voltammetric Electronic Tongue †

Hafsa El Youbi [1], Alassane Diouf [1,2], Benachir Bouchikhi [2] and Nezha El Bari [1,*]

1 Biosensors and Nanotechnology Group, Department of Biology, Faculty of Sciences, Moulay Ismaïl University of Meknes, B.P. 11201, Zitoune, Meknes 50003, Morocco; hafsa2017elyoubi@gmail.com (H.E.Y.); Alou20081@hotmail.fr (A.D.)

2 Sensor Electronic & Instrumentation Group, Department of Physics, Faculty of Sciences, Moulay Ismaïl University of Meknes, B.P. 11201, Zitoune, Meknes 50003, Morocco; benachir.bouchikhi@gmail.com

* Correspondence: n_elbari@hotmail.com

† Presented at the 1st International Electronic Conference on Chemical Sensors and Analytical Chemistry, 1–15 July 2021; Available online: https://csac2021.sciforum.net/.

Abstract: Codeine and diclofenac overdoses have been widely reported. Here, a biomimetic sensor (bi-MIP) was devised, and an electronic tongue was used to analyze water samples simultaneously containing both these drugs. The bi-MIP sensor limits of detection for diclofenac and codeine taken individually were 0.01 μg/mL and 0.16 μg/mL, respectively. Due to a cross-reactivity effect when using the bi-MIP sensor, the electronic tongue was shown to differentiate samples containing both analytes. The results confirm the feasibility of simultaneous detection of two target analytes via a bi-MIP sensor. Additionally, they demonstrate the ability of a multi-sensor to classify different water samples.

Keywords: drug analysis; molecularly imprinted polymer sensor; nanoparticles; electrochemical multi-sensor; chemometrics; water

Citation: El Youbi, H.; Diouf, A.; Bouchikhi, B.; El Bari, N. Simultaneous Sensing of Codeine and Diclofenac in Water Samples Using an Electrochemical Bi-MIP Sensor and a Voltammetric Electronic Tongue. *Chem. Proc.* **2021**, 5, 63. https://doi.org/10.3390/CSAC2021-10483

Academic Editor: Nicole Jaffrezic-Renault

Published: 30 June 2021

1. Introduction

Diclofenac (DCF) and codeine (COD) are drugs administered to treat certain human health problems. Here the focus is first on DCF, which is a non-steroidal anti-inflammatory drug (NSAID), widely prescribed for the treatment of a wide variety of conditions. It reduces the need for morphine after surgery and is effective against menstrual pain and endometriosis. Although DCF has outstanding medical features, it is sometimes misused and can, as a result, easily move into the synovial fluid. This unfortunately leads to a reduction in the secretion of prostaglandins [1]. In consequence, the consumer can experience many health problems [2].

The second study focus is on COD, which is an opiate used clinically for its analgesic, antitussive and antidiarrheal properties. However, it is said to be addictive and can cause psychological damage to the patient if abused. Extreme consumption of COD can even cause death [3]. For these reasons, the World Health Organization (WHO), the US Food and Drug Administration (FDA), and the European Medicines Agency (EMA), among other international organizations, have issued strict warnings about the adverse effects of COD [4].

Electrochemical methods are very good candidates for drug analysis [5]. This is attributed to their low cost, lower detection limits, wide range of potential windows, and ease of surface renewal.

Firstly, electrochemical devices based on molecularly imprinted polymers (MIPs) can be considered as good alternatives to conventional techniques. However, according to our literature research, the MIP strategy has not yet been exploited for the simultaneous detection of these two analytes. Currently, the immobilization of MIPs, as a sensing element on portable electrochemical transducers, such as screen-printed electrodes (SPEs), offers an interesting approach. A study has been reported for the detection of dopamine and uric acid using MIP technology.

Secondly, as drugs are usually released in wastewater, and wastewater treatment plants are not totally efficient, this work focuses on the analysis of mineral water samples with different concentrations of the drugs in question. When multiple targets are to be detected, it is appropriate to use various electrical interfaces, such as multi-sensor systems.

The following content of this study is devoted to the qualitative analysis of drugs in mineral water samples using a voltammetric electron tongue (VE-Tongue) combined with chemometric methods. When using the bi-MIP sensor, a cross-reactivity effect due to the presence of several compounds was encountered. To avoid it, qualitative analysis via VE-Tongue can help to classify/discriminate drug samples with different concentrations of the drugs in question.

Taking all these points into consideration, the primary objective of this paper was to report on the development of an electrochemical sensor based on molecularly imprinted polymers for the simultaneous detection of DCF and COD. Electrochemical techniques, such as electrochemical impedance spectroscopy (EIS), differential pulse voltammetry (DPV), and cyclic voltammetry (CV), were used to investigate the electrochemical behavior of the electrodes during the different steps of the bi-MIP sensor fabrication. Principal components analysis (PCA) was used to process the database from the VE-Tongue sensor array for the purpose of discriminating between water samples containing DCF and COD.

2. Materials and Methods

2.1. Samples

Five sets of mineral water samples were prepared for the electrochemical analysis:

Set 1: Mineral water sample used as reference sample which was not spiked;

Set 2: Mineral water samples spiked with different concentration of diclofenac (0.001, 0.01, 0.1, 1, 10, 100, 300, 500 μg/mL);

Set 3: Mineral water samples spiked with codeine at the same concentrations as described above;

Set 4: Mineral water samples spiked with diclofenac at the same concentrations as described above, but each containing 300 μg/mL codeine;

Set 5: Mineral water samples spiked with codeine at the same concentrations as described above, but each containing 300 μg/mL diclofenac.

2.2. Instrumentation and Electrochemical Techniques

Figure 1 shows the experimental setup used in this study. The five sets described above were studied using both detection systems (i.e., bi-MIP sensor and VE-Tongue).

The bi-MIP sensor was designed on a screen-printed gold electrode (Au-SPE).

The voltammetric electronic tongue (VE-Tongue) consisted of an array of 5 working electrodes made of gold, copper, glassy carbon, platinum, and palladium. A silver/silver chloride (Ag/AgCl) reference electrode and a platinum counter electrode completed the three-electrode configuration.

A computer interfaced to a potentiostat device was used for data acquisition. Using the potentiostat, electrochemical characterization techniques, including CV, DPV and EIS, were run.

These three established techniques were used for the electrochemical measurements. The CV was operated from −0.4 to 0.6 V at a scan rate of 30 mV/s. To investigate the surface properties of the bi-MIP sensor, the EIS was performed in an open circuit at a low AC potential of 10 mV amplitude and a frequency range of 0.1 to 50,000 Hz. The retention properties of the bi-MIP sensor were investigated using DPV over a potential range of

−0.2 to 0.3 V and a slew rate of 50 mV/s. All measurements were performed at room temperature (25 °C).

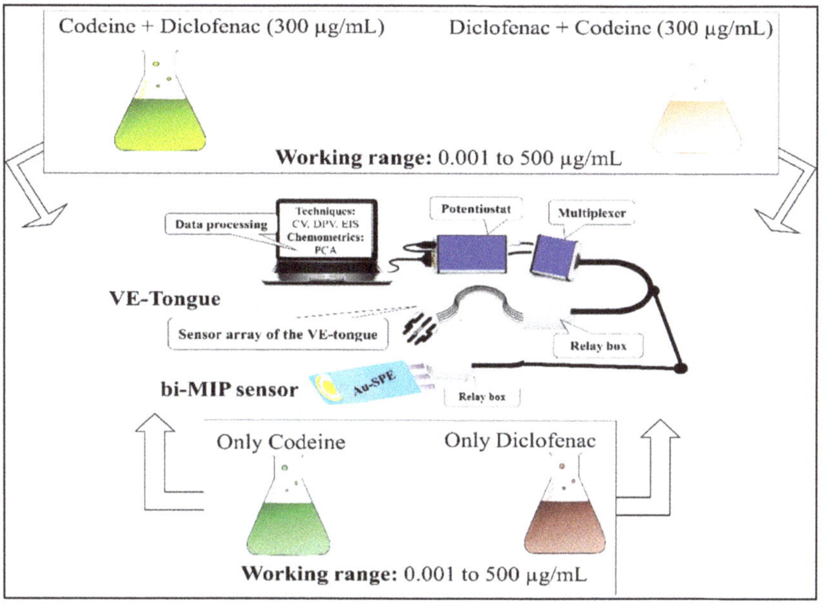

Figure 1. Graphical overview of the experimental setup.

2.3. Bi-MIP Sensor Preparation

Figure 2 illustrates the procedures for the bi-MIP sensor elaboration. Briefly, a layer of polyvinyl carboxylic chloride (PVC-COOH) was first assembled to modify the bare Au-SPE. Then, after activation of -COOH groups by 1-ethyl-3-(3-dimethylaminipropyl) carbodiimide (EDC) and N-hydrosuccinimide (NHS), a solution (1 mg/mL), containing simultaneously DCF and COD, was deposited on the modified electrode. After DCF and COD binding, a solution containing methacrylic acid, as the functional monomer, and silver nanoparticles (AgNPs) was immobilized. An extraction stage of template molecules followed to complete the fabrication of the bi-MIP sensor.

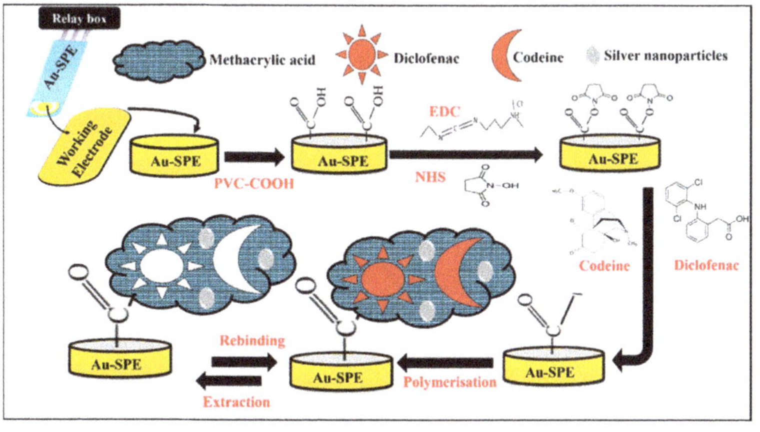

Figure 2. The development stages of the bi-MIP sensor.

2.4. Data Analysis

The multivariate responses of the VE-Tongue were processed by a known unsupervised method called PCA. This statistical technique reduces the dimensionality of the multivariate data while retaining maximum information on new variables called principal components (PCs) [6,7]. This allows for better visualization of the data and better interpretation of the analyzed samples.

3. Results and Discussion

3.1. Biomimetic Receptor Assembly

During the development of the biomimetic sensor, several immobilization procedures to form the sensitive layer were performed. After each step, the electrochemical behavior of the electrode was observed using a supporting electrolyte (PBS pH 7.4) containing electroactive species ($[Fe(CN)_6]^{4-/3-}$). For this purpose, the CV and EIS techniques were run in PsTRACE software. The results of these characterizations are presented in Figure 3. At each step of the sensor development, the electrochemical behavior of the electrode changed compared to the bare electrode. Moreover, the CV and EIS results were in good agreement.

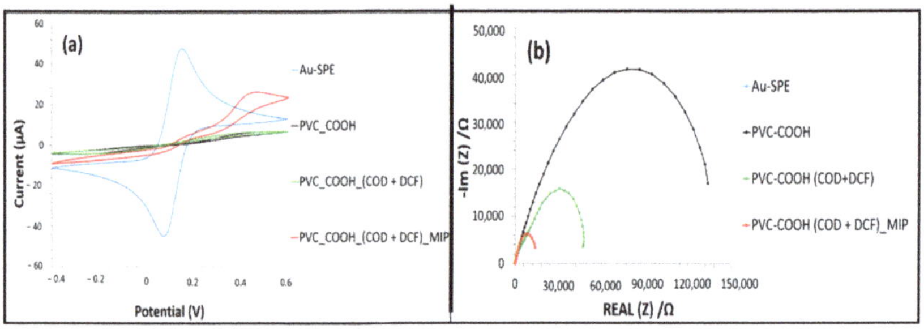

Figure 3. Electrochemical signals corresponding to the development stages of the bi-MIP sensor: (**a**) cyclic voltammograms, (**b**) *Nyquist* diagrams.

3.2. Bi-MIP Sensor Responses

In the first step, the analysis of DCF alone (set 2), at different concentrations on the bi-MIP sensor, was performed using the differential pulse voltammetry (DPV) technique. The calibration curves related to these responses are shown in Figure 4. A clear decrease in the amplitude of the voltammograms was observed as the concentration of DCF increased, expressed in the linear regression equation shown in Figure 4a. The equation is y = −0.083Log (C)—0.355 with a determination coefficient R^2 = 0.93. The calculated detection limit was 0.01 µg/mL using the formula described by *DIOUF* et al. [8].

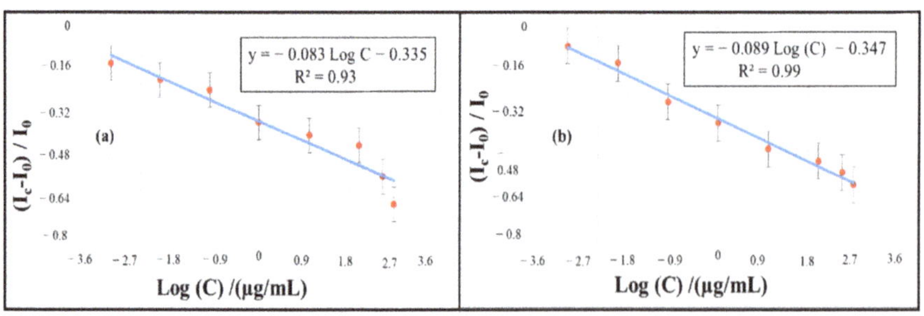

Figure 4. Calibration curves of the bi-MIP sensor with increasing concentrations of: (**a**) diclofenac (**b**) codeine, from 0.001 to 500 µg/mL.

Secondly, COD alone (set 3) was analyzed under the same conditions. The corresponding equation of the bi-MIP sensor responses (voltammograms) is shown in Figure 4b. Here, a similar trend to that of the DCF was obtained with a calibration equation of y = −0.089Log (C) − 0.347 with R^2 = 0.98. The limit of detection was 0.16 µg/mL.

When detecting the two analytes individually, it was found that the bi-MIP sensor had almost equivalent sensitivity. However, because of cross-reactivity, the results for the simultaneous detection of both analytes by the bi-MIP were not satisfactory. An electronic tongue was used to explore a potential strategy to address this.

3.3. PCA Analysis of the VE-Tongue Dataset

Due to cross-reactivity and limitations encountered with the bi-MIP sensor, measurement of samples containing both target analytes simultaneously was performed using the VE-Tongue. After data pre-processing, principal components analysis (PCA) was used to classify the samples from all sets. The results are presented in Figure 5, which shows the projections of the experimental results onto a two-dimensional (2D) space formed by the first two principal components; 78.90% of the total variance of the data was explained by the first two PCs indicating significant pattern separation.

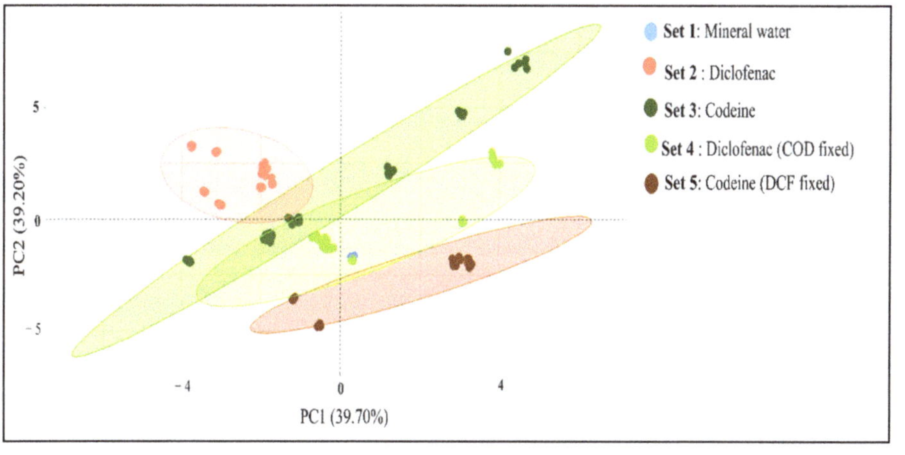

Figure 5. PCA plot showing the discrimination of the different sets using ΔI and Area as features. ΔI is the difference between the maximum current of the oxidation wave and the reduction wave. Area is the area of the VE-Tongue response (voltammogramme) using the trapezoidal method.

PCA was also applied to data after analysis of samples from set 4 and set 5, according to their concentrations.

Set 4 contained water samples with varying concentrations of DCF and a fixed concentration of COD (300 µg/mL) for each. As shown in Figure 6a, all samples in set 4 were well separated with only 85.87% of the total variance expressed by PC1 and PC2. In addition, the samples containing low and high concentrations of DCF clustered in the top right and bottom of the graph, respectively.

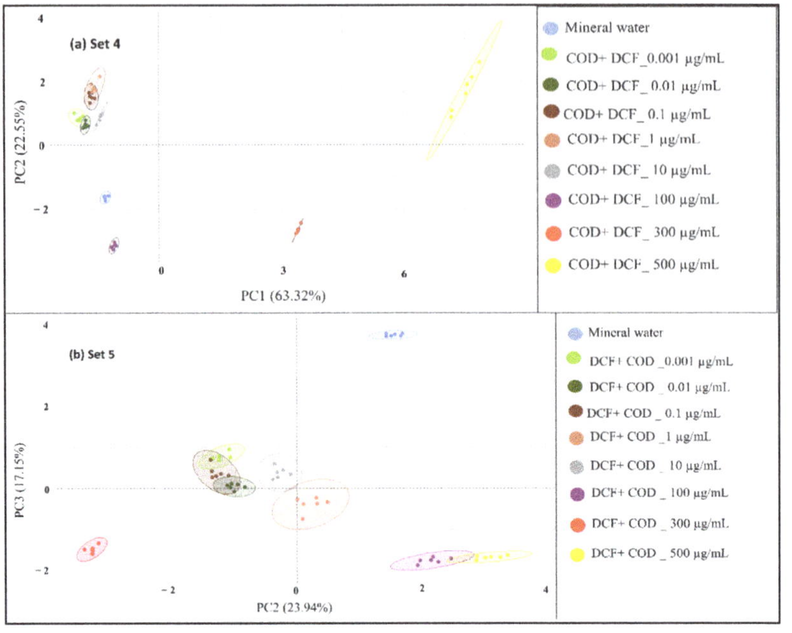

Figure 6. PCA plot showing the discrimination between set 1 and water samples of (**a**) set 4 and (**b**) set 5 at different concentrations using ΔI and Area as features.

In Figure 6b, the same trend is also observed for the analysis of samples in set 5. In this set, COD was varied but the concentration of DCF was maintained at 300 μg/mL. In the graph, the clean water sample and the spiked samples are well separated, with a score of 41.1% of the total variance, expressed as PC2 and PC3.

These results clearly show that the VE-Tongue was able to discriminate water samples containing several compounds at different concentrations.

4. Conclusions

In this study, a new bi-MIP-sensor-based electrochemical detection system for the detection of diclofenac and codeine was proposed. The principle of simultaneous detection was highlighted by using an electronic tongue combined with pattern recognition methods. The proposed analytical tools represent a breakthrough in water analysis.

Author Contributions: Conceptualization, N.E.B.; methodology, H.E.Y., N.E.B.; Software, B.B.; validation, N.E.B.; formal analysis, H.E.Y., N.E.B. investigation, N.E.B., H.E.Y.; resources, N.E.B.; data curation, N.E.B., H.E.Y.; writing—original draft preparation, H.E.Y., N.E.B., A.D.; writing—review and editing, H.E.Y., N.E.B.; visualization, supervision, N.E.B.; project administration, N.E.B.; funding acquisition, N.E.B.; Software: B.B. All authors have read and agreed to the published version of the manuscript.

Funding: This research received no external funding.

Institutional Review Board Statement: Not applicable.

Informed Consent Statement: Not applicable.

Data Availability Statement: Not applicable.

Acknowledgments: Authors gratefully acknowledge Moulay Ismaïl University of Meknes for financial support of the project "Research support".

Conflicts of Interest: The authors declare no conflict of interest.

References

1. Kormosh, Z.; Hunka, I.; Bazel, Y.; Laganovsky, A.; Mazurenko, I.; Kormosh, N. Determination of diclofenac in pharmaceuticals and urine samples using a membrane sensor based on the ion associate of diclofenac with Rhodamine B. *Open Chem.* **2007**, *5*, 813–823. [CrossRef]
2. Elbalkiny, H.T.; Yehia, A.M.; Safáa, M.R.; Elsaharty, Y.S. Potentiometric diclofenac detection in wastewater using functionalized nanoparticles. *Microchem. J.* **2019**, *145*, 90–95. [CrossRef]
3. Häkkinen, M.; Launiainen, T.; Vuori, E.; Ojanperä, I. Comparison of fatal poisonings by prescription opioids. *Forensic Sci. Int.* **2012**, *222*, 327–331. [CrossRef] [PubMed]
4. Tobias, J.D.; Green, T.P.; Coté, C.J. Codeine: Time to say "no". *Pediatrics* **2016**, *138*, e20162396. [CrossRef]
5. Gimenes, D.T.; Cunha, R.R.; de Carvalho Ribeiro, M.M.A.; Pereira, P.F.; Muñoz, R.A.A.; Richter, E.M. Two new electrochemical methods for fast and simultaneous determination of codeine and diclofenac. *Talanta* **2013**, *116*, 1026–1032. [CrossRef] [PubMed]
6. Haddi, Z. Conception et Développement d'un Système Multicapteurs en gaz et en Liquide pour la Sécurité Alimentaire. Ph.D. Thesis, Université Claude Bernard—Lyon I, Lyon, France, 2013.
7. Motia, S.; Bouchikhi, B.; El Bari, N. An electrochemical sensor based on molecularly imprinted polymer conjointly with a voltammetric electronic tongue for quantitative diphenyl phosphate detection in urine samples from cosmetic product users. *Sens. Actuators B-Chem.* **2021**, *332*, 129449. [CrossRef]
8. Diouf, A.; Bouchikhi, B.; El Bari, N. A nonenzymatic electrochemical glucose sensor based on molecularly imprinted polymer and its application in measuring saliva glucose. *Mater. Sci. Eng. C* **2019**, *98*, 1196–1209. [CrossRef]

chemistry proceedings

MDPI

Proceeding Paper

Electrochemical Detection of Fenthion Insecticide in Olive Oils by a Sensitive Non-Enzymatic Biomimetic Sensor Enhanced with Metal Nanoparticles †

Youssra Aghoutane [1,2], Nezha El Bari [1,*], Zoubida Laghrari [1] and Benachir Bouchikhi [2]

1 Biosensors and Nanotechnology Group, Department of Biology, Faculty of Sciences, Moulay Ismaïl University, B.P. 11201, Zitoune, Meknes 50000, Morocco; youssi.aghoutane@gmail.com (Y.A.); zlaghrari@gmail.com (Z.L.)

2 Sensor Electronic & Instrumentation Group, Department of Physics, Faculty of Sciences, Moulay Ismaïl University of Meknes, B.P. 11201, Zitoune, Meknes 50000, Morocco; benachir.bouchikhi@gmail.com

* Correspondence: n_elbari@hotmail.com

† Presented at the 1st International Electronic Conference on Chemical Sensors and Analytical Chemistry, 1–15 July 2021; Available online: https://csac2021.sciforum.net/.

Abstract: Fenthion, an organophosphate insecticide, is a cholinesterase inhibitor and is highly toxic. An electrochemical sensor based on molecularly imprinted polymer is developed here for its detection. For this purpose, 2-aminothiophenol mixed with gold nanoparticles was immobilized on screen-printed gold electrodes. The FEN pattern was then fixed before being covered with 2-aminothiophenol. Cyclic voltammetry, differential pulse voltammetry and electrochemical impedance spectroscopy methods were used for the electrochemical characterization. The low detection limit was 0.05 mg/Kg over a range of 0.01–17.3 μg/mL. The sensor was successfully applied for the determination of FEN in olive oil samples with high recovery values.

Keywords: fenthion; molecularly imprinted polymer (MIP); screen-printed gold electrode (Au-SPE); electrochemical sensor; olive oils

Citation: Aghoutane, Y.; Bari, N.E.; Laghrari, Z.; Bouchikhi, B. Electrochemical Detection of Fenthion Insecticide in Olive Oils by a Sensitive Non-Enzymatic Biomimetic Sensor Enhanced with Metal Nanoparticles. *Chem. Proc.* **2021**, *5*, 64. https://doi.org/10.3390/CSAC2021-10773

Academic Editor: Huangxian Ju

Published: 17 August 2021

1. Introduction

Olive oil production is located in the countries of the Mediterranean basin, specifically in Spain, Portugal, Italy, Greece, Turkey, Tunisia and Morocco [1]. However, production in other countries, such as Australia and the United States, is increasing.

Organophosphorus insecticides are the pesticides used in the largest quantities in olive groves to control pests. The most commonly used are those belonging to the class of organophosphate insecticides, because of their relatively low persistence under natural conditions, their ease of synthesis, their low cost and their high effectiveness in eradicating insects. Fat-soluble pesticides tend to concentrate in oils and toxic residues in lipids have been reported [2,3]. In addition, however, residues of POs in the environment present significant risks to the ecosystem and agricultural products; due to their lack of specificity, they affect the nervous system of non-target species, such as mammals, birds and aquatic fauna. Fenthion (FEN) is among the most commonly used pesticides [4]. The Codex Alimentarius Commission of the Food and Agriculture Organization of the United Nations (FAO) have set maximum residue limits (MRLs) for pesticides in olives and olive oil (e.g., 1 and 2 mg/kg for FEN) [5]. Many different detection methods have been used for the determination of residues of organophosphorus pesticides in olive oil. The most commonly used techniques are gas chromatography (GC) methods, which require the extraction of pesticides from samples. Reversed phase liquid chromatography–gas chromatography was also applied to olive oil [6]. These techniques are generally expensive, and require large quantities of samples and organic solvents as well as cleaning and preconcentration steps. Complementary analytical methods, such as enzymatic biosensors, which are based on the inhibition of acetylcholinesterase, were also developed for the detection of FEN [6]. However, enzymes have drawbacks such as difficulty in purification due to their instability, sensitivity and selectivity, which often depend on the nature of the materials. Immunosensing applications are also used for the FEN detection of organophosphorus insecticides, such as indirect enzyme-linked immunosorbent assay (ELISA) [7] and electrochemical immunosensor [8] applications. Their main limitations lie in the availability of antibodies and the fact that they cannot be used for the determination of low-molecular-weight hapten analytes [9,10].

Alternative methods based on molecularly imprinted polymers (MIPs) have also been adopted. They can mimic the functions of biological receptors but with fewer stability constraints. There are few applications of MIPs for the detection of FEN [11].

Due to the growing concern about the presence of pesticide residues in olive oil, it is necessary to implement procedures that can be applied in the field, with sensitive and selective detection.

The objective of this work was to develop a new low-cost, selective and highly sensitive electrochemical sensor, based on MIP, for the determination of FEN in olive oil samples.

This electrochemical sensor was fabricated by immobilizing a 2-aminothiophenol (2-ATP) complex mixed with gold nanoparticles (AuNPs) onto a screen-printed gold electrode (Au-SPE) via Au-S bonds. Then, the FEN template was bound onto Au-SPE/ATP-AuNPs before being coated with 2-ATP. The synthesis process of the electrochemical MIP sensor was straightforward. More generally, we believe that the results obtained open up new opportunities to detect other organophosphate insecticides in various food products and in the environment.

2. Materials and Methods

2.1. Reagents and Solutions

Fenthion (FEN), malathion (MAL), dimethoate (DMT), ethanol (99.8%), methanol, hydrochloric acid (HCL), acetonitrile, gold nanoparticles (AuNPs), potassium chloride, 2-aminothiophenol (2-ATP), phosphate-buffered saline (PBS) and ferri-ferrocyanide (K4[$Fe(CN)_6$], K3[$Fe(CN)_6$]), were all purchased from Sigma-Aldrich, Saint-Quenti-Fallavier, France. Ultra-pure water was used throughout the experiments.

2.2. Synthesis of MIP and NIP Materials on the Gold Electrode

As shown in Figure 1, the overall process for the preparation of the MIP sensor can be summarized by the following steps: First, the functionalization process was performed by immobilizing 0.1 M of 2-aminothiophenol (2-ATP), mixed with 1 mL of AuNPs, on the surface of the gold electrodes followed by incubation for 12 h at room temperature. In the second step, the FEN pattern was deposited on the modified surface and incubated for 2 h at room temperature. In the third step, the second layer containing 5 mM 2-ATP was electropolymerized with 0.1 M KCl and 0.05 M PBS (pH = 6.8) over the potential range (−0.2 V to 0.6 V) at a scan rate of 100 mV/s for 10 cycles. Finally, the printed pattern was extracted in HCl solution (0.5 M) for 20 min. Electrochemical detection of FEN by the MIP sensor was performed by placing 10 μL of each concentration of FEN on the working electrode for 30 min. The electrochemical characteristics of the stepwise MIP sensor fabrication were studied in a $[Fe(CN)_6]^{3-/4-}$ 5 mM solution containing PBS (pH = 7.2). The redox probe $[Fe(CN)_6]^{3-/4-}$ was chosen as a marker to study the changes at the electrode surface after each step of the sensor preparation. Similarly, a non-printed film (Au-SPE/NIP) was prepared using the same procedure, but without adding the template into the polymer solution. This was carried out in order to ensure that the observed effects during the MIP detection steps were only related to the printing characteristics.

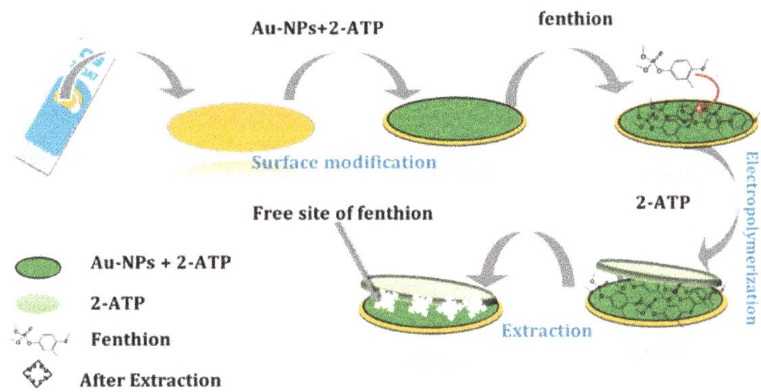

Figure 1. Representation of the experimental procedure of stepwise MIP sensor fabrication.

2.3. .Electrochemical Measurements

A portable instrument (PalmSens^3, Houten, The Netherlands) was used for performing the electrochemical measurements. The screen-printed gold electrodes (Au-SPE) consisted of a three-electrode system (purchased from Dropsens, Asturias, Spain), with a gold working electrode. (0.19 cm^2), a silver reference electrode and a gold counter electrode (0.54 cm^2). The following three electrochemical techniques, cyclic voltammetry (CV), differential pulse voltammetry (DPV) and electrochemical impedance spectroscopy (EIS), were applied. They were used during the development and detection phases to evaluate the sensor response by measuring changes in current and resistance. The DPV measurements were carried out by scanning the potential from −0.1 V to 0.2 V with a scan rate of 50 mV/s. The EIS was measured at a bias potential of 10 mV over a frequency range of 0.1 Hz to 50 kHz. The impedance data were appropriately adjusted using the Randles equivalent circuit.

2.4. *Analysis of Olive Oil Samples*

The prepared MIP sensor was applied to olive oil samples to detect the presence of FEN. For this purpose, the extract of a Moroccan extra virgin olive oil collected from fields in the province of Taounate, Morocco, supposed to be free of residues of the pesticide FEN, a contaminated oil collected from an olive field in the province of Ouarzazate in Morocco, and a commercial oil called Al Hora, were pretreated. The two olive oil samples were pre-treated as follows: 1 mL of the oil samples was macerated and mixed with 5 mL of methanol/water (4:1, *v*/*v*), for 1 h at room temperature. This allowed the remaining pesticide residues in the samples to be extracted. This solution was centrifuged for 15 min at 6000 rpm, which accelerated the transfer of the pesticide residues into the solution. Next, the solvent methanol was evaporated from the supernatant at 65 °C (boiling point of methanol), for 5 min and the extract was collected in an Eppendorf tube. Finally, we deposited a volume of 10 µL of this mixture on the surface of the MIP sensor.

3. Results and Discussion

3.1. *Electropolymerization of FEN Imprinted Film*

After the functionalization step of the Au-SPE electrode, by 2-ATP with Au-NPs, the prepared pre-electropolymerization complex, composed of a mixture of FEN with acetonitrile, was electropolymerized via reactions between FEN and the second layer containing 2-ATP.

In order to study the surface changes of Au-SPE (bare gold, after electropolymerization and extraction), the CV was used for further characterization in a $[Fe(CN)_6]^{3-/4-}$ 5 mM solution, over the potential range of −0.4 V to 0.6 V at a scan rate of 20 mV/s. As shown in Figure 2, the peak anode current (Ia) and potential (Ea) were recorded. As can be seen, the CV signal of the bare gold electrode was lower than the MIP current, indicating that FEN had been successfully trapped on the Au-SPE electrodes, preventing $[Fe(CN)_6]^{3-/4-}$ from diffusing onto the Au-SPE surface. In addition, the NIP sensor had a lower peak current due to the absence of FEN molecules in the polymer.

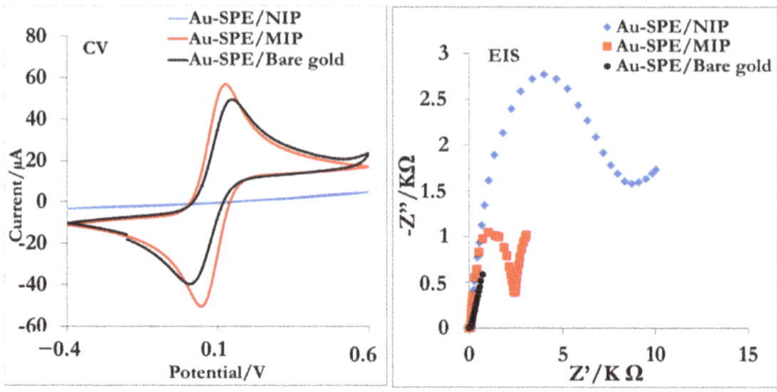

Figure 2. Cyclic voltammograms, Nyquist plots of 5 mM $[Fe(CN)_6]^{3-/4-}$ solution at bare Au-SPE, Au-SPE/MIP and Au-SPE/NIP.

Impedance spectroscopy was also used. As can be seen, the results obtained from the EIS and the CV are consistent with each other.

3.2. Molecular Recognition by MIP and NIP Sensors

To verify the retention capacity of the sensor for different concentrations of FEN, DPV and EIS techniques were used. Figure 3a shows the DPV and EIS responses of the modified electrode for the detection of FEN in the range of 0.01 µg/mL to 17.3 µg/mL. The $[Fe(CN)_6]^{3-/4-}$ was used as a mediator between the printed electrodes and the standard solutions.

In Figure 3a, it can be seen that the redox current peaks increased with increasing FEN concentrations. This increase may have been due to the conductivity of the film covering the surface of the Au-SPE. This confirms the binding of the FEN molecules that hindered the electron transport of the redox probe $[Fe(CN)_6]^{3-/4-}$.

The same results were observed for the EIS technique. The semicircles of the Nyquist diagrams decreased as the FEN concentrations increased. Thus, the increase in FEN concentration produced a decrease in charge transfer resistance. The deposition of FEN concentrations resulted in an increase in the overall conductivity of the electrode surface. This demonstrates that the FEN molecules were well captured by the MIP sites.

When analyzing the NIP data obtained under the same conditions, it was found that the current peaks varies slightly. This was probably due to the absence of FEN during electropolymerization that did not involve the creation of specific cavities (Figure 3b). The results of the NIP Nyquist plots show semi-circular patterns and negligible changes in resistance values (Rtc). This means that the NIP sensor was not specific to FEN molecules, confirming that the responses obtained by the MIP sensor were only due to the presence of FEN-specific cavities.

3.3. Calibration Curve and Detection Limit

After optimizing the manufacturing process parameters, the MIP sensor was used for synthetic FEN detection.

Figure 4 shows the calibration curve referring to the sensor responses to FEN exposure.

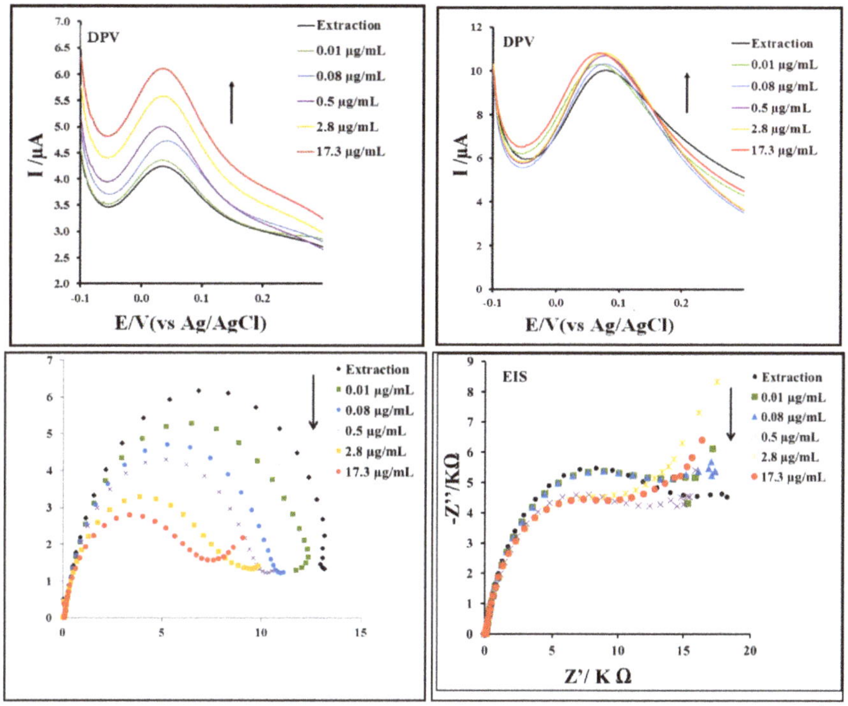

(a) (b)

Figure 3. Voltammograms, Nyquist plots of different FEN concentrations for (**a**) MIP and (**b**) NIP sensors.

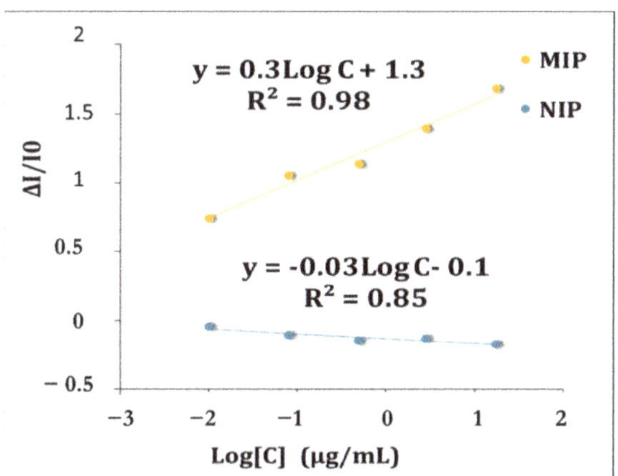

Figure 4. Calibration curves obtained by DPV for MIP and NIP sensors towards FEN.

The DPV technique was performed in a potential window of −0.1 to 0.2 V. The working range for synthetic detection was 0.01 to 17.3 µg/mL.

As a result, a satisfactory logarithmic relationship between the MIP sensor responses and FEN concentrations was achieved (Y = 0.3Log C + 1.3; R^2 = 0.98).

The limits of detection (LOD) and quantification (LOQ) were calculated using:

$$LOD/LOQ = k_{iLOD/LOQ} \times s/m, \quad (1)$$

where k_i corresponds to the signal/noise ratio, k = 3.3 for the LOD and k = 10 for the quantification limit (LOQ), s is the standard deviation of the intercept and m is the slope [12]. The LOD value was found to be 0.05 mg/Kg for the DPV measurements.

3.4. Selectivity of the MIP Sensor

To examine the selectivity of the MIP electrochemical sensor towards FEN, the interference of some similar molecular structures, including dimethoate (DMT) and malathion (MAL), was examined [11,13,14]. The interference test was carried out with satisfactory results. Figure 5 clearly shows that FEN's MIP sensor has much higher current responses compared to both analogues. It is, therefore, better suited for the selective detection of FEN.

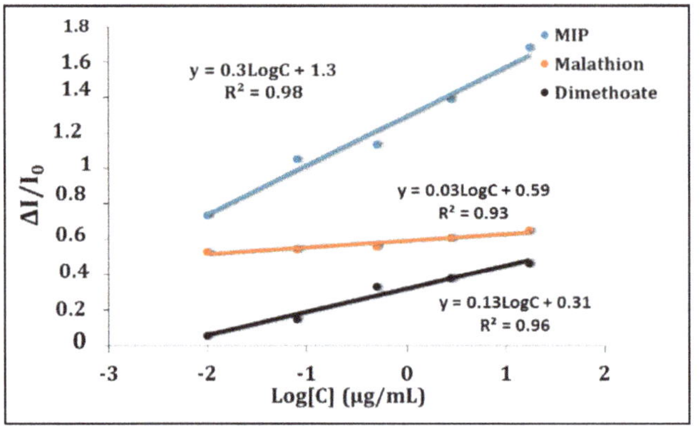

Figure 5. Calibration curves obtained by DPV for MIP towards FEN and interferences.

3.5. Analysis of Olive Oil Samples

The developed MIP sensor was tested for the determination of FEN in contaminated olive oil samples. Using the DPV technique, the responses of the MIP sensor were exploited. In fact, the difference between the maximum current values of the blank and the sample can be calculated using Equation (2) to obtain the FEN concentration in the real samples.

$$y = 0.3\text{Log } C + 1.3 \quad (2)$$

The results obtained are summarized in Table 1. These results show a satisfactory measurement accuracy of the MIP sensor with an acceptable RSD of 0.14% for olive oils. The measured FEN content of the olive oil samples was 0.25 pg/mL for the Al Horra commercial oil and 0.74 pg/mL for the Ouarzazate field oil.

Table 1. Detection of fen in olive oil samples.

Samples	Concentrations (mg/kg)	RSD (%) (n = 2)
Commercial oil Al Horra	0.3125	0.14
Ouarzazate field oil	0.925	0.20

Author Contributions: Conceptualization, N.E.B.; methodology, Y.A. and N.E.B.; software, B.B. and N.E.B.; validation, N.E.B.; formal analysis, Y.A. and N.E.B.; investigation, N.E.B. and Y.A.; resources, N.E.B.; data curation, N.E.B. and Y.A.; writing—original draft preparation, Y.A., N.E.B. and Z.L. All authors have read and agreed to the published version of the manuscript.

Funding: This research received no external funding.

Institutional Review Board Statement: Not applicable.

Informed Consent Statement: Not applicable.

Acknowledgments: The authors would like to express their thanks for the financial support received from the Moulay Ismail University "Scientific Research Promotion".

Conflicts of Interest: The authors declare no conflict of interest.

References

1. Garcés-García, M.; Brun, E.M.; Puchades, R.; Maquieira, Á. Immunochemical determination of four organophosphorus insecticide residues in olive oil using a rapid extraction process. *Anal. Chim. Acta* **2006**, *556*, 347–354. [CrossRef]
2. Lentza-Rizos, C. Monitoring pesticide residues in olive products: Organophosphorus insecticides in olives in oil. *J. AOAC Int.* **1994**, *77*, 1096–1100. [CrossRef]
3. Morchio, G.; de Andreis, R.; Verga, G.R. Indagine sul contenuto di composti fosforganici presenti negli oli vegetali e in particolare nell'olio di olive. *Riv. Ital. Sostanze Grasse* **1992**, *69*, 147–157.
4. Hiskia, A.E.; Atmajidou, M.E.; Tsipi, D.F. Determination of organophosphorus pesticide residues in Greek virgin olive oil by capillary gas chromatography. *J. Agric. Food Chem.* **1998**, *46*, 570–574. [CrossRef] [PubMed]
5. Sanchez, R.; Vazquez, A.; Andini, J.C.; Villén, J. Automated multiresidue analysis of pesticides in olive oil by on-line reversed-phase liquid chromatography–gas chromatography using the through oven transfer adsorption–desorption interface. *J. Chromatogr. A* **2004**, *1029*, 167–172. [CrossRef] [PubMed]
6. Wei, F.; Wang, Z.; Wang, J. Determination of Organophosphorus Pesticide Residues by an Acetylcholinesterase Biosensor in Vegetables and Fruits. *J. Food Sci.* **2007**, *28*, 229–231.
7. Zhanga, Q.; Sun, Q.; Hu, B.; Shen, Q.; Yang, G.; Liang, X.; Sun, X.; Liu, F. Development of a sensitive ELISA for the analysis of the organophosphorous insecticide fenthion in fruit samples. *Food Chem.* **2008**, *106*, 1278–1284. [CrossRef]
8. Cho, Y.; Cha, G.S.; Lee, Y.T.; Lee, H.S. A dipstick-type electrochemical immunosensor for the detection of the organophosphorus insecticide fenthion. *Food Sci. Biotechnol.* **2005**, *14*, 743–746.
9. Krämer, P.M.; Franke, A.; Zherdev, A.Z.; Yazynina, E.V.; Dzantiev, B.B. Comparison of two express immunotechniques with polyelectrolyte carriers, ELISA and FIIAA, for the analysis of atrazine. *Talanta* **2005**, *65*, 324–330. [CrossRef] [PubMed]
10. Ghindilis, A.L.; Atanasov, P.; Wilkins, M.; Wilkins, E. Immunosensors: Electrochemical sensing and other engineering approaches. *Biosens. Bioelectron.* **1998**, *13*, 113–131. [CrossRef]
11. Bakas, I.; Oujji, N.B.; Istamboulié, G.; Piletsky, S.; Piletska, E.; Ait-Addi, E.; Ait-Ichou, I.; Noguer, T.; Rouillon, R. Molecularly imprinted polymer cartridges coupled to high performance liquid chromatography (HPLC-UV) for simple and rapid analysis of fenthion in olive oil. *Talanta* **2014**, *125*, 313–318. [CrossRef] [PubMed]
12. Şengül, Ü. Comparing determination methods of detection and quantification limits for aflatoxin analysis in hazelnut. *J. Food Drug Anal.* **2016**, *24*, 56–62. [CrossRef] [PubMed]
13. Bakas, I.; Oujji, N.B.; Moczkoc, E.; Istambouliea, G.; Piletskyc, S.; Piletskac, E.; Ait-Ichoub, I.; Ait-Addi, E.; Noguer, T.; Rouillona, R. Molecular imprinting solid phase extraction for selective detection of methidathion in olive oil. *Anal. Chim. Acta* **2012**, *734*, 99–105. [CrossRef] [PubMed]
14. Capoferri, D.; Del Carlo, M.; Ntshongontshi, N.; Iwuoha, E.I.; Sergi, M.; Di Ottavio, F.; Compagnone, D. MIP-MEPS based sensing strategy for the selective assay of dimethoate, Application to wheat flour samples. *Talanta* **2017**, *174*, 599–604. [CrossRef] [PubMed]

chemistry proceedings

MDPI

Proceeding Paper

Spectroscopic Ellipsometry Detection of Prostate Cancer Bio-Marker PCA3 Using Specific Non-Labeled Aptamer: Comparison with Electrochemical Detection †

Sarra Takita [1,*], Alexei Nabok [1], David Smith [2] and Anna Lishchuk [3]

1 Material and Engineering Research Institute, City Campus, Sheffield Hallam University, Howard Street, Sheffield S1 1WB, UK; a.nabok@shu.ac.uk
2 Biomolecular Research Centre, City Campus, Sheffield Hallam University, Howard Street, Sheffield S1 1WB, UK; hwbds1@exchange.shu.ac.uk
3 Department of Chemistry, The Dainton Building, The University of Sheffield, Brook Hill, Sheffield S3 7HF, UK; a.tsargorodska@sheffield.ac.uk
* Correspondence: sarah.a.takita@gmail.com
† Presented at the 1st International Electronic Conference on Chemical Sensors and Analytical Chemistry, 1–15 July 2021; Available online: https://csac2021.sciforum.net/.

Abstract: The most common prostate cancer (PCa) diagnostics, which are based on detection of prostate-specific antigens (PSA) in blood, have specificity limitations often resulting in both false-positive and false-negative results; therefore, improvement in PCa diagnostics using more specific PCa biomarkers is of high importance. Studies have shown that the long noncoding RNA Prostate Cancer Antigen 3 (lncPCA3) that is over-expressed in the urine of prostate cancer patients is an ideal biomarker for non-invasive early diagnostics of PCa. Geno-sensors based on aptamer bioreceptors (aptasensors) offer cost- and time-effective, and precise diagnostic tools for detecting PCa biomarkers. In this study, we report on further developments of RNA-based aptasensors exploiting optical (spectroscopic ellipsometry) measurements in comparison with electrochemical (CV and IS) measurements published earlier. These sensors were made by immobilization of thiolated CG-3 RNA aptamers on the surface of gold. Instead of a redox-labelled aptamer used previously in electrochemical measurements, a non-labelled aptamer was used here in a combination with total internal reflection ellipsometry (TIRE) measurements. The results obtained by these two methods were compared. The method of TIRE is potentially highly sensitive and comparable in that respect with electrochemical methods capable of detection of PCA3 in sub-pM levels of concentration. The required selectivity is provided by the high affinity of PCA3-to-aptamer binding with K_D in the 10^{-9} M range. The spectroscopic ellipsometry measurements provided additional information on the processes of PCA3 to aptamer binding.

Keywords: aptamer; mRNA; PCA3; biosensor; TIRE

Citation: Takita, S.; Nabok, A.; Smith, D.; Lishchuk, A. Spectroscopic Ellipsometry Detection of Prostate Cancer Bio-Marker PCA3 Using Specific Non-Labeled Aptamer: Comparison with Electrochemical Detection. *Chem. Proc.* **2021**, *5*, 65. https://doi.org/10.3390/CSAC2021-10453

Academic Editor: Maria Emília de Sousa

Published: 30 June 2021

1. Introduction

Prostate cancer (PCa) is considered as one of the most common types of cancer worldwide, and is the second leading cause of mortality among men after lung cancer [1,2]. There are clinical challenges for PCa early-stage diagnosis related to the asymptomatic nature of the disease and the similarity of its symptoms to benign prostatitis [3]. Early diagnosis of PCa can reduce mortality rates and increase the opportunity for effective medical interventions, therefore, the development of reliable diagnostics of PCa is of high importance [4,5]. Current diagnostics of PCa is based on the detection of total serum prostate-specific antigens (PSA) in blood followed by (if PCa suspected) digital rectal examination and imaging studies [6,7]. However, the lack of specificity of PSA markers often leads to both false-positive and false-negative results of the PSA test [8]. Hence, identifying alternative specific prostate cancer biomarkers and developing methods for their detection in the early stage of the disease is of high importance nowadays [9,10]. Several PCa biomarkers have been identified as being over-expressed in prostate tumours [11]. The differential display code 3 (DD3) gene, also known as prostate cancer antigen 3 (PCA3), the long non-coding RNA (lncRNA) discovered in 1999 [12] has been widely accepted as one of the specific biomarkers for malignant PCa [13–15]. PCA3 levels can predict the prostatic biopsies' outcome, especially in combination with other PCa biomarkers, such as PSA, and can reduce the likelihood of false-positive results. The Prognesa® test based on simultaneous detection of PCA3 and PSA using quantitative nucleic acid amplification with high sensitivity and specificity is approved in USA [16]. However, such a test is time-consuming and expensive. The development of PCA3 biosensors for express, accurate and cost-effective diagnostics of PCa is a subject of high importance. Recent developments in biosensing technologies related to the use of aptamers, synthetic bioreceptors having specifically designed sequences of RNA or DNA oligonucleotides to provide the antibody-like function towards a wide range of analytes, leads to substantial progress in cancer diagnostics (including prostate cancer) [17,18]. The RNA-based aptamer (CG3 aptamer), with a high affinity towards 277-bases section of PCA3 transcript, has been developed recently [19]. The successful application of the CG3 aptamer functionalized with ferrocene at C5′ terminal and immobilized on the surface of gold screen-printed electrodes via thiol group at C3′ terminal for electrochemical in-vitro detection of PCA3 in low concentrations down to sub-pM range was reported for the first time in [20].

One of the most attractive optical biosensing technology developed in the last decade was the method of total internal reflection ellipsometry (TIRE) which is a combination of spectroscopic ellipsometry (SE) and surface plasmon resonance (SPR) [21]. The method of TIRE has a high sensitivity (10 times higher than conventional SRP) and thus is particularly attractive for detection of small molecules, such as mycotoxins, in concentrations down to ppt level [22,23]; it is also suitable for the study of adsorption kinetics and subsequent evaluation of the affinity of bioreceptors (antibodies and aptamers). This work is mostly focused on TIRE detection of PCA3 in direct assay with unlabelled CG3 aptamers immobilized on the surface of gold. The results are compared to our data of electrochemical detection of PCA3 using redox-labelled CG3 aptamer reported earlier in [20]. Our observations are a step towards the long-term aim of developing a novel, accurate, simple, and cost-effective diagnostic tool for the early detection of prostate cancer.

2. Materials and Methods

2.1. Chemicals

HEPES binding buffer (HBB) pH 7.2–7.6, sodium phosphate (Na_2HPO_4), potassium phosphate (KH_2PO_4), potassium chloride (KCl), magnesium chloride ($MgCl_2$), dithiothreitol (DTT), and sodium chloride (NaCl), were procured from Sigma-Aldrich (UK). All reagents were of analytical grade. The biological target, the 277 nt target analyte fragment of lncRNA PCA3, was purchased from Eurofins Genomics (Germany). The label-free CG-3 RNA-based aptamer (5′-AGUUUUUGCGUGUGCCCUUUUUGUCCCC-3′SH) for optical measurements was acquired from Sangon-Biotech, China. The same aptamer, but labelled with ferrocene at 5′, was used previously [20] for electrochemical detection of PCA3. Before immobilization, the stock solution of aptamer (100 μM) was diluted at desired concentration with PBB supplemented with 2 mM of DTT, then diluted aptamer solution was activated by thermocycling in PCR unit (Prime TC3600) by heating to 90 °C for 5 min and cooling down to 4 °C for 5 min. Immobilization of aptamers was carried out in 100 M HEPES buffer (pH 7.4) with the addition of 2.5 mM DTT and 3 mM $MgCl_2$. Optical measurements were performed in the detection buffer, e.g., 100 mM PBS (pH 7.4), prepared by dissolving 10 mM Na_2HPO_4, 1.56 mM KH_2PO_4, 2.5 mM KCl, and 135 mM NaCl in water. Milli-Q water was used for all preparations.

2.2. TIRE-Optical Bio-Transducer

The TIRE experimental set-up schematically shown on inset in Figure 1 is based on J.A. Woollam M2000 spectroscopic ellipsometer with the addition of a 68° glass prism (providing the light coupling at total internal reflection conditions) which was optically connected via index matching fluid with the gold coated glass slide. The PTFE cell of 0.2 mL in volume was sealed against the gold coated glass slide; the inlet and outlet tubes allow the injection of the required liquid reagents in the cell. The principles of total internal reflection ellipsometry (TIRE) measurements and data acquisition were described in detail previously in [21–24].

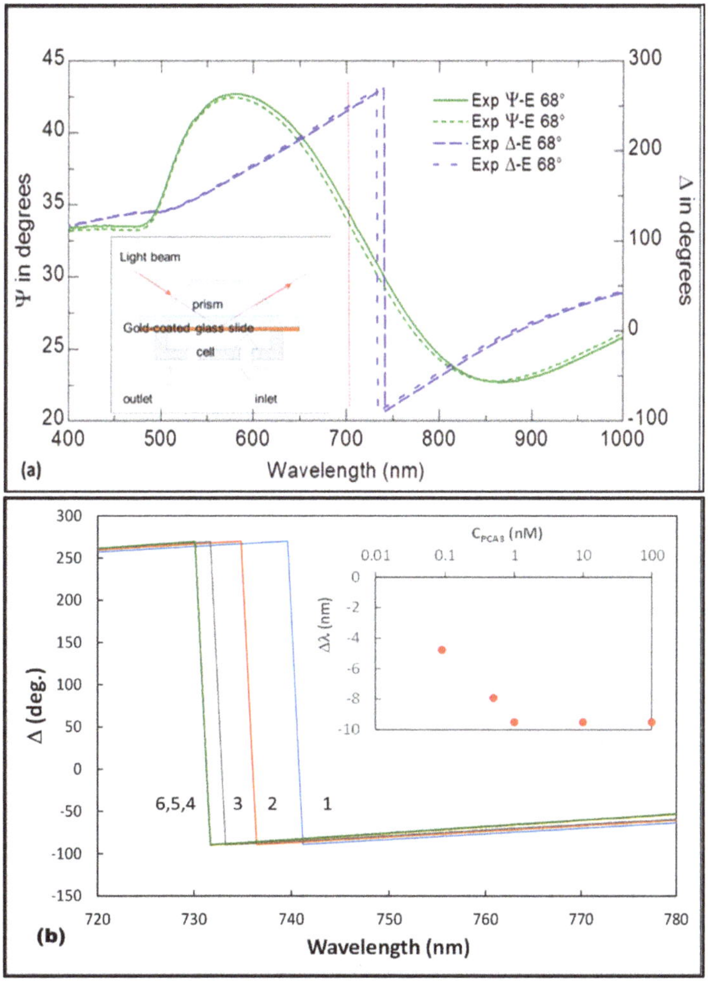

Figure 1. Results of TIRE spectra measurements: (**a**) TIRE spectra for Au layer with aptamers immobilized on the surface (solid line) and the same sample after binding 1 nM of PCA3 to aptamer (dotted line); (**b**) A series of TIRE D-spectra demonstrating the "blue" spectral shift caused by binding PCA3 of different concentrations: aptamer before exposure (1) and after exposure to PCA3 0.09 nM (2), 0.5 nM (3), 1 nM (4), 10nM (5), and 100 nM (6).

Standard microscopic glass slides were cleaned in hot piranha solution (3:1 mixture of H_2SO_4 and H_2O_2) for 10 min. followed by rinsing with deionized Milli-Q water and drying under a stream of nitrogen gas. Gold layers of about 20 to 25 nm in thickness were evaporated on glass slides using Edwards E306A metal evaporator unit; an intermediate layer of Cr (3 to 5 nm) was used to improve the adhesion of gold to glass. For TIRE measurements, gold coated glass slides functionalized with label-free aptamer were used.

The PCA3 solutions were prepared by diluting the original stock solution (100 μM) in PBS buffer to obtain the required concentrations of 0.09, 0.5, 1, 10, 100 nM. TIRE measurements were performed in a sequential adsorption manner (starting with the injection of lowest concentration of PCA3) and rinsing the cell after each adsorption step; the initial TIRE measurements of pure buffer were used as a reference. The TIRE setup allows two types of ellipsometric measurements: (i) single spectroscopic spectral scans performed in PBS after completing each stage of molecular adsorption, and (ii) dynamic measurements, e.g., recording of several spectroscopic scans during the binding of analytes (PCA3) to receptors (CG3 aptamer) which provides the information on the reaction's kinetics.

3. Results and Discussion

3.1. Results of TIRE Single Spectroscopic Measurements

Typical TIRE spectra of Ψ and Δ of Au/Cr layer on the glass slide functionalized with aptamers are shown in Figure 1. The maximum and minimum in the spectrum of the amplitude-related parameter Ψ correspond, respectively, to the conditions of total internal reflection (TIR) and surface plasmon resonance (SPR), while a sharp drop in a spectrum of phase related parameter Δ is a new quantity non-existent in traditional SPR. The position of such a phase drop is highly sensitive to changes in the optical density of a molecular layer adsorbed the surface of gold; the increase in the molecular layer thickness causes a "red" spectral shift, while thickness decrease causes a "blue" spectral shift of Δ spectrum.

As one ca see in Figure 1a, binding PCA3 from its 0.5 nM solution in PBS to its specific aptamer results in a blue shift of both Ψ and Δ spectra (dotted lines) as compared to the spectra of unperturbed aptamer layer (solid lines).

As shown in Figure 1b, the increase in concentration of PCA3 causes a progressive increase in the blue (negative) spectral shift until the saturation of binding sites, e.g., aptamers) occurs at concentrations larger than 1nM. The saturation of the TIRE sensor response at the level of about −9.5 nm shown as inset in Figure 1b corresponds to a decrease in the film thickness of around 2 nm. The negative control tests were carried out by adsorbing molecules having scrambled the sequence of PCA3, resulting in the "red" (positive) spectral shift of about 20 nm associated with the thickness increase of about 4 nm.

These results can be understood using the model shown in Figure 2 which schematically illustrates the process of PCA3 to aptamer binding during which the aptamer engulfs the target resulting in the thickness decrease. Contrarily, non-specific binding of the scrambled PCA3 to the aptamer results in the thickness increase. Interestingly, that prolonged exposure of aptamer to large concentrations of PCA3 (10 nM and 100 nM) also causes the red spectral shift which is due to the non-specific adsorption (or piling up) of PCA3 molecules. Another explanation of the saturation of the sensor response at relatively low concentrations of PCA3 is in high density of the immobilized aptamers which have no room for coiling around large target molecules of PCA3; optimization of the aptamer concentration is therefore required.

It is worth mentioning that the binding of 90 pM (the lowest concentration used) of PCA3 causes a substantial spectral shift of about 5 nm. Considering the high accuracy of ellipsometry measurements with the noise level of Ψ and Δ in the second decimal digits the actual limit of detection (LOD) could be much lower. In this work, however, the evaluation of LOD was not a priority.

Additional spectroscopic ellipsometry measurements were carried out on dried samples, in order to evaluate the thickness of the aptamer layer covalently bound on the surface of gold. The thicknesses of Au (17 nm to 20 nm) and Cr (5 nm to 7 nm) layers were evaluated first from the measurements on bare metal layers. The aptamer layer thickness of 2.5 nm was found. The thermocycling of gold coated slides with immobilized aptamer in

PCR unit (e.g., heating up to 95 °C and cooling down to 5 °C during 10 min.) has resulted in the aptamer layer thickness increase up to 4.5 nm. This shows that aptamer molecules tend to coil in a dry state, while thermocycling in a buffer solution containing Mg^{2+} ions stabilizes the aptamer structure in the original stretched form suitable for sensing. Such a procedure is recommended for "refreshing" the samples of aptamers immobilized on gold prior to conducting sensing tests.

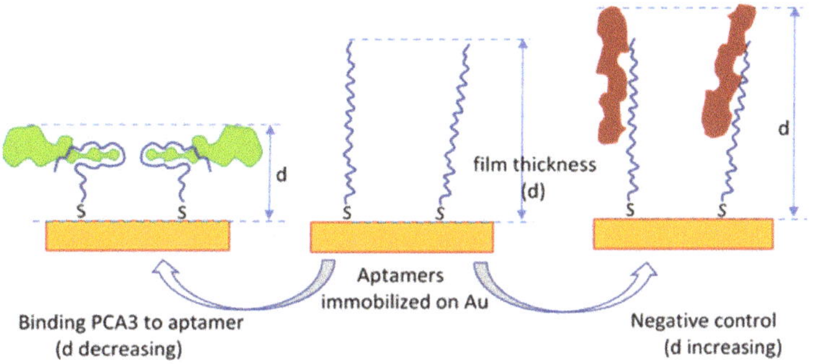

Figure 2. Schematic diagram of specific and non-specific binding of molecules to unlabeled aptamers immobilized on the surface of gold.

3.2. *TIRE Study of the Binding Kinetics*

TIRE spectral measurements were carried out (at certain time intervals) during binding PCA3 to aptamers immobilized on the surface of gold. The resulted massive data files can be processed by plotting the time dependencies of either Ψ or Δ at fixed wavelength typically selected on the left side of the resonance (see red dotted line in Figure 1a). A typical example of TIRE binding kinetics of PCA3 (0.5 nM) to aptamers is given in Figure 3a as the time dependence of Ψ at 700 nm. These data were fitted to the rising exponential function with the parameters of the equation given as an inset. The parameter of interest was the time constant (τ). Such measurements were carried out at different concentrations of PCA3 (C) and the characteristic time constants (τ) were evaluated at each concentration. According to the theory of molecular adsorption [20], the rates of adsorption and desorption (k_a and k_d) can be found, respectively, as the gradient and intercept of the following linear equation: $\frac{1}{\tau} = k_a C + k_d$, then the association and affinity constants (K_A and K_D) can be found as $K_A = k_a/k_d$, $K_D = 1/K_D$.

Linear dependence of $1/(\tau)$ vs. C given in Figure 3b in both logarithmic and linear coordinates yields the values of $K_A = 3.87 \times 10^{10}$ M^{-1} and $K_D = 2.58 \times 10^{-9}$ M which are very similar to those obtained earlier by the electrochemical method of CV [20]. Also, the TIRE experiments revealed anomalous kinetics at high concentrations of PCA3 when soon after reaching the saturation the response started to rise again. This is most likely associated with the non-specific adsorption of PCA3.

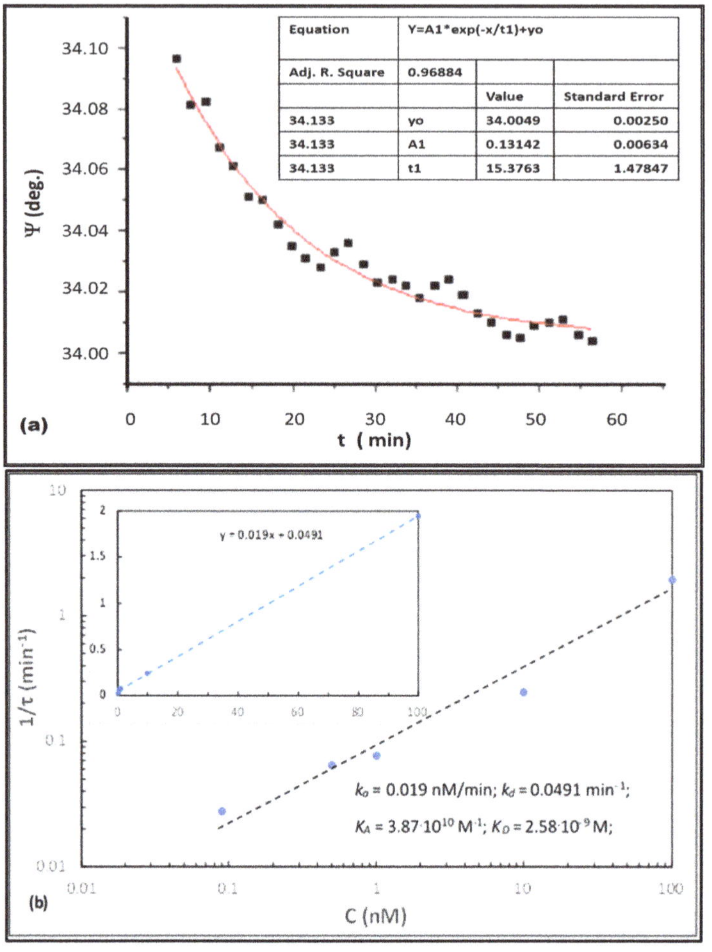

Figure 3. Evaluation of the PCA3 to aptamer binding affinity from dynamic TIRE measurements: (**a**) example of Ψ time dependence upon binding PCA3 (0.5 nM) to aptamer immobilized on the surface of Au; The values of Ψ at 700 nm were presented; (**b**) evaluation of K_A and K_D from the 1/τ(C) dependence given in logarithmic and linear scales.

4. Conclusions: Comparison of the Electrochemical and Optical Detection Strategies

The use of the TIRE method for detection of the lncRNA transcript PCA3 in direct assay with non-labelled aptamer immobilized on the surface proved to be promising. In contrast to electrochemical detection based on labelled aptamers, the optical detection using TIRE showed the increase in the molecular layer thickness caused by the non-specific binding of PCA3 which cannot be detected with the electrochemical method. In terms of sensitivity, the method of TIRE is potentially capable of detection of PCA3 in low concentrations (much lower than 0.09 nM used in this work) which could be comparable with the values of LOD reported earlier for electrochemical CV (0.35 pM–0.78 pM) and EIS (0.26 pM) methods [20]. More detailed optical study in a wider concentration range of PCA3 and optimization of aptamer concentration is currently underway. The high affinity of unlabeled aptamer towards PCA3 was confirmed by TIRE kinetics study which gave similar values K_A and K_D to those obtained previously from CV measurements for an aptamer labelled with ferrocene [20].

From the point of view of sensing, electrochemical methods are more attractive because of the low cost and simplicity of use, however the optical method of TIRE provides important complementary information on the thickness of molecular layers which allows for better understanding of the processes of aptamer-target interaction.

Author Contributions: Conceptualization A.N.; methodology, A.N. and A.L.; validation, S.T.; formal analysis, A.N.; investigation, S.T.; resources, S.T. and A.N.; writing—original draft preparation, A.N. and S.T.; writing—review and editing, A.N. and D.S.; supervision, A.N., A.L. and D.S. All authors have read and agreed to the published version of the manuscript.

Funding: This research received no external funding.

Informed Consent Statement: Not applicable.

Data Availability Statement: The data are not publicly available; The data files are stored on corresponding instruments and on personal computers.

Acknowledgments: We would like to acknowledge Material and Engineering Research Institute at Sheffield Hallam University for providing full access to its resources and materials.

Conflicts of Interest: The authors declare no conflict of interest.

References

1. Ferlay, J.; Soerjomataram, I.; Dikshit, R.; Eser, S.; Mathers, C.; Rebelo, M.; Parkin, D.M.; Forman, D.; Bray, F. Cancer incidence and mortality worldwide: Sources, methods and major patterns in GLOBOCAN 2012. *Int. J. Cancer* **2015**, *136*, E359–E386. [CrossRef]
2. Liss, M.A.; Santos, R.; Osann, K.; Lau, A.; Ahlering, T.E.; Ornstein, D.K. PCA3 molecular urine assay for prostate cancer: Association with pathologic features and impact of collection protocols. *World J. Urol.* **2011**, *29*, 683–688. [CrossRef]
3. Daniyal, M.; Siddiqui, Z.A.; Akram, M.; Asif, H.M. MINI-REVIEW Epidemiology, Etiology, Diagnosis and Treatment of Prostate Cancer. *Asian Pac. J. Cancer Prev.* **2014**, *15*, 9575–9578. [CrossRef]
4. Filella, X.; Foj, L. Prostate Cancer Detection and Prognosis: From Prostate Specific Antigen (PSA) to Exosomal Biomarkers. *Int. J. Mol. Sci.* **2016**, *17*, 1784. [CrossRef]
5. Miller, K.D.; Siegel, R.L.; Lin, C.C.; Mariotto, A.B.; Kramer, J.L.; Rowland, J.H.; Stein, K.D.; Alteri, R.; Jemal, A. Cancer treatment and survivorship statistics, 2016. *CA A Cancer J. Clin.* **2016**, *66*, 271–289. [CrossRef]
6. Salman, J.W.; Schoots, I.G.; Carlsson, S.; Jenster, G.; Roobol, M.J. Prostate Specific Antigen as a Tumor Marker in Prostate Cancer: Biochemical and Clinical Aspects. In *Advances in Experimental Medicine and Biology*; Springer: Dordrecht, The Netherlands, 2015; Volume 867, pp. 93–114. [CrossRef]
7. Buzzoni, C.; Auvinen, A.; Roobol, M.J.; Carlsson, S.; Moss, S.M.; Puliti, D.; de Koning, H.J.; Bangma, C.H.; Denis, L.J.; Kwiatkowski, M.; et al. Metastatic Prostate Cancer Incidence and Prostate-specific Antigen Testing: New Insights from the European Randomized Study of Screening for Prostate Cancer. *Eur. Urol.* **2015**, *68*, 885–890. [CrossRef] [PubMed]
8. Adhyam, M.; Gupta, A.K. A Review on the Clinical Utility of PSA in Cancer Prostate. *Indian J. Surg. Oncol.* **2012**, *3*, 120–129. [CrossRef]
9. Mistry, K.; Cable, G. Meta-Analysis of Prostate-Specific Antigen and Digital Rectal Examination as Screening Tests for Prostate Carcinoma. *J. Am. Board Fam. Med.* **2003**, *16*, 95–101. [CrossRef]
10. Altuwaijri, S. Role of Prostate Specific Antigen (PSA) in Pathogenesis of Prostate Cancer. *J. Cancer Ther.* **2012**, *3*, 331–336. [CrossRef]
11. Landers, K.A.; Burger, M.J.; Tebay, M.A.; Purdie, D.M.; Scells, B.; Samaratunga, H.; Lavin, M.F.; Gardiner, R.A. Use of multiple biomarkers for a molecular diagnosis of prostate cancer. *Int. J. Cancer* **2005**, *114*, 950–956. [CrossRef]
12. Bussemakers, M.J.; Van Bokhoven, A.; Verhaegh, G.W.; Smit, F.P.; Karthaus, H.F.; Schalken, J.A.; Debruyne, F.M.; Ru, N.; Isaacs, W.B. DD3: A new prostate-specific gene, highly overexpressed in prostate cancer. *Cancer Res.* **1999**, *59*, 5975–5979.
13. Chistiakov, D.A.; Myasoedova, V.A.; Grechko, A.V.; Melnichenko, A.A.; Orekhov, A.N. New biomarkers for diagnosis and prognosis of localized prostate cancer. *Semin. Cancer Biol.* **2018**, *52*, 9–16. [CrossRef] [PubMed]
14. Rönnau, C.G.H.; Verhaegh, G.W.; Luna-Velez, M.V.; Schalken, J.A. Noncoding RNAs as Novel Biomarkers in Prostate Cancer. *BioMed Res. Int.* **2014**, *2014*, 591703. [CrossRef]
15. Schalken, J.A.; Hessels, D.; Verhaegh, G. New targets for therapy in prostate cancer: Differential display code 3 (DD3PCA3), a highly prostate cancer–specific gene. *Urology* **2003**, *62*, 34–43. [CrossRef]
16. Nicholson, A.; Mahon, J.; Boland, A.; Beale, S.; Dwan, K.; Fleeman, N.; Hockenhull, J.; Dundar, Y. The clinical effectiveness and cost-effectiveness of the PROGENSA® prostate cancer antigen 3 assay and the Prostate Health Index in the diagnosis of prostate cancer: A systematic review and economic evaluation. *Health Technol. Assess.* **2015**, *19*, 1–192. [CrossRef] [PubMed]
17. Pasinszki, T.; Krebsz, M.; Tung, T.T.; Losic, D. Carbon Nanomaterial Based Biosensors for Non-Invasive Detection of Cancer and Disease Biomarkers for Clinical Diagnosis. *Sensors* **2017**, *17*, 1919. [CrossRef]

18. Topkaya, S.N.; Azimzadeh, M.; Ozsoz, M. Electrochemical Biosensors for Cancer Biomarkers Detection: Recent Advances and Challenges. *Electroanalysis* **2016**, *28*, 1402–1419. [CrossRef]
19. Marangoni, K.; Neves, A.F.; Rocha, R.M.; Faria, P.R.; Alves, P.T.; Souza, A.G.; Fujimura, P.T.; Santos, F.A.A.; Araújo, T.G.; Ward, L.S.; et al. Prostate-specific RNA aptamer: Promising nucleic acid antibody-like cancer detection. *Sci. Rep.* **2015**, *5*, 12090. [CrossRef]
20. Nabok, A.; Abu-Ali, H.; Takita, S.; Smith, D. Electrochemical Detection of Prostate Cancer Biomarker PCA3 Using Specific RNA-Based Aptamer Labelled with Ferrocene. *Chemosensors* **2021**, *9*, 59. [CrossRef]
21. Arwin, H.; Poksinski, M.; Johansen, K. Total internal reflection ellipsometry: Principles and applications. *Appl. Opt.* **2004**, *43*, 3028–3036. [CrossRef] [PubMed]
22. Nabok, A.; Tsargorodskaya, A.; Mustafa, M.; Szekacs, I.; Starodub, N.; Szekacs, A. Detection of low molecular weight toxins using an optical phase method of ellipsometry. *Sens. Actuators B Chem.* **2011**, *154*, 232–237. [CrossRef]
23. Al Rubaye, A.; Nabok, A.; Catanante, G.; Marty, J.-L.; Takacs, E.; Szekacs, A. Detection of ochratoxin A in aptamer assay using total internal reflection ellipsometry. *Sens. Actuators B Chem.* **2018**, *263*, 248–251. [CrossRef]
24. Al-Rubaye, A.G.; Nabok, A.; Catanante, G.; Marty, J.-L.; Takács, E.; Székács, A. Label-Free Optical Detection of Mycotoxins Using Specific Aptamers Immobilized on Gold Nanostructures. *Toxins* **2018**, *10*, 291. [CrossRef] [PubMed]

chemistry proceedings

MDPI

Proceeding Paper

Titanium Based Materials for High-Temperature Gas Sensor in Harsh Environment Application †

Roussin Lontio Fomekong [1,*] and Bilge Saruhan [2]

1 Higher Teacher Training College, University of Yaounde I, Yaounde P.O. Box 47, Cameroon
2 Department of High-Temperature and Functional Coatings, Institute of Materials Research, German Aerospace Center (DLR), 51147 Cologne, Germany; bilge.saruhan@dlr.de
* Correspondence: lonforou@yahoo.fr or roussin.lontio@univ-yaounde1.cm

† Presented at the 1st International Electronic Conference on Chemical Sensors and Analytical Chemistry, 1–15 July 2021; Available online: https://csac2021.sciforum.net/.

Abstract: Gas selective sensors that are capable of sensing at hot-gas environment are increasingly gaining importance for the monitoring of combustion and thermal processes releasing NO, NO_2 and H_2 containing hot gases at temperatures exceeding 600 °C. Despite some drawbacks, TiO_2 is capable of operating as a gas sensor above 500 °C. In this context, Ni-doped TiO_2, Co-doped TiO_2 and Rh-doped $BaTiO_3$ have been prepared by oxalate coprecipitation route and fully characterized. Co-doping of TiO_2 promotes p-type behavior exhibiting good sensing properties to NO_2 while Ni-doping displays the maintenance of n-type behavior and better H_2-sensing properties at 600 °C. Rh-doped $BaTiO_3$ shows excellent NO sensing properties at 900 °C.

Keywords: Ni-doped TiO_2; Co-doped TiO_2; Rh-doped TiO_2; coprecipitation; high-temperature gas sensor

check for updates

Citation: Fomekong, R.L.; Saruhan, B. Titanium Based Materials for High-Temperature Gas Sensor in Harsh Environment Application. *Chem. Proc.* **2021**, *5*, 66. https://doi.org/10.3390/CSAC2021-10480

Academic Editor: Huangxian Ju

Published: 30 June 2021

1. Introduction

High temperature gas sensors are mainly designed to solve gas detection and monitoring problems with a high operating temperature environment, such as gas turbine, nuclear power plants and automobile internal combustion engine emission [1]. As far as target gases are concerned NOx (NO_2, NO) and H_2 are among the most important. NOx is a severely toxic gas with a pungent odor arising mostly from the related human activities involving high temperature such as combustion of coal and oil at electric power plants, combustion of chemical plant and also in emissions from automotive and aircraft engines. NOx-emission leads to harmful effects on the environment and health. There is an urgent need to develop sensor control systems for exhaust emission gases to directly monitor NOx at temperatures in the range of 400–900 °C [2]. As far as hydrogen is concerned, it is the best candidate to replace the hydrocarbon-based fuels used in many combustion engines such as those in automobiles and aircraft, which are responsible for much of today's air pollution [3]. Hydrogen seems to be a green, renewable energy carrier that can help solve the problems of non-sustainable energy use (fossil fuels). However, the efficient application of hydrogen requires careful consideration of the relevant safety concern. In fact, its physico-chemical properties make hydrogen a highly explosive gas [3,4]. Moreover, as hydrogen is colorless, odorless and tasteless, the ability to detect a hydrogen leak by means of selective sensors is highly desired.

Chemiresistive gas sensors based on semiconductor metal oxides have been drawing more and more attention because of their advantages such as low cost, lightweight, fast response/recovery times and high compatibility with microelectronic processing. Graphene and its derivatives, as well as some organic semiconductor-based materials constitute also an interesting family of chemo-resistive gas sensor due to their large surface area, good electrical, thermal and mechanical properties [5,6].

Cost effective metal oxide-based gas sensors such as SnO_2, WO_3, ZnO, NiO or CuO operate mostly at temperatures below 400 °C [7–10]. There are only few reports in literature focusing on their gas sensing above 400 °C. TiO_2 is one of them to be capable of operating above 500 °C. The additional benefits of TiO_2 are non-toxicity, easy fabrication, and the good chemical stability [11]. However, TiO_2 is a high resistive n-type semiconductor with relatively poor conductivity for sensing oxidative gases such as NO_2. This disadvantage was previously reported to be overcome through addition of low valence dopant atoms which alter the electronic structure [12–15]. Another strategy is to use catalytically doped perovskite-based titanium compounds such as $BaTiO_3$. In this work, we report the synthesis of Co-doped TiO_2, Ni doped TiO_2 and Rh-doped $BaTiO_3$ by co-precipitation method and demonstrate gas sensing ability toward NO_2, NO and H_2 above 500 °C. Our results yield that Co-doping of TiO_2 promotes p-type behavior exhibiting good sensing properties to NO_2 while Ni-doping displays the maintenance of n-type behavior and better H_2-sensing properties at 600 °C. More interestingly, Rh-doped $BaTiO_3$ shows excellent NO sensing properties even at 900 °C.

2. Materials and Methods

2.1. Preparation of the Different Sensing Materials

Ni-doped TiO_2 nanoparticles were prepared using the co-precipitation synthesis route followed by calcination. The starting precursor solutions were first prepared by dissolving nickel acetate in acetic acid while pure ethanol was used separately to dissolve titanium iso-propoxide (TTIP). The adjustment between the previously prepared nickel and titanium solution has been performed so that the final obtained mix would contain 0.0, 0.5, 1.0 and 2.0 mol.% of nickel dopant in TiO_2 and were labeled as TN0, TN05, TN1 and TN2, respectively. The solutions were then mixed and stirred for 5 min. Oxalic acid was used as the precipitating agent. It was dissolved in absolute ethanol solution and poured progressively into the previously mixed solutions. In order to achieve a total precipitation, the resulting mixtures were stirred for 1 h at room temperature, followed by the filtration and drying of the obtained precipitate at 80 °C. A white powder was obtained for the undoped precursor and a yellowish colored powder for the Ni-doped precursor samples. The prepared precursor powders were then calcined in a muffle furnace under static air for 3 h at 700 °C to obtain the nano-particulate powders.

Co-doped TiO_2 nanoparticles were synthesized by employing the same processing route. The cobalt acetate and titanium iso-propoxide solutions were adjusted to obtain 0.5 and 1 mol.% of cobalt in TiO_2 and were labeled as 0.5Co-doped TiO_2 and 1Co-doped TiO_2, respectively. The so-obtained precipitate was filtered and dried in oven at 80 °C yielding a pink colored powder for the Co-doped samples.

The synthesis of Rh-doped $BaTiO_3$ (designated as BTR1-OX-900) was also prepared by coprecipitation by following the same procedure. The amount of the aqueous Rh-nitrate solution, barium acetate and titanium iso-propoxide were adjusted to yield perovskites with the following composition: $BaTi_{0.98}Rh_{0.02}O_3$. The as-prepared precursor powder was calcined in a ceramic combustion boat holder at 900 °C in a muffle furnace (5 °C min^{-1}) for one hour under static air. In order to activate this material, it was treated under 2% of hydrogen at 900 °C for two hours.

2.2. Materials Characterization

The XRD diffractograms of all the samples were obtained at room temperature with a D5000 Siemens Kristalloflex θ–2θ powder diffractometer which has a Bragg-Brentano geometry and equipped with Cu-Kα radiation (λ = 1.54178 Å) and a standard scintillation counter detector.

Bruker Senterra Raman spectrometer (from Bruker Optik GmbH, Ettlingen, Germany) was used to record all the Raman spectra at room temperature under 532 nm and 0.2 mW power laser excitation, which was focused on samples through a 50X objective (Olympus MPlan N 50X/0.75).

The particles' morphology was determined by Scanning Electron Microscopic (SEM) analysis and was carried out in a Zeiss Ultra 55 microscope.

2.3. Sensor Preparation

The as prepared materials (Co-doped TiO_2, Ni-doped TiO_2 and Rh-doped $BaTiO_2$) powders were deposited as thick films (~20 μm) using a simple drop-coating method on alumina substrates that were previously fitted with interdigitated electrodes. The sensor response for n-type semiconductors is defined by $(R_{gas}/R_{air} - 1) \times 100$ and $(R_{air}/R_{gas} - 1) \times 100$ for oxidizing and reducing gases, respectively, while for p-type semiconductor, $(R_{gas}/R_{air} - 1) \times 100$ and $(R_{air}/R_{gas} - 1) \times 100$ for reducing and oxidizing gases respectively. R is the electrical resistance of the sensor material in air (Rair) or in gas (Rgas)

3. Results and Discussion

The phase identification derived from powder XRD analysis is presented on Figure 1. As can be observed in Figure 1a for Ni-doped TiO_2, the results indicate that anatase is present in those from undoped to 1% Ni-doped TiO_2, while rutile is the only phase present in the 2% Ni-doped TiO_2 powders. The amount of anatase (JCPDS 21-1272) decreases as the amount of nickel increases while the amount of rutile (JCPDS 21-1276) follows the opposite trend. A trace amount of ilmenite, $NiTiO_3$ is observed at 32.5° in the TN1 and TN2 samples (JCPDS 33-0960). These results indicate that Ni promotes the transition of anatase to rutile. As far as Co-doped TiO_2 is concerned, the results depicted on Figure 1b reveal that, the undoped and the 0.5Co-doped TiO_2 samples showed pure TiO_2 consisted of its two polymorphs; anatase (majority) and rutile phases (minority). No other phase containing Co was observed. The 1Co-doped TiO_2 sample showed only single phase TiO_2 but this time the rutile polymorph was the major phase as anatase phase was in trace amount. The X-ray results indicate that the cobalt dopant promotes also the anatase-to-rutile phase conversion of TiO_2 but not progressively as nickel. The results of XRD analysis performed on the Rh-doped $BaTiO_3$ (before and after the activation) are presented in Figure 1c. As shown, the major phase is $BaTiO_3$ (according to JCPDS 075-0462) and a trace amount of $BaCO_3$ phase is observed. No evidence of an Rh phase was encountered indicating the substitution of Ti by Rh.

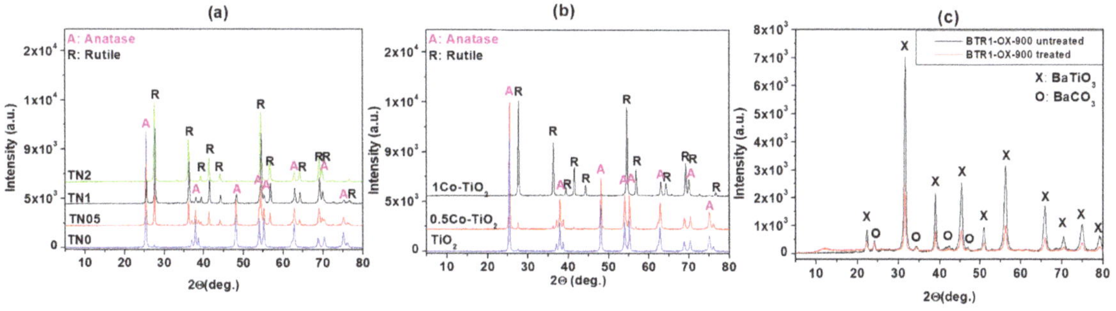

Figure 1. (**a**) Powder XRD patterns of Ni-doped TiO_2, (**b**) Co-doped TiO_2 and (**c**) Rh-doped $BaTiO_3$.

Figure 2 shows the morphology of the synthesized powders investigated by SEM. Concerning Ni-doped TiO_2 (Figure 2a), as far as the undoped powder sample is concerned, the microstructural investigation reveals spherical nanoparticles with sizes around 70 nm, which tend to agglomerate. While the TN05 sample shows more agglomerated spherical particles, the TN1 and TN2 samples show less agglomeration with the appearance of small pores and particle size reduction for the TN2 samples. Figure 2b shows the morphology of the Co-doped TiO_2 synthetized powders. The SEM investigation revealed the formation of spherical nanoparticles with sizes around 70 nm. As the sample 0.5Co-doped TiO_2 shows more agglomerated spherical particles, the sample 1Co-doped TiO_2 presented larger and well-faceted rhombohedral crystallites with less agglomeration. As SEM pictures display in Figure 2b the Rh-doped $BaTiO_3$ has the well-defined and homogeneously distributed spherical nanoparticles (~50 nm).

Figure 2. SEM images of (**a**) undoped and Ni-doped TiO_2 powders and (**b**) Co-doped TiO_2 and Rh-doped $BaTiO_3$.

The Raman spectra of the samples were obtained between the wavenumbers of 175–800 cm^{-1} and the results are presented in Figure 3. The results of Ni-doped TiO_2 Raman analysis (Figure 3a) show that the main signals came from TiO_2. The samples TN0 and TN05 show very strong Raman signals, with peaks at 196(E_g), 396($B1_g$), 517($A1_g$) and 638(B1g) cm^{-1} from the typical anatase TiO_2 phase [16]. A weak peak at 447 cm^{-1} (E_g) attributed to rutile is observed in TN05. Samples TN1 and TN2, which contain larger amounts of Ni, present the Raman signals corresponding to both the anatase and rutile (447(E_g), 612($A1_g$) cm^{-1}) phase [17]. In addition to the anatase and rutile phases, another set of Raman vibrations emerges in the TN1 and TN2 samples. The peaks at 244, 345 and 706 cm^{-1} are assigned to a trace amount of ilmenite, $NiTiO_3$. In the Figure 3b showing the results of Co-doped TiO_2, the Raman lines observed at 197, 390, 511, 637 cm^{-1} can be assigned to E_g, $B1_g$, $A1_g$, or $B1_g$ and E_g modes of anatase phase respectively. The spectra show that the peak intensities decrease drastically after doping, due certainly to the decrease of the amount of anatase and the formation of rutile as indicated by XRD. Moreover, the Raman spectra of 1Co-doped TiO_2 yielded a smaller shift towards lower wavelengths while new peaks (436 cm^{-1}) appeared indicating the presence of rutile polymorph, as mentioned in literature [17]. Raman spectra of Rh-doped $BaTiO_3$ are displayed in Figure 3c. It shows the peaks at 270, 308, 525 and 725 cm^{-1} which are assigned respectively to the A1(TO2), E(TO2), A1(TO3), and A1(LO3) of barium titanate modes of the room temperature P4mm phase.

Based on our previous results on undoped, Al and Cr-doped TiO_2, 600 °C was chosen as the optimum sensing temperature for Ni and Co doped TiO_2 in this work [12,13].

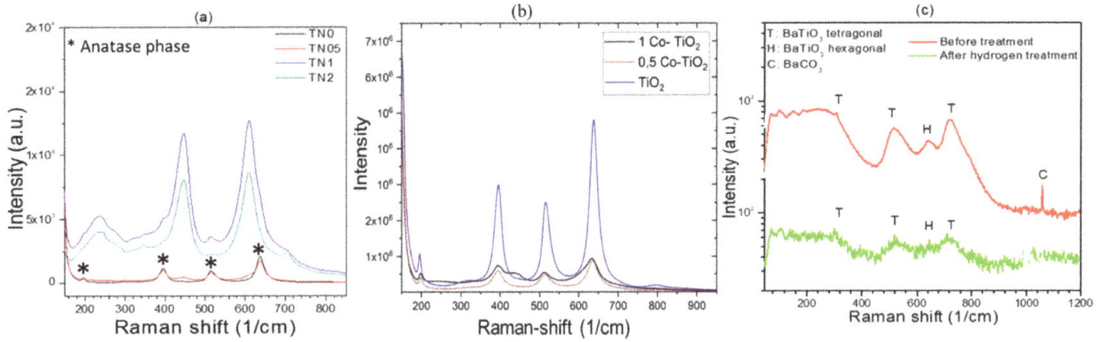

Figure 3. (**a**) Raman spectra of undoped and Ni-doped TiO_2, (**b**) Co-doped TiO_2 and (**c**) Rh-doped $BaTiO_3$.

The responses of the undoped TiO_2 and all Ni-doped TiO_2 towards 10,000 ppm H_2 in dry synthetic air at 600 °C are shown in Figure 4a. The sensor responses are 42, 72, 70 and 62% for TN0, TN05, TN1 and TN2, respectively. It can be observed that the sensor response increases greatly as Ni-content increases up to 0.5 mol.% and then decreases slowly with further increase in the Ni-content to 2.0 mol.%. This implies that the sensor reaches its maximum response of 72% with 0.5 mol.% of Ni dopant. It can be assumed that this enhancement of gas sensor response may be due to the formation of a n-n junction between the anatase (E_g = 3.2 eV) and rutile (E_g = 3.0 eV) phases. As revealed by XRD results, the TN05 sample contains almost the same amount of anatase and rutile phases (which is not the case with the other samples in this work), and thus, the highest amount of n-n junctions are expected to be present in this sample. This kind of junction effect has also been reported for other n-n junction systems such as ZnO-SnO_2 [18] and SnO_2-WO_3 [19]. The achievement of a great selectivity towards the target gas is a key parameter and a very important characteristic. Therefore, the responses of the TN05 gas sensor towards a variety of interference gases including NO_2, CO, and NO at 600 °C in dry synthetic air were explored to evaluate its selectivity. As observed in Figure 4b, the response of this sensor towards 600 ppm of H_2 (35%) is at least a factor of two higher than that towards 300 ppm of CO (12%), 300 ppm of NO_2 (11%) and 300 ppm of NO (7%). The sensor's H_2-response was a factor of two greater compared to that for 300 ppm of all the tested interfering gases, yielding the highest value and indicating a relatively high selectivity potential of the sensor towards H_2.

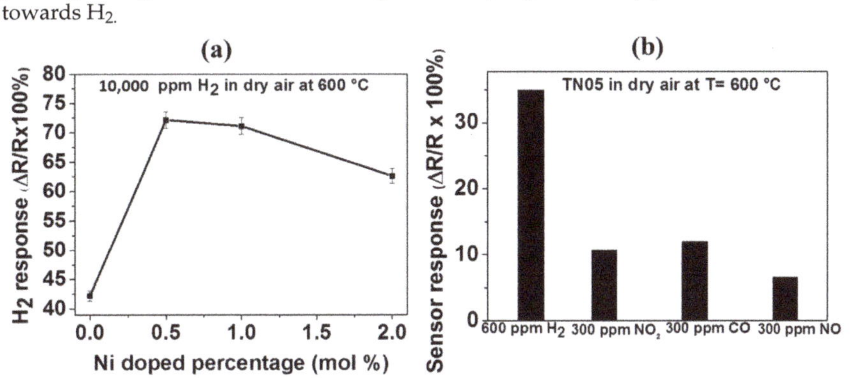

Figure 4. (**a**) Response of gas sensors based on undoped TiO_2 and Ni-doped TiO_2 to 10,000 ppm of hydrogen gas at the optimum operating temperature of 600 °C. (**b**) Response of gas sensor based on 0.5% Ni-doped TiO_2 to various gases including 600 ppm of H_2, 300 ppm of NO_2, NO and CO all in dry air at 600 °C.

As Figure 5a shows, the sensors yield higher response towards H_2 than NO_2. The H_2 sensor responses are 42, 23 and 33% for undoped, 0.5Co-doped TiO_2 and 1Co-doped TiO_2, respectively. The undoped sample shows the highest response toward H_2. The presence of Co-dopant seems to decrease the H_2-sensing performance of TiO_2 even though doping creates more oxygen vacancies in TiO_2. This behavior can be attributed to the increase of rutile polymorph content on doping with cobalt, as previously reported, rutile is the less active (in term of functional properties) polymorph of TiO_2 [20]. On the other hand, the 1Co-doped TiO_2 which contains predominantly rutile polymorph showed a higher response toward H_2 than 0.5Co-doped TiO_2. This discrepancy can be explained by alteration of the conductivity from n-type to p-type. In fact, the dynamic response of the sensors toward H_2 given in Figure 5b reveals that undoped and 0.5Co-doped TiO_2 exhibit n-type semi-conductivity (i.e., their electrical resistance decreases upon interaction with hydrogen) while the 1Co-doped TiO_2 yields p-type conductivity (its electrical resistance increases when reducing gas is introduced). The sensor responses measured toward the oxidizing gas NO_2 were 5, 3 and 8% for undoped, 0.5Co-doped TiO_2 and 1Co-doped TiO_2 respectively. In the case of NO_2 sensing, the 1Co-doped TiO_2 showed the highest response. This may be mainly due to the electronic alteration of TiO_2 from n to p-type semiconductor. Previous literature points out that this alteration can be utilized for the detection of oxidizing gas. Our current results confirm that the dominant factor for the gas sensing property of the Co-doped TiO_2 depend on the existing polymorphs as well as the nature of target gas (oxidizing or reducing). In the case of reducing gases, the type of polymorphs has more influence on the gas sensitivity than the type of electronic structure, while an opposite trend can be observed for oxidizing gases.

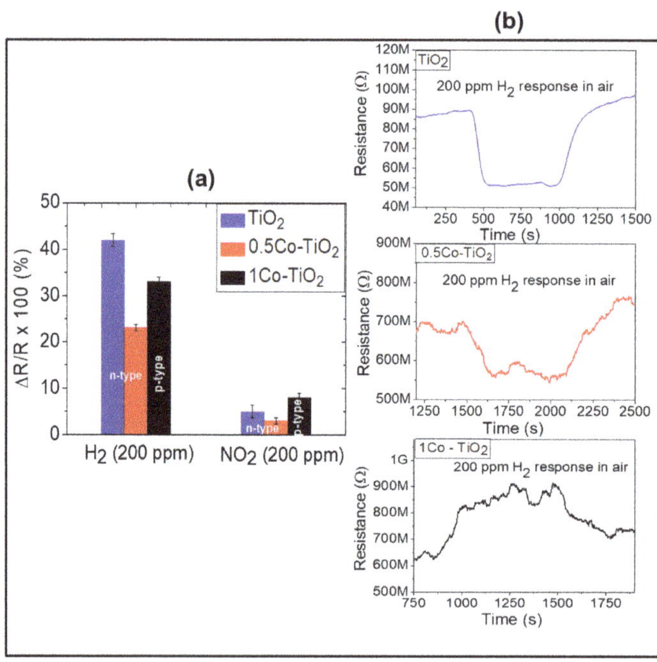

Figure 5. (**a**) Sensor response of undoped and Co-doped TiO_2 towards NO_2 and H_2 at 600 °C and (**b**) their dynamic responses towards H_2.

Figure 6a shows the sensor responses of the hydrogen treated Rh-doped $BaTiO_3$ towards 200 ppm of nitrogen oxide (NO) at a different operating temperature under dry and humid (10% of RH) synthetic air and the dynamic response at 700 and 900 °C, respectively. The sensor responses are 14, 2, 6, 7% in dry air, 6, 12, 16 and 18% in humid air at 600, 700, 800, and 900 °C, respectively. In dry air, the sensor response decreases from 14 to 7% in general with the increasing temperature, while in humid air, the sensor response increases from 6 to 18% as the temperature increases. The maximum sensor response is therefore obtained at 900 °C under humidity. This enhancement of sensing properties in the presence of humidity can be explained by the affinity between adsorbed hydroxyl group (generated after thermal decomposition of water) and the NO. In fact, at high temperatures, H_2O in water vapor decomposes, and hydroxyl is adsorbed on the sensing layer. As the temperature increases, more decomposition will occur, and more hydroxyl groups will be adsorbed on the surface, enhancing the NO sensor response. To the best of our knowledge, this is the first time that NO detection is reported at such a high temperature in the humid. Therefore, we have investigated intensely further gas sensing characteristics of this material under these extreme conditions (e.g., at 900 °C under humid air). Two other main products for fuel combustion are NO_2 and CO. Their presence in the exhaust gas stream at a high temperature can cause important hinderance for the NO gas sensing application. Our sensor's selectivity toward NO against CO and NO_2 at 900 °C under humid air was investigated. Figure 6b shows the different responses of the sensor to 200 ppm of NO, NO_2 and CO. The results indicate that at 900 °C, the response to 200 ppm of NO (18%) is higher than that of 200 ppm of NO_2 (8.7%) and 200 ppm of CO (8.4%). This implies that this sensor is at least twice as much sensitive to NO than NO_2 and CO. This good selectivity is ascribed to the catalytic effect of Rhodium-NPs on the oxidation of NO, which will promote and enhance the adsorption and the oxidation of NO preferentially. It is reported in the literature that Rhodium which is currently and often used in TWC, is a suitable catalyst for NO oxidation [21].

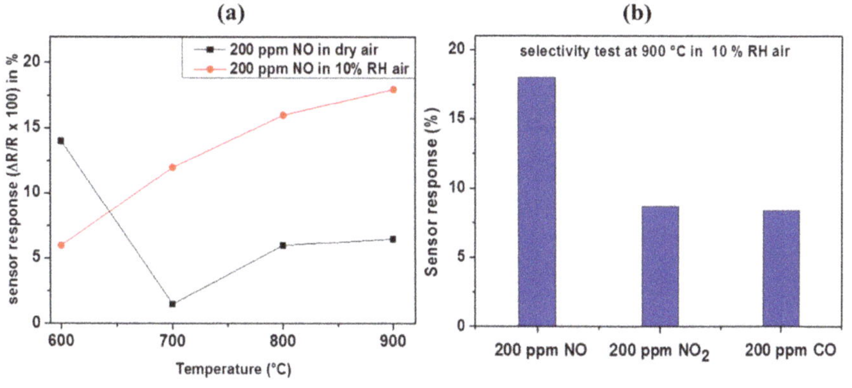

Figure 6. (**a**) Response of gas sensor based on activated Rh-doped $BaTiO_3$ to 200 ppm of NO in dry and humid air at different operating temperatures and (**b**) its selectivity investigation.

4. Conclusions

This paper reports the successful synthesis of Ni-doped TiO_2, Co-doped TiO_2 and Rh-doped $BaTiO_3$ nanoparticles by a facile co-precipitation route through the use of oxalic acid. Their NO_2, NO and hydrogen sensing properties at high temperatures (≤600 °C) were investigated. According to structural characterization the substitution of Ti^{4+} by dopant (Ni^{2+}, Co^{3+} and Rh^{3+}) was effective and creates more oxygen vacancies which promotes the anatase-to-rutile transformation in the case of Ni and Co doped TiO_2. Enhanced sensing properties with respect to H_2 were observed for 0.5% Ni-doped TiO_2 in comparison to undoped and 1 and 2% Ni-doped TiO_2. The sample 1Co-doped TiO_2 which reveals p-type conductive behavior yields an enhanced NO_2 response at 600 °C under air as carrier gas. With Rh-doped $BaTiO_3$, it was possible to detect NO at 900 °C under humid air with a good response (18% for 200 NO ppm) and good selectivity (twice as much sensitive to NO than CO and NO_2). Titanium based materials appear as a promising high temperature gas sensor in harsh environment.

Author Contributions: R.L.F.: conceptualization, data curation, formal analysis concerning Raman spectroscopy and SEM analysis, methodology and investigation concerning sensor testing, writing—original draft preparation; B.S.: supervision, project administration, funding acquisition, validation, visualization, writing—review and editing. All authors have read and agreed to the published version of the manuscript.

Funding: This research was funded by DAAD-DLR postdoc fellowship program, grant number 284.

Institutional Review Board Statement: Not applicable.

Informed Consent Statement: Not applicable.

Data Availability Statement: Data is contained within the article.

Acknowledgments: The authors thank Shujie You of the Luleå University of Technology in Sweden for the Raman spectroscopy measurements.

Conflicts of Interest: The authors declare no conflict of interest.

References

1. Liu, Y.; Parisi, J.; Sun, X.; Lei, Y. Solid-state gas sensors for high temperature applications—A review. *J. Mater. Chem. A* **2014**, *2*, 9919–9943. [CrossRef]
2. Resitoglu, I.A. NOx Pollutants from Diesel Vehicles and Trends in the Control Technologies. *IntechOpen* **2018**, *3*, 1–16. [CrossRef]
3. Gu, H.; Wang, Z.; Hu, Y. Hydrogen gas sensors based on semiconductor oxide nanostructures. *Sensors* **2012**, *12*, 5517–5550. [CrossRef]
4. Hermawan, A.; Asakura, Y.; Kobayashi, M.; Kakihana, M.; Yin, S. High temperature hydrogen gas sensing property of GaN prepared from α-GaOOH. *Sens. Actuators B Chem.* **2018**, *276*, 388–396. [CrossRef]
5. Singh, E.; Meyyappan, M.; Nalwa, H.S. Flexible Graphene-Based Wearable Gas and Chemical Sensors. *ACS Appl. Mater. Interfaces* **2017**, *9*, 34544–34586. [CrossRef]
6. Ali, S.; Gupta, A.; Shafiei, M.; Langford, S.J. Recent Advances in Perylene Diimide-Based Active Materials in Electrical Mode Gas Sensing. *Chemosensors* **2021**, *9*, 30. [CrossRef]
7. Dey, A. Semiconductor metal oxide gas sensors: A review. *Mater. Sci. Eng. B* **2018**, *229*, 206–217. [CrossRef]
8. Lontio Fomekong, R.; Tedjieukeng Kamta, H.M.; Ngolui Lambi, J.; Lahem, D.; Eloy, P.; Debliquy, M.; Delcorte, A. A sub-ppm level formaldehyde gas sensor based on Zn-doped NiO prepared by a co-precipitation route. *J. Alloys Compd.* **2018**, *731*, 1188–1196. [CrossRef]
9. Han, Y.; Liu, Y.; Su, C.; Chen, X.; Zeng, M.; Hu, N.; Su, Y.; Zhou, Z.; Wei, H.; Yang, Z. Sonochemical synthesis of hierarchical WO3 flower-like spheres for highly efficient triethylamine detection. *Sens. Actuators B Chem.* **2020**, *306*, 127536. [CrossRef]
10. Cho, H.-J.; Choi, S.-J.; Kim, N.-H.; Kim, I.-D. Porosity controlled 3D SnO_2 spheres via electrostatic spray: Selective acetone sensors. *Sens. Actuators B Chem.* **2020**, *304*, 127350. [CrossRef]
11. Chen, X.; Mao, S. Titanium Dioxide Nanomaterials: Synthesis, Properties, Modifications, and Applications. *Chem. Rev.* **2007**, *107*, 2891–2959. [CrossRef] [PubMed]
12. Lontio Fomekong, R.; Saruhan, B. Synthesis of Co^{3+} doped TiO_2 by co-precipitation route and its gas sensing properties. *Front. Mater.* **2019**, *6*, 252. [CrossRef]
13. Zhu, L.; Zhang, D.; Wang, Y.; Feng, C.; Zhou, J.; Liu, C.; Ruan, S. Xylene gas sensor based on Ni doped TiO_2 bowl-like submicron particles with enhanced sensing performance. *RSC Adv.* **2015**, *5*, 28105–28110. [CrossRef]
14. Gönüllü, Y.; Haidry, A.A.; Saruhan, B. Nanotubular Cr-doped TiO_2 for use as high-temperature NO_2 gas sensor. *Sens. Actuators B Chem.* **2015**, *217*, 78–87. [CrossRef]
15. Bayata, F.; Saruhan-Brings, B.; Ürgen, M. Hydrogen gas sensing properties of nanoporous Al-doped titania. *Sens. Actuators B Chem.* **2014**, *204*, 109–118. [CrossRef]
16. Chanda, A.; Rout, K.; Vasundhara, M.; Joshi, S.R.; Singh, J. Structural and magnetic study of undoped and cobalt doped TiO_2 nanoparticles. *RSC Adv.* **2018**, *8*, 912–916. [CrossRef]
17. Mazza, T.; Barborini, E.; Piseri, P.; Milani, P.; Cattaneo, D.; Li Bassi, A.; Bottani, C.E.; Ducati, C. Raman spectroscopy characterization of TiO_2 rutile nanocrystals. *Phys. Rev. B Condens. Matter.* **2007**, *75*, 045416. [CrossRef]
18. Chen, W.; Li, Q.; Xu, L.; Zeng, W. Gas sensing properties of ZnO-SnO_2 nanostructures. *J. Nanosci. Nanotechnol.* **2015**, *15*, 1245–1252. [CrossRef] [PubMed]
19. Van Toan, N.; Manh Hung, C.; Van Duy, N.; Duc Hoa, N.; Thanh Le, D.T.; Van Hieu, N. Bilayer SnO_2–WO_3 nanofilms for enhanced NH_3 gas sensing performance. *Mater. Sci. Eng. B* **2017**, *224*, 163–170. [CrossRef]
20. Luttrell, T.; Halpegamage, S.; Tao, J.; Kramer, A.; Sutter, E.; Batzill, M. Why is anatase a better photocatalyst than rutile? Model studies on epitaxial TiO_2 films. *Sci. Rep.* **2014**, *4*, 4043. [CrossRef] [PubMed]
21. Weiss, B.M.; Artioli, N.; Iglesia, E. Catalytic NO Oxidation Pathways and Redox Cycles on Dispersed Oxides of Rhodium and Cobalt. *ChemCatChem* **2012**, *4*, 1397–1404. [CrossRef]

chemistry proceedings

Proceeding Paper

Nutritional Composition of the Atlantic Seaweeds *Ulva rigida*, *Codium tomentosum*, *Palmaria palmata* and *Porphyra purpurea* †

Javier Echave [1], Catarina Lourenço-Lopes [1], Anxo Carreira-Casais [1], Franklin Chamorro [1], Maria Fraga-Corral [1,2], Paz Otero [1], Pascual Garcia-Perez [1], Sergio Baamonde [3], Fermín Fernández-Saa [3], Hui Cao [1], Jianbo Xiao [1], Miguel A. Prieto [1,2,*] and Jesus Simal-Gandara [1,*]

1 Nutrition and Bromatology Group, Faculty of Food Science and Technology, Ourense Campus, University of Vigo, E32004 Ourense, Spain; javier.echave@uvigo.es (J.E.); c.lopes@uvigo.es (C.L.-L.); anxocc@uvigo.es (A.C.-C.); franklin.noel.chamorro@uvigo.es (F.C.); mfraga@uvigo.es (M.F.-C.); paz.otero@uvigo.es (P.O.); pasgarcia@uvigo.es (P.G.-P.); hui.cao@uvigo.es (H.C.); jianboxiao@uvigo.es (J.X.)
2 Centro de Investigação de Montanha (CIMO-IPB), Campus de Santa Apolónia, 5300-252 Braganza, Portugal
3 Centro de Investigación e Innovación Tecnológico en Algas Marinas (CIITAM), Algas Atlánticas Algamar S.L., Polígono de Amoedo, Pazos de Borbén, E36840 Pontevedra, Spain; sergio.baamonde@algamar.com (S.B.); oficinas2@algamar.com (F.F.-S.)
* Correspondence: mprieto@uvigo.es (M.A.P.); jsimal@uvigo.es (J.S.-G.)
† Presented at the 1st International Electronic Conference on Chemical Sensors and Analytical Chemistry, 1–15 July 2021; Available online: https://csac2021.sciforum.net/.

Abstract: Macroalgae are regarded as a healthy food due to their composition and nutritional properties. In this work, nutritional composition of two green (*Ulva rigida*, *Codium tomentosum*) and two red (*Palmaria palmata*, *Porphyra purpurea*) edible seaweed was studied. Total lipids were measured gravimetrically as evaporated mass after petroleum-ether Soxhlet extraction of samples. In addition, fatty acid profile was determined by gas chromatography coupled to a flame ionization detector (GC-FID). Results showed that all studied species were accounted for very low levels of lipids (<1% dw), but levels of unsaturated fatty acids oleic, linoleic, and linolenic acids were present at high concentrations, with *P. palmata* displaying the highest quantities (>200 mg C18:1/g extract). In parallel, proteins were quantified following the macro-Kjeldahl method. In this analysis, red algae, especially *P. purpurea*, showed significant protein content up to 30% DW. Total organic acids were found by ultra-filtration liquid-chromatography coupled to an amperometry detector (UFLC-PAD) after an acid extraction, *P. purpurea* being the algae with the higher organic acid content (10.61% dw). Minerals were identified and quantified by inductively coupled plasma atomic emission spectroscopy (ICP-OES), suggesting that both algae groups are rich in K and Mg (>15 g/kg), but *U. rigida* also displayed a remarkable iron content (>1 g Fe/kg). Other detected minerals in minor concentrations were Ca, P or F. Altogether, results corroborate that these edible algae are a good source of nutrients in accordance with literature.

Keywords: macroalgae; nutrition; composition; chromatography; minerals

check for updates

Citation: Echave, J.; Lourenço-Lopes, C.; Carreira-Casais, A.; Chamorro, F.; Fraga-Corral, M.; Otero, P.; Garcia-Perez, P.; Baamonde, S.; Fernández-Saa, F.; Cao, H.; et al. Nutritional Composition of the Atlantic Seaweeds *Ulva rigida*, *Codium tomentosum*, *Palmaria palmata* and *Porphyra purpurea*. *Chem. Proc.* **2021**, *5*, 67. https://doi.org/10.3390/CSAC2021-10681

Academic Editor: Huangxian Ju

Published: 14 July 2021

1. Introduction

Seaweeds (macroalgae) are common ingredients in East Asian cuisine, of which many species such as wakame (*Undaria pinnatifida*), sweet kelp (*Saccharina latissima*) or nori (*Porphyra purpurea*) are used in different dishes. Seaweeds have long been recognized as healthy foods owing to their low caloric index and high content in dietary fiber, minerals and antioxidant molecules such as their cell wall polysaccharides [1,2] Besides, in recent years, there has been an increasing consumer interest in vegetarian food sources. In this context, algae could be a valuable alternative source of essential macronutrients. Indeed, seaweeds have been proposed as an alternative ingredient for the formulation of nutritional supplements that could cover various dietary needs [3]. One of the key nutritional components is protein and the amino acid composition of a food protein. It is known that red seaweeds account for a protein content between 20 and 47% of its dry weight (dw), while green algae generally contains about 9–26% and brown seaweeds 3 to 15% [4]. From a nutritional perspective, seaweed proteins are also valuable since their content in essential amino acids is generally higher (~50%) than legumes (~40%) [5]. Seaweeds have also been generally described to contain very low levels of lipids usually between 1 and 4% dw, but nonetheless rich in polyunsaturated fatty acids (PUFA) [6]. Considering their mineral composition, seaweeds tend to hold much higher content of potassium, magnesium or calcium than several terrestrial plants. However, they are also described to generally accumulate iodine in great amounts of which an excessive intake could be hazardous to thyroid function [7]. In some cases, hazardous levels of arsenic have also been reported, which requires monitoring and assessment upon consumption of certain species [8]. Nonetheless, the nutritional composition of several seaweed species considering their growing region remains to be described, especially considering traditional methods for determining proximate compositions. In this work, nutritional composition of edible seaweed *Ulva rigida* (UR), *Codium tomentosum* (CT), *Palmaria palmata* (PA) and *Porphyra purpurea* (PU) widely distributed in Atlantic shores was studied using standardized analytical methods.

2. Material and Methods

2.1. Sample Preparation

Algae samples were provided by Algas Atlánticas Algamar S.L company (www.algamar.es, accessed on 1 July 2021) located in Pontevedra, Spain. The algae samples were collected from the coasts of the Galician region, Pontevedra province (NW Spain), washed with distilled water, frozen at −80 °C and freeze-dried afterwards. The seaweed samples were then crushed and grinded to obtain a homogeneous matrix, which was stored at −20 °C for further analysis.

2.2. Proximate Composition

Proximate composition was studied following AOAC guidelines (1995) [9]. All thermogravimetric analyses were carried out with a SETSYS Evolution thermobalance (Setaram, Caluire-et-Cuire, France). Results are expressed as g per 100 g dw.

Humidity was determined by a gravimetric method. 1 g of sample was deposited in a previously dried (104 °C/24 h) ceramic crucible. Next, the crucibles with fresh sample were placed in an oven at 104 °C for 24 h. After that time, they were weighed again, and humidity was calculated as the weight difference.

2.2.1. Inorganic Material

To find the content of inorganic material (Ash), 250 mg were placed in a porcelain crucible (previously weighed) and the samples were incinerated at 600 ± 15 °C for 5 h. The crucible was then weighed with the resulting sample content. The obtained difference in weight was calculated as the ash value for each sample.

2.2.2. Protein Content

Protein content was determined according to the macro-Kjeldahl method [10]. Briefly, 500 mg of sample were placed in a Kjeldahl tube, then adding a catalytic tablet (Sigma Aldrich, St. Louis, MO, USA) and 20 mL of sulfuric acid. The tubes were placed in a digestive block and the temperature was gradually increased to 400 °C for 70 min. The tubes were then removed, allowed to briefly cool, and 25 mL of distilled H_2O was added. The nitrogen (N) converted to ammonia was measured with a macro-Kjeldahl distiller. The resulting N value was multiplied by a correction factor of 6.25 to obtain the estimate of the protein content, an extensively used correction factor for algae N-to-protein determinations [11].

2.2.3. Lipids

For total lipids determination, 3 g of sample were placed inside a paper cartridge. An extraction with petroleum ether was then conducted through a ST 243 SOXTEC Soxhlet extraction system (Foss, Hillerød, Denmark) at a constant temperature of 120 °C for 7 h. The resulting product was transferred to a ground test tube, previously weighed, and placed in the oven for evaporation of the solvent. After solvent evaporation, the tube was weighed again to obtain, by difference, the total lipids content.

2.2.4. Fiber and Hydrocarbons

Fiber was determined following the gravimetric AOAC method [9]. Briefly, 1 g of dried sample was sequentially treated with α-amylase from *Bacillus licheniformis* (pH 6, 30 min, 37 °C), protease from *Bacillus licheniformis* (pH 7.5, 30 min, 37 °C) and amyloglucosidase from *Aspergillus niger* (pH 4.5, 30 min, 40 °C). The obtained residue was precipitated with 4 times its volume in ethanol and filtered through a 0.45 μm paper syringe filter. The obtained difference in weight was calculated as total fiber.

Total hydrocarbons were calculated as the difference of the rest of the components, following Equation (1) and the results expressed as % (g/100 g dw) [12]:

$$\text{Hydrocarbons} = 100 - (\text{Lipids} + \text{Proteins} + \text{Ash} + \text{Fiber}) \quad (1)$$

2.3. *Organic Acids*

To determine the organic acid content, 1 g of each sample was weighed, and an extraction was carried out with 25 mL of 4.5% metaphosphoric acid, while stirring for 20 min. It was then filtered through paper and nylon (0.22 μm) to be able to work in ultra-fast liquid chromatography coupled to a photodiode array detector (UFLC-PAD).

The analysis was performed using a Shimadzu 20A series UFLC (Shimadzu, Kyoto, Japan) Separation was achieved on a SphereClone (Phenomenex, Torrance, CA, USA) reverse phase C_{18} column (5 μm, 250 × 4.6 mm) at 35 °C. Sulfuric acid 3.6 mM was used as mobile phase with a flow rate of 0.8 mL/min. Detection was carried out using wavelengths between 215 and 245 nm. Detected organic acids were quantified by comparison of the area of their peaks with calibration curves obtained by comparison to an ascorbic acid standard (Sigma Aldrich, St. Luois, MO, USA).

2.4. *Mineral Content*

For mineral detection and quantification, 2 g samples were subjected to a metaphosphoric acid digestion for 10 min and analyzed afterwards with optic emission spectrometry with inductively coupled plasma (ICP-OES) using an Optima 4300 DV instrument (Perkin Elmer, Whaltman, MA, USA) [9]. Briefly, quantification of minerals was determined following detection in specific wavelengths for Ca (317.9 nm), Mg, (285.2 nm), Cl, (134.7 nm), Fe (248.6 nm), Mn (279.4 nm), Zn (206.2 nm), K, (769.9 nm), I (178.2 nm), F, (685.6 nm), As (188.9 nm), P (213.6 nm) with RF power of 1450 W and at an argon plasma flow of 15 L/min. Results are expressed as g/kg dw.

2.5. *Fatty Acid Profile*

To carry out this determination, the product resulting from the lipids Soxhlet extraction was used and a derivatization process was carried out to obtain fatty acid methyl esters (FAME). 5 mL of reagent A (MeOH, H_2SO_4 and C_7H_8) was added in a 2:1:1 ratio and they were kept in a bath at 50 °C while stirring at 160 rpm for 12 h. Afterwards, 3 mL of distilled H_2O was added, then adding 3 mL of diethyl ether under vigorous and continuing stirring until a homogeneous sample was obtained. Later, the two phases separation was allowed to occur, and the supernatant was transferred to a vial with sodium sulfate. The contents of the vial were filtered through 0.22 μm nylon prior to their chromatographic analysis by gas chromatography coupled to an infrared detector (GC-FID). The GC system was an Agilent 7820A and an Agilent HP-88 (60 m, 250 μm × 0.25 μm) column was used (Agilent

Technologies, Santa Clara, CA, USA). Helium was used as carrier; 1µL of sample was injected. Oven temperature program started at 120 °C, increasing to 175 °C at 10 °C pre min. rate and hold for 10 min. Then, temperature was further increased to 220 °C at 3 °C increase per min. and kept at 220 °C for 5 min.

Different fatty acid levels were determined by comparing the relative retention times of the FAME peaks of the algae samples with respect to a commercial standard of FAME mix (Supelco 37 Component FAME MIX, Sigma Aldrich, St. Luois, MO, USA).

3. Results and Discussion

3.1. Proximate Composition

All studied species were accounted for very low levels of lipids (<1% dw), which could be due to the harvesting season or an incomplete extraction, as other studies report generally higher lipid content in some closely related species [7].

Green seaweeds displayed protein levels around 15% dw (Figure 1). Red algae, especially *P. purpurea*, showed a significant protein content, up to 30% dw. This is in contrast with other analyses reported on both *P. purpurea* and *P. palmata*, which displayed higher protein content [7,13]. Organic acids content was significantly heterogeneous, with only *P. purpurea* showing a high content (10.61% dw), half of which was determined as citrate. No organic acids were detected in *U. rigida*. Almost all the analyzed species showed more than 40% dw of insoluble fiber, with similar results to those generally reported in literature [5,13]. On the other hand, inorganic matter was somehow homogeneous for all seaweeds except for *C. tomentosum*, which showed a 37% dw ash content.

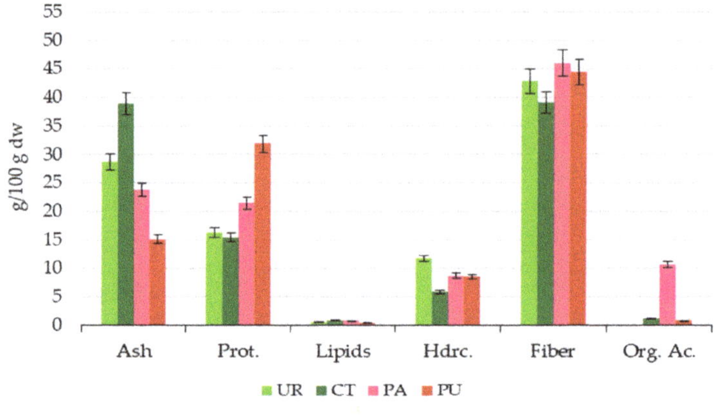

Figure 1. Proximate composition of the studied seaweed species (% dw). UR, *Ulva rigida;* CT, *Codium tomentosum;* PA, *Palmaria palmata;* PU, *Porphyra purpurea;* Prot, proteins; Hdrc., hydrocarbons; Org. Ac., organic acids.

3.2. Mineral Composition

The main minerals detected were Cl, K and Mg, of which *P. palmata* significantly outstand with as much as 100 g/kg dw of Cl and K, followed by *U. rigida* (Figure 2A). Indeed, the latter was accounted for the highest levels of Mg (22.9 g/kg dw), whereas *C. tomentosum* displayed the lowest mineral concentrations for Cl, Mg or K. *P. purpurea* displayed 33.5 g/kg dw of K and 12.1 g/kg dw of Cl, but the rest of its minerals were found at levels below 10 g/kg dw.

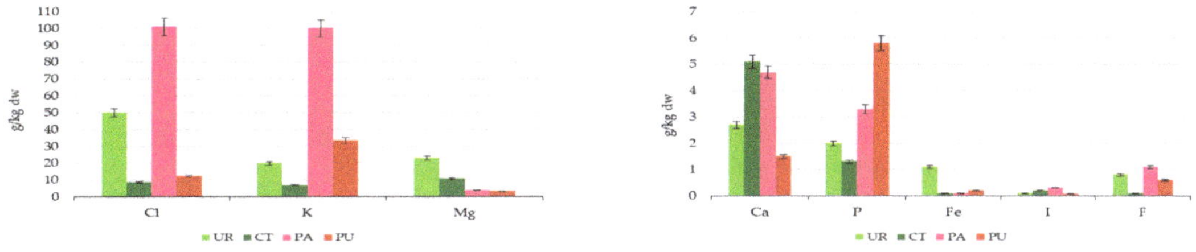

(A) **(B)**

Figure 2. Mineral composition of the studied seaweed species in: (**A**), in major concentrations (g/kg dw); and (**B**), in minor concentrations (g/kg dw). UR, *Ulva rigida*; CT, *Codium tomentosum*; PA, *Palmaria palmata*; PU, *Porphyra purpurea*.

Minerals detected at minor concentrations are displayed in Figure 2B. Ca levels for *C. tomentosum* and *P. palmata* were higher than those found in terrestrial plants. Conversely, out of the four studied species, *P. purpurea* displayed the highest P content, a feature that may be related to its higher protein content (Figure 1). It is worth noting the remarkably high Fe content of *U. rigida*, as it is several times higher than those reported in terrestrial plants. Considering this result, *U. rigida* may contain higher iron quantities than legumes, which are considered one of the main sources of this mineral [3]. On the other hand, iodine levels from the studied sampled species appear lower [14]. Although this could be due to experimental errors, it is plausible that it is related to the their growing region, since it seems that seaweeds from Galician waters accumulate less iodine [15]. It is also noteworthy that *C. tomentosum*, despite showing the highest ash content, did not display significantly high levels of any of the test minerals, except for Ca (5.1 g/kg dw). This could be due to the presence of other minerals or metals not tested in this work. On the other hand, other minerals and metals searched for like Mn, As and Zn were detected at trace amounts, below 0.01 g/kg dw (data not shown). F was also detected in all species, reaching up to 1.1 g/kg dw in *P. palmata*, although *C. tomentosum* had the lowest levels of it (0.1 g/kg dw).

3.3. Fatty Acid Profile

Regarding fatty acid profile (Figure 3), the proportion of PUFA was notably high, with *P. palmata* displaying the highest relative quantities. *P. palmata* levels of oleic acid were higher than 200 mg/g extract and more than 150 mg/g extract for palmitoleic acid. Moreover, *P. palmata* denoted more than 100 mg/g extract of eicosatetraenoic acid. *U. rigida* however, accounted for the most linoleic acid content (>150 mg/g extract).

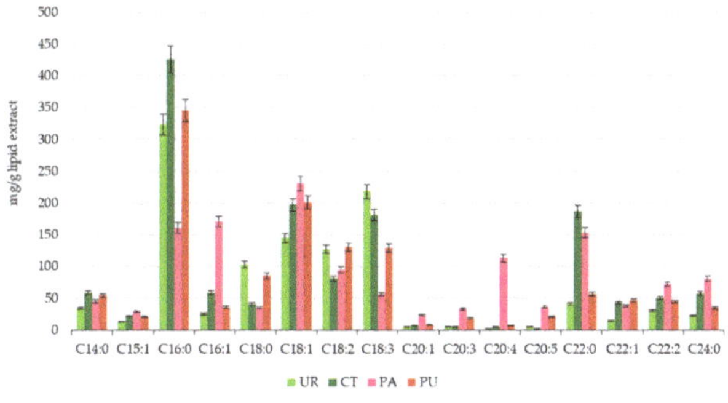

Figure 3. Fatty acid profile of the studied species. UR, *Ulva rigida*; CT, *Codium tomentosum*; PA, *Palmaria palmata*; PU, *Porphyra purpurea*.

In contrast, *P. purpurea* showed higher proportions of saturated fatty acids, foremost palmitic (>320 mg/g extract) and stearic acids (>70 mg/g extract). Nonetheless, *C. tomentosum* accounted for the highest proportion of saturated fatty acids, like palmitic acid (>400 mg/g extract) and behenic acid (>170 mg/g extract). Altogether, results corroborate that these edible algae are a good source of nutrients and analytical methods are suitable, in accordance with literature [7,16].

4. Conclusions

Seaweeds have been traditionally consumed for their nutritional value and availability. Moreover, in recent years they have been proposed as an alternative source of proteins and minerals, in contrast to terrestrial plants. It is of interest to accurately determine their nutritional composition to critically assess their value. In this work, standardized and recognized AOAC analytical methods for nutritional composition were employed to determine the major elements of four edible Atlantic seaweed species. Additionally, their mineral and fatty acid profile was investigated. Results unveiled that their protein and mineral content makes them notable sources of these nutrients, especially in red algae species. In these red seaweeds, protein content reached more than 20% for *P. palmata* and even more than 30% for *P. purpurea*. Whereas lipid content was particularly low (possibly due to specific environmental growing conditions), fatty acid analysis denoted high proportions of unsaturated fatty acids, i.e., oleic and linoleic acids in both *U. rigida* and *P. palmata*. Considering the reported results, these seaweed species growing in NW Spain could be a potential food and/or feed ingredient, especially owing to their high contents in minerals. Among the studied species, *P. palmata* stands out due to its PUFA and mineral composition, as well as its mineral content. *P. porphyra* on the other hand, was shown to be especially rich in proteins and P. Taking present results, seaweeds could be proposed as alternative supplementation ingredients for food and feed instead of animal or terrestrial plant sources; since not only they are rich in valuable nutrients, but also currently underexploited for this purpose. Further research should analyze other nutritional aspects from these widespread seaweed species in more depth. This would allow to assess with more accuracy the nutritional value of these Atlantic seaweeds.

Supplementary Materials: The poster presentation is available online at https://www.mdpi.com/article/10.3390/CSAC2021-10681/s1.

Author Contributions: Conceptualization, J.E., C.L.-L., A.C.-C., F.C., M.F.-C., P.O., P.G.-P., S.B., F.F.-S., H.C., J.X., J.S.-G. and M.A.P.; methodology, J.E., C.L.-L., A.C.-C. and F.C.; validation, M.F.-C., P.O., H.C. and P.G.-P.; formal analysis, M.F.-C., P.O., H.C. and P.G.-P.; investigation, S.B., F.F.-S., J.X., J.S.-G. and M.A.P.; writing—original draft preparation, J.E., C.L.-L., A.C.-C. and F.C.; writing—review and editing, M.F.-C., P.O., P.G.-P., H.C., J.X., J.S.-G. and M.A.P.; visualization, S.B., F.F.-S., J.X., J.S.-G. and M.A.P.; supervision, J.X., J.S.-G. and M.A.P. All authors have read and agreed to the published version of the manuscript.

Funding: The research leading to these results was supported by MICINN supporting the Ramón y Cajal grant for M.A. Prieto (RYC-2017-22891) and the FPU grant for A. Carreira-Casais (FPU2016/06135); by Xunta de Galicia for supporting the program EXCELENCIA-ED431F 2020/12, the post-doctoral grant of M. Fraga-Corral (ED481B-2019/096), the program BENEFICIOS DO CONSUMO DAS ESPECIES TINTORERA-CO-0019-2021 that supports the work of F. Chamorro and the program Grupos de Referencia Competitiva that supports the work of J. Echave (GRUPO AA1-GRC 2018); by the Bio Based Industries Joint Undertaking (JU) under grant agreement No 888003 UP4HEALTH Project (H2020-BBI-JTI-2019) that supports the work of P. Otero, P. Gar-cia-Perez and C. Lourenço-Lopes; and by Ibero-American Program on Science and Technology (CYTED—AQUA-CIBUS, P317RT0003).

Institutional Review Board Statement: Not applicable.

Informed Consent Statement: Not applicable.

Data Availability Statement: Not applicable.

Acknowledgments: The authors are grateful to Algamar (www.algamar.com, accessed on 1 July 2021) for their collaboration and algae material provision. The JU receives support from the European Union's Horizon 2020 research and innovation program and the Bio Based Industries Consortium. The project SYSTEMIC Knowledge hub on Nutrition and Food Security, has received funding from national research funding parties in Belgium (FWO), France (INRA), Germany (BLE), Italy (MIPAAF), Latvia (IZM), Norway (RCN), Portugal (FCT), and Spain (AEI) in a joint action of JPI HDHL, JPI-OCEANS and FACCE-JPI launched in 2019 under the ERA-NET ERA-HDHL (No 696295).

Conflicts of Interest: The authors declare no conflict of interest.

References

1. Martínez-Hernández, G.B.; Castillejo, N.; del M Carrión-Monteagudo, M.; Artés, F.; Artés-Hernández, F. Nutritional and bioactive compounds of commercialized algae powders used as food supplements. *Food Sci. Technol. Int.* **2018**, *24*, 172–182. [CrossRef] [PubMed]
2. Li, B.; Xu, H.; Wang, X.; Wan, Y.; Jiang, N.; Qi, H.; Liu, X. Antioxidant and antihyperlipidemic activities of high sulfate content purified polysaccharide from Ulva pertusa. *Int. J. Biol. Macromol.* **2020**, *146*, 756–762. [CrossRef] [PubMed]
3. Wells, M.L.; Potin, P.; Craigie, J.S.; Raven, J.A.; Merchant, S.S.; Helliwell, K.E.; Smith, A.G.; Camire, M.E.; Brawley, S.H. Algae as nutritional and functional food sources: Revisiting our understanding. *J. Appl. Phycol.* **2017**, *29*, 949–982. [CrossRef] [PubMed]
4. Pliego-Cortés, H.; Wijesekara, I.; Lang, M.; Bourgougnon, N.; Bedoux, G. Current knowledge and challenges in extraction, characterization and bioactivity of seaweed protein and seaweed-derived proteins. In *Advances in Botanical Research*; Elsevier Ltd.: Amsterdam, The Netherlands, 2020; Volume 95, pp. 289–326. ISBN 9780081027103.
5. Paiva, L.; Lima, E.; Patarra, R.F.; Neto, A.I.; Baptista, J. Edible Azorean macroalgae as source of rich nutrients with impact on human health. *Food Chem.* **2014**, *164*, 128–135. [CrossRef] [PubMed]
6. Paiva, L.; Lima, E.; Neto, A.I.; Marcone, M.; Baptista, J. Health-promoting ingredients from four selected Azorean macroalgae. *Food Res. Int.* **2016**, *89*, 432–438. [CrossRef] [PubMed]
7. Mæhre, H.K.; Malde, M.K.; Eilertsen, K.E.; Elvevoll, E.O. Characterization of protein, lipid and mineral contents in common Norwegian seaweeds and evaluation of their potential as food and feed. *J. Sci. Food Agric.* **2014**, *94*, 3281–3290. [CrossRef] [PubMed]
8. Mouritsen, O.G.; Dawczynski, C.; Duelund, L.; Jahreis, G.; Vetter, W.; Schröder, M. On the human consumption of the red seaweed dulse (*Palmaria palmata* (L.) Weber & Mohr). *J. Appl. Phycol.* **2013**, *25*, 1777–1791. [CrossRef]
9. Association of Official Chemistry (AOAC). *Official Methods of Analysis of the Association of Official Analytical Chemists*, 16th ed.; AOAC International: Arlington, TX, USA, 1995; Volume 1, ISBN 0935584544.
10. Bradstreet, R.B. Kjeldahl Method for Organic Nitrogen. *Anal. Chem.* **1954**, *26*, 185–187. [CrossRef]
11. Peinado, I.; Girón, J.; Koutsidis, G.; Ames, J.M. Chemical composition, antioxidant activity and sensory evaluation of five different species of brown edible seaweeds. *Food Res. Int.* **2014**, *66*, 36–44. [CrossRef]
12. Jayakody, M.M.; Vanniarachchy, M.P.G.; Wijesekara, W.L.I. Development and characterization of a seaweed snack using *Ulva fasciata*. *J. Food Sci. Technol.* **2021**, *58*, 1617–1622. [CrossRef] [PubMed]
13. Taboada, M.C.; Millán, R.; Miguez, M.I. Nutritional value of the marine algae wakame (Undaria pinnatifida) and nori (Porphyra purpurea) as food supplements. *J. Appl. Phycol.* **2013**, *25*, 1271–1276. [CrossRef]
14. Rubio, C.; Napoleone, G.; Luis-González, G.; Gutiérrez, A.J.; González-Weller, D.; Hardisson, A.; Revert, C. Metals in edible seaweed. *Chemosphere* **2017**, *173*, 572–579. [CrossRef] [PubMed]
15. Darias-Rosales, J.; Rubio, C.; Gutiérrez, Á.J.; Paz, S.; Hardisson, A. Risk assessment of iodine intake from the consumption of red seaweeds (*Palmaria palmata* and *Chondrus crispus*). *Environ. Sci. Pollut. Res.* **2020**, *27*, 45737–45741. [CrossRef] [PubMed]
16. Taboada, C.; Millan, R.; Miguez, I. Evaluation of marine algae Undaria pinnatifida and Porphyra purpurea as a food supplement: Composition, nutritional value and effect of intake on intestinal, hepatic and renal enzyme activities in rats. *J. Sci. Food Agric.* **2013**, *93*, 1863–1868. [CrossRef] [PubMed]

chemistry proceedings

MDPI

Proceeding Paper

Drift Compensation of the Electronic Nose in the Development of Instruments for Out-of-Laboratory Analysis †

Anastasiia Shuba [1,*], Tatiana Kuchmenko [1] and Dariya Menzhulina [2]

1 Department of Physical and Analytical Chemistry, Voronezh State University of Engineering Technologies, 394036 Voronezh, Russia; tak1907@mail.ru
2 Voronezh Children's Clinical Hospital Named after N.N. Burdenko, 394000 Voronezh, Russia; darinochka@mail.ru
* Correspondence: shuba1nastya@gmail.com; Tel.: +7-9204031622

† Presented at the 1st International Electronic Conference on Chemical Sensors and Analytical Chemistry, 1–15 July 2021; Available online: https://csac2021.sciforum.net/.

Abstract: A technique was developed to evaluate and compensate for the drift of eight mass-sensitive sensors in an open detection cell in order to estimate the influence of external factors (temperature, changes in the chemical composition of the background) on the out-of-laboratory analysis of biosamples. The daily internal standardization of the system is an effective way to compensate for the sensor signal drift when the sorption properties of sensitive coatings change during their long-term, intensive operation. In this study, distilled water was proposed as a standard for water matrix-based biosamples (blood, exhaled breath condensate, urine, etc.). Further, internal standardization was based on daily calculation of the specific sensor signals by dividing the sensor signals for the biosample according to the corresponding averaged values obtained from three to five standard measurements. The stability of the sensor array operation was estimated using the theory of statistical process control (exponentially weighted moving average control charts) based on the specific signal of the sensor array. The control limits for the statistical quantity of the central tendency for each sensor and the whole array, as well as the variations of the sensor signals, were determined. The average times required for signal and run lengths, for the purpose of statistically substantiated monitoring of the electronic nose's stability, were calculated. Based on an analysis of the tendency and variations in sensor signals during 3 months of operation, a technique was formulated to control the stability of the sensor array for the out-of-laboratory analysis of the biosamples. This approach was successfully verified by classifying the results of the analysis of the blood and water samples obtained for this period. The proposed technique can be introduced into the software algorithm of the electronic nose, which will improve decision-making during the long-term monitoring of health conditions in humans and animals.

Keywords: piezoelectric sensor array; drift; electronic nose; stability; statistical process control chart; blood

Citation: Shuba, A.; Kuchmenko, T.; Menzhulina, D. Drift Compensation of the Electronic Nose in the Development of Instruments for Out-of-Laboratory Analysis. *Chem. Proc.* **2022**, *5*, 68. https://doi.org/10.3390/CSAC2021-10464

Academic Editor: Ye Zhou

Published: 1 January 2022

1. Introduction

When designing a new methodology for the routine analysis of biological samples, significant attention is paid to reducing systematic errors linked to the changing properties of measuring devices. Control over external conditions (temperature and humidity) is determined by the nature of the working elements in the device. Thus, in gas analyzers of the sorption type, the signal is defined by the interaction of phases of modifiers with vapours and gases in the pre-sensory space; in reusable sensors, the stability of sorption surfaces, which defines the sorption–desorption processes, constitutes the main requirement. Since temperature changes the volatility of substances and the sorption properties of substances and sorbents, operating sorption-type sensors at a normal temperature and relative humidity is a critical factor. The development of approaches to ensure the stability of measurements during long-term operation is paramount in portable and mobile analytical systems [1] as this provides time minimization from the moment of sample selection to that of measurement, which is vital for biosample analysis.

It is well known that the drift of sensor signals influences classification solutions and the quantitative determination of components in a mixture [2–4]. The short-term drift of basic lines in sorption-type piezoelectric sensors is caused by malfunctions of electric schemes and an abrupt change in the external conditions of the experiment. As a rule, such mistakes are fixed immediately during the measurement and discarded as outliers. Essentially, a long-term drift relates to a change in the sensitive coatings of sensors during operation. Currently, two ways of decreasing the drift impact of a sensor array on analysis results exist: applying extremely stable sensitive sensor coatings [5–8] and mathematical algorithms of drift compensation [9–13]. These approaches increase accuracy while classifying samples. The mathematical algorithms based on multivariate analysis of the drift correction of sensor signals are difficult to implement in practice for out-of-laboratory analysis, since they require additional computational costs, software changes, or the application of additional programs along with the operator's modelling skills. The most critical component in mathematical processing implementation is valid information selection, obtained over a long period of time, for a single measurement but not for a data set. The goal of the investigation, the nature of the biosample, and the operating conditions are the factors that determine the choice of drift compensation approach.

In this study, we present a technique for drift compensation for the signal of mass-sensitive piezoelectric quartz sensors during operation in the open detection cell and frontal input of the gaseous phase over biosamples with significant water content. The technique consists of piezoelectric sensor signal correction in a portable e-nose as part of a lengthy routine analysis of water and blood samples, combined with a simple algorithm for assessing system instability, in order to increase the quality of measurement and decrease first- and second-order classification errors. This technique can be used for real sample analysis, including biological samples (blood, secretions, mucus, exhaled breath condensate, perspiration, urine, and saliva), with minor a modification of the device software for out-of-laboratory analysis.

2. Methods and Analysis

A gaseous phase analysis of distilled water and blood samples was conducted on the gas analyzer Diagnost-Bio-8 (LLC SenSino, Kursk, Russia) using frontal analyte input mode [14]. The sensor array consisted of eight piezoelectrical quartz BAW-type resonators with a 10.0 MHz basic oscillation frequency, with coatings from solid nanostructured sorbents ('living system' set): Sensor 1 and 8 are carboxylated carbon nanotubes of different masses, marked in the tables and text as MCNT1, MCNT2; Sensor 2 and 7 are phases of nitrate of zirconium oxide of different masses (Zr1, Zr2); Sensor 3 is dicyclohexano-18-crown-6 (DCH18C6); Sensor 4 and 5 are biohydroxyapatite phases of different masses (HA1, HA2); and Sensor 6 is polyethylene glycol succinate (PEGsc). This study describes sensor manufacturing [15]. Moreover, the work [16] presents the sorption characteristics of the sensor array from the 'living system' set. The basic lines of the sensors were observed to remain stable (± 1 Hz) during an 80 s blank measurement before the sample analysis.

The blood or water sample (volume: 0.5 cm^3) was placed on a glass petri plate and then covered by a detection cell of Diagnost-Bio-8. The measurement mode involved combined 80 s sorption and 120 s spontaneous desorption [14].

For the analysis of work stability and drift correction of the sensor array signal, a 3 month period (October–December 2019) of device operations (656 measurements) was chosen, during which 75 water samples and 31 blood samples of somatically healthy people, with indicators of general and biochemical analysis of blood within normal limits, were tested. As part of the training set of the samples, sensor array data for 64 samples (19 blood samples and 45 water samples) were selected for the first 1.5 months of operation (October–November). The sensor array data for 42 samples (12 blood samples and 30 water samples) were selected as the test set of samples for the next period of operation (November–December). During the experiment, external factors showed variation: the room temperature changed from 20 to 25 °C, the humidity was 45–55%, and the slight changes in room smell (background) per day corresponded to disinfection and ventilation schedule of the laboratory. Measurements were not performed during disinfection and ventilation and 90 min after these processes. Regarding the original sensor data for each sample, a special software determined maximum sensor responses ($\Delta F_{max,i}$, Hz), which corresponded to 80 s of measurement (sorption time).

We suggested the correction of the sensor signals in an array based on daily internal standardization. The distilled water was chosen as the standard. We calculated the specific signals $\overline{F}_i$ via the following formula:

$$\overline{F}_i = \frac{\Delta F_{max,i}(for\ sample)}{\Delta \overline{F}_{max,i}(for\ water)}, \quad (1)$$

where $\Delta F_{max,i}$, Hz are values of the original signal of i-th sensor during the biosample analysis and $\Delta \overline{F}_{max,i} \overline{F}_{max,i}$ is the average signal of the i-th sensor for three or five analyses of water samples on the same day.

For specific signals calculated for the water samples, we applied a statistical process control method, namely the exponentially weighted moving average (EWMA) control chart, for uni- and multivariate data [17], in order to assess the stability of the sensor arrays. We investigated the statistical quantity T_i^2 in the sensor arrays, for each sensor–parameter z_i, and exponentially weighted the mean square (EWMS) error s_i during 3 months of operation. Before the computation of the statistical criteria, the specific signals obtained were standardized by the average value equal to 1.00 for all the sensor signals. The standard deviation for each sensor was determined according to the data obtained for the first 25 days of operation.

As an alternative method of drift compensation for sensor signals, component correction by principal component analysis (CC-PCA) was employed in order to remove a part of the information. This described the sensor signal drift corresponding to the first principal component from the data matrix [18].

As variables for designing the classification models, we used original maximum sensor responses and specific signals. The results of the water and blood gaseous phase analysis by sensor array were classified into two classes—'water' and 'blood'—using linear discriminant analysis with preliminary processing by principal component analysis (PCA-LDA) with a significance level of 0.05. A CAMOSoftware Unscrambler (v.10.0.0, Oslo, Norway) was employed to assess the effectiveness of the drift compensation techniques.

3. Results and Discussion

The following are the significant factors determining the drift of sensor signals in the device with an open detection cell: the changes in the sorption properties of sensitive coatings during operation and the external factors of the experiment (temperature, humidity, air composition in the laboratory [(background)). Frequently, these factors act together, which makes it difficult to estimate and predict the contribution of an individual factor to the total signal change. The other internal factors, such as the malfunctions in the oscillating scheme of the device and the change in the features of utilized quartz and electrodes, and the problems linked to signal transmission into software for our devices, lead to no more than 5% variation in sensor signals [19]. The algorithm of multivariate data processing for the drift compensation of the sensor signals was limited by the pilot experiment conditions for the model design. It cannot always contain all the factor combinations influencing the system's signals.

A daily internal standardization may perform a simple accounting of the external factors that influence sensor signals during long-term operation. As the standard, we proposed to use the substance whose content was dominant in the analyzed samples. For biosamples, such as blood, urine, exhaled breath condensate and saliva, distilled water is an appropriate standard. During the 3 months of the experiment, the value of the original signals of the sensors in water vapour varied from 20 to 27%; moreover, the dispersions of the original sensor signals were non-homogeneous, according to Cochran's C-test. The application of the daily standardization reduced the variation to 12%, and the dispersions of all the signals became homogeneous. The average specific signals of all the sensors throughout the total operation period equalled 1.00. The implementation of the specific signals considers the changes in biosamples as well as the sorption properties of sensor coatings, depending on external factors, and thereby reduces the errors of sample classification. However, more accurate classification results could be obtained using information about the stability of sensors.

3.1. *Stability Assessment of Piezoelectric Sensor Array Using Specific Signals*

The application of multivariate EWMA control charts by specific signals helps to discover only abrupt changes in external conditions (caused by the measurement of a contaminated standard, which can be classified as operator error). Therefore, detailed information about stability sensor arrays describes EWMA control charts for individual signals of the sensor array (Figure 1).

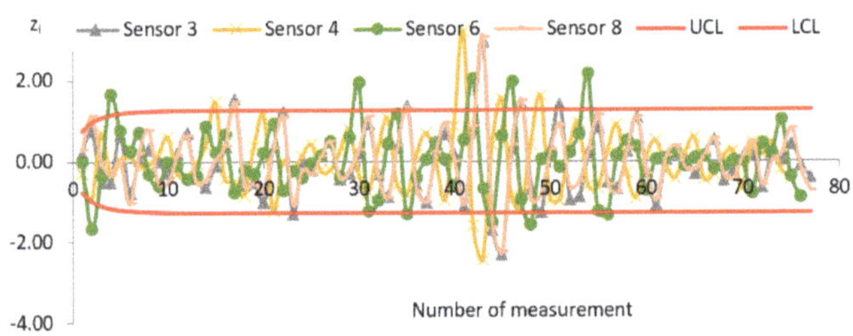

Figure 1. EWMA control chart for specific signals of four sensors from the array. The solid red lines indicate the upper and lower control limits for the third sensors in the array.

According to multiple excesses of control limits, the high instability of Sensor 6 with the PEGsc film was established because of the excellent water impact on the properties of polymer films. The control limits of the sensor with the DCH18C6 film (Sensor 3) featured the most significant values (Table 1). This is explained by the nature of the modifier, which consisted of solvated crown-ether macromolecules. This type of structure is less stable and varies more depending on external conditions.

Table 1. Maximum values of upper (+) and lower (−) control limits for parameter z_i for each sensor.

MCNT1	Zr1	DCH18C6	HA1	HA2	PEGsb	Zr2	MCNT2
±1.01	±1.24	±1.27	±1.25	±1.22	±1.06	±1.18	±1.22

The parameters z_i for the sensors with hypersensitive nanostructured films of carboxylated nanotubes (Sensor 8) and hydroxyapatite (Sensor 4) were almost always within the control limits. The exceptions were the day of synchronous change of the z_i parameters of all the sensors and the following three days (41–51 measurements); these exceptions were related to the changes in the coating properties during the sorption of the contaminated standard vapours and their subsequent reactivation by purging with laboratory air and repeated measurements of pure standard samples. The data from the biosample analysis obtained on these days may have been unreliable and may have contributed to error in the classification model.

To decide on the exclusion of the sensor signals from the classification data and to assess the sensor regeneration level after poisoning, an analysis of the changes to the signal variability was required. Here, the evaluation of the sample generalized variance for sensor array was not informative, since its values throughout the total studied period were either negative or extremely low (less than 10^{-60}). Therefore, we considered the EWMS errors s_i of specific sensor signals (Figure 2). On the day of the abrupt change in external conditions (measurements of the contaminated sample of the standard; measurement No. 41–43), variation for all the sensors and z_i exceeded the upper limit control (UCL) of 1.45. Therefore, in the case of the synchronous increase in control limits by parameters z_i and s_i for all the sensors, it was necessary to terminate measurements, verify compliance with the conditions for the measurement procedure and reactivate the sensor array's coatings. When calculating the EWMS for each subsequent day after conducting the reactivation of coatings, EWMS $s_0 = 1$ was accepted as the initial point for all the sensors. In the following eight days (measurement No. 44–65), the EWMS of all the sensors, except for Sensors 1 and 8, which featured carbon nanotube coatings, exceeded the UCL. This probably indicated the sorption quality changes in the hydrophilic modifiers following the sorption of contaminated standard vapor and their gradual recovery when measuring the pure standard or slight inertia of the surface reactivation process of nanostructured mineral coatings, such as hydroxyapatite.

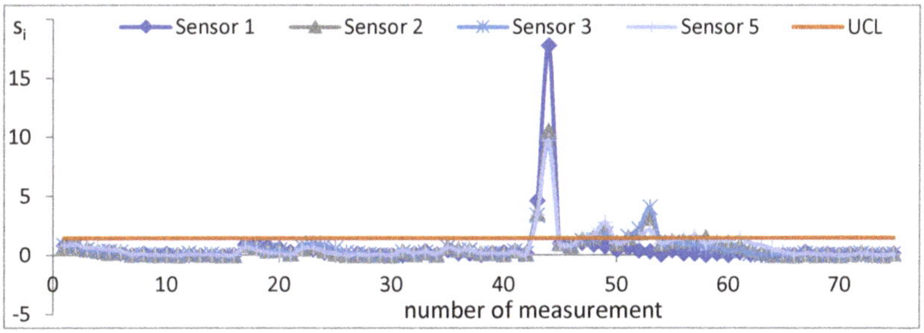

Figure 2. Control chart for EWMS errors of four sensors.

The average run length (ARL) and time to signal (ATS) were calculated according to EWMA theory. The ARL was equal to three measurements to detect three σ shifts in signals and the ATS was 2.5 h. After a long break or recovering of the sorption properties of coatings, the ARL was five measurements, whereas ATS was 1.5 h.

General recommendations for sensor data analysis of biosamples, accounting for their stability, were formulated as a scheme that could be implemented in the device's software. The purpose was to monitor the stability sensor array with minimal control by the operator, which is essential for out-of-laboratory analysis. The proposed approach was implemented for different forms biosample analysis, such as blood analysis.

3.2. *Blood and Water Sample Classification*

To ensure the orthogonality of the data for the discriminant analysis, we performed the principal component analysis of original and specific sensor signals. For further analysis, we used the first four principal components with 98% explained variance, which allowed us to consider all the information from the sensor array during the sample classification. In the first stage, we estimated the possibility of water and blood sample classification using original signals of sensors. The accuracy of this classification model for the training set of samples was 90.4% (Table 2).

The misclassification of blood samples as 'water' class was observed. This signifies that during the analysis of the gaseous phase over the blood samples, the sample matrix influences sensors and that over a period of time, differences within the gaseous composition in the samples become less noticeable for the sensors. The highest value for the accuracy and correctness of the blood sample classification belonged to the model constructed by specific sensor signals, which accounted for sensor stability information (Table 2). Consequently, the drift compensation of the sensor signals using specific signals with combination sensor stability monitoring by EWMA control charts could be applied, along with the routine sensor array operation in the laboratory and out-of-laboratory analyses. Furthermore, the correct classification of the samples considering the impact of the dominant component allows us to suggest that this approach can provide a more accurate classification of slight differences in the composition of gaseous phases of blood with pathologies.

Table 2. Accuracy and correctness of sample division into two classes by PCA-LDA models for original and specific sensor signals.

Variables	Accuracy of PCA-LDA Model for Training Set (%)		Correctness of Classification of Blood Samples from Test Set (%)	
	1	2	1	2
Original signals ($\Delta F_{max,i}$, Hz)	90.4	96.2	0	0
Specific signals ($\overline{F_i}$)	94.2	95.2	33.3	80.0
CC-PCA	90.4	96.2	8	10

1—for the model design, we used the whole data array; 2—the samples and sensor data were excluded according to the stability evaluation of sensor array operation.

4. Conclusions

We proposed a fast and efficient compensation method for sensor signal drift based on daily standardization (dividing sensor signals for biosamples into corresponding average signals for standard samples (measured thrice during the day)). Distilled water was suggested as a standard for blood samples and other biofluids (urine, perspiration, exhaled breath condensate, and saliva). EWMA control charts were applied for sensor array stability monitoring as an additional module in the device software for routine analysis. The coating of carboxylated carbon nanotubes of small mass (Sensor 8) was the most stable within the studied sensor array when measuring the water and blood samples based on the statistical quantities of the EWMA charts. The effectiveness of applying for drift compensation by daily standardization, with a combination of sensor stability monitoring, was proven by significantly improving the blood samples' classification accuracy. Similarly, it is possible to compensate for the drift of sensor signals when analysing other biosamples using appropriate standards. The additional parameters of sorption or features of sensor output curves after standardization could be used to improve the correctness of the classification during the long-term operation of the sensor array.

Author Contributions: Conceptualization, T.K.; methodology, T.K. and A.S.; validation, D.M. and A.S.; investigation, D.M.; data curation, A.S.; writing—original draft preparation, D.M.; writing—review and editing, A.S.; supervision, T.K.; project administration, T.K.; funding acquisition, T.K. All authors have read and agreed to the published version of the manuscript.

Funding: This research was partly funded by the Russian Science Foundation, grant number 18-76-10015.

Institutional Review Board Statement: The study was conducted according to the guidelines of the Declaration of Helsinki, and approved by Ethics Committee of Voronezh State University of Engineering Technologies (protocol code No. 2 from 21 February 2021).

Informed Consent Statement: Informed consent was obtained from all subjects involved in the study. Written informed consent has been obtained from the patients to publish this paper.

Conflicts of Interest: The authors declare no conflict of interest. The funders had no role in the design of the study; in the collection, analyses, or interpretation of data; in the writing of the manuscript, or in the decision to publish the results.

References

1. Di Carlo, S.; Falasconi, M. Drift Correction Methods for Gas Chemical Sensors in Artificial Olfaction Systems: Techniques and Challenges. In *Advances in Chemical Sensors*, 1st ed.; Wang, W., Ed.; IntechOpen: London, UK, 2011; pp. 305–326.
2. Ahmadou, D.; Laref, R.; Losson, E.; Siadat, M. Reduction of drift impact in gas sensor response to improve quantitative odor analysis. In Proceedings of the 2017 IEEE International Conference on Industrial Technology (ICIT), Toronto, ON, Canada, 22–25 March 2017; pp. 928–933.
3. Wang, Y.F.; Yin, Y.; Ge, F.; Yu, H.C. A drift correction method of E-nose data based on wavelet packet decomposition and no-load data: Case study on the robust identification of Chinese spirits. *Sens. Actuators B-Chem.* **2019**, *292*, 217–224. [CrossRef]
4. Yin, Y.; Bai, Y.; Ge, F.; Yu, H.C.; Liu, Y.H. Long-term robust identification potential of a wavelet packet decomposition based recursive drift correction of E-nose data for Chinese spirits. *Measurement* **2019**, *139*, 284–292. [CrossRef]
5. Ushimi, Y.; Ito, Y.; Horiuchi, H.; Kadota, M.; Nozaki, Y.; Hotta, Y.; Shiratori, S. Quartz Crystal Microbalance Sensor for NH_3 Gas with Compensation of Humidity Drift. *Electron. Commun. Jpn.* **2015**, *98*, 1–7. [CrossRef]
6. Tolentino, M.A.K.P.; Albano, D.R.B.; Sevilla, F.B. Piezoelectric sensor for ethylene based on silver(I)/polymer composite. *Sens. Actuators B-Chem.* **2018**, *254*, 299–306. [CrossRef]
7. Haghighi, E.; Zeinali, S. Nanoporous MIL-101(Cr) as a sensing layer coated on a quartz crystal microbalance (QCM) nanosensor to detect volatile organic compounds (VOCs). *RSC Adv.* **2019**, *9*, 24460–24470. [CrossRef]
8. Aleixandre, M.; Nakamoto, T. Study of Room Temperature Ionic Liquids as Gas Sensing Materials in Quartz Crystal Microbalances. *Sensors* **2020**, *20*, 4026. [CrossRef] [PubMed]
9. Liu, B.; Zeng, X.P.; Tian, F.C.; Zhang, S.Y.; Zhao, L.L. Domain Transfer Broad Learning System for Long-Term Drift Compensation in Electronic Nose Systems. *IEEE Access* **2019**, *7*, 143947–143959. [CrossRef]
10. Liu, T.; Chen, Y.B.; Li, D.Q.; Yang, T.; Cao, J.H.; Wu, M.Y. Drift Compensation for an Electronic Nose by Adaptive Subspace Learning. *IEEE Sens. J.* **2020**, *20*, 337–347. [CrossRef]
11. Liu, T.; Li, D.Q.; Chen, J.J.; Chen, Y.B.; Yang, T.; Cao, J.H. Active Learning on Dynamic Clustering for Drift Compensation in an Electronic Nose System. *Sensors* **2019**, *19*, 3601. [CrossRef] [PubMed]
12. Rehman, A.U.; Bermak, A. Heuristic Random Forests (HRF) for Drift Compensation in Electronic Nose Applications. *IEEE Sens. J.* **2019**, *19*, 1443–1453. [CrossRef]
13. Rehman, A.U.; Bermak, A.; Hamdi, M. Shuffled Frog-Leaping and Weighted Cosine Similarity for Drift Correction in Gas Sensors. *IEEE Sens. J.* **2019**, *19*, 12126–12136. [CrossRef]
14. Kuchmenko, T.; Shuba, A.; Umarkhanov, R.; Chernitskiy, A. Portable Electronic Nose for Analyzing the Smell of Nasal Secretions in Calves: Toward Noninvasive Diagnosis of Infectious Bronchopneumonia. *Vet. Sci.* **2021**, *8*, 74. [CrossRef] [PubMed]
15. Shuba, A.; Kuchmenko, T.; Umarkhanov, R.; Chernitskiy, A. Portable E-nose for Diagnostic of Inflammation and Diverse Variation in Health Status of Humans and Animals. In Proceedings of the Fifth International Conference on Advances in Sensors, Actuators, Metering and Sensing (ALLSENSORS 2020), Valencia, Spain,, 21–25 November 2020; pp. 56–62.
16. Kuchmenko, T.; Shuba, A.; Umarkhanov, R.; Lvova, L. The New Approach to a Pattern Recognition of Volatile Compounds: The Inflammation Markers in Nasal Mucus Swabs from Calves Using the Gas Sensor Array. *Chemosensors* **2021**, *9*, 116. [CrossRef]
17. Montgomery, D. *Introduction to Statistical Quality Control*, 6th ed.; Wiley: Jefferson City, MO, USA, 2009; pp. 419–424, 509–518.
18. Artursson, T.; Eklöv, T.; Lundström, I.; Mårtensson, P.; Sjöström, M.; Holmberg, M. Drift correction for gas sensors using multivariate methods. *J. Chemometr.* **2000**, *14*, 711–723. [CrossRef]
19. Kuchmenko, T.A.; Shuba, A.A.; Umarkhanov, R.U.; Drozdova, E.V.; Chernitskii, A.E. Application of a Piezoelectric Nose to Assessing the Respiratory System in Calves by Volatile Compounds. *J. Anal. Chem.* **2020**, *75*, 645–652. [CrossRef]

chemistry proceedings

Proceeding Paper

The Performance of Organophosphate Pesticides Determination Using Biosensor Based on Small Device Potentiometer as a Transducer †

Mashuni Mashuni [1,*], Halimahtussaddiyah Ritonga [1], Muhammad Jahiding [2], La Ode Ahmad Nur Ramadhan [1], Desy Kurniawati [1] and Fitri Handayani Hamid [1]

[1] Department of Chemistry, Halu Oleo University, Kendari 93132, Indonesia; halimahhalimah124@yahoo.co.id (H.R.); ramadhan305@gmail.com (L.O.A.N.R.); 081383720011nanda@gmail.com (D.K.); fitrihandayanihamid@gmail.com (F.H.H.)

[2] Department of Physics, Halu Oleo University, Kendari 93132, Indonesia; mjahiding2019@gmail.com

* Correspondence: mashuni2696@gmail.com

† Presented at the 1st International Electronic Conference on Chemical Sensors and Analytical Chemistry, 1–15 July 2021; Available online: https://csac2021.sciforum.net/.

Abstract: The need to control pesticide residues in foodstuffs in a fast and straightforward analysis for the field scale is required. Therefore, this research develops a transducer-based biosensor with a small device potentiometer (SDP) to produce a fast and accurate pesticide detection tool. The biosensor based on Au electrodes by immobilizing the acetylcholinesterase (AChE) enzyme coated membrane cellulose acetate (CA) 15% (*w*/*v*) cross-linked glutaraldehyde (GA) 25% (*v*/*v*) and SDP as a transducer that produces a potential value. The biosensor testing results on the organophosphate pesticide class, namely diazinon and profenofos, showed the sensitivity of 21.204 and 20.035 mV decade^{-1}, Limit of Detection (LoD) 10^{-7} mg L^{-1}, selectivity coefficient $-1 < \boldsymbol{K_{i,j}} < 1$ and accuracy of 99.497 and 94.765%, respectively. The results showed that the biosensor connected to an SDP transducer had an excellent performance in determining the presence of organophosphate pesticides.

Keywords: small device potentiometer; biosensor; acetylcholinesterase; organophosphate; pesticide; diazinon and profenofos

Citation: Mashuni, M.; Ritonga, H.; Jahiding, M.; Ramadhan, L.O.A.N.; Kurniawati, D.; Hamid, F.H. The Performance of Organophosphate Pesticides Determination Using Biosensor Based on Small Device Potentiometer as a Transducer. *Chem. Proc.* **2021**, *5*, 69. https://doi.org/10.3390/CSAC2021-10604

Academic Editor: Xin Wu

Published: 5 July 2021

1. Introduction

Organophosphate pesticides are a group of pesticides that contain a phosphate group. The organophosphate pesticides in agricultural and plantation processing systems are widely used to tackle pests and diseases that attack plants, the leading cause of declining crop yields. However, the use of pesticides harms the health of living things and the balance of the environment [1,2]. Analysis of pesticide residues from crop yields is necessary to ensure food safety. One alternative method of pesticide analysis is a biosensor [3,4].

In recent decades, biosensors have become a popular research area capable of identifying pesticide residues and other chemicals. Biosensors are "self-standing devices" that record physical, chemical or biological changes, convert them into measurable signals from the sample and monitor the analyte of interest [5,6]. The sensor contains a recognition element that allows a selective response to a specific analyte or group of analytes, minimizing interference from other sample components. Another significant sensor component is a transducer or detection device that produces a signal [7].

Electrochemical biosensors are a subclass of chemical sensors that combine sensitivity—such as low detection limit—electrochemical transducers with the high specificity of biological recognition processes. These devices contain biological recognition elements (enzymes, proteins, antibodies, nucleic acids, cells, tissues or receptors) that selectively react with the target analyte and generate an electrical signal related to the measured analyte concentration [5,8]. Enzyme-based electrochemical biosensors have advantages over conventional methods due to their excellent sensitivity, selectivity, mini size and fast response [5,9,10].

Maintaining the catalytic activity on the efficient immobilization of the acetylcholinesterase (AChE) enzyme is an essential consideration when developing electrochemical biosensors that can be used for practical applications. The highlight of the electrochemical biosensor is its unique ability to generate digital signals that can measure by converting the catalytic signal with the help of microfabricated electronics [3,11,12]. The electrochemical method using measurement tools is based on potentiometric [12], amperometric [13] or conductometric [14] biosensor. Potentiometric biosensors are suitable for measuring the response value of pesticide detection measurements [15].

Potentiometric biosensors are efficient for in-field analysis because they are more straightforward and ideal for real-time analysis [16]. The potentiometric detection system developed by Timur, S. and Telefoncu, A., 2004 [17], has the underlying principle of inhibition of AChE activity due to its properties in identifying organophosphate compounds. The enzyme was immobilized on the surface of the electrode with the help of a chitosan membrane [18]. Without a pre-concentration step, in both aqueous and organic media, detection of organophosphates without the requirement of trained personnel proved advantageous for the proposed portable biosensor. Pesticides were effectively detected in the range of 0.1–100 mM for parathion-methyl and methamidophos and 0.6–600 mM for Malathion [17]. However, in the presence of higher pesticide concentrations, only partial regeneration of the enzymatic activity was regenerated [15].

The combination of potentiometric-based AChE enzyme biosensors as transducers with analytical techniques has been widely reported in the literature as a suitable method. In this work, we report the development of a small device potentiometric (SDP) based biosensor as a transducer for the determination of pesticide organophosphates, based on the AChE enzyme immobilized on cellulose acetate (CA) and glutaraldehyde (GA) membrane-coated Au electrodes.

2. Materials and Methods

2.1. Meterials

Acetylcholinesterase (AChE, from electrophorus, Sigma-Aldrich, St. Louis, MO, USA, 1.17 mg with activity 425.94 units per mg (EC. 3.1.1.7)) in 9 mL PBS pH 8 and 1 mL KCl 10^{-1} M, cellulose acetate (CA, from Sigma-Aldrich, St. Louis, MO, USA, 15% *v*/*v* in acetone), glutaraldehyde (GA, from Sigma-Aldrich, St. Louis, MO, USA, 25% in H_2O), potassium chloride (KCl, from Merck, Darmstadt, Germany, 10^{-1} M in H_2O) and acetone (C_3H_6O, from Sigma-Aldrich, St. Louis, MO, USA, ≥99.5%). Phosphate buffer solutions (PBS) with values of pH 8.0 were prepared by mixing standard stock solutions of 0.2 M Na_2HPO_4 (99% purity) and 0.2 M NaH_2PO_4 (99% purity). Standard solution of acetylthiocholine chloride (ATCl, A5626 from Sigma-Aldrich, St. Louis, MO, USA) substrate with concentrations of 10^{-1}, 10^{-2} and 10^{-3} M in PBS solution. The pesticides used in the OP group are diazinon and profenofos was purchased from Merck, Darmstadt, Germany, as inhibitors, each made in concentrations of 10^{-9}, 10^{-8}, 10^{-7}, 10^{-6}, 10^{-5}, 10^{-4}, 10^{-3}, 10^{-2} and 10^{-1} mg L^{-1} in 5 mL ethanol (C_5H_6O, from Sigma-Aldrich, St. Louis, MO, USA, 99.8%) and H_2O.

2.2. Apparatus

The potentiometer is used as an experimental tool for measuring the potential value of analyte detection [19]. The working electrodes used are Au and a platinum (Pt) as a cathode of the electrolysis process and an Ag/AgCl as a reference electrode.

2.3. Electrolysis of Ag/AgCl

Reference electrodes of Ag/AgCl were carried out by electrolyzing Ag wire (anode) and Pt wire (cathode) in 0.1 M KCl solution for ±20 min. The length of time electrolysis will affect the thickness of AgCl on Ag wire, where the more extended the electrolysis process, the thicker it will be to a certain extent, and vice versa. Next, the Ag/AgCl wire that has been formed is then dried in the open air. Finally, Ag/AgCl wire that has been dried at room temperature is inserted into the electrode body as a comparison electrode for Ag/AgCl [20].

2.4. Preparation of Au Electrode Biosensor

The Au electrode tip was immersed in a 15% CA membrane solution. The CA membrane formed was rinsed with distilled water and then dipped in 25% GA solution for 6 h. Furthermore, the electrode was rinsed with distilled water and PB solutions pH 8, then an electrode membrane (Em) was formed. Then, Em was immersed in the AChE enzyme for 2×24 h at 4 °C. Before measuring the response to the biosensor, the components in the measurement, such as standard electrodes, coated wire type working electrodes, ATCl substrate and inhibitor solution, need to be left at room temperature for about 2 h until components are stable and produce a good response [20].

2.5. Measurement of the Potential Value Biosensor

Measurement of the potential value of the enzymatic biosensor electrodes with the pesticide inhibitors diazinon and profenofos in concentrations of 10^{-8}, 10^{-7}, 10^{-6}, 10^{-5}, 10^{-4}, 10^{-3}, 10^{-2} and 10^{-1} mg L^{-1} using a potentiometer. Em was immersed in PBS pH 8.0 for 10 min and then used ME to measure the potential value of 10^{-3} M ATCl substrate to obtain a constant value. The Em was removed and rinsed with distilled water, and then Em was immersed in a pesticide solution for 30 min, then removed and rinsed with PB solution pH 8.0 before being dipped again into the ATCl substrate solution. Furthermore, it made observations to obtain a constant potential value.

2.6. The performance Test of Biosensor

2.6.1. Sensitivity

The sensitivity value (Nernst factor) is determined using a graph of the relationship between the potential value and –log of inhibitor concentration. Then, we can see the linear equation from the chart to obtain the sensitivity range of the diazinon and profenofos pesticide electrode.

2.6.2. Limit of Detection (LoD)

LoD is the lowest limit of analyte concentration that can be measured by the instrument, which is statistically different from the blank. Determination of LoD was carried out by analyzing the potential response of a series of standard solutions of various pesticide concentrations of 10^{-8}, 10^{-7}, 10^{-6}, 10^{-5}, 10^{-4}, 10^{-3}, 10^{-2} and 10^{-1} mg L^{-1}. The analysis results obtained a linear equation of the calibration curve, y = ax + b, then the measurement of the potential value of the blank. The equation for the value of y at the detection limit is based on equation of Christian et al. (2014) [21]. They suggest calculation for the LoD = $3 \times$ (SD/S) based on the response's standard deviation, SD, and the slope or sensitivity, S, of the calibration curve at levels approaching the limit.

2.6.3. Selectivity

Selectivity is expressed as the degree of bias of the primary analyte analysis data with interference compared to the analyte analysis data without interference [21]. The sample analysis used a concentration of 10^{-5}, 10^{-4} and 10^{-3} mg L^{-1}. Measuring the potential value using a potentiometer and calculating the selectivity coefficient ($K_{i,j}$). The value of the electrode selectivity calculated based on the Nicolsky–Eisenman equation [22]. The potential of an ion-selective electrode in the presence of an interfering ion follows an equation:

$$E_{ISE} = k + \frac{S}{z} log\left(a_i + \sum_{j \neq i} K_{i,j}\, a_j^{z_i z_j}\right)$$

where S is the slope (theoretically 2.303RT/F) and k is the ion charge including all contributions independent of activities a and z is the ion charge. The subscript i stands for the primary pesticides and subscript j stands for the interfering pesticides.

2.6.4. Accuracy

Determination of the accuracy value is obtained by calculating the % recovery, where the sample used is mustard greens with the addition of inhibitor concentrations of 10^{-4}, 10^{-3} and 10^{-2} mg L^{-1}. Measurement of potential value using a potentiometer and calculated % recovery based on the equation of Christian et al. (2014) [21].

3. Results and Discussions

As shown in Figure 1, an analysis of the performance of the biosensor on the pesticide diazinon and profenofos using a small device potentiometric (SDP)-based biosensor as a transducer was carried out. SDP-based biosensor performance tests, including sensitivity, LoD, selectivity and accuracy, are essential parameters in biosensors.

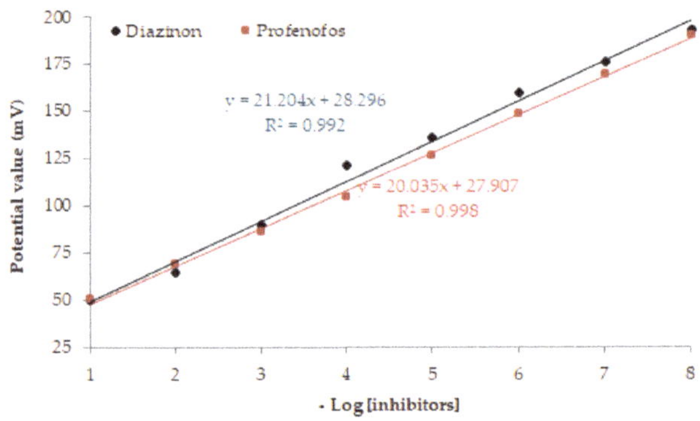

Figure 1. Graph of the relationship of −log [inhibitors] with the potential value of biosensor-based SDP.

The sensitivity, or Nernst factor, is one of the general parameters of biosensor performance testing, using SDP as a transducer indicated by the slope resulting from the calibration of the electrode potential response to the analyte's activity. Analysis using potentiometric is based on the potential change in each variation of ion concentration [16]. The feasibility of a tool used in detecting an analyte is seen from how much sensitivity is in the measurement process. So, in this study, the Nernst factor value was determined to see how well the sensitivity of the tool and the measurement range of an electrode was suitable for use as a pesticide detection tool. The results of the measurement of the potential value of the biosensor performance are presented in Table 1.

Figure 1 shows the sensitivity of the performance of SDP-based biosensors to the detection of pesticide diazinon and profenofos of 21.204 and 20.035 mV $decade^{-1}$, respectively. The sensitivity is the slope value of the linear regression equation from the graph of the relationship −log [inhibitor], with the potential value measured using a potentiometer. The value of the Nernst factor is more ideal if it is close to the value of 29.6 mV $decade^{-1}$ [23].

The characteristics of a biosensor are also determined by its ability to detect the concentration of an analyte. The smaller the concentration that can be seen, the better the biosensor features. Limit of Detection (LoD) is the minor analyte concentration that gives a sufficiently large signal and can be distinguished from the signal obtained from the blank with a 99% confidence level [15]. The optimum performance of a biosensor can be determined by its ability to detect the concentration of an analyte. The LoD was determined using the standard deviation of the intercept and the slope of the calibration line. Table 1 shows the LoD value of the SDP-based biosensor as a transducer, which is 10^{-7} mg L^{-1} instead of 10^{-8} mg L^{-1}, because the potential at that concentration is close to the potential blank value. So, the smaller the concentration that can be detected, the better the biosensor characteristics [24].

Table 1. Measurement of potential value, sensitivity value and LoD of biosensor.

Substrate Concentration (M)	Inhibitor Concentration (mg L^{-1})	Potential Value	
		Diazinon	Profenofos
10^{-3}	10^{-1}	50	50.7
	10^{-2}	64.8	69.3
	10^{-3}	89.5	86.1
	10^{-4}	121.1	104.2
	10^{-5}	135.7	126.5
	10^{-6}	159.9	148.7
	10^{-7}	175.8	169.4
	10^{-8}	192.9	189.6
Potential value of substrate (mV)		199.8	195.7
Potential value of blank (mV)		199.1	196,3
Potential value of 10^{-8} mg L^{-1} (mV)		195.5	194.6
Sensitivity (mV $decade^{-1}$)		21.204	21.035
Linear regression equation (R^2)		0.992	0.998
LoD (mg L^{-1})		10^{-7}	10^{-7}

The selectivity is carried out to determine the method's ability to measure the presence of pesticides carefully and thoroughly in interfering components. The ideal biosensor is expected to only respond to the primary analyte to be detected with selectivity coefficient $-1 < K_{i,j} < 1$. If the electrode is highly selective towards i rather than ion j, $K_{i,j} < 1$. Conversely, if the electrode is highly selective towards j, ion i, $K_{i,j} > 1$. Variations in the value of $K_{i,j}$ depend on the electrode's response and the component environment in solution. The selectivity coefficient value obtained is smaller than 1 (Table 2). Based on the overall value received, the range of low concentrations of interfering components is still within tolerance. The average selectivity coefficient value received still meets the specified selectivity value standard, greater than −1 and more minor than +1, except for the primary analyte analysis data, diazinon 10^{-5} mg L^{-1} and profenofos interfering 10^{-5} mg L^{-1}, then the average selective electrode for pesticide detection compared to interfering compounds [25,26].

Accuracy is a measure that shows the degree of closeness of the analysis results to the actual analyte content. Accuracy is expressed as the % recovery of the added analyte. In general, the acceptance criteria for accuracy (% recovery) are 80–110% [27]. Accuracy analysis is carried out using the recovery method by sample spiking or standard addition to the sample to be analyzed. The method is carried out by adding a certain amount of analyte with a certain concentration to the analyzed sample. The data in Table 3 show that the average % recovery of the SDP-based biosensor as a transducer has an accuracy rate of 99.497 and 94.765% for diazinon and profenofos pesticide detection, respectively. The percent recovery value obtained is in accordance with the required standard. See Supplementary Materials for details.

Table 2. Selectivity of biosensor-based SDP.

[Diazinon] (mg L^{-1})	[Profenofos] (mg L^{-1})	Potential Value (mV)				Selectivity	
		Diazinon		Profenofos			
		a_i (1)	a_j (1)	a_i (2)	a_j (2)	$K_{i,j}$ (1)	$K_{i,j}$ (2)
10^{-5}	0	160	0	126.5	0	0	0
	10^{-9}		159.5		127.9	−0.24	0.64
	10^{-8}		158.9		126.1	−0.53	−0.19
	10^{-7}		158.6		125.9	−0.67	−0.29
	10^{-6}		158.1		124.7	−0.91	−0.87
	10^{-5}		157.4		124.5	−1.26	−0.96
10^{-4}	0	131.1	0	104.2	0	0	0
	10^{-9}		130.5		106.3	−0.29	0.96
	10^{-8}		130.1		105.7	−0.48	0.68
	10^{-7}		129.8		104.8	−0.63	0.27
	10^{-6}		129.5		103.3	−0.77	−0.43
	10^{-5}		129.1		102.2	−0.97	−0.97
10^{-3}	0	90	0	80.1	0	0	0
	10^{-9}		89.8		82.2	−0.1	0.95
	10^{-8}		89		81.9	−0.49	0.81
	10^{-7}		88.7		80.5	−0.63	0.17
	10^{-6}		88.2		79.6	−0.88	−0.25
	10^{-5}		88		78.1	−0.98	−0.98

a_i is the concentration of primary pesticides, a_j is the concentration of interfering pesticides, $K_{i,j}$ is the selectivity coefficient, $_{(1)}$ for diazinon pesticide and $_{(2)}$ for profenofos pesticide.

Table 3. Accuracy of biosensor-based SDP.

[C′$_A$]	[C$_A$]	[C$_F$]	Potential Value (mV)						Accuracy, % Recovery	
			Diazinon			Profenofos			Diazinon	Profenofos
			[C′$_A$]	[C$_A$]	[C$_F$]	[C′$_A$]	[C$_A$]	[C$_F$]		
10^{-2}	10^{-3}	10^{-2}	64.8	131.1	79.1	80.4	118.6	76.9	79.123	76.899
10^{-3}		10^{-3}	90.0		84.8	118.6		99.2	84.790	99.232
10^{-4}		10^{-4}	131.1		107.7	130.8		108.2	107.690	108.165
Mean of % Recovery									99.497	94.765

C′$_A$ is the concentration of pesticides added, CA is the concentration of the sample, CF is the total concentration of the sample obtained from the measurement.

4. Conclusions

Based on the results and data obtained from the study of SDP-based biosensors as transducers in the detection of organophosphate pesticides, the sensitivity was 21.204 and 20.035 mV decade^{-1}, LoD 10^{-7} mg L^{-1}, selectivity coefficient $-1 < K_{i,j} < 1$ and accuracy of 99.497 and 94.765%. Thus, potentiometric biosensors with CA and GA membranes immobilized by AChE enzymes have good sensitivity, selectivity and accuracy in detecting the presence of organophosphate pesticides in a sample and LoD from tiny biosensors are effective for detecting at low scale and concentration.

Supplementary Materials: The following are available online at https://www.mdpi.com/article/10.3390/CSAC2021-10604/s1, presentation materials at the 1st International Electronic Conference on Chemical Sensors and Analytical Chemistry, 1–15 July 2021.

Author Contributions: Data curation, M.M., L.O.A.N.R. and D.K.; writing—original draft preparation, M.M. and F.H.H.; writing—review and editing, M.M., H.R. and F.H.H.; supervision, M.M., H.R. and M.J. All authors have read and agreed to the published version of the manuscript.

Funding: This research received funding from the Ministry of Education, Culture, Research, and Technology of the Republic of Indonesia.

Institutional Review Board Statement: Not applicable.

Informed Consent Statement: Not applicable.

Data Availability Statement: Data available on request due to restrictions, e.g., privacy or ethical.

Acknowledgments: The authors thank all the team for their contribution to this research, University of Halu Oleo (UHO) and the Ministry of Education, Culture, Research, and Technology of the Republic of Indonesia.

Conflicts of Interest: The authors declare no conflict of interest.

References

1. Zhang, C.; Jiang, Z.; Jin, M.; Du, P.; Chen, G.; Cui, X.; Zhang, Y.; Qin, G.; Yan, F.; Abd El-Aty, A.M.; et al. Fluorescence immunoassay for multiplex detection of organophosphate pesticides in agro-products based on signal amplification of gold nanoparticles and oligonucleotides. *Food Chem.* **2020**, *326*, 126813. [CrossRef] [PubMed]
2. Zhao, G.; Zhou, B.; Wang, X.; Shen, J.; Zhao, B. Detection of organophosphorus pesticides by nanogold/mercaptomethamidophos multi-residue electrochemical biosensor. *Food Chem.* **2021**, *354*. [CrossRef]
3. Singh, A.P.; Balayan, S.; Hooda, V.; Sarin, R.K.; Chauhan, N. Nano-interface driven electrochemical sensor for pesticides detection based on the acetylcholinesterase enzyme inhibition. *Int. J. Biol. Macromol.* **2020**, *164*, 3943–3952. [CrossRef] [PubMed]
4. Kalyani, N.; Goel, S.; Jaiswal, S. On-site sensing of pesticides using point-of-care biosensors: A review. *Environ. Chem. Lett.* **2021**, *19*, 345–354. [CrossRef]
5. Ronkainen, N.J.; Halsall, H.B.; Heineman, W.R. Electrochemical biosensors. *Chem. Soc. Rev.* **2010**, *39*, 1747–1763. [CrossRef]
6. Bacciu, A.; Arrigo, P.; Delogu, G.; Marceddu, S.; Monti, P.; Rocchitta, G.; Serra, P.A. A new perspective on using glycols in glutamate biosensor design: From stabilizing agents to a new containment net. *Chemosensors* **2020**, *8*, 23. [CrossRef]
7. Eggins, B.R. *Chemical Sensors and Biosensors*, 1st ed.; Johiley & Sons: Hoboken, NJ, USA, 2002.
8. Joseph, W. *Analytical Electrochemistry*, 3rd ed.; John Wiley & Sons: Hoboken, NJ, USA, 2006.
9. Kumar, H.; Kumari, N.; Sharma, R. Nanocomposites (conducting polymer and nanoparticles) based electrochemical biosensor for the detection of environment pollutant: Its issues and challenges. *Environ. Impact Assess. Rev.* **2020**, *85*, 106438. [CrossRef]
10. Karadurmus, L.; Kaya, S.I.; Ozkan, S.A. Recent advances of enzyme biosensors for pesticide detection in foods. *J. Food Meas. Charact.* **2021**, *15*, 4582–4595. [CrossRef]
11. Hu, H.; Yang, L. Development of enzymatic electrochemical biosensors for organophosphorus pesticide detection. *J. Environ. Sci. Health Part B Pestic. Food Contam. Agric. Wastes* **2020**, *56*, 168–180. [CrossRef]
12. Du, D.; Huang, X.; Cai, J.; Zhang, A. Amperometric detection of triazophos pesticide using acetylcholinesterase biosensor based on multiwall carbon nanotube-chitosan matrix. *Sens. Actuators B Chem.* **2007**, *127*, 531–535. [CrossRef]
13. Simonian, A.L.; Good, T.A.; Wang, S.S.; Wild, J.R. Nanoparticle-based optical biosensors for the direct detection of organophosphate chemical warfare agents and pesticides. *Anal. Chim. Acta* **2005**, *534*, 69–77. [CrossRef]
14. Hassani, S.; Momtaz, S.; Vakhshiteh, F.; Maghsoudi, A.S.; Ganjali, M.R.; Norouzi, P.; Abdollahi, M. Biosensors and their applications in detection of organophosphorus pesticides in the environment. *Arch. Toxicol.* **2017**, *91*, 109–130. [CrossRef] [PubMed]
15. Mashuni; Ritonga, H.; Hamid, F.H.; Widiyani, M.; Jahiding, M. Analysis of bio-oil effectiveness from coconut shells pyrolysis as biopesticide by potentiometric biosensor. *J. Phys.* **2021**, *1825*, 012095. [CrossRef]
16. Fernando, P.U.A.I.; Glasscott, M.W.; Pokrzywinski, K.; Fernando, B.M.; Kosgei, G.K.; Moores, L.C. Analytical Methods Incorporating Molecularly Imprinted Polymers (MIPs) for the Quantification of Microcystins: A Mini-Review. *Crit. Rev. Anal. Chem.* **2020**, *11*, 1–15. [CrossRef]
17. Timur, S.; Telefoncu, A. Acetylcholinesterase (AChE) electrodes based on gelatin and chitosan matrices for the pesticide detection. *Artif. Cells Blood Substit. Biotechnol.* **2004**, *32*, 427–442. [CrossRef]
18. Patel, H.; Rawtani, D.; Agrawal, Y.K. Trends in Food Science & Technology A newly emerging trend of chitosan-based sensing platform for the organophosphate pesticide detection using Acetylcholinesterase—A review. *Trends Food Sci. Technol.* **2019**, *85*, 78–91.
19. Mashuni; Ramadhan, L.O.A.N.; Jahiding, M.; Herniati. Analysis of diazinon pesticide using potentiometric biosensor based on enzyme immobilized cellulose acetate membrane in gold electrode. In *IOP Conference Series: Materials Science and Engineering*; IOP Publishing: Solo, Indonesia, 2016; Volume 107, pp. 1–7. [CrossRef]
20. Mashuni; Ramadhan, L.O.A.N.; Jahiding, M.; Syarfiah. Design of Pesticide Biosensor Using Glutaraldehyde Crosslinked-Cellulose Acetate Membrane in Gold Electrode. In *Journal of Chemical, Environmental & Biological Sciences*; Hakim, L., Ed.; Brawijaya University: Malang, Indonesia, 2016; Volume 4, pp. 147–151.
21. Christian, G.D.; Dasgupta, P.K.; Kevin, A.S. *Analytical Chemistry*, 7th ed.; Wiley: Haboken, NJ, USA, 2014; ISBN 9780470887578.

22. Choudhury, S.; Roy, S.; Bhattacharya, G.; Fishlock, S.; Deshmukh, S.; Bhowmick, S.; McLaughlign, J.; Roy, S.S. Potentiometric ion-selective sensors based on UV-ozone irradiated laser-induced graphene electrode. *Electrochim. Acta* **2021**, *387*, 138341. [CrossRef]
23. Mashuni; Syahrul, M.; Ahmad, A.; Wahab, A.W. Pengaruh Konsentrasi Glutaraldehid pada Komposisi Membran Elektroda Biosensor Pestisida Karbamat. *Indones. Chim. Acta* **2010**, *3*, 8–14.
24. Azizah, A.; Mulyasuryani, A. Sutrisno Biosensor Konduktometri Berbasis SPCE-Kitosan Untuk Mendeteksi Diazinon dan Malathion. *Kim. Stud. J.* **2014**, *1*, 50–56.
25. Ouyang, Q.; Wang, L.; Ahmad, W.; Rong, Y.; Li, H.; Hu, Y.; Chen, Q. A highly sensitive detection of carbendazim pesticide in food based on the upconversion-MnO2 luminescent resonance energy transfer biosensor. *Food Chem.* **2021**, *349*. [CrossRef]
26. Raymundo-Pereira, P.A.; Gomes, N.O.; Shimizu, F.M.; Machado, S.A.S.; Oliveira, O.N. Selective and sensitive multiplexed detection of pesticides in food samples using wearable, flexible glove-embedded non-enzymatic sensors. *Chem. Eng. J.* **2021**, *408*. [CrossRef]
27. Bozyiğit, G.D.; Ayyıldız, M.F.; Chormey, D.S.; Engin, G.O.; Bakırdere, S. Development of a sensitive and accurate method for the simultaneous determination of selected insecticides and herbicide in tap water and wastewater samples using vortex-assisted switchable solvent-based liquid-phase microextraction prior to determination. *Environ. Monit. Assess.* **2020**, *192*, 1–8. [CrossRef] [PubMed]

chemistry proceedings

MDPI

Proceeding Paper

Optimization of Bioactive Compounds with Antioxidant Activity of *Himanthalia elongata* by Microwave-Assisted Extraction Using Response Surface Methodology †

Franklin Chamorro [1], Lucia Cassani [1,2], Catarina Lourenço-Lopes [1], Anxo Carreira-Casais [1], Maria Carpena [1], Javier Echave [1], Sergio Baamonde [3], Fermin Fernández-Saa [3], Paz Otero [1], Pacual Garcia-Perez [1,4], Jesus Simal-Gandara [1,*] and Miguel Angel Prieto [1,4,5,*]

1 Nutrition and Bromatology Group, Faculty of Food Science and Technology, Ourense Campus, University of Vigo, E-32004 Ourense, Spain; franklin.noel.chamorro@uvigo.es (F.C.); lucia_cassani@hotmail.com (L.C.); c.lopes@uvigo.es (C.L.-L.); anxocc@uvigo.es (A.C.-C.); mcarpena@uvigo.es (M.C.); javier.echave@uvigo.es (J.E.); paz.otero@uvigo.es (P.O.); pasgarcia@uvigo.es (P.G.-P.)
2 Instituto de Investigaciones en Ciencia y Tecnología de Materiales (INTEMA, CCT-CONICET), Colón 10850, Mar del Plata 7600, Argentina
3 Centro de Investigación e Innovación Tecnológico en Algas Marinas (CIITAM), Algas Atlanticas Algamar S.L., Polígono de Amoedo, E-36840 Pazos de Borbén, Spain; sergio.baamonde@algamar.com (S.B.); oficinas2@algamar.com (F.F.-S.)
4 Department for Sustainable Food Process, Università Cattolica del Sacro Cuore, Via Emilia Parmense 84, 29122 Piacenza, Italy
5 Centro de Investigação de Montanha (CIMO-IPB), Campus de Santa Apolónia, E5300-253 Bragança, Portugal
* Correspondence: jsimal@uvigo.es (J.S.-G.); mprieto@uvigo.es (M.A.P.)

† Presented at the 1st International Electronic Conference on Chemical Sensors and Analytical Chemistry, 1–15 July; Available online: https://csac2021.sciforum.net/.

Citation: Chamorro, F.; Cassani, L.; Lourenço-Lopes, C.; Carreira-Casais, A.; Carpena, M.; Echave, J.; Baamonde, S.; Fernández-Saa, F.; Otero, P.; Garcia-Perez, P.; et al. Optimization of Bioactive Compounds with Antioxidant Activity of *Himanthalia elongata* by Microwave-Assisted Extraction Using Response Surface Methodology. *Chem. Proc.* **2021**, *5*, 70. https://doi.org/10.3390/CSAC2021-10478

Academic Editor: Ye Zhou

Published: 30 June 2021

Abstract: *Himanthalia elongata* is a brown alga used in applications in the food, pharmaceutical and nutraceutical industries due to its biological properties, such as antioxidant, anti-inflammatory, and antimicrobial, among others. These effects are attributed to the high content of nutrients and secondary metabolites, especially phenolic compounds. The objective of this study is to optimize the microwave-assisted extraction method to recover phenolic compounds and flavonoids, considering three extraction parameters: the concentration of ethanol in water, the extraction time and pressure. The total phenolic content and the total flavonoid content were evaluated, and two biological tests were performed to assess the antioxidant properties.

Keywords: macroalgae; microwave-assisted extraction; *Himanthalia elongata*; bioactive compounds; antioxidant

1. Introduction

Traditionally, algae have been used as food and for medicinal purposes, mainly in eastern countries. However, its popularity is increasing in western countries, due to the search for healthier and more natural products by consumers, including food, cosmetics, pharmaceutical products, etc. [1,2]. Numerous studies indicate the good nutritional value of algae: they provide proteins and essential amino acids, and they are rich in non-digestible carbohydrates and polyunsaturated fatty acids, vitamins, and minerals. Furthermore, they are a source of compounds with various biological activities (e.g., antioxidants, antivirals, antimicrobials, antifungals, etc.) [3–5], which has attracted the attention of researchers hoping to study them and develop new industrial applications [6–8]. The antioxidant activity of some species of algae has been attributed to the presence of phenolic compounds such as polyphenols, hydroquinones and flavonoids. *Himanthalia elongata* is a brown alga of the order Fucales, found mainly in the N-W Atlantic Ocean and the North Sea.

Its antioxidant properties have been described previously [4], and it is reported that the amount of polyphenolic content is higher than in other algae [9].

Bioactive compounds from algae were commonly extracted using organic solvents (methanol, ethanol, acetone) with application temperatures between 45 and 60 °C, for hours or days, which implies high energy and environmental costs [10]. In contrast, non-conventional or green extraction techniques, such as ultrasound-assisted extraction (EAU), high-pressure-assisted extraction (HPAE), microwave-assisted extraction (MAE), enzyme-assisted extraction (EAE), supercritical fluid extraction (SFE), pulsed electrified field extraction (PEF), pressurized-liquid-assisted extraction (PLE) and surfactant-assisted extraction (SAE), have proven to be a valid alternative in the recovery of bioactive compounds from algae [11–13]. Among them, microwave-assisted extraction (MAE) is an efficient and environmentally friendly technique, which reduces the extraction time and the amount of organic solvents and, in the best of cases, uses less polluting solvents such as water [10,13]. Different variables, such as the type of solvent, time and pressure, influence the recovery efficiency of bioactive compounds. Optimal extraction parameters can be estimated with statistical optimization methods. In this sense, the response surface methodology (RSM) uses quantitative data from an experimental design to solve the multivariate equation and maximize the results of the selected response variables. The objective of this study is to establish the most favorable conditions for MAE, in terms of the type of solvent, time and pressure required to produce extracts of *H. elongata* rich in bioactive compounds that present antioxidant activity.

2. Material and Methods

2.1. Sample Preparation

H. elongata samples were provided by the company Algas Atlánticas Algamar S.L located in Pontevedra, Spain. The algae were collected from the coasts of the province of Pontevedra, and they were washed with distilled water, frozen at −80 °C and later lyophilized. Next, the samples were crushed and ground to obtain a homogeneous matrix, which was stored at −20 °C until use.

2.2. Microwave-Assisted Extraction (MAE)

The process for obtaining bioactive compounds was carried out by MAE, using the multiwave-3000 equipment (Anton-Par). The extraction was carried out using 0.6 g of the lyophilized alga and 20 mL of solvent (solute/solvent ratio of 30 g/L). The variables studied were the ethanol concentration ($\%Et$), pressure (P) and time (t), as critical extraction parameters. Specifically, the $\%Et$ varied between 0 and 100% v/v, the P from 2 to 20 bar and t from 3 to 25 min. The power and frequency of the microwave were fixed for this batch of experiments and set at the maximum value of 1400 W for power and 2.45 GHz for frequency. Once the extraction was completed, the samples were placed in an ice bath for 5 min in order to rapidly lower the temperature and avoid degradation of the thermolabile compounds. Finally, the samples were centrifuged at 9000 rpm for 15 min and filtered to separate the supernatant from algae debris. These extracts were stored in a freezer at −80 °C.

In order to study the influence of MAE conditions ($\%Et$, P and t), the RMS was applied using circumscribed central composite design (CCCD), which allows one to identify the operating conditions for maximizing five response variables: extraction yield (EY), total phenolic content (TPC), total flavonoid content (TFC) and antioxidant activity of *H. elongata*. The interaction between the different variables generates a total of 28 experiments. The least squares regression method was used to fit the data obtained in the 28 experiments to a quadratic model shown in the following equation:

$$Y = b_0 + \sum_{i=1}^{n} b_i X_i + \sum_{\substack{i=1 \\ j>i}}^{n-1} \sum_{j=2}^{n} b_{ij} X_i X_j + \sum_{i=1}^{n} b_{ii} X_i^2 \quad (1)$$

where Y is the predicted responses (Y_1: EY, Y_2: TPC, Y_3: TFC, Y_4: DPPH assay, Y_5: ABTS assay), b_0 is the constant of the model, b_i is the linear coefficient, b_{ii} is the coefficient quadratic, b_{ij} is the coefficient of the interaction and X_i is the dimensionless coded value of the independent variables (X_1: $\%Et$, X_2: P and X_3: t).

2.3. Determination of Bioactive Compounds and Antioxidant Capacity

The *EY* was evaluated based on the dry weight (dw) obtained according to Equation (2).

$$EY\ (\%) = (P_2 - P_1)/P_0 \times 100 \quad (2)$$

where P_0 is the mass of lyophilized algae prior to extraction (mg), P_1 is the mass of the empty crucible (mg), P_2 is the mass of the dry extract in the crucible (mg).

The TPC was determined using the Folin-Ciocalteu reagent, while the TFC was evaluated according to the methodology proposed by Cassani et al. [14]. The results were expressed as mg of phloroglucinol equivalents (PGE)/g of dw and mg of quercetin equivalents (QE)/g of dw, respectively. Regarding the antioxidant capacity, it was determined using two assays: the diphenyl-2-picryl-hydrazyl radical (DPPH) and 2,2′-azino-bis(3-ethylbenzothiazoline-6-sulfonic acid) (ABTS) scavenging assays. The results of both assays are expressed in mg of scavenged compound/mL of extract.

3. Results and Discussions

The experimental results of the RSM of CCCD for the optimization of *H. elongata*' MAE for the five considered response variables are presented in Table 1.

Table 1. Experimental parameters of the optimization process.

	Independent Variables			Response Variables				
Run	***t* (min)**	***P* (bar)**	***Et* (%)**	***EY***	**TPC**	**TFC**	**DPPH**	**ABTS**
1	7.5	5.6	20.3	452	24.58	2.60	5.64	12.24
2	7.5	5.6	79.7	355	28.26	6.14	6.59	80.08
3	7.5	16.4	20.3	529	41.30	4.28	19.97	21.38
4	7.5	16.4	79.7	376	25.52	5.31	5.33	20.41
5	20.5	5.6	20.3	543	19.24	2.36	4.47	9.43
6	20.5	5.6	79.7	384	23.90	5.11	6.29	22.01
7	20.5	16.4	20.3	489	38.86	5.59	19.68	19.70
8	20.5	16.4	79.7	361	19.75	4.40	3.84	15.87
9	3	11	50	459	25.56	2.92	4.70	22.12
10	25	11	50	370	37.70	4.02	7.86	35.70
11	14	2	50	358	28.40	4.62	8.56	59.45
12	14	20	50	479	35.49	5.51	14.32	30.33
13	14	11	0	491	25.50	11.31	11.28	15.12
14	14	11	100	109	11.89	4.19	5.66	22.58
15	3	2	0	373	7.53	1.35	1.01	16.08
16	3	2	100	60	12.87	0.73	4.11	16.08
17	3	20	0	409	27.64	7.44	10.34	23.34
18	3	20	100	99	10.14	4.61	3.57	7.22
19	25	2	0	459	8.10	1.92	1.05	74.57
20	25	2	100	67	5.36	2.73	2.77	7.44
21	25	20	0	443	29.75	9.20	13.43	37.66
22	25	20	100	133	3.89	7.78	3.95	8.18
23	14	11	50	377	35.62	2.62	9.58	61.86
24	14	11	50	377	32.35	4.26	9.11	60.83
25	14	11	50	474	47.73	9.64	25.29	95.13
26	14	11	50	425	21.61	3.29	5.60	24.28
27	14	11	50	439	21.89	3.70	4.50	22.10
28	14	11	50	435	21.97	3.44	4.97	23.59

Abbreviations: extraction yield (*EY*), total phenolic content (TPC), total flavonoid content (TFC), antioxidant assays (DDPH and ABTS).

As can be observed, the five response variables were favored by different extraction conditions. Regarding *EY*, the most favorable conditions were 7.5 min, 16.4 bar and 20% *Et*. The TPC was favored by an extraction time of 14 min, 11 bars and a 50% *Et*. In contrast, TFC was favored under the same conditions of time and pressure, but differed with respect to TPC in the solvent, with 0% Et achieving the best results. Similar results have been reported previously [15]. On the other hand, higher TPC and TFC usually corresponded with higher antioxidant activity in ABTS and DPPH assays. In general terms, the time and pressure parameters with intermediate values favored the EY and the obtaining of TPC and TFC. On the other hand, the parameter with the greatest influence was the %*Et*, showing differences in obtaining bioactive compounds. This can be explained by considering the polarity of the solvent and the compounds [16].

In order to obtain a *H. elongata* extract rich in phenolic compounds and flavonoids, with the maximum antioxidant capacity, all the response variables were simultaneously optimized by means of RSM. The operational conditions that simultaneously optimize all the considered response variables are presented in Table 2. These optimal extraction conditions give rise to an *EY* of 502.28 ± 25.11 mg/g of dw, a TPC of 37.43 ± 3.74 mg PGE/g dw and a TFC of 9.93 ± 0.99 mg QE/g dw. Regarding the antioxidant assays, the radical elimination activity of DPPH and ABTS was 16.37 ± 0.82 and 65.77 ± 1.97 mg/mL, respectively (Table 2).

Table 2. Effect of *H. elongata* extract by MAE under optimal conditions on antioxidant activity.

Best Operating Conditions	**%*Et***	***P* (bar)**	***T* (min)**	
	0.00 ± 0.00	20.00 ± 0.50	16.01 ± 4.80	
EY	**TPC**	**TFC**	**Antioxidant Activity** DPPH	ABTS
502.28 ± 25.11	37.43 ± 3.74	9.93 ± 0.99	16.37 ± 0.82	65.77 ± 1.97

The optimized operating conditions are consistent with the study presented by Magnusson et al. [16], who obtained the best TPC using water as solvent and an extraction time between 3 and 15 min, but very high temperatures were required (160 °C). In this sense, Zhang et al. [17] stated, using terrestrial plants, that water is a solvent with good solubility and has an excellent ability to absorb microwave energy and lead to efficient heating of the sample. Regarding TPC, the results of previous studies are variable. For example, Jiménez-Escrig et al. [9] reported a similar TPC around 30 mg PGE/g dw; however, when using aqueous methanol (50%) and extraction times longer than 2 h, TPC was around 10 mg PGE/g dw when using water but also with longer extraction times (1 h). Fernández et al. [18] reported values of 18 mg gallic acid equivalents/g dw, but again they required the use of organic solvents and extraction times longer than one hour and with different steps. Therefore, our data present a rapid, simple and green method to effectively extract a different kind of biomolecule from *H. elongata*. Nevertheless, further microwave parameters, such as temperature, power and frequency need to be further analyzed to obtain the most efficient extraction method. Furthermore, it is noticeable that the differences observed between studies could be due to the great variability of the content and phytochemical profile of algae, which can be larger and affected by different climatic and intrinsic factors, such as season, age, geographical location and environmental conditions [19].

4. Conclusions

H. elongata is an alga species with reported antioxidant activity, which has been attributed to the presence of phenolic compounds and flavonoids. In this study, MAE resulted in a suitable technique to extract those compounds and obtain extracts with antioxidant activity. Furthermore, the RSM was a suitable statistical method to determine the optimal conditions that maximize the content of polyphenols and total flavonoids, the antioxidant capacity and the extraction performance using microwaves. According to the optimization results, the best operational conditions that allowed us to produce extracts rich in bioactive compounds and displayed significant antioxidant effects on DPPH and ABTS assays were 0% Et, 20.00 bar and an extraction time of 16 min. Considering the growing interest in algae compounds, this extract could be used in the development of functional food, cosmetic and pharmaceutical applications.

Author Contributions: Conceptualization, F.C., L.C., S.B. and F.F.-S.; methodology, F.C., L.C., A.C.-C. and C.L.-L.; software, M.C., P.O. and P.G.-P.; validation, M.C., P.O. and P.G.-P.; formal analysis, F.C., M.C., P.O. and P.G.-P.; investigation, F.C., L.C., A.C.-C. and C.L.-L.; writing—original draft preparation, F.C. and J.E.; writing—review and editing, F.C. and J.E.; visualization, J.S.-G. and M.A.P.; supervision, J.S.-G. and M.A.P. All authors have read and agreed to the published version of the manuscript.

Funding: The JU receives support from the European Union's Horizon 2020 research and innovation program and the Bio Based Industries Consortium. The project SYSTEMIC Knowledge hub on Nutrition and Food Security has received funding from national research funding parties in Belgium (FWO), France (INRA), Germany (BLE), Italy (MIPAAF), Latvia (IZM), Norway (RCN), Portugal (FCT), and Spain (AEI) in a joint action of JPI HDHL, JPI-OCEANS and FACCE-JPI launched in 2019 under the ERA-NET ERA-HDHL (n° 696295). The authors acknowledge the "Margarita Salas" grant awarded to Pascual Garcia-Perez, supported through the European Union by the "NextGenerationEU" program.

Data Availability Statement: Not applicable.

Acknowledgments: The research leading to these results was supported by MICINN supporting the Ramón y Cajal grant for M.A. Prieto (RYC-2017-22891) and the FPU grant for A. Carreira-Casais (FPU2016/06135); by Xunta de Galicia for supporting the program EXCELENCIA-(ED481B-2021/152), the post-doctoral grant of L. Cassani, the program BENEFICIOS DO CONSUMO DAS ESPECIES TINTORE-RA-(CO-0019-2021) that supports the work of F. Chamorro, the EcoChestnut Project (Erasmus+ KA202) that supports the work of J. Echave; by University of Vigo for supporting the predoctoral grant of M. Carpena (Uvigo-00VI 131H 6410211) and the European Union through the "NextGenerationEU" program supporting the "Margarita Salas" grant awarded to P. Garcia-Perez. The authors are grateful to Ibero-American Program on Science and Technology (CYTED—AQUA-CIBUS, P317RT0003), to the Bio Based Industries Joint Undertaking (JU) under grant agreement No 888003 UP4HEALTH Project (H2020-BBI-JTI-2019) that supports the work of P. Otero, and C. Lourenço-Lopes, and to AlgaMar enterprise for the collaboration and algae material provision.

Conflicts of Interest: The authors declare no conflict of interest.

References

1. Martínez-Hernández, G.B.; Castillejo, N.; Carrión-Monteagudo, M.D.M.; Artés, F.; Artés-Hernández, F. Nutritional and bioactive compounds of commercialized algae powders used as food supplements. *Food Sci. Technol. Int.* **2018**, *24*, 172–182. [CrossRef]
2. Lourenço-Lopes, C.; Fraga-Corral, M.; Jimenez-Lopez, C.; Pereira, A.G.; Garcia-Oliveira, P.; Carpena, M.; Prieto, M.A.; Simal-Gandara, J. Metabolites from macroalgae and its applications in the cosmetic industry: A circular economy approach. *Resources* **2020**, *9*, 101. [CrossRef]
3. De Quirós, A.R.B.; Frecha-Ferreiro, S.; Vidal-Pérez, A.M.; López-Hernández, J. Antioxidant compounds in edible brown seaweeds. *Eur. Food Res. Technol.* **2010**, *231*, 495–498. [CrossRef]
4. Rajauria, G.; Jaiswal, A.K.; Abu-Gannam, N.; Gupta, S. Antimicrobial, antioxidant and free radical-scavenging capacity of brown seaweed himanthalia elongata from western coast of Ireland. *J. Food Biochem.* **2013**, *37*, 322–335. [CrossRef]
5. Garcia-Oliveira, P.; Carreira-Casais, A.; Caleja, C.; Pereira, E.; Calhelha, R.C.; Sokovic, M.; Simal-Gandara, J.; Ferreira, I.C.F.R.; Prieto, M.A.; Barros, L. Macroalgae as an Alternative Source of Nutrients and Compounds with Bioactive Potential. *Proceedings* **2020**, *70*, 46. [CrossRef]
6. Leandro, A.; Pereira, L.; Gonçalves, A.M.M. Diverse applications of marine macroalgae. *Mar. Drugs* **2020**, *18*, 17. [CrossRef]
7. Wells, M.L.; Potin, P.; Craigie, J.S.; Raven, J.A.; Merchant, S.S.; Helliwell, K.E.; Smith, A.G.; Camire, M.E.; Brawley, S.H. Algae as nutritional and functional food sources: Revisiting our understanding. *J. Appl. Phycol.* **2017**, *29*, 949–982. [CrossRef]
8. Lorenzo, J.M.; Agregán, R.; Munekata, P.E.S.; Franco, D.; Carballo, J.; Şahin, S.; Lacomba, R.; Barba, F.J. Proximate composition and nutritional value of three macroalgae: *Ascophyllum nodosum*, *Fucus vesiculosus* and *Bifurcaria bifurcata*. *Mar. Drugs* **2017**, *15*, 360. [CrossRef]
9. Jiménez-Escrig, A.; Gómez-Ordóñez, E.; Rupérez, P. Brown and red seaweeds as potential sources of antioxidant nutraceuticals. *J. Appl. Phycol.* **2012**, *24*, 1123–1132. [CrossRef]

10. Ummat, V.; Sivagnanam, S.P.; Rajauria, G.; O'Donnell, C.; Tiwari, B.K. Advances in pre-treatment techniques and green extraction technologies for bioactives from seaweeds. *Trends Food Sci. Technol.* **2021**, *110*, 90–106. [CrossRef]
11. Lourenço-Lopes, C.; Otero, P.; Rodríguez, M.C.; Carreira-Casais, A.; Lourenço-Lopes, C.; Carpena, M.; Pereira, A.G.; Echave, J.; Soria-Lopez, A.; Chamorro, F.; et al. Application of Green Extraction Techniques for Natural Additives Production. *Food Addit.* **2021**. [CrossRef]
12. Bordoloi, A.; Goosen, N. *Green and Integrated Processing Approaches for the Recovery of High-Value Compounds from Brown Seaweeds*; Elsevier Ltd.: Vannes, France, 2020; Volume 95, ISBN 9780081027103.
13. Gomez, L.; Tiwari, B.; Garcia-Vaquero, M. *Emerging Extraction Techniques: Microwave-Assisted Extraction*; Elsevier Inc.: Amsterdam, The Netherlands, 2020; ISBN 9780128179437.
14. Cassani, L.; Tomadoni, B.; Ponce, A.; Agüero, M.V.; Moreira, M.R. Combined Use of Ultrasound and Vanillin to Improve Quality Parameters and Safety of Strawberry Juice Enriched with Prebiotic Fibers. *Food Bioprocess Technol.* **2017**, *10*, 1454–1465. [CrossRef]
15. Yuan, Y.; Zhang, J.; Fan, J.; Clark, J.; Shen, P.; Li, Y.; Zhang, C. Microwave assisted extraction of phenolic compounds from four economic brown macroalgae species and evaluation of their antioxidant activities and inhibitory effects on α-amylase, α-glucosidase, pancreatic lipase and tyrosinase. *Food Res. Int.* **2018**, *113*, 288–297. [CrossRef]
16. Magnusson, M.; Yuen, A.K.L.; Zhang, R.; Wright, J.T.; Taylor, R.B.; Maschmeyer, T.; de Nys, R. A comparative assessment of microwave assisted (MAE) and conventional solid-liquid (SLE) techniques for the extraction of phloroglucinol from brown seaweed. *Algal Res.* **2017**, *23*, 28–36. [CrossRef]
17. Zhang, B.; Yang, R.; Liu, C.Z. Microwave-assisted extraction of chlorogenic acid from flower buds of Lonicera japonica Thunb. *Sep. Purif. Technol.* **2008**, *62*, 480–483. [CrossRef]
18. Fernández-Segovia, I.; Lerma-García, M.J.; Fuentes, A.; Barat, J.M. Characterization of Spanish powdered seaweeds: Composition, antioxidant capacity and technological properties. *Food Res. Int.* **2018**, *111*, 212–219. [CrossRef]
19. Mekinić, I.G.; Skroza, D.; Šimat, V.; Hamed, I.; Čagalj, M.; Perković, Z.P. Phenolic content of brown algae (Pheophyceae) species: Extraction, identification, and quantification. *Biomolecules* **2019**, *9*, 244. [CrossRef]

chemistry proceedings

MDPI

Proceeding Paper

Semiconductor Oxide Gas Sensors: Correlation between Conduction Mechanisms and Their Sensing Performances †

Ambra Fioravanti [1,*], Sara Morandi [2] and Maria Cristina Carotta [1]

[1] Institute of Sciences and Technologies for Sustainable Energy and Mobility (CNR–STEMS), Via Canal Bianco 28, 44124 Ferrara, Italy; mariacristina.carotta@stems.cnr.it

[2] Department of Chemistry, University of Turin, Via P. Giuria 7, 10125 Torino, Italy; sara.morandi@unito.it

* Correspondence: ambra.fioravanti@stems.cnr.it; Tel.: +39-0532-735668

† Presented at the 1st International Electronic Conference on Chemical Sensors and Analytical Chemistry, 1–15 July 2021; Available online: https://csac2021.sciforum.net/.

Abstract: In this work, a variety of semiconducting oxides were prepared and principally characterized by means of spectroscopic techniques (absorbance FT-IR, diffuse reflectance UV-Vis-NIR) to shed light on the electronic properties and defects involved at the roots of gas sensing capabilities. The thick films were obtained by screen printing technology on which electrical characterization and gas sensing measurements were performed. From the cross analysis of the results, a description of the specific sensing mechanism for each material is proposed.

Keywords: thick film gas sensors; nanostructured semiconductor oxides; UV-Vis-NIR and FT-IR spectroscopies; electrical characterization; sensing mechanisms

Citation: Fioravanti, A.; Morandi, S.; Carotta, M.C. Semiconductor Oxide Gas Sensors: Correlation between Conduction Mechanisms and Their Sensing Performances. *Chem. Proc.* **2021**, *5*, 71. https://doi.org/10.3390/CSAC2021-10472

Academic Editor: Ye Zhou

Published: 30 June 2021

1. Introduction

The adsorption of a gas on the surface of a semiconducting oxide can induce a significant change in the electrical resistance of the material. This effect is at the basis of the development of chemiresistors for gas detection [1]. Due to their high sensitivity, tunable selectivity, easy production, small dimensions, and low cost, they are successfully used in a broad range of applications (pollutant monitoring, food quality control, industrial system control, and medical diagnosis) to detect a large number of gaseous compounds. Despite this, an increasing demand of gas sensors with high performances has been documented [2]. Many actions can be made to improve the sensing performances, such as the synthesis of nanostructures with a high specific surface area and the loading with noble metals, but the first issue is to understand the sensing mechanism of the materials and their sensing properties [3,4].

The IR and the UV-Vis spectroscopies are excellent experimental tools for investigating the electronic properties and surface chemistry of a large class of metal oxides used in the fabrication of solid state devices for gas sensing [5].

This work is aimed to determine the electronic properties for a variety of semiconducting oxides (single or combined, such as SnO_2 MoO_3, WO_3, ZnO, TiO_2, Ti-Sn, W-Sn, Mn-W mixed oxides, etc.) and to correlate them with the sensing mechanism and the sensor performances.

2. Materials and Methods

A selection of semiconducting oxides (single or combined, such as SnO_2 MoO_3, WO_3, ZnO, TiO_2, Ti-Sn, W-Sn, Mn-W mixed oxides, etc.) were prepared using wet chemistry methods [6,7]. The synthesis procedures were optimized to obtain nanopowders with a single phase, and a homogenous distribution in grain size (analyzed by X-ray diffraction and scanning electron microscopy, respectively) was presented.

The powders were characterized by means of spectroscopic techniques with the aim to evaluate the electronic properties and defects involved in the sensing processes. Absorbance IR spectra were run on a Perkin–Elmer System 2000 FT-IR spectrophotometer, equipped with a Hg–Cd–Te cryodetector, working in the range of wavenumbers 7800–580 cm^{-1}. Diffuse reflectance spectra in the UV-Vis-NIR region were run on a Varian Cary 5 spectrophotometer, working in the range of wavenumbers 53,000–4000 cm^{-1}. The spectroscopic responses were studied in the range from room temperature to 500 °C, both for the interaction with pure gases and for mixture gas/O_2 at different concentrations depending on the oxide [8,9].

For obtaining thick films for electrical characterization and gas sensing measurements, the functional materials were added to an organic vehicle together with a small percentage of glass frit. Then, they were deposited on alumina substrates with interdigitated Au contacts and a heating element, finally fired at 650 °C. The flow-through technique was used maintaining a flow rate of 0.5 L/min, using synthetic air as carrier gas in dry conditions for the: (i) conductance measurements vs. temperature (room temperature of 650 °C); (ii) surface potential barrier height measurements to determine the intergranular energy barrier (Schottky barrier) versus temperature, following the method of stimulated temperature conductance measurements, as described by Clifford and Tuma [8,10]; and (iii) dynamical responses obtained in the presence of a mixture of different gases and operating temperature from 350 to 550 °C. The sensor response was calculated as the ratio between the conductance in the presence of the gas test and the conductance in air.

Finally, a sensing mechanism was proposed for each material by combining the results of spectroscopic and electrical characterization.

3. Results and Discussion

In this section, the main findings of some single (SnO_2, WO_3, and TiO_2) and mixed (Ti-Sn and W-Sn) oxides, considering the carbon monoxide gas test, are summarized. More detailed characterizations are reported in [8,9].

The spectroscopic measurements on single oxides show that, on the one hand, SnO_2 is characterized by the presence of mono-ionized oxygen vacancies (evidence in the medium IR region); on the other hand, WO_3 is characterized by the presence of polarons, i.e., electrons trapped in deep levels in the band gap (evidenced in the Vis-NIR region). TiO_2 shows both mono-ionized oxygen vacancies and polarons. The amount of defects increases upon CO interaction at increasing temperatures.

As for mixed oxide $Sn_{1-x}Ti_xO_2$, the spectroscopic measurements in CO reveal two different behaviors: (i) samples with x = 0.1 shows absorption related to the photo-ionization of mono-ionized oxygen vacancies, as SnO_2 shows; and (ii) samples with $0.3 \leq x \leq 0.9$ show the increase in an absorption near to the VB–CB edge, as TiO_2 shows. Samples with x = 0.2 is a borderline sample, showing behavior in between those of SnO_2 and TiO_2. This behavior was confirmed by electrical measurements.

Combining the results of spectroscopic and electrical characterization, two detection mechanisms emerge depending on the kind of chemical reaction involved. The first occurs between ionosorbed oxygen atoms and CO, with a consequent electron transfer from surface to bulk, resulting in a conductance increase and a consequent barrier height decrease. The second occurs between surface lattice oxygen atoms and CO; the bond electrons of the surface lattice oxygen atoms do not contribute to the formation of the spatial charge region and the Schottky barrier. To conclude, the prevalent gas detection mechanism in the materials with $x < 0.3$ is that usually occurs through Schottky barrier modification, whilst $x \geq 0.3$ is based on bond electrons which, after the reaction, enter the conduction band without affecting the barrier height, but only bulk conductance.

Concerning the CO responses, the Ti-Sn solid solutions offer higher responses than those of pristine oxides, and the solution with x = 0.25 results in the best material to detect carbon monoxide.

The spectroscopic characterization of W-Sn mixed oxides highlighted the presence of polaron levels, with a position not affected by the Sn content, for mixed oxides with Sn molar content up to 33% (as for WO_3) and mono-ionized oxygen vacancies for the mixed oxide with Sn molar content of 89% (as for SnO_2).

The different positions of the defect levels in the band gap cause the formation of surface potential barriers significantly lower for WO_3 and for mixed oxides with Sn molar content up to 33% (WO_3-like samples) than for SnO_2 and for the mixed oxide with Sn molar content of 89% (SnO_2-like sample). This result allows to define the correlation between the electronic levels associated with the defects and the surface potential barriers in the air and in reducing atmospheres. In particular, the electrical measurements indicate that the changes in the already-low barriers of WO_3 and WO_3-like samples are almost negligible in the presence of a reducing gas, such as CO; otherwise, CO significantly decreases the barriers of SnO_2 and SnO_2-like sample. These results are completely in agreement with the low CO sensitivity of WO_3 and WO_3-like samples, and with the better CO sensitivity of SnO_2 and SnO_2-like sample.

4. Conclusions

In the electrical characterization, the main parameter typically measured is the conductance. All the operating characteristics of the sensors are derived from this measurement, considering the strength and the weakness of semiconductor sensors. On the one hand, it is simple and easily measured, but it is a second-order parameter that, although very sensitive to some reactions at the solid surface, is not a direct indicator of the exact processes taking place. For this reason, we investigated the behavior of different oxide materials by means of IR and UV-Vis spectroscopies to enlighten surface reactions and electronic properties and coupling the results to those of electrical characterization. We demonstrate the possibility to describe the processes involved in the detection mechanism with a method which can be applied to every functional material characterized towards every gas of interest.

Author Contributions: Conceptualization, characterizations, experiments conduction and data analysis A.F., S.M. and M.C.C.; writing—original draft preparation A.F. All authors have read and agreed to the published version of the manuscript.

Funding: This research received no external funding.

Data Availability Statement: Not applicable.

Conflicts of Interest: The authors declare no conflict of interest.

References

1. Neri, G. First fifty years of chemoresistive gas sensors. *Chemosensors* **2015**, *3*, 1. [CrossRef]
2. Fioravanti, A.; Carotta, M.C. Year 2020: A snapshot of the last progress in flexible printed gas sensors. *Appl. Sci.* **2020**, *10*, 1741. [CrossRef]
3. Galstyan, V. "Quantum dots: Perspectives in next-generation chemical gas sensors"—A Review. *Anal. Chim. Acta* **2021**, *152*, 238192. [CrossRef] [PubMed]
4. Dey, A. Semiconductor metal oxide gas sensors: A review. *Mater. Sci. Eng. B Solid-State Mater. Adv. Technol.* **2018**, *229*, 206–217. [CrossRef]
5. ViterIgor, R.; Barhoum, I. *Optical Spectroscopy for Characterization of Metal Oxide Nanofibers*; Springer: Cham, Switzerland, 2019; ISBN 9783319536545.
6. Nikam, A.V.; Prasad, B.L.V.; Kulkarni, A.A. Wet chemical synthesis of metal oxide nanoparticles: A review. *CrystEngComm* **2018**, *20*, 5091–5107. [CrossRef]
7. Fioravanti, A.; Marani, P.; Massarotti, G.P.; Lettieri, S.; Morandi, S.; Carotta, M.C. (Ti,Sn) solid solution-based gas sensors for new monitoring of hydraulic oil degradation. *Materials* **2021**, *14*, 605. [CrossRef] [PubMed]
8. Carotta, M.C.; Fioravanti, A.; Gherardi, S.; Malagù, C.; Sacerdoti, M.; Ghiotti, G.; Morandi, S. (Ti,Sn) solid solutions as functional materials for gas sensing. *Sens. Actuators B Chem.* **2014**, *194*, 195–205. [CrossRef]

9. Morandi, S.; Amodio, A.; Fioravanti, A.; Giacomino, A.; Mazzocchi, M.; Sacerdoti, M.; Carotta, M.C.; Skouloudis, A.N. Operational functionalities of air-quality W–Sn metal-oxide sensors correlating semiconductor defect levels and surface potential barriers. *Sci. Total Environ.* **2020**, *706*, 135731. [CrossRef] [PubMed]
10. Clifford, P.K.; Tuma, D.T. Characteristics of semiconductor gas sensors II. transient response to temperature change. *Sens. Actuators* **1982**, *3*, 255–281. [CrossRef]

chemistry proceedings

MDPI

Proceeding Paper

Cross-Sensitive Potentiometric Sensors Based on Anti-Crown $(C_6HgF_4)_3$ †

Ekaterina Yuskina [1], Kirill Tugashov [2], Vladimir B. Shur [2], Irina A. Tikhonova [2], Vasily Babain [1] and Dmitry Kirsanov [1,]*

[1] Institute of Chemistry, Saint Petersburg State University, Universitetskaya nab. 7/9, 199034 Saint Petersburg, Russia; lazyvelikaya@yandex.ru (E.Y.); babainv@mail.ru (V.B.)

[2] A.N.Nesmeyanov Institute of Organoelement Compounds, RAS, Vavilov Street 28, 119991 Moscow, Russia; kirill@ineos.ac.ru (K.T.); vbshur@ineos.ac.ru (V.B.S.); tikhonova@ineos.ac.ru (I.A.T.)

* Correspondence: d.kirsanov@gmail.com; Tel.: +7-921-333-1246

† Presented at the 1st International Electronic Conference on Chemical Sensors and Analytical Chemistry, 1–15 July; Available online: https://csac2021.sciforum.net/.

Abstract: In this work, we explore the possibility of using anti-crown ether $(C_6HgF_4)_3$ as a membrane-active component for potentiometric cross-sensitive sensors. Anti-crown ligands have already been employed as ionophores in plasticized polymeric membranes; however, the results of these studies are contradictory. In order to clarify the electrochemical sensitivity patterns of anti-crown-based sensors, we have studied plasticized polymeric membranes containing cation and anion-exchanging additives and various solvent-plasticizers. We explored the electrochemical sensitivity of these membranes in a wide variety of aqueous solutions of inorganic salts. Alkaline, alkaline-earth, and d-element salts with different anions were studied. It was found that the sensors based on anti-crown $(C_6HgF_4)_3$ exhibit cationic sensitivity, and no considerable anionic responses were observed.

Keywords: anti-crown; potentiometric sensors; plasticized polymeric membranes

check for updates

Citation: Yuskina, E.; Tugashov, K.; Shur, V.B.; Tikhonova, I.A.; Babain, V.; Kirsanov, D. Cross-Sensitive Potentiometric Sensors Based on Anti-Crown $(C_6HgF_4)_3$. *Chem. Proc.* **2021**, *5*, 72. https://doi.org/10.3390/CSAC2021-10424

Academic Editor: Xin Wu

Published: 30 June 2021

1. Introduction

In this work, we explore the possibility of using the so-called anti-crown compound $(C_6HgF_4)_3$ as a membrane-active component for potentiometric cross-sensitive sensors. Anti-crown ethers are macrocyclic organometallic compounds consisting of mercury atoms or tin atoms separated by carbon atoms, including fluorinated macrocycles and mercury-carborands.

The chemistry of anti-crowns is quite extensive and it is a rapidly grown field of research. Recently, it was reported that (o-$C_6F_6Hg)_3$ is readily coordinated with bromide and iodide anions to yield the following complexes: {o-$[(C_6F_6Hg)_3Br]^-$} and {o-$[(C_6F_6Hg)_3I]^-$}. When anti-crowns interact with halogens, they form polydecker wedge-shaped sandwiches: $[(\cdots(C_6F_6Hg)_3\cdots X\cdots)_n]^{n-}$, where X = chloride and bromide anions [1]. The ability of anti-crown ethers to coordinate with single-charge anions has attracted the attention of analytical chemists. Several attempts have been made to create sensors based on anti-crown ethers to detect anions. For example, mercury-carborand was used as an ionophore for the determination of chloride ions [2]. The authors made a PVC-plasticized membrane consisting of NPOE as a plasticizer, TDDMA-Cl as an anion exchange additive, and mercury-carborand-3 as an ionophore. It was experimentally shown that the sensors exhibit a near-Nernstian response when interacting with Cl^- in a wide range of concentrations, and detection limits were in the micromole range. It was also found that the presence of TDDMA-Cl in the membrane causes an increase in sensor sensitivity. Another study explored three mercury anti-crown ethers to develop sensors for the Me_4N^+ cation [3]. The authors of the article synthesized three different PVC-plasticized membranes with various plasticizers: bis(2-ethylhexyl) sebacate (DOS), dioctyl phthalate (DOP), and dibutyl phthalate (DBP).

Potassium tetrakis(4-chlorophenyl)-borate (KTpClPB) was used as a cation exchanger. A near-Nernstian response of 54.2 mV/dec towards the Me_4N^+ cation was observed when DOS was employed as a plasticizer. With other plasticizers, the responses were smaller. The electrode with the ionophore was more sensitive, with a response of 54.9 mV/dec. Figure 1 shows several typical anti-crown ether structures.

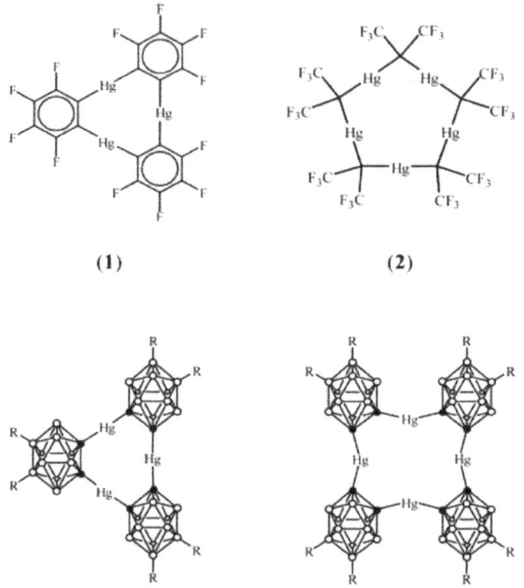

Figure 1. Structural formulas of typical anti-crown ethers based on fluorinated macrocycles and mercury-carborand: (**1**) Containing 3 mercury atoms; (**2**) containing 4 and 5 mercury atoms.

2. Experimental Part

2.1. Reagents

Three-mercury anti-crown ether was used as an ionophore in all membranes. The anti-crown was synthesized at A.N. Nesmeyanov Institute of Organoelement Compounds. The polymer matrix of the membranes was made of PVC (poly(vinylchloride)). O-nitrophenyloctyl ether (NPOE) and dioctyl sebacate (DOS) were used as a solvent-plasticizers. NaTpClPB was used as an anionic additive, and TDDMA-NO_3 was used as a cationic additive. All these substances were from SigmaAldrich, at Selectophore grade. We made three membranes of each type to assess repeatability of the results.

2.2. ISE Preparation and Potentiometric Measurements

The composition of the sensor membranes is shown in Table 1. The membranes were made by mixing the following components: 200 mg plasticizer, 109 mg PVC, 17 mg anti-crown ether, 30 mg NaTpClPB (membranes 2 and 5), and 20 mg TDDMA-NO_3 (membranes 3 and 6). The components where mixed with 5 mL THF using a magnetic stirrer until completely dissolved. After that, the mixture was transferred to a flat-bottom teflon beaker and left overnight for solvent evaporation. The disks (7 mm in diameter) were cut from the parent membrane and glued into electrode bodies. After the glue dried, the housing was connected to the electrodes and filled with 0.01 M NaCl solution. Finally, the electrodes were soaked for a day in a solution of sodium chloride of the same concentration. Between subsequent measurements, the sensors were stored in air. In total, we made 6 membranes and 18 sensors.

Table 1. Membrane compositions.

Sensor	Component	Weight (g)
1	PVC	0.1089
	DOS	0.2020
	Anti-crown	0.0170
2	PVC	0.1090
	DOS	0.2024
	Anti-crown	0.0171
	NaTpClPB	0.0030
3	PVC	0.1089
	DOS	0.2019
	Anti-crown	0.0171
	TDDMA-NO_3	0.0020
4	PVC	0.1089
	NPOE	0.2054
	Anti-crown	0.0172
5	PVC	0.1088
	NPOE	0.2026
	Anti-crown	0.0171
	NaTpClPB	0.0030
6	PVC	0.1089
	NPOE	0.2010
	Anti-crown	0.0046
	TDDMA-NO_3	0.0009

Solutions with a concentration of 1 M were prepared by weight, and less concentrated solutions were prepared by sequential volume dilution of the parent solution. All solutions were prepared with doubly distilled water.

The galvanic cell for the potentiometric measurements was the following:

Ag | AgCl, KCl(sat.) | analyzed solution | PVC membrane | NaCl, 0.01 M, AgCl | Ag

The reference electrode was a silver chloride electrode filled with a saturated solution of potassium chloride. A glass electrode was used to control the pH during the experiment. All measurements were carried out at room temperature.

3. Results and Discussion

3.1. Anion Sensitivity

To study the anion sensitivity of the sensors, a series of measurements was carried out in aqueous solutions of Li_2SO_4, NaCl, NaOAc, NaF, and $Ca(NO_3)_2$ in the concentration range of 10^{-6}–10^{-2} M. Based on the results of three parallel measurements, the slopes of the linear parts (10^{-5}–10^{-2} M) of the calibration curves were calculated. The sensor compositions 2 and 5 (Table 2) contain cation-exchange NaTpClPB, compositions 3 and 6-TDDMA-NO_3 contain an anion-exchange additive.

Table 2. Response characteristics of the electrodes to anions (±2 mV/dec).

Sensor	Cl^-	F^-	OAc^-	NO_3^-	SO_4^{2-}
1	5	9	8	4	6
2	16	18	14	19	11
3	−27	−1	−6	−3	−2
4	34	31	32	5	35
5	29	39	32	15	41
6	−41	−8	−10	−11	−3

As can be observed, the developed sensors in most cases do not demonstrate pronounced sensitivity to anions. The sensitivity values obtained for chloride for the sensors of composition 6 (−40.85 mV/dec with NPOE plasticizer) are the closest to the theoretical values. The sensors without any ion-exchanging additive showed cation sensitivity, thus suggesting that three mercury anti-crown ether promotes cation sensitivity.

3.2. Cation Sensitivity

We also studied the sensitivity of the electrodes to cations. A broad variety of inorganic cations was studied: Li^+, Na^+, Mg^{2+}, K^+, Ca^{2+}, Co^{2+}, Ni^{2+}, Cu^{2+}, Zn^{2+}, Sr^{2+}, Cd^{2+}, Cs^+, Pb^{2+}, La^{3+}, Eu^{3+}, and Lu^{3+}. The measurements were performed in the concentration range of 10^{-7}–10^{-3} M for lanthanides solutions; other salts were in the concentration range of 10^{-6}–10^{-2} M. The slopes were calculated for the concentration range of 10^{-5}–10^{-2} M (10^{-5}–10^{-3} M for La^{3+}, Eu^{3+}, and Lu^{3+}). The results are provided in Table 3.

Table 3. Cation sensitivities of the electrodes (±2 mV/dec).

Sensor	Li^+	Na^+	Mg^{2+}	K^+	Ca^{2+}	Co^{2+}	Ni^{2+}	Cu^{2+}
1	6	5	−1	23	4	−2	−3	8
2	11	16	9	44	19	12	11	10
3	−2	−27	−3	1	−3	−4	−3	2
4	35	34	4	39	5	15	10	13
5	41	29	10	44	15	15	12	14
6	−3	−41	−7	−6	−11	−2	−17	−10
	Zn^{2+}	Sr^{2+}	Cd^{2+}	Cs^+	Pb^{2+}	La^{3+}	Eu^{3+}	Lu^{3+}
1	−3	−3	1	36	2	−1	0	−1
2	9	13	8	51	19	1	2	2
3	−3	−5	0	12	−1	−1	−2	−4
4	13	13	11	42	23	3	3	2
5	12	14	13	47	17	0	0	1
6	−6	−4	−2	−12	−2	0	−1	−2

It can be seen that the sensors have pronounced cross-sensitivity to cations. The sensors of compositions 3 and 6 in most cases showed anion sensitivity, apparently due to the presence of the anion exchanger. The sensor with a cation-exchange additive based on the plasticizer DOS (composition 2) was more sensitive to all the cations than the sensor without an additive with the same plasticizer (composition 1). For example, for Cs^+ and Pb^{2+}, the slopes for sensor 1 were 35.8 mV/dec and 2.1 mV/dec, and the slopes for sensor 2 were 51.2 mV/dec and 18.7 mV/dec, respectively. Comparing sensors based on different plasticizers (sensors 2 and 5; plasticizers DOS and NPOE), we can say that for most of the cations (Li^+, Mg^{2+}, Ni^{2+}, Cu^{2+}, Sr^{2+}, and Cd2$^+$), the values of the slope of the electrode function are higher for the sensor based on o-nitrophenyloctyl ether (composition 5). For example, for Cd^{2+}, the slope for sensor 2 was 8.4 mV/dec, and the slope for sensor 5 was 12.9 mV/dec. At the same time, it is worth noting that the observed sensitivity values in most cases significantly differ from the theoretical values. The developed sensors did not show sensitivity to the La^{3+}, Eu^{3+}, and Lu^{3+} cations.

4. Conclusions

We studied the possibility of using a three-mercury anti-crown as an ionophore. Also, the electrochemical sensitivity of developed sensor membranes in solutions of inorganic anions and cations was studied. It was found that the presence of three-mercury anti-crown ether in the polymer plasticized membrane promoted the cation sensitivity of the sensors.

In general, the developed sensors may have potential in the development of potentiometric multi-sensor systems; however, further studies are needed to confirm that there is some benefit to using anti-crown compounds as ionophores, since the observed sensitivities followed the lipophilicity of the cations (the highest values were observed for cesium and lead).

Author Contributions: Conceptualization, D.K. and V.B.; methodology, E.Y., K.T., V.B.S. and I.A.T.; formal analysis, E.Y.; investigation, E.Y., K.T., V.B.S. and I.A.T.; resources, D.K.; writing—original draft preparation, E.Y.; writing—review and editing, D.K. and V.B.; visualization, E.Y.; supervision, D.K. and V.B. All authors have read and agreed to the published version of the manuscript.

Funding: This research received no external funding.

Conflicts of Interest: The authors declare no conflict of interest.

References

1. Tugashov, K.I.; Gribanyov, D.A.; Dolgushin, F.M.; Smol'yakov, A.F.; Peregudov, A.S.; Minacheva, M.K.; Tikhonova, I.A.; Shur, V.B. Coordination Chemistry of Anticrowns. Synthesis and Structures of Double-Decker Sandwich Complexes of the Three-Mercury Anticrown (o-$C_6F_4Hg)_3$ with Halide Anions Containinf and Not Containing Coordinated Dibromomethane Molecules. *Organometallic* **2016**, *3*, 2197–2206. [CrossRef]
2. Badr, I.H.; Martin Diaz, M.; Hawthorne, F.; Bachas, L.G. Mercuracarborand "Anti-crown Ether"-Based Chloride-Sensitive Liquid/Polymeric Membrane Electrodes. *Anal. Chem.* **1999**, *71*, 1371–1377. [CrossRef] [PubMed]
3. Yan, Z.; Zhou, Z.; Wu, Y.; Shur, V.B.; Tikhonova, I.A.; Shur, V.B. Cyclic Trimeric Perfluoro-o-Phenylenemercury as a New Ionophore for Quaternary Ammonium Cation-Selective Membrane Electrode. *Anal. Lett.* **2005**, *38*, 377–388. [CrossRef]

chemistry proceedings

MDPI

Proceeding Paper

An Optical Fiber Sensor for Hg^{2+} Detection Based on the LSPR of Silver and Gold Nanoparticles Embedded in a Polymeric Matrix as an Effective Sensing Material †

María Elena Martínez-Hernández [1,*], Xabier Sandua [2,3], Pedro J. Rivero [2,3], Javier Goicoechea [1,4] and Francisco J. Arregui [1,4]

1 Department of Electrical, Electronic and Communication Engineering, Public University of Navarre (UPNA), Edif. Los Tejos, Campus Arrosadía, 31006 Pamplona, Spain; javier.goico@unavarra.es (J.G.); parregui@unavarra.es (F.J.A.)
2 Institute for Advanced Materials and Mathematics (INAMAT2), Public University of Navarre (UPNA), 31006 Pamplona, Spain; xabier.sandua@unavarra.es (X.S.); pedrojose.rivero@unavarra.es (P.J.R.)
3 Engineering Department, Campus de Arrosadía S/N, Public University of Navarre (UPNA), 31006 Pamplona, Spain
4 Institute of Smart Cities (ISC), Public University of Navarre (UPNA), Campus Arrosadía, 31006 Pamplona, Spain
* Correspondence: mariaelena.martinez@unavarra.es

† Presented at the 1st International Electronic Conference on Chemical Sensors and Analytical Chemistry, 1–15 July 2021; Available online: https://csac2021.sciforum.net/.

Abstract: In this work, an optical fiber sensor based on the localized surface plasmon resonance (LSPR) phenomenon is presented as a powerful tool for the detection of heavy metals (Hg^{2+}). The resultant sensing film was fabricated using a nanofabrication process, known as layer-by-layer embedding (LbL-E) deposition technique. In this sense, both silver nanoparticles (AgNPs) and gold nanoparticles (AuNPs) were synthesized using a synthetic chemical protocol as a function of a strict control of three main parameters: polyelectrolyte concentration, loading agent, and reducing agent. The use of metallic nanostructures as sensing materials is of great interest because well-located absorption peaks associated with their LSPR are obtained at 420 nm (AgNPs) and 530 nm (AuNPs). Both plasmonic peaks provide a stable real-time reference that can be extracted from the spectral response of the optical fiber sensor, giving a reliable monitoring of the Hg^{2+} concentration.

Keywords: fiber optic sensor; gold nanoparticles; silver nanoparticles; localized surface plasmon resonance; layer-by-layer embedding; mercury; ppm

Citation: Martínez-Hernández, M.E.; Sandua, X.; Rivero, P.J.; Goicoechea, J.; Arregui, F.J. An Optical Fiber Sensor for Hg^{2+} Detection Based on the LSPR of Silver and Gold Nanoparticles Embedded in a Polymeric Matrix as an Effective Sensing Material. *Chem. Proc.* **2021**, *5*, 73. https://doi.org/10.3390/CSAC2021-10633

Academic Editor: Huangxian Ju

Published: 7 July 2021

1. Introduction

The presence of heavy metals in a human's daily life has become a concern due to their adverse health effects. Among all of them, mercury is receiving major attention because its presence is associated with serious problems, such as lung or nervous system damage, heart diseases, and even neurological and psychological symptoms [1]. Due to this, a wide variety of detection methods can be found in the bibliography, ranging from electrochemical sensors [2–4] to colorimetric sensors [5–7] and optical sensors [8–10]. This work focuses on optical fiber sensors based on the localized surface plasmon resonance (LSPR) phenomenon. It is well known that LSPR is an optical phenomenon that is generated, thanks to the interaction between the incident light and the electrons in the conduction band of the metal surface [11]. It has been demonstrated that the resultant amplitude and the plasmonic resonance energy can vary as a function of the geometry and the distance between the nanoparticles. Until now, LSPR optical fiber sensors for mercury ion detection mostly contain gold nanoparticles (AuNPs) as the main plasmonic sensing material, showing the interaction between gold and mercury as a change in the physical and chemical properties of the metallic nanoparticles [12,13]. The novelty of this work is the possibility of introducing two different metallic nanoparticles, AgNPs and AuNPs, into LbL films with the aim of obtaining two different LSPR sensing signals for the detection of mercury ions. This deposition technique makes it possible to obtain thin films with a good control in the resultant thickness in the nanometric range as a function of operational parameters, such as pH, ionic strength, and number of bilayers deposited [14–16]. An initial study is performed on glass slides in order to optimize the nanofabrication technique, and then sensing coating is implemented on an optical fiber. Finally, a change in the wavelength position of the LSPR band can be observed as a function of the concentration of the analyte. To sum up, this is the first time that an optical fiber sensor with a dual reference state is presented for mercury ion detection.

2. Methods

2.1. Materials

The polymeric matrix is composed of poly(allylamine hydrochloride) (PAH) (Mw ~15.000), which acts as a polycation, and poly(acrylic acid) (PAA) 35 wt% solution in water, which acts as a polyanion. In order to obtain AuNPs and AgNPs, gold(III) chloride trihydrate ($HAuCl_4 \cdot 3H_2O$) and silver nitrate ($AgNO_3$) were used as loading agents for the synthesis of metallic nanoparticles. Finally, dimethylamine borane complex (DMAB) was used as a reducing agent.

2.2. Chemical Process for the Synthesis of Metallic Nanoparticles

2.2.1. Gold Nanoparticle (AuNP) Synthesis

First, aqueous solutions of $HAuCl_4 \cdot 3H_2O$ (20 mL, 5 mM) and PAA (120 mL, 10 mM), which acts as a stabilizing agent, were mixed and stirred for a period of 2 h. After that, an aqueous solution of the reducing agent, DMAB (5 mL, 100 mM), was added to the previous solution, and the mixture was stirred for 24 h at room temperature. Finally, a color change from yellow to violet was obtained, indicating the synthesis of AuNPs. The combination of PAA and AuNPs is denoted as PAA-AuNPs. This colloidal dispersion solution (PAA-AuNPs) showed a spherical shape and nanometric range (10–20 nm) corroborated by transmission electron microscopy (TEM) [17].

2.2.2. Silver Nanoparticle (AgNPs) Synthesis

For the synthesis of AgNPs, first, aqueous solutions of $AgNO_3$ (20 mL, 10 mM) and PAH (120 mL, 10 mM), which acts as a stabilizing agent, were mixed and stirred for a period of 2 h. After that, an aqueous solution of the reducing agent, DMAB (5 mL, 100 mM), was added to the initial solution, and the mixture was stirred for 24 h at room temperature. Finally, a color change from transparent to orange was obtained, indicating the synthesis of AgNPs. The combination of PAH and AgNPs is denoted as PAH-AgNPs. The location of the LSPR absorption band clearly indicates the synthesis of AgNPs with a spherical shape and nanometric size [18].

2.3. Optical Characterization

The optical properties of the synthesized metallic nanoparticles were determined by using a Jasco V-630 spectrophotometer (Agilent, Santa Clara, CA, USA). Two different and well-separated absorption bands were obtained.

2.4. Layer-by-Layer Nanoassembly

The layer-by-layer nanoassembly technique was used for the fabrication of the thin films. In this work, the presence of PAH and PAA was used as the positively and negatively charged polyelectrolytes for the buildup of the polyelectrolyte structure film. In addition, as demonstrated in the previous section, these charged structures also play a key role in stabilizing the synthesized nanoparticles. More specifically, the polycationic solution PAH-capped AgNPs (PAH-AgNPs) and the polyanionic PAA-capped AuNPs (PAA-AuNPs) were used for the fabrication of the thin films.

2.5. Optical Fiber Detection Setup

A multimode optical fiber with a 200 μm core diameter with polymeric cladding, 0.39 NA (Thorlabs FT-200-EMT), was used for the fabrication of the optical fiber sensor. First, it was necessary to remove the acrylate cladding of a segment of approximately 2 cm of the optical fiber, and for this, a few drops of dry acetone and a blade were used, exposing the bare optical fiber core in its entire cylindrical section. This optical fiber segment was immersed for 5 min in piranha solution to eliminate the acetone that could remain. Temporary SMA connectors were used at the end of the optical fiber, exciting the sensor from one of the connectors with a halogen white source, and the other end collected the optical response with a CCD spectrometer (HR4000-UV Ocean Optics, Ocean Insight, FL, USA). A scheme of the deposition process and the optical setup is presented in Figure 1.

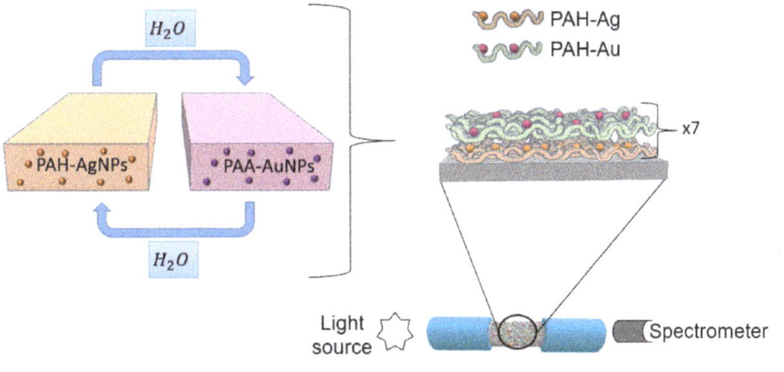

Figure 1. Schematic representation for the fabrication of the LbL films by using PAH-AgNPs as a polycation and PAA-AuNPs as a polyanion.

2.6. Mercury Sample Preparation

The mercury samples were prepared using mercury(II) chloride ($HgCl_2$). Every concentration of mercury was prepared with phosphate buffer (PB) solution for achieving a constant pH = 7.6. The Hg concentrations were varied from 50 to 1 and 0.1 ppm. An important aspect is that for each measurement, the fiber optical sensor was immersed in PB + DMAB buffer solution with the aim of obtaining a stable baseline for further mercury detection.

3. Results and Discussion

As an initial step, the nanofabrication process was performed on glass substrates, and then, the same procedure was extrapolated to the optical fiber for further chemical sensing. The selected pH for the fabrication of the whole process was 7.0 in the dipping polyelectrolytes. As can be observed in Figure 2, the sample for thickness coating with 10 bilayers showed a clear predominance of the LSPR in relation to AgNPs (plasmonic peak centered at 450 nm), without being able to identify the peak in relation to AuNPs. However, when the thickness coating was gradually increased up to a total thickness of 30 bilayers, both LSPR peaks can be clearly observed, which were centered at 420 nm (AgNPs) and 540 nm (AuNPs), although transparent films were still obtained, which were observed by the naked eye.

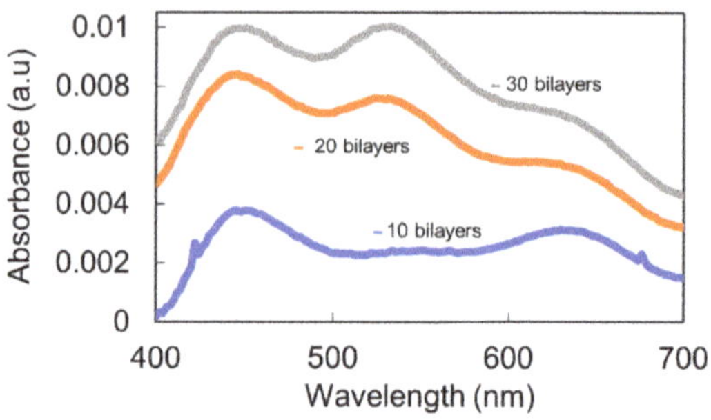

Figure 2. UV–VIS spectra of the LbL coatings based on PAH/AgNPs and PAA/AuNPs deposited on glass slides as a function of the thickness coating (10, 20, and 30 bilayers) for pH 7.0.

Once the presence of both LSPR peaks was demonstrated on glass slides, the next step was based on the deposition of this same thin film onto an optical fiber at the same pH value (87.0) in order to appreciate both absorption bands in the UV–VIS spectra. In Figure 3, it is demonstrated that by only a final thickness of 7 bilayers, it is possible to appreciate the LSPR of the AgNPs (centered at 420 nm) and AuNPs (centered at 540 nm), this sensing thin film being for the mercury ion detection.

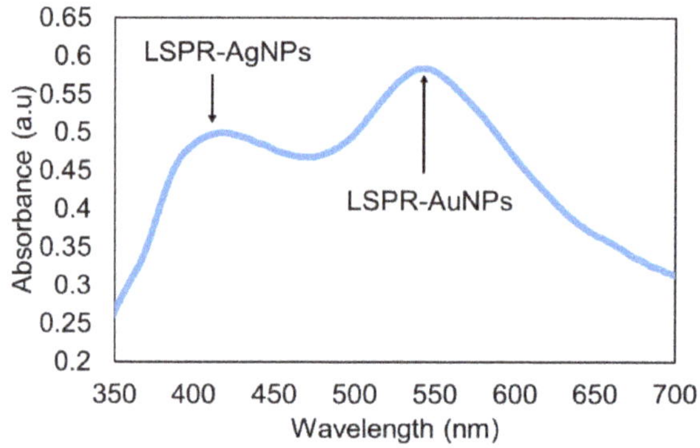

Figure 3. UV–VIS spectrum of the LbL coatings based on PAH/AgNPs and PAA/AuNPs deposited on an optical fiber for a thickness coating of 7 bilayers.

Detection of Mercury Ions with Fiber Optic Sensor

Once the thin film was fabricated, the optical fiber was immersed in the Buffer PB + DMAB solution for 1 h in order to have a stable baseline for the mercury detection stage. After that, the sensing film was immersed at a fixed mercury concentration of 50 ppm, and a very interesting result was that a clear wavelength shift of 23 nm was observed for LSPR (AuNPs), which remained stable in wavelength, compared with the wavelength shift observed in the AuNP LSPR band. According to this, LSPR-AuNPs are much more sensitive to the presence of Hg than LSPR-AgNPs (Figure 4). The UV–VIS spectra for the minimum (0.1 ppm) and maximum (50 ppm) H_2O_2 concentrations are presented in Figure 5 in order to have a better appreciation of the wavelength shift related to the LSPR-AuNPs, with the detection range being in the order of 100 ppb.

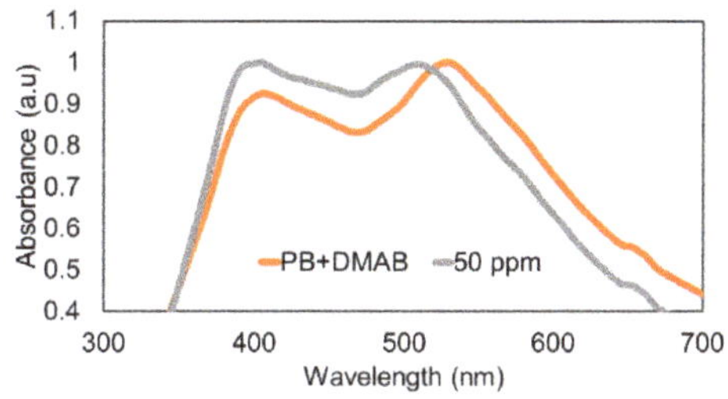

Figure 4. Wavelength shift of the LSPR absorption bands at 50 ppm of mercury concentration.

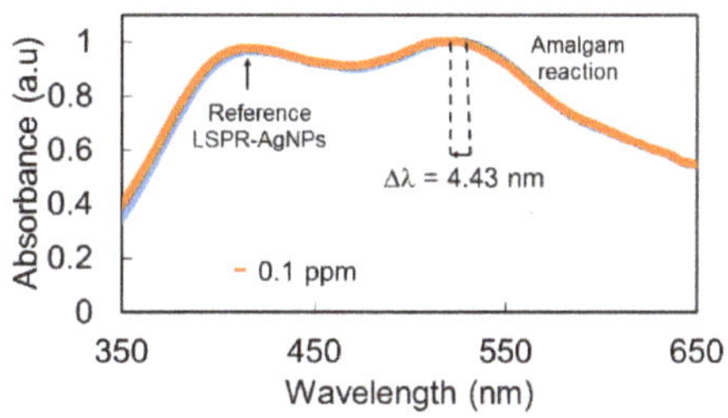

(**a**)

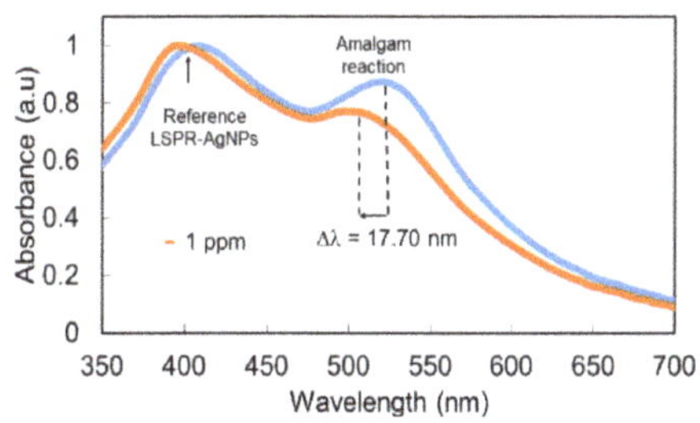

(**b**)

Figure 5. UV–VIS spectra for (**a**) the minimum concentration (0.1 ppm) and (**b**) the maximum Hg concentration (50 ppm).

The wavelength shift observed in the AuNP LSPR absorption band can be explained by the chemical reaction of the mercury present in the sample with the AuNPs. As previously reported [8], mercury ions, in the presence of DMAB of the stock buffered solution, are reduced to metallic mercury, which is known to show a high affinity for gold to form an amalgam [13]. Mercury reacts with AuNPs, changing their surface chemistry, so it is possible that their effective diameter in terms of LSPR resonances is reduced, which explains the blue shift in the LSPR band maximum. The higher reaction affinity of mercury towards gold compared with silver makes the stability of the AgNP absorption band possible, which can be used as a quite stable wavelength reference.

Different sensors were fabricated with the same sensing materials in order to detect a particular mercury concentration. Although both LSPR bands experimented changes in the presence of mercury ions, it is clearly visible that the LSPR band corresponding to AuNPs showed a greater blue shift in comparison with the LSPR of AgNPs. Finally, the dynamic response of the LSPR band inherent in AuNPs is presented in Figure 6 for different mercury concentrations (0.1, 1, and 50 ppm).

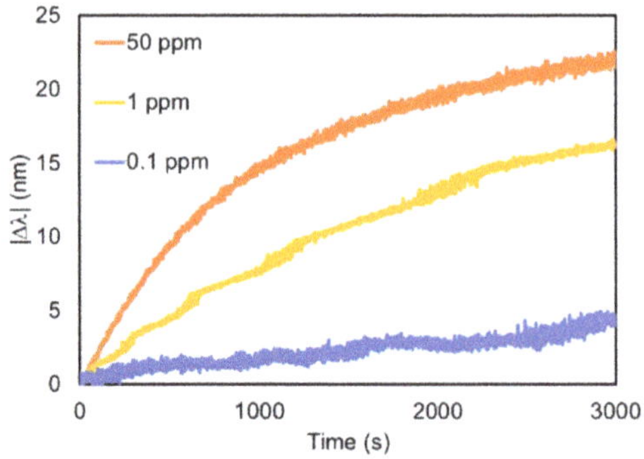

Figure 6. Dynamic response of the optical fiber sensors for the LSPR (AuNPs) to different Hg concentrations, ranging from 50 to 0.1 ppm.

4. Conclusions

In this work, a fiber optic sensor based on two different LSPR sensing signals for the detection of Hg^{2+} was presented. The metallic nanoparticles were incorporated into the sensing films by using the layer-by-layer nanoassembly technique. The sensors were exposed to different Hg^{2+} concentrations, with the wavelength response of the AuNP LSPR greater than that of the AgNP LSPR. Finally, this resultant sensing material can be extrapolated for the detection of different heavy metals in environmental applications.

Author Contributions: Conceptualization and methodology, M.E.M.-H., X.S., P.J.R., J.G. and F.J.A.; investigation and validation, M.E.M.-H. and X.S.; writing—original draft preparation, M.E.M.-H., X.S. and P.J.R.; writing—review and editing, P.J.R., J.G. and F.J.A.; supervision, P.J.R., J.G. and F.J.A.; project administration and funding acquisition, F.J.A. All authors have read and agreed to the published version of the manuscript.

Funding: This work has been supported by the Spanish State Research Agency (AEI) through the project PID2019-106070RB-I00 and the Public University of Navarre with a PhD research grant.

Data Availability Statement: Not applicable.

Conflicts of Interest: The authors declare no conflict of interest.

References

1. Järup, L. Hazards of heavy metal contamination. *Br. Med. Bull.* **2003**, *68*, 167–182. [CrossRef] [PubMed]
2. Jiang, B.; Yu, L.; Li, F.; Xie, J. A dual functional electrochemical "on-off" switch sensor for the detection of mercury(II) and melamine. *Sens. Actuators B Chem.* **2015**, *212*, 446–450. [CrossRef]
3. Zhang, Y.; Kang, S.; Zhang, Q.; Grigholm, B.; Kaspari, S.; You, Q.; Qin, D.; Mayewski, P.A.; Cong, Z.; Huang, J.; et al. A 500year atmospheric dust deposition retrieved from a Mt. Geladaindong ice core in the central Tibetan Plateau. *Atmos. Res.* **2015**, *166*, 1–9. [CrossRef]
4. Zhang, Y.; Ma, H.; Wu, D.; Li, Y.; Du, B.; Wei, Q. Label-free immunosensor based on Au@Ag2S nanoparticles/magnetic chitosan matrix for sensitive determination of ractopamine. *J. Electroanal. Chem.* **2015**, *741*, 14–19. [CrossRef]
5. Lee, J.S.; Han, M.S.; Mirkin, C.A. Colorimetric detection of mercuric ion (Hg2+) in aqueous media using DNA-functionalized gold nanoparticles. *Angew. Chem.-Int. Ed.* **2007**, *46*, 4093–4096. [CrossRef] [PubMed]
6. Coronado, E.; Galán-Mascarós, J.R.; Martí-Gastaldo, C.; Palomares, E.; Durrant, J.R.; Villar, R.; Gratzel, M.; Nazeeruddin, M.K. Reversible colorimetric probes for mercury sensing. *J. Am. Chem. Soc.* **2005**, *127*, 12351–12356. [CrossRef]
7. Chen, X.; Nam, S.W.; Jou, M.J.; Kim, Y.; Kim, S.J.; Park, S.; Yoon, J. Hg^{2+} selective fluorescent and colorimetric sensor: Its crystal structure and application to bioimaging. *Org. Lett.* **2008**, *10*, 5235–5328. [CrossRef]
8. Martínez-Hernández, M.E.; Goicoechea, J.; Arregui, F.J. Hg2+ optical fiber sensor based on LSPR generated by gold nanoparticles embedded in LBL nano-assembled coatings. *Sensors* **2019**, *19*, 4906. [CrossRef] [PubMed]
9. Cao, J.; Sun, T.; Grattan, K.T.V. Gold nanorod-based localized surface plasmon resonance biosensors: A review. *Sens. Actuators B Chem.* **2014**, *195*, 332–351. [CrossRef]
10. James, J.Z.; Lucas, D.; Koshland, C.P. Gold nanoparticle films as sensitive and reusable elemental mercury sensors. *Environ. Sci. Technol.* **2012**, *46*, 9557–9562. [CrossRef] [PubMed]

11. Caucheteur, C.; Guo, T.; Albert, J. Review of plasmonic fiber optic biochemical sensors: Improving the limit of detection. *Anal. Bioanal. Chem.* **2015**, *407*, 3883–3897. [CrossRef] [PubMed]
12. Sadani, K.; Nag, P.; Mukherji, S. LSPR based optical fiber sensor with chitosan capped gold nanoparticles on BSA for trace detection of Hg (II) in water, soil and food samples. *Biosens. Bioelectron.* **2019**, *134*, 90–96. [CrossRef] [PubMed]
13. Schopf, C.; Martín, A.; Schmidt, M.; Iacopino, D. Investigation of Au-Hg amalgam formation on substrate-immobilized individual Au nanorods. *J. Mater. Chem. C* **2015**, *3*, 8865–8872. [CrossRef]
14. Shiratori, S.S.; Rubner, M.F. pH-dependent thickness behavior of sequentially adsorbed layers of weak polyelectrolytes. *Macromolecules* **2000**, *33*, 4213–4219. [CrossRef]
15. Decher, G.; Eckle, M.; Schmitt, J.; Struth, B. Layer-by-layer assembled multicomposite films. *Curr. Opin. Colloid Interface Sci.* **1998**, *3*, 32–39. [CrossRef]
16. Rivero, P.J.; Goicoechea, J.; Matias, I.R.; Arregui, F.J. A comparative study of two different approaches for the incorporation of silver nanoparticles into layer-by-layer films. *Nanoscale Res. Lett.* **2014**, *9*, 1–11. [CrossRef] [PubMed]
17. Goicoechea, J.; Rivero, P.J.; Sada, S.; Arregui, F.J. Self-Referenced Optical Fiber Sensor for Hydrogen Peroxide Detection based on LSPR of Metallic Nanoparticles in Layer-by-Layer Films. *Sensors* **2019**, *19*, 3872. [CrossRef] [PubMed]
18. Rivero, P.J.; Ibañez, E.; Goicoechea, J.; Urrutia, A.; Matias, I.R.; Arregui, F.J. A self-referenced optical colorimetric sensor based on silver and gold nanoparticles for quantitative determination of hydrogen peroxide. *Sens. Actuators B Chem.* **2017**, *251*, 624–631. [CrossRef]

chemistry proceedings

MDPI

Proceeding Paper

Development of Graphene-Doped TiO_2-Nanotube Array-Based MIM-Structured Sensors and Its Application for Methanol Sensing at Room Temperature †

Teena Gakhar * and **Arnab Hazra**

Department of Electrical and Electronics Engineering, Birla Institute of Technology and Science BITS, Pilani 333031, India; arnabhazra2013@gmail.com
* Correspondence: gakharteena16@gmail.com

† Presented at the 1st International Electronic Conference on Chemical Sensors and Analytical Chemistry, 1–15 July 2021; Available online: https://csac2021.sciforum.net/.

Abstract: This work concerns the development of a good quality graphene doped TiO_2 nanotube array sensor for efficient detection of methanol. A pure and graphene doped TiO_2 nanotube array was synthesized by electrochemical anodization. Morphological, structural and optical characterizations were performed to study the samples. Both the nanotube samples were produced in Au/TiO_2 nanotube/Ti type MIM-structured devices. Pure and graphene-doped TiO_2 nanotubes offered a response magnitude of 20% and 28% to 100 ppm of methanol at room temperature, respectively. Response/Recovery time was fast for the graphene doped TiO_2 nanotube array (34 s/40 s) compared to a pure TiO_2 nanotube array (116 s/576 s) at room temperature. This study confirmed the notable enhancement in methanol sensing due to the formation of local heterojunctions between graphene and TiO_2 in the hybrid sample.

Keywords: methanol sensing; graphene doping; electrochemical anodization

Citation: Gakhar, T.; Hazra, A. Development of Graphene-Doped TiO_2-Nanotube Array-Based MIM-Structured Sensors and Its Application for Methanol Sensing at Room Temperature. *Chem. Proc.* **2021**, 5, 74. https://doi.org/10.3390/CSAC2021-10620

Academic Editor: Ye Zhou

Published: 6 July 2021

1. Introduction

Methanol is one of the essential organic solvents having numerous applications in the production of dyes, drugs, perfumes and colors. Moreover, it is extensively utilized in automobile fuel, wastewater denitrification and electricity generation [1]. Methanol is an extremely toxic VOC which is disastrous to human health. Repeated exposure to methanol vapors causes many problems to human beings, such as blindness, acidosis, headaches, blurred vision, shortness of breath and dizziness. Skin contact with methanol results in dermatitis or scaling and eye contact results in vision destruction [2]. With all these concerns, there is a high demand for the development of methanol sensors which are reliable, stable, and sensitive as well as able to perform at low temperatures.

Different materials like metal oxide semiconductors, polymers, carbon nanostructures, metal nanoparticles, and nanocomposites have been extensively utilized by different researchers for chemical sensing. Solid state sensors based on semiconducting metal oxides have achieved a lot in the field of chemical sensing due to their exceptional properties [3]. TiO_2 is an efficient semiconducting metal oxide which can be synthesized in different nanoforms (nanotubes, nanorods, nanoparticles and nanospheres, etc.) for different applications like photocatalysis [4], chemical sensing [5], and wastewater purification [6]. In the field of vapor sensing, the 1D-TiO_2 nanotube performs very well due to its ideal properties such as uniformity, stability and one dimensional electron flow [7]. Different researchers have applied different techniques such as the formation of TiO_2-based hybrid to improvise the performance of TiO_2-based sensors.

Two-dimensional graphene offers advanced opportunities to develop hybrids with amazing electronic catalytic behavior. The flat monolayer of graphene offers unique properties such as high surface-to-volume ratio, excessive mobility and good electrical conductivity [8]. These properties make graphene an ideal candidate to support or form hybrid with metal oxide semiconductors having high catalytic properties [9]. Some reports have been published demonstrating the sensing performance of a graphene-TiO_2-based hybrid. Fan and group described the hydrothermal production of a TiO_2-graphene nanocomposite and its implementation in electrochemical sensing. They showed electrochemical sensing of dopamine with excellent sensitivity and selectivity [10]. Ye and co-workers reported room temperature ammonia sensing by an rGO-TiO_2 hybrid. They produced the hybrid by a simple hydrothermal method [11]. Galstyan and group reported the production of an rGO-TiO_2 nanotube hybrid for hydrogen sensing. They showed the impact of GO concentration on the response of TiO_2 nanotubes [12].

In this current work, a highly aligned and uniform graphene-doped TiO_2 nanotube array was synthesized by way of electrochemical anodization for efficient detection of methanol vapors. A pure TiO_2 nanotube array and graphene-doped TiO_2 nanotube array were produced by way of electrochemical anodization. Both samples were examined and analyzed through various characterization techniques which confirmed the presence of graphene in the graphene-doped TiO_2 nanotube array. Metal-insulator-metal (MIM)-structured sensors were produced by using both pure and graphene-doped TiO_2 nanotubes. Graphene-doped TiO_2 nanotubes showed a sensitivity of 28% with quite a fast response and recovery time of 34 s and 40 s towards 100 ppm of methanol. A pure TiO_2 nanotube array, however, showed a sensitivity of 20% with relatively slow response/recovery time (116 s/576 s) in the same conditions.

2. Experimental Details

A highly ordered and oriented pure TiO_2 nanotube array and graphene-doped TiO_2 nanotube array were synthesized by electrochemical anodization route. Two-electrode anodic oxidation was performed for 120 min under 40 V potential where Ti foil was used as the anode and graphite was used as the cathode. The electrolyte was made up of 0.5 wt% of NH_4F, 10% vol of DI water and ethylene glycol. The method to synthesize the TiO_2 nanotube array was described in detail in our previous reports [13].

High purity graphene oxide suspension was used to prepare 0.2 wt% graphene oxide (GO) aqueous solution. Then an electrolyte was prepared with 0.5 wt% NH_4F, 10 vol% of GO aqueous solution and ethylene glycol for the preparation of a graphene-doped TiO_2 nanotube array. Again, the anodization was performed for 120 min by applying a constant voltage of 40 V. Due to the constant availability of GO in the electrolyte, graphene was doped uniformly in the TiO_2 nanotubes. Both the pure TiO_2 nanotube array and graphene-doped TiO_2 nanotube array were annealed for 3 h at 450 °C in ambient air. Annealing made the nanotubes more robust and stable and hence more reliable to use. The flow chart describing the steps for the synthesis of pure TiO_2 nanotube array and graphene doped TiO_2 nanotube array is represented in Figure 1.

The morphology of the produced samples was analysed by FESEM. The crystallographic structure of both the samples was examined via X-ray diffraction spectroscopy. Raman spectroscopy was performed for both samples which confirmed the doping of graphene in graphene-doped TiO_2 nanotube array (GO-TiO_2).

To produce the MIM structure for the sensors, Au was deposited on top of TiO_2 nanotube/Ti and GO-TiO_2 nanotube/Ti samples by electron beam evaporation.. 100 nm thick deposited Au was considered as the top electrode and Ti was considered as the bottom electrode. Both samples were enveloped in Cu mask to ensure 1*1mm^2 Au top electrode. A part from the corner of the TiO_2 nanotubes was etched with hydro fluoric acid to induce Ti as the bottom electrode.

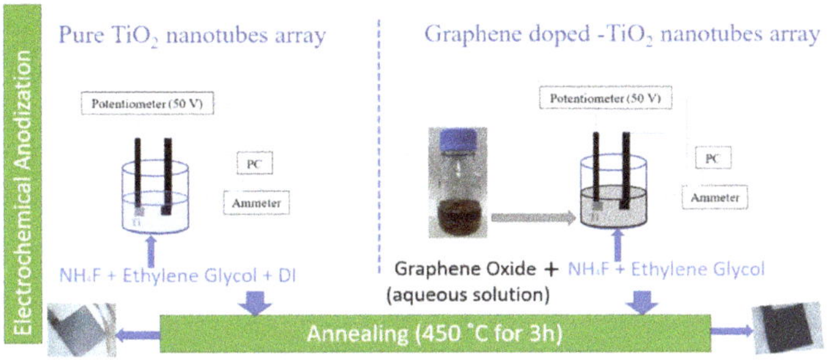

Figure 1. Flow chart describing the synthesis procedure of pure TiO_2 nanotube array and graphene-doped TiO_2 nanotube array were synthesized.

The produced sensors were tested against the methanol vapours. The sensors were examined at room temperature. The sensor setup with their properties has been discussed previously [14]. Resistance in the ambient air (R_a) and in the exposure of the reducing vapours (R_g), methanol was observed. The response magnitude is calculated as $[(R_a - R_g)/R_a] * 100$. The response time and recovery time for both sensors is defined as 90% of maximum change of the resistance when exposed to methanol vapours and exposed to synthetic air for the removal of vapors, respectively.

3. Results and Discussion

3.1. Material Characterization

FESEM confirmed the formation of highly ordered and uniform nanotubes in both the samples (Figure 2). Highly aligned nanotubes were formed with an approximate average outer diameter of 110 nm and length of 1 μm in both the pure TiO_2 nanotube array and graphene doped TiO_2 nanotube array. Graphene does not hamper the original morphology of TiO_2 nanotubes (Figure 2c,d). As graphene was uniformly doped inside the nanotubes, it was hard to observe the graphene with scanning electron microscopy. The chemical composition was studied through the EDS spectra, where the evidence of carbon is clearly visible in a GO-doped TiO_2 nanotube array (Figure 2f).

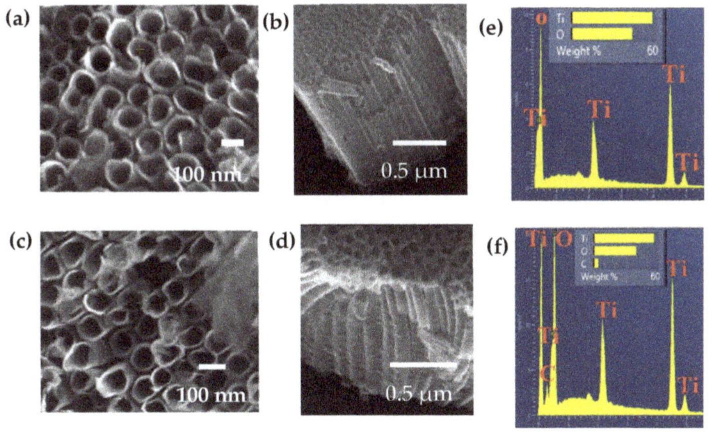

Figure 2. FESEM Image of Pure TiO_2 nanotube array (**a**) Top view, (**b**) Side view (**e**) EDS spectra and Graphene-doped TiO_2 nanotube array (**c**) Top view, (**d**) Side view (**f**) EDS spectra.

The sharp intensity peak at 25.3° in both the samples is attributed to the anatase crystallinity of TiO_2 nanotubes (Figure 3a). A low intensity peak at 52° corresponds to the anatase crystallinity A (105) in both the samples. A small peak at 54.1°, present only in the pure TiO_2 nanotube array, corresponds to A (201) and clearly shows the presence of more anatase in the pure TiO_2 nanotube array. The peaks labelled as T arise due to the use of a Titanium substrate in both the samples. T peak intensity is high in the pure TiO_2 nanotube array and less in the graphene-doped TiO_2 nanotube array in comparison to the A (101) peak.

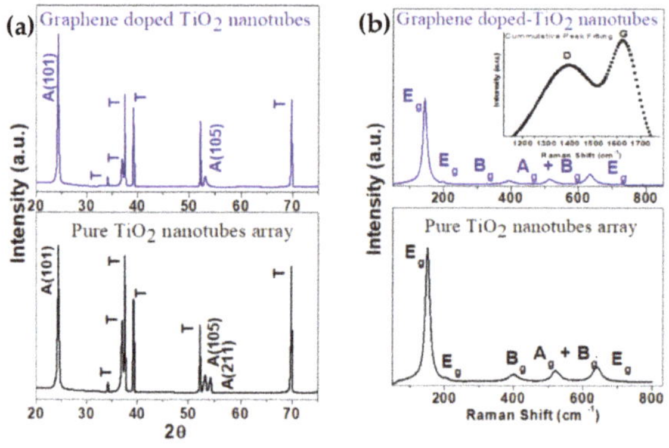

Figure 3. Pure TiO_2 nanotube array and Graphene doped TiO_2 nanotube array (**a**) XRD Spectra, (**b**) Raman spectra.

The Raman spectra of pure TiO_2 nanotube array and graphene-doped TiO_2 nanotube array is represented in Figure 3b. The presence of pure anatase is determined by six active modes E_g (144 cm^{-1}), E_g(197 cm^{-1}), B_g (399 cm^{-1}) A_g + B_g (516 cm^{-1}) and E_g (639 cm^{-1}) present in both the samples [13]. The sharp intensity peak at 144 cm^{-1} determines the formation of Ti-O in the anatase phase of TiO_2. The presence of graphene is authenticated by the sharp peaks at 1348 cm^{-1} (D band) and 1596 cm^{-1} (G band) in the graphene-doped TiO_2 nanotube array [15]. The active modes of the anatase TiO_2 nanotube and graphene were present at their corresponding positions even after the uniform doping of graphene.

3.2. Methanol Sensing

The two MIM-structure based sensors were examined against the reducing vapours, methanol, at room temperature. The resistance of the pure TiO_2 nanotube array and graphene-doped TiO_2 nanotube array was 90 MΩ and 30 MΩ, respectively. The reduced resistance (increased conductance) of graphene-doped TiO_2 nanotube array sensor clearly defines the incorporation of graphene inside a TiO_2 nanotube. Both the sensors were subjected to 100 ppm of methanol at room temperature. The response magnitude of pure the TiO_2 nanotube array and graphene-doped TiO_2 nanotube array was 20% and 28%, respectively. The pure TiO_2 nanotube array had a response time and recovery time of 116 s and 576 s, respectively. Moreover, there was the improvement in response time (34 s) and recovery time (40 s) in the case of the graphene-doped TiO_2 nanotube array (Figure 4a).

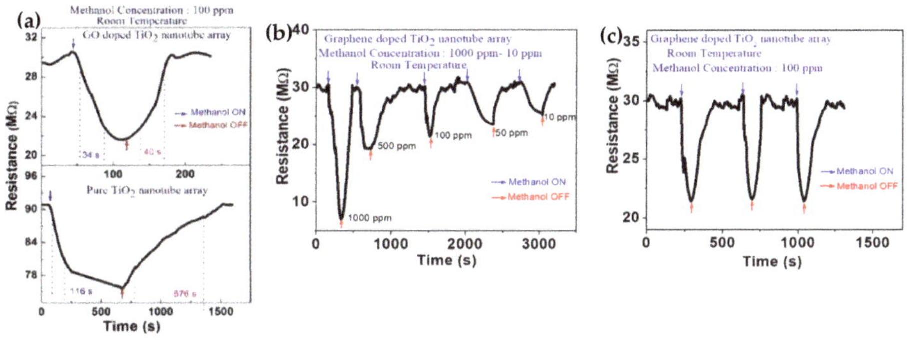

Figure 4. (**a**)Transient behavior of pure TiO_2 nanotube array sensor and graphene-doped TiO_2 nanotube array sensor in 100 ppm methanol at RT with measured response time and recovery time; Graphene-doped TiO_2 nanotube array sensor; (**b**) Transient behavior from methanol concentration range of: −1000 ppm to 10 ppm; (**c**) Repeated cycles in 100 ppm methanol at RT.

A transient was measured within a concentration range of 1000 ppm to 10 ppm for a graphene-doped TiO_2 nanotube array sensor (Figure 4b). A good response magnitude was obtained in 1000 ppm methanol (77.7%) and an average response magnitude was obtained in 10 ppm methanol (19%) at room temperature. The graphene-doped TiO_2 nanotube array sensor exhibited a stable baseline resistance with highly repeatable transient behavior at room temperature (Figure 4c). The graphene-doped TiO_2 nanotube array sensor response was improvised with short response time and recovery time due to the incorporation of graphene inside TiO_2 nanotubes.

3.3. *Methanol Sensing Mechanism*

The large surface area and two dimensional structure of graphene enhanced the sensing performance of graphene-doped TiO_2 nanotube array sensor. This increased conductance of graphene doped TiO_2 nanotube array sensor can be attributed to the large carrier mobility and high electrical conductance of graphene. The uniform doping of graphene inside the TiO_2 nanotubes improved the sensing parameters of graphene-doped TiO_2 nanotube array and enabled room temperature sensing.

An energy band diagram of both junctions was sketched by considering the work function of GO $q\varphi_{GO} \sim 4.5$ eV [16] and anatase n-TiO_2 $q\varphi_{TiO2} \sim 5.1$ eV [17]. An energy band gap of 3.59 eV for pure GO and 3.2 eV for pure TiO_2 (S_0) were estimated from a literature survey. On the formation of a heterojunction between TiO_2 and GO, electrons are transferred to TiO_2 and get accumulated on the TiO_2 surface.

$$O_2(\text{gas}) \rightarrow O_2(\text{adsorbed}) \quad (1)$$

$$O_2(\text{adsrorbed}) + e^- \rightarrow O_2^- \ (\text{adsorbed}) \quad (2)$$

$$O_2^- + e^- \rightarrow 2O^-(\text{adsorbed}) \quad (3)$$

Surface adsorption of oxygen groups ($O_2{}^-$, O^-, O^{2-}) reduces the electron concentration (Equations (1)–(3)) and increases the width of the depletion region, resulting in the formation of built-in potential on the surface of the graphene-doped TiO_2 nanotube array sensor as represented in Figure 5. Upon exposure to the methanol vapors, the trapped electron oxygen groups are released back to the surface of the graphene-doped TiO_2 nanotube array sensor, lowering the built-in potential.

$$CH_3OH + O^- \rightarrow HCOH + H_2O + e^- \quad (4)$$

$$CH_3OH + O_2^- \rightarrow HCOOH + H_2O + e^- \quad (5)$$

When methanol vapor react with the oxygen species it gets oxidised into formaldehyde, then to formic acid and then it releases electrons to the conduction band which, in turn, reduces the resistance of the sensor in exposure to methanol vapors (Equations (4) and (5)) [18].

Formation of depletion region across the TiO_2 and GO junction plays an important role for improving the sensor response. Uniform doping of graphene on the TiO_2 surface is the main reason for enhancing the change of current in-between air and VOC ambient that eventually shows high sensitivity towards methanol by the graphene doped TiO_2 nanotube sensor at room temperature with quick response time and recovery time.

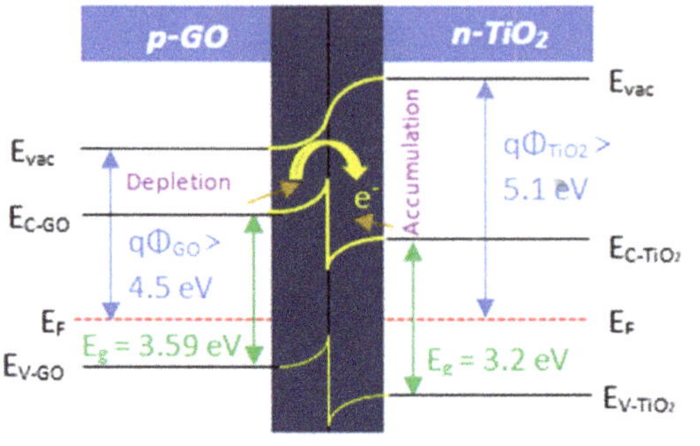

Figure 5. Heterojunction formed between *p*-type GO and n-TiO_2 nanotubes with electron depletion in GO and electron accumulation in TiO_2.

4. Conclusions

In this work, electrochemical anodization was applied to develop pure TiO_2 nanotube array and graphene-doped TiO_2 nanotube array. Graphene was doped in the TiO_2 nanotubes without hampering the original morphology of the nanotubes. Morphological characterization confirmed the formation of highly aligned and uniform nanotubes and structural characterization confirmed the anatase crystallinity of TiO_2 nanotubes in both the samples. The evidence of graphene in the hybrid nanotubes was authenticated by the D and G peaks in the Raman spectra. The pure and graphene-doped TiO_2 nanotube array sensor was produced in MIM structure where Au was considered as the top electrode and Ti was considered as the bottom electrode. Pure TiO_2 nanotube array showed a response magnitude of 20% with slow response time (116 s) and recovery time (576 s) to 100 ppm methanol at room temperature. Graphene-doped TiO_2 nanotube array showed a better response magnitude of 28% with a quick response time (34 s) and recovery time (40 s) to 100 ppm of methanol at room temperature. Also, lower detection limit till 10 ppm with good response magnitude (19%) towards methanol was achieved with the graphene-doped TiO_2 nanotube array sensor at room temperature. A significant improvement in methanol sensing was achieved by the formation of localized heterojunctions between graphene and TiO_2 in the hybrid sample.

Supplementary Materials: The following are available online at https://www.mdpi.com/article/10.3390/CSAC2021-10620/s1.

Author Contributions: Experiment, analysis and writing by T.G.; A.H. designed the research. Review editing and validation was also done by A.H. All authors have read and agreed to the published version of the manuscript.

Funding: This work was supported in part by Department of Biotechnology grant (Letter No. BT/PR28727/NNT/28/1569/ 2018), SPARC grant (SPARC/2018-2019/P1394/SL), and CSIR-SRF Direct (1431120/2k19/1) fellowship Govt. of India.

Institutional Review Board Statement: The study was conducted according to the guidelines and Deceleration of BITS Pilani.

Informed Consent Statement: Not applicable.

Data Availability Statement: Not applicable.

Conflicts of Interest: There are no conflicts to declare.

References

1. Bindra, P.; Hazra, A. Impedance behavior of n-type TiO_2 nanotubes porous layer in reducing vapor ambient. *Vacuum* **2018**, *152*, 78–83. [CrossRef]
2. Mirzaei, A.; Leonardi, S.G.; Neri, G. Detection of hazardous volatile organic compounds (VOCs) by metal oxide nanostructures-based gas sensors: A review. *Ceram. Int.* **2016**, *42*, 15119–15141. [CrossRef]
3. Bindra, P.; Gangopadhyay, S.; Hazra, A. Au/TiO_2 Nanotubes/Ti-based solid-state vapor sensor: Efficient sensing in resistive and capacitive modes. *IEEE Trans. Electron. Devices* **2018**, *65*, 1918–1924. [CrossRef]
4. Zhang, J.; Xu, Q.; Feng, Z.; Li, M.; Li, C. Importance of the relationship between surface phases and photocatalytic activity of TiO_2. *Angew. Chem.* **2008**, *120*, 1790–1793. [CrossRef]
5. Liu, Y.; Li, J.; Qiu, X.; Burda, C. Novel TiO_2 nanocatalysts for wastewater purification: Tapping energy from the sun. *Water Sci. Technol.* **2006**, *54*, 47–54. [CrossRef] [PubMed]
6. Hazra, A.; Bhowmik, B.; Dutta, K.; Chattopadhyay, P.P.; Bhattacharyya, P. Stoichiometry, length, and wall thickness optimization of TiO_2 nanotube array for efficient alcohol sensing. *ACS Appl. Mater. Interfaces* **2015**, *7*, 9336–9348. [CrossRef] [PubMed]
7. Hazra, A.; Bhattacharyya, P. Tailoring of the gas sensing performance of TiO_2 nanotubes by 1-D vertical electron transport technique. *IEEE Trans. Electron. Devices* **2014**, *61*, 3483–3489. [CrossRef]
8. Geim, A.K.; Novoselov, K.S. *The Rise of Graphene*; Manchester Centre for Mesoscience and Nanotechnology, University of Manchester: Manchester, UK, 2009.
9. Williams, G.; Seger, B.; Kamat, P.V. TiO_2-graphene nanocomposites. UV-assisted photocatalytic reduction of graphene oxide. *ACS Nano* **2008**, *2*, 1487–1491. [CrossRef] [PubMed]
10. Fan, Y.; Lu, H.T.; Liu, J.H.; Yang, C.P.; Jing, Q.S.; Zhang, Y.X.; Yang, X.K.; Huang, K.J. Hydrothermal preparation and electrochemical sensing properties of TiO_2–graphene nanocomposite. *Colloids Surf. B Biointerfaces* **2011**, *83*, 78–82. [CrossRef] [PubMed]
11. Ye, Z.; Tai, H.; Guo, R.; Yuan, Z.; Liu, C.; Su, Y.; Chen, Z.; Jiang, Y. Excellent ammonia sensing performance of gas sensor based on graphene/titanium dioxide hybrid with improved morphology. *Appl. Surf. Sci.* **2017**, *419*, 84–90. [CrossRef]
12. Galstyan, V.; Ponzoni, A.; Kholmanov, I.; Natile, M.M.; Comini, E.; Nematov, S.; Sberveglieri, G. Reduced graphene oxide–TiO_2 nanotube composite: Comprehensive study for gas-sensing applications. *ACS Appl. Nano Mater.* **2018**, *1*, 7098–7105. [CrossRef]
13. Gakhar, T.; Hazra, A. Oxygen vacancy modulation of titania nanotubes by cathodic polarization and chemical reduction routes for efficient detection of volatile organic compounds. *Nanoscale* **2020**, *12*, 9082–9093. [CrossRef] [PubMed]
14. Bindra, P.; Hazra, A. Selective detection of organic vapors using TiO_2 nanotubes based single sensor at room temperature. *Sens. Actuators B Chem.* **2019**, *290*, 684–690. [CrossRef]
15. Malard, L.M.; Pimenta, M.A.; Dresselhaus, G.; Dresselhaus, M.S. Raman spectroscopy in graphene. *Phys. Rep.* **2009**, *473*, 51–87. [CrossRef]
16. Singhal, A.V.; Charaya, H.; Lahiri, I. General synthesis of graphene-supported bicomponent metal monoxides as alternative high-performance Li-ion anodes to binary spinel oxides. *Crit. Rev. Solid State Mater. Sci.* **2017**, *42*, 1–28.
17. Scanlon, D.O.; Dunnill, C.W.; Buckeridge, J.; Shevlin, S.A.; Logsdail, A.J.; Woodley, S.M.; Catlow, C.R.A.; Powell, M.J.; Palgrave, R.G.; Parkin, I.P.; et al. Band alignment of rutile and anatase TiO_2. *Nat. Mater.* **2013**, *12*, 789. [CrossRef] [PubMed]
18. Sahay, P.P.; Nath, R.K. Al-doped ZnO thin films as methanol sensors. *Sens. Actuators B Chem.* **2008**, *134*, 654–659. [CrossRef]

chemistry proceedings

MDPI

Proceeding Paper

Morphological Effects in SnO_2 Chemiresistors for Ethanol Detection: A Systematic Statistical Analysis of Results Published in the Last 5 Years †

Andrea Ponzoni

National Institute of Optics of the National Research Council (CNR-INO), Unit of Brescia, 25123 Brescia, Italy; andrea.ponzoni@ino.cnr.it

† Presented at the 1st International Electronic Conference on Chemical Sensors and Analytical Chemistry, 1–15 July 2021; Available online: https://csac2021.sciforum.net/.

Abstract: SnO_2 is one of the most studied materials in gas sensing. Among the many strategies adopted to optimize its sensing properties, the fine tuning of the morphology in nanoparticles, nanowires, and nanosheets, as well as their eventual hierarchical organization, has become an active field of research. In this work, results published in the literature over the last five years are systematically analyzed focusing on response intensities recorded with chemiresistors based on pure SnO_2 for ethanol detection in dry air. Results indicate that no morphology clearly outperforms others, while a few individual sensors emerge as remarkable outliers with respect to the whole dataset.

Keywords: chemiresistors; SnO_2; ethanol; nanoparticles; nanorods; nanosheets

Citation: Ponzoni, A. Morphological Effects in SnO_2 Chemiresistors for Ethanol Detection: A Systematic Statistical Analysis of Results Published in the Last 5 Years. *Chem. Proc.* **2021**, *5*, 75. https://doi.org/10.3390/CSAC2021-10474

Academic Editor: Ye Zhou

Published: 30 June 2021

1. Introduction

Chemiresistors based on semiconducting metal oxides are among the most popular gas sensing devices. Their success comes from their high sensitivity to a broad range of chemicals, their reduced size and power consumption, and their suitability for mass production at relatively reduced costs. To optimize the sensing layer, the fine control of the morphology, both at the level of individual nanostructures and at the level of their hierarchical assembly, has been reported as very effective [1,2].

In this work, with the aim to have a more general and reliable picture of the state of the art, results published in the literature in the last five years are systematically analyzed, focusing on response intensities recorded with chemiresistors based on pure SnO_2 for ethanol detection in dry air, as the case example. In particular, we chose to focus on SnO_2 because it is the most studied material among semiconducting metal oxides. Similarly, we chose ethanol as target gas because it is widely used as a test gas for the development of innovative materials (morphologies) and it is a key component in many applications [3].

2. Materials and Methods

This work considers the responses to ethanol reported for chemiresistors based on pure SnO_2 in the period from January 2015 to July 2020. In order to have a common background between all the considered responses, only dry air tests have been taken into account.

The morphology of the SnO_2 layer is described at two different levels: at the level of individual nanostructures and the level of their eventual hierarchical assembly.

Concerning the shape of individual crystallites composing the sensing layer, it has been categorized as follows:

- Nanorods: elongated nanostructures with a high aspect-ratio, and surfaces identified by well-defined crystalline planes;
- Nanoparticles: spherical nanostructures, such as those used in thick films;

- Nanosheets: thin nanostructures extending in two dimensions.

3. Results

As an example of the shape of elementary nanostructures widely investigated in the literature, Figure 1 reports the SEM images for two SnO_2 layers composed by a disordered network of nanowires (Figure 1a), and by a disordered network of nanoparticles (Figure 1b) [1]. Therefore, some nanoparticles are distributed over the substrate individually, while others are distributed in μm-sized grains as a consequence of aggregation often observed in nanoparticle-based layers [1].

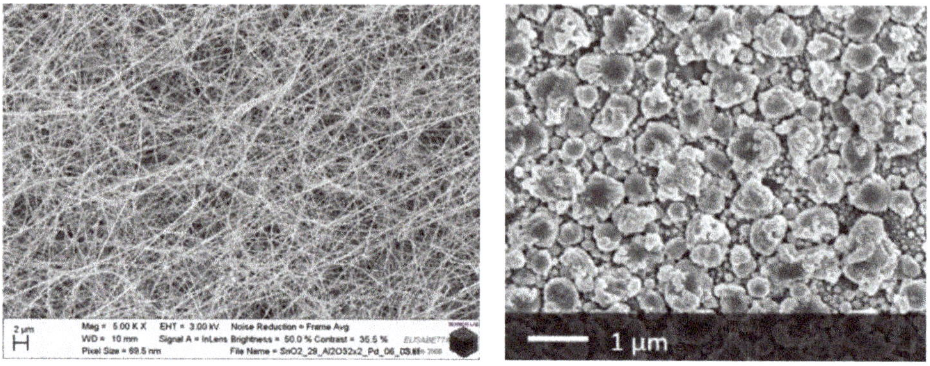

(a) (b)

Figure 1. Examples of two different morphologies investigated in the literature for SnO_2-based chemiresistors. (**a**) Film composed by a disordered network of SnO_2 nanowires; (**b**) film composed by a disordered network of SnO_2 nanoparticles, which are distributed either individually or in μm-sized aggregates. Reprinted from [1].

Boxplots resuming the responses to 10 ppm and to 300 ppm of ethanol reported in literature are shown in Figure 2a,b, respectively, grouping the results by nanostructure morphologies, namely nanorods, nanoparticles, and nanosheets.

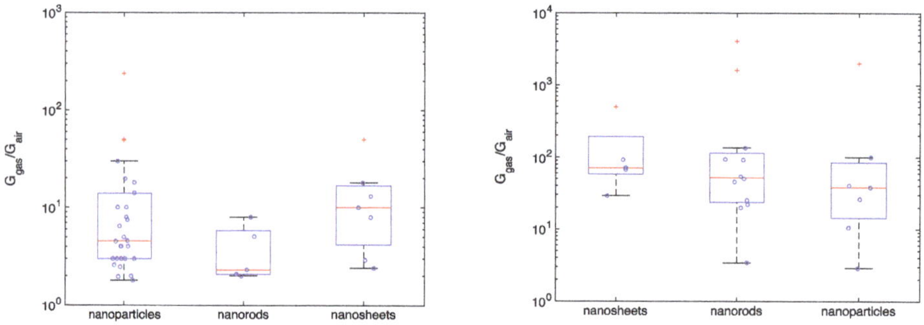

(a) (b)

Figure 2. Boxplots resuming the statistics of the response intensities of SnO_2 chemiresistors grouped by crystallite shape. (**a**) Statistics recorded vs. 10 ppm of ethanol; (**b**) statistics recorded vs. 300 ppm of ethanol.

The statistical parameters describing these distributions are reported in Tables 1 and 2 for data shown in Figure 2a,b, respectively.

Table 1. Statistics of data shown in Figure 1a (responses to 10 ppm of ethanol).

	Nanoparticles	Nanorods	Nanosheets
Number of samples [Refs.]	30 [4–24]	5 [25–29]	7 [22,30–34]
Number of outliers	4	0	1
G_{gas}/G_{air}, Q1	3	2.075	4.175
G_{gas}/G_{air}, Q2 (median)	4.55	2.3	10
G_{gas}/G_{air}, Q3	14	5.825	16.75
G_{gas}/G_{air}, whisker low	1.8	2	2.4
G_{gas}/G_{air}, whisker up	30	8	18
p-value median test, nanorods	NaN	0.68	0.18
p-value median test, nanoparticles	0.68	NaN	0.08
p-value median test, nanosheets	0.18	0.08	NaN

Table 2. Statistics of data shown in Figure 1b (responses to 300 ppm of ethanol).

	Nanosheets	Nanorods	Nanoparticles
Number of samples [Refs.]	5 [30,35–38]	12 [35–41]	7 [35,37,40,42–45]
Number of outliers	1	2	1
G_{gas}/G_{air}, Q1	58.25	23.5	14.375
G_{gas}/G_{air}, Q2 (median)	71	52	38
G_{gas}/G_{air}, Q3	193	115	85
G_{gas}/G_{air}, whisker low	29	3.4	2.9
G_{gas}/G_{air}, whisker up	93	135	100
p-value median test, nanorods	NaN	0.5	0.08
p-value median test, nanoparticles	0.5	NaN	0.21
p-value median test, nanosheets	0.08	0.21	NaN

Statistical parameters reported in these tables are: the number of samples considered in each category (morphology of elementary nanostructures); the number of outliers identified for each category; the values of the 1st, 2nd, and 3rd quartiles (Q1, Q2, and Q3) of the response amplitude G_{gas}/G_{air}; and the values of the upper and lower whiskers. The p-value of the median test comparing the median response of morphologies two by two are also reported in order to have a statistical check about the similarity and dissimilarity between median responses of the different morphologies.

4. Discussion

The distributions of the response intensities shown in Figure 2 depend on the gas concentration. This is partially due to the fact that different authors often tested their sensors against different ethanol concentration so there is no a complete overlap between concentration used in different articles. In other words, the sensors whose response is shown in Figure 2a are not exactly the same sensors whose response is shown in Figure 2b. Nonetheless, despite these differences, a common feature is that no morphology clearly performs better than other morphologies. Median tests reported in Tables 1 and 2 feature a p-value that is larger than 0.05 in all situations. This means that there is no clear evidence to reject the null hypothesis, i.e., there is no clear evidence to reject the hypothesis that the couple of morphologies under the test are not distinguishable. The same is observed for other concentrations and also considers the eventual hierarchical organization of the individual nanostructures into assemblies, such as hollow spheres, fibers, hollow fibers, etc. [46]. On the other hand, some materials emerge as outliers with respect to all morphologies. In Figure 2a, there are five outliers: four are the responses from layers composed by nanoparticles, namely [4–7] with response intensities of about 236, 50, 49, and 50 (to 10 ppm of ethanol), and one composed by nanosheets [33] featuring a response $G_{gas}/G_{air} \approx 50$. As a reference, the median responses to this ethanol concentration are around 4.55, 2.3, and 10 for nanoparticles, nanorods, and nanosheets, respectively. Concerning the concentration of 300 ppm, four outliers emerges: the nanoparticles synthesized by [45], and two types of nanorods and the nanosheets developed by [38]. These materials feature responses of about 2000, 4070, 1609, and 495, compared with the median responses of 71, 52, and 38 for nanosheets, nanorods, and nanoparticles, respectively.

These results are arguably due to the longer tradition of the synthesis of nanoparticles with respect to those of nanowires and nanosheets. Such a longer experience may reasonably imply a more developed capability to effectively combine the many parameters underlying the sensing mechanism, which may counterbalance the advantages arising from the fine morphological tuning inherent in the more recent nanostructures.

Funding: His research was funded by Regione Lombardia and Fondazione Cariplo through the project EMPATIA@LECCO.

Institutional Review Board Statement: Not applicable.

Informed Consent Statement: Not applicable.

Data Availability Statement: All data are in the reported figures and in the cited references.

Conflicts of Interest: The authors declare no conflict of interest.

References

1. Ponzoni, A.; Comini, E.; Concina, I.; Ferroni, M.; Falasconi, M.; Gobbi, E.; Sberveglieri, V.; Sberveglieri, G. Nanostructured Metal Oxide Gas Sensors, a Survey of Applications Carried out at SENSOR Lab, Brescia (Italy) in the Security and Food Quality Fields. *Sensors* **2012**, *12*, 17023–17045. [CrossRef] [PubMed]
2. Lee, J.-H. Gas Sensors Using Hierarchical and Hollow Oxide Nanostructures: Overview. *Sens. Actuators B Chem.* **2009**, *140*, 319–336. [CrossRef]
3. Palma, S.I.C.J.; Traguedo, A.P.; Porteira, A.R.; Frias, M.J.; Gamboa, H.; Roque, A.C.A. Machine Learning for the Meta-Analyses of Microbial Pathogens' Volatile Signatures. *Sci. Rep.* **2018**, *8*, 3360. [CrossRef] [PubMed]
4. Lee, S.-H.; Galstyan, V.; Ponzoni, A.; Gonzalo-Juan, I.; Riedel, R.; Dourges, M.-A.; Nicolas, Y.; Toupance, T. Finely Tuned SnO_2 Nanoparticles for Efficient Detection of Reducing and Oxidizing Gases: The Influence of Alkali Metal Cation on Gas-Sensing Properties. *ACS Appl. Mater. Interfaces* **2018**, *10*, 10173–10184. [CrossRef] [PubMed]
5. Tricoli, A.; Pratsinis, S.E. Dispersed Nanoelectrode Devices. *Nat. Nanotechnol.* **2010**, *5*, 54–60. [CrossRef] [PubMed]
6. Li, H.; Chu, S.; Ma, Q.; Li, H.; Che, Q.; Wang, J.; Wang, G.; Yang, P. Multilevel Effective Heterojunctions Based on SnO_2/ZnO 1D Fibrous Hierarchical Structure with Unique Interface Electronic Effects. *ACS Appl. Mater. Interfaces* **2019**, *11*, 31551–31561. [CrossRef] [PubMed]
7. Tofighi, G.; Degler, D.; Junker, B.; Müller, S.; Lichtenberg, H.; Wang, W.; Weimar, U.; Barsan, N.; Grunwaldt, J.-D. Microfluidically Synthesized Au, Pd and AuPd Nanoparticles Supported on SnO_2 for Gas Sensing Applications. *Sens. Actuators B Chem.* **2019**, *292*, 48–56. [CrossRef]
8. Wang, T.; Jiang, B.; Yu, Q.; Kou, X.; Sun, P.; Liu, F.; Lu, H.; Yan, X.; Lu, G. Realizing the Control of Electronic Energy Level Structure and Gas-Sensing Selectivity over Heteroatom-Doped $In2O_3$ Spheres with an Inverse Opal Microstructure. *ACS Appl. Mater. Interfaces* **2019**, *11*, 9600–9611. [CrossRef] [PubMed]
9. Zhang, Y.; He, X.; Li, J.; Miao, Z.; Huang, F. Fabrication and Ethanol-Sensing Properties of Micro Gas Sensor Based on Electrospun SnO_2 Nanofibers. *Sens. Actuators B Chem.* **2008**, *132*, 67–73. [CrossRef]
10. Tricoli, A.; Righettoni, M.; Pratsinis, S.E. Minimal Cross-Sensitivity to Humidity during Ethanol Detection by SnO_2–TiO_2 Solid Solutions. *Nanotechnology* **2009**, *20*, 315502. [CrossRef]
11. Kassem, O.; Saadaoui, M.; Rieu, M.; Viricelle, J.-P. A Novel Approach to a Fully Inkjet Printed SnO_2-Based Gas Sensor on a Flexible Foil. *J. Mater. Chem. C* **2019**, *7*, 12343–12353. [CrossRef]
12. Kotchasak, N.; Wisitsoraat, A.; Tuantranont, A.; Phanichphant, S.; Yordsri, V.; Liewhiran, C. Highly Sensitive and Selective Detection of Ethanol Vapor Using Flame-Spray-Made CeOx-Doped SnO_2 Nanoparticulate Thick Films. *Sens. Actuators B Chem.* **2018**, *255*, 8–21. [CrossRef]
13. Li, X.; Peng, K.; Dou, Y.; Chen, J.; Zhang, Y.; An, G. Facile Synthesis of Wormhole-Like Mesoporous Tin Oxide via Evaporation-Induced Self-Assembly and the Enhanced Gas-Sensing Properties. *Nanoscale Res. Lett.* **2018**, *13*, 14. [CrossRef]
14. Li, R.; Chen, S.; Lou, Z.; Li, L.; Huang, T.; Song, Y.; Chen, D.; Shen, G. Fabrication of Porous SnO_2 Nanowires Gas Sensors with Enhanced Sensitivity. *Sens. Actuators B Chem.* **2017**, *252*, 79–85. [CrossRef]
15. Xue, N.; Zhang, Q.; Zhang, S.; Zong, P.; Yang, F. Highly Sensitive and Selective Hydrogen Gas Sensor Using the Mesoporous SnO2 Modified Layers. *Sensors* **2017**, *17*, 2351. [CrossRef]
16. Zito, C.A.; Perfecto, T.M.; Volanti, D.P. Impact of Reduced Graphene Oxide on the Ethanol Sensing Performance of Hollow SnO_2 Nanoparticles under Humid Atmosphere. *Sens. Actuators B Chem.* **2017**, *244*, 466–474. [CrossRef]

17. Li, S.-H.; Meng, F.-F.; Chu, Z.; Luo, T.; Peng, F.-M.; Jin, Z. Mesoporous SnO_2 Nanowires: Synthesis and Ethanol Sensing Properties. *Adv. Condens. Matter Phys.* **2017**, *2017*, 1–6. [CrossRef]
18. Palla Papavlu, A.; Mattle, T.; Temmel, S.; Lehmann, U.; Hintennach, A.; Grisel, A.; Wokaun, A.; Lippert, T. Highly Sensitive SnO_2 Sensor via Reactive Laser-Induced Transfer. *Sci. Rep.* **2016**, *6*, 25144. [CrossRef]
19. Liu, J.; Dai, M.; Wang, T.; Sun, P.; Liang, X.; Lu, G.; Shimanoe, K.; Yamazoe, N. Enhanced Gas Sensing Properties of SnO_2 Hollow Spheres Decorated with CeO_2 Nanoparticles Heterostructure Composite Materials. *ACS Appl. Mater. Interfaces* **2016**, *8*, 6669–6677. [CrossRef]
20. Naik, A.; Parkin, I.; Binions, R. Gas Sensing Studies of an N-n Hetero-Junction Array Based on SnO_2 and ZnO Composites. *Chemosensors* **2016**, *4*, 3. [CrossRef]
21. Tan, W.; Yu, Q.; Ruan, X.; Huang, X. Design of SnO_2-Based Highly Sensitive Ethanol Gas Sensor Based on Quasi Molecular-Cluster Imprinting Mechanism. *Sens. Actuators B Chem.* **2015**, *212*, 47–54. [CrossRef]
22. Lou, Z.; Wang, L.; Wang, R.; Fei, T.; Zhang, T. Synthesis and Ethanol Sensing Properties of SnO_2 Nanosheets via a Simple Hydrothermal Route. *Solid-State Electron.* **2012**, *76*, 91–94. [CrossRef]
23. Francioso, L.; De Pascali, C.; Creti, P.; Radogna, A.V.; Capone, S.; Taurino, A.; Epifani, M.; Baldacchini, C.; Bizzarri, A.R.; Siciliano, P.A. Nanogap Sensors Decorated with SnO_2 Nanoparticles Enable Low-Temperature Detection of Volatile Organic Compounds. *ACS Appl. Nano Mater.* **2020**, *3*, 3337–3346. [CrossRef]
24. Zhang, L.; Tong, R.; Ge, W.; Guo, R.; Shirsath, S.E.; Zhu, J. Facile One-Step Hydrothermal Synthesis of SnO_2 Microspheres with Oxygen Vacancies for Superior Ethanol Sensor. *J. Alloy Compd.* **2020**, *814*, 152266. [CrossRef]
25. Van Hieu, N.; Kim, H.-R.; Ju, B.-K.; Lee, J.-H. Enhanced Performance of SnO_2 Nanowires Ethanol Sensor by Functionalizing with La_2O_3. *Sens. Actuators B Chem.* **2008**, *133*, 228–234. [CrossRef]
26. Nguyen, K.; Hung, C.M.; Ngoc, T.M.; Thanh Le, D.T.; Nguyen, D.H.; Nguyen Van, D.; Nguyen Van, H. Low-Temperature Prototype Hydrogen Sensors Using Pd-Decorated SnO_2 Nanowires for Exhaled Breath Applications. *Sens. Actuators B Chem.* **2017**, *253*, 156–163. [CrossRef]
27. Choi, K.S.; Park, S.; Chang, S.-P. Enhanced Ethanol Sensing Properties Based on SnO_2 Nanowires Coated with Fe_2O_3 Nanoparticles. *Sens. Actuators B Chem.* **2017**, *238*, 871–879. [CrossRef]
28. Wang, Q.; Yao, N.; An, D.; Li, Y.; Zou, Y.; Lian, X.; Tong, X. Enhanced Gas Sensing Properties of Hierarchical SnO_2 Nanoflower Assembled from Nanorods via a One-Pot Template-Free Hydrothermal Method. *Ceram. Int.* **2016**, *42*, 15889–15896. [CrossRef]
29. Cai, Z.; Park, S. Enhancement Mechanisms of Ethanol-Sensing Properties Based on Cr_2O_3 Nanoparticle-Anchored SnO_2 Nanowires. *J. Mater. Res. Technol.* **2020**, *9*, 271–281. [CrossRef]
30. Cui, Y.; Zhang, M.; Li, X.; Wang, B.; Wang, R. Investigation on Synthesis and Excellent Gas-Sensing Properties of Hierarchical Au-Loaded SnO_2 Nanoflowers. *J. Mater. Res.* **2019**, *34*, 2944–2954. [CrossRef]
31. Wang, B.; Sun, L.; Wang, Y. Template-Free Synthesis of Nanosheets-Assembled SnO_2 Hollow Spheres for Enhanced Ethanol Gas Sensing. *Mater. Lett.* **2018**, *218*, 290–294. [CrossRef]
32. Zito, C.A.; Perfecto, T.M.; Volanti, D.P. Palladium-Loaded Hierarchical Flower-like Tin Dioxide Structure as Chemosensor Exhibiting High Ethanol Response in Humid Conditions. *Adv. Mater. Interfaces* **2017**, *4*, 1700847. [CrossRef]
33. Zhou, Q.; Chen, W.; Li, J.; Tang, C.; Zhang, H. Nanosheet-Assembled Flower-like SnO_2 Hierarchical Structures with Enhanced Gas-Sensing Performance. *Mater. Lett.* **2015**, *161*, 499–502. [CrossRef]
34. Xu, M.-H.; Cai, F.-S.; Yin, J.; Yuan, Z.-H.; Bie, L.-J. Facile Synthesis of Highly Ethanol-Sensitive SnO_2 Nanosheets Using Homogeneous Precipitation Method. *Sens. Actuators B Chem.* **2010**, *145*, 875–878. [CrossRef]
35. Cao, S.; Zeng, W.; Zhu, Z.; Peng, X. Synthesis of SnO_2 Nanostructures from 1D to 3D via a Facile Hydrothermal Method and Their Gas Sensing Properties. *J. Mater. Sci. Mater. Electron.* **2015**, *26*, 1820–1826. [CrossRef]
36. Zeng, W.; Zhang, H.; Li, Y.; Chen, W.; Wang, Z. Hydrothermal Synthesis of Hierarchical Flower-like SnO_2 Nanostructures with Enhanced Ethanol Gas Sensing Properties. *Mater. Res. Bull.* **2014**, *57*, 91–96. [CrossRef]
37. Zeng, W.; He, Q.; Pan, K.; Wang, Y. Synthesis of Multifarious Hierarchical Flower-like SnO_2 and Their Gas-Sensing Properties. *Phys. E Low-Dimens. Syst. Nanostructures* **2013**, *54*, 313–318. [CrossRef]
38. Firooz, A.A.; Mahjoub, A.R.; Khodadadi, A.A. Highly Sensitive CO and Ethanol Nanoflower-like SnO_2 Sensor among Various Morphologies Obtained by Using Single and Mixed Ionic Surfactant Templates. *Sens. Actuators B Chem.* **2009**, *141*, 89–96. [CrossRef]
39. Zhang, B.; Fu, W.; Meng, X.; Ruan, A.; Su, P.; Yang, H. Enhanced Ethanol Sensing Properties Based on Spherical-Coral-like SnO_2 Nanorods Decorated with α-Fe_2O_3 Nanocrystallites. *Sens. Actuators B Chem.* **2018**, *261*, 505–514. [CrossRef]
40. Hoa, L.T.; Cuong, N.D.; Hoa, T.T.; Khieu, D.Q.; Long, H.T.; Quang, D.T.; Hoa, N.D.; Hieu, N.V. Synthesis, Characterization, and Comparative Gas Sensing Properties of Tin Dioxide Nanoflowers and Porous Nanospheres. *Ceram. Int.* **2015**, *41*, 14819–14825. [CrossRef]
41. Wang, X.; Han, X.; Xie, S.; Kuang, Q.; Jiang, Y.; Zhang, S.; Mu, X.; Chen, G.; Xie, Z.; Zheng, L. Controlled Synthesis and Enhanced Catalytic and Gas-Sensing Properties of Tin Dioxide Nanoparticles with Exposed High-Energy Facets. *Chem. Eur. J.* **2012**, *18*, 2283–2289. [CrossRef]
42. Punginsang, M.; Wisitsoraat, A.; Sriprachuabwong, C.; Phokharatkul, D.; Tuantranont, A.; Phanichphant, S.; Liewhiran, C. Roles of Cobalt Doping on Ethanol-Sensing Mechanisms of Flame-Spray-Made SnO_2 Nanoparticles—electrolytically Exfoliated Graphene Interfaces. *Appl. Surf. Sci.* **2017**, *425*, 351–366. [CrossRef]

43. Asgari, M.; Saboor, F.H.; Mortazavi, Y.; Khodadadi, A.A. SnO_2 Decorated SiO_2 Chemical Sensors: Enhanced Sensing Performance toward Ethanol and Acetone. *Mater. Sci. Semicond. Process.* **2017**, *68*, 87–96. [CrossRef]
44. Li, M.; Zhu, H.; Cheng, J.; Zhao, M.; Yan, W. Synthesis and Improved Ethanol Sensing Performance of CuO/SnO_2 Based Hollow Microspheres. *J. Porous Mater.* **2017**, *24*, 507–518. [CrossRef]
45. NaderiNasrabadi, M.; Mortazavi, Y.; Khodadadi, A.A. Highly Sensitive and Selective Gd_2O_3-Doped SnO_2 Ethanol Sensors Synthesized by a High Temperature and Pressure Solvothermal Method in a Microreactor. *Sens. Actuators B Chem.* **2016**, *230*, 130–139. [CrossRef]
46. Ponzoni, A. Morphological Effects in SnO_2 Chemiresistors for Ethanol Detection: A Review in Terms of Central Performances and Outliers. *Sensors* **2021**, *21*, 29. [CrossRef]

chemistry proceedings

MDPI

Proceeding Paper

Review of the Recent Advances in Nano-Biosensors and Technologies for Healthcare Applications †

Maha Wajeeh Aqra [1] and Amall Ahmed Ramanathan [2,*]

[1] National Center for Research and Development, Amman 11941, Jordan; maha.aqra@ncrd.gov.jo
[2] Department of Physics, University of Jordan, Amman 11942, Jordan
* Correspondence: amallahmad@gmail.com

† Presented at the 1st International Electronic Conference on Chemical Sensors and Analytical Chemistry, 1–15 July 2021; Available online: https://csac2021.sciforum.net/.

Abstract: The growing human population and the discovery of new diseases and emerging pandemics have increased the need for healthcare treatments and medications with innovative designs. The emergence of nanotechnology provides a platform for novel diagnostic and therapeutic in vivo non-invasive detection and treatment of ailments. It is now the era of the Internet of things (IoT), and data acquisition and interpretation from various parts of the human body in real time is possible with interconnected sensors and information transfer devices. Miniaturization, low power consumption and price with compatibility to existing network circuits are essential requirements in the IoT. Biosensors made of nanostructured materials are the ideal choice due to the unique structural, chemical and electronic properties of these materials with the advantage of a large surface-to-volume ratio, which makes them very successful for use as sensors for the detection of diseases, drug carriers, filters, fillers and reaction catalysts in healthcare applications. In this paper, we reviewed the recent progress made in the research and applications of biosensors in health and preventive medicine. The focus of the paper is biosensors made of nanostructured layered materials such as graphene and its structural analogs molybdenum disulphide (MoS_2) and boron nitride (BN). We discussed and highlighted the present capabilities of the different nano-forms of these materials in the detection and analysis of diseases. Their efficiencies in terms of the detection limit, the sensitivity and the adaptability to different environments were be discussed. In addition, the challenges and future perspectives of using nano-biosensors to develop efficient diagnostic, therapeutic and cost-effective monitoring devices with smart technologies were explored.

Keywords: electronic tongues and noses; 2D materials; nanopores; preventive medicine; non-invasive

check for updates

Citation: Aqra, M.W.; Ramanathan, A.A. Review of the Recent Advances in Nano-Biosensors and Technologies for Healthcare Applications. *Chem. Proc.* **2022**, *5*, 76. https://doi.org/10.3390/CSAC2021-10473

Academic Editor: Xin Wu

Published: 30 June 2021

1. Introduction

The detection of biological molecules, ions or species of interest (analyte) through the measurement and analysis of signals proportional to the concentration of the analyte is the basic function of a biosensor. The biological/chemical information needs to be transformed into readable outputs through the transducer. Biosensors used in the detection and prevention of diseases need to be non-invasive, highly selective, flexible and sensitive [1–3]. In addition in order to acquire and interpret signals from different parts of the body with interconnected or multifunctional sensors, the sensor design needs to be innovative and compatible with smart technologies that can transfer data with a high speed and accuracy [4,5]. Moreover, several constraints such as biocompatibility, reliability, stability, comfort, convenience, miniaturization and costs need to be considered [6]. The last decade has seen tremendous research on Two dimensional materials such as grapheme (Gr), graphene oxide (GrO) and molybdenum disulphide (MoS_2) in different nano-forms for sensing applications in the healthcare, environment and other sectors [7–14].

Gr as the first 2D material discovered with its one-atomic-layer honeycomb structure has remarkable electronic, mechanical and optical properties and has seen a multitude of applications [15–17]. Gr analogs MoS_2 and boron nitride (BN) also have a honeycomb lattice and a layered structure that allow for the easy fabrication of 2D and other nanostructures due to the weak inter-layer van der Waal interactions.

A lot of research has been going on in the area of Gr and beyond Gr nanomaterials (NMs) during the past decade, and it is necessary to put into perspective and highlight the progress of Gr, MoS_2 and BN nanostructures in biosensing for the healthcare sector. This is a rapidly changing and highly researched field with new discoveries and innovation and requires the frequent updates of progresses and challenges. This motivated us to present a focused review with the literature survey of the recent developments (last five years) on Gr and its structural analogs MoS_2 and BN nanostructures in the detection and analysis of diseases in terms of efficiency, detection limits, sensitivity and adaptability to different environments. We discussed and highlighted the present capabilities of the different nano-forms of these materials. In addition, the challenges and future perspectives of using nano-biosensors to develop efficient diagnostic, therapeutic and cost-effective monitoring devices with smart technologies for healthcare and preventive medicine were explored. The article was arranged under the main headings: Introduction, Nano-Biosensors, Smart Technologies and Challenges or Opportunities.

2. Nano-Biosensors

2.1. Biosensor Types

Biosensors are two component devices consisting of a receptor and a transducer. The receptor is a biological recognition element which could be an enzyme, micro-organism, tissue, antibody or nucleic acid. The transducer converts the physiochemical change due to the interaction of the analyte with the receptor into an analytical output signal, which is coupled to an appropriate data-processing system. A schematic diagram of the process is shown in Figure 1.

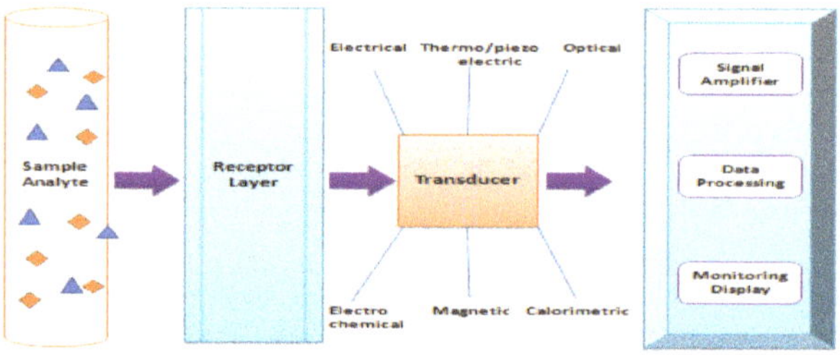

Figure 1. Schematics of a biosensor unit.

Electrical, optical, electrochemical, micromechanical, calorimetric, magnetic, thermoelectric and piezoelectric transducers can be employed in biosensors, and the choice depends on the sensing environment and needs. Materials have been researched widely by the materials science community for use to fabricate the best-suited biosensor.

2.2. Nanostructured Materials for Biosensing

Gr and its analogs such as MoS_2 and BN NMs have been the best materials for biosensing so far. The unique layered and honeycomb structure of these materials allows for the easy synthesis of monolayers (MLs), bilayers (BLs), nano-flakes, nanotubes and hetrostructures with a wide range of bandgaps and a diverse variety of optoelectronic properties. In addition, due to the weak inter-layer van der Waals forces, one can intercalate with atoms of different species and functionalize them easily to obtain the desired properties at will; moreover, NMs have the advantage of a large surface-to-volume ratio, which is important in the efficient immobilization of receptors on the surface of the NMs for good sensor performance [18]. All these factors make them prime candidates for use as biosensors in healthcare applications. In Figure 2, we give a graphical representation of NMs for biosensing that best describes the scope of this review.

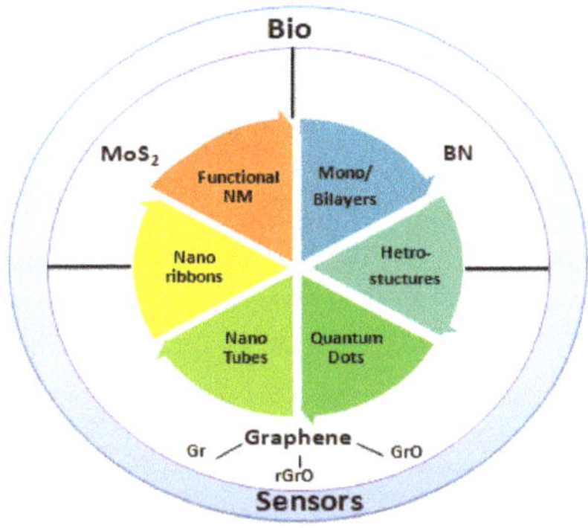

Figure 2. Graphical representation of nanostructured materials for biosensing.

2.2.1. Gr Nano-Biosensors

A Gr layer has a hexagonal symmetry with a honeycomb structure, and in-plane C atoms are bonded by strong covalent sp^2 bonds with the nearest neighbours and an out-of-plane delocalised π bond, as shown in Figure 3a. It is the delocalised π electrons that are responsible for the extremely high room temperature mobility of 15,000–200,000 cm V^{-1} s^{-1} [19]. Moreover, Gr has an excellent mechanical strength on account of the strong covalent bonding and is optically transparent and highly flexible [20,21].

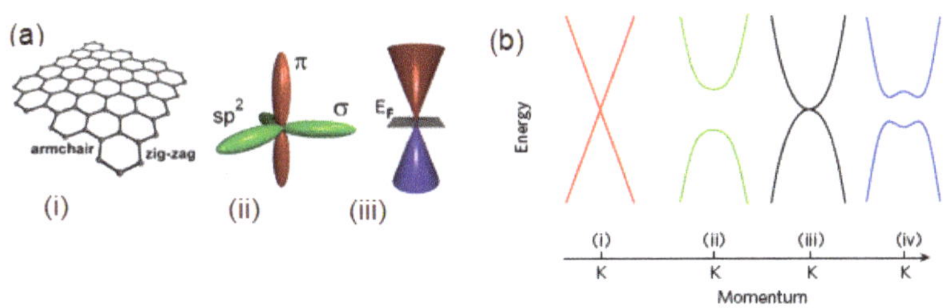

Figure 3. (**a**) Graphene geometry (**i**), bonding (**ii**) and the related band diagram (**iii**) [19]. (**b**) Schematic diagram showing the Dirac Fermi cone (**i**), the modification of the band by chemical or geometry restrictive doping (**ii**), the modification of the band by bilayer graphene (**iii**) and, finally, the modification of the bands in doped bilayer graphene (**iv**) [22].

The high electrical and thermal conductivity, mechanical strength, flexibility, optical transparency and ultrathin feature (one-atom thickness) of Gr are ideal characteristics for sensing applications. The sensor selectivity plays a very important role in its design, and this is very closely related to NM sensors characteristics, so selectivity can only be improved by fine-tuning the NM properties. The NM interacts with target bio-molecules by either a physisorption or chemisorption process. Physisorption, although fast, is a non-covalent bonding reaction and is not preferred as the bio-molecules do not bind completely, thereby affecting the sensitivity. Chemisorption can be brought about by the presence of defects, vacancies, doping and chemical functionalization, all of which increase the reactivity and enhance the selectivity to the target species. Figure 3b depicts the band structure changes of Gr by changing the geometry, thickness and doping mechanisms. The GrO 2D material produced by the oxidation of Gr is semiconducting and has a finite gap as compared to Gr. It has the advantage of being stable in water and other solvents and can be easily functionalized. Reduced graphene oxide (rGrO) is obtained by the removal of the oxygen functional groups and has the advantages of Gr and GrO, which include being conducting and having chemically active defect sites. The bandgap engineering and chemical functionalizing of Gr through the use of Gr derivatives such as GrO and rGrO and composites have proved to work well as sensors (including wearable sensors and implantable devices) for human health monitoring, as reported in Table 1. The body temperature is an important indicator of abnormal body functions, and its measurement is one of the first lines of action in suspicious cases. We have seen ample evidence of this aspect during the current COVID-19 pandemic. It is also linked to our biological clock and can be used to monitor an individual's sleep patterns, which is important in determining the overall health and mental fitness. Table 1 gives a summary of various Gr-based sensors along with body functions tested, the mechanisms of sensing, the sensitivities and the ranges when available and the references.

Table 1. Summary of the details of graphene (Gr)-based sensors in health monitoring.

NM	Body Function	Sensing Mechanism	Sensitivity	Range	Reference
Freestanding single reduced graphene oxide (rGrO); 3D Gr–PDS composite	Body temperature	Resistance-based	-	- -	[23] [24]
Gr/PDMS; Gr	Body movements	Piezo-capacitive strain; textile strain	0.24 kPa^{-1} 0.0078 kPa^{-1}	0–10 kPa 10–100 kPa	[25] [26]
Inkjet-printed Gr	Heart rate	Electronic	-	-	[27]
	Wrist pulse	Strain; pressure	-	-	[28] [29]
Gr–rubber composite; rGrO	Body movements + respiration rate	Strain	- -	- -	[30] [31]
Gr porous network	Blood pressure	Pressure + strain	-	-	[32]
3D nano-implant; Nano-hybrid fiber	Blood glucose Sweat glucose	Electrode Electrocatalytic	- -	- -	[33] [34]
3D Gr scaffold	ECG	Implant	-	-	[35]
Gr; Porous Gr	EMG	Electronic skin	- -	- -	[36] [37]
3D Gr scaffold; Gr	EEG	Implant Electronic skin	- -	- -	[35] [36]

2.2.2. MoS_2 and BN Nano-Biosensors

Similar to Gr, 2D MoS_2 and hBN materials with a honeycomb structure have all the advantages of Gr for sensing mentioned in the previous section. These van der Waal structures exhibit unique optical and electronic properties that make them very appealing for biosensing [38]. Moreover, they have the added advantage of bandgaps unlike Gr which has a zero bandgap; this improves the sensitivity of sensor devices made of these materials, especially in sensors.

MoS_2 is a prototype of a class of materials termed transition metal dichalcogenides (TMDs) and has markedly anisotropic properties, as seen from its electrical resistivity among other properties. The resistivity in a direction perpendicular to the planes is about 1000 times greater than in the parallel direction. Unlike Gr which is one-atom thick, an ML of MoS_2 has three atomic layers sulfur–molybdenum–sulfur. The physical properties of MoS_2 change markedly at the nanoscale. The bulk material has an indirect bandgap of ~1.2 eV, while the ML material has a direct and broader bandgap of ~1.8 eV [39]. Hence, it shows thickness-dependent bandgap properties, allowing for the production of tuneable optoelectronic devices with diversified spectral operation.

The electronic and optical properties of Gr and MoS_2 are complemented by those of hBN, which is an insulator with a large indirect bandgap value of ~5.95 eV [40] in the bulk form and in the ML limit crossover to a direct-bandgap material with a gap of 6.1 eV [41]. The sensing mechanisms of these materials could be electrical-based sensing, through charge transfer which alters the resistance or optical sensing where due to the charge transfer the surface plasmon resonance (SPR) gets modified and can be detected; or biomolecules are detected by their spectral fingerprints.

The hybrid structures of Gr, MoS_2 and BN have also been highly researched to increase the scope of the biosensing capabilities of these NMs. This is the topic of the next section.

2.2.3. Hetrostructures

2D Gr, GrO, rGrO, MoS_2 and hBN can all be used like Lego blocks to build interesting hetrostructures by mixing and matching for the increased selectivity and sensitivity of the nano-biosensensors. This process of electrostatic doping by the stacking of these van der Waal structures can be used to obtain unique and tuneable electronic properties. Figure 4 shows a graphical representation of a hetrostructure that can be made with the basic single layers of Gr, hBN and MoS_2.

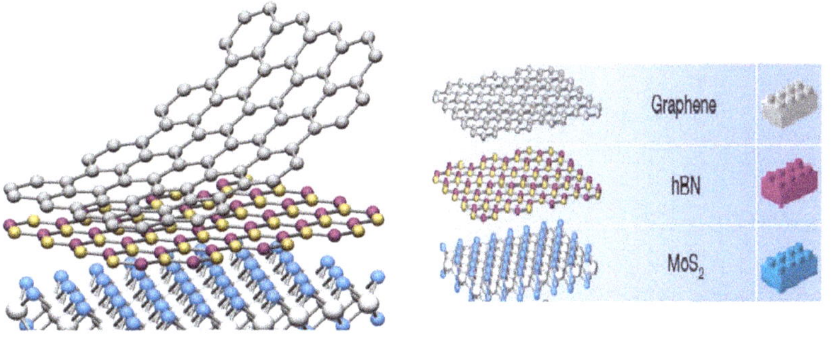

Figure 4. Graphical representation of possible hetrostructures that can be made by stacking multiple van der Waal layered structures in different orderings. Adapted from [42].

Hetrostructures, although highly desirable, require careful considerations of the lattice mismatch, the misalignment of layers and the introduction of unforeseen defects during the deposition and the epitaxial growth. Hexagonal BN is an insulating analogue of graphite with a small lattice mismatch (~1.8%), so it is an ideal substrate for graphene and a key building block in many van der Waals hetrostructures. Gr–hBN-integrated devices have been recently used for DNA sequencing by current modulation [43] and distinguishing nucleotides in DNA [44]. An SPR-based biosensor consisting of Gr/hBN hybrid structures for the detection of biomolecules was reported in 2019 [45]. The SPR technique is also used in a biosensor consisting of a MoS_2/Gr hybrid structure with Au, as a substrate, used to detect biomolecules using SPR [46]. Again in 2017, an angle-based SPR biosensor made of a MoS_2/Al film/MoS_2/Gr heterostructure was used to detect biomolecules [47]. In Table 2, we summarized various nano-biosensors made of MoS_2, hBN, Gr, Gr derivatives and hetrostructures of these NMs in the recent years. Table 2 gives the NMs used, the species detected, the sensing mechanisms, the sensitivities, the detection ranges, publications and the years of publications.

Table 2. Summary of nano-biosensors with references and the years of research.

Nanomaterial (NM)	Analyte	Sensing Mechanism	Detection Limit	Range	Reference + Year
MoS_2	DNA	Fluorescence quenching	500 pM	0–50 nM	[48]; 2014
MoS_2/Gr	Acetaminophen	Electrochemical	20 nM	0.1–100 µM	[49]; 2013
Gr/MoS_2	DNA hybridization	Photoluminescence	1 attomolar		[50]; 2014
MoS_2/Gr on Au	Biomolecule	Surface plasmon resonance (SPR)		10−6 RIU	[46]; 2015
MoS_2/Gr–Al hybrid	Biomolecule	Angle-based SPR	190.83° RIU^{-1}		[47]; 2017
Gr	PMMA, PVP	IR transmission spectroscopy	-	-	[51]; 2014
Gr	ssDNA	Phase-based SPR	1 attomolar	-	[52]; 2015
Gr	Glucose	FET	0.5 µM	-	[53]; 2015
Gr	Carcinoembryonic antigen (CEA)	FET	100 pg mL^{-1}		[54]; 2016
Gr	Protein	Acoustic Gr plasmons	-	-	[55]; 2017
Multichannel Gr	DNA	FET	10 pM	-	[56]; 2017
rGrO + trityl organic radical	Xanthine	Electrode-based	0.52 nM	-	[57]; 2017
GrO	hCG	Angle-based SPR	0.06 mM	-	[58]; 2017
hBN	Dopamine	Neurotransmittor	10 µM	-	[59]; 2016
hBN	CBP	IR vibrational spectroscopy	-	-	[60]; 2018
Gr/hBN	DNA sequencing	Current Modulation	-	-	[44]; 2017
Gr/hBN	DNA sequencing	Current Modulation	-	-	[43]; 2019
Gr/hBN	Biomolecule	SPR	4.207 µm RIU^{-1}		[45]; 2019

3. Smart Technologies

The early-stage detection and prevention of chronic and fatal diseases requires continuous monitoring. Data acquisition and interpretation from various parts of the human body in real time is possible with interconnected sensors and information transfer devices in today's era of the Internet of things (IoT). The unprecedented advancements in electronics and sensor technologies coupled with Big Data and AI offer exciting opportunities in the field of smart and sustainable healthcare. The stage is now set to shift from old medical procedures and protocols and adapt smart integrated medical testing with nano-devices for diagnosis and therapeutics [61,62]. We need to discard costly and bulky equipment and old fashioned laboratories and embrace wearable and miniaturised sensors that use interstitial fluid (ISF), instead of blood, to detect minute changes in biomarkers with sweat, tears and breath analysis that contain a wealth of information about body malfunctions [1–3]. Wireless, powerless nano-devices made of biocompatible materials that can be worn on the skin (patches, tattoos, watches, etc.), in textiles, the eye, mouth, teeth (miniaturised implants) and other innovative means using non-invasive probes are the need of the day. Electronic nose, tongues and skin are the new innovative smart technologies that are the future of healthcare monitoring and preventive medicine [63–67].

4. Challenges or Opportunities

A challenge, limitation or drawback is an opportunity for improvement, change in strategy or chance for innovation. Although the nano-biosensors research shows that considerable improvements in healthcare monitoring can be made, commercial products are few and from small companies [68]. Before large-scale and widespread manufacturing of 2D and other nanostructured devices for health-related applications can be realized, uniformity and controlled synthesis is necessary to rule out the device-to-device variability. This is crucial for large-scale commercialization, and the challenge has been met as indicated by the recent research and publications addressing this issue [69,70].

In addition, in vivo and point-of-care diagnostics require biocompatibility and toxicity issues to be addressed. The precise control of NMs properties and biocompatibility is required, especially in the local biological environment, where the devices are to be used with a thorough understanding of complex physiochemical interactions. The recent years have seen tremendous work in this direction with good progresses [71–73].

Author Contributions: A.A.R. conceived, designed and wrote the project manuscript; M.W.A. contributed towards literature survey, curation of data and writing. All authors have read and agreed to the published version of the manuscript.

Funding: This research received no external funding.

Institutional Review Board Statement: Not applicable.

Informed Consent Statement: Not applicable.

Data Availability Statement: Not applicable.

Conflicts of Interest: The authors have no conflict of interests to declare.

References

1. Tricoli, A.; Nasiri, N.; De, S. Wearable and Miniaturized Sensor Technologies for Personalized and Preventive Medicine. *Adv. Funct. Mater.* **2017**, *27*, 1605271. [CrossRef]
2. Wang, X.; Liu, Z.; Zhang, T. Flexible sensing electronic for wearable/attachable health monitoring. *Adv. Sci.* **2017**, *13*, 1602790. [CrossRef]
3. Yao, S.; Swetha, P.; Zhu, Y. Nanomaterial-enabled wearable sensors for healthcare. *Adv. Healthc. Mater.* **2017**, *7*, 1700889. [CrossRef] [PubMed]
4. Li, X.; Yang, J. First-principles design of spintronics materials. *Natl. Sci. Rev.* **2016**, *3*, 365–381. [CrossRef]
5. Ramanathan, A.A.; Khalifeh, J.M. Electronic, magnetic and optical properties of XScO3 (X = Mo, W) perovskites. *PeerJ Mater. Sci.* **2021**, *3*, e15. [CrossRef]
6. Pantelopoulos, A.; Bourbakis, N.G. Prognosis—A wearable health monitoring system for people at risk: Methodology and modeling. *IEEE Trans. Inf. Technol. Biomed.* **2010**, *14*, 613–621. [CrossRef] [PubMed]
7. Lu, C.; Huang, P.J.J.; Liu, B.; Ying, Y.; Liu, J. Comparison of Graphene Oxide and Reduced Graphene Oxide for DNA Adsorption and Sensing. *Langmuir* **2016**, *32*, 10776–10783. [CrossRef] [PubMed]
8. Hu, K.; Tsukruk, V.V. Tuning the Electronic Properties of Robust Bio-Bond Graphene Papers by Spontaneous Electrochemical Reduction: From Insulators to Flexible Semi-Metals. *Chem. Mater.* **2015**, *27*, 6717–6729. [CrossRef]
9. Chong, Y.; Ge, C.; Yang, Z.; Garate, J.A.; Gu, Z.; Weber, J.K.; Liu, J.; Zhou, R. Reduced Cytotoxicity of Graphene Nanosheets Mediated by Blood-Protein Coating. *ACS Nano* **2015**, *9*, 5713–5724. [CrossRef] [PubMed]
10. Ramanathan, A.A. Defect Functionalization of MoS_2 nanostructures as toxic gas sensors: A review. *IOP Conf. Ser. Mater. Sci. Eng.* **2018**, *305*, 012001. [CrossRef]
11. Wang, Y.H.; Huang, K.J.; Wu, X. Recent advances in transition-metal dichalcogenides based electrochemical biosensors: A review. *Biosens. Bioelectron.* **2017**, *97*, 305. [CrossRef] [PubMed]
12. Chen, Y.; Tan, C.; Zhang, H.; Wang, L. Two-Dimensional Graphene Analogues for Biomedical Applications. *Chem. Soc. Rev.* **2015**, *44*, 2681–2701. [CrossRef] [PubMed]
13. Ji, D.K.; Ménard-Moyon, C.; Bianco, A. Physically-triggered nanosystems based on two-dimensional materials for cancer theranostics. *Adv. Drug Deliv. Rev.* **2019**, *138*, 211–232. [CrossRef] [PubMed]
14. Ramanathan, A.A.; Aqra, M.W.; Al-Rawajfeh, A.E. Recent advances in 2D-nanopores for desalination. *Environ. Chem. Lett.* **2018**, *16*, 1217–1231. [CrossRef]
15. Yang, T.; Lin, H.; Loh, K.P.; Jia, B. Fundamental Transport Mechanisms and Advancements of Graphene Oxide Membranes for Molecular Separation. *Chem. Mater.* **2019**, *31*, 1829–1846. [CrossRef]
16. Li, M.; Liu, C.; Zhao, H.; An, H.; Cao, H.; Zhang, Y.; Fan, Z. Tuning Sulfur Doping in Graphene for Highly Sensitive Dopamine Biosensors. *Carbon* **2015**, *86*, 197–206. [CrossRef]
17. Aqra, M.W.; Ramanathan, A.A. Graphene and related 2D materials for desalination: A review of recent patents. *Jordan J. Phys.* **2020**, *13*, 233–242.
18. Szunerits, S.; Boukherroub, R. Graphene-based biosensors. *Interface Focus* **2018**, *8*, 20160132. [CrossRef] [PubMed]
19. Lemme, M.C. Current Status of Graphene Transistors. *Solid State Phenom.* **2010**, *156–158*, 499–509. [CrossRef]
20. Xiang, L.; Ma, S.Y.; Wang, F.; Zhang, K. Nano indentation models and young's modulus of few-layer graphene: A molecular dynamics simulation study. *J. Phys. D Appl. Phys.* **2015**, *48*, 395305. [CrossRef]
21. Yang, H.; Xue, T.; Li, F.; Liu, W.; Song, Y. Graphene: Diversified flexible 2D material for wearable vital signs monitoring. *Adv. Mater. Technol.* **2018**, *4*, 1800574. [CrossRef]
22. Schwierz, F. Graphene Transistors. *Nat. Nanotechnol.* **2010**, *5*, 487–496. [CrossRef] [PubMed]

23. Trung, T.Q.; Le, H.S.; Dang, T.M.L.; Ju, S.; Park, S.Y.; Lee, N.-E. Freestanding, fiber-based, wearable temperature sensor with tunable thermal index for healthcare monitoring. *Adv. Healthc. Mater.* **2018**, *7*, 1800074. [CrossRef]
24. Wang, Z.; Gao, W.; Zhang, Q.; Zheng, K.; Xu, J.; Xu, W.; Shang, E.; Jiang, J.; Zhang, J.; Liu, Y. 3D printed graphene/ polydimethylsiloxane composites for stretchable and strain insensitive temperature sensors. *ACS Appl. Mater. Interfaces* **2018**, *11*, 1344–1352. [CrossRef] [PubMed]
25. Kou, H.; Zhang, L.; Tan, Q.; Liu, G.; Lv, W.; Lu, F.; Dong, H.; Xiong, J. Wireless flexible pressure sensor based on micro-patterned Graphene/PDMS composite. *Sens. Actuators A Phys.* **2018**, *277*, 150–156. [CrossRef]
26. Yang, Z.; Pang, Y.; Han, X.; Yang, Y.; Ling, J.; Jian, M.; Zhang, Y.; Yang, Y.; Ren, T.L. Graphene textile strain sensor with negative resistance variation for human motion detection. *ACS Nano* **2018**, *12*, 9134–9141. [CrossRef]
27. Karim, N.; Afroj, S.; Malandraki, A.; Butterworth, S.; Beach, C.; Rigout, M.; Novoselov, K.S.; Casson, A.J.; Yeates, S.G. All inkjet-printed graphene-based conductive patterns for wearable e-textile applications. *J. Mater. Chem. C* **2017**, *5*, 11640–11648. [CrossRef]
28. Yang, T.; Jiang, X.; Zhong, Y.; Zhao, X.; Lin, S.; Li, J.; Li, X.; Xu, J.; Li, Z.; Zhu, H. A wearable and highly sensitive graphene strain sensor for precise home-based pulse wave monitoring. *ACS Sens.* **2017**, *2*, 967–974. [CrossRef] [PubMed]
29. Pang, Y.; Zhang, K.; Yang, Z.; Jiang, S.; Ju, Z.; Li, Y.; Wang, X.; Wang, D.; Jian, M.; Zhang, Y.; et al. Epidermis microstructure inspired graphene pressure sensor with random distributed spinosum for high sensitivity and large linearity. *ACS Nano* **2018**, *12*, 2346–2354. [CrossRef] [PubMed]
30. Boland, C.S.; Khan, U.; Backes, C.; O'Neill, A.; McCauley, J.; Duane, S.; Shanker, R.; Liu, Y.; Jurewicz, I.; Dalton, A.B.; et al. Sensitive, high-strain, high-rate bodily motion sensors based on graphene–rubber composites. *ACS Nano* **2014**, *8*, 8819–8830. [CrossRef] [PubMed]
31. Xu, M.; Qi, J.; Li, F.; Zhang, Y. Highly stretchable strain sensors with reduced graphene oxide sensing liquids for wearable electronics. *Nanoscale* **2018**, *10*, 5264–5271. [CrossRef]
32. Pang, Y.; Tian, H.; Tao, L.; Li, Y.; Wang, X.; Deng, N.; Yang, Y.; Ren, T.L. Flexible, highly sensitive, and wearable pressure and strain sensors with graphene porous network structure. *ACS Appl. Mater. Interfaces* **2016**, *8*, 26458–26462. [CrossRef] [PubMed]
33. Pu, Z.; Tu, J.; Han, R.; Zhang, X.; Wu, J.; Fang, C.; Wu, H.; Zhang, X.; Yu, H.; Li, D. A flexible enzyme-electrode sensor with cylindrical working electrode modified with a 3D nanostructure for implantable continuous glucose monitoring. *Lab Chip* **2018**, *18*, 3570–3577. [CrossRef]
34. Toi, P.T.; Trung, T.Q.; Dang, T.M.L.; Bae, C.W.; Lee, N.E. Highly electrocatalytic, durable, and stretchable nano hybrid fiber for on body sweat glucose detection. *ACS Appl. Mater. Interfaces* **2019**, *11*, 10707–10717. [CrossRef]
35. Ameri, S.K.; Singh, P.K.; D'Angelo, R.; Stoppel, W.; Black, L.; Sonkusale, S.R. Three dimensional graphene scaffold for cardiac tissue engineering and in-situ electrical recording. In Proceedings of the 2016 38th Annual International Conference of the IEEE Engineering in Medicine and Biology Society (EMBC), Orlando, FL, USA, 16–20 August 2016; pp. 4201–4203. [CrossRef]
36. Yun, Y.J.; Ju, J.; Lee, J.H.; Moon, S.-H.; Park, S.-J.; Kim, Y.H.; Hong, W.G.; Ha, D.H.; Jang, H.; Lee, G.H.; et al. Highly elastic graphene-based electronics toward electronic skin. *Adv. Funct. Mater.* **2017**, *27*, 1701510–1701513. [CrossRef]
37. Sun, B.; McCay, R.N.; Goswami, S.; Xu, Y.; Zhang, C.; Ling, Y.; Lin, J.; Yan, Z. Gas-permeable, multifunctional on-skin electronics based on laser-induced porous graphene and sugar-templated elastomer sponges. *Adv. Mater.* **2018**, *30*, 1804327–1804328. [CrossRef] [PubMed]
38. Gupta, A.; Sakthivel, T.; Seal, S. Recent development in 2D materials beyond graphene. *Prog. Mater. Sci.* **2015**, *73*, 44–126. [CrossRef]
39. Kadantsev, E.S.; Hawrylak, P. Electronic structure of a single MoS2 monolayer. *Solid State Commun.* **2012**, *152*, 909–913. [CrossRef]
40. Cassabois, G.; Valvin, P.; Gil, B. Hexagonal boron nitride is an indirect bandgap semiconductor. *Nat. Photon* **2016**, *10*, 262–266. [CrossRef]
41. Elias, C.; Valvin, P.; Pelini, T.; Summerfield, A.; Mellor, C.J.; Cheng, T.S.; Eaves, L.; Foxon, C.T.; Beton, P.H.; Novikov, S.V.; et al. Direct band-gap crossover in epitaxial monolayer boron nitride. *Nat. Commun.* **2019**, *10*, 2639. [CrossRef]
42. Geim, A.K.; Grigorieva, I.V. Van der Waals heterostructures. *Nature* **2013**, *499*, 419–425. [CrossRef] [PubMed]
43. Panahi, S.F.K.S.; Namiranian, A.; Jamaati, M. Graphene-hBN hybrid nanogap for boosting DNA nucleobases recognition sensitivity. *ChemNanoMat* **2019**, *5*, 488–498. [CrossRef]
44. de Souza, F.A.; Amorim, R.G.; Scopel, W.L.; Scheicher, R.H. Electrical detection of nucleotides via nanopores in a hybrid graphene/h-BN sheet. *Nanoscale* **2017**, *9*, 2207–2212. [CrossRef] [PubMed]
45. Jiang, H.; Choudhury, S.; Kudyshev, Z.A.; Wang, D.; Prokopeva, L.J.; Xiao, P.; Jiang, Y.; Kildishev, A.V. Enhancing sensitivity to ambient refractive index with tunable few-layer graphene/hBN nanoribbons. *Photonics Res.* **2019**, *7*, 815–822. [CrossRef]
46. Zeng, S.; Hu, S.; Xia, J.; Anderson, T.; Dinh, X.-Q.; Meng, X.-M.; Coquet, P.; Yong, K.-T. Graphene-MoS_2 hybrid nanostructures enhanced surface plasmon resonance biosensors. *Sens. Actuators B* **2015**, *207*, 801–810. [CrossRef]
47. Wu, L.; Jia, Y.; Jiang, L.; Guo, J.; Dai, X.; Xiang, Y.; Fan, D. Sensitivity improved SPR biosensor based on the MoS_2/graphene–aluminum hybrid structure. *J. Lightwave Technol.* **2016**, *35*, 82–87. [CrossRef]
48. Huang, Y.; Shi, Y.; Yang, H.Y.; Ai, Y. A novel single-layered MoS_2 nanosheet based microfluidic biosensor for ultrasensitive detection of DNA. *Nanoscale* **2015**, *7*, 2245–2249. [CrossRef] [PubMed]
49. Huang, K.-J.; Wang, L.; Li, J.; Liu, Y.-M. Electrochemical sensing based on layered MoS_2-graphene composites. *Sens. Actuators B* **2013**, *178*, 671–677. [CrossRef]

50. Loan, P.T.K.; Zhang, W.; Lin, C.-T.; Wei, K.-H.; Li, L.-J.; Chen, C.-H. Graphene/MoS_2 heterostructures for ultrasensitive detection of DNA hybridization. *Adv. Mater.* **2014**, *26*, 4838–4844. [CrossRef]
51. Li, Y.; Yan, H.; Farmer, D.B.; Meng, X.; Zhu, W.; Osgood, R.M.; Heinz, T.F.; Avouris, P. Graphene plasmon enhanced vibrational sensing of surface-adsorbed layers. *Nano Lett.* **2014**, *14*, 1573–1577. [CrossRef]
52. Zeng, S.; Sreekanth, K.V.; Shang, J.; Yu, T.; Chen, C.K.; Yin, F.; Baillargeat, D.; Coquet, P.; Ho, H.P.; Kabashin, A.V.; et al. Graphene-gold metasurface architectures for ultrasensitive plasmonic biosensing. *Adv. Mater.* **2015**, *27*, 6163–6169. [CrossRef]
53. Zhang, M.; Liao, C.; Mak, C.H.; You, P.; Mak, C.L.; Yan, F. Highly sensitive glucose sensors based on enzyme-modified whole-graphene solution-gated transistors. *Sci. Rep.* **2015**, *5*, 8311. [CrossRef] [PubMed]
54. Zhou, L.; Mao, H.; Wu, C.; Tang, L.; Wu, Z.; Sun, H.; Zhang, H.; Zhou, H.; Jia, C.; Jin, Q.; et al. Label-free graphene biosensor targeting cancer molecules based on non-covalent modification *Biosens. Bioelectron.* **2017**, *87*, 701–707. [CrossRef] [PubMed]
55. Chen, S.; Autore, M.; Li, J.; Li, P.; Alonso-Gonzalez, P.; Yang, Z.; Martin-Moreno, L.; Hillenbrand, R.; Nikitin, A.Y. Acoustic graphene plasmon nanoresonators for field-enhanced infrared molecular spectroscopy. *ACS Photonics* **2017**, *4*, 3089–3097. [CrossRef]
56. Xu, S.; Zhan, J.; Man, B.; Jiang, S.; Yue, W.; Gao, S.; Guo, C.; Liu, H.; Li, Z.; Wang, J.; et al. Real-time reliable determination of binding kinetics of DNA hybridization using a multi-channel graphene biosensor. *Nat. Commun.* **2017**, *8*, 14902. [CrossRef] [PubMed]
57. Seber, G.; Muñoz, J.; Sandoval, S.; Rovira, C.; Tobias, G.; Mas-Torrent, M.; Crivillers, N. Synergistic exploitation of the superoxide scavenger properties of reduced graphene oxide and a trityl organic radical for the impedimetric sensing of xanthine. *Adv. Mater. Interface* **2018**, *5*, 1701072. [CrossRef]
58. Chiu, N.-F.; Kuo, C.-T.; Lin, T.-L.; Chang, C.-C.; Chen, C.-Y. Ultra-high sensitivity of the non-immunological affinity of graphene oxide-peptide-based surface plasmon resonance biosensors to detect human chorionic gonadotropin. *Biosens. Bioelectron.* **2017**, *94*, 351–357. [CrossRef] [PubMed]
59. Nurunnabi, M.; Nafiujjaman, M.; Lee, S.-J.; Park, I.-K.; Huh, K.M.; Lee, Y.-k. Preparation of ultra-thin hexagonal boron nitride nanoplates for cancer cell imaging and neurotransmitter sensing. *Chem. Commun.* **2016**, *52*, 6146–6149. [CrossRef] [PubMed]
60. Autore, M.; Li, P.; Dolado, I.; Alfaro-Mozaz, F.J.; Esteban, R.; Atxabal, A.; Casanova, F.; Hueso, L.E.; Alonso-González, P.; Aizpurua, J.; et al. Boron nitride nanoresonators for phonon-enhanced molecular vibrational spectroscopy at the strong coupling limit. *Light Sci. Appl.* **2018**, *7*, 17172. [CrossRef] [PubMed]
61. Kostarelos, K.; Vincent, M.; Hebert, C.; Garrido, J.A. Graphene in the Design and Engineering of Next-Generation Neural Interfaces. *Adv. Mater.* **2017**, *29*, 1700909. [CrossRef] [PubMed]
62. Sazonov, E. *Wearable Sensors: Fundamentals, Implementation and Applications*; Technology & Engineering; Academic Press: Cambridge, MA, USA, 2020; 660p.
63. Belizário, J.E.; Faintuch, J.; Malpartida, M.G. Breath Biopsy and Discovery of Exclusive Volatile Organic Compounds for Diagnosis of Infectious Diseases. *Front. Cell. Infect. Microbiol.* **2021**, *10*, 564194. [CrossRef]
64. Scarlata, S.; Finamore, P.; Meszaros, M.; Dragonieri, S.; Bikov, A. The Role of Electronic Noses in Phenotyping Patients with Chronic Obstructive Pulmonary Disease. *Biosensors* **2020**, *10*, 171. [CrossRef] [PubMed]
65. Al Ramahi, R.; Zaid, A.N.; Abu-Khalaf, N. Evaluating the potential use of electronic tongue in early identification and diagnosis of bacterial infections. *Infect. Drug Resist.* **2019**, *12*, 2445–2451. [CrossRef] [PubMed]
66. Yang, J.C.; Mun, J.; Kwon, S.Y.; Park, S.; Bao, Z.; Park, S. Electronic Skin: Recent Progress and Future Prospects for Skin-Attachable Devices for Health Monitoring, Robotics, and Prosthetics. *Adv. Mater.* **2019**, *31*, 1904765. [CrossRef]
67. Behera, B.; Joshi, R.; Anil Vishnu, G.K.; Bhalerao, S.; Pandya, H.J. Electronic nose: A non-invasive technology for breath analysis of diabetes and lung cancer patients. *J. Breath Res.* **2019**, *13*, 024001–024023. [CrossRef] [PubMed]
68. Briggs, N.; Subramanian, S.; Lin, Z.; Li, X.; Zhang, X.; Zhang, K.; Xiao, K.; Geohegan, D.; Wallace, R.; Chen, L.Q.; et al. A Roadmap for Electronic Grade 2Dimensional Materials. *2D Mater.* **2019**, *6*, 022001. [CrossRef]
69. Magda, G.Z.; Peto, J.; Dobrik, G.; Hwang, C.; Biro, L.P.; Tapaszto, L. Exfoliation of Large-Area Transition Metal Chalcogenide Single Layers. *Sci. Rep.* **2015**, *5*, 14714. [CrossRef] [PubMed]
70. Ramanathan, A.A. and Aqra, M.W. An Overview of the Green Road to the Synthesis of Nanoparticles. *J. Mater. Sci. Res. Rev.* **2019**, *2*, 1–11.
71. Wang, S.; Li, K.; Chen, Y.; Chen, H.; Ma, M.; Feng, J.; Zhao, Q.; Shi, J. Biocompatible PEGylated MoS2 nanosheets: Controllable bottom-up synthesis and highly efficient photothermal regression of tumor. *Biomaterials* **2015**, *39*, 206–217. [CrossRef] [PubMed]
72. Zielińska, A.; Costa, B.; Ferreira, M.V.; Miguéis, D.; Louros, J.; Durazzo, A.; Lucarini, M.; Eder, P.; Chaud, M.V.; Morsink, M.; et al. Nanotoxicology and Nanosafety: Safety-By-Design and Testing at a Glance. *Int. J. Environ. Res. Public Health* **2020**, *17*, 4657. [CrossRef]
73. Ramanathan, A.A. Toxicity of nanoparticles_challenges and opportunities. *Appl. Microsc.* **2019**, *49*, 2. [CrossRef] [PubMed]

chemistry proceedings

MDPI

Proceeding Paper

Feasibility of Total White Blood Cells Counts by Visible-Near Infrared Spectroscopy †

Teresa Guerra Barroso [1], Lénio Ribeiro [2,3], Hugo Gregório [2], Filipe Santos [4] and Rui Costa Martins [4,*]

1 TOXRUN—Toxicology Research Unit, University Institute of Health Sciences, CESPU, CRL, 4585-116 Gandra, Portugal; teresa.barroso@gmail.com
2 Veterinary Hospital Center (CHV), R. Manuel Pinto de Azevedo 118, 4100-320 Porto, Portugal; lenioribeirovet@gmail.com (L.R.); hugogregvet@hotmail.com (H.G.)
3 Department of Veterinary Medicine, Universidade Lusófona de Humanidades e Tecnologias, Campo Grande 376, 1749-024 Lisboa, Portugal
4 INESC TEC, Institute for Systems and Computer Engineering, Technology and Science, Campus da FEUP, Rua Dr. Roberto Frias, 4200-465 Porto, Portugal; fbsantos@inesctec.pt
* Correspondence: rmcm@inesctec.pt
† Presented at the 1st International Electronic Conference on Chemical Sensors and Analytical Chemistry, 1–15 July 2021; Available online: https://csac2021.sciforum.net/.

Abstract: Total white blood cells (WBC) count is an important indication for infection diagnosis, in both human and veterinary medicine. State-of-the-art WBC counts are performed by flow cytometry combined with light scattering or impedance measurements, in the clinical analysis laboratory. These technologies are complex and difficult to be miniaturized into a portable point-of-care (POC) system. Spectroscopy is one of the most powerful technologies for POC miniaturization due to its capacity to analyze low sample quantities, little to no sample preparation, and 'real-time' results. WBC is in the proportion of 1:1000 to red blood cells (RBC), and the latter dominate visible-near infrared (Vis-NIR) information due to their large quantities and hemoglobin absorbance. WBC are difficult to be detected by traditional spectral analysis because their information is contained within the interference of hemoglobin bands. Herein, we perform a feasibility study for the direct detection of WBC counts in canine blood by Vis-NIR spectroscopy for veterinary applications, benchmarking current chemometrics techniques with self-learning artificial intelligence—a new advanced method for high-accuracy quantification from spectral information. Results show that total WBC counts can be detected by Vis-NIR spectroscopy to an average detection limit of 7.8×10^9 cells/L, with an R^2 of 0.9880 between impedance flow cytometry analysis and spectral quantification. This result opens new possibilities for reagent-less POC technology in infection diagnosis. As WBC counts in dogs range from 5 to 45 $\times 10^9$ cells/L, the detection limit obtained in this research allows concluding that the combined use of spectroscopy with this SL-AI new algorithm is a step towards the existence of portable and miniaturized Spectral POC hemogram analysis.

Keywords: point-of-care; spectroscopy; white blood cells; artificial intelligence

Citation: Barroso, T.G.; Ribeiro, L.; Gregório, H.; Santos, F.; Martins, R.C. Feasibility of Total White Blood Cells Counts by Visible-Near Infrared Spectroscopy. *Chem. Proc.* **2021**, *5*, 77. https://doi.org/10.3390/CSAC2021-10434

Published: 30 June 2021

1. Introduction

Total white blood cell (WBC) count is one of the most requested hematology parameters because of its broad diagnostic value, including for infection and leukemia. Leukocytosis and leukopenia, which are abnormal values (high/low, respectively) in WBC counts, are more frequently associated with neutrophil changes, although other leukocytes and neoplastic cells can also cause fluctuations. Neutrophilia is usually related to inflammation, and neutropenia to greater peripheral use or reduced bone marrow production [1].

Most common methods for WBC differential are based on electrical impedance, laser light scattering, radio frequency conductivity, and/or flow cytometry [2] (Figure 1). The basic principles of operation for automated hematology analyzers are based on cell size affecting directly impedance and scattering angle. This approach has disadvantages for WBC differential, because cell sizes for each type of leukocyte are highly dependent on the development stage and differentiation, leading to inaccurate counts in current automated equipment [3]. Despite laser scattering technology provides better accuracy than impedance technology, the latter is widely adopted in Veterinary Medicine. Impedance counting is a cheaper technology and the best hematology practices recommend that blood smear microscope counts are performed on abnormal cases [4].

Spectroscopy is one of the leading technologies for the development of reagent-less point-of-care (POC) devices [5,6], capable of providing comprehensive clinical information from a single drop of blood (<10 μL), with little or no sample preparation and real-time results.

Visible short-wave near-infrared (Vis-SWNIR) spectroscopy is an information-rich technology that carries both physical and chemical information, where the information about blood cells and constituents is distributed across the different wavelengths. Dominant spectral information in blood comes from highly absorbent constituents in the Vis-SWNIR region, such as hemoglobin present in red blood cells (RBC) and bilirubin in serum.

WBC is present in significantly lower quantities than RBC (∼1:1000), being considerably more difficult to be detected because the information about WBC is a small interference effect on the hemoglobin bands. State-of-the-art chemometrics and artificial intelligence technologies are unable to deal with small-scale interference and non-dominant spectral information sample constituents with good accuracy [6]. Such may lead to non-causal correlation in spectroscopy quantification, where the quantification is not obtained by direct relationship to the spectral absorbance bands, but rather by intrinsic correlations of the dataset [7], which may lead to erroneous diagnosis [6].

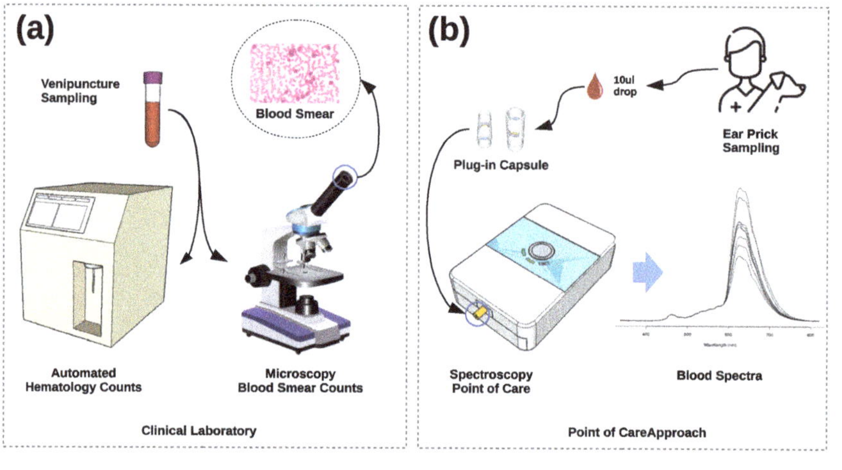

Figure 1. Total white blood cell counts: (**a**) current laboratory methods—automated cell counting using electric impedance or laser scattering, and manual smear count at the microscope by trained hematologist; and (**b**) Point-of-care approach—single blood drop spectroscopy counts using artificial intelligence.

In this research, we study the capacity of WBC quantification by Vis-SWNIR spectroscopy and a new algorithm based on Self-Learning Artificial Intelligence [6]. This new approach isolates spectral interference by searching consistent covariance between WBC and spectral features—the covariance mode (CovM). CovM is a set of samples that allow the direct relationship between spectral features and WBC, by sharing the same latent structure information [6]. Ideally, the relationship between WBC and spectral features is given by a single eigenvector or latent variable (LV), allowing to unscramble spectral interference in complex samples such as blood.

Herein, we provide a feasibility study on using Vis-SWNIR spectroscopy for the quantification and diagnosis of WBC, by providing a benchmark between a common chemometrics technique—partial least squares (PLS), and our new methodology (SL-AI).

2. Materials and Methods

2.1. Hemogram Analysis

Blood samples from daily clinical practice were collected from the jugular vein by qualified personnel using standardized venipuncture procedures at the Centro Hospitalar Veterinário do Porto into EDTA tubes. The sample was measured to WBC by a Beckman–Coulter capillary impedance [8] Mindray BC-2800-vet auto-hematology analyzer (Mindray, Shenzen, China), and a drop of blood (10 μL) was used for spectroscopy measurement.

2.2. Spectroscopy

Blood spectra were recorded using a POC prototype (INESC TEC, Porto, Portugal) using a 4500 K power LED as light source, and an USB-based miniaturized spectrometer (Ocean Insight STS-vis, Orlando, FL, USA) with an optical configuration and plug-in capsule system according to [5]. LED temperature and spectrometer integration times were automatically managed to maintain result consistency. Three replicates measurements were made for each blood sample.

2.3. Chemometrics

Spectral records were subjected to scattering correction (Mie and Rayleigh) before modeling. A feasibility benchmark is performed between PLS and SL-AI methods. PLS maximizes the global covariance between spectral features and WBC, by determining the orthogonal eigenvectors of the covariance matrix. The relationship between WBC and signal features is derived by the latent variables (LV), at each deflation. The number of LV is determined by cross-validation at the minimum value of the predicted residuals sum of squares (PRESS) [9].

SL-AI searches for stable covariance in spectral datasets, finding covariance modes (CovM). CovM is a group of samples that hold the same interference information characteristics, carrying proportionality between WBC and spectral features. Ideally, the CovM relationship between WBC and spectral features is given by a single eigenvector or latent variable (LV). The CovM is validated by leave one-out cross-validation [6].

3. Results and Discussion

PLS attains a correlation of 0.5687 and a SE of 11.60×10^9 cell/L (Table 1). PLS analysis demonstrates that there is a significant correlation between spectral features and WBC, and the small-scale interference of WBC is present in the spectra records. PLS model is obtained with 5 LV. Such means that the interference information about WBC in the blood Vis-SWNIR spectra is present in a significant number of differentiated covariance modes, where the non-dominant spectral interference can be related to WBC. PLS collapses the 5 LV into a single linear coefficient, which relates the WBC to the recorded spectra, leading to an averaged representation of all covariance modes present in the dataset. Such results in a high SE and MAPE of 44.62%. The PLS model is unable to estimate WBC values above 45.00×10^9 cell/L, misdiagnosing severe infection cases (Figure 2).

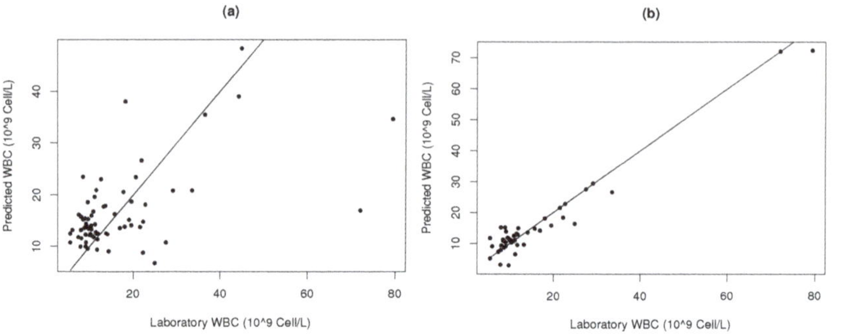

Figure 2. Total white blood cell counts spectral quantification: (**a**) PLS and (**b**) SL-AI.

The minimal total error criteria established by the American Society for Veterinary Clinical Pathology (ASVCP) for WBC is 20%. PLS shows to be unable to provide the necessary accuracy for WBC spectral POC technology.

SL-AI has a significantly higher correlation (R = 0.9733), a SE of 2.16 $\times$ 10^9 cell/L, and a MAPE of 20.00%. SL-AI covariance modes are obtained with 3 LV (Table 1). Results show that the different covariance modes (CovM) hold spectral interference proportional to WBC. Such demonstrates that it is possible to search non-dominant spectral interference from WBC and correlate it to total WBC count (Figure 3).

SL-AI CovM relationships are obtained with 3 LV. This is an indication that interference with other constituents and WBC differential population are incorporated in total WBC count, and that this higher complexity is not completely unscrambled in the dataset. In ideal conditions, CovM is obtained with a single LV (one eigenvector), directly relating the constituent concentration to spectral interference. The results show that non-dominant WBC spectral interference information has high complexity, which can be attributed to complex immune response, where differentiated cell types act at different stages and levels of infection or inflammation. The LV number re-assures the need for further studies, in order to investigate the source of non-dominant spectral interference attributed to WBC. Results may be improved by:

i. Larger datase—more data can help to complement the information of consistent CovM, allowing detection of single LV CovM;
ii. Feature space optimization—optimize the search for a feature space that better discriminates the small variation of WBC interference (e.g., Fourier or Wavelets decomposition).

Despite the limitations shown in this feasibility study, WBC quantification using Vis-SWNIR spectroscopy in conjunction with the new SL-AI algorithm can attain a total error estimate of 20%. Such result is following the ASVCP total allowable error for WBC in dog blood [4], but is above the 15% total allowable error in humans defined by CLIA [10].

Table 1. WBC quantification benchmarks for PLS and SL-AI.

Method	SE	LV	R^2	MAPE (%)	$R_{Pearson}$
PLS	11.06	5	0.3234	44.62	0.5687
SL-AI	2.16	3	0.9473	20.00	0.9733

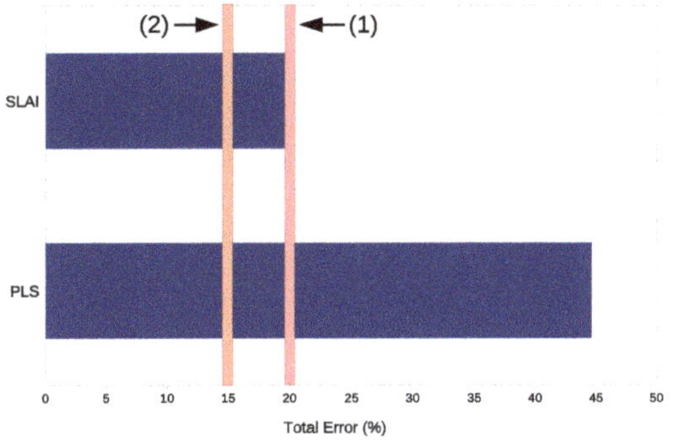

Figure 3. Percentage total error for PLS and SL-AI predictions: (1) ASVCP acceptable error limit (20%) and (2) CLIA acceptable error limit (15%).

4. Conclusions

This feasibility study has shown that low intensity, non-dominant, and multi-scale interferent spectral information is possible to be accessed by unscrambling information with the CovM principle included in our SL-AI method. The smaller quantities of WBC and corresponding interference with dominant constituents, such as erythrocytes, hemoglobin, and bilirubin, are detectable in each CovM. The results allow us to conclude that a spectral POC in the Vis-SWNIR for measuring WBC is achievable, for the application in both veterinary and human medicine.

Author Contributions: T.G.B., L.R. and H.G.: Investigation, methodology, validation, writing—review and editing; F.S.: investigation, hardware and firmware; R.C.M.: conceptualization, software and hardware, funding acquisition, writing—original draft, resources and formal analysis, project administration. All authors have read and agreed to the published version of the manuscript.

Funding: Rui Costa Martins acknowledges Fundação para a Ciência e Tecnologia (FCT) research contract grant (CEEIND/017801/2018).

Data Availability Statement: Not applicable.

Conflicts of Interest: The authors declare that they have no known competing financial interests or personal relationships that could have appeared to influence the work reported in this paper.

References

1. Burton, A.G.; Jandrey, K.E. Leukocytosis and Leukopenia. In *Textbook of Small Animal Emergency Medicine*; Drobatz, K.J., Hopper, K., Rozanski, E., Silverstein, D.C., Eds.; John Wiley & Sons, Inc.: New York, NY, USA, 2018. [CrossRef]
2. Chabot-Richards, D.S.; George, T.I. White blood cell counts: Reference methodology. *Clin. Lab Med.* **2015**, *35*, 11–24. [CrossRef] [PubMed]
3. Rishniw, M.; Pion, P.D. Evaluation of Performance of Veterinary In-Clinic Hematology Analyzers. *Vet. Clin. Path.* **2016**, *45*, 604–614. [CrossRef] [PubMed]
4. Nabity, M.B.; Harr, K.E.; Camus, M.S.; Flatland, B.; Vap, L.M. ASVCP guidelines: Allowable total error hematology. *Vet. Clin. Pathol.* **2018**, *47*, 9–21. [CrossRef]
5. Martins, R.C.; Sousa, N.J.; Osorio, R. Optical System for Parameter Characterization of an Element of Body Fluid or Tissue. U.S. Patent 10209178B2, 19 February 2019.
6. Martins, R.C. Big Data Self-Learning Artificial Intelligence Methodology for the Accurate Quantification and Classification of Spectral Information under Complex Variability and Multi-Scale Interference. WO Patent WO/2018/060967, 5 April 2018.
7. Pasquini, C. Near infrared spectroscopy: A mature analytical technique with new perspectives—A review. *Anal. Chim. Acta* **2018**, *1026*, 8–36. [CrossRef] [PubMed]
8. Brown, M.D.; Carl, T. Wittwer. Flow cytometry: Principles and clinical applications in hematology. *Clin. Chem.* **2000**, *46*, 1221–1229. [CrossRef] [PubMed]
9. Geladi, P.; Kowalsky, B. Partial least squares regression: A tutorial. *Anal. Acta* **1988**, *185*, 1–17. [CrossRef]
10. Westgard, J. CLIA Requirements for Analytical Chemistry. 2021. Available online: https://www.westgard.com/clia.htm (accessed on 1 May 2021).

chemistry proceedings

MDPI

Proceeding Paper

Visible–Near-Infrared Platelets Count: Towards Thrombocytosis Point-of-Care Diagnosis †

Teresa Guerra Barroso [1], Lénio Ribeiro [2,3], Hugo Gregório [2], Filipe Santos [4] and Rui Costa Martins [4,*]

1 TOXRUN—Toxicology Research Unit, University Institute of Health Sciences, CESPU, CRL, 4585-116 Gandra, Portugal; teresa.barroso@gmail.com
2 Veterinary Hospital Center (CHV), R. Manuel Pinto de Azevedo 118, 4100-320 Porto, Portugal; lenioribeirovet@gmail.com (L.R.); hugogregvet@hotmail.com (H.G.)
3 Department of Veterinary Medicine, Universidade Lusófona de Humanidades e Tecnologias, Campo Grande 376, 1749-024 Lisboa, Portugal
4 INESC TEC, Institute for Systems and Computer Engineering, Technology and Science, Campus da FEUP, Rua Dr. Roberto Frias, 4200-465 Porto, Portugal; fbsantos@inesctec.pt
* Correspondence: rmcm@inesctec.pt
† Presented at the 1st International Electronic Conference on Chemical Sensors and Analytical Chemistry, 1–15 July 2021; Available online: https://csac2021.sciforum.net/.

Abstract: Thrombocytosis is a disorder with an excessive number of platelets in the blood, where total platelet counts (TPC) are crucial for diagnosis. This condition predisposes to blood vessels clotting and diseases such as stroke or heart attack. TPC is generally performed at the laboratory by flow cytometry with laser scattering or impedance detection. Due to the limited capacity of automated hematology in performing TPC quantification, a manual microscopy count is a very common quality assurance measure undertaken by clinical pathologists. Monitoring coagulation risk is key in many health conditions, and point-of-care platforms would simplify this procedure by taking platelet counts to the bedside. Spectroscopy has high potential for reagent-less point-of-care miniaturized technologies. However, platelets are difficult to detect in blood by standard spectroscopy analysis, due to their small size, low number when compared to red blood cells, and low spectral contrast to hemoglobin. In this exploratory research, we show that it is possible to perform TPC by advanced spectroscopy analysis, using a new processing methodology based on self-learning artificial intelligence. The results show that TPC can be measured by visible–near-infrared spectroscopy above the standard error limit of 61.19×10^9 cells/L (R^2 = 0.7016), tested within the data range of 53×10^9 to 860×10^9 cells/L of dog blood. These results open the possibility for using spectroscopy as a diagnostic technology for the detection of high levels of platelets directly in whole blood, towards the rapid diagnosis of thrombocytosis and stroke prevention.

Keywords: point-of-care; spectroscopy; platelets; artificial intelligence

Citation: Barroso, T.G.; Ribeiro, L.; Gregório, H.; Santos, F.; Martins, R.C. Visible–Near-Infrared Platelets Count: Towards Thrombocytosis Point-of-Care Diagnosis. *Chem. Proc.* **2021**, *5*, 78. https://doi.org/10.3390/CSAC2021-10435

Academic Editor: Manel del Valle

Published: 30 June 2021

1. Introduction

Platelets (PLT) are the smallest cells in the blood, being responsible for coagulation and blood vessel repair. The PLT counts reference interval in dogs is 300 to 500×10^9 cell/L. High PLT counts is a condition known as thrombocytosis, being attributed to abnormal bone marrow production or an ongoing condition such as anemia or inflammation [1]. Thrombocytosis can result in blood clots, leading to life-threatening or impairing conditions such as heart attack or stroke [2]. Automated PLT counts are mostly performed by flow cytometry, electric impedance (Coulter principle), or laser-scattering technologies [3]. However, these methods are prone to erroneous PLT counts, because of changes in cell size and morphology, due to blood clotting, activation, aggregation, or even post-sampling artifacts. This limits scattering angle and impedance detection, leading to misidentification as larger cells, such as erythrocytes or leucocytes. Laser scattering is significantly more accurate than electric impedance, but the latter is cheaper and has a higher implementation in veterinary medicine. Veterinary doctors make use of blood-smear PLT manual counts for ensuring results quality in abnormal (low or high) values [4].

Visible-shortwave–near-infrared (Vis–NIR) spectroscopy has a high potential for the development of point-of-care (POC) without the need for reagents or complex sample preparation. The developed Vis–SWNIR POC system (Figure 1b) records the blood spectra of a single drop of blood (<10 µL) to provide a significant number of clinical analysis parameters with real-time results [5].

Visible-short-wave–near-infrared (Vis–SWNIR) spectroscopy is an information-rich technology that carries both physical and chemical information, where the information about blood cells and constituents is distributed across the different wavelengths. Dominant spectral information in blood comes from highly absorbent constituents in the Vis–SWNIR region, such as hemoglobin present in red blood cells (RBC) and bilirubin in blood serum.

Platelets are present in significantly lower values than red blood cells (RBC) (Figure 1a). The PLT reference interval in dogs is 300 to 500 $\times$ 10^9 cells/L and RBC is 5500 to 8500 $\times$ 10^9 cells/L, being at approximately 1:18 ratio to RBC, which makes the detection difficult:

i. Smaller size of PLT with the significantly lower area and volume for light absorbance, resulting in low sensitivity in the spectral signal;
ii. High interference between PLT and RBC and hemoglobin and bilirubin, which leads to the existence of significantly different characteristic interferences;
iii. High variance of PLT morphology—which can vary from small platelets to activated platelets with branches and clotted cells.

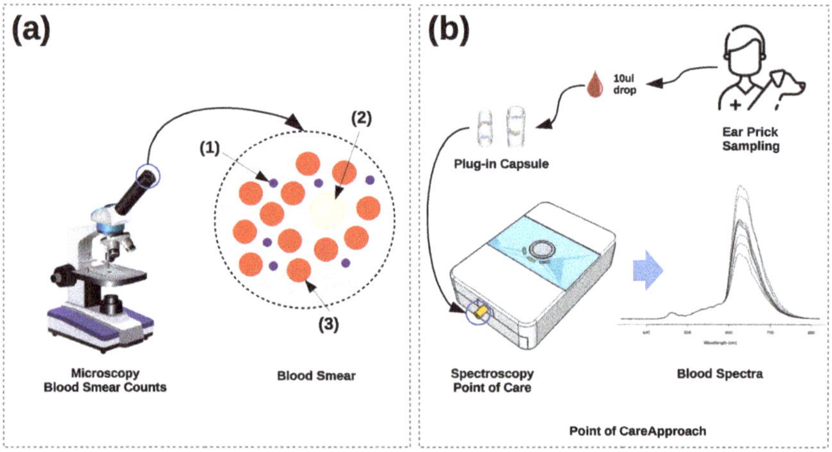

Figure 1. Platelets cell counts: (**a**) manual smear count at the microscope by trained hematologist demonstrating the proportionality between (1) platelets, (2) white blood cells, and (3) red blood cells and (**b**) point-of-care approach—single-blood-drop spectroscopy counts using artificial intelligence.

PLT counts are difficult to obtain, even by microscopy methods, exhibiting high variability. Herein, we explore the capacity of Vis–SWNIR and self-learning artificial intelligence (SL-AI) for PLT quantification [5]. This new approach isolates spectral interference by searching consistent covariance between PLT and spectral features, which belong to a covariance mode (CovM). CovM is a set of samples that can hold a direct relationship between spectral features and PLT counts, by sharing a common latent structure [5]. Ideally, PLT counts are related to spectral-interference features by a single latent variable (LV) or eigenvector. This allows unscrambling the interference of PLT concerning the other blood constituents. This research provides a feasibility benchmark between the widely used chemometrics partial least squares (PLS) method and the SL-AI method.

2. Materials and Methods

2.1. Hemogram Analysis

Dog blood samples from routine clinical practice were collected by qualified personnel by standard venipuncture, at the Centro Hopitalar Veterinário do Porto. No animal experimentation was involved or any additional procedure. Samples used in this study are remnants from already necessary routine clinical analysis medical practice. Dataset is anonymized. PLT was determined by Beckman–Coulter capillary impedance using a Mindray B-2800 vet auto-hematology analyzer (Mindray, Shenzen, China).

2.2. Spectroscopy

Blood spectra were recorded using a POC prototype (INESC TEC, Porto, Portugal) using a 4500 K power LED as a light source and a USB-based miniaturized spectrometer (Ocean Insight STS-vis, Orlando, FL, USA), with an optical configuration and plug-in capsule system according to [6]. LED temperature and spectrometer integration times were automatically managed to maintain result consistency. Three replicate measurements were made for each blood sample.

2.3. Chemometrics

Spectral records were subjected to scattering correction (Mie and Rayleigh) before modeling. A feasibility benchmark was performed between PLS and SL-AI methods. PLS maximizes the global covariance between spectral features and PLT, by determining the orthogonal eigenvectors of the covariance matrix. The relationship between PLT and signal features is derived by the latent variables (LV), at each deflation. The number of LV is determined by cross-validation at the minimum value of the predicted residuals sum of squares (PRESS) [7].

SL-AI searches for stable covariance in spectral datasets, finding covariance modes (CovM). CovM is a group of samples that contains the same interference information characteristics, holding proportionality between PLT and spectral features. Ideally, the relationship between PLT and spectral features is given by a single eigenvector or latent variable (LV). The CovM is validated by leave-one-out cross-validation [5].

3. Results and Discussion

The PLS model attained a correlation of 0.2613 with a very poor R^2 (0.068) and a corresponding high SE of 175.99×10^9 cells/L. The PLS analysis shows that the correlation between spectral features and PLT counts is highly unstable and non-linear. Such is because PLT is present in much fewer quantities than other blood constituents (Figure 1), as well as, due to the small size and high interference with the other major blood constituents (e.g., RBC, hemoglobin, and bilirubin). Another indication of non-linearity is that the PLS algorithm attains the optimum prediction error with two LV, resulting in a non-significant model (Figure 2a). The PLS is unable to increase the number of LV because the information about PLT is scattered in significantly different interference modes that cannot be collapsed into a linear oblique projection model [5,7]. PLS cannot be used in a POC as it does not attain a MAPE similar to 25%—the total allowable error established by the American Society for Veterinary Clinical Pathology (ASVCP) for PLT counts [8].

SL-AI presented a significant correlation of 0.8376, an SE of 61.19×10^9 cell/L, and a MAPE of 24.67%, with R^2 of 0.7016 (Table 1). SL-AI covariance modes (CovM) were obtained with 1 to 3 LV. This means that, although statistically valid relationships were obtained for each CovM, some of these were integrating more than one type of interference. Under ideal conditions, all CovM should have only one LV, directly relating PLT counts and spectral interference.

Table 1. PLS and SL-AI benchmark results.

Method	SE	LV	R^2	MAPE (%)	$R_{Pearson}$
PLS	175.99	2	0.068	88.89	0.2613
SL-AI	61.19	1-3	0.7016	24.67	0.8376

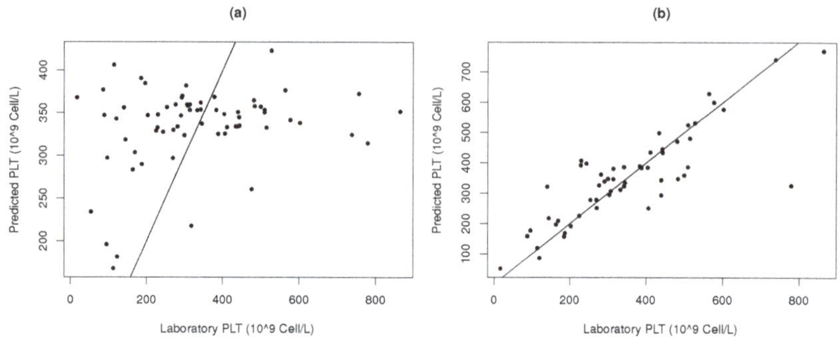

Figure 2. Total platelet counts spectral quantification: (**a**) PLS and (**b**) SL-AI.

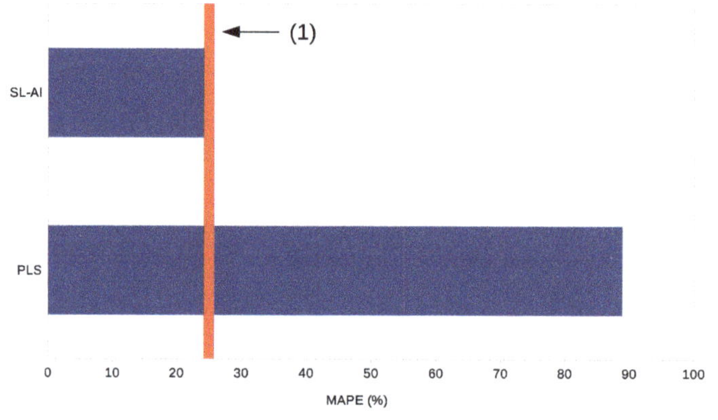

Figure 3. Percentage total error for PLS and SL-AI predictions: (1) ASVCP acceptable error limit (25%).

The results also show that non-dominant spectral information and low-scale spectral variation is unscrambled by the CovM principle. The number of LV can be attributed to the high diversity of PLT morphology present in dog blood (non-activated, activated, and clotted PLT) and the particular conditions of the tested blood, with correspondence in the major constituents.

Despite the limitations shown in this feasibility study, PLT quantification using Vis–SWNIR spectroscopy in conjunction with the new SL-AI algorithm can attain a total error estimate of 25%. Such a result is following the ASVCP total allowable error for PLT in dog blood [8] (Figure 3).

Vis–SWNIR POC technology based on SL-AI has shown high potential for PLT quantification and thrombocytosis diagnosis. The results presented for dog blood are within the acceptable error defined by the ASVCP of 25% [8]. The presented results also allow extending the potential application to both human and other animal species in further studies.

4. Conclusions

This feasibility study showed that low intensity, non-dominant, and multi-scale interferent spectral information is possible to be accessed, by unscrambling information with the CovM principle included in the SL-AI method. The small variations in the spectral signal that contain information about PLT cannot be modeled by PLS. SL-AI can unscramble PLT interference information based on the CovM principle, allowing the quantification of PLT. Future studies, with more samples, may provide better insights on the full potential of the developed POC technology in both veterinary and human medicine.

Author Contributions: T.G.B., L.R. and H.G.: investigation, methodology, validation, writing—review and editing; F.S.: investigation, hardware, and firmware; R.C.M.: conceptualization, software and hardware, funding acquisition, writing—original draft, resources and formal analysis, and project administration. All authors have read and agreed to the published version of the manuscript.

Funding: Rui Costa Martins acknowledges Fundação para a Ciência e Tecnologia (FCT) research contract grant (CEEIND/017801/2018).

Data Availability Statement: Not applicable.

Conflicts of Interest: The authors declare that they have no known competing financial interests or personal relationships that could have appeared to influence the work reported in this article.

References

1. Woolcock, A.; Keenan, A.; Cheung, C.; Christian, J.; Moore, G. Thrombocytosis in 715 Dogs (2011–2015). *J. Vet. Intern. Med.* **2017**, *31*, 1691–1699. [CrossRef] [PubMed]
2. Soucy-Giguère, M.C.; Turgeon, P.Y.; Sènèchal, M. What cardiologists should know about essential thrombocythemia and acute myocardial infarction: Report of two cases and advanced heart failure therapies considerations. *Int. Med Case Rep. J.* **2019**, *12*, 253–259. [CrossRef] [PubMed]
3. Baccini, V.; Geneviéve, F.; Jacqmin, H.; Chatelain, B.; Girard, S.; Wuilleme, S.; Vedrenne, A.; Guiheneuf, E.; Toussaint-Hacquard, M.; Everaere, F.; et al. Platelet Counting: Ugly Traps and Good Advice. Proposals from the French-Speaking Cellular Hematology Group (GFHC). *J. Clin. Med.* **2020**, *9*, 808. [CrossRef] [PubMed]
4. Arnold, J.E.; Camus, M.S.; Freeman, K.P.; Giori, L.; Hooijberg, E.H.; Jeffery, U.; Korchia, J.; Meindel, M.J.; Moore, A.R.; Sisson, S.C.; et al. ASVCP Guidelines: Principles of Quality Assurance and Standards for Veterinary Clinical Pathology (version 3.0). *Vet. Clin. Pathol.* **2019**, *48*, 542–618. [CrossRef] [PubMed]
5. Martins, R.C. Big Data Self-Learning Artificial Intelligence Methodology for the Accurate Quantification and Classification of Spectral Information Under Complex Variability and Multi-Scale Interference. Patent WO/2018/060967, 5 April 2018.
6. Martins, R.C.; Sousa, N.J.; Osorio, R. Optical System for Parameter Characterization of an Element of Body Fluid or Tissue. U.S. Patent 10209178B2, 19 February 2019.
7. Geladi, P.; Kowalski, B.R. Partial least squares regression: A tutorial. *Anal. Acta* **1988**, *185*, 1–17. [CrossRef]
8. Nabity, M.B.; Harr, K.E.; Camus, M.S.; Flatl, B.; Vap, L.M. ASVCP guidelines: Allowable total error hematology. *Vet. Clin. Pathol.* **2018**, *47*, 9–21. [CrossRef] [PubMed]

chemistry proceedings

MDPI

Proceeding Paper

Development of an Integrated In-Vehicle Driver Breath Ethanol System Based on α-Fe_2O_3 Sensing Material †

Roberto Di Chio *, Monica Galtieri, Nicola Donato and Giovanni Neri

Department of Engineering, University of Messina, Contrada Di Dio, 98166 Messina, Italy; monicagaltieri94@gmail.com (M.G.); nicola.donato@unime.it (N.D.); giovanni.neri@unime.it (G.N.)

* Correspondence: rdichio@unime.it

† Presented at the 1st International Electronic Conference on Chemical Sensors and Analytical Chemistry, 1–15 July 2021; Available online: https://csac2021.sciforum.net/.

Abstract: Alcohol abuse is the dominant cause of fatal car accidents (about 25% of all road deaths in Europe). The large-scale implementation of systems aimed at the realization of in-vehicle driver breath ethanol detection is therefore in high demand. For this reason, we devoted our attention to the design of an inexpensive and reliable breath alcohol sensor for use in an Advanced Driver Assistance System (ADAS). The main challenge in the development of this sensor is related to the complexity of breath composition and its high humidity content, coupled with the high dilution of breath reaching the sensor. In this work, a simple α-Fe_2O_3 film-based sensor was developed and validated in laboratory tests. Tests were also performed by placing the ethanol sensor within the casing of the upper steering column of a car to simulate real driving conditions. Using an array provided with the developed ethanol sensor and humidity, temperature and CO_2 sensors, it was possible to differentiate the signal of a driver's breath before and after alcohol consumption.

Keywords: gas-sensing; ethanol; iron oxide; sensing materials; ADAS

Citation: Di Chio, R.; Galtieri, M.; Donato, N.; Neri, G. Development of an Integrated In-Vehicle Driver Breath Ethanol System Based on α-Fe_2O_3 Sensing Material. *Chem. Proc.* **2021**, *5*, 79. https://doi.org/10.3390/CSAC2021-10476

Academic Editor: Huangxian Ju

Published: 30 June 2021

1. Introduction

Advanced Driver Assistance Systems (ADASs) are intelligent systems that assist the driver in a variety of ways [1]. They may be used to provide useful traffic information but may also be used to evaluate whether or not the driver is in physical condition to drive. Among other driver-related risk factors (e.g., drug intake or altered emotional state), alcohol abuse remains the dominant cause of fatal car accidents (about 25% of all road deaths in Europe). It is well known that too much alcohol in the blood leads to various serious effects on human health and the condition of drivers [2,3]. Alcohol interferes with the brain, affecting the way that it looks and works and reducing movement coordination. Further, alcohol can slow reflexes, slow eye muscle function, and alter visual perception. These conditions are very critical for car drivers, so maintaining an acceptable blood alcohol level is necessary in order to limit car accident risks.

Based on these concerns, we initiated research activity with the main objective of developing an in-vehicle driver breath ethanol detection system [4]. To facilitate the large-scale implementation of these systems, the design of inexpensive, reliable and easily fabricated sensors is required. Conductometric sensors apply very well to this scope, as they possess all of the required characteristics [5]. Many examples of ethanol sensors have been developed and show a remarkable sensing capacity [6–8]. In particular, we have shown that α-Fe_2O_3 is an ideal candidate as a sensing material to be used in breath ethanol conductometric sensors [9,10].

Based on previous work, in this research, the α-Fe_2O_3 material was employed for fabricating conductometric gas sensors to be used for breath ethanol detection in ADASs. Preliminary laboratory tests were performed to validate the fabricated sensors and optimize the operating conditions. Then, tests were performed by placing the ethanol sensor within the casing of the upper steering column of a car to simulate the driving position. The main challenge in the development of this system is related to the complexity of breath composition and its high humidity content, coupled with the high dilution of breath reaching the sensor. For this reason, it was necessary to install the ethanol sensor in an array that also contains humidity, temperature and CO_2 sensors (the latter breath component is employed as an internal standard). Through the simultaneous use of these three sensors, it was possible to differentiate the signal of a driver's breath before and after alcohol beverage consumption.

2. Materials and Methods

2.1. Material Preparation

For the synthesis of α-Fe_2O_3 material, a simple Pechini sol–gel process was employed [6,7]. This method is based on the polymerization of metallic citrate by ethylene glycol. Iron nitrate ($Fe(NO_3)_3 \cdot 9H_2O$), citric acid ($C_6H_8O_7$ H_2O), poly(vinylpyrrolidone) and ethylene glycol ($C_2H_6O_2$) were purchased from Merck. All of the chemicals were used as received and without further purification. Double-distilled water was used to prepare precursor solutions.

First, the appropriate amount of $Fe(NO_3)_3 \cdot 9H_2O$ was dissolved in distilled water at 70 °C for 1 h under magnetic stirring to produce a 0.5 M Fe^{+3} solution. Then, this solution was mixed with PVP solution with a molar ratio of [PVP]/[Fe^{+3}] = 1. On the other hand, citric acid was dissolved in distilled water at 70 °C for 30 min. Afterwards, the citric acid solution was added slowly to the Fe^{+3}/PVP solution with stirring. The citric acid to Fe^{+3} molar ratio was 2. Then the esterification agent, i.e., ethylene glycol (EG), was added with a molar ratio of [citric acid]/[EG] = 2 while stirring and heating the solution. The final solution was refluxed at 100 °C for 2h. The clear yellow-colored precursor solution obtained was dried at 120 °C for 12 h to obtain the precursor powders. Finally, the amorphous powders were calcined at 550 °C in air for 3 h using a muffle furnace to obtain iron oxide nanoparticles.

2.2. Sensor Preparation and Sensing Tests

Sensor devices were fabricated by the spray-coating method as follows. An appropriate volume of the α-Fe_2O_3 suspension was sprayed on alumina substrates (3 × 6 mm) supplied with interdigitated Pt electrodes and a heating element on the backside. The prepared sensors were dried at room temperature and then heat-treated at 400 °C to obtain a mechanically stable sensing layer. The structure of the fabricated ethanol sensor is shown in Figure 1.

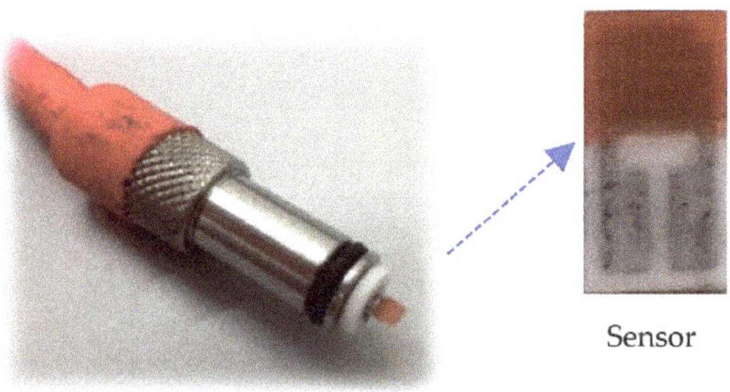

Figure 1. Photo of the fabricated ethanol sensor.

Measurements were performed under both a dry and wet (50% relative humidity) air stream of 100 mL/min in total, and the sensor resistance data were collected in four-point mode using an Agilent 34970A multimeter. Electrical measurements were carried out at a working temperature of 300 °C. Laboratory sensing tests were performed in a lab apparatus that allows operation at controlled temperature and the performance of resistance measurements while varying the ethanol concentration from 12.5 to 400 ppm.

The gas response was defined as the ratio R_{air}/R_{gas}, where R_{air} represents the electrical resistance of the sensor in dry air, and R_{gas} is the electrical resistance of the sensor at different ethanol concentrations. Response time, t_{res}, was defined as the time required for the sensor resistance to reach 90% of the equilibrium value after ethanol is injected, and recovery time, t_{rec}, was taken as the time necessary for the sensor resistance to reach 90% of the baseline value in air.

3. Results

3.1. Laboratory Sensing Tests

The characteristics of the developed α-Fe_2O_3 sensor were first evaluated in laboratory tests. Based on the preliminary results, a temperature of 300 °C was selected as the operating temperature. Figure 2a shows the sensor behavior versus ethanol concentration, which ranged from 400 to 12.5 ppm at this temperature. A reversible variation in the resistance was observed with the concentration of ethanol. As usually verified for metal oxide-based conductometric sensors, response and recovery times are dependent on the alcohol concentration. This is also the case for our sensor. Response (8–15 s) and recovery (120–45 s) times were observed for ethanol concentrations ranging from 400 to 12.5 ppm. At the intermediate concentration of 100 ppm, the sensor showed a noticeable reversible response (see Figure 2b) with a fast response and recovery (about 10 s and 60 s, respectively).

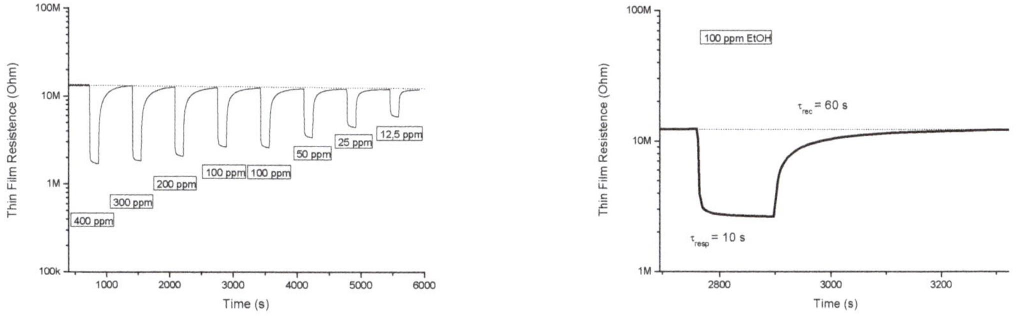

(a) (b)

Figure 2. (**a**) Response of the sensor to a variable concentration of ethanol in dry air at 300 °C; (**b**) response of the sensor to an ethanol pulse of 100 ppm. The measured response and recovery times are reported.

From the above test, the calibration curve shown in Figure 3 was obtained. Plotting the data in a log–log graph, a high linear correlation between the sensor resistance and the ethanol concentration is observed. The same graph also shows the calibration curve for the same sensor obtained in conditions of higher relative humidity (50% RH). Breath is highly saturated with water vapor; therefore, the sensor performance must not be influenced by changes in the humidity level [11–13]. Interestingly, the sensor signal that we collected in different humidity conditions appears to be independent of this variable.

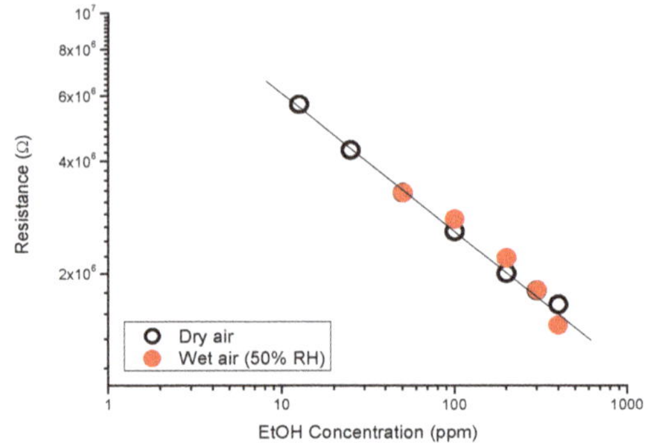

Figure 3. Calibration curve at 300 °C for the α-Fe_2O_3 sensor in *dry* and *wet* conditions.

3.2. Ethanol Sensor Implementation in ADASs

Then, the research work continued with the installation of the ethanol sensor in the casing of the upper steering column of a car to simulate real driving conditions (see Figure 4). A diagram of the designed and constructed module consisting of the ethanol sensor used in this research was reported in a previous paper [14]. Humidity, temperature and CO_2 sensors were also installed. The detected CO_2 concentration was used to account for the dilution of the breath sample. A suitable chamber was therefore designed and built to contain the sensor array.

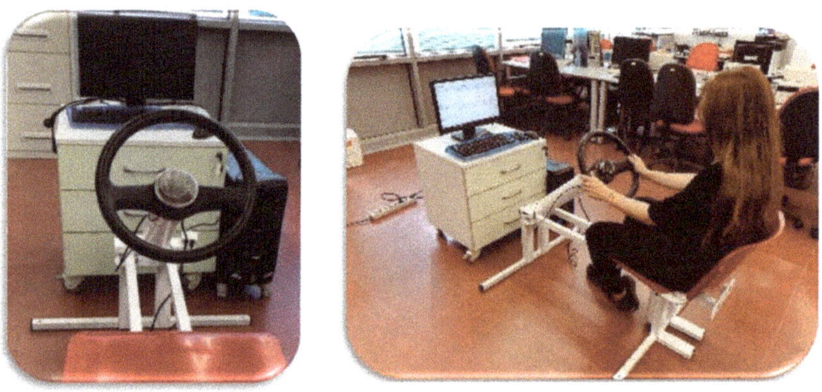

Figure 4. Pictures showing the location of the ethanol sensor inside the casing of an upper steering column (**left**) and the position of the driver during the test, simulating real driving conditions (**right**).

After installing the sensors, some preliminary tests to validate their correct functioning were carried out, especially to verify if the breath of the driver can be well detected by the sensor array when it is located at a distance of 30–50 cm from the driver's mouth. Indeed, in the conditions adopted, breath is diluted with ambient air by a factor as high as 5–10 [15,16].

One subject (male, 70 kg) was used for the test. He was allowed to drink two glasses (50 mL in total) of a commercial alcoholic beverage (44% of ethanol in *v*/*v*) in less than 15 min, for a total weight of 17.5 g of ingested alcohol (0.25 g of alcohol per kilogram of body mass). Measurements were performed before drinking and repeated every 30 min until the breath alcohol level returned to the background level (approximatively 3 h).

The graphs reported in Figure 5 show the signals coming from the ethanol, humidity, temperature and CO_2 sensors, recorded when the driver was in different conditions, i.e., before drinking alcoholic beverages and therefore in the absence of alcohol in the breath (white zone, left column) and subsequently after drinking an alcoholic beverage and thus in the presence of alcohol in the breath (red zone, right column).

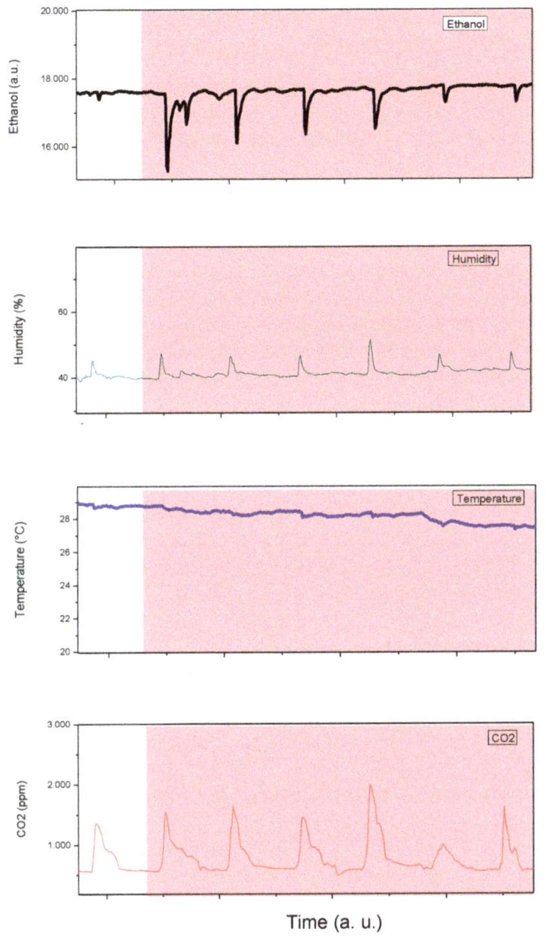

Figure 5. Signals from the ethanol, humidity, temperature and CO_2 sensors, recorded before drinking alcoholic beverages (white zone, left column) and subsequently after drinking an alcoholic beverage (red zone, right column).

By analyzing data coming from the ethanol sensor, we can see that after the alcoholic beverage is consumed, the signal of the sensor undergoes a quick increase, reaching a maximum after about 30 min (see Figure 6). Subsequently, the signal of the ethanol sensor tends to decrease, as expected by considering the well-known dynamic process of ethanol absorption, metabolism and elimination from the body after its ingestion [17].

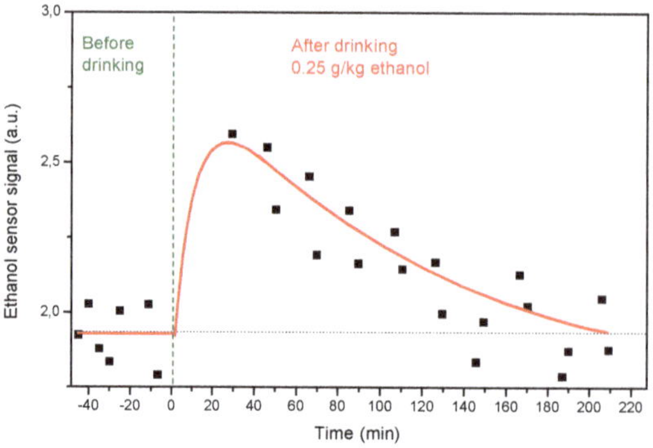

Figure 6. Response from the ethanol sensor during the entire duration of breath test.

The measurements carried out demonstrate that the designed and built sensor module correctly fulfills its functions and is thus able to monitor the level of ethanol in the driver's breath in real time.

4. Conclusions

An in-vehicle driver breath ethanol detection system was realized by using a simple α-Fe_2O_3 film-based conductometric sensor for detecting breath ethanol. Using an array provided with the developed ethanol sensor and humidity, temperature and CO_2 sensors, it was possible to differentiate the signal of a driver's breath before and after alcohol consumption, thus demonstrating that the developed sensor module can monitor the level of ethanol in the driver's breath in real time.

Author Contributions: Conceptualization, G.N. and N.D.; investigation, R.D.C. and M.G.; data curation, R.D.C. and M.G.; writing—review and editing, G.N. and R.D.C. All authors have read and agreed to the published version of the manuscript.

Funding: This research was funded by MUR (project code ARS01_00459) under the framework of *PON*—National Operational Programme for Research and Competitiveness 2014–2020.

Institutional Review Board Statement: Not applicable.

Informed Consent Statement: Informed consent was obtained from all subjects involved in the study.

Data Availability Statement: Not applicable.

Conflicts of Interest: The authors declare no conflict of interest.

References

1. Kala, R. Advanced Driver Assistance Systems. In *On-Road Intelligent Vehicles*; Butterworth Heinemann Publisher: Oxford, UK, 2016; pp. 59–82.
2. Liu, Y.-C.; Ho, C.H. Effects of different blood alcohol concentrations and post-alcohol impairment on driving behavior and task performance. *Traffic Inj. Prev.* **2010**, *11*, 334–341. [CrossRef] [PubMed]
3. Zhao, X.; Zhang, X.; Rong, J. Study of the Effects of Alcohol on Drivers and Driving Performance on Straight Road. *Modeling Simul. Transp. Eng.* **2014**, *2014*, 607652. [CrossRef]
4. Zaouk, B. Driver Alcohol Detection System for Safety. In Proceedings of the Transportation Research Board 90th Annual Meeting, Washington, DC, USA, 23–27 January 2011.
5. Neri, G. First fifty years of chemoresistive gas sensors. *Chemosensors* **2015**, *3*, 1–20. [CrossRef]
6. Chen, X.; Deng, N.; Zhang, X.; Li, J.; Yang, Y.; Hong, B.; Jin, D.; Peng, X.; Wang, X.; Ge, H.; et al. Cerium-doped indium oxide nanosphere arrays with enhanced ethanol-sensing properties. *J. Nanopart. Res.* **2019**, *21*, 77. [CrossRef]
7. Guo, W.; Wang, Z. Composite of ZnO spheres and functionalized SnO_2 nanofibers with an enhanced ethanol gas sensing properties. *Mater. Lett.* **2016**, *169*, 246–249. [CrossRef]
8. Wang, Y.; Liu, L.; Meng, C.; Zhou, Y.; Gao, Z.; Li, X.; Cao, X.; Xu, L.; Zhu, W. A novel ethanol gas sensor based on $TiO_2/Ag_{0.35}V_2O_5$ branched nanoheterostructures. *Sci. Rep.* **2016**, *6*, 33092. [CrossRef] [PubMed]

9. Mirzaei, A.; Janghorban, K.; Hashemi, B.; Bonyani, M.; Leonardi, S.G.; Neri, G. Highly stable and selective ethanol sensor based on α-Fe_2O_3 nanoparticles prepared by Pechini sol–gel method. *Ceram. Int.* **2016**, *42*, 6136–6144. [CrossRef]
10. Mirzaei, A.; Bonyani, M.; Torkian, S.; Feizpour, M.; Bonavita, A.; Leonardi, S.G.; Neri, G. A comparative study on the electrical and gas sensing properties of thick films prepared with synthesized nano-sized and commercial micro-sized Fe_2O_3 powders. *Process. Appl. Ceram.* **2017**, *11*, 265–274. [CrossRef]
11. Yan, W.; Zeng, X.; Liu, H.; Guo, C.; Ling, M.; Zhou, H. Highly reliable and selective ethanol sensor based on α-Fe_2O_3 nanorhombs working in realistic environments. *Chin. Phys. B* **2019**, *28*, 106801. [CrossRef]
12. Liu, Y.; Yao, S.; Yang, Q.; Sun, P.; Gao, Y.; Liang, X.; Liu, F.; Lua, G. Highly sensitive and humidity-independent ethanol sensors based on In_2O_3 nanoflower/SnO_2 nanoparticle composites. *RSC Adv.* **2015**, *5*, 52252–52258. [CrossRef]
13. Niu, G.; Zhao, C.; Gong, H.; Yang, Z.; Leng, X.; Wang, F. NiO nanoparticle-decorated SnO2 nanosheets for ethanol sensing with enhanced moisture resistance. *Microsyst. Nanoeng.* **2019**, *5*, 21. [CrossRef] [PubMed]
14. Lombardo, L.; Grassini, S.; Parvis, M.; Donato, N.; Gulino, A. Ethanol breath measuring system. In Proceedings of the 2020 IEEE International Symposium on Medical Measurements and Applications (MeMeA), Bari, Italy, 1 June–1 July 2020; pp. 1–6.
15. Kaisdotter Andersson, A.; Hök, B.; Karlsson, A.; Pettersson, H. Unobtrusive Breath Alcohol Testing. In Proceedings of the 20th International Conference on Alcohol, Drugs and Traffic Safety, Brisbane, Australia, 25–28 August 2013.
16. Ljungblad, J.; Hök, B.; Allalou, A.; Pettersson, H. Passive in-vehicle driver breath alcohol detection using advanced sensor signal acquisition and fusion. *Traffic Inj. Prev.* **2017**, *18* (Suppl. 1), S31–S36. [CrossRef] [PubMed]
17. Pizon, A.F.; Becker, C.E.; Bikin, D. The Clinical Significance of Variations in Ethanol Toxicokinetics. *J. Med. Toxicol.* **2007**, *3*, 63–72. [CrossRef] [PubMed]

chemistry proceedings

MDPI

Proceeding Paper

Developing an Electrochemical Biosensor for the Detection of Hemagglutinin Protein of Influenza A Virus Subtype H1N1 in Artificial Saliva †

Carlos Torres-Méndez [1,*], Jayendra Ellamathy [1], Maria Ines Mascarenhas [2], Yifan Liu [1], Georgia-Vasiliki Gkountana [2,*], Patrizia Kühne [2], Javier Sebastián [1], Ivana Jovanovic [2], David Bern [1], Sharmilee Nandi [2], Maike Lüftner [2], Viktoria Langwallner [1], Maria Lysandrou [2], Sam Taylor [1], Klara Martinovic [2], Abdul-Raouf Atif [1], Ehsan Manouchehri Doulabi [2], Masood Kamali-Moghaddam [2,*] and Gemma Mestres [1,*]

1 Division of Biomedical Engineering, Department of Materials Science and Engineering, Uppsala University, Box 35, 751 03 Uppsala, Sweden; jayendra.ellamathy.0855@student.uu.se (J.E.); yifan.liu.5161@student.uu.se (Y.L.); javierenrique.sebastianalonso.8237@student.uu.se (J.S.); david.bern.5596@student.uu.se (D.B.); Viktoria.Langwallner.2522@student.uu.se (V.L.); samtamlyn.taylor.3541@student.uu.se (S.T.); abdul-raouf.atif@angstrom.uu.se (A.-R.A.)

2 Genetics and Pathology, Science for Life Laboratory, Department of Immunology, Uppsala University, 751 08 Uppsala, Sweden; mariaines.berrojoromeyromascarenhas.2772@student.uu.se (M.I.M.); patrizia.kuhne.3958@student.uu.se (P.K.); ivana.jovanovic.1181@student.uu.se (I.J.); sharmilee.nandi.8806@student.uu.se (S.N.); Maike.Luftner.8985@student.uu.se (M.L.); maria.lysandrou.6833@student.uu.se (M.L.); klara.martinovic.2708@student.uu.se (K.M.); ehsan.manouchehri@igp.uu.se (E.M.D.)

* Correspondence: carlosenrique.torresmendez.5581@student.uu.se (C.T.-M.); georgiavasiliki.gkountana.3705@student.uu.se (G.-V.G.); masood.kamali@igp.uu.se (M.K.-M.); gemma.mestres@angstrom.uu.se (G.M.)

† Presented at the 1st International Electronic Conference on Chemical Sensors and Analytical Chemistry, 1–15 July 2021; Available online: https://csac2021.sciforum.net/.

Citation: Torres-Méndez, C.; Ellamathy, J.; Mascarenhas, M.I.; Liu, Y.; Gkountana, G.-V.; Kühne, P.; Sebastián, J.; Jovanovic, I.; Bern, D.; Nandi, S.; et al. Developing an Electrochemical Biosensor for the Detection of Hemagglutinin Protein of Influenza A Virus Subtype H1N1 in Artificial Saliva. *Chem. Proc.* **2021**, 5, 80. https://doi.org/10.3390/CSAC2021-10477

Academic Editor: Maria Emília de Sousa

Published: 30 June 2021

Publisher's Note: MDPI stays neutral with regard to jurisdictional claims in published maps and institutional affiliations.

Abstract: Influenza A virus belongs to the Orthomyxoviridae family and, to date, is one of the most important pathogens causing acute respiratory infections, such as the recent pandemic of 2009. Hemagglutinin (HA) is one of the surface proteins of the virus that allow it to interact with cellular molecules. Due to the fact that it is the most abundant protein in the virus capsule, it is the best target in the detection of the Influenza A H1N1 virus through biosensing devices. Our aim is to develop an electrochemical biosensor to detect H1 by modifying carbon screen-printed electrodes (CSPE) with gold nanoparticles and to add further functionalization with monoclonal antibodies that are specific to this protein. The electrodes were characterized by the means of cyclic voltammetry, differential pulse voltammetry and electrochemical impedance spectroscopy. Our preliminary results suggest that the selected monoclonal antibodies have acceptable affinity and bind effectively to the H1 protein and that the electrodes have a wide potential window in the presence of $[Fe(CN)_6]^{3-/4-}$. In the future, we will continue to develop this biosensor in hope that it will be commercialized and be common in medical procedures during flu seasons and future influenza pandemics.

Keywords: influenza virus; voltammetry; screen-printed electrodes; hemagglutinin/HA protein; thiol chemistry

1. Introduction

In 2009, a novel H1N1 influenza A virus caused a pandemic leading to the death of 151,700–575,400 people worldwide according to the estimates of the Centers for Disease Control and Prevention (CDC) of the United States [1,2]. H1N1 influenza is a subtype of influenza A virus that was previously detected in swine, which causes upper and, in some cases, lower respiratory tract infections in its host [1]. Influenza A virus causes one of the most common respiratory diseases globally, seasonal flu, and together with Influenza B, C, and D, is a part of the Orthomyxoviridae virus family. Moreover, influenza A virus belongs to the single-stranded RNA viruses. It has a segmented genome that encodes several viral proteins that are important for the pathogenesis of the virus [3,4]. Two of these proteins are important for detecting the virus in human specimens; these are hemagglutinin (HA) and neuraminidase (NA), which are the surface proteins of the virus involved in host invasion [5]. HA is the major protein of H1N1 and it is the protein with which the virus binds to the host's cells and invades them, while NA helps in the viral spreading from cell to cell [5].

So far, most of the detection methods for the influenza A virus are characterized by a long detection time, expensive instruments and reagents, and the need for trained technicians, thus creating an inconvenience for both the patients and the healthcare workers [1,6]. The development of sensitive and rapid detection methods, such as biosensors, is now the focus of many research groups and could be a great solution to the aforementioned problem. A lot of different biorecognition elements can be used for the detection of an analyte. However, antibodies seem to be the most widely used type among these elements.

Antibodies are specialized, Y-shaped proteins that identify pathogens by selectively binding to their membranes [7]. Due to their high specificity and sensitivity, antibodies are ideal biorecognition elements for biosensors [8]. Other biorecognition elements commonly used in biosensors include enzymes, nucleic acids, aptamers and molecular-imprinted polymers [9]. The focus of this paper is the development of an electrochemical antibody-based biosensor for the detection of the influenza A surface protein H1.

2. Materials and Methods

2.1. Reagents and Materials

HA H1N1 protein, mouse monoclonal antibodies (mAbs) and rabbit polyclonal antibodies (pAb) were purchased from Sinobiological (Frankfurt, Germany). Secondary goat anti-Rabbit IgG antibodies (Alexa Fluor 568) were purchased from Thermo Fisher (Waltham, MA, USA). Chloroauric acid ($HAuCl_4$), Sulphuric acid (H_2SO_4), 4 aminothiophenol (4-ATP), ethanol, potassium hexacyanoferrate (II) trihydrate and Potassium hexacyanoferrate (III) were purchased from Sigma Aldrich (Darmstadt, Germany). Carbon screen-printed electrodes (CSPE) were provided by Zimmer & Peacock (Horten, Norway).

2.2. Electrochemical Measurements

The EmStat Pico Module potentiostat controlled using the PSTrace 5.8 computer software was employed for all cyclic voltammetry (CV), electrochemical impedance spectroscopy (EIS), chronoamperometric electrodeposition and differential pulse voltammetry (DPV) experiments. The CSPE was used as a three-electrode cell system comprising a carbon working electrode (WE), a carbon counter electrode (CE) and an Ag/AgCl reference electrode (RE). EIS measurements were made at 6 mV ac amplitude in the frequency range of 5.0 mHz to 50 kHz and the equivalent circuit models were fitted using PSTrace software.

2.3. Electrodeposition of Gold Nanoparticles on CSPE

A modified method from the literature was employed [10], an aqueous solution containing 2 mM $HAuCl_4$ and 0.5 M H_2SO_4 was used to cover the CSPE, a chronoamperometric method using a constant potential of −0.25 V for 60 s was used to deposit gold nanoparticles on top of the CSPE, and the electrode was washed with abundant deionized water, left to dry at room temperature and identified as AuNP-CSPE.

2.4. Modification of AuNP-CSPE Electrodes with 4-ATP

A previously reported method was adapted [11], in a typical experiment, and the working electrode was covered with 10 μL of 10 mM 4-ATP solution in ethanol at room temperature (22 °C) for 15 min. Nonspecifically adsorbed molecules were flushed off by careful rinsing with ethanol and deionized water. The electrode was dried under a stream of nitrogen and identified as NH_2-AuNP-CSPE. The amine functionality in the electrode could be used later to form an amide bond [12] and immobilize the mouse monoclonal antibodies against the HA H1N1 protein.

2.5. Testing of mAb Specificity and Sensitivity

The enzyme-linked immunosorbent assay (ELISA) was used for this purpose. The protocol used for this indirect sandwich ELISA assay was in accordance with the mAb provider [13].

3. Results and Discussion

3.1. Electrodeposition of Gold Nanoparticles

The CSPEs offered a reasonable potential window to study the redox reaction of the $[Fe(CN)_6]^{3-/4-}$ system and showed a symmetric shape, and the distance between the oxidation and the reduction peaks was of 727 mV (Figure 1); this value was much higher than the prediction of the Nerst equation for single electron transfer reactions and was attributed to a drop in potential due to the resistance of the carbon material [14]. When the CSPEs were modified with gold nanoparticles, the reversibility of the $[Fe(CN)_6]^{3-/4-}$ redox system increased as the distance between the oxidation and reduction peaks was 280 mV on the voltammogram; this was attributed to the increase in the surface area of the electrode and to the high conductivity of metallic gold nanoparticles.

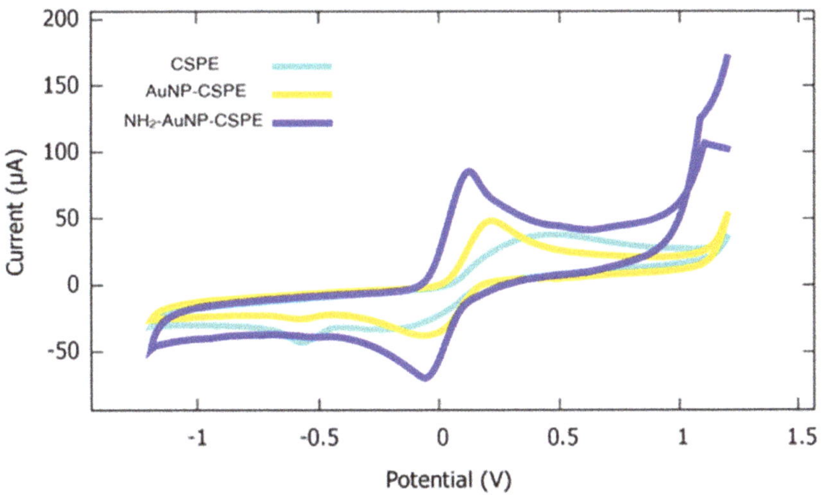

Figure 1. Cyclic voltammogram of CSPE, AuNP-CSPE and NH_2-NP-CSPE in the presence of $[Fe(CN)_6]^{3-/4-}$ obtained at a scan rate of 100 mV/s.

3.2. Electrodeposition Length

Further study into the gold electrodeposition process as a function of time (Figure 2) showed that longer reaction times than one minute do not increase either the current response of the electrode or the reversibility of the system.

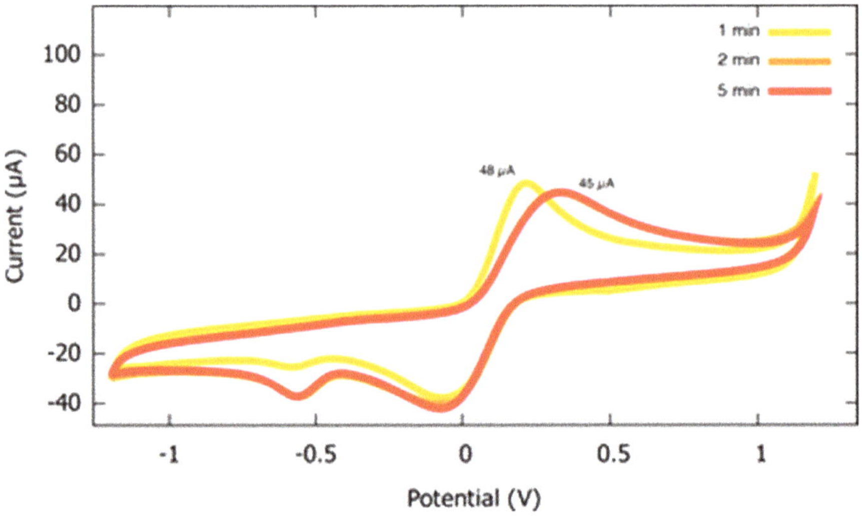

Figure 2. Cyclic voltammogram of modified AuNP-CSPEs using different electrodeposition times; the experiment was conducted in the presence of $[Fe(CN)_6]^{3-/4-}$ at a scan rate of 100 mV/s.

3.3. *Characterization of NH_2-AuNP-CSPE*

The cyclic voltammogram of NH_2-AuNP-CSPE showed promising results due to the functionalization of the nanoparticles with the 4-ATP linker molecule (Figure 1), as the reversibility of the system increased and the electron transfer process for the reduction and oxidation reactions of $[Fe(CN)_6]^{3-/4-}$ was facilitated on these modified electrodes due to the π delocalized system of the linker molecule. The electrochemical impedance spectroscopy tests indicated a decrease in impedance after the modification of the electrodes with gold nanoparticles and 4-ATP linker (Figure 3). The Nyquist plot of the bare CSPE can be fitted to an equivalent circuit for a simple electron-transfer reaction and NH_2-AuNP-CSPE can be fitted to the classical Randles equivalent circuit comprising the ohmic resistance of the electrolyte solution (R_Ω) and the charge transfer resistance (R_{CT}), in series with the Warburg impedance element (diffusion controlled impedance) and in parallel with a double layer capacitance (C_{DL}) (Figure 4). The modified NH_2-AuNP-CSPE showed a significantly smaller R_{CT} and is, therefore, highly conductive compared to the bare CSPE.

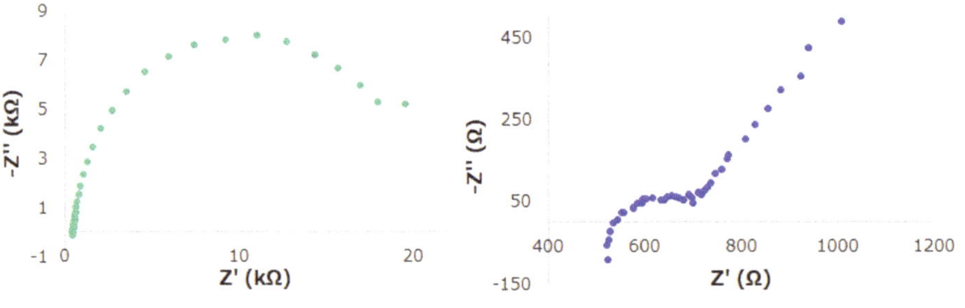

(**A**) (**B**)

Figure 3. Nyquist plot of CSPE (**A**) and NH_2-AuNP-CSPE (**B**) using the frequency range of 5.0 mHz to 50 kHz.

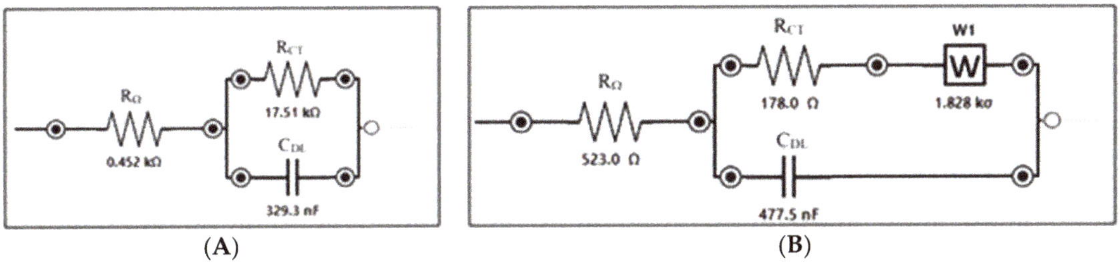

Figure 4. Fitting of CSPE to equivalent circuit for a simple electron transfer (**A**) and fitting of the NH_2-AuNP-CSPE Randles equivalent circuit (**B**).

3.4. Effect of the $[Fe(CN)_6]^{3-/4-}$ Concentration

The effect of the $[Fe(CN)_6]^{3-/4-}$ concentration over the current response was evaluated by the means of CV and DPV (Figure 5), with a higher concentration of the electroactive species in the solution implying that more $[Fe(CN)_6]^{3-/4}$ molecules could reach the surface of the NH_2-AuNP-CSPE to undergo oxidation and reduction, respectively. In the experiment, higher current responses were observed at high concentrations of $[Fe(CN)_6]^{3-/4}$. This experiment provided a visual basis for the expected effect on the final design of the biosensor, where the NH_2-AuNP-CSPE is going to be coupled to monoclonal antibodies against the H1 protein and a blocking effect over $[Fe(CN)_6]^{3-/4}$ (lowering the current response) could take place once the H1 protein is bound to the antibody-modified electrode.

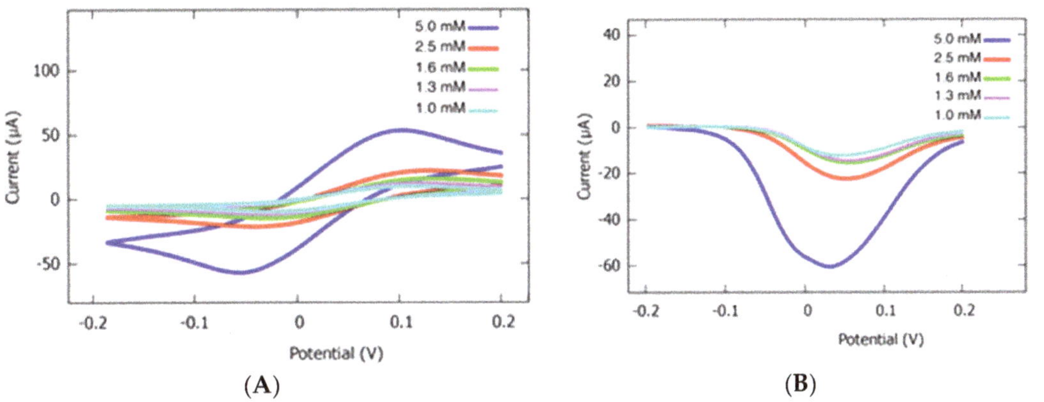

Figure 5. Cyclic voltammogram (**A**) and differential pulse voltammogram (**B**) for NH_2-AuNP-CSPE as a function of the $[Fe(CN)_6]^{3-/4-}$ concentration.

3.5. mAb Characterization

To characterize the specificity of the monoclonal antibody targeting the H1 protein, an indirect sandwich ELISA was conducted. It was shown that the antibody can detect the H1 protein specifically (Figure 6). The approach was tested for different H1 concentrations and the lowest detectable concentration was 10 ng/mL.

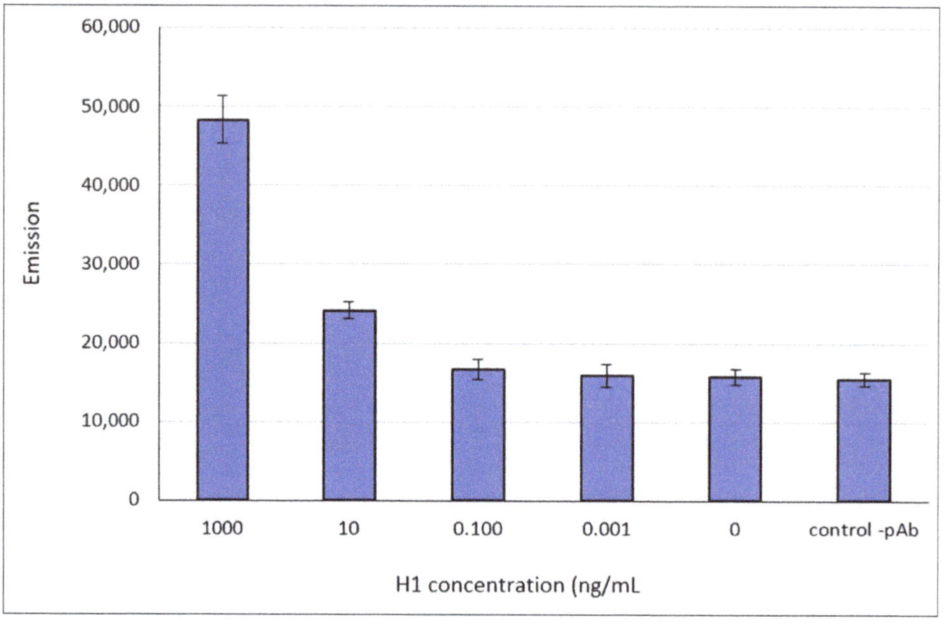

Figure 6. Indirect sandwich ELISA results using 1:500 dilution of mAb, 1:1000 dilution of pAb and different H1 concentrations as shown in the *x* axis; the *y* axis shows the emission measured after the addition of the secondary antibody. Experiments were conducted twice with replicates of 2–6.

4. Conclusions

Cost-effective CSPE were modified with gold nanoparticles and 4-ATP, and the electrodes were characterized by means of electrochemical methods in the presence of $[Fe(CN)_6]^{3-/4-}$ complexes. The redox system $[Fe(CN)_6]^{3-/4-}$ was found to be more reversible in the modified electrodes AuNP-CSPE and NH_2-AuNP-CSPE than in the bare CSPE. The modified NH_2-AuNP-CSPE showed a decrease in impedance compared to CSPE, indicating that the electron transfer process is more favorable in the modified electrode. With the indirect sandwich ELISA, it was shown that the monoclonal antibody specifically targets the HA H1N1 protein and can be further used in the biosensor setup. The amine functionality on the modified electrodes can be exploited to couple mouse monoclonal antibodies against the HA H1N1 protein in future work within this project. In addition to this, aims for future work include the detection of the HA H1N1 protein in artificial saliva using DPV, the establishment of a protocol using bovine serum albumin (BSA) to avoid non-specific binding, and the determination of the sensibility and detection limits of the biosensor.

Author Contributions: Conceptualization, data curation, formal analysis and visualization, C.T.-M., S.N., P.K., J.S. and V.L.; methodology, A.-R.A. and E.M.D.; software and validation, J.E., Y.L. and S.T.; investigation, I.J., M.L. (Maike Lüftner), K.M. and M.L. (Maria Lysandrou); resources, D.B.; writing—original draft preparation, C.T.-M., G.-V.G., P.K. and M.I.M.; writing—review and editing, C.T.-M. and P.K.; supervision, project administration and funding acquisition M.K.-M. and G.M. All authors have read and agreed to the published version of the manuscript.

Funding: This work was supported by the Disciplinary Domain of Medical and Pharmacy and Disciplinary Domain of Science and Technology at Uppsala University, Swedish Research Council under grant 2020-02258 and Swedish Institute's scholarship to C. Torres-Méndez.

Institutional Review Board Statement: Not applicable.

Informed Consent Statement: Not applicable.

Data Availability Statement: All data are included herein and no data deposited in external repositories.

Acknowledgments: We are grateful to SensUs organization for all the support. We are also grateful to Quentin Palomar-Marchand for the technical advice provided for the electrochemical measurements and Zimmer & Peacock for providing the carbon screen-printed electrodes used in this study.

Conflicts of Interest: The authors declare no competing interest.

References

1. Patel, M.; Dennis, A.; Flutter, C.; Khan, Z. Pandemic (H1N1) 2009 influenza. *Br. J. Anaesth.* **2010**, *104*, 128–142. [CrossRef] [PubMed]
2. 2009 H1N1 Pandemic. Centers for Disease Control and Prevention. Available online: https://www.cdc.gov/flu/pandemic-resources/2009-h1n1-pandemic.html (accessed on 11 June 2019).
3. Krammer, F.; Smith, G.J.D.; Fouchier, R.A.M.; Peiris, M.; Kedzierska, K.; Doherty, P.C.; Palese, P.; Shaw, M.L.; Treanor, J.; Webster, R.G.; et al. Influenza. Nature Reviews. *Dis. Primers* **2018**, *4*, 3. [CrossRef] [PubMed]
4. Jilani, T.N.; Jamil, R.T.; Siddiqui, A.H. H1N1 Influenza. In *StatPearls*; StatPearls Publishing: Treasure Island, FL, USA, 2021.
5. Sriwilaijaroen, N.; Suzuki, Y. Molecular basis of the structure and function of H1 hemagglutinin of influenza virus. *Proc. Jpn. Acad. Ser. B* **2012**, *88*, 226–249. [CrossRef] [PubMed]
6. Ravina, R.; Dalal, A.; Mohan, H.; Prasad, M.; Pundir, C. Detection methods for influenza A H1N1 virus with special reference to biosensors: A review. *Biosci. Rep.* **2020**, *40*, BSR20193852. [CrossRef] [PubMed]
7. Antibodies Use in Biosensors. Available online: News-Medical.Net (accessed on 20 January 2021).
8. Sharma, S.; Byrne, H.; O'Kennedy, R.J. Antibodies and antibody-derived analytical biosensors. *Essays Biochem.* **2016**, *60*, 9–18. [PubMed]
9. Chambers, J.P.; Arulanandam, B.P.; Matta, L.L.; Weis, A.; Valdes, J.J. Biosensor recognition elements. *Curr. Issues Mol. Biol.* **2008**, *10*, 1–12. [PubMed]
10. Wang, Y.C.; Cokeliler, D.; Gunasekaran, S. Reduced graphene oxide/carbon nanotube/gold nanopar-ticles nanocomposite functionalized screen-printed electrode for sensitive electrochemical detection of endocrine disruptor bisphenol A. *Electroanalysis* **2015**, *27*, 2527–2536. [CrossRef]
11. Valerio, E.; Abrantes, L.M.; Viana, A.S. 4-Aminothiophenol Self-Assembled Monolayer for the De-velopment of a DNA Biosensor Aiming the Detection of Cylindrospermopsin Producing Cyanobacteria. *Electro-Anal. Int. J. Devoted Fundam. Pract. Asp. Electroanal.* **2008**, *20*, 2467–2474.
12. Rezki, M.; Septiani, N.L.W.; Iqbal, M.; Harimurti, S.; Sambegoro, P.; Adhika, D.R.; Yuliarto, B. Amine-functionalized Cu-MOF Nanospheres towards Label-free Hepatitis B Surface Antigen Electrochemical Immunosensors. *J. Mater. Chem. B* **2021**, *9*, 5711–5721. [CrossRef] [PubMed]
13. Sandwich ELISA Protocol | Sino Biological. (n.d.) Available online: https://www.sinobiological.com/category/sandwich-elisa-protocol (accessed on 13 June 2021).
14. Damiati, S.; Haslam, C.; Sopstad, S.; Peacock, M.; Whitley, T.; Davey, P.; Awan, S. Sensitivity Comparison of Macro- and Micro-Electrochemical Biosensors for Human Chorionic Gonadotropin (hCG) Biomarker Detection. *IEEE Access* **2019**, *7*, 94048–94058. [CrossRef]

chemistry proceedings

MDPI

Abstract

Determination of Chemical Oxygen Demand (COD) Using Nanoparticle-Modified Voltammetric Sensors and Electronic Tongue Principles †

Qing Wang and Manel del Valle *

Group of Sensors and Biosensors, Department of Chemistry, Universitat Autònoma de Barcelona, Bellaterra, 08193 Barcelona, Spain; arieswq@163.com

* Correspondence: manel.delvalle@uab.cat

† Presented at the 1st International Electronic Conference on Chemical Sensors and Analytical Chemistry, 1–15 July 2021; Available online: https://csac2021.sciforum.net/.

Abstract: Chemical Oxygen Demand (COD) is a widely used parameter in analyzing and controlling the degree of pollution in water. COD is defined as the amount of molecular oxygen (in milligrams of O_2) required to decompose all the organic compounds in 1 L of aqueous solution to carbon dioxide and water. There are many methods reported for COD determination, such as the conventional dichromate titration method. Electro-oxidizing the organic contaminants to completely transform them into CO_2 and H_2O using sensors is considered the best method for COD estimation. Increasing attention has been paid to electrochemical methods because they are highly sensitive, time-saving, low-cost, and easy to operate. In this sense, copper electrodes have been reported based on the fact that copper in alkaline media acts as a powerful electrocatalyst for the oxidation of aminoacids and carbohydrates, which are believed to be the major culprits for organic pollution. Cyclic voltammetry was the technique used to obtain the voltammetric responses. It is common for different organic compounds to show different cyclic voltammogram shapes and current intensities in different concentrations. In this work, four kinds of electrodes modified with copper (Cu)/copper oxide (CuO)/nickel copper alloy (Ni Cu alloy) nanoparticles were studied for COD analysis; this was done by employing the cyclic voltammetry technique, which involved a Nafion film-covered electrodeposited CuO/Cu nanoparticle electrode (**E1**), a Cu nanoparticle–graphite composite electrode (**E2**), a CuO nanoparticle–graphite composite electrode (**E3**), and a Ni Cu alloy nanoparticle–graphite composite electrode (**E4**). The COD values were determined via the plotted calibration of COD values vs. the current intensity. Glucose, glycine, potassium hydrogen phthalate (KHP), and ethylene glycol—which show different reducibilities—were chosen as the standard substances to play the role of organic contaminants with different degradation difficulties. From the obtained cyclic voltammograms, we can see that glucose is very easily oxidized by those four electrodes, with electrode **E1** displaying the best performance, with a linear range of 19.2~1120.8 mg/L and limit of detection of 27.5 mg/L (calculated based on the formula $3\sigma/k$). In contrast, it is very difficult for the compound KHP to be oxidized by these four electrodes. Nevertheless, the obtained voltammetric profiles presented different shapes with the tested organic compounds, suggesting these four electrodes can compose an electronic tongue array for multivariate analysis. As a result, the main component of river samples—whose degradation could be easy or difficult—can be evaluated via the PCA technique. This evaluation is very helpful for the accuracy of COD detection. The resulting sensor-based method demonstrates great potential not only for estimating the precise value of COD, but for predicting the difficulty of its degradation; this represents a simple, fast, and clean methodology, which is perfectly suited to the present demands of green techniques.

Keywords: Chemical Oxygen Demand; copper (oxide) nanoparticle electrodes; nickel copper alloy nanoparticle electrode; cyclic voltammetry; electronic tongue

Citation: Wang, Q.; del Valle, M. Determination of Chemical Oxygen Demand (COD) Using Nanoparticle-Modified Voltammetric Sensors and Electronic Tongue Principles. *Chem. Proc.* **2021**, *5*, 81. https://doi.org/10.3390/CSAC2021-10442

Academic Editor: Ye Zhou

Published: 30 June 2021

Supplementary Materials: The poster presentation is available online at https://www.mdpi.com/article/10.3390/CSAC2021-10442/s1.

Author Contributions: Conceptualization, Q.W. and M.d.V.; methodology, M.d.V.; writing—original draft preparation, Q.W.; writing—review and editing, M.d.V.; supervision, M.d.V.; Funding acquisition, M.d.V. All authors have read and agreed to the published version of the manuscript.

Funding: This research was funded by the Spanish Ministry of Science and Innovation, MCINN (Madrid) through project PID2019-107102RB-C21C. Q.W. acknowledges the concession of a PhD grant of the Chinese Scholarship Council (China). M.d.V. thanks the support from Generalitat de Catalunya through the program ICREA Academia.

Institutional Review Board Statement: No applicable.

Informed Consent Statement: No applicable.

Data Availability Statement: The data presented in this study are available in this article and supporting Information.

Conflicts of Interest: The authors declare no conflict of interest.

chemistry proceedings

MDPI

Proceeding Paper

New Half Metal Perovskite $NbScO_3$ for Spintronic Sensing Applications †

Amall Ahmed Ramanathan

Department of Physics, The University of Jordan, Amman 11942, Jordan; amallahmad@gmail.com

† Presented at the 1st International Electronic Conference on Chemical Sensors and Analytical Chemistry, 1–15 July 2021; Available online: https://csac2021.sciforum.net/.

Abstract: Half-metallic ferromagnetic (HMF) materials demonstrate 100% spin polarization at the Fermi level, making them promising candidates for spintronic sensing applications. In this work, the full potential linearized augmented plane wave (FP-LAPW) density functional theory (DFT) method is used to calculate the electro-magnetic properties of the transition metal perovskite $NbScO_3$ using the generalized gradient approximation (GGA) and the modified Becke-Johnson (mBJ) approximation for the exchange correlations. The electronic band structures for the two spin orientations using GGA, predict $NbScO_3$ to be an HMF with an integer magnetic moment of 2.0 μB and hence a promising candidate for spintronics. The new half metal perovskite shows metallic behavior in the majority spin and semiconducting in the minority spin channel with a direct $\Gamma-\Gamma$ band gap of 1.870 eV. The integer magnetic moment of 2.0 μB is also preserved with mBJ exchange potential. The band structure, however, shows indirect gaps $R-\Gamma$ and $X-\Gamma$ of 2.023 eV and 0.780 eV in the minority and majority channels, respectively indicating $NbScO_3$ to be a magnetic semiconductor. The results indicate the suitability of $NbScO_3$ for spintronics as the necessary conditions are satisfied.

Keywords: half metal; band structure; spintronics; sensors; information technology; perovskite

check for updates

Citation: Ramanathan, A.A. New Half Metal Perovskite $NbScO_3$ for Spintronic Sensing Applications. *Chem. Proc.* **2021**, *5*, 82. https://doi.org/10.3390/CSAC2021-10628

Academic Editor: Ye Zhou

Published: 7 July 2021

Publisher's Note: MDPI stays neutral with regard to jurisdictional claims in published maps and institutional affiliations.

1. Introduction

The rapid technological advancements in the last decade call for smart and sustainable lifestyle management, with sensors playing a vital role [1–4]. Electron spin is fast becoming a very useful tool in sensing devices that are based on spintronics. Spintronics is a science in which the electron spin instead of the charge is used as the information carrier, providing the advantages of low energy consumption, high speed data processing and circuit integration density [5–7]. Among today's various proposed information transfer methodologies like molecular/nano electronics and quantum technologies, spintronics stands out due to the fact that it is compatible with conventional electronics making it easy to extend the existing well known electronic techniques to spintronic circuits. HMFs, due to their exceptional electronic structure, satisfy the needs for spintronic applications. The electrons of one spin direction behave as metals and those of the other spin direction act as semiconductors. Recently, quite a few new perovskites have been predicted to be half-metals [8–10].

Transition metals (TM) are of special interest, and a variety of interesting magnetic properties have been identified, as seen from recent research results. Depending upon the local environment non-magnetic materials have become magnetic due to their presence [11–13]. TM perovskites have piqued the interest of the scientific community due the intriguing nature of the TM ion interplay with the oxide or halide ion [14,15] with the great possibilities of different electronic and magnetic properties.

Unlike the majority of previous research, wherein the TM occupies the B site, in this work we switched the sites, and the TM Niobium occupies the A site with some very interesting magneto-electronic results. The purpose of the paper is to give the essential and accurate theoretical characterization using DFT_FP-LAPW of the perovskite $NbScO_3$ which is being investigated for the first time for potential use in spintronics and sensing.

2. Method and Materials

The full-potential linearized-augmented plane wave (FP-LAPW) method, as implemented in the WIEN2k [16] package, is used to calculate the spin polarized ground states of the cubic perovskite $NbScO_3$ within the DFT [17,18] formalism. The Perdew, Burke and Ernzerhof (PBE) [19] generalized gradient approximation (GGA) is used to calculate the optimized structures for a $10 \times 10 \times 10$ k-point grid. The optimized lattice constant value is then used to evaluate the electronic and magnetic properties with the more accurate mBJ exchange correlation of Trans Blaha [20] at a denser k-point grid of $15 \times 15 \times 15$. K_{max}, which provides the magnitude of the largest K vector in the plane-wave expansion is set to 8. The muffin-tin radii were set to 1.60 a.u for Sc and O atoms and 2.7 a.u for Nb. The tetrahedron method [21] with 120 k points in the irreducible Brillouin zone is employed for integrations within the self-consistency cycle (SCF). The convergence tolerance thresholds for SCF is less than 10^{-4} Ry for energy and 10^{-4} for electron charges.

3. Results and Discussion

3.1. Structural and Electronic

The cubic $NbScO_3$ perovskite has the space group Pm-3m (#221) symmetry, and the atoms occupy the positions 1a (0, 0,0), 1b ($\frac{1}{2}$, $\frac{1}{2}$,$\frac{1}{2}$) and 3c (0,$\frac{1}{2}$, $\frac{1}{2}$) sites of Wyckoff coordinates for Nb, Sc and O atoms, respectively, as depicted in Figure 1 in an inset image.

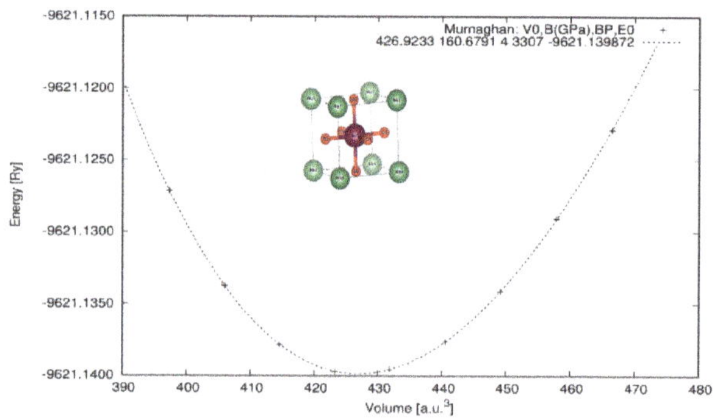

Figure 1. The energy-volume optimization and Murnaghan fit for the perovskite $NbScO_3$.

The lattice constants are optimized using the Murnaghan equation of state [22] with only volume optimization as the structure is cubic is required. The energy vs. volume optimization gives the minimum equilibrium energy state lattice parameter and is presented in Figure 1. The optimized values of the lattice constant and bulk modulus obtained are 3.985 Å and 160.679 GPa, respectively. This lattice constant value of $NbScO_3$ is used alongside the GGA-PBE and mBJ exchange correlations at denser grids to calculate the electronic band structures along the high symmetry points. The GGA band structure is shown in Figure 2 for both the spin Dn (minority) and spin Up (majority) orientations.

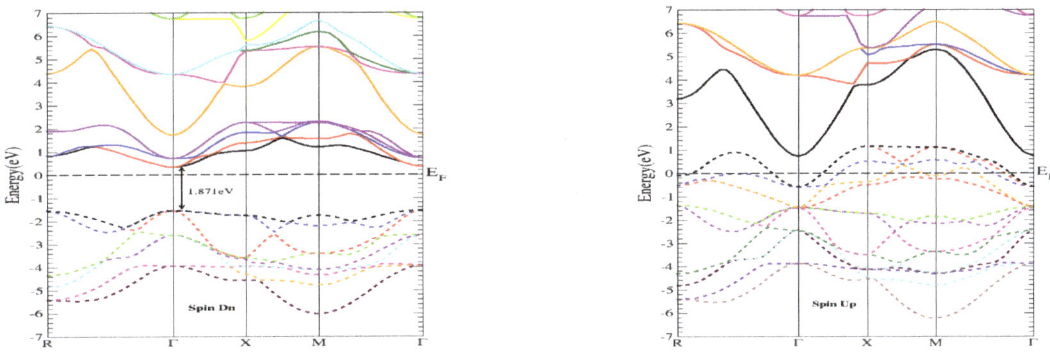

(**a**) (**b**)

Figure 2. The $NbScO_3$ electronic band structures for (**a**) the minority (Spin Dn) and (**b**) the majority (Spin Up) channels with PBE-GGA.

We see from the figure that $NbScO_3$ shows typical semi-conducting behavior in the minority spin with a direct Γ−Γ gap of 1.87 eV, and that it is metallic in the majority spin resulting in a half metal ferromagnetic behavior (HMF). HMFs have 100% spin polarization and can intrinsically provide single-spin channel electrons, which are very useful in spintronics.

The band structure of $NbScO_3$ with the mBJ exchange potential on the other hand shows it to be a magnetic semiconductor as seen from the Figure 3 plots for the spin Dn and spin Up states.

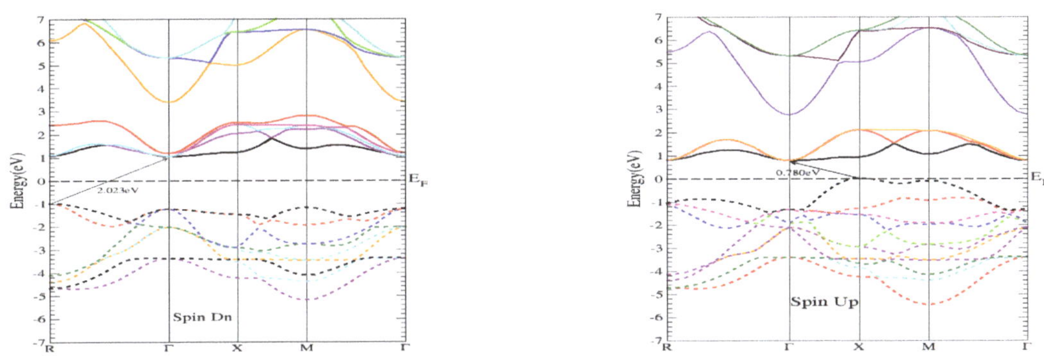

(**a**) (**b**)

Figure 3. The $NbScO_3$ electronic band structures for (**a**) the minority (Spin Dn) and (**b**) the majority (Spin Up) channels with mBJ.

The minority spin and majority spin have indirect R−Γ and X−Γ gaps of 2.02 and 0.78 eV, respectively. Magnetic semiconductors combine the advantages of both magnets and semiconductors, and form the basis for spintronics. Magnetic semiconductors can be used for spin generation, injection, and spin manipulation and detection. Since the mBJ exchange potential provides very reliable and accurate band structures in comparison to that of GGA or hybrid functionals, the correct behavior of $NbScO_3$ would be a magnetic semiconductor.

The half metal gap or spin flip energy E_{HM} is defined here as the minimum energy required to flip a minority-spin electron from the valance band maximum to the majority spin Fermi level. The predicted band gaps and E_{HM} for the GGA and mBJ exchange potential are listed in Table 1.

Table 1. The electronic band gaps in the minority and majority spin channels for the $NbScO_3$ perovskite.

$NbScO_3$ Perovskite	Band Gap (eV)				E_{HM} (eV)
	Minority (Spin Dn)		Majority (Spin Up)		
GGA-PBE	Γ−Γ	1.871	metallic	No gap	1.547
mBJ	R−Γ	2.023	X−Γ	0.780	0.979

3.2. Magnetic

The band-structure plots in the previous section have clearly shown the magnetic nature of $NbScO_3$, and to fully understand the origin and hybridization of the atomic orbitals, the total and partial density of states (TDOS/PDOS) are calculated using the mBJ exchange potential. These are depicted in Figure 4.

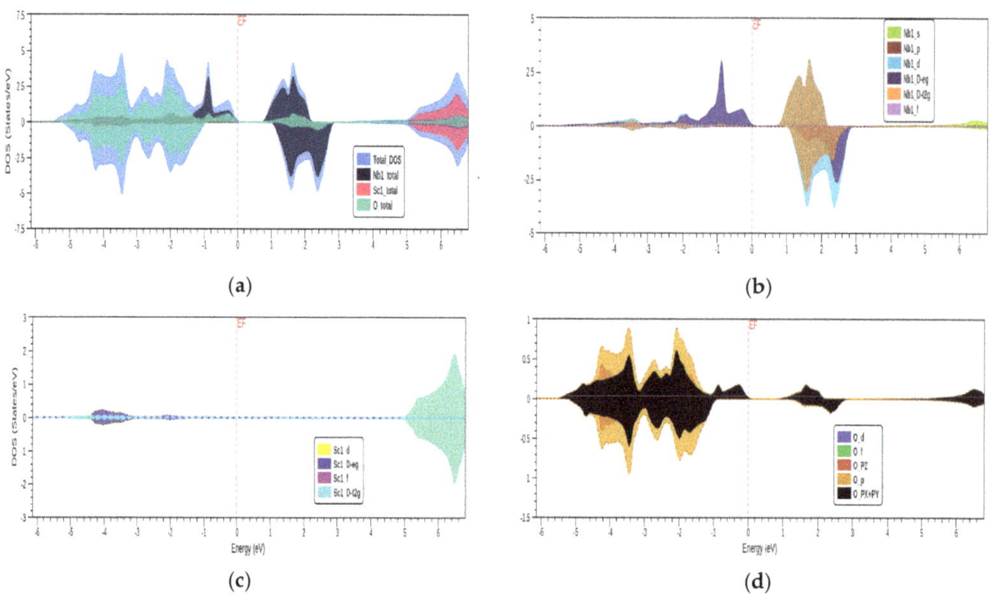

Figure 4. The $NbScO_3$ electronic DOS in both minority (bottom part) and majority (top part) spin channels (**a**) the TDOS for the compound and the atom constituents (**b–d**) show the PDOS in the different orbitals for the Nb atom, the Sc atom and the oxygen atom, respectively.

We can clearly observe the semiconducting and magnetic nature of $NbScO_3$from the plots. The minority channel shows the wide band gap, with no states at E_F (Fermi-energy), whereas the majority channel has the valence band edge at the Fermi energy and one can say that $NbScO_3$ is an intrinsic magnetic semiconductor. The role of Sc in the magnetism is negligible as indicated by the spin polarized PDOS plot for Sc. The main contribution to the magnetism comes from the 'd' orbital of Nb and the 'Px' and 'Py' orbitals of O_2. Additionally, the significant difference in the majority and minority TDOS can be clearly seen; resulting in the total integer magnetic moment of 2 μB typical of HMF as given in Table 2. The table also lists the atom-wise and interstitial magnetic moments of $NbScO_3$ with the GGA and mBJ exchange correlations.

Table 2. The total and atom projected magnetic moments of the $NbScO_3$ perovskite in units of μ_B.

Exchange	Nb	Sc	O	Intertitial	Total Moment
GGA_PBE	1.8484	0.0126	−0.0761	0.3673	2.0000
mBJ	1.7095	−0.0007	0.0460	0.1533	2.0000

4. Conclusions

In conclusion, the FP-LAWP investigation of the perovskite $NbScO_3$ has predicted the system to be a HMF and an intrinsic magnetic semiconductor using the GGA and mBJ exchange potentials, respectively. Moreover, the considerable size of the bandgap and magnetic moment obtained confirms the feasibility of $NbScO_3$ for spintronic applications. In addition, the large value of E_{HM} supports the robustness of this system for spintronics and sensing applications, whereby a significant role is played by the spin of the electrons in the sensor design.

Funding: This research no external funding.

Institutional Review Board Statement: Not applicable.

Informed Consent Statement: Not applicable.

Data Availability Statement: Data can be provided upon reasonable request.

Conflicts of Interest: The authors declare no conflict of interest.

References

1. Tricoli, A.; Nasiri, N.; De, S. Wearable and miniaturized sensor technologies for personalized and preventive medicine. *Adv. Funct. Mater.* **2017**, *27*, 1605271. [CrossRef]
2. Aqra, M.W.; Ramanathan, A.A. Review of the Recent Advances in Nano-Biosensors and Technologies for Healthcare Applications. *Chem. Proc.* **2021**, *5*, 76. [CrossRef]
3. Behera, B.; Joshi, R.; Vishnu, G.K.A.; Bhalerao, S.; Pandya, H.J. Electronic nose: A non-invasive technology for breath analysis of diabetes and lung cancer patients. *J. Breath Res.* **2019**, *13*, 024001. [CrossRef] [PubMed]
4. Ramanathan, A.A. Defect Functionalization of MoS_2 nanostructures as toxic gas sensors: A review. *IOP Conf. Ser. Mater. Sci. Eng.* **2018**, *305*, 012001. [CrossRef]
5. Ramanathan, A.A. First Principles Investigation of the Optoelectronic Properties of Molybdenum Dinitride for Optical Sensing Applications. *Chem. Proc.* **2021**, *5*, 27. [CrossRef]
6. Wolf, S.A.; Awschalom, D.D.; Buhrman, R.A.; Daughton, J.M.; von Molnár, V.S.; Roukes, M.L.; Chtchelkanova, A.Y.; Treger, D.M. Spintronics: A spin-based electronics vision for the future. *Science* **2001**, *294*, 1488–1495. [CrossRef]
7. Li, X.; Yang, J. First-principles design of spintronics materials. *Natl. Sci. Rev.* **2016**, *3*, 365–381. [CrossRef]
8. Khandy, S.A.; Gupta, D.C. Structural, elastic and magneto-electronic properties of half-metallic BaNpO3 perovskite. *Mater. Chem. Phys.* **2017**, *198*, 380–385. [CrossRef]
9. He, M.S. Half-metallic double perovskites Sr2CrWO6 and Sr2FeReO6 materials for spintronics applications. *Adv. Tissue Eng. Regen. Med. Open Access* **2018**, *4*, 1–6. [CrossRef]
10. Ramanathan, A.A.; Khalifeh, J.M. Electronic, magnetic and optical properties of XScO3 (X = Mo, W) perovskites. *PeerJ Mater. Sci.* **2021**, *3*, e15. [CrossRef]
11. Babar, Z.U.D.; Fatheema, J.; Arif, N.; Anwar, M.S.; Gul, S.; Iqbal, M.; Rizwan, S. Magnetic phase transition from paramagnetic in Nb2AlC-MAX to superconductivity-like dia-magnetic in Nb2C-MXene: An experimental and computational analysis. *RSC Adv.* **2020**, *10*, 25669–25678. [CrossRef]
12. Ramanathan, A.A.; Khalifeh, J.M.; Hamad, B.A. Evidence of surface magnetism in the V/Nb(0 0 1) system: A total energy pseudopotential calculation. *Surf. Sci.* **2008**, *602*, 607. [CrossRef]
13. Ramanathan, A.; Khalifeh, J.; Hamad, B. Structure and magnetism of the V/Ta(001) surface: A DFT calculation. *J. Magn. Magn. Mater.* **2009**, *321*, 3804–3807. [CrossRef]
14. Tomiyasu, K.; Okamoto, J.; Huang, H.Y.; Chen, Z.Y.; Sinaga, E.P.; Wu, W.B.; Chu, Y.Y.; Singh, A.; Wang, R.-P.; de Groot, F.M.F.; et al. Coulomb Correlations Intertwined with Spin and Orbital Excitations in LaCoO3. *Phys. Rev. Lett.* **2017**, *119*, 196402. [CrossRef] [PubMed]
15. Julian Gebhardt, J.; Rappe, A.M. Transition metal inverse-hybrid perovskites. *Mater. Chem. A* **2018**, *6*, 14560–14565. [CrossRef]
16. Blaha, P.; Schwarz, K.; Tran, F.; Laskowski, R.; Madsen, G.K.; Marks, L.D. WIEN2k: An APW+ lo program for calculating the properties of solids. *J. Chem. Phys.* **2020**, *152*, 074101. [CrossRef]
17. Ramanathan, A.A. A DFT calculation of Nb and Ta (001) Surface Properties. *JMP* **2013**, *4*, 432–437. [CrossRef]

18. Ramanathan, A.A. The Magnetism of a Ni Monolayer on Vanadium: Structure and Exchange Correlation Effects. *IEEE Trans. Nanotechnol.* **2020**, *19*, 11–16. [CrossRef]
19. Perdew, J.P.; Burke, K.; Ernzerhof, M. Generalized gradient approximation made simple. *Phys. Rev. Lett.* **1996**, *77*, 3865–3868. [CrossRef]
20. Tran, F.; Blaha, P. Accurate Band Gaps of Semiconductors and Insulators with a Semilocal Exchange-Correlation Potential. *Phys. Rev. Lett.* **2009**, *102*, 226401. [CrossRef]
21. Blöchl, P.E.; Jepsen, O.; Andersen, O.K. Improved tetrahedron method for Brillouin-zone integrations. *Phys. Rev. B* **1994**, *49*, 16223. [CrossRef] [PubMed]
22. Murnaghan, F.D. The Compressibility of Media under Extreme Pressures. *Proc. Natl. Acad. Sci. USA* **1944**, *30*, 244–247. [CrossRef] [PubMed]

chemistry proceedings

MDPI

Proceeding Paper

Group 14 Metallafluorenes for Lipid Structure Detection and Cellular Imaging †

Helena J. Spikes, Shelby J. Jarrett-Noland, Stephan M. Germann, Wendy Olivas, Janet Braddock-Wilking and Cynthia M. Dupureur *

Department of Chemistry & Biochemistry, University of Missouri St. Louis, St. Louis, MO 63121, USA; hjs7c5@mail.umsl.edu (H.J.S.); sjdzd@umsystem.edu (S.J.J.-N.); smg8v5@mail.umsl.edu (S.M.G.); olivasw@umsl.edu (W.O.); wilkingj@umsl.edu (J.B.-W.)

* Correspondence: cdup@umsl.edu; Tel.: +1-314-516-4392

† Presented at the 1st International Electronic Conference on Chemical Sensors and Analytical Chemistry, 1–15 July 2021; Available online: https://csac2021.sciforum.net/.

Abstract: Fluorescent compounds have been shown to be useful in probing lipid dynamics, and there is ongoing interest in nontoxic, photostable, and sensitive dyes. Recently, we evaluated a number of 2,7-disubstituted-alkynyl(aryl)-3,6-dimethoxy-9,9-diphenyl sila- and germafluorenes for their potential as cellular fluorescent probes. These compounds exhibit remarkable quantum yields in hydrophobic environments and dramatic increases in emission intensity in the presence of surfactants. Here, we show that they exhibit significant emission enhancements in the presence of small unilamellar vesicles and are nontoxic to *E. coli*, *S. aureus*, and *S. cerevisiae.* Furthermore, they luminesce in *S. cerevisiae* cells with strong photostability and colocalize with the lipid droplet stain Nile Red, demonstrating their promise as lipid probes.

Keywords: fluorescence; lipid; metallafluorene

check for updates

Citation: Spikes, H.J.; Jarrett-Noland, S.J.; Germann, S.M.; Olivas, W.; Braddock-Wilking, J.; Dupureur, C.M. Group 14 Metallafluorenes for Lipid Structure Detection and Cellular Imaging. *Chem. Proc.* **2021**, *5*, 83. https://doi.org/10.3390/CSAC2021-10455

Academic Editor: Nicole Jaffrezic-Renault

Published: 30 June 2021

1. Introduction

The introductory understanding of the role of biological membranes is that they are barriers and are used to regulate transport and for energy processes. Only fairly recently has there been a developing understanding that membranes are dynamic in their lipid composition and properties, and that these local differences participate in cellular processes in a profound way and have been linked to disease states [1,2], including cellular stress [3].

Fluorescence spectroscopy is an accessible and powerful tool for the sensing of molecules and their behaviors, including binding interactions, conformational changes, and catalytic activities, in both *in vitro* and *in vivo* via cellular imaging [4]. Due to their ability to respond to changes in molecular environment, molecules that exhibit intramolecular charge transfer (ICT) or excited-state intramolecular proton transfer (ESIPT) are particularly attractive as probes of lipids, their interactions, and dynamics [5,6].

Probes commonly used for such purposes include Nile Red, dansyl, NBD [7], and F2N12S [6]. These vary with respect to relevant properties such as the excitation and extinction wavelength, extinction coefficient, working concentration (sensitivity), photostability, and quantum yield, all of which can impact their utility. When this is coupled with the rapidly expanding research area, it is no surprise that a call for more probes to meet expanding needs is prominently articulated [5].

Recently, we evaluated a small library of sila- and germafluorenes (metallafluorenes or MFs) containing alkynyl(aryl) substituents at the 2,7-position ([8,9]; Figure 1) for their potential as fluorescent probes of surfactants. These compounds are soluble and luminescent in aqueous solution and exhibit high quantum yields and dramatic emission enhancements in the presence of various surfactants (5–25-fold) [10]. These results suggest that MFs could have biological applications. Here, we examine the sensitivity, toxicity, and photostability of MFs toward lipids both *in vitro* and *in vivo* and demonstrate the potential of these compounds as lipid probes. Indeed, they are sensitive to DOPC small unilamellar vesicles (SUVs) with significant fluorescence enhancements. These dyes show no toxicity to Gram-positive bacteria, Gram-negative bacteria, and yeast cells and demonstrate high photostability. When compared to the commercially available lipid droplet dye Nile Red, these MFs show strong colocalization with more punctate staining, demonstrating their potential as lipid probes.

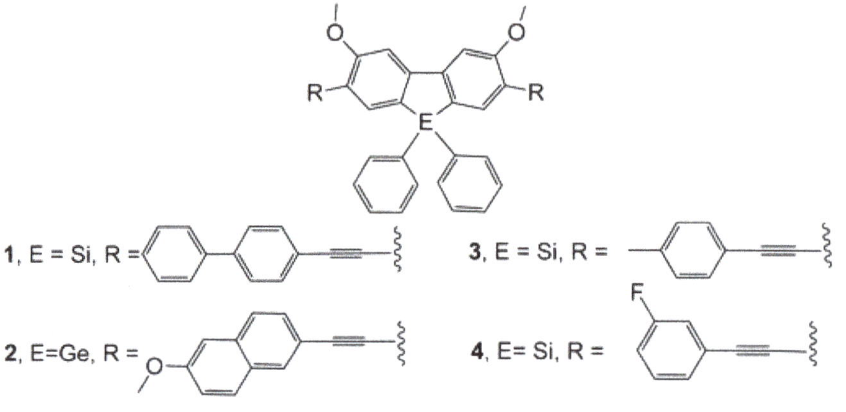

Figure 1. Structures of 2,7-disubstituted sila- and germafluorenes used in this study. **1** silicon based 4-ethynyl-1,1′-biphenyl substitutent; **2** germanium based 2-ethynyl-6-methoxynaphthalene substitutent; **3** silicon based 4-ethynyltoluene substitutent; **4** silicon based 1-ethynyl-3-fluorobenzene substitutent.

2. Materials and Methods

2.1. Materials

Phospholipids were purchased from Avanti Polar Lipids (Alabaster, AL, USA). Two-hundred-proof ethanol was purchased from Decon Labs (King of Prussia, PA, USA). DMSO and Nile Red were obtained from Millipore Sigma (Milkwaukee, WI, USA). p-Xylene was obtained from ThermoFisher (Waltham, MA, USA). All chemicals were of reagent grade and were used as received without further purification. Compounds **1–4** were synthesized as previously described using an appropriate alkynyl(aryl) precursor in a palladium-catalyzed Sonagashira cross-coupling reaction [8,9] and dispensed from stocks in DMSO as previously described [10].

2.2. Preparation of Small Unilamellar Vesicles (SUVs)

At 25 °C, a stock concentration of 4.2 mM DOPC was prepared by drying under inert gas and then resuspended in 10 mM Tris buffer. After 30 min, DOPC was sonicated for 27 min at 25 °C until cloudy. The DOPC-SUV solution was then passed through an Avanti Mini Extruder eleven times to make uniformly sized 0.1 µm DOPC-SUVs at 25 °C. DOPC-SUVs were then diluted to 0.1 mM in a quartz cuvette for fluorescence measurements [11,12].

2.3. Spectroscopy

Absorbance spectra were recorded on a Shimadzu 1800 (Kyoto, Japan) with slits (bandpass) set to 1 nm. Emission spectra were collected in an acid-washed quartz cuvette on a Fluorolog-3 (SPEX) spectrofluorimeter (Horiba Scientific, Piscataway, NJ, USA). The temperature was maintained at 25 °C with a thermostatted cell holder equipped with a magnetic stirrer. Emission spectra were collected with the indicated excitation wavelength and slits (bandpass). MF photostability in xylene was observed at the indicated emission maximum.

2.4. *Microbial Toxicity*

Culture tubes containing LB media or YPD media were inoculated with *Escherichia coli* (Gram-negative), *Staphylococcus aureus* (Gram-positive), or *Saccharomyces cerevisiae*, respectively. Compounds **1–4** were added such that the final DMSO concentration was 2–10% and the MF at its solubility limit in the media. The tubes were incubated at either 37 °C (bacteria) or 30 °C (yeast) overnight and visually inspected for growth.

2.5. *Confocal Laser Scanning Microscopy*

Samples were prepared by smearing a small amount of cells onto a glass microscope slide and heat-fixed by passing the slide through a flame no more than 5 times. Then, **1–4** or NR was applied to heat-fixed cells at 15 μM and incubated at room temperature for 15 min for MFs and 10 min for Nile Red. Slides were then rinsed with 2–3 mL of deionized water, topped with coverslips, and sealed with clear nail polish. Cells were imaged with a Zeiss LSM 900 (Zeiss, Oberkochen, Germany) confocal microscope with an excitation wavelength of 405 nm. For photostability, the sample was illuminated with 1% laser power and images collected periodically.

3. Results and Discussion

3.1. *Spectroscopic Studies*

To assess their sensitivity to a biologically relevant membrane, the emission spectra of **1–4** were compared in the absence and presence of DOPC-SUVs. As shown in Figure 2, fold-enhancements range from two- to sevenfold, with **1** and **2** showing the most dramatic changes.

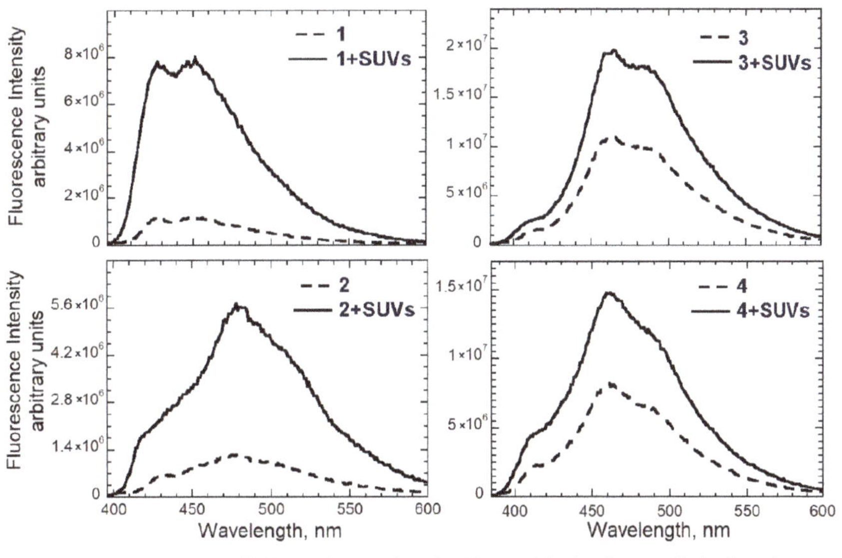

Figure 2. Emission spectra of MFs **1–4** (as numbered in Figure 1) in the absence (dashed) and presence (solid) of 0.1 mM DOPC-SUVs. Conditions: 1 μM MF, 0.1 mM DOPC, 10 mM Tris pH 8, 25 °C. The excitation wavelength was 387 nm and the slits (bandpass) set to 1.0 nm. Three minute incubation.

The photostability of these MFs was initially probed by observing the emission signal as a function of time in xylene, which is used to mimic the interior of membranes [13]. As summarized in Figure 3, these signals are remarkably stable over two hours of continuous excitation. Together, the responsiveness to SUVs and photostability in xylene indicate promise for MFs as probes of lipids *in vivo*.

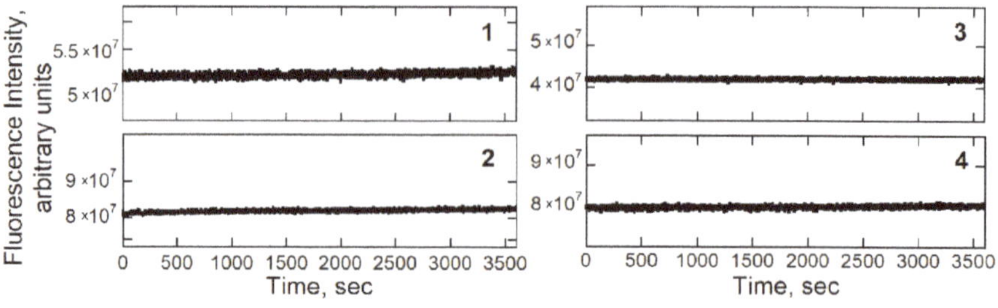

Figure 3. Photostability *in vitro*. Compounds **1–4** (as defined in Figure 1) were diluted into p-xylene and excited continuously. Conditions: 1 μM MF, 0.1–0.4% DMSO. **1**: excitation at 387 nm, slits 1 nm; **2**: excitation at 390 nm, slits 0.8 nm; **3**: excitation at 376 nm, slits 0.9 nm; **4**: excitation at 376 nm, slits 0.9 nm.

3.2. Microbial Toxicity Studies

To assess their potential for use in cellular imaging, MFs were screened for toxicity against microorganisms. For yeast, *E. coli* and *S. aureus*, no inhibition of growth was observed at the MF solubility limit in media (at least 50 μM).

3.3. Imaging of S. cerevisiae with Metallafluorenes

To determine if these MFs can be used to stain cells, **1–4** were introduced to yeast cells and subsequently imaged using confocal microscopy. Figure 4 illustrates that in all cases, the MF emission intensity is visible inside fixed yeast cells.

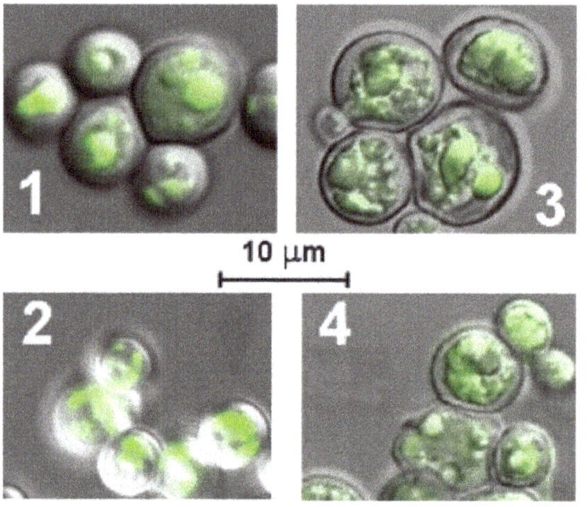

Figure 4. Confocal imaging of MFs **1–4** in yeast cells. Conditions: 15 μM MF as indicated, 63×. The excitation wavelength was 405 nm and scan range 400–600 nm. Numbers refer to MFs as defined in Figure 1.

To assess the MF photostability in yeast cells, excitation was applied and fluorescence was observed as a function of time. As summarized in Figure 5 for **1** and **2**, fluorescence persisted for over 2 min, with **2** showing greater photostability. See Supplemental Figure S1 for photostability studies of **3** and **4**.

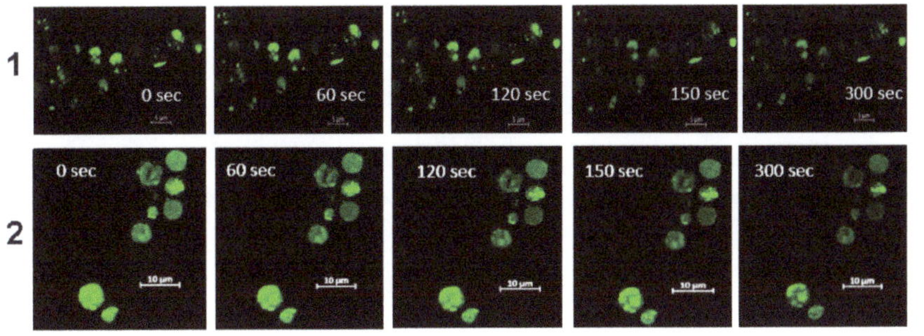

Figure 5. Photostability of **1** and **2** in yeast cells. See Methods for details. *S. cerevisiae* were stained for 15 min with 15 μM **1** or **2** and then imaged periodically during continuous excitation. Magnification is 63×. Numbers at left refer to compounds as defined in Figure 1.

Finally, to determine where these MF localize in yeast, we costained with Nile Red, a well-known lipid droplet stain [14]. As shown in Figure 6, **1** yields more punctate images and colocalizes with this probe, demonstrating clear specificity for *S. cereviseiae* organelles, including the vacuole and possibly lipid granules. See Supplemental Figure S2 for a colocalization study of **2** and **4**.

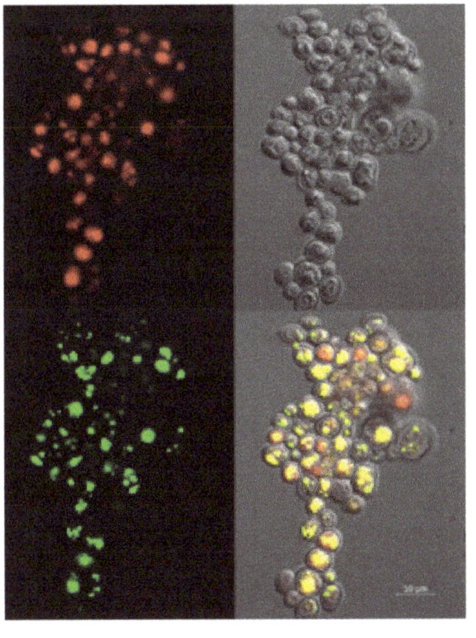

Figure 6. MF **1** colocalizes in yeast with Nile Red. Red, Nile Red; green, **1**; top right, transmitted light; bottom right, yellow indicates colocalization. 15 μM probe at 63× magnification.

4. Conclusions

We show here that these metallafluorenes have good photostability and are sensitive to lipid structures *in vitro*, demonstrating impressive fold enhancements in the presence of SUVs. Furthermore, they are non-toxic to cells and can enter cells and colocalize with Nile Red, a lipid probe. In addition, the higher extinction coefficients of MFs and competitive quantum yields [10] make them more sensitive. All of these observations bode well for the application of MFs as lipid probes both *in vitro* and *in vivo*. The synthetic scaffolding of these MFs provides convenient tuning of desired properties by changing the 2,7 substituent.

This feature facilitates designs that incorporate optimal solubility, emission spectra, dipole moment, and solvatochromism for specific applications.

5. Patents

WO/2020/210416; PCT International Patent Application No.: PCT/US2020/027355.

Supplementary Materials: The following are available online at https://www.mdpi.com/article/10.3390/CSAC2021-10455/s1, Figure S1: Photostability of 3 and 4 in Yeast Cells, Figure S2: 2 and 4 Colocalize with Nile Red in Yeast.

Author Contributions: Conceptualization, C.M.D., H.J.S. and S.J.J.-N.; investigation, H.J.S. and S.J.J.-N.; resources, C.M.D., J.B.-W., S.M.G. and W.O.; data curation, S.J.J.-N.; writing—original draft preparation, C.M.D. and H.J.S.; writing—review and editing, C.M.D. and S.J.J.-N.; supervision, C.M.D.; project administration, C.M.D.; funding acquisition, J.B.-W. and C.M.D.; H.J.S. and S.J.J.-N. contributed equally to the preparation of this manuscript. All authors have read and agreed to the published version of the manuscript.

Funding: This work was supported by the National Science Foundation (CHE-1362431 to J.B.-W.).

Data Availability Statement: The data presented in this study are available on request from the corresponding author.

Conflicts of Interest: The authors declare no conflict of interest.

References

1. Levental, I.; Levental, K.R.; Heberle, F.A. Lipid Rafts: Controversies Resolved, Mysteries Remain. *Trends Cell Biol.* **2020**, *30*, 341–353. [CrossRef] [PubMed]
2. Carravilla, P.; Nieva, J.L.; Eggeling, C. Fluorescence Microscopy of the HIV-1 Envelope. *Viruses* **2020**, *12*, 348. [CrossRef] [PubMed]
3. Ashoka, A.H.; Ashokkumar, P.; Kovtun, Y.P.; Klymchenko, A.S. Solvatochromic Near-Infrared Probe for Polarity Mapping of Biomembranes and Lipid Droplets in Cells under Stress. *J. Phys. Chem. Lett.* **2019**, *10*, 2414–2421. [CrossRef]
4. Sedgwick, A.C.; Wu, L.; Han, H.H.; Bull, S.D.; He, X.P.; James, T.D.; Sessler, J.L.; Tang, B.Z.; Tian, H.; Yoon, J. Excited-state intramolecular proton-transfer (ESIPT) based fluorescence sensors and imaging agents. *Chem. Soc. Rev.* **2018**, *47*, 8842–8880. [CrossRef] [PubMed]
5. Klymchenko, A.S.; Kreder, R. Fluorescent probes for lipid rafts: From model membranes to living cells. *Chem. Biol.* **2014**, *21*, 97–113. [CrossRef] [PubMed]
6. Shynkar, V.V.; Klymchenko, A.S.; Kunzelmann, C.; Duportail, G.; Muller, C.D.; Demchenko, A.P.; Freyssinet, J.M.; Mely, Y. Fluorescent biomembrane probe for ratiometric detection of apoptosis. *J. Am. Chem. Soc.* **2007**, *129*, 2187–2193. [CrossRef]
7. Klymchenko, A.S. Solvatochromic and Fluorogenic Dyes as Environment-Sensitive Probes: Design and Biological Applications. *Acc. Chem. Res.* **2017**, *50*, 366–375. [CrossRef] [PubMed]
8. Hammerstroem, D.W.; Braddock-Wilking, J.; Rath, N.P. Synthesis and characterization of luminescent 2,7-disubstituted silafluorenes. *J. Organomet. Chem.* **2016**, *813*, 110–118. [CrossRef]
9. Hammerstroem, D.W.; Braddock-Wilking, J.; Rath, N.P. Luminescent 2,7-disubstituted germafluorenes. *J. Organomet. Chem.* **2017**, *830*, 196–202. [CrossRef]
10. Spikes, H.J.; Jarrett-Noland, S.J.; Germann, S.M.; Braddock-Wilking, J.; Dupureur, C.M. Group 14 Metallafluorenes as Sensitive Luminescent Probes of Surfactants in Aqueous Solution. *J. Fluoresc.* **2021**, *31*, 961–969. [CrossRef] [PubMed]
11. Amaro, M.; Filipe, H.A.; Ramalho, J.P.; Hof, M.; Loura, L.M.S. Fluorescence of nitrobenzoxadiazole (NBD)-labeled lipids in model membranes is connected not to lipid mobility but to probe location. *Phys. Chem. Chem. Phys.* **2016**, *18*, 7042–7054. [CrossRef] [PubMed]
12. Mazeres, S.; Joly, E.; Lopez, A.; Tardin, C. Characterization of M-laurdan, a versatile probe to explore order in lipid membranes. *F1000Research* **2014**, *3*, 172. [CrossRef] [PubMed]
13. Shaya, J.; Collot, M.; Benailly, F.; Mahmoud, N.; Mely, Y.; Michel, B.Y.; Klymchenko, A.S.; Burger, A. Turn-on Fluorene Push-Pull Probes with High Brightness and Photostability for Visualizing Lipid Order in Biomembranes. *ACS Chem. Biol.* **2017**, *12*, 3022–3030. [CrossRef] [PubMed]
14. Greenspan, P.; Mayer, E.P.; Fowler, S.D. Nile red: A selective fluorescent stain for intracellular lipid droplets. *J. Cell Biol.* **1985**, *100*, 965–973. [CrossRef] [PubMed]

chemistry proceedings

MDPI

Proceeding Paper

QSPR Modelling of Potentiometric HCO_3^-/Cl^- Selectivity for Polymeric Membrane Sensors †

Nadezhda Vladimirova *, Julia Ashina and Dmitry Kirsanov

Institute of Chemistry, Saint Petersburg State University, 198504 Saint Petersburg, Russia; y.ashina@spbu.ru (J.A.); d.kirsanov@gmail.com (D.K.)

* Correspondence: hada96@mail.ru

† Presented at the 1st International Electronic Conference on Chemical Sensors and Analytical Chemistry, 1–15 July 2021; Available online: https://csac2021.sciforum.net/.

Abstract: Since the development process of new sensors is long and tedious, it would be very helpful to develop a model that could predict sensor properties based on active compound structure without the actual synthesis and characterization of the corresponding sensors. In this work, the model for the prediction of logK (HCO_3^-/Cl^-) was constructed based on 40 ligand structures suggested in the literature for carbonate sensing. Substructural molecular fragments (SMF) were used to describe the structure of compounds, where fragments were considered as sequences of bonds and atoms. The projection on latent structures (PLS) method was used to calculate the regression model.

Keywords: membrane sensor; carbonate ionophore; QSPR; PLS

Citation: Vladimirova, N.; Ashina, J.; Kirsanov, D. QSPR Modelling of Potentiometric HCO_3^-/Cl^- Selectivity for Polymeric Membrane Sensors. *Chem. Proc.* **2021**, *5*, 84. https://doi.org/10.3390/CSAC2021-10621

Academic Editor: Ye Zhou

Published: 6 July 2021

1. Introduction

Polymeric membrane electrodes offer numerous advantages, and their properties can be tuned in a wide range by the modification of membrane composition. However, the process of the selection of an appropriate ligand to construct the sensor for a particular task is time-consuming and requires ligand synthesis, sensor preparation, and characterization. It would be very helpful for researchers to make a model that allows the prediction of sensor properties of an electrode based on the structure of the employed ionophore.

Using the computational chemistry, different characteristics of chemical compounds can be predicted. Quantitative relations between physical or chemical properties of organic compounds and their chemical structures can be set with the QSPR (quantitative structure property relationship). This methodology is widely applied nowadays, e.g., in pharmaceutical investigations; specifically, a search for new drugs [1]. There are QSPR models for various materials, such as nanomaterials [2], catalysts [3], and ionic liquids [4]. Recently, an application of QSPR for predicting the sensor properties of membrane electrodes was suggested [5]. It was possible to relate the structure of the organic ligand with the selectivity constant of the corresponding membrane sensor for Ca^{2+}/Mg^{2+}.

This study aims to expand this approach to anion-selective sensors, where ligand selection is much more challenging than in the case of metal cations, due to the wide variability of the geometries of inorganic ligands, and their hydrolysis in the case of weak acids.

Among anions, there are ones with important biological and industrial roles. An example of such anions is carbonate. Therefore, it is a prospective object for predicting selectivity to carbonate against chloride anions by means of QSPR.

2. Materials and Methods

2.1. Dataset

The dataset of 40 samples was composed with literature sources and experimental data. A summary table with compositions of all samples and literature sources can be found in the supplementary materials. Whereas the number of anionic (especially carbonate) ionophores is small due to the reasons that we have already discussed earlier, there are a few articles about carbonate ionophores. A great part of the data was extracted from IUPAC review "POTENTIOMETRIC SELECTIVITY COEFFICIENTS OF ION-SELECTIVE ELECTRODES", with a summary table of anions existing in 2002 [6]. Considering the significant shortage of carbonate ionophores, we had to add in the table ionophores with Cl^-/HCO_3^- selectivity that were converted to HCO_3^-/Cl^- selectivity according to the Nikolsky–Eisenman equation:

$$E = E_I^0 + \frac{RT}{z_I F}\ln(a_I + \sum K_{IJ} a_J^{\frac{z_I}{z_J}}). \quad (1)$$

All of the structures of ionophores are available in Table S1 in Supplementary Materials.

We also made sure that all these data were obtained in the narrow range of pH (7.0–8.6) for understanding which particular ionic form prevailed in an examined solution.

All the membranes described in the literature are based on polyvinylchloride (PVC) as a polymer, and one of three plasticizers: nitrophenyl octyl ether (NPOE), dioctyl adipate (DOA) and dioctyl sebacate (DOS), which is also known as bis(2-ethylhexyl) sebacate (BEHS). Ion-exchanger is tridodecylmethylammonium (TDMA) with Cl^- or NO_3-counter-ion.

Due to the small number of carbonate ionophores examined, we had to compile the resulting database with membranes that differ not just in ionophores, but plasticizers as well. We took different plasticizers into account and added their dielectric constant (also known as relative permittivity) as a descriptor for adjusting our model and making it more comprehensive.

The selectivity coefficients of these ionophores varied from −5.8 to 6.2 on the logarithmic scale. The average selectivity logK (HCO_3^-/Cl^-) was −1.425, and the median value was −2.6.

2.2. Descriptors

We used substructural molecular fragments (SMF) for encoding molecular structures in a matrix. A molecular structure can be described with this method by dividing a molecule into all possible fragments and writing the number of these fragments into the matrix. These SMFs were obtained by using "ISIDA QSPR" software [7]. There are two approaches for obtaining a molecule's SMF in ISIDA: sequences of atoms and/or bonds (topological path) and selected ("augmented") atom (atom-centered fragments) with its environment that can be atoms, bonds, or both (Figure 1). In this work, atom and bond sequences were applied. A molecule was represented as a graph and its descriptors were, consequently, subgraphs.

Hence, ISIDA SMF descriptors are numbers of fragments (or subgraphs) in a molecule with each element of the descriptor associated with one of the detected possible fragments. Only the shortest paths from one atom to the other were used. It should be noted that the length of sequences is limited. The minimal and the maximal lengths are 2 and 15, respectively.

2.3. Projection on Latent Structures (PLS) Modelling

In order to relate molecular descriptors of ligands with selectivity coefficients of the corresponding sensors, we employed the PLS regression algorithm. PLS regression searches for a set of components are known as latent vectors that perform a synchronous decomposition of X and Y, with the clause that these components explain, as much as possible, the covariance between X and Y. Data matrix size was 40 × 1855, where 40 is the number of samples (ligands) and 1855 is the number of descriptors.

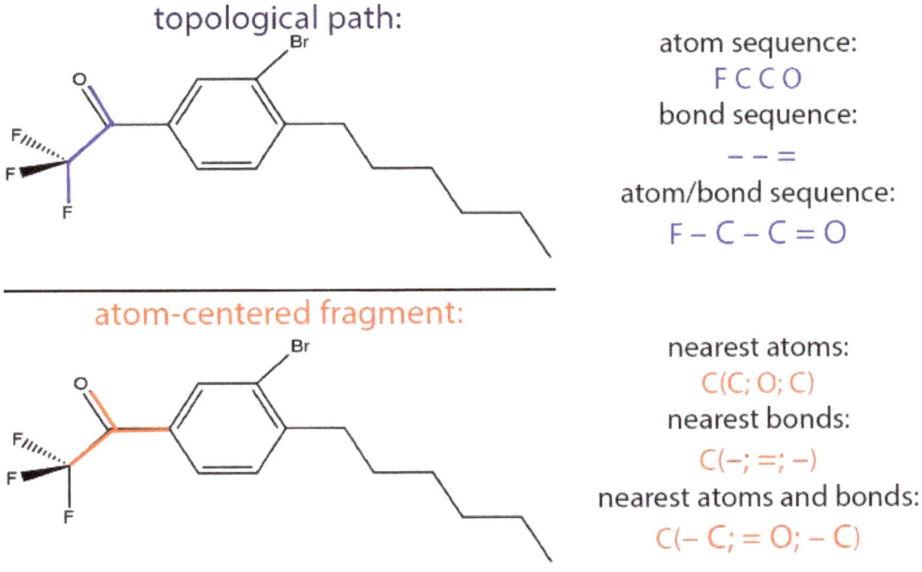

Figure 1. Two approaches to obtaining ISIDA SMF: topological path and atom-centered fragments.

3. Results and Discussion

The molecular descriptors obtained for the chosen ligands were calculated with ISIDA QSPR software [7]. The 40 ionophore structures and their properties, specifically substructural molecular fragments (SMF) and the permittivity of membranes, were used as descriptors. The PLS model relating the descriptors with selectivity was evaluated according to the following parameters: root mean square error (RMSE) and squared determination coefficient (R^2). The results of QSPR modelling are shown in Figure 2. Each point in the graph corresponds to an item in the database, whereas straight lines represent the resulting models. Blue and red colours correspond to training and test samples, respectively.

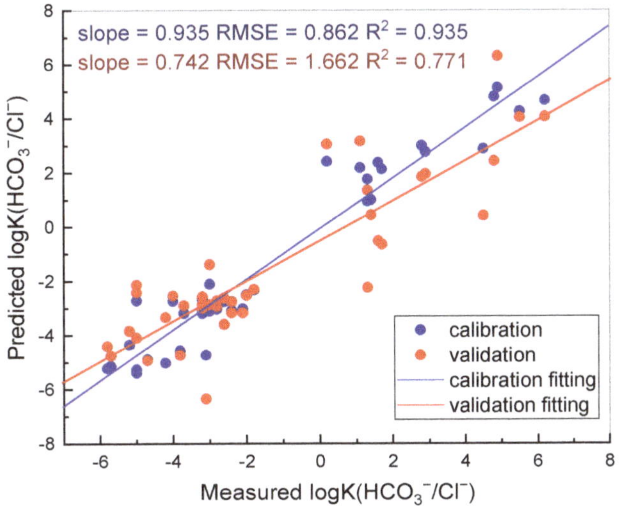

Figure 2. "Measured vs. predicted" plot for the QSPR model for predicting the selectivity of membrane sensors.

It can be seen that the derived model allows for a semi-quantitative estimation of the selectivity coefficients, based on the ligand structure. As far as each compound is an individual molecule, which consists of a variety of molecular fragments, regression coefficients allow evaluating the contribution of each fragment (represented as an independent variable in the matrix) in the selectivity of the sensors. The largest coefficients signify the variables (in this case, SMF, which are the encoded fragments of molecules) with the most important impact. The fragments with the largest contribution in the absolute value of the selectivity logK (HCO_3^-/Cl^-) of potentiometric membrane sensors are shown in Figure 3.

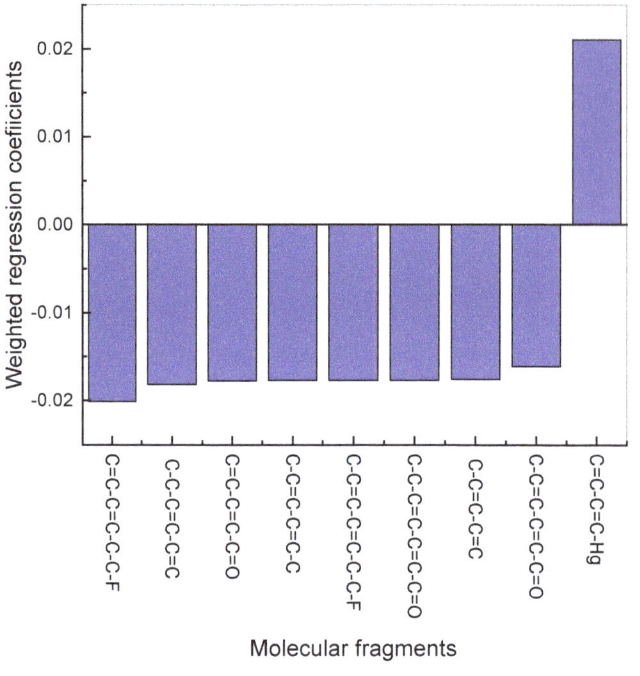

Figure 3. Fragments with the largest contribution in logK (HCO_3^-/Cl^-) values, where '-' is a single bond and '=' is a double bond.

As follows from the graph, the fragment C=C-C=C-C-C-F makes the largest negative contribution, and it is part of a longer fragment C-C=C-C=C-C-C-F with a smaller contribution. The shortest fragment with negative contribution C-C=C-C=C is included in the remaining fragments with the largest negative contribution. The fragment with positive contribution C=C-C=C-Hg contains mercury in its composition. These observations provide valuable information for the further design of the ligands with required selectivity.

4. Conclusions

Despite some problems with anions' ionophores that are described more fully in the introduction, we were able to collect the database that allowed making a QSPR model while satisfying RMSE and R^2 for a relatively small amount of data. We found the fragments with the highest impact in selectivity logK (HCO_3^-/Cl^-) of potentiometric membrane sensors, and we believe that this will help in the future search for the new ionophores. It appears that semi-quantitative prediction of sensor selectivity is possible based on the ligand structure.

Supplementary Materials: The following are available online at https://www.mdpi.com/article/10.3390/CSAC2021-10621/s1, Table S1: QSPR modelling of potentiometric HCO_3^-/Cl^- selectivity for polymeric membrane sensors supporting information.

Author Contributions: Conceptualization, D.K.; methodology, D.K.; formal analysis, D.K. and N.V.; investigation, N.V.; writing—original draft preparation, N.V.; writing—review and editing, D.K. and J.A.; visualization, N.V.; supervision, D.K.; funding acquisition, J.A. All authors have read and agreed to the published version of the manuscript.

Funding: This research was funded by RFBR, grant number 20-33-70272.

Conflicts of Interest: The authors declare no conflict of interest. The funders had no role in the design of the study; in the collection, analyses, or interpretation of data; in the writing of the manuscript, or in the decision to publish the results.

References

1. Kapetanovic, I.M. Computer-aided drug discovery and development (CADDD): In silico-chemico-biological approach. *Chem.-Biol. Interact.* **2008**, *171*, 165–176. [CrossRef] [PubMed]
2. Ruoff, R.; Tse, D.; Malhotra, R.; Lorents, D. Solubility of C_{60} in a Variety of Solvents. *J. Phys. Chem.* **1993**, *97*, 3379–3383. [CrossRef]
3. Cruz, V.; Ramos, J.; Munoz-Escalona, A.; Lafuente, P.; Pena, B.; Martinez-Salazar, J. 3D-QSAR analysis of metallocene-based catalysts used in ethylene polymerization. *Polymer* **2004**, *45*, 2061–2072. [CrossRef]
4. Trohalaki, S.; Pachter, R.; Drake, G.W.; Hawkins, T. Quantitative Structure-Property Relationships for Melting Points and Densities of Ionic Liquids. *Energy Fuels* **2005**, *19*, 279–284. [CrossRef]
5. Martynko, E.; Solov'Ev, V.; Varnek, A.; Legin, A.; Kirsanov, D. QSPR Modeling of Potentiometric Mg^{2+}/Ca^{2+} Selectivity for PVC-plasticized Sensor Membranes. *Electroanalysis* **2020**, *32*, 792–798. [CrossRef]
6. Umezawa, Y.; Umezawa, K.; Bühlmann, P.; Hamada, N.; Aoki, H.; Nakanishi, J.; Sato, M.; Xiao, K.P.; Nishimura, Y. Potentiometric selectivity coefficients of ion-selective electrodes. Part II. Inorganic anions (IUPAC Technical Report). *Pure Appl. Chem.* **2002**, *74*, 923–994. [CrossRef]
7. Varnek, A. ISIDA (In Silico Design and Data Analysis) Program. Available online: http://infochim.u-strasbg.fr/recherche/isida/index.php (accessed on 7 June 2012).

chemistry proceedings

MDPI

Proceeding Paper

Room Temperature Hydrogen Sensing Based on Tapered Optical Fiber Coated with Polyaniline (PANI) †

Mohammed Majeed Alkhabet [1,2], Saad Hayatu Girei [1,3], Abdul Hadi Ismail [4], Suriati Paiman [5], Norhana Arsad [6], Mohd Adzir Mahdi [1] and Mohd Hanif Yaacob [1,*]

1 Wireless and Photonics Network Research Centre, Faculty of Engineering, Universiti Putra Malaysia (UPM), Serdang 43400, Selangor, Malaysia; mohammed.alkhabet@gmail.com (M.M.A.); gireisaad3@gmail.com (S.H.G.); mam@upm.edu.my (M.A.M.)
2 Department of Computer Engineering Techniques, Al-Rasheed University College, Baghdad 10011, Iraq
3 Department of Computer Engineering, Federal Polytechnic Mubi, Mubi 652105, Adamawa State, Nigeria
4 Institute of Advanced Technology (ITMA), Universiti Putra Malaysia, Seri Kembangan 43400, Selangor, Malaysia; ab.hadismail@gmail.com
5 Department of Physics, Faculty of Science, Universiti Putra Malaysia, Serdang 43400, Selangor, Malaysia; suriati@upm.edu.my
6 Department of Electrical, Electronic and Systems Engineering, Faculty of Engineering and Built Environment, Universiti Kebangsaan Malaysia (UPM), Bangi 43600, Selangor, Malaysia; noa@ukm.edu.my
* Correspondence: hanif@upm.edu.my; Tel.: +60-16-328-9044
† Presented at the 1st International Electronic Conference on Chemical Sensors and Analytical Chemistry, 1–15 July 2021; Available online: https://csac2021.sciforum.net/.

Abstract: This work demonstrates a hydrogen (H_2) sensor at room temperature made of tapered optical fibers coated with a polyaniline (PANI) nanofiber. A transducing platform was constructed using a multimode optical fiber (MMF) with a 125 µm cladding and a 62.5 µm core diameter. In order to enhance the light evanescent field surrounding the fiber, the fibers were tapered from 125 µm in diameter to 20 µm in diameter with 10 mm waist and coated PANI using the drop casting technique. Various characterization techniques, such as field emission scanning electron microscopy (FESEM), energy dispersive X-ray (EDX), differential X-ray (XRD), and atomic force microscopy, have been used to establish the PANI's properties. When H_2 is subtracted, the optical properties of the PANI layer change, resulting in a change in light absorption. The fabricated sensor was tested by exposing it to H_2 at different concentration from 0.125% to 1%. In this case, the sensitivity, response, and recovery times were 15.928/vol%, 110 s, and 160 s, respectively. The improved hydrogen sensor holds great promise for environmental and industrial applications due to its ability to operate at room temperature.

Keywords: hydrogen (H_2); tapered optical fiber; polyaniline (PANI); drop casting technique

Citation: Alkhabet, M.M.; Girei, S.H.; Ismail, A.H.; Paiman, S.; Arsad, N.; Mahdi, M.A.; Yaacob, M.H. Room Temperature Hydrogen Sensing Based on Tapered Optical Fiber Coated with Polyaniline (PANI). *Chem. Proc.* **2021**, *5*, 85. https://doi.org/10.3390/CSAC2021-10415

Academic Editor: Nicole Jaffrezic-Renault

Published: 30 June 2021

1. Introduction

Hydrogen (H_2), due to its high fuel efficiency, abundance, uncontaminated character, and sustainability, is one of the possible solutions to the impending energy crisis [1]. It is also a reliable gas for misdiagnosis in power transformers [2,3]. H_2 has also been used in other sectors, such as aerospace engineering, mineral refineries, oil exploration, chemical processing, cryogenic refrigeration, and many more [4]. However, the high diffusion coefficient (0.16 cm^2/s in air), low ignition energy (0.018 mJ), wide explosion concentration range (4–75%), and high heat of combustion (285.8 kJ/mol) convert them to gases, which are explosive and potentially dangerous for use, transport, and storage [5].

On the other hand, optical sensors rely on optical fibers, which have unique characteristics such as lightness, small size, electromagnetic interference resistance, instability, and stiffness in harsh environments [6]. Due to their peculiar properties, optical fibers are perfect candidates for detection in harsh environments [7].

Recently, the scientific community has made significant progress on the synthesis and application of conducting polymers [8–10]. One of the promising materials for gas sensing is polyaniline (PANI), which among polymers has peak environmental stability. It is currently known as the only conducting polymer that is stable in air [11]. As a sensor, PANI is particularly helpful because it is sensitive at room temperature [12,13], and it can be applied to detect a range of gases in combination with additional nanomaterials [14].

Several optical hydrogen sensors using PANI as an energy transformer have been reported in recent years. Most of them rely on fiber gratings (FBGs) [15] and plastic optical fiber [16]. To make fiber optic sensors sensitive to their surroundings, most of them must be functionalized [17]. In this research, PANI coated with dissolved, tapered optical fibers is used to detect hydrogen gas.

2. Experiments

2.1. Tapered Optical Fiber Fabrication

The H_2 gas sensor was developed using a multimodal tapered fiber optic (MMF) with cladding and core dimensions of 125 m and 62.5 m, respectively, as a transducing platform. For reduction, the Vytran glass processing system (Vytran GPX-3400, Thorlabs, Inc., Trenton, NJ, USA) was employed. The machine operates on a heating and pulling principle, with a graphite filament acting as a heating source to create the desired tapered profile geometry. The MMF had a 125 mm cladding diameter that tapered to a 20 mm waist diameter, a 10 mm waist length, and a 5 mm tapering top. The picture of the tapered optical fiber created with the tapered area is shown in Figure 1.

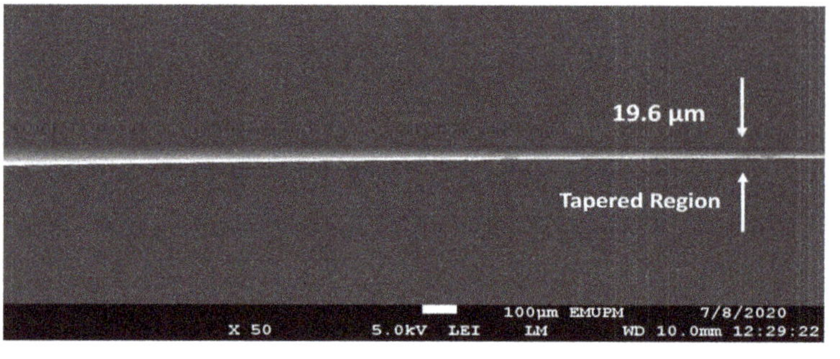

Figure 1. Scanning electron microscopy (SEM) micrograph of a tapered multimode optical fiber's transition area (MMF).

2.2. PANI Functionalization on Tapered Optical Fiber

Aniline was purified by distillation before polymerization. The purified aniline (0.16 M) was instantly dissolved in pre-prepared 0.05 M perchloric acid ($HClO_4$) [18,19] (Merck, 70–72%) to avoid atmospheric oxidation. In another volumetric flask, 0.16 M of ammonium peroxodisulfate was also dissolved in $HClO_4$, and both mixtures were rested overnight. Ammonium peroxodisulfate solution was carefully added to the aniline solution with continuous stirring at room condition. The oxidative polymerization reaction of the mixture was left stirring for 24 h. The obtained PANI (dark green precipitate) was filtered and washed with ethanol (EtOH) until a colorless supernatant liquid can be observed in order to minimize the amount of unreacted monomers and oligomers [20]. It is then dried at 60 °C until a constant weight is obtained. A drop casting technique was used to coat the tapered optical fibers. To guarantee full evaporation of the aqueous medium, a drop of mixture (about 20 µL) was placed into the base of the tapered optical fiber by using a micropipette, and the sample was heated in the oven at 80 °C for 40 min.

A light source (tungsten halogen, HL-2000, Ocean Optics, Dunedin, FL, USA) with a coverage wavelength of 360 nm to 2500 nm and a spectrophotometer (USB 4000, Ocean Optics) comprised the experimental phase of the gas optical sensing system. It can detect anything between 200 and 1100 times per second. The optical gas detection system's experimental setup includes a customized gas chamber for measuring the optical absorption spectrum. The PANI-coated sensor was placed in a close gas unit and purged with a centrifuge at a gas flow rate of 200 sccm from a computer managed mass flow controller. Figure 2 depicts the H_2 sensor's experimental setup.

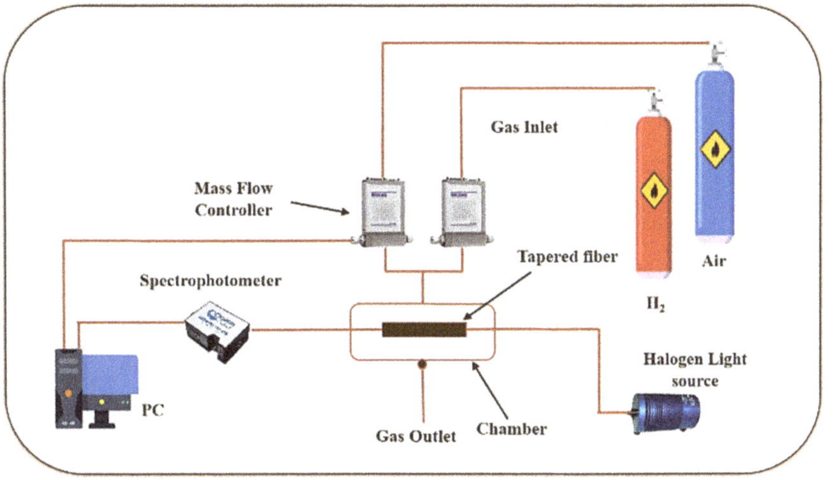

Figure 2. H_2 sensor experimental setup.

2.3. Materials Characterization

FESEM (JSM-7600F, Musashino, Akishima, Tokyo, 8558, Japan) was used to examine the films' morphology, while EDX analysis was used to assess their original composition. XRD examination indicated material identification, crystallization, and the PANI phase transition (APD 2000). FESEM images of PANI nanoparticles are shown in Figure 3. The PANI is mostly made up of uneven grains and chips with sharp edges, as can be observed. Furthermore, the structure seems to be completely porous, creating very small polyaniline particles that can expand the liquid–solid interface [21,22].

Figure 4a shows the XRD pattern of the PANI, which shows an amorphous nature in a partly crystalline condition with a diffraction peak at 22.34°. (200). Due to the repeating of benzenoid and quinoid rings in the PANI chains, this pattern displays poor conductive polymer crystallinity.

Figure 4b shows the UV–vis spectra of PANI structures. Spectroscopic scanning was performed at a wavelength range from 300 to 1000 nm. The UV–vis spectrum is useful for measuring the amount of conjugation, and for this reason, conductive polyaniline samples exhibit a broad absorbance, referred to as a free carrier tail, at wave lengths greater than about 800 nm. As the conjugation length becomes longer, the peak shifts to broader wavelengths and becomes extremely large [23].

Figure 4c shows the EDX pattern of PANI, which revealed that the key elements in PANI film are C, N, O, and Si, as observed by their peaks. A silica fiber was employed as the substrate shared the silicon (Si) peak.

The atomic force microscope (AFM) can verify the average surface roughness and thicknesses of PANI. A 10 × 10 μm section of the boundary area was scanned for AFM analysis. The average surface roughness values of the PANI were ≈23.4 nm, as shown in Figure 5a,b. As part of this study, the thicknesses of the PANI coatings were measured. As shown in Figure 5c, measurements were taken by surrounding parts of fiber with aluminum tape and then assessing the thickness differences between coated and uncoated fiber. The average thickness of the PANI coatings was 690 nm.

Figure 3. FESEM micrograph of Polyaniline (PANI).

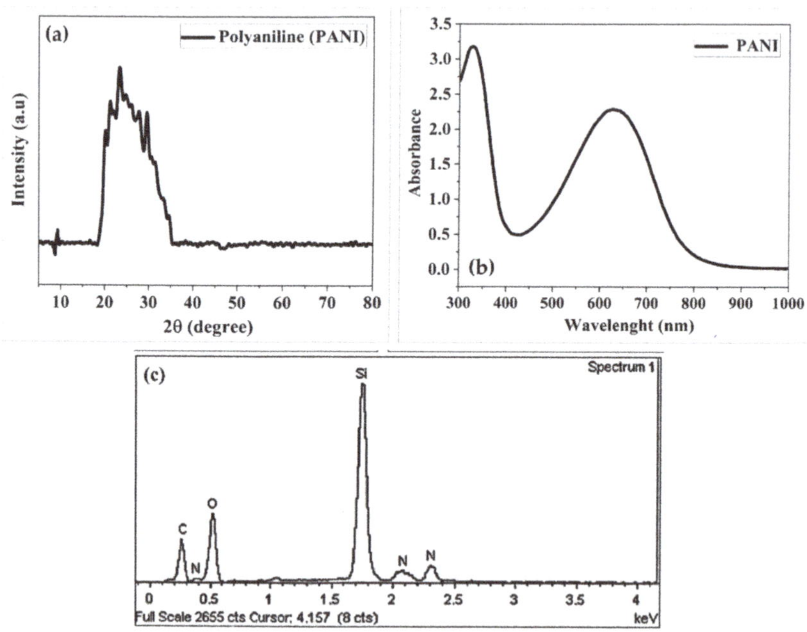

Figure 4. (**a**) XRD pattern of PANI; (**b**) shows the UV–vis spectra of PANI structures; and (**c**) shows the EDX measurement of PANI.

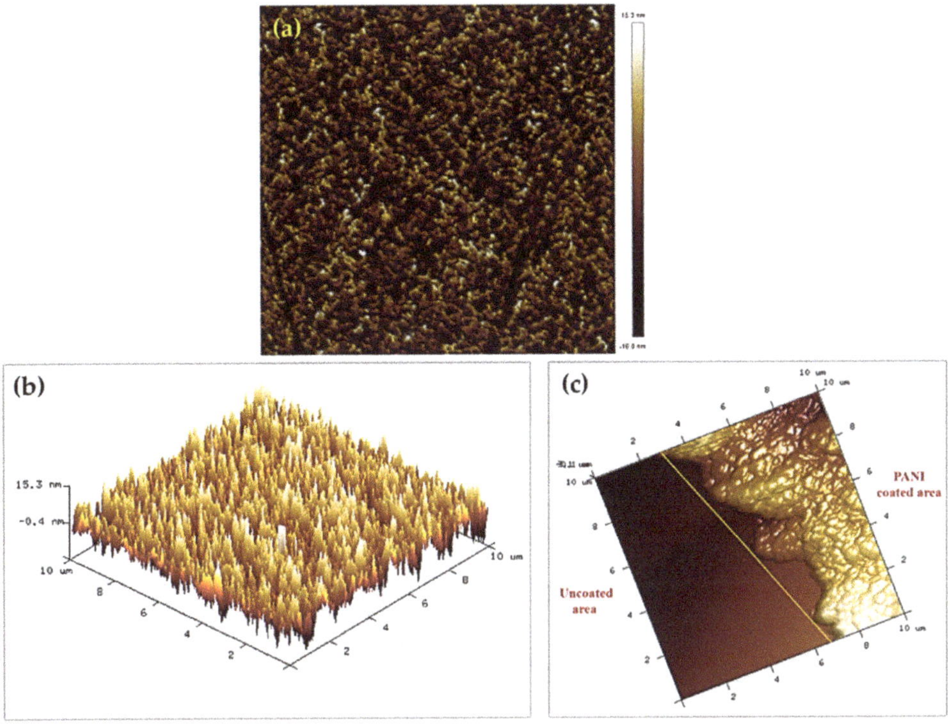

Figure 5. Two-dimensional topography AFM images (**a**) PANI; (**b**) 3D topography AFM images of PANI; and (**c**) 3D topography AFM images of the boundary area between the uncoated and coated fiber for PANI sensing layer.

3. Results and Discussion

The absorption spectra of the developed sensor coated with PANI relative to synthetic air and 1.00% H_2 at room temperature were observed. As shown in Figure 6a, the PANI-coated sensor shows notable changes in absorption, particularly in the wavelength range of 550 to 850 nm. The sensor performance of the PANI-coated sensor was monitored in terms of cumulative absorbance, which is the product of a combination of response curves over a particular wavelength range. Figure 6b shows the dynamic response of PANI-coated based sensors to H_2 concentrations in air ranging from 0.125% to 1.00% at room temperature. The response and recovery times of the developed PANI-coated based sensor were 110 s and 160 s, respectively. The absorbance change was approximately 4% at 0.125% H_2 and 19% higher with 1.00% H_2. Compared with the results of previous studies [24,25], the PANI-coated based sensor showed greater H_2 absorbance and recovery, as well as advanced compromise differences.

The repeatability of the PANI-coated based sensor was tested by exposing it to three cycles of 1.00% H_2. Overall, the PANI-coated based sensor showed a strong and stable absorbance response as well as high repeatability towards H_2.

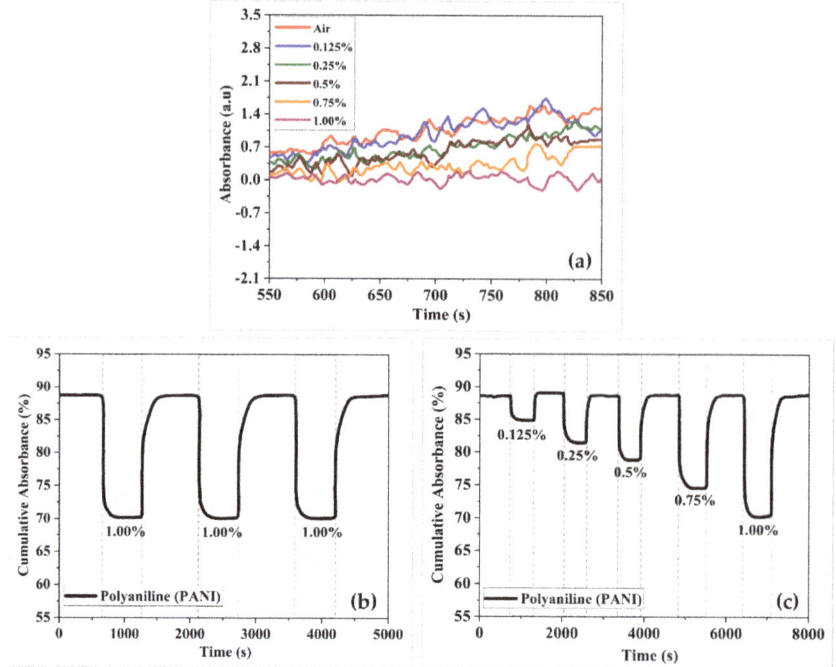

Figure 6. (**a**) Absorbance versus optical wavelength, (**b**) dynamic absorbance curves, and (**c**) repeatability of PANI-coated based sensor towards H_2.

Figure 7a shows the absorption versus H_2 concentration for PANI-coated based sensors. The PANI-coated based sensors had a sensitivity of 15.928/vol% and a linearity slope of 98%. When measuring gas-sensing properties, selectivity is an important key to consider. As shown in Figure 7b, the sensor's absorption properties toward ammonia (NH_3) and methane gas (CH_4) at a concentration of 1.00% were investigated. The PANI-coated based sensor had a very high NH_3 absorption response but a substantially lower response for the other gases. Furthermore, the adsorption of PANI based materials was highly selective for polar molecules such as NH_3, whereas sensitivity was low for non-polar molecules such as H_2 and CH_4 [26].

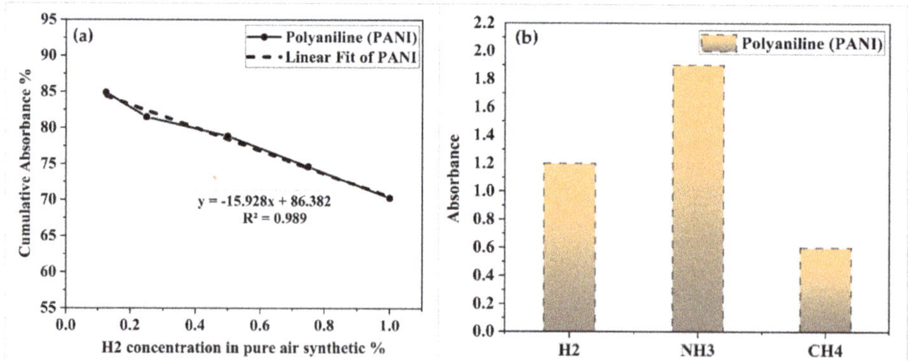

Figure 7. (**a**) Absorbance changes at different H_2 concentration for PANI-coated based sensor and (**b**) the selectivity of PANI-coated based sensor.

4. H_2 Mechanism of PANI Coated on Tarped Optical Fiber

The H_2 sensor with PANI mechanism consists of two parts, as illustrated in Figure 8. The first is the physical absorption of gas molecules in the PANI, which causes changes in the refractive index of the optical fiber's surface, which in turn will result in changes in the amount of light transmitted in the fiber. A higher RI would allow more light to escape, lowering the intensity of light detected by the spectrophotometer. The charge transfer between the adsorbent and the PANI molecules is the second step. The charge transfer from the electron donating H_2 gas to PANI changes the surface chemistry of the sensor layer as it is absorbed into the walls and sides of PANI. This involves changes in the sensor layer's optical properties, as the light that traveled through the fiber is absorbed by environmental changes, causing spectrum shifts.

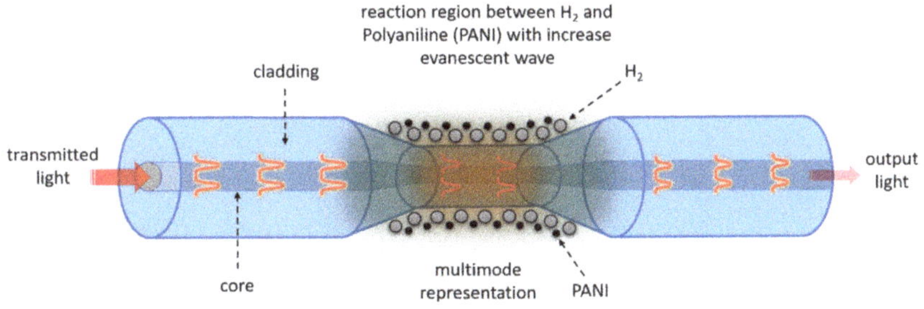

Figure 8. Hydrogen PANI-sensing mechanism.

5. Conclusions

By using a drop casting technique, this study proved that optical fiber sensors could be fabricated from Polyaniline (PANI). The response of the established sensor to various concentrations of H_2 gas at room temperature was used to measure its efficiency. When exposed to 1.00 % H_2 in synthetic air, the PANI-coated based sensor enhanced its ab-sorption response by 19%, according to these findings. The selectivity investigation indicated that the PANI-based optical sensor responded strongly towards ammonia, methane, and hydtogen chemicals. The findings suggest that an affordable and easy methodology may be utilized to enhance an effective, accurate, and repeatable H_2 sensor in real-world atmospheric conditions.

Supplementary Materials: The poster presentation is available online at https://www.mdpi.com/article/10.3390/CSAC2021-10415/s1.

Author Contributions: Conceptualization, M.M.A., and M.H.Y.; methodology, M.M.A., S.H.G., A.H.I., and M.H.Y.; writing—original draft preparation, M.M.A.; review and editing, M.H.Y., M.A.M., S.P., and N.A. All authors have read and agreed to the published version of the manuscript.

Funding: Universiti Putra Malaysia funded this research, grant number GP-IPS/2019/9674900.

Institutional Review Board Statement: Not applicable.

Informed Consent Statement: Not applicable.

Data Availability Statement: Not applicable.

Conflicts of Interest: The authors declare no conflict of interest.

References

1. Hübert, T.; Detjens, M.; Tiebe, C. Response Time Measurement of Hydrogen Sensors. In Proceedings of the 7th International Conference on Hydrogen Safety, Hamburg, Germany, 11–13 September 2017; pp. 1–11.
2. Bychkov, A.L.; Korobeynikov, S.M.; Ryzhkina, A.Y. Determination of the Hydrogen Diffusion Coefficient in Transformer Oil. *Tech. Phys.* **2011**, *56*, 421–422. [CrossRef]

3. Islam, M.M.; Lee, G.; Hettiwatte, S.N. A Review of Condition Monitoring Techniques and Diagnostic Tests for Lifetime Estimation of Power Transformers. *Electr. Eng.* **2018**, *100*, 581–605. [CrossRef]
4. Staffell, I.; Scamman, D.; Velazquez Abad, A.; Balcombe, P.; Dodds, P.E.; Ekins, P.; Shah, N.; Ward, K.R. The Role of Hydrogen and Fuel Cells in the Global Energy System. *Energy Environ. Sci.* **2019**, *12*, 463–491. [CrossRef]
5. Singh, A.P.; Kumar, D.; Agarwal, A.K. Particulate Characteristics of Laser Ignited Hydrogen Enriched Compressed Natural Gas Engine. *Int. J. Hydrogen Energy* **2020**, *45*, 18021–18031. [CrossRef]
6. Pospíšilová, M.; Kuncová, G.; Trögl, J. Fiber-Optic Chemical Sensors and Fiber-Optic Bio-Sensors. *Sensors* **2015**, *15*, 25208–25259. [CrossRef] [PubMed]
7. Rajan, G. *Optical Fiber Sensors: Advanced Techniques and Applications*; CRC Press: Boca Raton, FL, USA, 2015; ISBN 1482228297.
8. Zou, Y.; Pisciotta, J.; Baskakov, I.V. Nanostructured Polypyrrole-Coated Anode for Sun-Powered Microbial Fuel Cells. *Bioelectrochemistry* **2010**, *79*, 50–56. [CrossRef] [PubMed]
9. Wang, M.; Wang, D.N.; Yang, M.; Cheng, J.; Li, J. In-Line Mach-Zehnder Interferometer and FBG with Pd Film for Simultaneous Hydrogen and Temperature Detection. *Sens. Actuators B Chem.* **2014**, *202*, 893–896. [CrossRef]
10. Yu, L.; Gan, M.; Ma, L.; Huang, H.; Hu, H.; Li, Y.; Tu, Y.; Ge, C.; Yang, F.; Yan, J. Facile Synthesis of MnO2/Polyaniline Nanorod Arrays Based on Graphene and Its Electrochemical Performance. *Synth. Met.* **2014**, *198*, 167–174. [CrossRef]
11. Cichosz, S.; Masek, A.; Zaborski, M. Polymer-Based Sensors: A Review. *Polym. Test.* **2018**, *67*, 342–348. [CrossRef]
12. Zhang, Y.; Wen, F.; Jiang, Y.; Wang, L.; Zhou, C.; Wang, H. Layer-by-Layer Construction of Caterpillar-like Reduced Graphene Oxide–Poly (Aniline-Co-o-Aminophenol)–Pd Nanofiber on Glassy Carbon Electrode and Its Application as a Bromate Sensor. *Electrochim. Acta* **2014**, *115*, 504–510. [CrossRef]
13. Zheng, Z.; Du, Y.; Feng, Q.; Wang, Z.; Wang, C. Facile Method to Prepare Pd/Graphene–Polyaniline Nanocomposite and Used as New Electrode Material for Electrochemical Sensing. *J. Mol. Catal. A Chem.* **2012**, *353*, 80–86. [CrossRef]
14. Nazemi, H.; Joseph, A.; Park, J.; Emadi, A. Advanced Micro-and Nano-Gas Sensor Technology: A Review. *Sensors* **2019**, *19*, 1285. [CrossRef] [PubMed]
15. Ahad, I.Z.M.; Harun, S.W.; Gan, S.N.; Phang, S.W. Application of Fiber Bragg Grating Sensor Coated with Polyaniline as an Optical Sensor for Chloroform Detection. *Polym. Polym. Compos.* **2017**, *25*, 555–562. [CrossRef]
16. Khalaf, A.L.; Mohamad, F.S.; Arasu, P.T.; Shabaneh, A.A.; Abdul Rahman, N.; Lim, H.N.; Paiman, S.; Yusof, N.A.; Mahdi, M.A.; Yaacob, M.H. Modified Plastic Optical Fiber Coated Graphene/Polyaniline Nanocompositefor Ammonia Sensing. In Proceedings of the 2016 IEEE 6th International Conference on Photonics, ICP 2016, Kuching, Malaysia, 14–16 March 2016; pp. 31–33. [CrossRef]
17. Alkhabet, M.M.; Girei, S.H.; Paiman, S.; Arsad, N.; Mahdi, M.A.; Yaacob, M.H. Highly Sensitive Hydrogen Sensor Based on Palladium-Coated Tapered Optical Fiber at Room Temperature. *Eng. Proc.* **2020**, *2*, 8.
18. Andika, A.I.W.; Hafizah, M.A.E.; Manaf, A. Electrical Conductivity and Microwave Characteristics of HCl and HClO4-Doped Polyaniline Synthesized through Chemical Oxidative Continuous Polymerization Process at Various Polymerization Temperatures. In *Proceedings of the AIP Conference Proceedings*; AIP Publishing LLC: Depok, Indonesia, 2018; Volume 2023, p. 20027.
19. Mogre, P. Synthesis and Characterization Studies of Polyaniline Nano Fibres. *Adv. Mater. Proc.* **2018**, 3, 178–180. [CrossRef]
20. Deng, J.; Wang, X.; Guo, J.; Liu, P. Effect of the Oxidant/Monomer Ratio and the Washing Post-Treatment on Electrochemical Properties of Conductive Polymers. *Ind. Eng. Chem. Res.* **2014**, *53*, 13680–13689. [CrossRef]
21. Jarad, A.N. Synthesis and Characterization of Conductive Polyaniline Using 24 Hours Chemical Oxidative Process for Organic Solar Cells. Ph.D. Thesis, Universiti Sains Malaysia, George Town, Malaysia, 2017.
22. Al-Daghman, A.N.J. Optical Characteristics of Conductive Polymer Polyaniline PANI-EB. *Int. J. Res. Appl. Sci. Eng. Technol.* **2018**, *6*, 1555–1560. [CrossRef]
23. Karaoğlan, N.; Bindal, C. Synthesis and Optical Characterization of Benzene Sulfonic Acid Doped Polyaniline. *Eng. Sci. Technol. Int. J.* **2018**, *21*, 1152–1158. [CrossRef]
24. Arasu, P.T.; Noor, A.S.M.; Khalaf, A.L.; Yaacob, M.H. Highly Sensitive Plastic Optical Fiber with Palladium Sensing Layer for Detection of Hydrogen Gas. In Proceedings of the 2016 IEEE Region 10 Symposium, TENSYMP 2016, Bali, Indonesia, 9–11 May 2016; pp. 390–393. [CrossRef]
25. González-Sierra, N.E.; Gómez-Pavón, L.d.C.; Pérez-Sánchez, G.F.; Luis-Ramos, A.; Zaca-Morán, P.; Muñoz-Pacheco, J.M.; Chávez-Ramírez, F. Tapered Optical Fiber Functionalized with Palladium Nanoparticles by Drop Casting and Laser Radiation for H2 and Volatile Organic Compounds Sensing Purposes. *Sensors* **2017**, *17*, 2039. [CrossRef] [PubMed]
26. Li, Z.F.; Blum, F.D.; Bertino, M.F.; Kim, C.S. Amplified Response and Enhanced Selectivity of Metal-PANI Fiber Composite Based Vapor Sensors. *Sens. Actuators B Chem.* **2012**, *161*, 390–395. [CrossRef]

chemistry proceedings

MDPI

Proceeding Paper

Aqueous Medium Fluoride Anion Sensing by Fluorophore Encapsulated UiO-66 Type Zirconium Metal–Organic Framework †

Rana Dalapati and Ling Zang *

Department of Materials Science and Engineering, Nano Institute of Utah, University of Utah, Salt Lake City, UT 84112, USA; rana.dalapati@utah.edu
* Correspondence: lzang@eng.utah.edu

† Presented at the 1st International Electronic Conference on Chemical Sensors and Analytical Chemistry, 1–15 July 2021; Available online: https://csac2021.sciforum.net/.

Abstract: A well-known fluorophore molecule, pyrene was encapsulated into a stable metal organic framework by in situ encapsulation method. The existing metal-organic framework (MOF) called UiO-66 (UiO = University of Oslo) served as host material for pyrene fluorophore. The fluorescence of pyrene was quenched after encapsulation inside the porous host. Recovery of quenched fluorescence was accomplished by anion induced host dissolution, followed by the release of the fluorophore molecule. Using this anion induced dissolution, a selective sensing of fluoride anion in pure aqueous was achieved.

Keywords: metal-organic framework; pyrene; fluoride anion sensing

Citation: Dalapati, R.; Zang, L. Aqueous Medium Fluoride Anion Sensing by Fluorophore Encapsulated UiO-66 Type Zirconium Metal–Organic Framework. *Chem. Proc.* **2021**, *5*, 86. https://doi.org/10.3390/CSAC2021-10551

Academic Editor: Ye Zhou

Published: 1 July 2021

1. Introduction

For the past few decades, a tremendous effort has been dedicated by the scientific community towards the development of a modest strategy for the selective and precise sensing of anions [1], as they perform a key role in biological systems, health, and environment [2]. Among the anions present in biological systems, the smallest fluoride anion has drawn significant attention, due to its biological and environmental impact [3]. Currently, the presence of fluoride in drinking water and commercial household products is the emerging concern for public health [4]. Although fluoride is considered as a micronutrient [5], the excess uptake of fluoride can cause fluorosis [6], and even chronic renal failure [7]. As such, there is an urgent necessity for selective and precise determination of fluoride anions in fluoride-contaminated water.

Metal-organic framework (MOF), a new class of porous materials, have received tremendous attention for their potential applications in gas storage [8], chemical separation [9], catalysis [10], and drug delivery [11]. The UiO-66 framework (UiO = University of Oslo) is one typical Zr-MOF, constructed with $Zr_6O_4(OH)_4$ clusters and 1,4-benzenedicarboxylate, BDC linkers [12]. With a higher surface area, thermal resistivity, and an exceptional structural stability in water, this becomes an ideal molecular host material [13]. The triggered release of a guest molecule by host dissolution is one of the efficient strategies for molecular recognition [14]. Recently Bein et al. reported fluoride sensing by using the hybrid composite of the metal-organic framework NH_2-MIL-101(Al) and fluorescein [15].

Herein, we report a selective and precise sensing of fluoride ions in a pure aqueous medium by a fluoride triggered release of pyrene fluorophore from Zirconium based MOF, UiO-66 [16]. First, in one step we have synthesized pyrene encapsulated UiO-66. Where the Zr-O or μ_3-oxo bond of UiO-66 framework acts as a reactive probe and the pyrene molecule acts as a signal transductor. Upon encapsulating, the inside of the pore of the framework fluorescence of pyrene was found to be completely quenched. The addition of the fluoride ion provoked the decomposition of the host UiO-66 and the released pyrene provides a turn-on fluorescence.

2. Methods

All the starting materials were of reagent grade and used as they were received from the commercial suppliers.

2.1. Synthesis

Syntheses of the pyrene containing UiO-66 framework were performed as reported by Biswas et al. with modification [17]. In brief, a mixture of ZrCl4 (72.24 mg, 0.31 mmol), Benzene-1,4-dicarboxylic acid (H2BDC) (0.31 mmol), pyrene (10 mg, 0.05 mmol), and formic acid (1.2 mL, 3.18 mmol) in Dimethylacetamide (DMA) (3 mL) was placed in a Pyrex tube. The tube was sealed and heated in a preheated heating block to 150 °C for 24 h. The reaction mixture was then cooled to room temperature. Finally, the precipitate was collected by filtration, washed with acetone, and dried in an air oven (60 °C).

2.2. Fluorescence Titration Measurement

For fluorescence titration measurements, a stock solution of pyrene@UiO-66 (1 mg/mL) was diluted in water (final concentration of 99 μg/mL) in a quartz glass cuvette at room temperature. A 4 mm solution of different anions was used. All the titration fluorescence emission was monitored using an excitation wavelength of 337 nm.

3. Results and Discussion

3.1. Material Characterization

Pyrene encapsulated UiO-66 (pyrene@UiO-66) was synthesized by a single step in situ encapsulation method. Initially, a certain amount of pyrene was added with $ZrCl_4$ and H_2BDC during the synthesis process. The pyrene@UiO-66 material was well characterized by various instrumental techniques, such as X-ray powder diffraction (XRPD), Fourier-transform infrared spectroscopy (FT-IR), and N_2 sorption analysis.

XRPD experiments showed that UiO-66 and pyrene@UiO-66 possessed very similar XRPD patterns (Figure 1a). The similarities between the patterns simulated and found during the experiments of XRPD confirmed the formation of pure UiO-66.

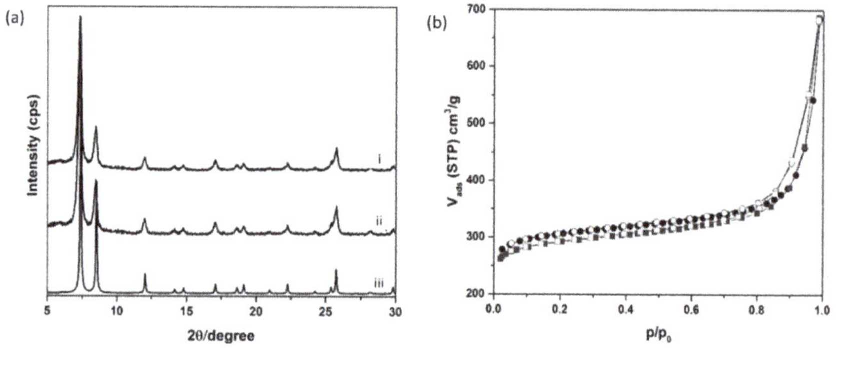

Figure 1. (**a**) XRPD pattern of (i) pyrene@UiO-66; (ii) only UiO-66; (iii) simulated pattern of UiO-66 MOF. (**b**) Nitrogen adsorption (filled symbols) and desorption (empty symbols) isotherms of UiO-66 (circle) and pyrene@UiO-66 (square) collected at −196 °C.

The N_2 sorption isotherms of pyrene@UiO-66 (Figure 1b) exhibited an insufficient decrease in the surface area, compared to UiO-66, which indicated the successful encapsulation of the pyrene into the pore framework of UiO-66. Pyrene encapsulated UiO-66 showed a surface area of 898 cm^3/g, which was lower than the guest-free UiO-66 material (955 cm^3/g).

3.2. Anion Sensing Experiment

The fluorescence emission spectra of pyrene@UiO-66 in water was recorded upon the gradual addition of sodium (Na^+) salts of various anions (F^-, Cl^-, Br^-, I^-, NO_2^-, NO_3^-, AcO^-, $S_2O_3^{2-}$, HSO_3^-, SO_4^{2-}, HSO_4^-, SO_3^{2-}, ClO_4^-, SCN^-, and HCO_3^-). Figure 2 shows the "turn-on" response of F^- anion towards pyrene@UiO-66 in water. There were almost no changes observed in the fluorescence emission spectra for all other anions. The bar plot in Figure 3a summarized the selectivity of pyrene@UiO-66 towards F^- anion over all other anions.

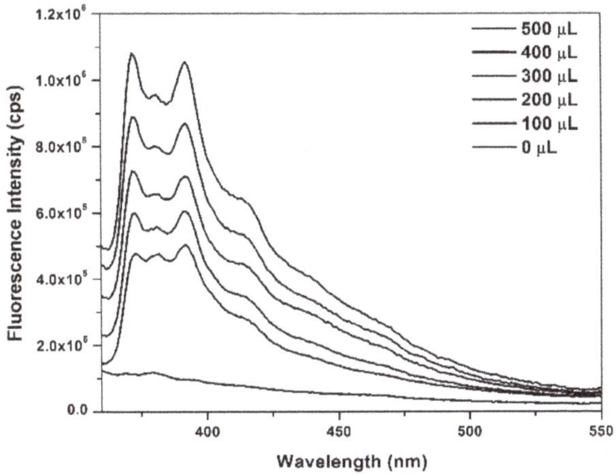

Figure 2. Change in fluorescence intensity with the gradual addition of the F^- solution to a suspension of pyrene@UiO-66 in aqueous medium.

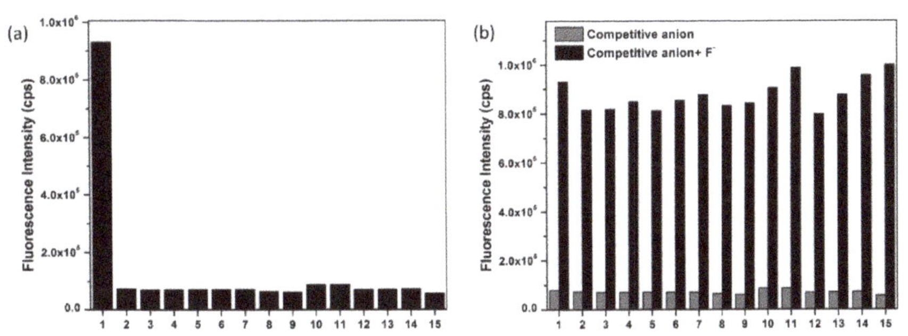

Figure 3. (**a**) Change in the fluorescence intensity of pyrene@UiO-66 upon the incremental addition of different anions. (**b**) Change in the fluorescence intensity of pyrene@UiO-66 upon the addition of F^- solution in the absence and presence of different anions. (F^- (1), Cl^- (2), Br^- (3), I^- (4), NO_2^- (5), NO_3^- (6), AcO^- (7), $S_2O_3^{2-}$ (8), HSO_3^- (9), SO_4^{2-} (10), HSO_4^- (11), SO_3^{2-} (12), ClO_4^- (13), SCN^- (14), and HCO_3^- (15)).

To examine the sensitivity of the pyrene@UiO-66 sensor material towards fluoride ions, even in the presence of other known interfering ions generally present in water, competitive experiments were performed by monitoring the fluorescence emission intensity of pyrene@UiO-66 in the absence and presence of other anions. During these experiments, solutions of interfering anions were added first to a water of pyrene@UiO-66, followed by the addition of the F^- anion. The change of the fluorescence intensity of pyrene@UiO-66 upon the addition of the F^- anion, in absence and presence of other interfering anions, are displayed in Figure 3b. In all cases, the interfering anions did not show any interference in their sensing of the F^- anion.

3.3. Mechanism for Anion Sensing

Until now, few mechanisms have been proposed for anion sensing via metal-organic framework. Some include: (1) anion induced coordination to metal-oxygen cluster [18], (2) hydrogen bonding formation with solvated framework [19], and (3) anion induced structural decomposition [15]. To understand the mechanism of fluoride sensing, XRPD and FT-IR measurement was carried out. To check the fluoride induced UiO-66 framework decomposition, MOF material was soaked in the fluoride anion solution. From Figure 4a it was shown that after the fluoride treatment, the characteristic diffraction peak for UiO-66 framework vanished, which confirmed the collapse of the framework in presence of fluoride. However, no change was observed in XRPD pattern after treatment with other anions in aqueous solution (data not shown).

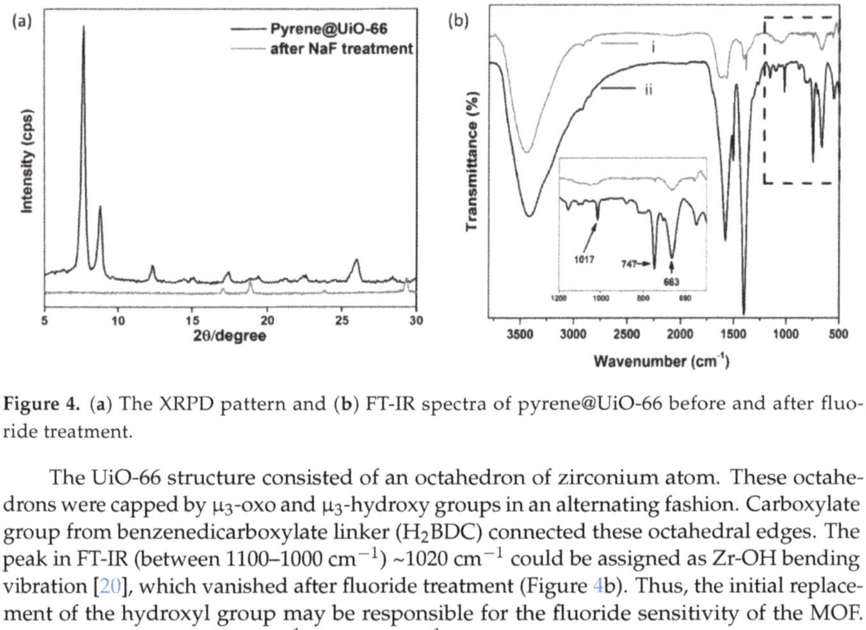

Figure 4. (**a**) The XRPD pattern and (**b**) FT-IR spectra of pyrene@UiO-66 before and after fluoride treatment.

The UiO-66 structure consisted of an octahedron of zirconium atom. These octahedrons were capped by μ_3-oxo and μ_3-hydroxy groups in an alternating fashion. Carboxylate group from benzenedicarboxylate linker (H_2BDC) connected these octahedral edges. The peak in FT-IR (between 1100–1000 cm^{-1}) ~1020 cm^{-1} could be assigned as Zr-OH bending vibration [20], which vanished after fluoride treatment (Figure 4b). Thus, the initial replacement of the hydroxyl group may be responsible for the fluoride sensitivity of the MOF. A peak in FT-IR near 747 cm^{-1} and 663 cm^{-1} was responsible for a Zr-μ_3-oxo bond that almost disappeared, which also suggested fragmentation of the μ3-oxo bond. Observation suggested that fluorescence enhancement occurred via zirconium and fluoride coordination, which led to the release of pyrene from the framework host.

4. Conclusions

We have demonstrated that fluoride induced UiO-66 framework decomposition can be successfully used as a selective sensing probe for the same. Although this system is not reversible in nature, the simple one-step synthesis protocol, high stability, and low toxicity makes this material a promising candidate for fluoride sensing in an aqueous medium.

Author Contributions: Conceptualization, L.Z. and R.D.; methodology, R.D.; formal analysis, R.D.; investigation, R.D.; resources, R.D.; data curation, R.D.; writing—original draft preparation, R.D.; writing—review and editing, R.D.; visualization, R.D.; supervision, L.Z.; project administration, R.D.; funding acquisition, L.Z. All authors have read and agreed to the published version of the manuscript.

Funding: The authors are highly grateful to University of Utah for instrumental facilities and other support.

Institutional Review Board Statement: Not applicable.

Informed Consent Statement: Not applicable.

Acknowledgments: The support from the University of Utah is acknowledged.

Conflicts of Interest: The authors declare no conflict of interest.

References

1. Busschaert, N.; Caltagirone, C.; van Rossom, W.; Gale, P.A. Applications of Supramolecular Anion Recognition. *Chem. Rev.* **2015**, *115*, 8038–8155. [CrossRef] [PubMed]
2. Gale, P.A.; Caltagirone, C. Anion sensing by small molecules and molecular ensembles. *Chem. Soc. Rev.* **2015**, *44*, 4212–4227. [CrossRef] [PubMed]
3. Zhou, Y.; Zhang, J.F.; Yoon, J. Fluorescence and Colorimetric Chemosensors for Fluoride-Ion Detection. *Chem. Rev.* **2014**, *114*, 5511–5571. [CrossRef] [PubMed]
4. Jagtap, S.; Yenkie, M.K.; Labhsetwar, N.; Rayalu, S. Fluoride in Drinking Water and Defluoridation of Water. *Chem. Rev.* **2012**, *112*, 2454–2466. [CrossRef] [PubMed]
5. Nielsen, F.H. Micronutrients in parenteral nutrition: boron, silicon, and fluoride. *GASTAB* **2009**, *137*, S55–S60. [CrossRef] [PubMed]
6. Burgmaier, G.M.; Schulze, I.; Attin, T. Fluoride uptake and development of artificial erosions in bleached and fluoridated enamel in vitro. *J. Oral Rehabil.* **2002**, *29*, 799–804. [CrossRef] [PubMed]
7. Chandrajith, R.; Dissanayake, C.; Ariyarathna, T.; Herath, H.; Padmasiri, J. Dose-dependent Na and Ca in fluoride-rich drinking water—Another major cause of chronic renal failure in tropical arid regions. *Sci. Total Environ.* **2011**, *409*, 671–675. [CrossRef] [PubMed]
8. Alezi, D.; Belmabkhout, Y.; Suyetin, M.; Bhatt, P.M.; Weseliński, Ł.J.; Solovyeva, V.; Adil, K.; Spanopoulos, I.; Trikalitis, P.N.; Emwas, A.-H. MOF Crystal Chemistry Paving the Way to Gas Storage Needs: Aluminum-Based soc-MOF for CH_4, O_2, and CO_2 Storage. *J. Am. Chem. Soc.* **2015**, *137*, 13308–13318. [CrossRef] [PubMed]
9. Stock, N.; Biswas, S. Synthesis of Metal-Organic Frameworks (MOFs): Routes to Various MOF Topologies, Morphologies, and Composites. *Chem. Rev.* **2011**, *112*, 933–969. [CrossRef] [PubMed]
10. Liu, J.; Chen, L.; Cui, H.; Zhang, J.; Zhang, L.; Su, C.-Y. Applications of metal–organic frameworks in heterogeneous supramolecular catalysis. *Chem. Soc. Rev.* **2014**, *43*, 6011–6061. [CrossRef] [PubMed]
11. Horcajada, P.; Chalati, T.; Serre, C.; Gillet, B.; Sebrie, C.; Baati, T.; Eubank, J.F.; Heurtaux, D.; Clayette, P.; Kreuz, C. Porous metal–organic-framework nanoscale carriers as a potential platform for drug delivery and imaging. *Nat. Mater.* **2010**, *9*, 172–178. [CrossRef] [PubMed]
12. Cavka, J.H.; Jakobsen, S.; Olsbye, U.; Guillou, N.; Lamberti, C.; Bordiga, S.; Lillerud, K.P. A New Zirconium Inorganic Building Brick Forming Metal Organic Frameworks with Exceptional Stability. *J. Am. Chem. Soc.* **2008**, *130*, 13850–13851. [CrossRef] [PubMed]
13. Liu, X.; Demir, N.K.; Wu, Z.; Li, K. Highly Water-Stable Zirconium Metal–Organic Framework UiO-66 Membranes Supported on Alumina Hollow Fibers for Desalination. *J. Am. Chem. Soc.* **2015**, *137*, 6999–7002. [CrossRef] [PubMed]
14. Ma, X.; Zhao, Y. Biomedical Applications of Supramolecular Systems Based on Host–Guest Interactions. *Chem. Rev.* **2014**, *115*, 7794–7839. [CrossRef] [PubMed]
15. Hinterholzinger, F.M.; Rühle, B.; Wuttke, S.; Karaghiosoff, K.; Bein, T. Highly sensitive and selective fluoride detection in water through fluorophore release from a metal-organic framework. *Sci. Rep.* **2013**, *3*, 2562. [CrossRef] [PubMed]
16. Huang, Y.; Qin, W.; Li, Z.; Li, Y. Enhanced stability and CO_2 affinity of a UiO-66 type metal–organic framework decorated with dimethyl groups. *Dalton Trans.* **2012**, *41*, 9283–9285. [CrossRef] [PubMed]
17. Biswas, S.; der Voort, P.V. A General Strategy for the Synthesis of Functionalised UiO-66 Frameworks: Characterisation, Stability and CO_2 Adsorption Properties. *Eur. J. Inorg. Chem.* **2013**, *2013*, 2154–2160. [CrossRef]
18. Yang, J.; Dai, Y.; Zhu, X.; Wang, Z.; Li, Y.; Zhuang, Q.; Shi, J.; Gu, J. Metal–organic frameworks with inherent recognition sites for selective phosphate sensing through their coordination-induced fluorescence enhancement effect. *J. Mater. Chem. A* **2015**, *3*, 7445–7452. [CrossRef]
19. Chen, B.; Wang, L.; Zapata, F.; Qian, G.; Lobkovsky, E.B. A Luminescent Microporous Metal−Organic Framework for the Recognition and Sensing of Anions. *J. Am. Chem. Soc.* **2008**, *130*, 6718–6719. [CrossRef] [PubMed]
20. Wang, C.; Liu, X.; Chen, J.P.; Li, K. Superior removal of arsenic from water with zirconium metal-organic framework UiO-66. *Sci. Rep.* **2015**, *5*, 16613. [CrossRef] [PubMed]

chemistry proceedings

MDPI

Proceeding Paper

Customized Screen-Printed Electrodes Based on Ag-Nanoseeds for Enhanced Electroanalytical Response towards Cd(II), Pb(II) and As(V) in Aqueous Samples †

Karina Torres-Rivero [1,2,*], Clara Pérez-Ràfols [3], Julio Bastos-Arrieta [4], Núria Serrano [3,5], Vicenç Martí [1,2,6] and Antonio Florido [1,2]

1 Departament d'Enginyeria Química, Escola d'Enginyeria de Barcelona Est (EEBE), Universitat Politècnica de Catalunya, BarcelonaTEch (UPC), Av. Eduard Maristany 16, 08019 Barcelona, Spain; vicens.marti@upc.edu (V.M.); antonio.florido@upc.edu (A.F.)
2 Barcelona Research Center for Multiscale Science and Engineering, Av. Eduard Maristany 16, 08019 Barcelona, Spain
3 Departament d'Enginyeria Química i Química Analítica, Facultat de Química, Universitat de Barcelona, Martí i Franquès 1-11, 08028 Barcelona, Spain; claraperezrafols@ub.edu (C.P.-R.); nuria.serrano@ub.edu (N.S.)
4 Grup de Biotecnologia Molecular i Industrial, Universitat Politècnica de Catalunya, Rambla Sant Nebridi 22, Edifici Gaia TR14, 08222 Terrassa, Spain; julio.bastos@upc.edu
5 Institut de Recerca de l'Aigua (IdRA), Universitat de Barcelona, Martí i Franquès 1-11, 08028 Barcelona, Spain
6 Eurecat-Centre Tecnològic de Catalunya, Water Air and Soil Unit, Plaça de la Ciència 2, 08243 Manresa, Spain
* Correspondence: karina.torres.rivero@upc.edu

† Presented at the 1st International Electronic Conference on Chemical Sensors and Analytical Chemistry, 1–15 July 2021; Available online: https://csac2021.sciforum.net/.

Abstract: Electrochemical analysis based on screen-printed electrodes (SPEs) represents a great alternative to conventional analytical methods such as ICP-MS or LC-MS due to their portability, sensitivity, selectivity, and cost-effectiveness. In addition, the functionalization of SPEs with nanomaterials has been reported to provide an enhanced analytical performance. In this regard, silver nanoparticles (AgNPs) were synthesized and appropriately characterized, showing spherical silver nanoseeds (Ag-NS) with a diameter of 12.20 ± 0.04 nm. Using the drop-casting methodology, the synthesized AgNPs were used to modify screen-printed carbon nanofiber electrodes (SPCNFEs). Ag-NS deposition onto the electrode surface was confirmed by scanning electron microscopy (SEM). Furthermore, the analytical response of the modified electrodes (Ag-NS-SPCNFE) was evaluated for the determination of trace Pb(II), Cd(II), and As(V) using differential pulse anodic stripping voltammetry (DPASV), obtaining detection limits of 3.3, 3.7, and 2.6 μg L^{-1}, for Pb(II), Cd(II) and As(V), respectively. Finally, Ag-NS-SPCNFE was tested towards the determination of As(V) in a spiked tap water sample, showing a good agreement with concentrations determined by ICP-MS.

Keywords: screen-printed electrodes; Ag nanoparticles; anodic stripping voltammetry; lead determination; cadmium determination; arsenic determination

check for updates

Citation: Torres-Rivero, K.; Pérez-Ràfols, C.; Bastos-Arrieta, J.; Serrano, N.; Martí, V.; Florido, A. Customized Screen-Printed Electrodes Based on Ag-Nanoseeds for Enhanced Electroanalytical Response towards Cd(II), Pb(II) and As(V) in Aqueous Samples. *Chem. Proc.* **2021**, *5*, 87. https://doi.org/10.3390/CSAC2021-10469

Academic Editor: Ye Zhou

Published: 30 June 2021

Publisher's Note: MDPI stays neutral with regard to jurisdictional claims in published maps and institutional affiliations.

1. Introduction

Water contamination caused by heavy metal ions (HMIs) is a concerning issue due to their high toxicity, non-biodegradability, bioaccumulation, and adverse health effects in humans [1]. In particular, for As, Cd, and Pb, the World Health Organization (WHO) has established the maximum allowed concentration in drinking water as 10 μg L^{-1}, 3 μg L^{-1}, and 10 μg L^{-1}, respectively [2]. The determination of these low concentration levels requires very sensitive analytical techniques, such as flameless atomic adsorption spectroscopy (FAAS) [3], inductively coupled plasma mass spectrometry (ICP-MS) [4,5], and hydride generation atomic fluorescence spectrometry (HG-AFS) [6]. However, these analytical techniques require expensive equipment and highly trained personnel, extended analysis time, and high operating costs.

In contrast, electrochemical techniques and anodic stripping voltammetry (ASV) allow relatively fast determination of trace HMIs with easy-handling and low-cost equipment [7,8]. In particular, electrochemical sensors represent a versatile tool for monitoring different samples in the environmental field. In addition, the literature has reported how modifying their surface with nanomaterials enhanced the electrochemical reactivity and sensitivity to specific analytes [9], allowing lower detection limits and higher sensitivity for stripping techniques [10–12].

The use of metallic nanoparticles (MNPs) to modify screen-printed electrodes (SPEs) can reduce the electron transfer resistance at the electrode surface, decreasing the electron transfer limited process and consequently catalyzing the electrode's response at low analyte concentrations [13–15]. Nanoparticles exhibit higher reactive surface influenced by the exposed atoms disposition resulting in more electrocatalytically active sites [16].

MNPs-modified sensors have been reported to allow the detection of arsenic [17,18], lead [19,20], and cadmium [21,22] at the level of a few $\mu g\ L^{-1}$, fulfilling the WHO guidelines for drinking-water quality [2]. These sensors were based on screen-printing technology, offering significant advantages over conventional voltammetric sensors such as low-cost, disposable character, portability, and commercial availability [1,9]. Thus, in this work, the voltammetric determination of HMIs, based on the use of carbon-nanofiber-based screen-printed electrodes (SPCNFEs) modified with silver nanoparticles (Ag-NPs), is proposed. Ag-NPs were synthesized in the shape of silver nanoseeds (Ag-NS), and the resulting modified electrodes were microscopically and analytically characterized for the determination of As(V), Pb(II), and Cd(II) by means of differential pulse anodic stripping voltammetry (DPASV). In addition, the applicability to real sample analysis was demonstrated through the direct determination of As(V) in spiked tap water samples.

2. Materials and Methods

2.1. Apparatus

DPASV measurements were carried out with either a Multi Autolab/M204 Modular Multi Potentiostat/Galvanostat or an Autolab PGSTAT204, attached to a Metrohm 663 VA Stand, as well as a personal computer with NOVA 2.1 software package to control the potentiostat and perform the required data treatment. All the electrochemical instrumentations and software were acquired from Metrohm (Herisau, Switzerland).

A Crison Basic 20 pH meter (Hach Lange Spain, L'Hospitalet de Llobregat, Spain) was used for pH measurements.

Ag-NS and SPCNFE modified with Ag-NS electrodes were characterized using a JEM-2010 transmission electron microscope (TEM) from JEOL (Tokyo, Japan) and a Gemini scanning electron microscope (SEM) from ZEISS® (Jena, Germany), respectively. Size distribution histograms were calculated using the ImageJ version 1.51 m software by the National Institute of Health (NIH, Bethesda, MD, USA).

ICP-MS measurements were carried out by means of inductively coupled plasma mass spectrometer model 7800 by Agilent Technologies (Santa Clara, CA, USA).

Commercial SPCNFEs, including working (4 mm disk), counter, and reference electrodes, were purchased from Dropsens (Llanera, Spain).

2.2. Preparation of Working Electrodes

The working electrode (WE) was a SPCNFE modified with silver nanoseeds (Ag-NS-SPCNFE). Ag-NS were first synthesized following a seed-mediated methodology, combining aqueous trisodium citrate (5 mL, 2.5 mmol L^{-1}), aqueous poly sodium styrene-sulfonate (PSSS) (0.25 mL, 500 mg L^{-1}), aqueous sodium borohydride ($NaBH_4$) (0.3 mL, 10 mmol L^{-1}) freshly prepared, followed by the addition of aqueous silver nitrate ($AgNO_3$) (5 mL, 0.5 mmol L^{-1}) using a syringe pump at a rate of 2 mL min^{-1} under continuous stirring [12,23]. Then, SPCNFEs were modified by drop-casting, dropping 40 µL of Ag-NS onto the working electrode, and evaporating the solvent at 50 °C for 30 min.

2.3. Electrochemical Measurements

DPASV measurements of Pb(II) and Cd(II) were carried out at a deposition potential (E_d) of −1.4 V, applied under stirring conditions during a deposition time (t_d) of 180 s in 0.1 mol L^{-1} acetate buffer (pH 4.5) and scanning the potential from −1.4 to 0.0 V. For As(V) determination, the experimental conditions used were E_d of −1.3 V and t_d of 120 s in 0.01 mol L^{-1} HCl pH 2 with a potential scan from −1.3 to −0.65 V.

A step potential of 5 mV, a pulse time of 50 ms, and a pulse amplitude of 50 mV were employed in all cases. All experiments were performed at room temperature (22 ± 1 °C) and without oxygen removal.

For real sample analysis, tap water samples were collected from the local water distribution network in Barcelona (Spain) and spiked with 20 µg L^{-1} of As(V). Prior to electrochemical analysis, water samples were diluted and acidified with 0.01 mol L^{-1} of HCl (pH 2.0), resulting in a final concentration of 10 µg L^{-1} of As(V). Sample analysis was carried out by means of the standard addition method, performing four successive As(V) additions from a standard solution of 1 mg L^{-1}. DPASV measurements were recorded under the above-mentioned electrochemical conditions.

3. Results

3.1. Microscopic Characterization

Ag-NS synthesis was microscopically confirmed by both SEM (Figure 1a) and TEM (Figure 1b). As it can be deduced from the TEM image, most Ag-NS presented a spherical shape. On the other hand, SEM images were used to calculate the corresponding size distribution histogram (Figure 1c), which was computed from 400 Ag-NSs. The obtained results show that the synthesized Ag-NS presented an average diameter of 12.2 ± 0.4 nm. These structures are in good agreement with the reported shapes of Ag-NPs [24,25].

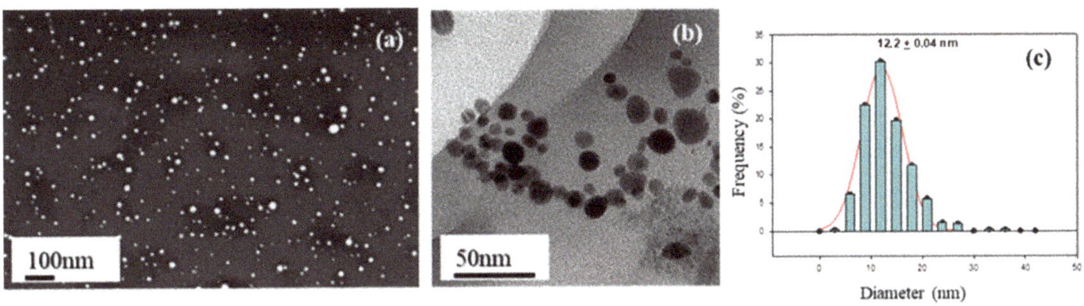

Figure 1. Microscopic characterization of Ag-Nanoseeds (**a**) SEM micrograph. (**b**) TEM micrograph and (**c**) corresponding size distribution histogram [15].

SEM micrographs were also obtained for a bare SPCNFE (Figure 2a) and an Ag-NS-SPCNFE (Figure 2b) to assess the modification of SPCNFEs by drop-casting. Compared to the non-modified carbon nanofiber surface of the bare electrode, Ag-NS can be spotted as white dots deposited onto the carbon nanofibers in the modified electrode (Ag-NS-SPCNFE), thus confirming the successful modification of the working electrode.

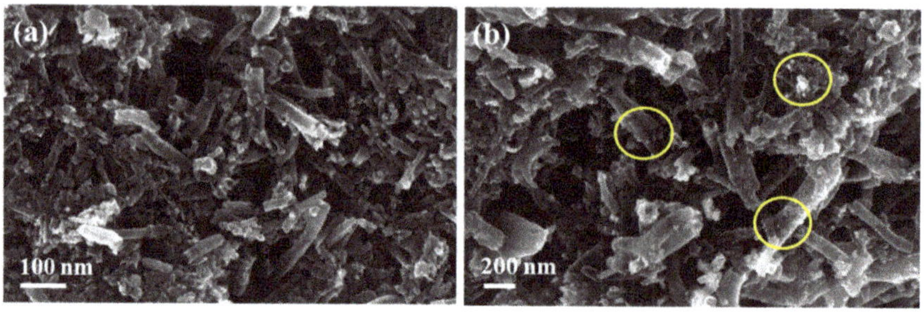

Figure 2. SEM micrographs for (**a**) Bare SPCNFE, and (**b**) Ag-NS-SPCNFE modified using the drop-casting methodology [12,15].

3.2. Electrochemical Characterization

DPASV measurements were carried out in solutions containing either Pb(II), Cd(II), or As(V). Well-defined voltammetric peaks were obtained in all cases, with peak potentials of ca. −0.65 V, −0.75 V, and −1.0 V for Pb(II), Cd(II), and As(V), respectively (see Figure 3a).

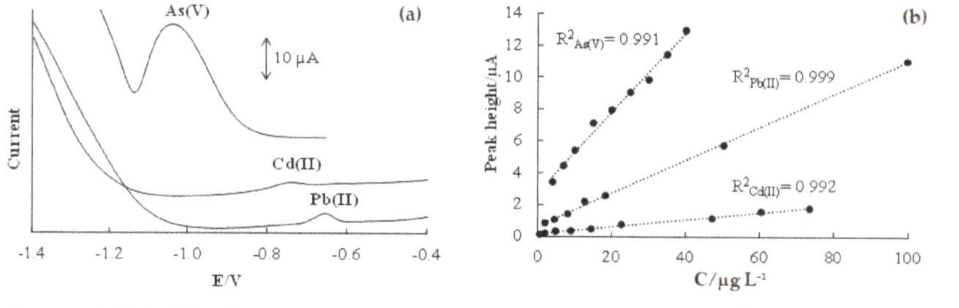

Figure 3. (**a**) DPASV voltammograms of Pb(II), Cd(II), and As(V) at 25 μg L^{-1} and (**b**) their calibration plots at the previously mentioned conditions using Ag-NS-SPCNFE.

Individual calibration curves of Pb(II), Cd(II), and As(V) were obtained by DPASV by increasing metal ion concentration in the ranges 1.9 to 150.0 μg L^{-1}, 0.6 to 120.6 μg L^{-1}, and 1.0 to 50.1 μg L^{-1}, respectively. The obtained data were used to calculate the corresponding analytical parameters (i.e., sensitivity, limit of detection (LOD), limit of quantification (LOQ), and linear range), which are displayed in Table 1.

Table 1. Calibration data for the individual determination of Pb(II), Cd(II), and As(V) using Ag-NS-SPCNFE and the corresponding buffer and DPASV parameters (see the experimental section for more details).

Analyte	LOD (μg L^{-1})	Linear Range (μg L^{-1}) [a]	R^2	Sensitivity (nA μg^{-1} L) [b]
Pb(II)	3.3	11.00–99.6	0.999	103 (1)
Cd(II)	3.7	12.2–73.4	0.992	22 (1)
As(V)	2.6	8.9–40.0	0.991	260 (10)

[a] The lowest value of the linear range corresponds to the LOQ. [b] The standard deviations are expressed in parentheses.

From this data, LODs and LOQs were calculated by using the Miller and Miller procedure [26,27].

As shown in Table 1, good linear response between the peak heights and the concentration of the different analytes was achieved using the Ag-NS-SPCNFE. LODs were at μg L^{-1} levels in all cases, and similar or even lower to other LODs reported in the literature. For example, LODs of 3.30 and 4.43 μg L^{-1} for Pb(II) and Cd(II), respectively, were reported using a graphene/polyaniline/polystyrene (G/PANI/PS) nanoporous fiber-modified screen-printed carbon electrode [28]. Additionally, the obtained LOD for As(V) is considerably lower than that reported using boron-doped diamond electrodes and ASV (12 μg L^{-1}) [29]. However, it is important to mention that Ag-NS-SPCNFE for the determination of As(V) presented a more restricted linear range in which the highest value is limited to a lower concentration value (until 40.0 μg L^{-1}), compared to the one reached by Nagaoka et al. (until 100 μg L^{-1}) [29].

In terms of sensitivities (nA $μg^{-1}$ L), which were calculated as the slope of the calibration curves, Ag-NS-SPCNFE exhibited higher sensitivity toward As(V) (260 nA $μg^{-1}$ L). In the case of Pb(II) and Cd(II), the sensitivities were significantly lower (103 and 22 nA $μg^{-1}$ L, respectively).

3.3. *Application to the Analysis of Spiked Tap Water*

The applicability of Ag-NS-SPCNFE for real sample determination was evaluated through the determination of As(V) in a spiked tap water sample. The determination of As(V) was performed in triplicate by the standard addition calibration method. Representative voltammograms are shown in Figure 4. As it can be observed, a well-shaped As(V) peak and a good correlation between peak area and concentration were acquired. Sample concentration calculated by extrapolation was 10.04 μg L^{-1} (SD:0.37 μg L^{-1}), which is in good agreement with values obtained by ICP-MS (10.7 μg L^{-1}, SD:0.20 μg L^{-1}), as an analytical reference technique. These results confirm the suitability of Ag-NS-SPCNFE for the analysis of real samples.

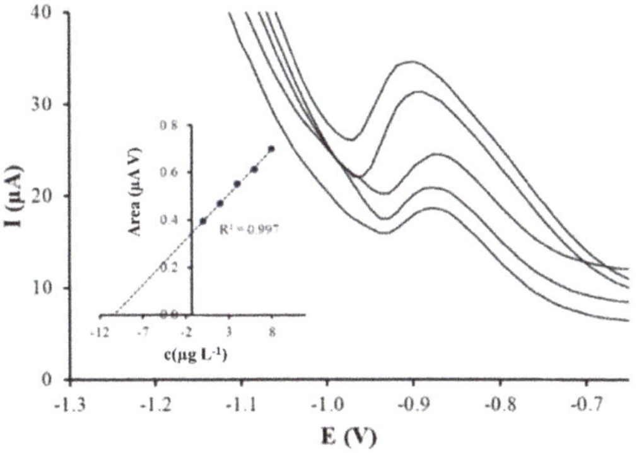

Figure 4. DPASV measurements of As(V) in a spiked tap water sample on Ag-NS-SPCNFE at pH 2.0 applying an E_d of −1.30 V and a t_d of 120 s. Inset: As(V) standard addition plot [12].

4. Conclusions

In this work, a DPASV method for the determination of trace Pb(II), Cd(II), and As(V) based on the modification of SPCNFE with Ag-NS has been proposed. The Ag-NSs were synthesized, microscopically characterized, and used for the modification of SPCNFEs.

The analytical performance of the modified electrode was evaluated for the three studied analytes. It was demonstrated that Ag-NS-SPCNFE is suitable for determining Pb(II), Cd(II), and As(V) at low μg L^{-1} levels, showing wider linear ranges for Pb(II) and Cd(II) but lower sensitivities as compared to As(V). Regarding previous studies of Pb(II), Cd(II), and As(V) determination, the LODs achieved in this investigation are equal or lower than other LODs previously reported.

The suitability of Ag-NS-SPCNFE for real sample analysis was demonstrated for the determination of As(V) in spiked water samples, achieving comparable results to those obtained by ICP-MS measurements with good reproducibility.

Author Contributions: K.T.-R., J.B.-A. and A.F. carried out the synthesis and microscopic characterization of silver nanoparticles. K.T.-R., C.P.-R. and N.S. carried out the modification of screen-printed electrodes with nanoparticles, the voltammetry measurements, and the data treatment. All authors contributed to the writing, revision, and critical discussion of the results presented in the final version of the manuscript. A.F., V.M. and N.S. were responsible for the student supervision. A.F. and V.M. dealt with the project administration and funding acquisition. All authors have read and agreed to the published version of the manuscript.

Funding: This research has been funded by Ministerio de Ciencia, Innovación y Universidades, and European Union Funds for Regional Development (FEDER), projects CTM2015-68859-C2-2-R and CGL2017-87216-C4-3-R, as well as by the Generalitat de Catalunya (Project 2017SGR312 and 2017SGR-311).

Data Availability Statement: Not applicable.

Conflicts of Interest: The authors declare no conflict of interest.

References

1. Barton, J.; García, M.B.G.; Santos, D.H.; Fanjul-Bolado, P.; Ribotti, A.; McCaul, M.; Diamond, D.; Magni, P. Screen-printed electrodes for environmental monitoring of heavy metal ions: A review. *Microchim. Acta* **2016**, *183*, 503–517. [CrossRef]
2. WHO. *Guidelines for Drinking-Water Quality*, 4th ed.; WHO, Ed.; World Health Organization: Geneva, Switzerland, 2017; ISBN 978-92-4-154995-0.
3. Daşbaşi, T.; Saçmaci, Ş.; Çankaya, N.; Soykan, C. A new synthesis, characterization and application chelating resin for determination of some trace metals in honey samples by FAAS. *Food Chem.* **2016**, *203*, 283–291. [CrossRef] [PubMed]
4. Azimi, S.; Es'haghi, Z. A Magnetized Nanoparticle Based Solid-Phase Extraction Procedure Followed by Inductively Coupled Plasma Atomic Emission Spectrometry to Determine Arsenic, Lead and Cadmium in Water, Milk, Indian Rice and Red Tea. *Bull. Environ. Contam. Toxicol.* **2017**, *98*, 830–836. [CrossRef] [PubMed]
5. Xu, Y.; Zhou, J.; Wang, G.; Zhou, J.; Tao, G. Determination of trace amounts of lead, arsenic, nickel and cobalt in high-purity iron oxide pigment by inductively coupled plasma atomic emission spectrometry after iron matrix removal with extractant-contained resin. *Anal. Chim. Acta* **2007**, *584*, 204–209. [CrossRef] [PubMed]
6. Shirkhanloo, H.; Mousavi, H.Z.; Rouhollahi, A. Speciation and determination of trace amount of inorganic arsenic in water, environmental and biological samples. *J. Chin. Chem. Soc.* **2011**, *58*, 623–628. [CrossRef]
7. Zhang, X.; Zhang, Y.; Ding, D.; Zhao, J.; Liu, J.; Yang, W.; Qu, K. On-site determination of Pb2+and Cd2+in seawater by double stripping voltammetry with bismuth-modified working electrodes. *Microchem. J.* **2016**, *126*, 280–286. [CrossRef]
8. Liu, Z.G.; Huang, X.J. Voltammetric determination of inorganic arsenic. *TrAC-Trends Anal. Chem.* **2014**, *60*, 25–35. [CrossRef]
9. Torres-Rivero, K.; Florido, A.; Bastos-Arrieta, J. Recent Trends in the Improvement of the Electrochemical Response of Screen-Printed Electrodes by Their Modification with Shaped Metal Nanoparticles. *Sensors* **2021**, *21*, 2596. [CrossRef]
10. Zhang, L.; Song, W.; Shi, L.; Li, Y.; Long, Y. High Sensitive On-Site Cadmium Sensor Based on AuNPs Amalgam Modified Screen-Printed. *IEEE Sen. J.* **2010**, *10*, 1583–1588. [CrossRef]
11. Li, D.; Li, J.; Jia, X.; Wang, E. Gold nanoparticles decorated carbon fiber mat as a novel sensing platform for sensitive detection of Hg(II). *Electrochem. Commun.* **2014**, *42*, 30–33. [CrossRef]
12. Torres-Rivero, K.; Pérez-Ràfols, C.; Bastos-Arrieta, J.; Florido, A.; Martí, V.; Serrano, N. Direct As(V) determination using screen-printed electrodes modified with silver nanoparticles. *Nanomaterials* **2020**, *10*, 1280. [CrossRef]
13. Guo, Z.; Luo, X.K.; Li, Y.H.; Zhao, Q.N.; Li, M.M.; Zhao, Y.T.; Sun, T.S.; Ma, C. Simultaneous determination of trace Cd(II), Pb(II) and Cu(II) by differential pulse anodic stripping voltammetry using a reduced graphene oxide-chitosan/poly-L-lysine nanocomposite modified glassy carbon electrode. *J. Colloid Interface Sci.* **2017**, *490*, 11–22. [CrossRef]
14. Priya, T.; Dhanalakshmi, N.; Thennarasu, S.; Thinakaran, N. A novel voltammetric sensor for the simultaneous detection of Cd2+and Pb2+using graphene oxide/κ-carrageenan/L-cysteine nanocomposite. *Carbohydr. Polym.* **2018**, *182*, 199–206. [CrossRef]
15. Torres-Rivero, K.; Torralba-Cadena, L.; Espriu-Gascon, A.; Casas, I.; Bastos-Arrieta, J.; Florido, A. Strategies for Surface Modification with Ag-Shaped Nanoparticles: Electrocatalytic Enhancement of Screen-Printed Electrodes for the Detection of Heavy Metals. *Sensors* **2019**, *19*, 4249. [CrossRef]
16. Murray, R.W. Nanoelectrochemistry: Metal Nanoparticles, Nanoelectrodes, and Nanopores. *Chem. Rev.* **2008**, *108*, 2688–2720. [CrossRef]
17. Sanllorente-Méndez, S.; Domínguez-Renedo, O.; Arcos-Martínez, M.J. Determination of arsenic(III) using platinum nanoparticle-modified screen-printed carbon-based electrodes. *Electroanalysis* **2009**, *21*, 635–639. [CrossRef]
18. Khairy, M.; Kampouris, D.K.; Kadara, R.O.; Banks, C.E. Gold Nanoparticle Modified Screen Printed Electrodes for the Trace Sensing of Arsenic(III) in the Presence of Copper(II). *Electroanalysis* **2010**, *22*, 2496–2501. [CrossRef]
19. Wan, H.; Sun, Q.; Li, H.; Sun, F.; Hu, N.; Wang, P. Screen-printed gold electrode with gold nanoparticles modification for simultaneous electrochemical determination of lead and copper. *Sens. Actuators B Chem.* **2015**, *209*, 336–342. [CrossRef]

20. Tukur, S.; Azah Yusof, N.; Hajian, R. Gold Nanoparticles Modified Screen Printed Electrode for Determination of Pb (II) Ion Using Linear Sweep Anodic Stripping Voltammetry. *IEEE Sens. J.* **2014**, *15*, 2780–2784. [CrossRef]
21. Rico, M.Á.G.; Olivares-Marín, M.; Gil, E.P. Modification of carbon screen-printed electrodes by adsorption of chemically synthesized Bi nanoparticles for the voltammetric stripping detection of Zn(II), Cd(II) and Pb(II). *Talanta* **2009**, *80*, 631–635. [CrossRef] [PubMed]
22. Wang, H.; Zhao, G.; Yin, Y.; Wang, Z.; Liu, G. Screen-Printed Electrode Modified by Bismuth /Fe_3O_4 Nanoparticle/Ionic Liquid Composite Using Internal Standard Normalization for Accurate Determination of Cd(II) in Soil. *Sensors* **2017**, *18*, 6. [CrossRef] [PubMed]
23. Pérez-Ràfols, C.; Bastos-Arrieta, J.; Serrano, N.; Díaz-Cruz, J.M.; Ariño, C.; de Pablo, J.; Esteban, M. Ag nanoparticles drop-casting modification of screen-printed electrodes for the simultaneous voltammetric determination of Cu(II) and Pb(II). *Sensors* **2017**, *17*, 1458. [CrossRef] [PubMed]
24. Aherne, D.; Ledwith, D.M.; Gara, M.; Kelly, J.M. Optical properties and growth aspects of silver nanoprisms produced by a highly reproducible and rapid synthesis at room temperature. *Adv. Funct. Mater.* **2008**, *18*, 2005–2016. [CrossRef]
25. Aherne, D.; Cara, M.; Kelly, J.M.; Gun'Ko, Y.K. From Ag nanoprisms to triangular AuAg nanoboxes. *Adv. Funct. Mater.* **2010**, *20*, 1329–1338. [CrossRef]
26. Miller, J.N.; Miller, J. *Métodos de Calibración en Análisis Instrumental: Regresión y Correlación*, 4th ed.; Capella, I., Ed.; Prentice Hall: Madrid, Spain, 2002.
27. Muñoz, J.; Bastos-Arrieta, J.; Muñoz, M.; Muraviev, D.; Céspedes, F.; Baeza, M. CdS quantum dots as a scattering nanomaterial of carbon nanotubes in polymeric nanocomposite sensors for microelectrode array behavior. *J. Mater. Sci.* **2016**, *51*, 1610–1619. [CrossRef]
28. Promphet, N.; Rattanarat, P.; Rangkupan, R.; Chailapakul, O.; Rodthongkum, N. An electrochemical sensor based on graphene/polyaniline/polystyrene nanoporous fibers modified electrode for simultaneous determination of lead and cadmium. *Sens. Actuators B Chem.* **2015**, *207*, 526–534. [CrossRef]
29. Nagaoka, Y.; Ivandini, T.A.; Yamada, D.; Fujita, S.; Yamanuki, M.; Einaga, Y. Selective detection of As(V) with high sensitivity by as-deposited boron-doped diamond electrodes. *Chem. Lett.* **2010**, *39*, 1055–1057. [CrossRef]

chemistry proceedings

MDPI

Proceeding Paper

Hydroponics Monitoring through UV-Vis Spectroscopy and Artificial Intelligence: Quantification of Nitrogen, Phosphorous and Potassium †

Aníbal Filipe Silva [1,2], Klara Löfkvist [3], Mikael Gilbertsson [4], Erik Van Os [5], Geert Franken [5], Jos Balendonck [5], Tatiana M. Pinho [6], José Boaventura-Cunha [6,7], Luís Coelho [2], Pedro Jorge [1,2] and Rui Costa Martins [2,*]

1 Faculty of Sciences of the University of Porto, Rua do Campo Alegre, s/n, 4169-007 Porto, Portugal; afilipe.silva@fc.up.pt (A.F.S.); pedro.jorge@inesctec.pt (P.J.)
2 Centre for Applied Photonics, INESC TEC, Rua Dr. Roberto Frias, 4200-465 Porto, Portugal; lcoelho@inesctec.pt
3 HIR Skåne AB Borgeby Slottsväg 11, 237 91 Bjärred, Sweden; klara.lofkvist@hushallningssallskapet.se
4 RISE Food and Agriculture, Scheelevägen 17, 223 70 Lund, Sweden; mikael.gilbertsson@ri.se
5 Wageningen University and Research, Droevendaalsesteeg 1, 6708 PB Wageningen, The Netherlands; erik.vanos@wur.nl (E.V.O.); geertcfranken@gmail.com (G.F.); jos.balendonck@wur.nl (J.B.)
6 Centre for Robotics in Industry and Intelligent Systems, INESC TEC, Rua Dr. Roberto Frias, 4200-465 Porto, Portugal; tatiana.m.pinho@inesctec.pt (T.M.P.); jboavent@utad.pt (J.B.-C.)
7 School of Science and Technology, University of Trás-os-Montes and Alto Douro, 5000-801 Vila Real, Portugal
* Correspondence: rui.c.martins@inesctec.pt; Tel.: +351-22-040-2301

† Presented at the 1st International Electronic Conference on Chemical Sensors and Analytical Chemistry, 1–15 July 2021; Available online: https://csac2021.sciforum.net/.

check for updates

Citation: Silva, A.F.; Löfkvist, K.; Gilbertsson, M.; Os, E.V.; Franken, G.; Balendonck, J.; Pinho, T.M.; Boaventura-Cunha, J.; Coelho, L.; Jorge, P.; et al. Hydroponics Monitoring through UV-Vis Spectroscopy and Artificial Intelligence: Quantification of Nitrogen, Phosphorous and Potassium. *Chem. Proc.* **2021**, *5*, 88. https://doi.org/10.3390/CSAC2021-10448

Academic Editor: Nicole Jaffrezic-Renault

Published: 30 June 2021

Publisher's Note: MDPI stays neutral with regard to jurisdictional claims in published maps and institutional affiliations.

Abstract: In hydroponic cultivation, monitoring and quantification of nutrients is of paramount importance. Precision agriculture has an urgent need for measuring fertilization and plant nutrient uptake. Reliable, robust and accurate sensors for measuring nitrogen (N), phosphorus (P) and potassium (K) are regarded as critical in this process. It is vital to understand nutrients' interference; thusly, a Hoagland fertilizer solution-based orthogonal experimental design was deployed. Concentration ranges were varied in a target analyte-independent style, as follows: [N] = [103.17–554.85] ppm; [P] = [15.06–515.35] ppm; [K] = [113.78–516.45] ppm, by dilution from individual stock solutions. Quantitative results for N and K, and qualitative results for P were obtained.

Keywords: nitrogen; phosphorus; potassium; Hoagland; nutrient; spectrophotometry; interferences

1. Introduction

Fertilizer usage represents an important part of traditional agriculture and crop yield. In a world of growing food (and other agricultural products) demand—estimates indicate up to 50% increase in the 2012–2050 time frame [1]—fertilizer (ab)use is seldomly a go-to solution for crop yield increase. Additionally, although growth rates for arable land are expected to increase within a sustainable manner, if an arable land loss scenario due to climate changing conditions is taken into account [2], further conflicts and competition might arise between protected lands, agricultural exploitation and human expansion.

Considering these concerns—well reflected by the United Nation's 2030 Agenda for Sustainable Development [3]—and also motivated to provide a solution for sustainable agriculture, our group has undertaken the task to develop a technology that is able to help farmers ensure that their crops' needs are being met, through their fertilization procedures. Knowing what is being fed to the crops and what is being taken up, it is possible to reduce water/fertilizer consumption to an optimal level, reducing the operational costs, whilst allowing crops to develop at their optimal speed, towards a bigger crop turnover.

Spectroscopy is, among others, one of the most well-established techniques for chemical identification and quantification. Several chemical determination methodologies that rely on spectroscopy exist (e.g., ICP-AES, LIBS, FTIR, GC-VUV); nevertheless, the following limitations also exist: the sample must be responsive to electromagnetic radiation (absorption/emission); linearity outside the Beer–Bouguer–Lambert Law [4–6] can sometimes be problematic, or the simple fact that spectroscopy is a molecular-level information tool, which can add entropy to the analysis by providing a wider range of information than the one desired. If pure compounds are analysed, little or no interference exists; when more complex mixtures are targeted, interferences might play a key role on the successful outcome. In such cases, in order to obtain an accurate and reliable measurement, interferences have to be taken into account. Chemometrics presents itself as a putative solution to, by employing varying complexity mathematical calculations, together with statistics and algorithms, allow the extraction of relevant information from the superimposed and—sometimes—latent data. Linearity-based models are unable to solve the interference pattern between any constituents present; this is the case for either interferents and non-interferents (target analytes), due to the fact that light has a wave-like nature and, hence, the sample information might suffer from constructive or destructive interferences [7]. Nevertheless, new chemometrics methodologies that encompass interferences already allow critical developments to be achieved, e.g., on health-related point-of-care analysis [8].

In hydroponics, most of these interferences can be attributed to the fertilizers. Fertilizers are mixtures of several different nutrients, mostly in their inorganic salt form (e.g., $MgSO_4$, $CaCO_3$, $FeCl_3$) whilst some might be in aqueous solutions (e.g., Mo, Ba, B). In complex mixtures, some signals might superimpose over others, causing a concentration misevaluation, or resulting in a continuous spectrum of overlapping signals.

This study aims to provide insight on the interferences within a complex matrix orthogonal design consisting of 8^3 independent concentration Hoagland solution samples. The performed assay further complements on our previous findings [9] on the feasibility of information extraction of highly constrained samples, by using an advanced algorithm—self-learning artificial intelligence (SLAI)—in order to find the adequate co-variance modes for accurate model prediction.

2. Materials and Methods

Hoagland solutions were chosen as a matrix due to their widely accepted status among the agronomical community as being a good model for complex nutrient solutions in hydroponics. Stock Hoagland solutions are composed as described by Table 1.

Table 1. Half-strength Hoagland solution individual component concentration.

Element	Ionic Form	C (ppm)	C (mM)
	Macronutrients		
Potassium (K)	K^+	117.29	3.00
Calcium (Ca)	Ca^{2+}	80.16	2.00
Magnesium (Mg)	Mg^{2+}	24.31	1.00
Nitrogen (N)	Total N	105.06	7.50
	$N(NH_4^+)$	7.01	0.50
	$N(NO_3^-)$	98.05	7.00
Phosphorus (P)	Total P (HPO_4^{2-}; H_2PO^4) *	15.49	0.50
Sulphur (S)	SO_4^{2-}	32.08	1.00

Table 1. *Cont.*

Element	Ionic Form	C (ppm)	C (mM)
	Micronutrients		
Chlorine (Cl)	Cl^{-}	0.324	9.15
Boron (B)	$BO_3{}^{3-}$	0.250	23.12
Manganese (Mn)	* Mn^{2+}	0.251	4.57
Zinc (Zn)	Zn^{2+}	0.025	0.38
Copper (Cu)	Cu^{2+}	0.0102	0.16
Molybdenum (Mo)	MoO^{4-}	0.0053	0.06
Iron (Fe)	Fe^{2+}; Fe^{3+} **	2.500	44.77
Sodium (Na)	Na^{+}	1.029	44.77

The ratio of ionic forms depends on * the pH and ** also on the O_2 level within the nutrient solution.

Three stock solutions for N, P and K were freshly and individually prepared in order for it to be possible to vary each target element (N, P or K) independently. Each stock solution comprised of ionic elements already present in the base matrix, as follows: N—$NaNO_3/NH_3$; P—H_3PO_4; K—KCl. The ratio of $NaNO_3/NH_3$ of the stock spiking solutions was the same as the ratio on the Hoagland solution (≈93:7).

Final concentrations of all samples (matrix + individual spikes) were corrected taking into account any variations derived from the preparation of fresh stock solutions each day during the execution of the assay.

The tested final concentration ranges, for the target analytes, were as described in Table 2.

Table 2. Tested concentration ranges for N, P and K within the design matrix.

	Concentration (ppm)			Concentration (mM)		
	N	P	K	N	P	K
Minimum	103.17	15.06	113.78	7.37	0.49	2.91
Maximum	554.85	515.35	516.45	39.61	16.64	13.21

The designed orthogonal matrix was composed by 8^3 samples, each one with an independent N, P and K concentration level. At each corresponding level, the N, P and K corresponding spike was added to the matrix and stirred for 10 s in order to attain full homogeneity. Afterwards, the pH value was registered (Crison GLP 21, Crison Instruments, SA, Barcelona, Spain) and the sample pumped into a custom-built flow cell for spectral data acquisition. Each sample had a final total volume of 30 mL (Hoagland + [N + P + K] spikes). After data acquisition, the sample was discarded and the system flushed with deionised water.

Data acquisition was performed with an in-house LabView-based developed software (National Instruments™ Corp., Austin, TX, USA) for pumping control and data acquisition.

Sample irradiation was performed with a D_2 light source (FiberLight® D_2 HighPower DTM 10/50S, Heraeus Noblelight GmbH, Hanau, Germany) whereas the detection was performed by a miniaturized spectrometer with a 190–650 nm range (STS-UV-L-50-400, Ocean Insight, Inc., Orlando, FL, USA). Individual components were assembled with custom-length 600 nm UV-Vis optical fibres (in-house customization), as depicted in Figure 1.

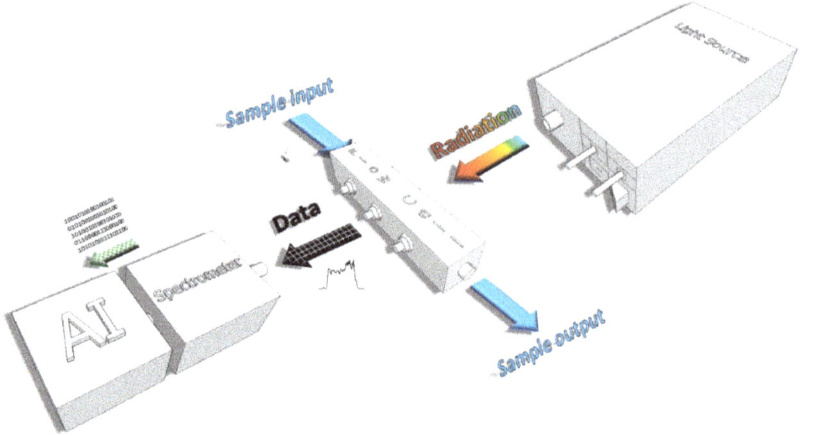

Figure 1. Scheme for individual components arrangement, within the developed prototype.

3. Results

The obtained results from the execution of this matrix were compiled and are depicted on Figure 2.

The collection of spectral data and cross correlation with the concentration information for each solution was performed. Spectroscopy signals were processed accordingly to [7]. Nevertheless, using advanced signal processing it is possible to train the system to recognize and extract the information from the relevant features, incorporating multi-scale interference into the NPK quantification models.

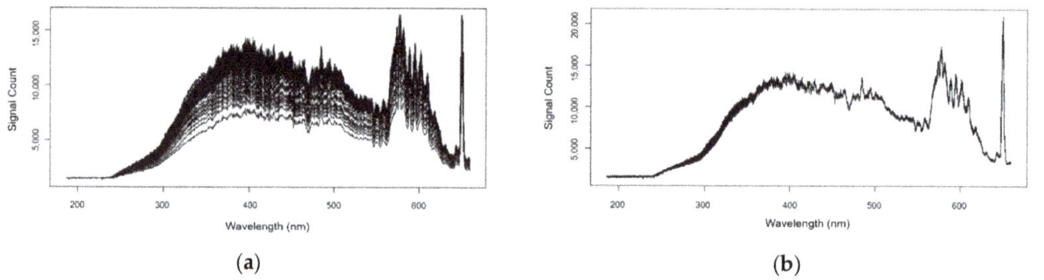

Figure 2. Compiled spectra (8^3 samples) without (**a**) and with scatter correction (**b**).

The correlation of the different levels among the NPK nutrients of the matrix design, can be represented as displayed by Figure 3a whereas Figure 3b shows the corresponding recorded spectra in the UV-Vis region (*circa* 200–650 nm) of the factorial design samples. As expected, most of the systematic spectral variation occurs at ≈250 to 450 nm, and, to a lesser extent, to 500 nm. This figure provides evidence that information about P and K is present, because, even to the naked eye, one can observe that there are more spectral patterns in the region of ≈250 to 450 nm than the expected nitrogen levels of the experimental design; that is a good indication that the interferences between all the constituents are being registered on the spectra.

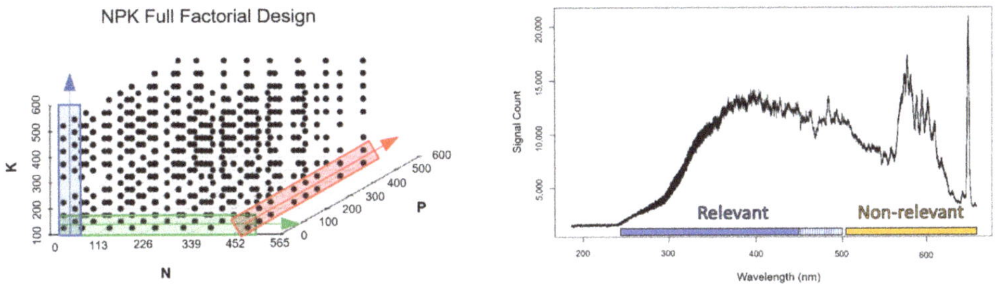

(a) (b)

Figure 3. Sample distribution of the NPK full factorial design (**a**), whole matrix data (with scatter correction) of the relevant wavelengths of the obtained spectra, where most of the relevant NPK-bearing information is contained (**b**).

The principal component analysis (PCA) (please refer to Figure 4a) scores plot of the corresponding experimental design spectra is shown, where the different colours represent the different levels of total nitrogen. The main variance present in the spectra corresponds to the nitrogen absorbance, where the first principal component is highly correlated to the nitrogen content. It is also possible to see that the K-level information is embedded inside each N-concentration level. Analysis of the second component allows to unveil that information of K-level also carries the information of the different P-levels of the sample matrix (please refer to Figure 4b).

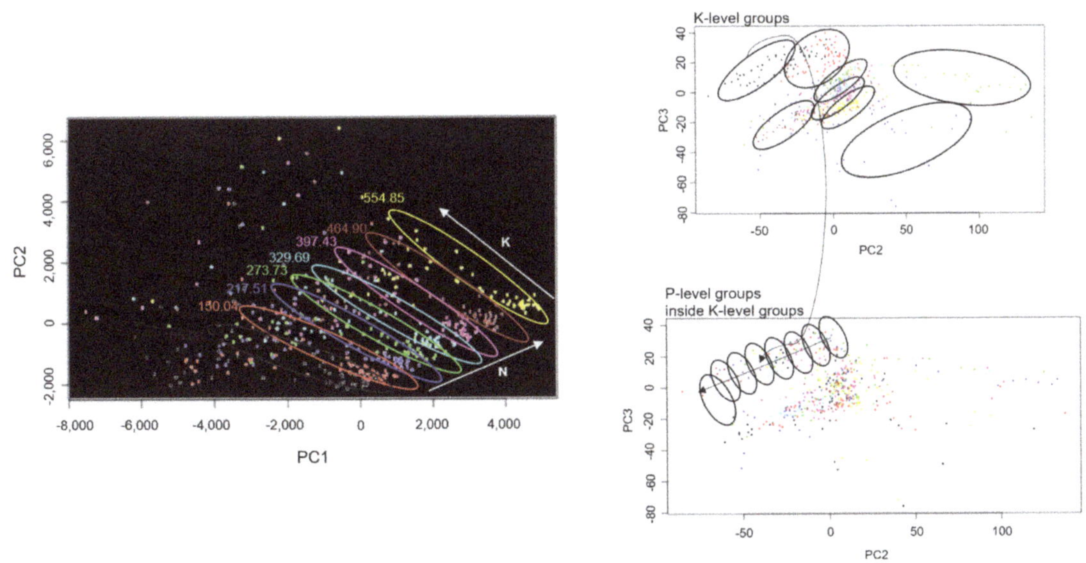

(a) (b)

Figure 4. Principal components analysis (**a**) and the information of K-levels within the N-levels (**b**, left-hand side) whereas (**b**, right-hand side) demonstrates the information of P embedded within K-level groups.

Using the data obtained from the executed matrix, it was possible to train the self-learning AI of the system in order to quantify N and K with 6.7% (0.997) and 3.8% (0.987), respectively, and to obtain qualitative results for P, as shown in Figure 5.

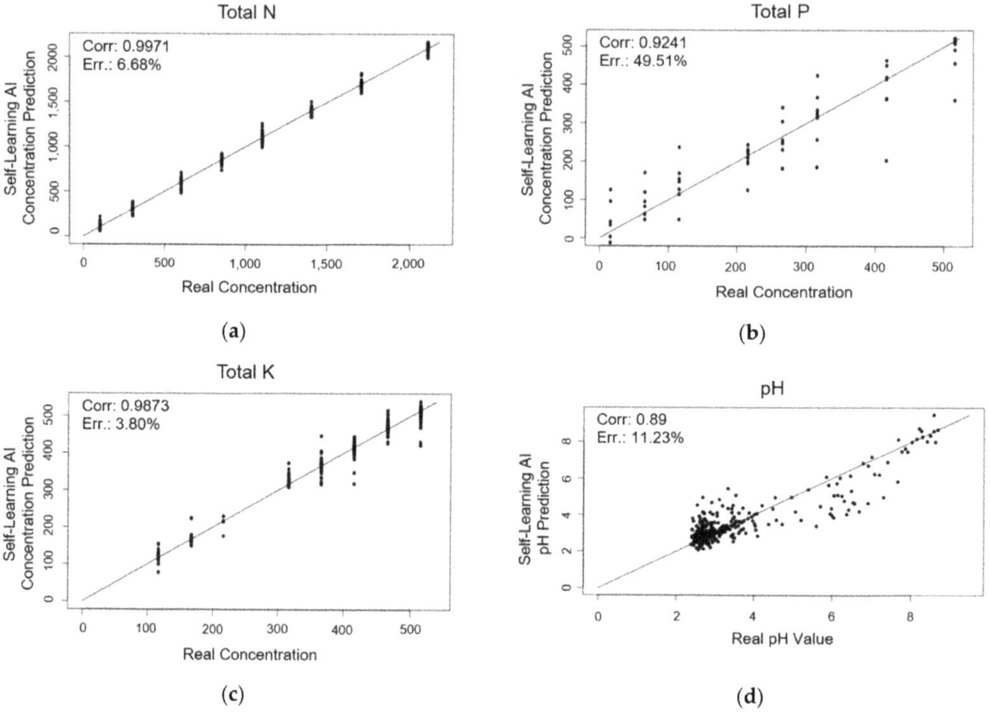

Figure 5. Matrix results for total N (**a**), total P (**b**), total K (**c**) and also pH (**d**). Quantitative results for N and K, whereas qualitative results for P are possible to be inferred.

4. Conclusions

A NPK spectroscopy-based, AI-supported by a robust self-learning artificial intelligence was developed in order to be able to cope with increasing interference complexity of fertilizer solutions in greenhouses. The obtained results allow to be inferred that the current system's performance is adequate for Hoagland solutions, which are used in research and high-end hydroponic systems.

The assembled system aimed to keep a good balance between cost–benefit, without relinquishing reliability, robustness and accuracy; this objective has been successfully attained.

Further analysis of the results—not within the scope of this manuscript—as well as of unpublished data, allows further developments to be implemented to the system/prototype, in order to enhance its robustness and accuracy.

Author Contributions: Conceptualization, A.F.S. and R.C.M.; methodology, A.F.S. and R.C.M.; software, R.C.M.; validation, A.F.S. and R.C.M.; formal analysis, A.F.S. and R.C.M.; investigation, A.F.S.; resources, R.C.M., L.C. and P.J.; data curation, A.F.S.; writing—original draft preparation, A.F.S.; writing—review and editing, A.F.S., K.L., M.G., E.V.O., G.F., J.B., T.M.P., J.B.-C., L.C., P.J. and R.C.M.; visualization, A.F.S., R.C.M.; supervision, R.C.M. and P.J.; project administration, J.B.-C.; funding acquisition, J.B.-C. All authors have read and agreed to the published version of the manuscript.

Funding: This work was funded by the ERA-NET Cofund WaterWorks2015 Call, within the frame of the collaborative international consortium AGRINUPES. This ERA-NET is an integral part of the 2016 Joint Activities developed by the Water Challenges for a Changing World Joint Programme Initiative (Water JPI/002/2015). R.M. acknowledges Fundação para a Ciência e Tecnologia (FCT) research contract grant (CEEIND/017801/2018). A.F.S. gratefully acknowledges the financial support provided by FCT (Portugal's Foundation for Science and Technology) within grant (DFA/BD/9136/2020).

Institutional Review Board Statement: Not applicable.

Informed Consent Statement: Not applicable.

Data Availability Statement: Not applicable.

Conflicts of Interest: The authors declare no conflict of interest.

References

1. FAO. *The Future of Food and Agriculture—Trends and Challenges*; FAO: Rome, Italy, 2017.
2. FAO. *The Future of Food and Agriculture—Alternative Pathways to 2050*; FAO: Rome, Italy, 2018.
3. United Nations. *Transforming Our World: The 2030 Agenda for Sustainable Development*; United Nations: San Francisco, CA, USA, 2015.
4. Lambert, J.H. *Photometry, or, on the Measure and Gradations of Light Intensity, Colors, and Shade*; Eberhardt Klett: Augsburg, Germany, 1760.
5. Bouguer, P. *Optics Essay on the Attenuation of Light*; Claude Jombert: Paris, France, 1729.
6. Beer, A. Determination of the absorption of red light in colored liquids. *Ann. Phys. Chem.* **1852**, *162*, 11.
7. Martins, R.C. Big Data Self-Learning Artificial Intelligence Methodology for the Accurate Quantification and Classification of Spectral Information under Complex Variability and Multi-Scale Interference. WO Patent WO2018060967, 5 April 2018.
8. Barroso, T.G.; Ribeiro, L.; Gregório, H.; Santos, F.; Martins, R.C. Point-of-care Vis-SWNIR spectroscopy towards reagent-less hemogram analysis. *Sens. Actuators B Chem.* **2021**, *343*, 130138. [CrossRef]
9. Monteiro-Silva, F.; Jorge, P.A.S.; Martins, R.C. Optical Sensing of Nitrogen, Phosphorus and Potassium: A Spectrophotometrical Approach Toward Smart Nutrient Deployment. *Chemosensors* **2019**, *7*, 51. [CrossRef]

chemistry proceedings

MDPI

Abstract

Detection of Indoor Air Pollutants Using Reactive Sputtering/GLAD of Tin Oxide Thin Films †

Achraf El Mohajir [1], Jean-Baptiste Sanchez [1,*], Mohammad Arab Pour Yazdi [1], Olivier Heintz [2] and Nicolas Martin [1]

1 Institut FEMTO-ST, UMR 6174 CNRS, Université de Bourgogne-Franche-Comté, 15B, Avenue des Montboucons, CEDEX, 25030 Besançon, France; achraf.elmohajir@femto-st.fr (A.E.M.); mohammad.arabpourya@femto-st.fr (M.A.P.Y.); nicolas.martin@femto-st.fr (N.M.)
2 Laboratoire Interdisciplinaire Carnot de Bourgogne, UMR 6303 CNRS, Université de Bourgogne-Franche Comté, 9 Avenue Alain Savary, BP 47870, 21078 Dijon, France; olivier.heintz@u-bourgogne.fr
* Correspondence: jbsanche@univ-fcomte.fr
† Presented at the 1st International Electronic Conference on Chemical Sensors and Analytical Chemistry, 1–15 July 2021; Available online: https://csac2021.sciforum.net/.

Abstract: Indoor air quality is a topic of major importance for public health. Among the numerous chemical compounds that can be found in indoor air, BTEX (i.e., benzene, toluene, ethylbenzene, and xylene) is considered one of the most toxic volatile organic compounds (VOCs). The present contribution is focused on the use of an original approach to produce nanostructured materials based on tin oxide with unexplored features, especially for gas sensors. In this work, we combine two physical vapor deposition techniques based, first, on a pulsing injection of the reactive gas during the deposition and second focused on the glancing angle deposition (GLAD) technique, which enables the structuring of various architectures. These active layers are deposited on a micro-hotplate to produce micro-chemical gas sensors for the detection of BTEX. Here, we have demonstrated the utility of using the GLAD deposition technique and the role of sputtering pressure in obtaining porous sensitive thin films. In particular, we established relationships between deposition parameters and gas sensing performances.

Keywords: BTEX gas sensor; GLAD method; sensitivity; porous surface

Citation: El Mohajir, A.; Sanchez, J.-B.; Arab Pour Yazdi, M.; Heintz, O.; Martin, N. Detection of Indoor Air Pollutants Using Reactive Sputtering/GLAD of Tin Oxide Thin Films. *Chem. Proc.* **2021**, *5*, 89. https://doi.org/10.3390/CSAC2021-10548

Academic Editor: Elisabetta Comini

Published: 1 July 2021

Publisher's Note: MDPI stays neutral with regard to jurisdictional claims in published maps and institutional affiliations.

Supplementary Materials: The following are available online at https://www.mdpi.com/article/10.3390/CSAC2021-10548/s1.

chemistry proceedings

Proceeding Paper

Influence of the Type and Amount of Plasticizer on the Sensory Properties of Microspheres Sensitive to Lipophilic Ions †

Aleksandra Kalinowska, Patrycja Matusiak, Sandra Skorupska, Ilona Grabowska-Jadach and Patrycja Ciosek-Skibińska *

Chair of Medical Biotechnology, Faculty of Chemistry, Warsaw University of Technology, Noakowskiego 3, 00-664 Warsaw, Poland; aleksandra.kalinowska3.dokt@pw.edu.pl (A.K.); patrycja.matusiak08@gmail.com (P.M.); sskorupska@ch.pw.edu.pl (S.S.); ilona@ch.pw.edu.pl (I.G.-J.)

* Correspondence: pciosek@ch.pw.edu.pl

† Presented at the 1st International Electronic Conference on Chemical Sensors and Analytical Chemistry, 1–15 July 2021; Available online: https://csac2021.sciforum.net/.

Abstract: Working parameters of chemical sensors, such as selectivity and sensitivity, can be adjusted by optimizing components of chemosensitive layers, including type and amount of plasticizer in the case of PVC membranes in optodes. Plasticizers are also used in the process of creating micro/nanospheres that are incorporated with chemical indicators to form micro/nano-scale optodes. This study investigated the influence of the type of plasticizer (polar o-NPOE and non-polar DOS) on the optical response of microspheres that are sensitive to lipophilic ions. Moreover, the amount of plasticizer was also adjusted in order to obtain satisfactory sensitivity in the widest linear range. The chemosensory response of the developed microspheres was studied with the use of spectrophotometry and spectrofluorimetry, while size of the optodes was estimated by confocal microscopy.

Keywords: plasticizer; microspheres; optode; chemosensors

Citation: Kalinowska, A.; Matusiak, P.; Skorupska, S.; Grabowska-Jadach, I.; Ciosek-Skibińska, P. Influence of the Type and Amount of Plasticizer on the Sensory Properties of Microspheres Sensitive to Lipophilic Ions. *Chem. Proc.* **2021**, *5*, 90. https://doi.org/10.3390/CSAC2021-10487

Academic Editor: Huangxian Ju

Published: 30 June 2021

1. Introduction

For many years, ion-selective electrodes (ISE) have been considered important analytical chemistry tools. They are widely used both in environmental analysis and in medical diagnostics [1]. Chemosensitive membranes of the ISEs whose signal is potentiometric can be easily modified by the incorporation of chromoionophores, which gives the possibility for optical transduction. On the other hand, polymer scaffolds of such optode membrane can be replaced by surfactants, which leads to the fabrication of suspension of micro- or nano-sized optodes.

Various achievements in the field of micellization of dyes can be applied for the fabrication of micro- and nano-optodes. Micelles are micro- or nanometer-sized particles that exist solution in equilibrium with the particles or ions. They can be incorporated with additional lipophilic components forming nano/microspheres, aggregates of surfactant molecules with a hydrophilic part being in contact with outer water solution, and a hydrophobic part creating inner lipophilic microenvironment [2]. Micelles are widely used in medicine for drug delivery and increase water solubility and pharmaceuticals' bioavailability. In recent years, the huge potential of micellar systems for various applications in analytical chemistry has been noticed [3].

Working parameters of chemical sensors, such as selectivity and sensitivity, can be adjusted by optimizing components of chemosensitive membranes, including type and amount of plasticizer [4]. A plasticizer is a substance allowing for increasing polymer flexibility, as well as susceptibility to its further processing. In membranes of ISEs and optodes, it forms a lipophilic environment in which receptors and additional components responsible for sensing are suspended. A plasticizer in the simplest sense is considered an organic solvent. There are many types of such compounds, such as animal fats, petroleum fractions, and all kinds of plant extracts. Plasticizers can be used in the process of creating micro/nanospheres incorporated with chemical indicators and stabilized by surfactants. As environmental conditions change, such systems may change their optical properties, becoming optodes in the micro/nanoscale [3,5].

This study aimed to develop two microsphere types sensitive towards lipophilic ions: anion-sensitive (AS) and cation-sensitive (CS). Each of these systems includes a chromoionophore, an ion exchanger, a surfactant, and a plasticizer. In this work we were focused on selecting an appropriate plasticizer, as well as adjusting its quantity. Changes in chemosensory properties were observed in absorbance and fluorescence mode, while confocal microscopy observations were applied to verify the microspheres' size.

2. Experimental

2.1. Chemicals

Sodium phosphate monohydrate, disodium phosphate dodecahydrate, Tris-HCl, Pluronic F127 were supplied by Sigma-Merck (Poznań, Poland). Milli-Q water was used for preparation of all aqueous solutions, including phosphate buffer pH 7.4 and Tris-HCl buffer pH 9.0. Plasticizers (2-Nitrophenyl octyl ether, o-NPOE, Bis(2-ethylhexyl) sebacate, DOS), lipophilic salts (Tridodecylmethylammonium chloride, TDMAC, Potassium tetrakis [3,5-bis(trifluoromethyl)phenyl]borate, KTFPB), chromoionophores I and XI were obtained from Fluka (Selectophore). Freshly distilled tetrahydrofuran, THF (Fluka) was used as a solvent for the microspheres' components. All chemicals were used as received.

2.2. Preparation of Microspheres Suspensions

Two types of optode microspheres: with generic anion-sensitivity (AS) and with generic cation-sensitivity (CS), were prepared (Figure 1). The composition of the membranes and proposed mechanism of optical signal generation are based on the literature data on optodes [6–10]. AS microspheres contained Chromoionophore XI, TDMAC, o-NPOE or DOS, and Pluronic (F-127), whereas CS microspheres contained Chromoionophore I, KTFPB, o-NPOE or DOS, and Pluronic (F-127). Each of the components included in the given microsphere was weighed and dissolved in 1.5 mL of THF. To ensure that all ingredients dissolve well, the vial was placed in an ultrasonic bath for 5 min. 0.5 mL portions of THF solutions were pipetted to 4.5 mL of deionized water on a vortex. The last step was to remove the solvent by passing compressed air through the solution (the process was carried out for 1 h). Clear particle suspensions were obtained that were applied for further measurements using microtiter plates. For this purpose, 100 µL of the prepared microsphere suspension were pipetted to each well and 100 µL of appropriate analyte solution was added. We examined the chemosensory response in the presence of 0.1 M NaOH, 0.1 M HCl, and calibration solutions appropriate for each type of microspheres (1 µM-0.1 M $NaClO_4$ and 1 µM-0.1 M NH_4NO_3 for AS and CS optodes, respectively). When protonation degree in AS was determined, microspheres suspensions were two times diluted (50 µL of optode cocktail + 50 µL of deionized water) to avoid recording signals out of range by microtiter plate reader.

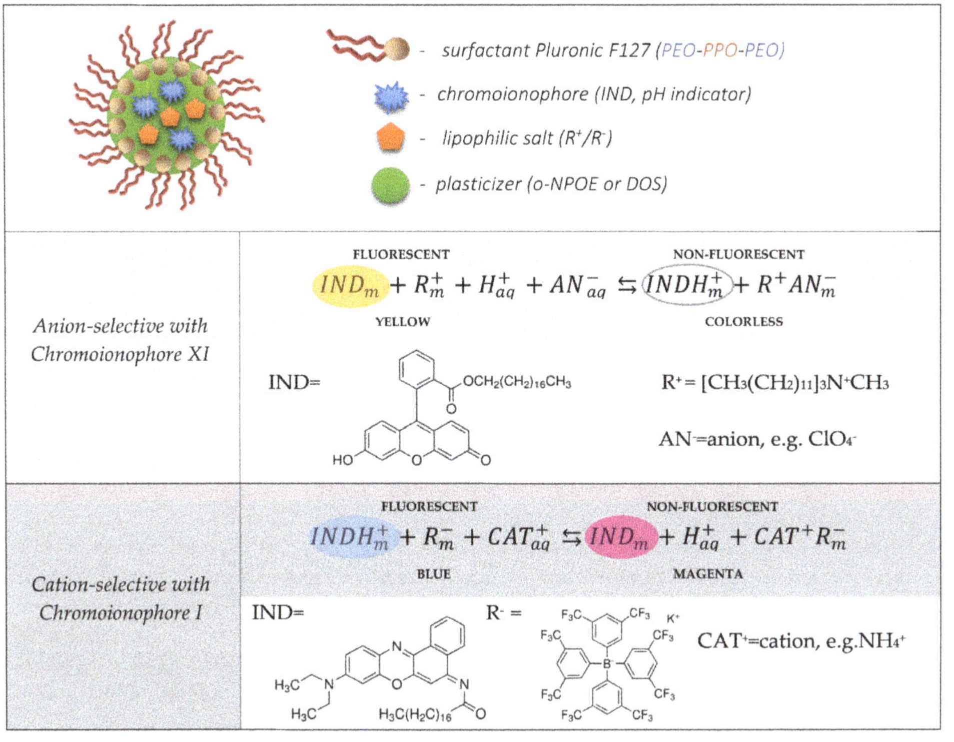

Figure 1. Schematic representation of the prepared microspheres, their composition, and the mechanisms of target analyte recognition with optical signal generation. The mechanisms by which the spheres operate are based on the extraction of the analyte from the aqueous medium (aq subscript) into the lipophilic interior of the microsphere (m subscript). Each type of microsphere solution is based on a plasticizer emulsion stabilized with a Pluronic non-ionic surfactant having hydrophobic and hydrophilic domains (PPO and PEO, respectively). The plasticizer droplets contain all components needed to extract target analytes (lipophilic salt) and generate a signal (IND) [6–10].

2.3. Examination of the Optical Properties of Microspheres

Both spectrophotometric and spectrofluorimetric measurements were applied to study the chemosensory properties of the obtained microspheres. They were tested using Synergy 2 Multi-Mode Reader (BioTek Instruments, Inc., Winooski, VT, USA). In order to examine the sizes of the obtained microspheres, a confocal microscope Fluoview FV10i (Olympus, Japanese) was used. The following parameters were used to observe the samples: λ_{ex} = 473 nm, λ_{em} = 490–590 nm for AS microspheres and λ_{ex} = 635 nm, λ_{em} = 660–760 nm for CS microspheres. Measurements of microspheres were made using the FV10i-SW software. Samples were observed using Cellview™ Cell Culture Dish (Greiner Bio One, Germany) with a glass bottom and four-compartments. Three. Hundred microliters of suspension of microspheres were placed in each compartment.

3. Results and Discussion

3.1. Spectrophotometric Measurements of Chemosensory Properties

The microspheres prepared for this work, both AS and CS, contain an ion exchanger in their composition, whose task is to exchange ions between the aqueous and organic phases according to equations given in Figure 1 (exchange of lipophilic cations to protons in the case of CS microspheres, coextraction of protons with lipophilic anions in the case of AS microspheres). This processes ensure the change in the protonation degree of the chromoionophores and thus cause a change in the optical properties of the produced system. The sensory response of four independent replicates of microspheres suspensions was tested by recording UV-Vis spectra in the presence of model lipophilic ions—perchlorate ions in the case of AS system, and ammonium ions in the case of CS system (Figures 2 and 3, respectively). The tested solutions were buffered to make the change of the chromoionophore spectrum independent of pH, in order to ensure, that the spectral change was influenced only by the change of the target ion concentration. The spectra were gradually changing following change in ion concentration and thus protonation degree of chromoionophore (represented by arrows in Figures 2a and 3a). Calibration curves were determined for the obtained UV-Vis spectra Figures 2b and 3b). First, signal based on baseline correction was calculated (ratio of absorbances obtained for 454 nm and 380 nm, K = A454/A380, and then only change related to this signal in smallest concentration was presented ($K - K_{C=1\mu M}$) for more precise comparison of the two systems using various plasticizers (Figure 2b). The obtained results clearly show that in the AS optode system, the non-polar plasticizer DOS allowed for obtaining responses with a greater degree of sensitivity and enables the determination of anions in almost the entire concentration range.

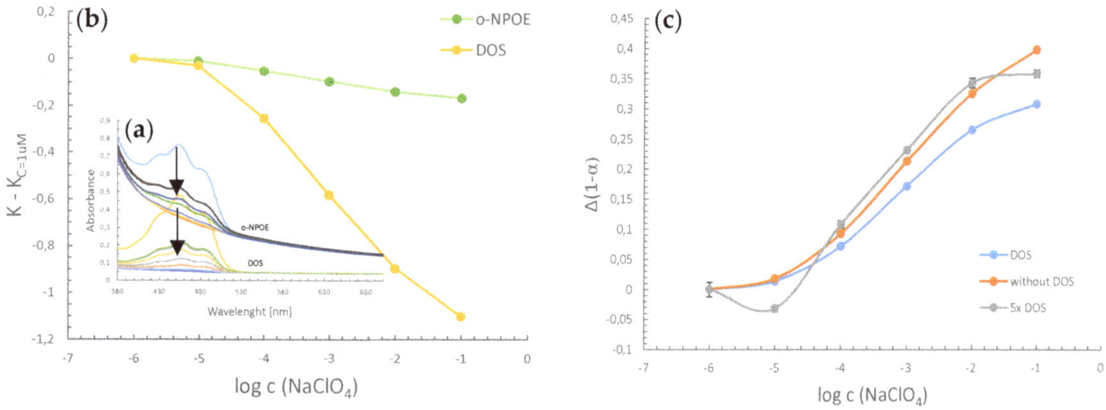

Figure 2. Sensory properties of AS microspheres: (**a**) Absorbance spectrum of microspheres with various plasticizers obtained in the presence of 0.1 M NaOH (deprotonated form of chromoionophore), in 0.1 M HCl (protonated form of chromoionophore), in 0.01 M PBS pH = 7.4 (control sample) and when $NaClO_4$ was added (calibration solutions in concentration range 1 µM–0.1 M); (**b**) calibration curves with signal based on change of baseline correction for K = A454/A380; (**c**) signals stated as change in the protonation degree of chromoionophore for microspheres with various amount of DOS. All calibration solutions were buffered with 0.01 M phosphate buffer pH 7.4. Points of calibration curves were determined as mean ± SD; $n = 4$.

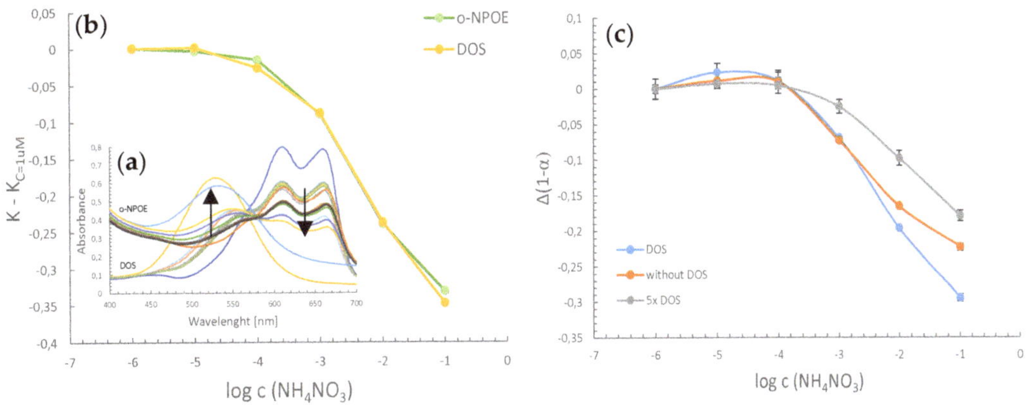

Figure 3. Spectrophotometric sensory properties of CS microspheres: (**a**) Absorbance spectrum of microspheres with various plasticizers obtained in the presence of 0.1 M NaOH (deprotonated form of chromoionophore), in 0.1 M HCl (protonated form of chromoionophore), in 0.01M Tris-HCl pH = 9 (control sample) and when NH_4NO_3 was added (calibration solutions in concentration range 1 µM–0.1 M); (**b**) Calibration curves with signal based on change of baseline correction for K = A610/A570; (**c**) signals stated as change in the protonation degree of chromoionophore for microspheres with various amount of DOS. All calibration solutions were buffered with 0.01 M Tris-HCl buffer pH 9.0. Points of calibration curves were determined as mean ± SD; $n = 4$.

In the case of CS microspheres, baseline correction was calculated using absorbances at 610 nm and 570 nm, and resulting signal was the change of K = A454/A380 value (Figure 3b). For CS microspheres based on DOS and o-NPOE, there was no significant difference in calibration curves—they were comparable. However, in the case of DOS-based optodes, a slightly more comprehensive linear range was observed.

Thus, a non-polar plasticizer DOS was chosen for further experiments, as the better for both microsphere types. The next phase of the research involved the study of the influence of plasticizer amount on chemosensory properties. Three formulations for AS optodes were prepared that was based on various plasticizer content. Again, UV-Vis spectra were recorded as previously, and calibration curves were determined (Figure 2c). Signals were presented as change in protonation degree, 1-α, of chromoionophore (change of 1-α related to this value obtained for the smallest concentration of the analyte, for clarity of presentation). The observation of these calibration curves revealed that the different amount of plasticizer did not significantly affect the sensory response of the system. The microspheres without plasticizers and with its five-fold amount exhibited slightly better sensitivity compared to the system with the standard amount of plasticizer. Additionally, without the addition of DOS, slightly better quantification range was noticed (10 µM–0.1 M). The narrowest linear range was observed for optode system with the highest amount of DOS. The same procedure was repeated for CS microsphere optodes. Respective calibration curves for three studied systems with various amounts of plasticizer are presented in Figure 3c. The best results in terms of sensitivity and quantification range were obtained for the standard amount of the plasticizer.

3.2. Spectrofluorimetric Measurements of Chemosensory Properties

In addition to measuring the absorbance, chemosensory properties were also tested in the fluorescence mode. Similarly, as in the case of absorbance, a gradual change in the spectrum of AS and CS microspheres, influenced by the growing concentration of target ions was observed (respective emission spectra and calibration curves presented in Figure 4). In the case of both AS and CS microspheres, DOS was recognized as the better plasticizer, which enables the determination of analytes in a wider concentration range with high sensitivity. Observation of protonation degree (data not shown) revealed that the widest linear range was obtained when the standard amount of DOS was applied, while its five-fold amount caused increase in the sensitivity.

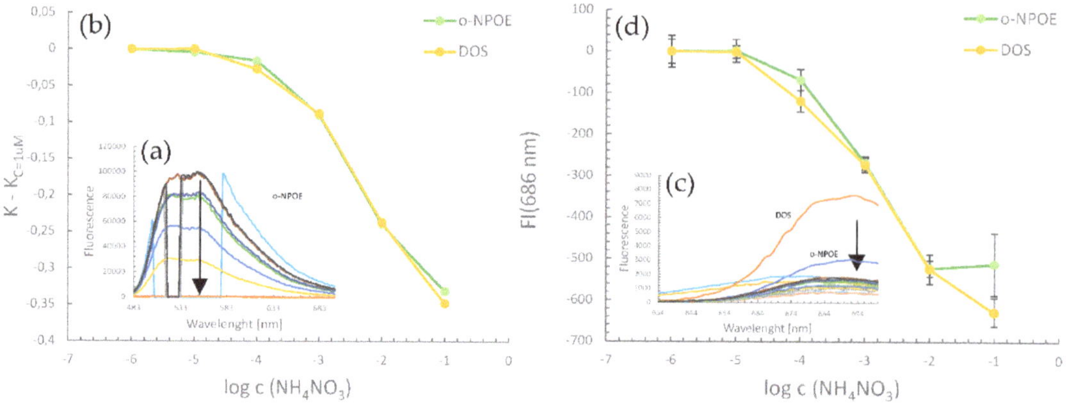

Figure 4. Spectrofluorimetric sensory properties of AS (a, b) and CS (**c**,**d**) microspheres with various plasticizers. Calibration curves (**b**,**d**) obtained on the basis of fluorescence emission spectra (**a**,**c**). All calibration solutions were buffered (0.01 M phosphate buffer pH 7.4 for AS microspheres, 0.01 M Tris-HCl buffer pH 9.0 for CS microspheres). Points of calibration curves were determined as means; $n = 4$.

3.3. Confocal Microscope

Figures 5 and 6 show pictures of microspheres taken with the use of a confocal microscope. The analyzes confirmed the formation of spherical microspheres in all experimental samples regardless of the amount of plasticizer added. A slight spread of the size of the microspheres was observed. Their diameters oscillated between 2–10 μm in the case of both types of microspheres.

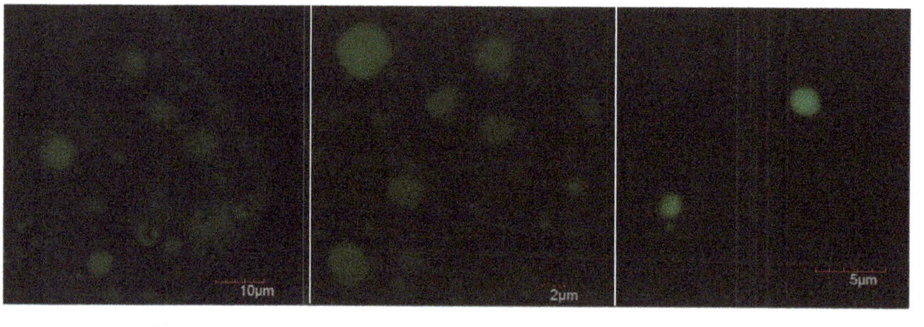

(a) (b) (c)

Figure 5. Confocal microscope pictures of AS microspheres: (**a**) without plasticizer; (**b**) with a standard amount of plasticizer; (**c**) with its five-fold amount.

Based on the observation of the fluorescence intensity of the population of microspheres, it was found that they are homogeneous in terms of fluorescence intensity, which proves the incorporation of the chromoionophore. Moreover, the chromoionophores are evenly distributed in both AS and CS microspheres.

(**a**) (**b**) (**c**)

Figure 6. Confocal microscope pictures of CS microspheres: (**a**) without plasticizer; (**b**) with a standard amount of plasticizer; (**c**) with its five-fold amount.

4. Conclusions

The presented results show that in the case of both AS and CS microspheres used in this work, the type and amount of plasticizer change chemosensory properties of the micro-optodes. For both types of microspheres, non-polar plasticizer DOS allowed for obtaining better results compared to the polar plasticizer in the composition. Thanks to DOS, it was possible to determine perchlorate and ammonium ions in a wide concentration range and the obtained sensitivity was higher compared to microspheres with o-NPOE content. The plasticizer amount should also be carefully adjusted to obtain satisfactory chemosensory characteristics. Generally, in the case of microspheres studied in this work, the standard amount led to the best results. All these observations were confirmed both in spectrophotometric and spectrofluorimetric measurements. The obtained results from the confocal microscope confirm the formation of spherical microspheres in size range of 2–10 μm. No influence of the amount of added plasticizer on the shape and size of the microspheres was observed. In summary, the use of an appropriate amount of a non-polar plasticizer allows for obtaining chemosensory microspheres sensitive to lipophilic ions in a broad quantification range from 10 μM to 0.1 M in both absorbance and fluorescence intensity mode.

Supplementary Materials: The following are available online at https://www.mdpi.com/article/10.3390/CSAC2021-10487/s1.

Author Contributions: Conceptualization, P.C.-S. and I.G.-J.; methodology, design of experiments, A.K. and S.S; investigation, P.M., A.K. and S.S.; Original draft preparation, A.K. and S.S.; resources, supervision, project administration, funding acquisition, review and editing of the manuscript were completed by P.C.-S. All authors have read and agreed to the published version of the manuscript.

Funding: National Science Centre (Poland): UMO-2018/30/E/ST4/00481.

Institutional Review Board Statement: Not applicable.

Informed Consent Statement: Not applicable.

Data Availability Statement: Not applicable.

Acknowledgments: This work was financially supported by National Science Centre (Poland) within the framework of the SONATA BIS project No. UMO-2018/30/E/ST4/00481. Aleksandra Kalinowska acknowledges financial support from IDUB project (Scholarship Plus program).

Conflicts of Interest: The authors declare no conflict of interest.

References

1. Wygladacz, K.; Radu, A.; Xu, C.; Qin, Y.; Bakker, E. Fiber-Optic Microsensor Array Based on Fluorescent Bulk Optode Microspheres for the Trace Analysis of Silver Ions. *Anal. Chem.* **2005**, *77*, 4706–4712. [CrossRef] [PubMed]

2. Xie, X.; Bakker, E. Ion selective optodes: From the bulk to the nanoscale. *Anal. Chem.* **2015**, *407*, 3899–3910. [CrossRef] [PubMed]
3. Wang, C.X.; Wu, B.; Zhou, W.; Wang, Q.; Yu, H.; Deng, K.; Li, J.M.; Zhuo, R.X.; Huang, S.W. Turn-on fluorescent probe-encapsulated micelle as colloidally stable nanochemosensor for highly selective detection of Al^{3+} in aqueous solution and living cell imaging. *Sens. Actuators B. Chem.* **2018**, *271*, 225–238. [CrossRef]
4. Bakker, E.; Bühlmann, P.; Pretsch, E. Carrier-Based Ion-Selective Electrodes and Bulk Optodes. 1. General Characteristics. *Chem. Rev.* **1997**, *97*, 3083–3132. [CrossRef] [PubMed]
5. Wypych, G. *Handbook of Plasticizers*, 3rd ed.; ChemTech Publishing: Toronto, ON, Canada, 2017.
6. Zeng, H.H.; Wang, K.M.; Li, D.; Yu, R.Q. Development of an alcohol optode membrane based on fluorescence enhancement of fluorescein derivatives. *Talanta* **1994**, *41*, 969–975. [CrossRef]
7. Ortuño, J.A.; Albero, M.I.; García, M.S.; Sánchez-Pedreño, C.; García, M.I.; Expósito, R. Flow-through bulk optode for the fluorimetric determination of perchlorate. *Talanta* **2003**, *60*, 563–569. [CrossRef]
8. Chan, W.H.; Lee, A.W.M.; Wang, K. Design of a primary amine-selective optode membrane based on a lipophilic hexaester of calix[6]arene. *Analyst* **1994**, *119*, 2809–2812. [CrossRef]
9. Chan, W.H.; Lee, A.W.M.; Lee, C.M.; Yau, K.W.; Wang, K. Design and characterization of sodium-selective optode membranes based on the lipophilic tetraester of calix[4]arene. *Analyst* **1995**, *120*, 1963–1967. [CrossRef]
10. Du, X.; Zhu, C.; Xie, X. Thermochromic Ion-Exchange Micelles Containing H^+ Chromoionophores. *Langmuir* **2017**, *33*, 5910–5914. [CrossRef] [PubMed]

chemistry proceedings

MDPI

Abstract

Intrinsically Coloured Red Aromatic Polyamides †

Patricia Peredo-Guzmán *, Miriam Trigo-López *, Saúl Vallejos *, Félix García * and José Miguel García *

Chemistry Department, Faculty of Science, Burgos University, Plaza Misael Bañuelos s/n. 09001 Burgos, Spain
* Correspondence: pdperedo@ubu.es (P.P.-G.); mtrigo@ubu.es (M.T.-L.); svallejos@ubu.es (S.V.); fegarcia@ubu.es (F.G.); jmiguel@ubu.es (J.M.G.)

† Presented at the 1st International Electronic Conference on Chemical Sensors and Analytical Chemistry, 1–15 July 2021; Available online: https://csac2021.sciforum.net/.

Abstract: Aromatic polyamides or aramids are materials with exceptional thermal and mechanical properties. For this reason, they are considered high-performance materials with many applications in fields such as civil security (bullet-proof body armour or fire, chemical, and saw protection suits), transport (automotive and aerospace), and civil engineering, among many others. The remarkable properties arise from the high cohesive energy due to their chemical structure, including the rigidity of the main chain due to the wholly aromatic structure conjugated with the amide groups, the high average bond energy, and a strong and highly directional interchain hydrogen bonds between the amide moieties. Although the natural yellowish colour of the fibres is used, generally, most of the applications require coloured fibres. However, aramid fibres have poor dyeing properties for the same reasons that make them thermally and mechanically resistant, and traditional dyeing methods, such as dope dyeing, are inefficient and aggressive, which impairs the fibres' properties. The ideal colour fastness of fibres is achieved by intrinsically, inherently, or self-coloured polymers by introducing a dye motif or chromophore monomer in the chemical structure of the polymer. In addition, the colour hue can be controlled by tuning the chromophore monomer molar content in the final composition. In previous research, we successfully obtained inherently blue-coloured aramids, with blue chromophore motifs unable to migrate and evenly distribute along the polymer chain and maintain their high-performance properties, and our aim now is to obtain red-coloured aramids prepared in the same fashion.

Keywords: red aromatic polyamides; chromophore monomer; self-coloured polymers; high-performance polymers

Citation: Peredo-Guzmán, P.; Trigo-López, M.; Vallejos, S.; García, F.; García, J.M. Intrinsically Coloured Red Aromatic Polyamides. *Chem. Proc.* **2021**, *5*, 91. https://doi.org/10.3390/CSAC2021-10421

Academic Editor: Nicole Jaffrezic-Renault

Published: 30 June 2021

Supplementary Materials: The following supporting information can be downloaded at: https://www.mdpi.com/article/10.3390/CSAC2021-10421/s1, Poster S1: Intrinsically coloured red aromatic polyamides.

Author Contributions: Conceptualization, M.T.-L. and F.G.; methodology, S.V.; formal analysis, P.P.-G.; investigation, P.P.-G. and M.T.-L.; resources, J.M.G.; data curation, P.P.-G.; writing—original draft preparation, P.P.-G. and M.T.-L.; writing—review and editing, P.P.-G. and M.T.-L.; visualization P.P.-G.; supervision, F.G. and S.V.; project administration, J.M.G.; funding acquisition, M.T.-L. All authors have read and agreed to the published version of the manuscript.

Funding: This research was funded by the Spanish Ministerio de Ciencia e Innovación, grant number PID2019-108583RJ-I00/AEI/10.13039/501100011033.

Institutional Review Board Statement: Not applicable.

Informed Consent Statement: Not applicable.

Conflicts of Interest: The authors declare no conflict of interest.

MDPI
St. Alban-Anlage 66
4052 Basel
Switzerland
Tel. +41 61 683 77 34
Fax +41 61 302 89 18
www.mdpi.com

Chemistry Proceedings Editorial Office
E-mail: chemproc@mdpi.com
www.mdpi.com/journal/chemproc

www.ingramcontent.com/pod-product-compliance
Lightning Source LLC
Chambersburg PA
CBHW061454220926
51850CB00004B/89
9783036541150